Selected Titles in This Subseries

40 **M. V. Karasev, Editor,** Coherent Transform, Quantization, and Poisson Geometry (TRANS2/187)

39 **A. Khovanskiĭ, A. Varchenko, and V. Vassiliev, Editors,** Geometry of Differential Equations (TRANS2/186)

38 **B. Feigin and V. Vassiliev, Editors,** Topics in Quantum Groups and Finite-Type Invariants (Mathematics at the Independent University of Moscow) (TRANS2/185)

37 **Peter Kuchment and Vladimir Lin, Editors,** Voronezh Winter Mathematical Schools (Dedicated to Selim Krein) (TRANS2/184)

36 **V. E. Zakharov, Editor,** Nonlinear Waves and Weak Turbulence (TRANS2/182)

35 **G. I. Olshanski, Editor,** Kirillov's Seminar on Representation Theory (TRANS2/181)

34 **A. Khovanskiĭ, A. Varchenko, and V. Vassiliev, Editors,** Topics in Singularity Theory (TRANS2/180)

33 **V. M. Buchstaber and S. P. Novikov, Editors,** Solitons, Geometry, and Topology: On the Crossroad (TRANS2/179)

32 **R. L. Dobrushin, R. A. Minlos, M. A. Shubin, and A. M. Vershik, Editors,** Topics in Statistical and Theoretical Physics (F. A. Berezin Memorial Volume) (TRANS2/177)

31 **R. L. Dobrushin, R. A. Minlos, M. A. Shubin, and A. M. Vershik, Editors,** Contemporary Mathematical Physics (F. A. Berezin Memorial Volume) (TRANS2/175)

30 **A. A. Bolibruch, A. S. Merkur'ev, and N. Yu. Netsvetaev, Editors,** Mathematics in St. Petersburg (TRANS2/174)

29 **V. Kharlamov, A. Korchagin, G. Polotovskiĭ, and O. Viro, Editors,** Topology of Real Algebraic Varieties and Related Topics (TRANS2/173)

28 **L. A. Bunimovich, B. M. Gurevich, and Ya. B. Pesin, Editors,** Sinai's Moscow Seminar on Dynamical Systems (TRANS2/171)

27 **S. P. Novikov, Editor,** Topics in Topology and Mathematical Physics (TRANS2/170)

26 **S. G. Gindikin and E. B. Vinberg, Editors,** Lie Groups and Lie Algebras: E. B. Dynkin's Seminar (TRANS2/169)

25 **V. V. Kozlov, Editor,** Dynamical Systems in Classical Mechanics (TRANS2/168)

24 **V. V. Lychagin, Editor,** The Interplay between Differential Geometry and Differential Equations (TRANS2/167)

23 **Yu. Ilyashenko and S. Yakovenko, Editors,** Concerning the Hilbert 16th Problem (TRANS2/165)

22 **N. N. Uraltseva, Editor,** Nonlinear Evolution Equations (TRANS2/164)

Published Earlier as Advances in Soviet Mathematics

21 **V. I. Arnold, Editor,** Singularities and bifurcations, 1994

20 **R. L. Dobrushin, Editor,** Probability contributions to statistical mechanics, 1994

19 **V. A. Marchenko, Editor,** Spectral operator theory and related topics, 1994

18 **Oleg Viro, Editor,** Topology of manifolds and varieties, 1994

17 **Dmitry Fuchs, Editor,** Unconventional Lie algebras, 1993

16 **Sergei Gelfand and Simon Gindikin, Editors,** I. M. Gelfand seminar, Parts 1 and 2, 1993

15 **A. T. Fomenko, Editor,** Minimal surfaces, 1993

14 **Yu. S. Il'yashenko, Editor,** Nonlinear Stokes phenomena, 1992

13 **V. P. Maslov and S. N. Samborskiĭ, Editors,** Idempotent analysis, 1992

12 **R. Z. Khasminskiĭ, Editor,** Topics in nonparametric estimation, 1992

11 **B. Ya. Levin, Editor,** Entire and subharmonic functions, 1992

(Continued in the back of this publication)

Coherent Transform, Quantization, and Poisson Geometry

American Mathematical Society

TRANSLATIONS

Series 2 • Volume 187

Advances in the Mathematical Sciences — 40

(*Formerly Advances in Soviet Mathematics*)

Coherent Transform, Quantization, and Poisson Geometry

M. V. Karasev
Editor

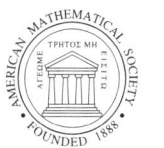

American Mathematical Society
Providence, Rhode Island

ADVANCES IN THE MATHEMATICAL SCIENCES
EDITORIAL COMMITTEE

V. I. ARNOLD
S. G. GINDIKIN
V. P. MASLOV

1991 *Mathematics Subject Classification.* Primary 58F05, 81R30, 78H05, 81Sxx; Secondary 33Cxx, 51H15, 53C05.

ABSTRACT. This book consists of three long articles devoted to advanced problems in quantization theory, as well as in symplectic and Poisson geometry. The main topics are: coherent states and irreducible representations for algebras with non-Lie permutation relations, Hamilton dynamics and quantization over stable isotropic submanifolds, and infinitesimal tensor complexes over degenerate symplectic leaves in Poisson manifolds. The articles include many examples (in particular, examples from physics) and complete proofs. The book will be of interest to researchers and graduate students working in these areas of mathematics and mathematical physics.

Library of Congress Card Number 91-640741
ISBN 0-8218-1178-9
ISSN 0065-9290

Copying and reprinting. Material in this book may be reproduced by any means for educational and scientific purposes without fee or permission with the exception of reproduction by services that collect fees for delivery of documents and provided that the customary acknowledgment of the source is given. This consent does not extend to other kinds of copying for general distribution, for advertising or promotional purposes, or for resale. Requests for permission for commercial use of material should be addressed to the Assistant to the Publisher, American Mathematical Society, P. O. Box 6248, Providence, Rhode Island 02940-6248. Requests can also be made by e-mail to reprint-permission@ams.org.

Excluded from these provisions is material in articles for which the author holds copyright. In such cases, requests for permission to use or reprint should be addressed directly to the author(s). (Copyright ownership is indicated in the notice in the lower right-hand corner of the first page of each article.)

© 1998 by the American Mathematical Society. All rights reserved.
The American Mathematical Society retains all rights
except those granted to the United States Government.
Printed in the United States of America.

∞ The paper used in this book is acid-free and falls within the guidelines
established to ensure permanence and durability.
Visit the AMS home page at URL: http://www.ams.org/

10 9 8 7 6 5 4 3 2 1   03 02 01 00 99 98

Contents

Preface ... ix

Non-Lie Permutation Representations, Coherent States, and Quantum Embedding
MIKHAIL KARASEV AND ELENA NOVIKOVA ... 1

Adapted Connections, Hamilton Dynamics, Geometric Phases, and Quantization over Isotropic Submanifolds
MIKHAIL KARASEV AND YURII VOROBJEV ... 203

Infinitesimal Poisson Cohomology
VLADIMIR ITSKOV, MIKHAIL KARASEV, AND YURII VOROBJEV ... 327

Preface

The present collection conrains three paper on the quantization theory and symplectic and Poisson geometry.

In the paper "*Non-Lie permutation relations, coherent states, and quantum embedding*" by Mikhail Karasev and Elena Novikova, irreducible representations for some interesting classes of algebras with non-Lie permutation relations are constructed. The approach used in construction these represenations is related to the theory of coherent states. The solution of the well-known problem of quantum embedding of surfaces (symplectic leaves) is presented. The relationship between quantum surfaces (of revolution, of birevolution, etc.), irreducible representations of non-Lie algebras and the theory of (generalized) hypergeometric functions is investigated. Explicit formulas for $\hbar$-expansions of the Weyl and Wick quantum products are found in geometric terms. Meromorphic coherent states are used to establish a duality between Wick and anti-Wick symbols. Numerous examples of non-Lie permutation relations, in particular, some examples of quadratic algebras that are important in physics, are considered in details.

The paper "*Adapted connections, Hamilton dynamics, geometric phases, and quantization over isotropic submanifolds*" by Mikhail Karasev and Yurii Vorobjev is devoted to symplectic geometry and Hamilton dynamics in a neighborhood of a stable isotropic submanifold (for example, a torus). The approach in the paper is based on the notion of adapted symplectic connection. The results generalize and develop a variety of classical theorems related to the case of a stable trajectory. Flat coisotropic extensions, Darboux atlases, Lagrangian inflations, Floquet multipliers in the symplectic setting, and other structures on isotropic submanifolds are introduced and studied. The method of adapted connections is also applied to the quantization problem and to the spectral geometry. Analogs and generalizations of the Gelfand–Lidskii index, the Johnson–Moser rotation number, and the Maslov class are defined for isotropic submanifolds of dimension more than 1 and less than $\frac{1}{2}$(phase space dimension). These invariants are combined together in a quantization condition.

In the paper "*Infinitesimal Poisson cohomology*" by Vladimir Itskov, Mikhail Karasev, and Yurii Vorobjev infinitesimal geometry of a degenerate symplectic leaf of a Poisson manifold is studied. Using linear connection and the Lie algebroid structure in the corresponding normal bundle, the infinitesimal Poisson cohomology is studied in detail, and applications to the problem of deformations of Poisson brackets are considered.

©1998 American Mathematical Society

The articles in the book continue and substantially develop the authors' previous results about nonlinear Poisson brackets, Hamilton dynamics, and quantization.

The collection can be viewed as a summary of some new ideas and approaches suggested by participants of our research seminar in the last five years. This seminar takes place in the Quantum Dynamical Systems Laboratory at the Applied Mathematics Department of the Moscow Institute of Electronics and Mathematics (MIEM). The participants of the seminar, who are my co-authors in this collection, graduated from MIEM in different years. For example, Yurii Vorobjev graduated from a special group initiated by V. P. Maslov in 1972. Students of this group attended lectures by V. Arnold, Yu. Manin, and other outstanding mathematicians. Our collaboration with Yurii started back then. Elena Novikova recently completed graduate studies (very successfully) and received the PhD degree. Vladimir Itskov is now a graduate student; he is certainly highly talented.

Discussions with my students Mikhail Kozlov and Alexander Pereskokov, who also are participants of the seminar, contributed to the common work. Unfortunately, for various reasons, this collection does not contain their papers.

The friendly scientific atmosphere at the Chair of Applied Mathematics greatly facilitates the collection. The help offered by M. A. Shishkova in translating and typesetting the articles in this collection was invaluable.

Mikhail Karasev

Non-Lie Permutation Relations, Coherent States, and Quantum Embedding

Mikhail Karasev and Elena Novikova

ABSTRACT. For classes of algebras generated by non-Lie permutation relations, we develop a method for constructing irreducible representations (in Hilbert spaces of antiholomorphic distributions), as well as coherent states, reproducing kernels, and Wick products. Relations between these irreducible representations and the theory of (basic) hypergeometric functions are investigated. For representations of higher-dimensional algebras, the reproducing kernels (quantum metrics) are generated by the Horn bihypergeometric functions and by "twisted" hypergeometric functions of several complex variables. Explicit power series for Weyl and Wick products, as well as for the reproducing measure, in terms of the quantum $*$-tensor and the quantum Kähler metric are given. Meromorphic coherent states are employed to establish a mutual duality of Wick and anti-Wick products. A number of examples, in particular, quadratic algebras related to the hydrogen atom and to the Zeeman effect and the degenerate Sklyanin algebra, are considered in detail (including Jacobi polynomials, Bessel functions, the Euler summation formula, etc.). The problem of correspondence between irreducible representations of a quantum algebra and symplectic leaves of the Poisson algebra is stated as a problem of quantum embedding. The solution of this problem is obtained, and quantum operators of complex structure are described. Quantum surfaces of revolution and birevolution, as well as some other quantum hypersurfaces, are constructed. Formulas for semiclassical embeddings are also given. In this paper we pay special attention to details, examples, and complete proofs.

Contents

INTRODUCTION	2
PART I. COHERENT STATES AND QUANTUM GEOMETRY	5
§1. Surfaces of revolution and representations of basic nonlinear commutation relations	5
§2. Representations by differential and q-differential operators. Relationship to the theory of hypergeometric functions	25
§3. Quantum embeddings of surfaces	36
§4. Multidimensional generalizations. Surfaces of birevolution	58
§5. Hypersurfaces and twisted hypergeometric functions of several variables	79

1991 *Mathematics Subject Classification.* Primary 81Sxx, 81R30; Secondary 33Cxx, 58F05.

This research was partially supported by the Russian Foundation for Basic Research under grant no. 96-01-01426.

§6. Quantum tensors and explicit asymptotic expansions for Weyl
and Wick products — 108
PART II. EXAMPLES OF QUANTUM EMBEDDINGS AND IRREDUCIBLE REPRESENTATIONS — 122
§1. Simplest Lie algebras — 122
§2. Quadratic algebra related to the Zeeman effect — 130
§3. Quadratic Faddeev–Sklyanin algebra — 144
§4. Weakly nonlinear algebras and representations with two vacuum vectors — 153
§5. Eight-dimensional quadratic algebra (quantum embedding of $T^*\mathbb{S}^3$) and Bessel coherent states — 164
§6. Reduction of coherent states — 179
REFERENCES — 197

Introduction

Interest in studying algebras with non-Lie permutation relations (the term "non-Lie" means different from the standard case of Lie algebras) is mostly based on profound examples in the q-deformation theory and in the "quantum inverse scattering problem."[1] Another very general way in which non-Lie permutation relations appear is the quantum version of the Marsden–Weinstein–Lie–Cartan reduction.[2]

There are several directions in the study of non-Lie permutation relations and quantization of general nonlinear Poisson brackets. The *first direction* deals with computation of the product in the enveloping algebra (or the convolution on its dual) via the operators of regular representation (or via the "generalized shift" or Delsarte operators).[3] In fact, this is a way to construct a quantum groupoid for given non-Lie relations, as well as to construct objects related to the noncommutative geometry due to Connes [**C**].

The *second direction* is based on the theory of deformation quantization proposed in [**BFFLS**$_{1,2}$, **FLS**, **Ve**]. It is a fundamental basis for the quantum group theory that related the quantum Yang–Baxter equation and the quadratic Faddeev–Zamolodchikov algebras. About new developments of the deformation quantization theory, see, for instance, [**CFS, FS, Ka**$_{17}$**, OMY, OMMY, Ri**$_{1,2}$**, We**$_2$].

The *third direction* of study is the "semiclassical" approach that makes it possible to derive all essential objects related to a non-Lie algebra and its quantum groupoid, but only approximately, with an accuracy $O(\hbar^\infty)$ with respect to the Planck "constant" $\hbar \to 0$; see [**Ka**$_{4-6}$**, KaM**$_{1,2}$] and detailed proofs in [**KaM**$_4$]. This is the theory of asymptotic quantization.

The *fourth direction* is based on the notion of coherent states, initiated in the earliest times of quantum mechanics by Erwin Schrödinger [**Sch**] and Werner Heisenberg [**H**], in optics by Glauber [**Gl**$_{1,2}$] (where the name "coherent states"

---

[1] See the background in [**Bi, Dr, FRT, GFa, J, KuR, Sk**], as well as some interesting examples in [**GLZh, GZhL, M**].

[2] See, for instance, [**Ka**$_{5-7,9,10}$**, KaN**$_{3,4}$] and [**KaM**$_4$] for more detail.

[3] For the basic facts about the generalized shift operators, see [**Le**]; for applications to non-Lie relations, see the general formulas in [**Ka**$_{2-4,17}$**, KaM**$_3$] and the review of this topic in the book [**KaM**$_4$].

was first introduced), and defined in general form by Klauder [**Kl**$_{1,2}$] and Berezin [**B**$_1$]. Of course, the ideas of von Neumann and Mackey regarding the projection representation of quantum mechanics were at the basis.

In fact, a version of coherent states was also developed in the theory of holomorphic functions under the name "reproducing kernels." See [**Aro, Brg, Bgm, BC**].

From the viewpoint of representation theory, the main property of coherent states is the possibility of writing out an irreducible representation of a given algebra in terms of differential operators acting on the space of parameters of the states. For the Heisenberg algebra, this fact was understood already by Fock [**Fo**$_{1,2}$] and Dirac [**D**] (see also in [**Coo**]), and was generalized for a wide class of Lie algebras [**O, P**$_{1,2}$] and for certain q-analogs (see, for instance, in [**ArC, Ku, Mac**] and also relations with p-adic analysis in [**VVZ**]). In the framework of a general quantization process, coherent states first were used in [**B**$_{1-4}$, **Kl**$_2$] and were deeply involved into this theory [**AAn, AAG, BMS, BC, DaG, DaGM, DeB, Gi**$_{1,2}$, **Kb**$_1$, **Kl**$_3$, **KlS, Mo, Od**$_{1,2}$, **OPr, Si, Sl, Sp, UU, VoB, Zh, ZhFG**], especially in the context of the fundamental geometric quantization due to Kostant and Souriau and the representation theory [**AEm, Bl**$_{1,2}$, **BKo**$_{1,2}$, **CaGR, GGP-S, HNØ**$_{1,2}$, **Hu**$_{1,2}$, **K**$_2$, **Ka**$_{8-16}$, **KaKz**$_{1,2}$, **KaV**$_{1,2}$, **R**$_{1-4}$, **Tu**]. This list of references is very far from being complete but shows the variety of approaches to this area. We also stress that there are many crucial questions in this theory that were posed back in the 70s and that still remain open.

In the present paper, we are interested in two questions. First, how to derive the coherent state and the corresponding irreducible representations for algebras with non-Lie permutation relations. Second, how to relate these quantum representations to classical symplectic leaves in the Poisson manifold,[4] or more precisely, how to define and construct a quantum embedding of leaves and quantum complex structures on them.

Simple basic classes of non-Lie permutation relations[5] are considered in § I.1.1, then they are generalized in §§ I.1.4, I.1.5, in §§ I.4, I.5, and in §§ II.4, II.5. For all these non-Lie algebras, we describe the set of irreducible Hermitian representations with a vacuum vector. This set can be identified with a subset $\mathcal{R} \subset \mathbb{R}^k$; points from $\mathcal{R}$ are analogs of highest weights in the case of Lie algebras. In order to define $\mathcal{R}$, we introduce a *basic function* $\mathcal{F}_a$ (or basic matrix in § I.6) and consider properties of its *roots*. To each $a \in \mathcal{R}$, we assign a Hilbert space of the representation, which is a space of antiholomorphic distributions over $\mathbb{R}^{2d}$. In general, these distributions are not regular functions (this is the difference from all standard approaches in "holomorphic quantization" theories).

Then we construct coherent states and reproducing kernels. They are also defined, in general, only as distributions.

---

[4]In the case of Lie algebras, the leaves are orbits of the coadjoint representation of a Lie group. In this case, the orbit method [**K**$_1$] gives the correspondence between leaves and representations via the character formula.

[5]For instance, these relations include the cases considered in [**M, Ka**$_{2-4}$, **KaM**$_{1,3}$] and in [**KaM**$_4$, Appendix II]; we also refer to a series of interesting physical papers [**AzE, BD, ChEK, Das, DQ**$_{1,2}$, **GrN, Ja, KQ, MOd, MMP, OKK, Q, Ro, W**], where versions of the basic relations appeared and representations were described (by methods that differ from ours). Our approach follows [**KaN**$_{1,3}$].

The operators representing a non-Lie algebra are realized as pseudodifferential operators acting on distributions. Under certain conditions, these operators are differential, and their orders are independent of a. In fact, we obtain not just one but many equivalent representations corresponding to the same $a \in \mathcal{R}$. The number of such representations, roughly speaking, coincides with the number of partitions of all roots of the basic function $\mathcal{F}_a$ into two groups.

It is remarkable that *these roots are parameters of hypergeometric functions* through which one can derive reproducing kernels and coherent states. Here we give the total classification and consider (in Part II) a series of examples in detail.[6]

Under some other conditions (see § I.2.4), the basic function $\mathcal{F}_a$ is a q-polynomial, the corresponding reproducing kernels and coherent states are derived in terms of basic q-hypergeometric functions, and the operators representing the non-Lie permutation relations are q-differential operators.[7] In Part II we investigate, among other things, two examples of quadratic algebras (of Sklyanin type), and describe all their irreducible representations via differential and q-differential operators.

In § I.4 we assign some reproducing kernels to the Horn *hypergeometric functions of two variables*; in § I.5 we generalize this construction by introducing a matrix factorial and a *matrix (twisted) version of hypergeometric functions*.

These multivariable twisted generalizations correspond to higher dimensional quantum algebras, whose structural functions are replaced by *structural matrices*, and so the basic function is replaced by a *basic matrix*. Since the basic matrix is factored by matrix factors, the Pochhammer symbols or shifted factorials (which appear in the definition of the usual hypergeometric functions) should be replaced by matrix factorials.

In addition to this generalization, in § I.5.6 a *universal exponential form* of coherent states, and thus of reproducing kernels, is described and applied to a "nonscalar" generalization of the original algebra.

In § I.3 we consider the problem of correspondence between irreducible representations and symplectic leaves, and introduce a definition of *quantum embedding of leaves*. Then, following [**Ka**$_{17}$], we describe the solution of this problem by constructing $*$-products on the entire space $\mathbb{R}^{k+2d}$ and on the symplectic leaf $\Omega = \Omega^{2d}$ (classically embedded into $\mathbb{R}^{k+2d}$), as well as by an additional procedure of *quantum restriction* on a leaf. We conclude with the following general statement [**Ka**$_{17}$]: "Quantization commutes with quantum restriction."

We also make another interesting general observation: quantum restriction of complex structure coincides with "classical" complex structure on a leaf.[8]

Examples of the quantum restriction (or quantum embedding) are given in Part II. First we examine the simplest case of the Lie algebras $\mathbf{su}(2)$ and $\mathbf{su}(1)$. Then we investigate two remarkable examples of quadratic 4-dimensional algebras in detail. In conclusion, in §§5, 6 of Part II, we consider a complicated example of quantum embedding of $T^*\mathbb{S}^3$ into $\mathbb{R}^8$ by means of special Bessel coherent

---

[6]In [**KaN**$_1$], we observed a relationship between algebras with non-Lie (quadratic) commutation relations and hypergeometric functions. For further development of this particular example and of other examples related to the Zeeman effect, see [**KaN**$_{2-4}$], as well as Part II of the present paper.

[7]See also the important papers [**Ku**, **Go**, **Od**$_2$, **Sp**] and the references therein.

[8]A variety of complex structures on the leaf arises as a result of different partitions into two groups of all roots of the basic function $\mathcal{F}_a$. In the polynomial case, different complex structures correspond to different hypergeometric series with one and the same total set of parameters.

states. These states were first derived in [**Ka**$_{10}$] by using the *reduction procedure* (see § II.6) with respect to a symmetry group corresponding to the Kustaanheimo regularization of the Kepler problem.[9] We show that these states are assigned to an irreducible representation of an interesting 8-dimensional quadratic algebra. The related quadratic Poisson algebra has two Casimir functions determining the classical embedding of $T^*\mathbb{S}^3$ into $\mathbb{R}^8$.

Other examples, as well as many-step reduction of coherent states: ["Bessel" of order 0] → ["Bessel" of order m] → [hypergeometric polynomial (Jacobi)], are given in [**KaN**$_{3,4}$].

In § I.3, in the framework of the Berezin–Klauder–Wick quantization procedure, we also describe *a duality between Wick and anti-Wick symbols* (or covariant and contravariant symbols). A pair of dual resolutions of unity are written over the symplectic (complex) groupoid $\Omega \times \Omega$. It is shown that both Wick and anti-Wick products can be derived in terms of a pair of probability operators that correspond to a pair of quantum Kähler forms over Ω and to the same reproducing measure. The first two terms of the semiquantum approximation for the quantum Kähler potential and for the reproducing measure are described in terms of the classical symplectic and Ricci forms on the leaf Ω. This provides us with simple formulas describing the semiquantum restriction on a symplectic leaf (or the semiquantum embedding of the leaf).

In § I.6 we follow certain results from [**Ka**$_{17}$] and demonstrate an *explicit procedure for calculating the $\hbar$-expansion of the Weyl star-product*, as well as of the *Wick product*, in geometric terms (via quantum tensors or quantum metrics). We also derive an explicit $\hbar$-expansion for the reproducing measure[10] and find a sequence of first cohomology classes that are potential obstructions to the global existence of the reproducing measure over not simply connected Kähler manifolds.

Part I. Coherent States and Quantum Geometry

§1. Surfaces of revolution and representations of basic nonlinear commutation relations

In the standard case of Lie algebras (that is, of linear commutation relations), the background of the representation theory is based on the simplest low-dimensional examples in which the coadjoint orbits are submanifolds in the classical "Greek" geometry: planes, spheres, cones, hyperboloids. In this section we consider a nonlinear generalization of such simplest Lie algebras and develop the representation theory by using the notion of coherent states. Algebras with nonlinear commutation relations are again associated with "Greek" surfaces and, at the same time, with important physical models. There also are remarkable relationships to general hypergeometric functions and their q-analogs, as well as to quantum geometry; we consider them in §2 and §3.

1.1. Hilbert spaces of antiholomorphic distributions. The following basic nonlinear relations are considered:

(1.1)
$$[\mathbf{C}, \mathbf{B}] = f(\mathbf{A}),$$
$$\mathbf{CA} = \varphi(\mathbf{A})\mathbf{C}, \qquad \mathbf{AB} = \mathbf{B}\varphi(\mathbf{A}),$$

---

[9] Note that these states differ from those obtained in Rawnsley's classical paper [**R**$_1$].

[10] The existence of such a formal expansion was proved in [**Kb**$_2$].

where $\mathbf{A} = (\mathbf{A}_1, \ldots, \mathbf{A}_k)$ is a set of commuting generators

(1.1a) $$[\mathbf{A}_\mu, \mathbf{A}_\nu] = 0, \qquad \mu, \nu = 1, \ldots, k.$$

The second and the third relations in (1.1) are vector-valued, that is, $\mathbf{A}$ and $\varphi(\mathbf{A})$ are vector-valued. The function f in (1.1) is scalar.

Here we do not specify the properties of the functions f and φ_μ; in general, they are not just polynomial but can contain, say, exponential functions. Now, for simplicity, the reader may assume that all f and φ_μ are polynomial.

Suppose that we have a representation of the algebra (1.1) in a Hilbert space. A unit vector $\mathfrak{P}_0$ (from the common domain of all generators) is called a *vacuum vector* if it is an eigenvector of all $\mathbf{A}_\mu$, and the annihilation operator $\mathbf{C}$ vanishes at $\mathfrak{P}_0$, that is,

(1.2) $$\mathbf{A}\mathfrak{P}_0 = a\mathfrak{P}_0, \qquad \mathbf{C}\mathfrak{P}_0 = 0,$$
$$a = (a_1, \ldots, a_k), \qquad \|\mathfrak{P}_0\| = 1.$$

We consider only *Hermitian representations*

$$\mathbf{B}^* = \mathbf{C}, \qquad \mathbf{A}_\mu^* = \mathbf{A}_\mu, \qquad \mu = 1, \ldots, k.$$

Thus, the vector a in (1.2) is real, $a \in \mathbb{R}^k$, and the structural functions f and φ in (1.1) are also real, $f \colon \mathbb{R}^k \to \mathbb{R}$, $\varphi \colon \mathbb{R}^k \to \mathbb{R}^k$. Let us introduce the basic function

(1.3) $$\mathcal{F}_a \colon \mathbb{Z}_+ \to \mathbb{R}, \qquad \mathcal{F}_a(n) \stackrel{\text{def}}{=} \frac{1}{n+1} \sum_{j=0}^n f(\mathcal{A}_a(j)),$$

where

(1.4) $$\mathcal{A}_a(n) \stackrel{\text{def}}{=} \underbrace{\varphi(\ldots (\varphi(\varphi(a))))}_{n} \qquad \text{is a composition of } n \text{ mappings } \varphi.$$

Define a subset $\mathcal{R} \subset \mathbb{R}^k$ as follows: a vector a belongs to $\mathcal{R}$ if

$$\mathcal{F}_a(n) > 0 \qquad \text{for all } n \in \mathbb{Z}_+,$$

or if there exists an integer $N \in \mathbb{Z}_+$ such that

(1.5) $$\mathcal{F}_a(n) > 0 \quad \text{for} \quad 0 \leq n < N, \qquad \text{and} \qquad \mathcal{F}_a(N) = 0.$$

In the first case (if $\mathcal{F}_a$ is positive everywhere on $\mathbb{Z}_+$), we set $N = \infty$.

Note that if $f(a) = 0$, then $a \in \mathcal{R}$ and $N = 0$, and if $f(a) < 0$, then $a \notin \mathcal{R}$. Also note that the number $N \leq \infty$ depends on a. So, sometimes it would be better to use a more precise notation $N = N_a$.

For each $a \in \mathcal{R}$, we split the function $\mathcal{F}_a$ into two factors (real or complex):

(1.6) $$\mathcal{F}_a(n) = \mathcal{B}_a(n)\mathcal{C}_a(n), \qquad n \geq 0,$$

with the single condition

(1.6a) $$\mathcal{B}_a(N) = 0$$

in the case of (1.5). Note that in this case, both factors $\mathcal{B}_a(n)$ and $\mathcal{C}_a(n)$ do not vanish for $0 \leq n \leq N - 1$.

Then we define positive numbers

(1.7)
$$s_0(a) = 1,$$
$$s_n(a) = \frac{n!\,\mathcal{F}_a(n-1)\ldots\mathcal{F}_a(0)}{|\mathcal{B}_a(n-1)|^2\ldots|\mathcal{B}_a(0)|^2} \quad \text{for} \quad 1 \leq n \leq N,$$
$$s_n(a) = \infty \quad \text{for} \quad n \geq N+1,$$

where $N = N_a$ (if it exists) is taken from (1.5).

Using the numbers $s_n(a)$, we introduce the Hilbert space of distributions over $\mathbb{R}^2$. These distributions are antiholomorphic, that is, they are annihilated by the operator $\partial/\partial z$, where z is the standard complex coordinate on the plane $\mathbb{R}^2$. Namely, let us consider distributions g over $\mathbb{R}^2$ that can be represented by power series

$$g(\bar{z}) = \sum_{n=0}^{\infty} g_n \bar{z}^n.$$

Here $\bar{z}$ denotes the conjugate complex coordinate. By $\mathcal{P}_{s(a)}$ we denote the space of all such series that satisfy the condition

(1.8)
$$\|g\|_{\mathcal{P}_{s(a)}} \stackrel{\text{def}}{=} \left(\sum_{n=0}^{\infty} s_n(a) |g_n|^2 \right)^{1/2} < \infty.$$

The inner product in $\mathcal{P}_{s(a)}$ is defined by

(1.8a)
$$(g', g)_{\mathcal{P}_{s(a)}} \stackrel{\text{def}}{=} \sum_{n=0}^{\infty} s_n(a) \overline{g'_n} g_n.$$

In the case of (1.5), the space $\mathcal{P}_{s(a)}$ is finite-dimensional and consists of all polynomials of degree $\leq N$. In any case, the set of polynomials is dense in $\mathcal{P}_{s(a)}$, the unit function 1 belongs to $\mathcal{P}_{s(a)}$, and the basis of monomials $\{\bar{z}^n\}$ is generated by applying to 1 powers of the multiplication operator that acts in $\mathcal{P}_{s(a)}$ as follows:

(1.9)
$$\bar{z} \colon \mathcal{P}_{s(a)} \to \mathcal{P}_{s(a)}, \quad \bar{z}\left(\sum_{n=0}^{N} g_n \bar{z}^n \right) \stackrel{\text{def}}{=} \sum_{n=1}^{N} g_{n-1} \bar{z}^n.$$

Here $N \leq \infty$, and N is finite in the case of (1.5).

Obviously, different factorizations (1.6) generate isomorphic Hilbert spaces; namely, if $\mathcal{P}_{s(a)}$ and $\mathcal{P}_{s'(a)}$ are such two spaces, then an isomorphism between them is given by the mapping

(1.10)
$$\sum_n g_n \bar{z}^n \to \sum_n g'_n \bar{z}^n, \quad g'_n = \sqrt{\frac{s_n(a)}{s'_n(a)}} g_n.$$

Now let H be an abstract Hilbert space, let $\mathcal{P}$ be a Hilbert space of antiholomorphic distributions, and let there exist a monomorphism $j \colon \mathcal{P} \to H$. The "integral kernel" $\mathfrak{P}$ of this monomorphism is an H-valued holomorphic distribution such that

$$\bigl(\mathfrak{p}, j(g)\bigr)_H = \bigl((\mathfrak{P}, \mathfrak{p})_H, g\bigr)_{\mathcal{P}} \quad \forall g \in \mathcal{P}, \quad \mathfrak{p} \in H.$$

DEFINITION 1.1. *The monomorphism* $\mathcal{P} \to H$ *is called a* coherent transform, *and its integral kernel* $\mathfrak{P} \in H \otimes \overline{\mathcal{P}}$ *is called a* family of generalized coherent states.

The coherent transform is a unitary map; its inverse is given by
$$j^{-1}(\mathfrak{p}) = (\mathfrak{P}, \mathfrak{p})_H, \qquad \forall \mathfrak{p} \in H.$$

If $\mathcal{P} = \mathcal{P}_s$, then the family $\mathfrak{P} = \{\mathfrak{P}_z\}$ can be represented as a power series in z: $\mathfrak{P}_z = \sum_{n \geq 0} \frac{z^n}{\sqrt{s_n}} e_n$ with respect to any orthonormal system $e_n \in H$, $e_0 = j(1)$.

Each representation of the algebra (1.1) with a vacuum vector $\mathfrak{P}_0$ naturally generates a system $\{e_n\}$ that is the system of eigenvectors for the operators $\mathbf{A}_1, \ldots, \mathbf{A}_k$ with $e_0 = \mathfrak{P}_0$. This system can be used for constructing the coherent transform (that is, the intertwining map) and realizing the representation of the algebra (1.1) in the space $\mathcal{P} = \mathcal{P}_{s(a)}$.

1.2. Irreducible representations and coherent states. Now we describe irreducible representations of the algebra (1.1) and objects related to them.

THEOREM 1.1. (a) *There is a one-to-one correspondence between the set $\mathcal{R}$ and the set of irreducible Hermitian representations of the algebra* (1.1) *that possess a vacuum vector* (1.2). *A representation corresponding to* $a \in \mathcal{R}$ *is finite-dimensional if and only if there exists an integer* $N = N_a$ *satisfying* (1.5); *in this case, the number* $N + 1$ *is the dimension of the representation.*

(b) *For each* $a \in \mathcal{R}$ *and for each factorization* (1.6), *the operators*

$$(1.11) \qquad \overset{\circ}{B} = \bar{z}\,\mathcal{B}_a\!\left(\bar{z}\frac{d}{d\bar{z}}\right), \qquad \overset{\circ}{C} = \mathcal{C}_a\!\left(\bar{z}\frac{d}{d\bar{z}}\right)\frac{d}{d\bar{z}}, \qquad \overset{\circ}{A} = \mathcal{A}_a\!\left(\bar{z}\frac{d}{d\bar{z}}\right)$$

represent the algebra (1.1), *and moreover, this representation is irreducible Hermitian and possesses the vacuum* 1 *in the Hilbert space* $\mathcal{P}_{s(a)}$.

(c) *Representations* (1.11) *assigned to different vectors* $a \in \mathcal{R}$ *are not equivalent, but for each chosen* $a \in \mathcal{R}$, *representations assigned to different factorizations* (1.6) *are equivalent under the isomorphism* (1.10).

(d) *An abstract Hermitian representation of the algebra* (1.1) *in a Hilbert space* H_a *with the vacuum vector* $\mathfrak{P}_0 = \mathfrak{P}_0(a)$, *satisfying* (1.2) *for some* $a \in \mathbb{R}^k$, *can be intertwined with the representation* (1.11) *by means of the following generalized coherent states*:

$$(1.12) \qquad \mathfrak{P}_z = \mathfrak{P}_0 + \sum_{n \geq 1} \frac{1}{n!\,\overline{\mathcal{C}}_a(n-1)\ldots\overline{\mathcal{C}}_a(0)}(z\mathbf{B})^n \mathfrak{P}_0.$$

In the case of (1.5), the summation over n in (1.12) automatically stops at the number $n = N$ (since $\mathbf{B}^{N+1}\mathfrak{P}_0 = 0$). The generalized "reproducing kernel," corresponding to the states (1.12),

$$\mathcal{K}_{s(a)}(\overline{w}, z) = (\mathfrak{P}_w, \mathfrak{P}_z)_{H_a}$$

is the kernel of the identity operator in $\mathcal{P}_{s(a)}$ and corresponds to the following distribution over $\mathbb{R}^2 \times \mathbb{R}^2$:

$$(1.13) \qquad \mathcal{K}_{s(a)}(\overline{w}, z) = \sum_{n \geq 0} \frac{(\overline{w}z)^n}{s_n(a)}.$$

This distribution satisfies the equations

(1.14)
$$\mathcal{C}_a\left(\overline{w}\frac{\partial}{\partial\overline{w}}\right)\frac{\partial}{\partial\overline{w}}\mathcal{K}_{s(a)} = z\overline{\mathcal{B}}_a\left(z\frac{\partial}{\partial z}\right)\mathcal{K}_{s(a)},$$
$$\mathcal{A}_a\left(\overline{w}\frac{\partial}{\partial\overline{w}}\right)\mathcal{K}_{s(a)} = \mathcal{A}_a\left(z\frac{\partial}{\partial z}\right)\mathcal{K}_{s(a)}.$$

To prove this theorem, we need the following lemma.

LEMMA 1.1. *Suppose that an Hermitian representation of the algebra* (1.1) *in a Hilbert space H_a possesses the vacuum vector $\mathfrak{P}_0 = \mathfrak{P}_0(a)$* (1.2). *Then*
(a) $\mathbf{CB}^m\mathfrak{P}_0 = m\mathcal{F}_a(m-1)\mathbf{B}^{m-1}\mathfrak{P}_0$, $m \geq 1$;
(b) $\|\mathbf{B}^m\mathfrak{P}_0\|^2_{H_a} = m!\,\mathcal{F}_a(m-1)\ldots\mathcal{F}_a(0)$, $m \geq 1$;
(c) $(\mathbf{B}^m\mathfrak{P}_0, \mathbf{B}^{m'}\mathfrak{P}_0)_{H_a} = 0$ *if* $m \neq m'$.

PROOF. (a) The general quasicommutation formulas (see [**KaM**$_4$, Appendix I]) and the last relation in (1.1) imply

(1.15) $$f(\mathbf{A})\mathbf{B}^n = \mathbf{B}^n f\big(\underbrace{\varphi(\varphi(\ldots\varphi(\mathbf{A})\ldots))}_{n}\big).$$

So, it follows from the first relation in (1.1) that

$$\mathbf{CB}^m = \mathbf{B}^m\mathbf{C} + \sum_{n=0}^{m-1}\mathbf{B}^{m-1-n}[\mathbf{C},\mathbf{B}]\mathbf{B}^n = \mathbf{B}^m\mathbf{C} + \sum_{n=0}^{m-1}\mathbf{B}^{m-1-n}f(\mathbf{A})\mathbf{B}^n$$
$$= \mathbf{B}^m\mathbf{C} + \mathbf{B}^{m-1}\sum_{n=0}^{m-1}f\big(\underbrace{\varphi(\varphi(\ldots\varphi(\mathbf{A})\ldots))}_{n}\big).$$

Applying this equality to the vacuum vector $\mathfrak{P}_0$ and taking the equations (1.2) into account, we obtain
$$\mathbf{CB}^m\mathfrak{P}_0 = \mathbf{B}^{m-1}\sum_{n=0}^{m-1}f\big(\mathcal{A}_a(n)\big)\mathfrak{P}_0,$$
which coincides with equality (a) in the lemma.

(b) Let us apply (a) step by step m times:
$$(\mathbf{B}^m\mathfrak{P}_0, \mathbf{B}^m\mathfrak{P}_0)_{H_a} = (\mathbf{CB}^m\mathfrak{P}_0, \mathbf{B}^{m-1}\mathfrak{P}_0)_{H_a}$$
$$= m\mathcal{F}_a(m-1)(\mathbf{B}^{m-1}\mathfrak{P}_0, \mathbf{B}^{m-1}\mathfrak{P}_0)_{H_a}$$
$$= \cdots = \big(m\cdot\mathcal{F}_a(m-1)\big)\ldots\big(1\cdot\mathcal{F}_a(0)\big)(\mathfrak{P}_0,\mathfrak{P}_0)_{H_a}.$$

This is equality (b) in the lemma.

(c) Let $m' > m$. Then
$$(\mathbf{B}^m\mathfrak{P}_0, \mathbf{B}^{m'}\mathfrak{P}_0)_{H_a} = (\mathbf{C}^{m+1}\mathbf{B}^m\mathfrak{P}_0, \mathbf{B}^{m'-m-1}\mathfrak{P}_0)_{H_a}.$$

Thus, it suffices to prove the equality

(1.16) $$\mathbf{C}^{m+1}\mathbf{B}^m\mathfrak{P}_0 = 0.$$

For $m = 0$ this follows from (1.2). Suppose that (1.16) holds for all numbers up to $m-1$, in particular, $\mathbf{C}^m\mathbf{B}^{m-1}\mathfrak{P}_0 = 0$. Then (a) implies
$$\mathbf{C}^{m+1}\mathbf{B}^m\mathfrak{P}_0 = \mathbf{C}^m\mathbf{CB}^m\mathfrak{P}_0 = m\mathcal{F}_a(m-1)\mathbf{C}^m\mathbf{B}^{m-1}\mathfrak{P}_0 = 0.$$

So, (1.16) holds, and equality (c) in Lemma 1.1 is proved. □

PROOF OF THEOREM 1.1. Suppose that a representation of the algebra (1.1) in a Hilbert space H_a is irreducible, Hermitian, and has the vacuum vector $\mathfrak{P}_0 = \mathfrak{P}_0(a)$. It follows from Lemma 1.1(a) and from (1.15) that the span of the vectors $\{\mathbf{B}^m \mathfrak{P}_0 \mid m \in \mathbb{Z}_+\}$ is invariant under the representation, and thus the closure of this span coincides with H_a.

Next, applying equality (b) in Lemma 1.1 several times, we can obtain the following statement.

First we set $m = 1$. Then $\|\mathbf{B}\mathfrak{P}_0\|_{H_a}^2 = \mathcal{F}_a(0) \equiv f(a)$, and so one has the following alternative: either $\mathcal{F}_a(0) > 0$, or $\mathbf{B}\mathfrak{P}_0 = 0$ and $\mathcal{F}_a(0) = 0$. In the latter case, the space
$$H_a = \overline{\text{span}}\{\mathbf{B}^n \mathfrak{P}_0\}$$
is, in fact, one-dimensional and coincides with the linear space generated by $\mathfrak{P}_0$, and we see that (1.5) is satisfied.

If $\mathcal{F}_a(0) > 0$, then we set $m = 2$ and obtain $\|\mathbf{B}^2 \mathfrak{P}_0\|_{H_a}^2 = 2\mathcal{F}_a(1)\mathcal{F}_a(0)$. So, either $\mathcal{F}_a(1) > 0$ or $\mathbf{B}^2 \mathfrak{P}_0 = 0$ and $\mathcal{F}_a(1) = 0$. In the latter case, we obtain (1.5) for $N = 1$, and $\dim H_a = N + 1 = 2$.

If $\mathcal{F}_a(1) > 0$, then we set $m = 3$ in equality (b) in Lemma 1.1, and so on. Thus, we have proved that the function $\mathcal{F}_a$ is positive until it vanishes, that is, $a \in \mathcal{R}$ and $\dim H_a = N + 1$ in the case of (1.5). This is a part of statement (a) in Theorem 1.1.

Now let us take some $a \in \mathcal{R}$ and construct operators (1.11). The rest of statement (a), as well as statement (b) in Theorem 1.1, is a consequence of the following lemma.

LEMMA 1.2. *Let $\mathcal{A}$, $\mathcal{B}$, and $\mathcal{C}$ be functions over $\mathbb{Z}_+$. Then the operators*

$$(1.17) \qquad \overset{\circ}{B} = \bar{z}\mathcal{B}\Big(\bar{z}\frac{d}{d\bar{z}}\Big), \qquad \overset{\circ}{C} = \mathcal{C}\Big(\bar{z}\frac{d}{d\bar{z}}\Big)\frac{d}{d\bar{z}}, \qquad \overset{\circ}{A} = \mathcal{A}\Big(\bar{z}\frac{d}{d\bar{z}}\Big)$$

are well defined on all polynomials and, in particular, on the unit function 1. These operators generate an irreducible Hermitian representation of the algebra (1.1) in a Hilbert space of distributions with the vacuum 1 if and only if the functions $\mathcal{A}$, $\mathcal{B}$, and $\mathcal{C}$ satisfy the following conditions:
- *the vector $a = \mathcal{A}(0)$ belongs to the set $\mathcal{R}$, that is, $a \in \mathcal{R}$;*
- *$\mathcal{A}$ coincides with the functions $\mathcal{A}_a$ given by (1.4);*
- *the product $\mathcal{BC}$ coincides with the function $\mathcal{F}_a$ given by (1.3);*
- *the function $\mathcal{B}$ satisfies equality (1.6a).*

In this case, the Hilbert space of the representation automatically coincides with $\mathcal{P}_{s(a)}$, where the sequence $s(a)$ is given by (1.7).

PROOF. Let $\mathcal{B}(n) \neq 0$ for $0 \leq n \leq N - 1$. Then the formula $\overset{\circ}{B}(\bar{z}^n) = \mathcal{B}(n)\bar{z}^{n+1}$ implies that all monomials $\bar{z}^n$ ($0 \leq n \leq N$) must belong to any space that is invariant with respect to operators (1.17) and contains the function 1. On the other hand, if $\mathcal{B}(N) = 0$ for some $N \geq 0$, then the span of $\{\bar{z}^n \mid 0 \leq n \leq N\}$ is such an invariant space.

Now let us suppose that $N \geq 1$ and take some n, $0 \leq n \leq N - 1$. The formula $\overset{\circ}{C}(\bar{z}^{n+1}) = (n+1)\mathcal{C}(n)\bar{z}^n$ implies

$$\mathcal{B}(n)(\bar{z}^{n+1}, \bar{z}^{n+1}) = \big(\bar{z}^{n+1}, \overset{\circ}{B}(\bar{z}^n)\big) = \big(\overset{\circ}{C}(\bar{z}^{n+1}), \bar{z}^n\big) = (n+1)\overline{\mathcal{C}}(n)(\bar{z}^n, \bar{z}^n).$$

Here the brackets $(\cdot, \cdot)$ denote an inner product in the Hilbert space of distributions. If we denote $s_n = (\bar{z}^n, \bar{z}^n)$ then the last formula reads

$$\mathcal{B}(n) s_{n+1} = (n+1)\overline{\mathcal{C}}(n) s_n$$

or

(1.18) $$s_{n+1} = \frac{(n+1)(\mathcal{BC})(n)}{|\mathcal{B}(n)|^2} s_n.$$

Since $\{s_n\}$ is a positive sequence, we have

(1.19) $$(\mathcal{BC})(n) > 0 \quad \text{for} \quad 0 \leq n \leq N-1.$$

Now let us try to substitute the operators (1.17) into the relations (1.1). We derive

$$\overset{\circ}{\mathcal{C}}\overset{\circ}{\mathcal{B}} = \mathcal{C}\left(\bar{z}\frac{d}{d\bar{z}}\right)\left(\bar{z}\frac{d}{d\bar{z}}+1\right)\mathcal{B}\left(\bar{z}\frac{d}{d\bar{z}}\right),$$

$$\overset{\circ}{\mathcal{B}}\overset{\circ}{\mathcal{C}} = \mathcal{B}\left(\bar{z}\frac{d}{d\bar{z}}-1\right)\bar{z}\frac{d}{d\bar{z}}\mathcal{C}\left(\bar{z}\frac{d}{d\bar{z}}-1\right),$$

$$\overset{\circ}{\mathcal{C}}\overset{\circ}{\mathcal{A}} = \mathcal{C}\left(\bar{z}\frac{d}{d\bar{z}}\right)\mathcal{A}\left(\bar{z}\frac{d}{d\bar{z}}+1\right)\frac{d}{d\bar{z}},$$

$$\overset{\circ}{\mathcal{A}}\overset{\circ}{\mathcal{B}} = \bar{z}\,\mathcal{A}\left(\bar{z}\frac{d}{d\bar{z}}+1\right)\mathcal{B}\left(\bar{z}\frac{d}{d\bar{z}}\right).$$

Hence, the first relation in (1.1) is equivalent to

(1.20) $$(n+1)(\mathcal{BC})(n) - n(\mathcal{BC})(n-1) = f\bigl(\mathcal{A}(n)\bigr),$$

while the second and the third relations are equivalent to

$$\mathcal{C}(n)\bigl(\mathcal{A}(n+1) - \varphi(\mathcal{A}(n))\bigr) = 0, \qquad \mathcal{B}(n)\bigl(\mathcal{A}(n+1) - \varphi(\mathcal{A}(n))\bigr) = 0$$

for any $0 \leq n \leq N$. Note that for $n=0$, (1.20) reads $(\mathcal{BC})(0) = f\bigl(\mathcal{A}(0)\bigr)$.

It follows from (1.19) that $\mathcal{B}(n)$ and $\mathcal{C}(n)$ are nonzero for $0 \leq n \leq N-1$. So, the equalities

(1.21) $$\mathcal{A}(n+1) = \varphi\bigl(\mathcal{A}(n)\bigr), \qquad n=0,\ldots, N-1,$$

are equivalent to the second and the third relations in (1.1). It follows from (1.20) and (1.21) that the conditions

$$\mathcal{BC} = \mathcal{F}_a, \qquad \mathcal{A} = \mathcal{A}_a, \qquad \text{where} \quad a = \mathcal{A}(0),$$

are necessary and sufficient for the operators (1.17) to satisfy (1.1). Then, (1.19) implies that $a \in \mathcal{R}$, and (1.18) implies $s = s_n(a)$. Lemma 1.2 is proved. $\square$

Statement (c) in Theorem 1.1 can be proved as follows. First, we note that the operator $\overset{\circ}{C}$ (1.11) annihilates only a constant function. Indeed, if $\overset{\circ}{C}$ annihilates a function $g(\bar{z}) = \sum_{0 \leq n \leq N} g_n \bar{z}^n$, then $\sum_{n=0}^{N-1}(n+1)\mathcal{C}_a(n) g_{n+1} \bar{z}^n \equiv 0$, and therefore $\mathcal{C}_a(n) g_{n+1} = 0$ for $0 \leq n \leq N-1$. But $\mathcal{C}_a(n) \neq 0$ for $\leq n \leq N-1$, and therefore $g_{n+1} = 0$. Thus, $g(\bar{z}) = g_0 \equiv \text{const}$.

If $a, a' \in \mathcal{R}$ and the representations (1.11) corresponding to a and a' are equivalent, then there is an isomorphism $U\colon \mathcal{P}_{s(a)} \to \mathcal{P}_{s(a')}$ such that
$$\overset{\circ}{C}_{a'} = U^{-1}\overset{\circ}{C}_a U, \qquad \overset{\circ}{A}_{a'} = U^{-1}\overset{\circ}{A}_a U.$$
Since $\overset{\circ}{C}_{a'} 1 = 0$, we have $\overset{\circ}{C}_a U(1) = 0$, and therefore $U(1) = \text{const}$. Thus,
$$a' = \overset{\circ}{A}_{a'} 1 = U^{-1}\overset{\circ}{A}_a U(1) = \overset{\circ}{A}_a 1 = a.$$

Now let us prove (d). The intertwining map $j\colon \mathcal{P}_{s(a)} \to H_a$ and its inverse (that is, the coherent transform) can be performed as follows:

(1.22) $\qquad j\colon g \mapsto \mathfrak{p}, \qquad (\mathfrak{p}', \mathfrak{p})_{H_a} = \big((\mathfrak{P}, \mathfrak{p}')_{H_a}, g\big)_{\mathcal{P}_{s(a)}}, \qquad \forall \mathfrak{p}' \in H_a,$

(1.23) $\qquad j^{-1}\colon \mathfrak{p} \mapsto g, \qquad g = (\mathfrak{P}, \mathfrak{p})_{H_a}.$

Here g is a distribution in the space $\mathcal{P}_{s(a)}$ and $\mathfrak{p}$ is a vector in H_a; the coherent state $\mathfrak{P} = \mathfrak{P}_z$ is a power series in z with coefficients from the Hilbert space H_a:

(1.24) $$\mathfrak{P}_z = \sum_{0 \le m,n \le N} \gamma_{m,n} z^m \mathbf{B}^n \mathfrak{P}_0.$$

Here $N \le \infty$ and N is finite in the case of (1.5). The intertwining property means that the distributions $\overset{\circ}{B}g$, $\overset{\circ}{C}g$, $\overset{\circ}{A}g$ with $\overset{\circ}{B}$, $\overset{\circ}{C}$, and $\overset{\circ}{A}$ given by (1.11), correspond via (1.22) to the vectors $\mathbf{B}\mathfrak{p}$, $\mathbf{C}\mathfrak{p}$, $\mathbf{A}\mathfrak{p}$, where $\mathbf{B}$, $\mathbf{C}$, and $\mathbf{A}$ are the generators of the representation of the algebra (1.1) in the space H_a.

The correspondence $\overset{\circ}{B}g \to \mathbf{B}\mathfrak{p}$ is equivalent to the chain of equations

(1.25) $\qquad \gamma_{m+1,n+1}(m+1)\overline{\mathcal{C}}_a(m) = \gamma_{m,n}, \qquad 0 \le m, n \le N-1,$

with the "boundary conditions"

(1.25a) $\qquad \begin{cases} \gamma_{m,0} = 0, & 1 \le m \le N, \\ \gamma_{N,n} = 0, & 0 \le n \le N-1. \end{cases}$

The correspondence $\overset{\circ}{C}g \to \mathbf{C}\mathfrak{p}$ is equivalent to the chain of equations

(1.25b) $\qquad \gamma_{m+1,n+1}(n+1)\mathcal{F}_a(n) = \gamma_{m,n}\overline{\mathcal{B}}_a(m), \qquad 0 \le m,n \le N-1,$

with the "boundary conditions"

(1.25c) $\qquad \begin{cases} \gamma_{0,n} = 0, & 1 \le n \le N, \\ \gamma_{m,N} = 0, & 0 \le m \le N-1. \end{cases}$

In both these equivalences, we have used the orthogonality
$$(\overline{z}^m, \overline{z}^n)_{\mathcal{P}_{s(a)}} = 0, \qquad m \ne n,$$
which follows from Lemma 1.1(c) (applied to the Hermitian representation (1.11) in the space $\mathcal{P}_{s(a)}$ with the vacuum vector 1). Lemma 1.1(a) was also used in the second equivalence.

The last correspondence $\overset{\circ}{A}g \to \mathbf{A}\mathfrak{p}$ is equivalent to

(1.26) $\qquad \big(\mathcal{A}(m) - \mathcal{A}(n)\big)\gamma_{m,n} = 0, \qquad 0 \le m, n \le N,$

where we used (1.15).

The correspondence of vacuum vectors $1 \to \mathfrak{P}_0$ implies

(1.25d)
$$\gamma_{0,0} = 1.$$

It follows from the "boundary conditions" (1.25a), (1.25c) and from equations (1.25), (1.25b) that
$$\gamma_{m,n} = 0 \quad \text{for } m \neq n.$$

In particular, conditions (1.26) are satisfied, and equations (1.25) and (1.25b) are equivalent. By (1.25) and (1.25d), we have

(1.27) $$\gamma_{m,n} = \frac{1}{n!\,\overline{\mathcal{C}}_a(n-1)\ldots\overline{\mathcal{C}}_a(0)}, \qquad 1 \leq n \leq N,$$

on the diagonal $m = n$.

Thus, the coherent state (1.24) looks like (1.12).

One can also easily verify that the map (1.23) is inverse to (1.22). Indeed, (1.22) is given by the relation

$$g = \sum_{n=0}^{N} g_n \overline{z}^n \to \mathfrak{p} = \sum_{n=0}^{N} g_n s_n(a) \gamma_{n,n} \mathbf{B}^n \mathfrak{P}_0,$$

and (1.23) is given by the relation

$$\mathfrak{p} \to g = \sum_{n=0}^{N} (\mathbf{B}^n \mathfrak{P}_0, \mathfrak{p}) \overline{\gamma}_{n,n} \overline{z}^n.$$

These maps are mutually inverse if and only if

$$|\gamma_{n,n}|^2 s_n(a) = \|\mathbf{B}^n \mathfrak{P}_0\|^{-2}, \qquad 0 \leq n \leq N.$$

Lemma 1.1(b) and formulas (1.7) and (1.27) prove these identities, and hence, equality (1.13). Equations (1.14) follow from the intertwining property of the coherent state (1.12). Theorem 1.1 is thereby proved. $\square$

COROLLARY 1.1. *The orthonormal bases*
$$\left\{ \frac{\overline{z}^n}{\sqrt{s_n(a)}} \,\Big|\, n \geq 0 \right\} \qquad in \quad \mathcal{P}_{s(a)}$$

and
$$\{\mathfrak{P}_0\} \cup \left\{ \frac{1}{\sqrt{s_n(a)}\mathcal{B}_a(n-1)\ldots\mathcal{B}_a(0)} \mathbf{B}^n \mathfrak{P}_0 \,\Big|\, n \geq 1 \right\} \qquad in \quad H_a$$

correspond to each other under the coherent transform (1.22), (1.23). *They are eigenbases for the commuting operators* $\overset{\circ}{A}_\mu$ (*in the space* $\mathcal{P}_{s(a)}$) *or* $\mathbf{A}_\mu$ (*in the space* H_a), $\mu = 1, \ldots, k$; *the corresponding eigenvalues are equal to* $\mathcal{A}_a(n)_\mu$, $0 \leq n \leq N$.

1.3. Casimir elements and regular relations. Now let us more carefully consider the influence of properties of the functions f and φ from (1.1) on irreducible representations of this algebra. A variety of particular cases of the relations (1.1) have been considered in [**M, Ka**$_{2-4}$**, KaM**$_{1,3}$**, Ro, ChEK, W, Das, DQ, Ja**] and elsewhere.

LEMMA 1.3. (a) *If a function $\varkappa$ on $\mathbb{R}^k$ is preserved by the mapping $\varphi\colon \mathbb{R}^k \to \mathbb{R}^k$, then $\varkappa(\mathbf{A})$ is an element of the center of the algebra* (1.1) *(that is, a Casimir element).*

(b) *If a function ρ on $\mathbb{R}^k$ satisfies the equation*

$$\rho(\varphi(A)) - \rho(A) = f(A), \qquad A \in \mathbb{R}^k, \tag{1.28}$$

then

$$\mathbf{K} = \mathbf{BC} - \rho(\mathbf{A}) \tag{1.29}$$

is a Casimir element of algebra (1.1).

PROOF. (a) In view of (1.1),

$$\varkappa(\mathbf{A})\mathbf{B} = \mathbf{B}\varkappa(\varphi(\mathbf{A})), \qquad \mathbf{C}\varkappa(\mathbf{A}) = \varkappa(\varphi(\mathbf{A}))\mathbf{C}. \tag{1.30}$$

Thus, if $\varkappa(\varphi(A)) = \varkappa(A)$, then $\varkappa(\mathbf{A})$ commutes with $\mathbf{B}$ and $\mathbf{C}$.

(b) Similarly to (1.30), we have $\mathbf{BCA} = \mathbf{B}\varphi(\mathbf{A})\mathbf{C} = \mathbf{ABC}$. So, the element $\mathbf{K}$ (1.29) commutes with $\mathbf{A}$. It follows from (1.1) that

$$[\mathbf{K},\mathbf{B}] = \mathbf{B}[\mathbf{C},\mathbf{B}] - [\rho(\mathbf{A}),\mathbf{B}] = \mathbf{B}f(\mathbf{A}) - \rho(\mathbf{A})\mathbf{B} + \mathbf{B}\rho(\mathbf{A})$$
$$= \mathbf{B}\big(f(\mathbf{A}) - \rho(\varphi(\mathbf{A})) + \rho(\mathbf{A})\big).$$

Thus, under condition (1.28), we have $[\mathbf{K},\mathbf{B}] = 0$. Similarly, $[\mathbf{K},\mathbf{C}] = 0$. The lemma is proved. $\square$

LEMMA 1.4. (a) *Suppose that the mapping φ is the shift by unit time along trajectories of a vector field $v = \sum_{\mu=1}^{k} v_\mu(A)\frac{\partial}{\partial A_\mu}$ on $\mathbb{R}^k$. Moreover, suppose that these trajectories fiber $\mathbb{R}^k$ (or fiber a certain φ-invariant domain that includes the set $\mathcal{R}$). Then there are independent real smooth functions $\varkappa_1,\ldots,\varkappa_{k-1}$ on $\mathbb{R}^k$ that are preserved by φ, and therefore, there are independent Casimir elements $\varkappa_1(\mathbf{A}),\ldots,\varkappa_{k-1}(\mathbf{A})$ of the algebra* (1.1).

(b) *In addition to* (a), *suppose that f is a smooth tempered function on $\mathbb{R}^k$, the field v is also tempered, and the Jacobi matrix Dv is nilpotent on the entire $\mathbb{R}^k$. Then there exists a real smooth solution ρ of equation* (1.28), *and therefore, there is a Casimir element of type* (1.29).

PROOF. (a) The functions $\varkappa_1,\ldots,\varkappa_{k-1}$ are just functions on the base of the fibration by the trajectories.

(b) Equation (1.28) along the trajectories reads

$$\rho(\tau+1) - \rho(\tau) = f(\tau). \tag{1.28a}$$

Here τ is the time along the trajectories, and $f(\tau)$ is the value of f on a trajectory. The conditions imposed on the field v imply that the growth of a trajectory as $\tau \to \infty$ is at most polynomial. So, $f(\tau)$ is tempered with respect to τ, and the solution of (1.28a) can be calculated as follows:

$$\rho(\tau) = \sum_{m=-\infty}^{\infty} \int \frac{e^{i\omega\tau} - e^{i\,2\pi m\tau}}{e^{i\omega} - 1} e_m(\omega)\,d\mu(\omega) \qquad \text{if} \quad f(\tau) = \int e^{i\omega\tau}\,d\mu(\omega).$$

Here $\{e_m\}$ is a partition of unity such that $e_m \in C_0^\infty(\mathbb{R})$ and $e_m(2\pi s) = \delta_{m,s}$ (the Kronecker symbol) for any $m,s \in \mathbb{Z}$. The lemma is proved. $\square$

DEFINITION 1.2. The permutation relations (1.1) are said to be *regular* if on $\mathbb{R}^k$ there are independent real smooth φ-invariant functions $\varkappa_1, \ldots, \varkappa_{k-1}$, as well as a real smooth solution ρ of the difference equation (1.28).

Lemma 1.4 provides sufficient conditions for the regularity of relations (1.1).

Note that if we are interested in Hermitian representations of (1.1) that possess a vacuum vector, then it suffices to know the function ρ only on the spectrum of the operators $\mathbf{A}_\mu$, that is, on the subset $\mathcal{R}^+ = \{\mathcal{A}_a(n) \mid a \in \mathcal{R}, 0 \leq n \leq N\}$. But from (1.28) we obtain

$$(1.28b) \qquad \rho\big(\mathcal{A}_a(n)\big) = \rho(a) + \sum_{j=0}^{n-1} f\big(\mathcal{A}_a(j)\big), \qquad n \geq 1.$$

By $\mathcal{R}_0$ we denote the subset of those $a \in \mathcal{R}^+$ that cannot be represented as $a = \mathcal{A}_{a'}(m)$ for some $a' \in \mathcal{R}$, $m \geq 1$. Then one can arbitrarily define the function ρ on the subset $\mathcal{R}_0$ and extend this function on $\mathcal{R}^+$ using (1.28b). The same is true of the functions $\varkappa_j$. One can arbitrarily define them on $\mathcal{R}_0$ and extend to $\mathcal{R}^+$ using $\varkappa_j\big(\mathcal{A}_a(n)\big) = \varkappa_j(a)$, $a \in \mathcal{R}_0$. In this sense, there is no problem of the existence of Casimir elements for (1.1). In Definition 1.2, the crucial assumption is the smoothness of the functions $\varkappa_j$ and ρ.

Now, using the Hermitian generators

$$(1.31) \qquad \mathbf{S}_1 = \frac{1}{2}(\mathbf{B} + \mathbf{C}), \qquad \mathbf{S}_2 = \frac{i}{2}(\mathbf{B} - \mathbf{C}),$$

let us write the Casimir element $\mathbf{K}$ as follows:

$$\mathbf{K} = \mathbf{S}_1^2 + \mathbf{S}_2^2 - \rho(\mathbf{A}) - \frac{1}{2}f(\mathbf{A}) = \mathbf{S}_1^2 + \mathbf{S}_2^2 - \frac{1}{2}\big(\rho(\mathbf{A}) + \rho(\varphi(\mathbf{A}))\big).$$

In any irreducible representation, the element $\mathbf{K}$ is a scalar. If there is a vacuum vector (1.2), this scalar has the form

$$(1.32) \qquad \mathbf{K} = -\rho(a) \cdot \mathbf{I}, \qquad a \in \mathcal{R},$$

and therefore,

$$(1.32a) \qquad \mathbf{BC} = \rho(\mathbf{A}) - \rho(a) \cdot \mathbf{I}$$

or

$$(1.32b) \qquad \mathbf{S}_1^2 + \mathbf{S}_2^2 = \frac{1}{2}\big(\rho(\mathbf{A}) + \rho(\varphi(\mathbf{A}))\big) - \rho(a) \cdot \mathbf{I}.$$

The last formula together with

$$(1.33) \qquad \varkappa_j(\mathbf{A}) = \varkappa_j(a) \cdot \mathbf{I}, \qquad j = 1, \ldots, k-1,$$

can be read as equations for a symplectic leaf corresponding to the irreducible representation of (1.1) assigned to $a \in \mathcal{R}$. This leaf is a surface embedded into the space $\mathbb{R}^{k+2}$ with the classical coordinates $S_1, S_2, A_1, \ldots, A_k$. The regularity of relations (1.1) precisely means that such an embedding exists.

In our case, in view of (1.32b), each leaf is a *surface of revolution*, that is, it is invariant under rotations in the plane (S_1, S_2). Different choices of the function ρ determine a variety of revolution surfaces: spheres, cones, hyperboloids, etc.

So, the algebras (1.1) (the enveloping algebras) can be considered as quantizations of such families of surfaces of revolution.

It is also useful to reformulate the construction of irreducible representations of algebra (1.1) in terms of the function ρ.

In view of (1.28), the basic object $\mathcal{F}_a$ (1.3), which we have already used, can be derived as

$$(1.34) \qquad \mathcal{F}_a(n) = \frac{1}{n+1}\Big(\rho\big(\mathcal{A}_a(n+1)\big) - \rho(a)\Big).$$

Thus, one has the following statement.

LEMMA 1.5. *The condition $a \in \mathcal{R}$ is equivalent to the following one: either*

$$(1.35) \qquad \rho\big(\mathcal{A}_a(n)\big) > \rho(a) \qquad \text{for any} \quad n \in \mathbb{Z}_+, \quad n \geq 1,$$

or there exists an integer $N \in \mathbb{Z}_+$ such that

$$(1.35a) \qquad \begin{cases} \rho\big(\mathcal{A}_a(n)\big) > \rho(a) & \text{for} \quad 1 \leq n \leq N, \\ \rho\big(\mathcal{A}_a(N+1)\big) = \rho(a). \end{cases}$$

The latter situation corresponds to the case of (1.5).

Instead of the factorization (1.6), one can consider a slightly different procedure. For $a \in \mathcal{R}$ and for any $A \in \mathbb{R}^k$, let us decompose the difference

$$(1.36) \qquad \rho(A) - \rho(a) = \mathcal{D}_a(A)\,\mathcal{E}_a(A)$$

into two smooth factors $\mathcal{D}_a$ and $\mathcal{E}_a$, and suppose that

$$(1.37) \qquad \mathcal{E}_a(a) = 0,$$

as well as

$$(1.38) \qquad \mathcal{D}_a\big(\mathcal{A}_a(N+1)\big) = 0$$

(the latter equality corresponds to (1.35a)).

Let us choose

$$(1.39) \qquad \mathcal{B}_a(n) = \mathcal{D}_a\big(\mathcal{A}_a(n+1)\big), \qquad \mathcal{C}_a(n) = \frac{1}{n+1}\mathcal{E}_a\big(\mathcal{A}_a(n+1)\big).$$

Then, in view of (1.36), (1.34), and (1.38), conditions (1.6), (1.6a) hold, and we arrive at the following statement.

THEOREM 1.2. *Suppose that the function ρ is assigned to f by (1.28), and $a \in \mathcal{R}$. Then:*

• *Each decomposition (1.36) generates a Hilbert space of antiholomorphic distributions $g(\bar{z}) = \sum_{n=0}^{N} g_n \bar{z}^n$ with the inner product*

$$(1.40) \qquad (g', g) \stackrel{\text{def}}{=} \bar{g}'_0 g_0 + \sum_{n=1}^{N} X_a\big(\mathcal{A}_a(n)\big) \ldots X_a\big(\mathcal{A}_a(1)\big) \bar{g}'_n g_n,$$

where

$$(1.41) \qquad \begin{aligned} X_a(a) &\stackrel{\text{def}}{=} 0, \\ X_a(A) &\stackrel{\text{def}}{=} \mathcal{D}_a(A)^{-1} \cdot \overline{\mathcal{E}_a(A)} \qquad \text{while} \quad \rho(A) \neq \rho(a), \end{aligned}$$

and the number $N \in \mathbb{Z}_+ \cup \{\infty\}$ is taken from (1.35a).

- *In this Hilbert space the irreducible Hermitian representation of the algebra (1.1) is given by the operators*

(1.42) $$\overset{\circ}{\mathbf{B}} = \mathcal{D}_a(\overset{\circ}{A}) \cdot \overline{z}, \qquad \overset{\circ}{\mathbf{C}} = \frac{1}{\overline{z}} \cdot \mathcal{E}_a(\overset{\circ}{A}), \qquad \overset{\circ}{\mathbf{A}} = \mathcal{A}_a\left(\overline{z}\frac{d}{d\overline{z}}\right).$$

- *The corresponding coherent states are the following:*

(1.43)
$$\mathfrak{P}_z = \mathfrak{P}_0 + \sum_{n=1}^{N} \frac{1}{\overline{\mathcal{E}}_a(\mathcal{A}_a(n))\ldots\overline{\mathcal{E}}_a(\mathcal{A}_a(1))}(z\mathbf{B})^n\mathfrak{P}_0$$
$$= \begin{cases} \left[\mathbf{I} - (z\mathbf{B}^+)^{N+1}\right](\mathbf{I} - z\mathbf{B}^+)^{-1}\mathfrak{P}_0, & \text{if } N < \infty, \\ (\mathbf{I} - z\mathbf{B}^+)^{-1}\mathfrak{P}_0, & \text{if } N = \infty, \end{cases}$$

where $\mathfrak{P}_0$ is the vacuum vector (1.2); in the last "resolvent form," the operator $\mathbf{B}^+$ is defined by the formula

(1.44) $$\mathbf{B}^+ \overset{\text{def}}{=} \overline{\mathcal{E}}_a(\mathbf{A})^{-1}\mathbf{B}.$$

- *The reproducing kernel* $\mathcal{K}(\overline{w}, z) = (\mathfrak{P}_w, \mathfrak{P}_z)$ *can also be written in the "resolvent form"*

(1.45)
$$\mathcal{K}(\overline{w}, z) = 1 + \sum_{n \geq 1} \frac{(\overline{w}z)^n}{X_a(\mathcal{A}_a(n))\ldots X_a(\mathcal{A}_a(1))}$$
$$= \left(I - X_a(\overset{\circ}{A})^{-1} \cdot x\right)^{-1} 1(x) \bigg|_{x = \overline{w}z};$$

here $\overset{\circ}{A} = \mathcal{A}_a\left(x\frac{d}{dx}\right)$, and all formulas are written under the assumption that $N = \infty$ (for $N < \infty$, the same version as in (1.43) should be used).

- *The operators of complex structure corresponding to the irreducible representation of the algebra (1.1) are given by the following formulas:*

(1.46) $$\widehat{\overline{z}} = \mathcal{D}_a(\mathbf{A})^{-1}\mathbf{B}, \qquad \widehat{z} \equiv \widehat{\overline{z}}^* = \overline{\mathcal{D}}_a(\varphi(\mathbf{A}))^{-1}\mathbf{C}, \qquad |\widehat{z}|^2 \equiv \widehat{\overline{z}}\widehat{z} = X_a(\mathbf{A}).$$

PROOF. Formula (1.40) for the inner product and the first formula in (1.43) for the coherent states follow from (1.8) and (1.12) if one takes into account that

$$\frac{s_n(a)}{s_{n-1}(a)} \equiv \frac{n\mathcal{F}_a(n-1)}{|\mathcal{B}_a(n-1)|^2} = \frac{\overline{\mathcal{E}}_a(\mathcal{A}_a(n))}{\mathcal{D}_a(\mathcal{A}_a(n))} \equiv X_a(\mathcal{A}_a(n))$$

(see (1.6), (1.39)). In view of (1.15), we also have

$$\frac{1}{\overline{\mathcal{E}}_a(\mathcal{A}_a(n))\ldots\overline{\mathcal{E}}_a(\mathcal{A}_a(1))}(z\mathbf{B})^n\mathfrak{P}_0 = \frac{1}{\overline{\mathcal{E}}_a\big(\underbrace{\varphi(\ldots\varphi(a)\ldots)}_{n}\big)\cdot\ldots\cdot\overline{\mathcal{E}}(\varphi(a))}(z\mathbf{B})^n\mathfrak{P}_0$$
$$= z^n \underbrace{(\overline{\mathcal{E}}_a(\mathbf{A})^{-1}\mathbf{B})\ldots(\overline{\mathcal{E}}_a(\mathbf{A})^{-1}\mathbf{B})}_{n}\mathfrak{P}_0 = (z\mathbf{B}^+)^n\mathfrak{P}_0.$$

So, the first formula in (1.43) reads

$$\mathfrak{P}_z = \sum_{n=0}^{N}(z\mathbf{B}^+)^n\mathfrak{P}_0,$$

which implies the second formula in (1.43).

The resolvent formula (1.45) is a version of (1.43) since $\mathcal{K}(\cdot, z)$ is itself a coherent state (with values in the space of antiholomorphic distributions). In the antiholomorphic representation (1.42), in view of (1.41), the operator (1.44) is given by the relations
$$\overset{\circ}{B}{}^+ = \overline{\mathcal{E}}_a(\overset{\circ}{A})^{-1}\overset{\circ}{B} = \overline{\mathcal{E}}_a(\overset{\circ}{A})^{-1}\mathcal{D}_a(\overset{\circ}{A})\overline{z} = X_a(\overset{\circ}{A})^{-1}\overline{z}.$$

Now let us prove the last statement of the theorem. By definition, the operator of complex structure in the antiholomorphic representation (1.11), (1.42) is just the multiplication operator $\overline{z}$ (1.9).

In the space H_a of an abstract representation, *the operator $\widehat{\overline{z}}$ of the complex structure corresponds to the multiplication operator $\overline{z}$ (1.9) via the coherent transform j (1.22), (1.23).* Thus, $\widehat{\overline{z}}$ is defined as follows:
$$(\mathfrak{p}', \widehat{\overline{z}}\mathfrak{p})_{H_a} = \big((\mathfrak{P}, \mathfrak{p}')_{H_a}, \overline{z}g\big)_{\mathcal{P}_{s(a)}}, \qquad \text{if} \quad \mathfrak{p} = j(g), \quad g \in \mathcal{P}_{s(a)}, \quad \mathfrak{p}' \in H_a.$$

This is equivalent to

(1.47)
$$\widehat{\overline{z}}\mathbf{B}^n \mathfrak{P}_0 = \mathcal{B}_a(n)^{-1} \cdot \mathbf{B}^{n+1} \mathfrak{P}_0, \qquad 0 \le n \le N-1,$$
$$\widehat{\overline{z}}\mathbf{B}^N \mathfrak{P}_0 = 0.$$

By (1.39) and (1.15), we have
$$\mathcal{B}_a(n)^{-1}\mathbf{B}^{n+1}\mathfrak{P}_0 = \mathbf{B}^{n+1}\mathcal{D}_a\big(\mathcal{A}_a(n+1)\big)^{-1}\mathfrak{P}_0$$
$$= \mathbf{B}^{n+1}\mathcal{D}_a\big(\underbrace{\varphi(\ldots\varphi(\mathbf{A})\ldots)}_{n+1}\big)^{-1}\mathfrak{P}_0 = \mathcal{D}_a(\mathbf{A})^{-1}\mathbf{B}^{n+1}\mathfrak{P}_0.$$

So, the operator $\widehat{\overline{z}} = \mathcal{D}_a(\mathbf{A})^{-1}\mathbf{B}$ satisfies the first equation in (1.47). The second equation is also satisfied since $\mathbf{B}^{N+1}\mathfrak{P}_0 = 0$.

The formula for the operator $\widehat{z}$ can be derived from the formula for $\widehat{\overline{z}}$ by using conjugation and the permutation formula (1.15).

The formula for the modulus operator $|\widehat{z}|^2$ follows from the previous formulas for $\widehat{\overline{z}}$, $\widehat{z}$, and from (1.32a) and (1.36). The theorem is proved. $\square$

Note that the formulas (1.46) for operators of complex structure were obtained in a special situation (when the function $\mathcal{D}_a$ is defined from the factorization (1.36)). In fact, we really need only the second relation (1.39), that is,
$$\mathcal{B}_a(n) = \mathcal{D}_a\big(\mathcal{A}_a(n+1)\big).$$

If one assumes that there exists a function $\mathcal{D}_a$ with this last property (independently and out of any factorization (1.36)), then it suffices to prove formulas (1.46) for $\widehat{\overline{z}}$ and $\widehat{z}$.

In Part II, §2, we shall describe an example of such type.

Respectively, for each irreducible representation (1.11), the corresponding complex structure can be determined each time, since the function $\mathcal{D}_a$ with the above-mentioned property exists.

It is also important to note that in (1.36) it suffices to determine both factors $\mathcal{D}_a(A)$ and $\mathcal{E}_a(A)$ only for those A that satisfy the relations $\varkappa_j(A) = \varkappa_j(a)$, $j = 1, \ldots, k$. We shall exploit this fact in Part II, §3.

1.4. Floquet generalization of the basic algebra.
In this section we consider the following generalization of the nonlinear relations (1.1) (see also [W, DQ, Ja, MMP]):

$$\tag{1.48} \begin{aligned} \mathbf{CB} &= \mathbf{B}\,\omega(\mathbf{A})\,\mathbf{C} + f(\mathbf{A}), \\ \mathbf{CA} &= \varphi(\mathbf{A})\,\mathbf{C}, \qquad \mathbf{AB} = \mathbf{B}\,\varphi(\mathbf{A}), \end{aligned}$$

where $\mathbf{A} = (\mathbf{A}_1, \ldots, \mathbf{A}_k)$ is a set of commuting elements (see (1.1a)). As previously, $\mathbf{A}$ and $\varphi(\mathbf{A})$ are vector-valued, the functions ω and f in (1.48) are scalar, and the functions ω, f, and φ are real:

$$\omega\colon \mathbb{R}^k \to \mathbb{R}, \qquad f\colon \mathbb{R}^k \to \mathbb{R}, \qquad \varphi\colon \mathbb{R}^k \to \mathbb{R}^k.$$

LEMMA 1.6. (a) *If a function $\varkappa$ on $\mathbb{R}^k$ is preserved by the mapping $\varphi\colon \mathbb{R}^k \to \mathbb{R}^k$, then $\varkappa(\mathbf{A})$ is an element of the center of the algebra (1.48).*

(b) *If functions σ and ρ on $\mathbb{R}^k$ satisfy the equations*

$$\tag{1.49} \begin{aligned} \sigma\big(\varphi(A)\big)\omega(A) &= \sigma(A), \\ \rho\big(\varphi(A)\big) - \rho(A) &= \sigma(A) f(A), \qquad A \in \mathbb{R}^k, \end{aligned}$$

then $\mathbf{K} = \mathbf{B}\sigma(\mathbf{A})\mathbf{C} - \rho(\mathbf{A})$ is a Casimir element of the algebra (1.48).

The proof of this lemma is similar to that of Lemma 1.3, which is a special case of Lemma 1.6.

REMARK 1.1. Equation (1.49) means that the function $1/\omega$ plays the role of the *Floquet multiplier* with respect to the mapping φ, and the function σ is the corresponding *Floquet solution*.

Here we distinguish the following two special cases: the function ω is everywhere strictly positive and the function ω is everywhere strictly negative. In both cases, we can make the change of variables

$$\widetilde{\mathbf{B}} \stackrel{\text{def}}{=} \mathbf{B}\big(\widetilde{\sigma}(\mathbf{A})\big)^{1/2}, \qquad \widetilde{\mathbf{C}} \stackrel{\text{def}}{=} \big(\widetilde{\sigma}(\mathbf{A})\big)^{1/2}\mathbf{C},$$

where $\widetilde{\sigma}$ is a positive Floquet solution of the equation $\widetilde{\sigma}\big(\varphi(A)\big)|\omega(A)| = \widetilde{\sigma}(A)$. This change of variables reduces relations (1.48) either to the basic algebra (in the first case)

$$[\widetilde{\mathbf{C}}, \widetilde{\mathbf{B}}] = \widetilde{f}(\mathbf{A}), \qquad \widetilde{\mathbf{C}}\mathbf{A} = \varphi(\mathbf{A})\widetilde{\mathbf{C}}, \qquad \mathbf{A}\widetilde{\mathbf{B}} = \widetilde{\mathbf{B}}\varphi(\mathbf{A}),$$

or to the *anticommutative version* of the basic algebra (in the second case)

$$[\widetilde{\mathbf{C}}, \widetilde{\mathbf{B}}]_+ = \widetilde{f}(\mathbf{A}), \qquad \widetilde{\mathbf{C}}\mathbf{A} = \varphi(\mathbf{A})\widetilde{\mathbf{C}}, \qquad \mathbf{A}\widetilde{\mathbf{B}} = \widetilde{\mathbf{B}}\varphi(\mathbf{A}),$$

where $[\,,\,]_+$ is the anticommutator and (in both cases) $\widetilde{f}(A) \stackrel{\text{def}}{=} \widetilde{\sigma}(A) f(A)$.

However, we are chiefly interested in the general situation in which the sign of ω varies and ω controls both the commutator and the anticommutator. In this situation, by using a change of variables of the cited type, it is impossible to reduce algebra (1.48) either to the commutation relations or to the anticommutation relations.

In the following, we construct irreducible representations in the general case, that is, without the assumption about the existence of the Floquet solution $\widetilde{\sigma}$.

Let us introduce the functions

(1.50) $\mathcal{F}_a^\omega : \mathbb{Z}_+ \to \mathbb{R}$,

$$\mathcal{F}_a^\omega(n) \stackrel{\text{def}}{=} \begin{cases} f(a), & n = 0, \\ \frac{1}{n+1}\left(f(\mathcal{A}_a(n)) + \sum_{j=0}^{n-1} f(\mathcal{A}_a(j)) \prod_{\ell=j}^{n-1} \omega(\mathcal{A}_a(\ell))\right), & n \geq 1, \end{cases}$$

where $\mathcal{A}_a(n)$ is the composition (1.4) of n mappings φ.

Now we define a subset $\mathcal{R} \subset \mathbb{R}^k$ as follows: a vector a belongs to $\mathcal{R}$ if $\mathcal{F}_a^\omega(n) > 0$ for all $n \in \mathbb{Z}_+$, or if there exists an integer $N \in \mathbb{Z}_+$ such that

(1.51) $\qquad \mathcal{F}_a^\omega(n) > 0 \quad \text{for } 0 \leq n < N, \quad \text{and} \quad \mathcal{F}_a^\omega(N) = 0.$

In the first case, we set $N = \infty$.

Note that if $f(a) = 0$, then $a \in \mathcal{R}$ and $N = 0$; but if $f(a) < 0$, then $a \notin \mathcal{R}$.

For each $a \in \mathcal{R}$, we decompose the function $\mathcal{F}_a^\omega$ into two factors (real or complex):

(1.52) $\qquad \mathcal{F}_a^\omega(n) = \mathcal{L}_a(n) \mathcal{M}_a(n), \quad n \geq 0,$

with the single condition $\mathcal{L}_a(N) = 0$ in the case (1.51). Note that $\mathcal{L}_a(n) \neq 0$ and $\mathcal{M}_a(n) \neq 0$ for $0 \leq n \leq N-1$.

Then we define positive numbers

$$t_0(a) = 1,$$
$$t_n(a) = \frac{n! \mathcal{F}_a^\omega(n-1) \ldots \mathcal{F}_a^\omega(0)}{|\mathcal{L}_a(n-1)|^2 \ldots |\mathcal{L}_a(0)|^2} \quad \text{for} \quad 1 \leq n \leq N,$$
$$t_n(a) = \infty \quad \text{for} \quad n \geq N+1,$$

where N (if it exists) is taken from (1.51).

By using the numbers $t_n(a)$, we introduce the Hilbert space $\mathcal{P}_{t(a)}$ of distributions on $\mathbb{R}^2$.

Now we describe irreducible representations of the algebra (1.48) and objects related to them.

THEOREM 1.3. (a) *There is a one-to-one correspondence between the set $\mathcal{R}$ and the set of irreducible Hermitian representations of the algebra (1.48) that possess the vacuum vector (1.2). Such a representation is finite-dimensional if and only if there exists an integer N satisfying the property (1.51); in this case, the number $N+1$ is the dimension of the representation.*

(b) *For each $a \in \mathcal{R}$ and for each factorization (1.52), the operators*

(1.53) $\qquad \overset{\circ}{B} = \bar{z} \mathcal{L}_a\left(z \frac{d}{d\bar{z}}\right), \quad \overset{\circ}{C} = \mathcal{M}_a\left(\bar{z} \frac{d}{d\bar{z}}\right) \frac{d}{d\bar{z}}, \quad \overset{\circ}{A} = \mathcal{A}_a\left(\bar{z} \frac{d}{d\bar{z}}\right)$

represent the algebra (1.48), (1.1a), and moreover, this representation is irreducible Hermitian and possesses the vacuum vector 1 in the Hilbert space $\mathcal{P}_{t(a)}$.

(c) *All other statements of Theorem 1.1 hold for the permutation relations (1.48), where the functions $\mathcal{B}_a$, $\mathcal{C}_a$, and the numbers $s_n(a)$ are replaced by $\mathcal{L}_a$, $\mathcal{M}_a$, and $t_n(a)$ (respectively).*

The proof of this theorem is similar to that of Theorem 1.1; we just need the following analog of Lemma 1.1.

LEMMA 1.7. *Let* $\mathfrak{P}_0 = \mathfrak{P}_0(a)$ (1.2) *be the vacuum vector of an Hermitian representation of the algebra* (1.48) *in a Hilbert space* H_a. *Then the following equalities hold*:

$$\begin{aligned}\mathbf{C}\mathbf{B}^m\mathfrak{P}_0 &= m\mathcal{F}_a^\omega(m-1)\,\mathbf{B}^{m-1}\mathfrak{P}_0, & m &\geq 1; \\ \|\mathbf{B}^m\mathfrak{P}_0\|_{H_a}^2 &= m!\,\mathcal{F}_a^\omega(m-1)\ldots\mathcal{F}_a^\omega(0), & m &\geq 1; \\ (\mathbf{B}^m\mathfrak{P}_0, \mathbf{B}^{m'}\mathfrak{P}_0)_{H_a} &= 0 & \text{if } m &\neq m'.\end{aligned}$$ (1.54)

1.5. Further generalization of the basic algebra. Now let us consider more general relations:

(1.55)
$$\begin{aligned}\mathbf{C}\mathbf{B} &= \mathbf{B}\,\omega(\mathbf{A})\,\mathbf{C} + f(\mathbf{A}), \\ \mathbf{C}\mathbf{A} &= \varphi(\mathbf{A})\,\mathbf{C} + \psi(\mathbf{A}), \qquad \mathbf{A}\mathbf{B} = \mathbf{B}\,\varphi(\mathbf{A}) + \psi(\mathbf{A}),\end{aligned}$$

where $\mathbf{A} = (\mathbf{A}_1, \ldots, \mathbf{A}_k)$ is a set of commuting elements and φ and ψ are vector-valued functions. We still assume that ω, f, and φ, as well as ψ, are real functions.

There is the following generalized Jacobi rule for algebras with abstract relations (see [**M**] or [**KaM**$_4$, Appendix II]): the identity $F(\overset{3}{\mathbf{B}}, \overset{2}{\mathbf{A}}, \overset{1}{\mathbf{C}}) = 0$, once satisfied for a polynomial F, must imply $F \equiv 0$. This condition can be formulated in another way: the Poincaré–Birkhoff–Witt property holds for given permutation relations.

In the case of (1.55), this generalized Jacobi rule implies

(1.56)
$$\begin{aligned}&(a) & \big(\varphi_\mu(A) - A_\mu\big)\psi_\nu(A) &= \big(\varphi_\nu(A) - A_\nu\big)\psi_\mu(A), \\ &(b) & \psi_\mu\big(\varphi(A)\big) &= \omega(A)\,\langle\psi(A), \delta\varphi_\mu(\varphi(A), A)\rangle.\end{aligned}$$

Here the vector-valued difference derivative δ is defined by the formula

$$\delta F(A, A') \stackrel{\text{def}}{=} \int_0^1 \frac{\partial F}{\partial A}\big(\tau A + (1-\tau)A'\big)\,d\tau,$$

and $\langle\cdot,\cdot\rangle$ denotes the inner product of vectors in $\mathbb{R}^k$.

Note that we have derived (1.56)(b) by using the commutation formula

(1.57) $$\mathbf{C}F(\mathbf{A}) = F\big(\varphi(\mathbf{A})\big)\mathbf{C} + \langle\psi(\mathbf{A}), \delta F(\varphi(\mathbf{A}), \mathbf{A})\rangle$$

that follows from the second relation in (1.55) (see [**KaM**$_4$, Appendix I]).

In what follows, we assume that the structural functions satisfy conditions (1.56). Then (1.56)(a) implies that the ratio $\psi_\mu(A)/(\varphi_\mu(A) - A_\mu)$ is independent of μ; we denote this ratio by

(1.58) $$\theta(A) \stackrel{\text{def}}{=} \frac{\psi_\mu(A)}{\varphi_\mu(A) - A_\mu}.$$

The denominator may vanish, and thus, some problems may arise.

Note that if $\theta(A)$ is finite at some point $A \in \mathbb{R}^k$, then by (1.56)(b) and (1.58), we have

$$\begin{aligned}\theta\big(\varphi(A)\big) &= \frac{\psi_\mu\big(\varphi(A)\big)}{\varphi_\mu\big(\varphi(A)\big) - \varphi_\mu(A)} = \frac{\omega(A)\,\langle\psi(A), \delta\varphi_\mu(\varphi(A), A)\rangle}{\varphi_\mu\big(\varphi(A)\big) - \varphi_\mu(A)} \\ &= \omega(A)\theta(A)\frac{\langle\varphi(A) - A, \delta\varphi_\mu(\varphi(A), A)\rangle}{\varphi_\mu\big(\varphi(A)\big) - \varphi_\mu(A)} = \omega(A)\theta(A),\end{aligned}$$

and hence, $\theta\big(\underbrace{\varphi(\ldots(\varphi(A))\ldots)}_{n}\big)$ is finite for any positive integer n:

$$\theta\big(\underbrace{\varphi(\ldots(\varphi(A))\ldots)}_{n}\big) = \theta(A)\prod_{j=0}^{n-1}\omega\big(\underbrace{\varphi(\ldots(\varphi(A))\ldots)}_{j}\big). \tag{1.59}$$

Now let us describe irreducible representations of the algebra (1.55) that possess a vacuum vector $\mathfrak{P}_0$ (1.2) such that the eigenvalues $a = (a_1, \ldots, a_k)$ of the operators $\mathbf{A}$ on $\mathfrak{P}_0$ satisfy the condition that $\theta(a)$ (1.58) is finite.

Consider the function
$$\widetilde{\mathcal{F}}_a^\omega : \mathbb{Z}_+ \to \mathbb{R},$$
$$\widetilde{\mathcal{F}}_a^\omega(n) \stackrel{\mathrm{def}}{=} \begin{cases} \widetilde{f}(a), & n = 0, \\ \frac{1}{n+1}\Big(\widetilde{f}(\mathcal{A}_a(n)) + \sum_{j=0}^{n-1}\widetilde{f}(\mathcal{A}_a(j))\prod_{\ell=j}^{n-1}\omega(\mathcal{A}_a(\ell))\Big), & n \geq 1, \end{cases}$$
where $\mathcal{A}_a$ is defined in (1.4) and
$$\widetilde{f}(A) \stackrel{\mathrm{def}}{=} f(A) + \big(\theta(A)\big)^2\big(\omega(A) - 1\big). \tag{1.60}$$

We define the subset $\widetilde{\mathcal{R}} \subset \mathbb{R}^k$ as follows: $a \in \widetilde{\mathcal{R}}$ if $|\theta(a)| < \infty$ and either
$$\widetilde{\mathcal{F}}_a^\omega(n) > 0 \qquad \text{for all} \quad n \in \mathbb{Z}_+$$
or
$$\widetilde{\mathcal{F}}_a^\omega(n) > 0 \quad \text{for} \quad 0 \leq n < N, \quad \text{and} \quad \widetilde{\mathcal{F}}_a^\omega(N) = 0. \tag{1.61}$$

Note that if $|\theta(a)| < \infty$ and $\widetilde{f}(a) = 0$, then $a \in \widetilde{\mathcal{R}}$ and $N = 0$.

For each $a \in \widetilde{\mathcal{R}}$, we split the function $\widetilde{\mathcal{F}}_a^\omega$ into two factors:
$$\widetilde{\mathcal{F}}_a^\omega(n) = \widetilde{\mathcal{L}}_a(n)\widetilde{\mathcal{M}}_a(n), \qquad n \geq 0, \tag{1.62}$$
with the single condition $\widetilde{\mathcal{L}}_a(N) = 0$ in the case of (1.61). Note that $\widetilde{\mathcal{L}}_a(N) \neq 0$ and $\widetilde{\mathcal{M}}_a(N) \neq 0$ for all $n < N$.

Then we define positive numbers
$$\widetilde{t}_0(a) = 1,$$
$$\widetilde{t}_n(a) = \frac{n!\,\widetilde{\mathcal{F}}_a^\omega(n-1)\ldots\widetilde{\mathcal{F}}_a^\omega(0)}{|\widetilde{\mathcal{L}}_a(n-1)|^2\ldots|\widetilde{\mathcal{L}}_a(0)|^2} \qquad \text{for} \quad 1 \leq n \leq N, \tag{1.63}$$
$$\widetilde{t}_n(a) = \infty \qquad \text{for} \quad n \geq N+1,$$
where N (if it exists) is taken from (1.61).

By using the numbers $\widetilde{t}_n(a)$, we introduce the Hilbert space $\mathcal{P}_{\widetilde{t}(a)}$ of antiholomorphic distributions over $\mathbb{R}^2$.

THEOREM 1.4. (a) *There is a one-to-one correspondence between the set $\widetilde{\mathcal{R}}$ and the set of irreducible Hermitian representations of algebra* (1.55) *that possess the vacuum vector* (1.2) *and satisfy the condition* $|\theta(a)| < \infty$, *where θ is given by* (1.58). *Such a representation is finite-dimensional if and only if there exists an integer N satisfying* (1.61); *in this case, the number $N+1$ is the dimension of the representation.*

(b) *For each $a \in \widetilde{\mathcal{R}}$ and for each factorization (1.62), the operators*

$$
\begin{aligned}
\overset{\circ}{B} &= \overline{z}\,\widetilde{\mathcal{L}}_a\!\left(\overline{z}\frac{d}{d\overline{z}}\right) - \theta(a)\,\chi_a\!\left(\overline{z}\frac{d}{d\overline{z}}\right), \\
\overset{\circ}{C} &= \widetilde{\mathcal{M}}_a\!\left(\overline{z}\frac{d}{d\overline{z}}\right)\frac{d}{d\overline{z}} - \theta(a)\,\chi_a\!\left(\overline{z}\frac{d}{d\overline{z}}\right), \\
\overset{\circ}{A} &= \mathcal{A}_a\!\left(\overline{z}\frac{d}{d\overline{z}}\right),
\end{aligned}
\tag{1.64}
$$

where

$$
\chi_a(n) \overset{\text{def}}{=} \begin{cases} 1, & n=0, \\ \prod_{j=0}^{n-1} \omega(\mathcal{A}_a(j)), & n \geq 1, \end{cases}
$$

represent the algebra (1.55), (1.1a), and moreover, this representation is irreducible Hermitian and possesses the vacuum 1 in the Hilbert space $\mathcal{P}_{\widetilde{t}(a)}$.

(c) *Representations assigned to different vectors $a \in \widetilde{\mathcal{R}}$ are not equivalent, but for each chosen $a \in \widetilde{\mathcal{R}}$, representations assigned to different factorizations (1.62) are equivalent.*

(d) *An abstract Hermitian representation of algebra (1.55) in a Hilbert space H_a with the vacuum vector $\mathfrak{P}_0 = \mathfrak{P}_0(a)$, satisfying (1.2) for some $a \in \mathbb{R}^k$, can be intertwined with the representation (1.64) by means of the following generalized coherent states:*

$$
\mathfrak{P}_z = \mathfrak{P}_0 + \sum_{n=1}^{N} \frac{1}{n!\,\widetilde{\mathcal{M}}_a(n-1)\ldots\widetilde{\mathcal{M}}_a(0)}\, z^n \prod_{j=0}^{n-1}(\mathbf{B}+\theta(a)\chi_a(j))\mathfrak{P}_0.
\tag{1.65}
$$

The generalized "reproducing kernel" corresponding to the states (1.65) is given by the following distribution over $\mathbb{R}^2 \times \mathbb{R}^2$:

$$
\mathcal{K}_{\widetilde{t}(a)}(\overline{w},z) = \sum_{n=0}^{N} \frac{(\overline{w}z)^n}{\widetilde{t}_n(a)}.
$$

The proof of this theorem is similar to that of Theorem 1.1, and we just need the following analog of Lemma 1.1.

LEMMA 1.8. *Let $\mathfrak{P}_0 = \mathfrak{P}_0(a)$ be the vacuum vector (1.2) of an Hermitian representation of the algebra (1.55) in a Hilbert space H_a, and let $|\theta(a)| < \infty$. Then the following equalities hold:*

(a) $\quad (\mathbf{C}+\theta(a))(\mathbf{B}+\theta(a))\mathfrak{P}_0 = \widetilde{\mathcal{F}}_a^\omega(0)\mathfrak{P}_0,$

$$
(\mathbf{C}+\theta(a)\chi_a(m-1)) \prod_{j=0}^{m-1}(\mathbf{B}+\theta(a)\chi_a(j))\mathfrak{P}_0
$$
$$
= m\widetilde{\mathcal{F}}_a^\omega(m-1) \prod_{j=0}^{m-2}(\mathbf{B}+\theta(a)\chi_a(j))\mathfrak{P}_0, \quad m \geq 2;
$$

(b) $\quad \left\| \prod_{j=0}^{m-1}(\mathbf{B}+\theta(a)\chi_a(j))\mathfrak{P}_0 \right\|_{H_a}^2 = m!\,\widetilde{\mathcal{F}}_a^\omega(m-1)\ldots\widetilde{\mathcal{F}}_a^\omega(0), \quad m \geq 1;$

(c) $\left(\prod_{j=0}^{m-1} (\mathbf{B} + \theta(a)\chi_a(j))\mathfrak{P}_0, \prod_{j=0}^{m'-1} (\mathbf{B} + \theta(a)\chi_a(j))\mathfrak{P}_0 \right)_{H_a} = 0, \quad m \neq m'.$

COROLLARY 1.2. *The following orthonormal eigenbasis of operators $\overset{\circ}{A}$:*

$$\left\{ \frac{\overline{z}^n}{\sqrt{\tilde{t}_n(a)}} \,\Big|\, n \geq 0 \right\} \quad in \quad \mathcal{P}_{\tilde{t}(a)}$$

corresponds to the following orthonormal eigenbasis of operators $\mathbf{A}_\mu$:
(1.66)
$$\{\mathfrak{P}_0\} \cup \left\{ \frac{1}{\sqrt{\tilde{t}_n(a)}\widetilde{\mathcal{L}}_a(n-1)\ldots\widetilde{\mathcal{L}}_a(0)} \prod_{j=0}^{n-1} (\mathbf{B} + \theta(a)\chi_a(j))\mathfrak{P}_0 \,\Big|\, n \geq 1 \right\} \quad in \; H_a$$

under the coherent transform (1.23).

REMARK 1.2. If $|\theta(a)| < \infty$, then it follows from (1.59) that the function $\theta(A)$ is defined on the entire spectrum $\{\mathcal{A}_a(n) \mid 0 \leq n \leq N\}$ of the operators $\mathbf{A}$. In this case, using the translation

(1.67) $$\widetilde{\mathbf{B}} = \mathbf{B} + \theta(\mathbf{A}), \qquad \widetilde{\mathbf{C}} = \mathbf{C} + \theta(\mathbf{A}),$$

we can formally reduce (1.55) to the form (1.48):

$$\widetilde{\mathbf{C}}\widetilde{\mathbf{B}} = \widetilde{\mathbf{B}}\omega(\mathbf{A})\widetilde{\mathbf{C}} + \widetilde{f}(\mathbf{A}), \qquad \widetilde{\mathbf{C}}\mathbf{A} = \varphi(\mathbf{A})\widetilde{\mathbf{C}}, \qquad \mathbf{A}\widetilde{\mathbf{B}} = \widetilde{\mathbf{B}}\varphi(\mathbf{A}),$$

where $\widetilde{f}$ is the function (1.60).

Indeed, the definition of θ implies

$$\widetilde{\mathbf{C}}\mathbf{A} = \varphi(\mathbf{A})\mathbf{C} + \psi(\mathbf{A}) + \theta(\mathbf{A})\mathbf{A} = \varphi(\mathbf{A})\widetilde{\mathbf{C}}.$$

Then, using (1.57) and its conjugate, we obtain

$$\widetilde{\mathbf{C}}\widetilde{\mathbf{B}} - \widetilde{\mathbf{B}}\omega(\mathbf{A})\widetilde{\mathbf{C}} = \mathbf{C}\mathbf{B} - \mathbf{B}\omega(\mathbf{A})\mathbf{C} + \big(\theta(\varphi(\mathbf{A})) - \theta(\mathbf{A})\omega(\mathbf{A})\big)\mathbf{C}$$
$$+ \mathbf{B}\big(\theta(\varphi(\mathbf{A})) - \theta(\mathbf{A})\omega(\mathbf{A})\big) + 2\langle\psi(\mathbf{A}), \delta\theta(\varphi(\mathbf{A}), \mathbf{A})\rangle + \big(\theta(\mathbf{A})\big)^2\big(1 - \omega(\mathbf{A})\big).$$

By (1.58) and (1.59), this and the relation

$$\langle\psi(A), \delta\theta(\varphi(A), A)\rangle = \theta(A)\langle\varphi(A) - A, \delta\theta(\varphi(A), A)\rangle$$
$$= \theta(A)\big(\theta(\varphi(A)) - \theta(A)\big) = \big(\theta(A)\big)^2\big(\omega(A) - 1\big)$$

imply

$$\widetilde{\mathbf{C}}\widetilde{\mathbf{B}} - \widetilde{\mathbf{B}}\omega(\mathbf{A})\widetilde{\mathbf{C}} = f(\mathbf{A}) + \big(\theta(\mathbf{A})\big)^2\big(\omega(\mathbf{A}) - 1\big) = \widetilde{f}(\mathbf{A}).$$

Note that the translation (1.67) introduces the "new" orthogonal basis

$$\widetilde{\mathbf{B}}^n \mathfrak{P}_0 = \big(\mathbf{B} + \theta(\mathbf{A})\big)^n \mathfrak{P}_0 = \prod_{j=0}^{n-1} (\mathbf{B} + \theta(a)\chi_a(j))\mathfrak{P}_0$$

instead of the "old" basis $\mathbf{B}^n \mathfrak{P}_0$.

We conclude this section by a lemma similar to Lemma 1.3.

LEMMA 1.9. (a) *If the function $\varkappa$ on $\mathbb{R}^k$ satisfies the condition $\varkappa(\varphi(A)) = \varkappa(A)$, then the operator $\varkappa(\mathbf{A})$ is a Casimir element for relations (1.55).*

(b) *Suppose that the function ω has no zeros on the spectrum of the operators $\mathbf{A}$, and there exist functions σ and ρ such that the following conditions are satisfied:*

(1.68) $\qquad \sigma(\varphi(A))\omega(A) = \sigma(A), \qquad \rho(\varphi(A)) - \rho(A) = \sigma(A)f(A).$

Then the operator

$$\mathbf{K} \stackrel{\text{def}}{=} \mathbf{B}\sigma(\mathbf{A})\mathbf{C} + \mathbf{B}\theta(\mathbf{A})\sigma(\mathbf{A}) + \theta(\mathbf{A})\sigma(\mathbf{A})\mathbf{C} - \rho(\mathbf{A})$$

is a Casimir element in the algebra (1.55).

PROOF. Statement (a) follows from (1.57):

$$\begin{aligned}\mathbf{C}\varkappa(\mathbf{A}) &= \varkappa(\varphi(\mathbf{A}))\mathbf{C} + \langle\psi(\mathbf{A}), \delta\varkappa(\varphi(\mathbf{A}), \mathbf{A})\rangle \\ &= \varkappa(\mathbf{A})\mathbf{C} + \theta(\mathbf{A})\langle\varphi(\mathbf{A}) - \mathbf{A}, \delta\varkappa(\varphi(\mathbf{A}), \mathbf{A})\rangle \\ &= \varkappa(\mathbf{A})\mathbf{C} + \theta(\mathbf{A})\big(\varkappa(\varphi(\mathbf{A})) - \varkappa(\mathbf{A})\big) = \varkappa(\mathbf{A})\mathbf{C}.\end{aligned}$$

Let us prove statement (b). The operator $\mathbf{K}$ commutes with all operators $\mathbf{A}$ independently of the choice of σ and ρ:

$$\begin{aligned}[\mathbf{K}, \mathbf{A}_\mu] = \mathbf{B}\sigma(\mathbf{A})\Big(\big(\mathbf{A}_\mu - \varphi_\mu(\mathbf{A})\big)\theta(\mathbf{A}) + \psi_\mu(\mathbf{A})\Big) \\ + \Big(-\psi_\mu(\mathbf{A}) + (\varphi_\mu(\mathbf{A}) - \mathbf{A}_\mu)\theta(\mathbf{A})\Big)\sigma(\mathbf{A})\mathbf{C} = 0.\end{aligned}$$

Using (1.57), we obtain

$$\begin{aligned}[\mathbf{C}, \mathbf{K}] = {}& \mathbf{B}\Big(\omega(\mathbf{A})\sigma(\varphi(\mathbf{A})) - \sigma(\mathbf{A})\Big)\mathbf{C}^2 \\ & + \mathbf{B}\Big(\omega(\mathbf{A})\langle\psi(\mathbf{A}), \delta\sigma(\varphi(\mathbf{A}), \mathbf{A})\rangle + \omega(\mathbf{A})\sigma(\varphi(\mathbf{A}))\theta(\varphi(\mathbf{A})) - \sigma(\mathbf{A})\theta(\mathbf{A})\Big)\mathbf{C} \\ & + \mathbf{B}\omega(\mathbf{A})\langle\psi(\mathbf{A}), \delta(\sigma\theta)(\varphi(\mathbf{A}), \mathbf{A})\rangle + \Big(\sigma(\varphi(\mathbf{A}))\theta(\varphi(\mathbf{A})) - \sigma(\mathbf{A})\theta(\mathbf{A})\Big)\mathbf{C}^2 \\ & + \Big(f(\mathbf{A})\sigma(\mathbf{A}) + \langle\psi(\mathbf{A}), \delta(\sigma\theta)(\varphi(\mathbf{A}), \mathbf{A})\rangle - \rho(\varphi(\mathbf{A})) + \rho(\mathbf{A})\Big)\mathbf{C} \\ & + f(\mathbf{A})\sigma(\mathbf{A})\theta(\mathbf{A}) - \langle\psi(\mathbf{A}), \delta\rho(\varphi(\mathbf{A}), \mathbf{A})\rangle = 0,\end{aligned}$$

since, according to (1.59) and (1.68), the coefficients (that is, functions of $\mathbf{A}$) in this formula are equal to zero for all powers $\mathbf{B}^\ell \ldots \mathbf{C}^m$. $\square$

§2. Representations by differential and q-differential operators. Relationship to the theory of hypergeometric functions

Till now, the functional coefficients f and φ in relations (1.1) have been more or less arbitrary. Accordingly, the representation operators (1.11) or (1.42) are, in general, pseudodifferential operators. In this section we consider special important situations in which the operators (1.11) are differential or q-differential.

2.1. Representation by differential operators.

LEMMA 2.1. *Suppose that the following conditions are satisfied:*
(a) *the function $f\colon \mathbb{R}^k \to \mathbb{R}$ is a polynomial;*
(b) *the components of the mapping $\varphi\colon \mathbb{R}^k \to \mathbb{R}^k$ are polynomials;*

(c) *the Jacobi matrix $D\varphi$ is upper triangular, and its diagonal elements are equal to 1;*

(d) *all fixed points of the mapping φ on the complex space $\mathbb{C}^k$ are nondegenerate and the polynomial f vanishes at these points.*

Then the relations (1.1) are regular, and for any $A \in \mathbb{R}^k$ the basic functions $\mathcal{F}_A$ (1.3) and $\mathcal{A}_A$ (1.4) are polynomial over $\mathbb{Z}_+$. Moreover, there exist independent polynomial functions $\varkappa_1, \ldots, \varkappa_{k-1}$ and a polynomial ρ over $\mathbb{R}^k$ that determine the Casimir elements of the algebra (1.1) (see Lemma 1.3).

In particular, conditions (b) and (c) are satisfied if φ is the shift by unit time along trajectories of a vector field $v = \sum_{\mu=1}^{k} v_\mu(A) \frac{\partial}{\partial A_\mu}$ on $\mathbb{R}^k$, where all the v_μ are polynomial and the Jacobi matrix $Dv = ((\partial v_\mu / \partial A_\nu))$ is nilpotent for each $A \in \mathbb{R}^k$.

PROOF. For simplicity, let us consider the case $k = 2$. Condition (c) implies

(2.1)
$$\varphi_2(A) = A_2 + \beta_2, \qquad \text{where } \beta_2 = \text{const};$$
$$\varphi_1(A) = A_1 + \beta_1(A_2), \qquad \text{where } \beta_1 \text{ is a function of a single variable } A_2.$$

Hence, for any $n \in \mathbb{Z}_+$, $n \geq 1$, we have

(2.2)
$$\bigl(\mathcal{A}_A(n)\bigr)_2 = \bigl(\underbrace{\varphi \circ \cdots \circ \varphi}_{n}(A)\bigr)_2 = A_2 + n\beta_2,$$
$$\bigl(\mathcal{A}_A(n)\bigr)_1 = \bigl(\underbrace{\varphi \circ \cdots \circ \varphi}_{n}(A)\bigr)_1 = A_1 + \sum_{m=0}^{n-1} \beta_1(A_2 + m\beta_2).$$

In view of (b), the function β_1 is a polynomial. So, the sum in the second formula in (2.2) is reduced to a linear combination of several sums of the type

(2.3)
$$\sum_{m=0}^{n-1} m^\ell = n\mathcal{P}_\ell(n),$$

where $\mathcal{P}_\ell$ is a (well-known) polynomial of degree ℓ for each $\ell \in \mathbb{Z}_+$. Thus, $\mathcal{A}_A(n)$ is a polynomial in $n \in \mathbb{Z}_+$, and formulas (1.3), (2.3) and condition (a) imply that $\mathcal{F}_A(n)$ is also a polynomial in n.

If $\beta_2 = 0$, then any function $\varkappa_1 = \varkappa_1(A_2)$ depending only on the variable A_2 is φ-invariant.

If $\beta_2 \neq 0$, then the function $\varkappa_1$ invariant with respect to the mapping φ (2.1) can be found in the form

$$\varkappa_1(A) = A_1 - \sum_{\ell=0}^{L} \varkappa_{1,\ell} A_2^{\ell+1},$$

where L is the degree of the polynomial $\beta_1 = \sum_{\ell=0}^{L} \beta_{1,\ell} A_2^\ell$. From the equation

(2.4)
$$\sum_{\ell=0}^{L} \varkappa_{1,\ell} \bigl[(A_2 + \beta_2)^{\ell+1} - A_2^{\ell+1}\bigr] = \sum_{\ell=0}^{L} \beta_{1,\ell} A_2^\ell,$$

one can recurrently (step by step, starting with $\ell = L$) obtain the coefficients $\varkappa_{1,\ell}$ in terms of the known numbers $\beta_{1,0}, \ldots, \beta_{1,L}$.

For the function ρ, we have the equation (see (1.28))

(2.5)
$$\rho\bigl(A_1 + \beta_1(A_2), A_2 + \beta_2\bigr) - \rho(A_1, A_2) = f(A_1, A_2).$$

Here again there are two possibilities.

If $\beta_2 = 0$, then the solution of (2.5) is obtained by the same formula as in Lemma 1.4, that is,

$$\rho(A_1, A_2) = \int \frac{e^{i\omega A_1} - 1}{e^{i\omega \beta_1(A_2)} - 1} d\mu_{A_2}(\omega) \quad \text{if} \quad f(A_1, A_2) = \int e^{i\omega A_1} d\mu_{A_2}(\omega).$$

Note that since $f(A_1, A_2)$ is a polynomial, its Fourier transform with respect to the variable A_1 (that is, the measure $d\mu_{A_2}(\omega)$) is just a sum of derivatives of $\delta(\omega)$ with coefficients polynomial in A_2. Moreover, in view of condition (d) in the case $\beta_2 = 0$, the polynomial $f(A_1, A_2)$ is divisible by the polynomial $\beta_1(A_2)$, and so, the above formula for $\rho(A_1, A_2)$ actually represents a polynomial in both A_1 and A_2.

Now let $\beta_2 \neq 0$. The right-hand side of (2.5) is a polynomial

$$f = \sum_{k_1, k_2 = 0}^{M} f_{k_1, k_2} A_1^{k_1} A_2^{k_2}.$$

If we look for the solution ρ in the form

$$\rho = \sum_{k_1 = 0}^{M} \sum_{k_2 = 0}^{N_{k_1}} \rho_{k_1, k_2} A_1^{k_1} A_2^{k_2 + 1},$$

then we obtain the following relation for the coefficients ρ_{k_1, k_2}:

$$(2.6) \quad \sum_{\ell_2 = 0}^{N_{k_1}} \rho_{k_1, \ell_2} \left[(A_2 + \beta_2)^{\ell_2 + 1} - A_2^{\ell_2 + 1} \right]$$
$$+ \sum_{\ell_1 > k_1}^{M} \sum_{\ell_2 = 0}^{N_{\ell_1}} \rho_{\ell_1, \ell_2} \beta_1(A_2)^{\ell_1 - k_1} C_{\ell_1}^{k_1} (A_2 + \beta_2)^{\ell_2 + 1} = \sum_{k_2 = 0}^{M} f_{k_1, k_2} A_2^{k_2},$$

where the C_ℓ^k are the binomial coefficients.

First, in (2.6), we take $k_1 = M$ and set $N_M = M$. Then we obtain the following equation, similar to (2.4), for ρ_{M, ℓ_2}:

$$\sum_{\ell_2 = 0}^{M} \rho_{M, \ell_2} \left[(A_2 + \beta_2)^{\ell_2 + 1} - A_2^{\ell_2 + 1} \right] = \sum_{k_2 = 0}^{M} f_{M, k_2} A_2^{k_2}.$$

From this equation, starting from $\ell_2 = M$, we recurrently derive all coefficients ρ_{M, ℓ_2} in terms of the known numbers $f_{M, 0}, \ldots, f_{M, M}$.

Once this is completed, we take $k_1 = M - 1$ in (2.6), and hence obtain the equation for ρ_{M-1, ℓ_2}:

$$\sum_{\ell_2 = 0}^{N_{M-1}} \rho_{M-1, \ell_2} \left[(A_2 + \beta_2)^{\ell_2 + 1} - A_2^{\ell_2 + 1} \right]$$
$$= \sum_{k_2 = 0}^{M} f_{M-1, k_2} A_2^{k_2} - M \beta_1(A_2) \sum_{\ell_2 = 0}^{M} \rho_{M, \ell_2} (A_2 + \beta_2)^{\ell_2 + 1}.$$

We set $N_{M-1} = M + 1 + M_1$, where M_1 is the degree of the polynomial β_1 in the variable A_2. Since the coefficients ρ_{M, ℓ_2} are already known, the last equation is, in fact, of type (2.4), and thus can be solved by just the same method as previously. So, all coefficients ρ_{M-1,ℓ_2} have been calculated.

In a similar way, we find all ρ_{k_1,k_2} from (2.6). The lemma is proved. □

Now, in expansions (1.6) or (1.36), one can assume that the factors $\mathcal{B}_a(n)$ and $\mathcal{C}_n(n)$ are polynomial in $n \in \mathbb{Z}_+$ or that the factors $\mathcal{D}_a(A)$ and $\mathcal{E}_a(A)$ are polynomial in $A \in \mathbb{R}^k$.

THEOREM 2.1. *Under assumptions* (a), (b), (c) *of Lemma 2.1, there exist irreducible representations of the algebra* (1.1) *by antiholomorphic differential operators with polynomial coefficients (given by* (1.11) *or* (1.42)*). The orders of these operators, as well as the degrees of their coefficients, are independent of the parameter $a \in \mathcal{R}$.*

2.2. Relations with hypergeometric series. Recall that hypergeometric series are defined as follows (see [**BE**]):

$$(2.7) \qquad {}_pF_r(b;c;z) \stackrel{\text{def}}{=} \sum_{n=0}^{\infty} \frac{(b)_n}{(c)_n} \cdot \frac{z^n}{n!},$$

where z is a variable, $b = (b_1, \ldots, b_p)$ and $c = (c_1, \ldots, c_r)$ are parameters, $(-c_j) \notin \mathbb{Z}_+$, $p, r \in \mathbb{Z}_+$, and the symbol $(b)_n$ denotes a shifted factorial

$$(b)_0 = 1, \qquad (b)_n \stackrel{\text{def}}{=} \prod_{j=1}^{p} b_j(b_j + 1) \ldots (b_j + n - 1), \quad n = 1, 2, \ldots.$$

Note that in general the series (2.7) does not converge and should be considered as a distribution in the variable z.

In addition to a complete hypergeometric series, it is useful to have its finite jet. The polynomial

$$(2.7a) \qquad {}_pf_r(b;c;z;N) = \sum_{n=0}^{N} \frac{(b)_n}{(c)_n} \cdot \frac{z^n}{n!}$$

is called a *hypergeometric jet of degree N*.

Now let us consider the function $\mathcal{F}_a$ (1.3). Under the assumptions (a), (b), (c) of Lemma 2.1, this function is a polynomial; thus, $\mathcal{B}_a$ and $\mathcal{C}_a$ in (1.6) can be chosen as polynomials

$$(2.8) \qquad \mathcal{B}_a(n) = \beta \prod_{j=1}^{p} \bigl(n - b_j(a)\bigr), \qquad \mathcal{C}_a(n) = \gamma \prod_{j=1}^{r} \bigl(n - c_j(a)\bigr).$$

Here by $b_j = b_j(a)$ we denote the (complex) roots of $\mathcal{B}_a$, and by $c_j = c_j(a)$ the roots of $\mathcal{C}_a$. The numbers $p = p(a)$ and $r = r(a)$ are the degrees of the polynomials $\mathcal{B}_a$ and $\mathcal{C}_a$. The sum $p+r$ is fixed and is the degree of the polynomial $\mathcal{F}_a$, which is uniquely defined by the structural polynomials f and φ from (1.1). The degree $p+r$ is independent of $a \in \mathcal{R}$ at generic points, but it can fall down on some exceptional surfaces.

THEOREM 2.2. *Under the assumptions* (a), (b), (c) *of Lemma* 2.1, *the coherent states* $\mathfrak{P}_z$ *and the reproducing kernel* $\mathcal{K}_{s(a)}$, *corresponding to the polynomial antiholomorphic representations in Theorem* 2.1, *can be derived via hypergeometric series and hypergeometric jets as follows.*

- *In the case* $N = \infty$, *that is, if the polynomials* $\mathcal{B}_a$ *and* $\mathcal{C}_a$ *have no roots in* $\mathbb{Z}_+$,

$$
(2.9) \quad \begin{aligned}
\mathfrak{P}_z &= \sum_{n=0}^{\infty} \frac{1}{n!(-\bar{c})_n} \left(\frac{z\mathbf{B}}{\bar{\gamma}}\right)^n \mathfrak{P}_0 \equiv {}_0F_r\left(-\bar{c}; \frac{z\mathbf{B}}{\bar{\gamma}}\right)\mathfrak{P}_0, \\
\mathcal{K}_{s(a)}(\overline{w}, z) &= \sum_{n=0}^{\infty} \frac{(-b)_n}{n!(-\bar{c})_n} \left(\frac{\beta\,\overline{w}z}{\bar{\gamma}}\right)^n \equiv {}_pF_r\left(-b; -\bar{c}; \frac{\beta\,\overline{w}z}{\bar{\gamma}}\right).
\end{aligned}
$$

- *In the case* $N < \infty$ (1.5), *that is, if the polynomial* $\mathcal{B}_a$ *has a root in* $\mathbb{Z}_+$,

$$
(2.10) \quad \begin{aligned}
\mathfrak{P}_z &= {}_0f_r\left(-\bar{c}; \frac{z\mathbf{B}}{\bar{\gamma}}; N\right)\mathfrak{P}_0, \\
\mathcal{K}_{s(a)}(\overline{w}, z) &= {}_pf_r\left(-b; -\bar{c}; \frac{\beta\,\overline{w}z}{\bar{\gamma}}; N\right).
\end{aligned}
$$

- *In the case* $N < \infty$, *the reproducing kernel* $\mathcal{K}_{s(a)}$ *can also be derived by the second formula* (2.9), *and the coherent state* $\mathfrak{P}_z$ *can also be derived by using the hypergeometric jet* ${}_0f_r$ *of any degree* N' *such that*

$$(2.11) \quad N \leq N' \leq N^+,$$

where N^+ *is a minimal root of* $\mathcal{C}_a$ *in* $\mathbb{Z}_+$ (*provided that it exists*). *If* $\mathcal{C}_a$ *does not have any roots in* $\mathbb{Z}_+$, *then the jet* ${}_0f_r$ *in* (2.10) *can be of any arbitrary degree* $N' \geq N$, *in particular, of degree* $N' = \infty$; *that is, the first formula* (2.9) *also holds.*

Note that in the finite-dimensional case $N < \infty$, the reproducing kernel $\mathcal{K}_{s(a)}$ is a hypergeometric jet, but, at the same time, it can be called a "hypergeometric polynomial" of degree N (following the definition from [**BE**, vol. I, Sect. 2.1.1, formula (4)]). In particular, if $p = 2$ and $r = 1$, then $\mathcal{K}_{s(a)}$ is a Jacobi polynomial (see examples in Part II).

To conclude this subsection, let us consider the following question: *what is a class of polynomials* $\mathcal{F}$ *whose roots* b_j, c_j *determine the hypergeometric representation* (2.9) *of the reproducing kernel for some permutation relations of type* (1.1)?

We briefly describe the answer to this question.

Let $\varphi\colon \mathbb{R}^k \to \mathbb{R}^k$ be a polynomial mapping whose Jacobi matrix $D\varphi$ is upper triangular and whose spectrum consists of the point $\{1\}$ only. Let $\mathcal{F}_A(n)$ be a polynomial both in $n \in \mathbb{Z}_+$ and in $A \in \mathbb{R}^k$. We say that this polynomial is *compatible* with the mapping φ if the following three conditions are satisfied:

(i) the degree of $\mathcal{F}_A(n)$ with respect to n does not depend on $A \in \mathbb{R}^k$;

(ii) the identity

$$(n + n' + 2)\,\mathcal{F}_A(n + n' + 1) = (n+1)\,\mathcal{F}_A(n) + (n'+1)\,\mathcal{F}_{\varphi^{n+1}(A)}(n')$$

holds for any $n, n' \in \mathbb{Z}_+$ (here $\varphi^k(A) = \underbrace{\varphi \circ \cdots \circ \varphi}_{k}(A)$);

(iii) there exists $A \in \mathbb{R}^k$ such that $\mathcal{F}_A(0) \geq 0$.

Then we can define a set

$$\mathcal{R} = \{a \in \mathbb{R}^k \mid \mathcal{F}_A(n) > 0 \quad \text{either for all } n \in \mathbb{Z}_+,$$
$$\text{or for } 0 \leq n \leq N-1 \text{ and } \mathcal{F}_A(N) = 0\}.$$

This set is not empty. For each $a \in \mathcal{R}$, we are in the situation of Theorems 2.1 and 2.2. In this case, the polynomial $f(A) \stackrel{\text{def}}{=} \mathcal{F}_A(0)$ is the right-hand side of the relation $[\mathbf{C}, \mathbf{B}] = f(\mathbf{A})$ in (1.1). The roots of the polynomial

$$\mathcal{F}_A(n) = \beta \prod_{j=1}^{p} \left(n - b_j(A)\right) \cdot \gamma \prod_{j=1}^{r} \left(n - c_j(A)\right)$$

are the parameters of the hypergeometric reproducing kernel (2.9) that corresponds to the irreducible representation of algebra (1.1). Thus, *the polynomials in questions are the polynomials compatible with mappings* $\mathbb{R}^k \to \mathbb{R}^k$.

2.3. Representations by q-differential operators. Now we drop assumption (c) in Lemma 2.1. Namely, let us consider relations (1.1) in which the structural mapping $\varphi \colon \mathbb{R}^k \to \mathbb{R}^k$ acts by contractions or dilations. For simplicity, we assume that φ is a linear mapping and its spectrum can be written as follows:

$$(2.12) \qquad \operatorname{spec}(\varphi) = \{q^{m_1}, \ldots, q^{m_k}\}, \qquad m_\mu \in \mathbb{Z},$$

where q is some positive number. We also assume that φ has no Jordan blocks. Then after a linear change of variables A_μ in (1.1), we can reduce the situation to the case of a diagonal mapping φ:

$$(2.12a) \qquad \varphi(A)_\mu = q^{m_\mu} A_\mu, \qquad \mu = 1, \ldots, k.$$

In this case, irreducible representations of the algebra (1.1) can be constructed via q-differential operators, and the coherent states, as well as the reproducing kernel, can be derived by using q-hypergeometric series (or "basic hypergeometric series").

Let us introduce a *q-dilation (contraction)* operator acting on antiholomorphic distributions as follows:

$$(2.13) \qquad (I^{(q)}g)(\bar{z}) \stackrel{\text{def}}{=} g(q\bar{z}), \qquad I^{(q)} = q^{\bar{z}\,d/d\bar{z}}.$$

Let us also introduce the operator of the *q-derivative*:

$$(2.13a) \qquad \delta^{(q)} = \frac{1}{\bar{z}(1-q)}(I - I^{(q)}), \qquad (\delta^{(q)}g)(\bar{z}) = \frac{g(\bar{z}) - g(q\bar{z})}{\bar{z} - q\bar{z}}.$$

The operators $\bar{z}$, $\delta^{(q)}$, and $I^{(q)}$ generate q-analogs of the Heisenberg algebra:

$$\delta^{(q)} \circ \bar{z} - \bar{z} \circ \delta^{(q)} = I^{(q)} \quad \text{or} \quad \delta^{(q)} \circ \bar{z} - q\bar{z} \circ \delta^{(q)} = I,$$
$$(2.13b) \qquad \delta^{(q)} \circ I^{(q)} - q I^{(q)} \circ \delta^{(q)} = 0,$$
$$I^{(q)} \circ \bar{z} - q\bar{z} \circ I^{(q)} = 0.$$

One also has

$$(2.14) \qquad \delta^{(q^{-1})} = (I^{(q)})^{-1} \circ \delta^{(q)}.$$

DEFINITION 2.1. An operator is called a *q-differential operator* if it can be represented by a polynomial function of $\delta^{(q)}$ and $\delta^{(q^{-1})}$ with coefficients that are polynomial functions of $\bar{z}$.

In view of (2.13) and (2.14), q-differential operators can also be described as polynomial functions of the operators $\bar{z}$, $\delta^{(q)}$, $I^{(q)}$, and $(I^{(q)})^{-1}$.

Note that if $q \to 1$, then the q-derivative $\delta^{(q)}$ approaches the ordinary derivative $\frac{d}{dz}$, but the operator $\frac{d}{dz}$ itself is not a q-differential operator.

We are interested in conditions that guarantee the realization of algebra (1.1) by q-differential operators.

Let us consider the structural function f from (1.1). Suppose that f is a polynomial in A:

$$(2.15) \qquad f(A) = \sum_{|\alpha|=0}^{M} f_\alpha A^\alpha.$$

Here $\alpha = (\alpha_1, \ldots, \alpha_k)$ are multi-indices, $\alpha_\mu \in \mathbb{Z}_+$. Then

$$(2.15\text{a}) \qquad f(\varphi(A)) = \sum_{|\alpha|=0}^{M} f_\alpha q^{m \cdot \alpha} A^\alpha,$$

where $m = (m_1, \ldots, m_k)$ is an integer-valued vector from (2.12). From (2.15) we see that there is a φ-stable part of the polynomial f, that is,

$$f_*(A) \stackrel{\text{def}}{=} \sum_{m \cdot \alpha = 0} f_\alpha A^\alpha, \qquad f_*(\varphi(A)) = f_*(A).$$

By $\mathcal{R}_*$ we denote the zero set of this φ-stable part:

$$(2.16) \qquad \mathcal{R}_* = \{A \in \mathbb{R}^k \mid f_*(A) = 0\}.$$

LEMMA 2.2. *Suppose that*
(a) *f is a polynomial, and*
(b) *φ is linear and of form (2.12a).*
Then if $a \in \mathcal{R}_*$, the function $\mathcal{F}_a$ (1.3) has the form

$$(2.17) \qquad \mathcal{F}_a(n) = \frac{1 - q^{n+1}}{(n+1)(1-q)} \cdot \frac{\Gamma_a(q^n)}{q^{nL}},$$

where $L \in \mathbb{Z}$, Γ_a is a polynomial, $\Gamma_a(1) = f(a)$, and $\Gamma_a(0) \neq 0$ if $f(a) \neq 0$.

PROOF. By (2.15) and (2.16), we have

$$f\bigl(\underbrace{\varphi(\ldots \varphi(a) \ldots)}_{n}\bigr) = \sum_{m \cdot \alpha \neq 0} f_\alpha (q^{m \cdot \alpha})^n a^\alpha,$$

and thus,

$$\mathcal{F}_a(n) = \sum_{m \cdot \alpha \neq 0} f_\alpha a^\alpha \frac{1 - (q^{n+1})^{m \cdot \alpha}}{(1 - q^{m \cdot \alpha})(n+1)}.$$

So, formula (2.17) is satisfied with

$$\Gamma_a(x) = x^L \sum_{m\cdot\alpha>0} \frac{1+qx+(qx)^2+\cdots+(qx)^{m\cdot\alpha-1}}{1+q+q^2+\cdots+q^{m\cdot\alpha-1}} f_\alpha a^\alpha$$
$$+ \sum_{m\cdot\alpha<0} x^{L-|m\cdot\alpha|} \frac{1+qx+(qx)^2+\cdots+(qx)^{|m\cdot\alpha|-1}}{1+q+q^2+\cdots+q^{|m\cdot\alpha|-1}} f_\alpha a^\alpha.$$

This is an explicit formula for the polynomial Γ_a, and the number L is determined as follows: either $L \geq 1$ is the maximal integer such that $\sum_{m\cdot\alpha=-L} f_\alpha a^\alpha \neq 0$, or if there are no such integers, then $L \leq 0$ is the maximal integer such that

$$\sum_{m\cdot\alpha \geq 1-L} \frac{f_\alpha a^\alpha}{1+q+\cdots+q^{m\cdot\alpha-1}} \neq 0. \quad \square$$

The results of §1 imply the following theorem.

THEOREM 2.3. *Suppose that* $f(A) = \sum_{|\alpha|=0}^M f_\alpha A^\alpha$ *is a polynomial on* $\mathbb{R}^k$ *and* m_μ, $\mu = 1,\ldots,k$, *are integer numbers. Then for the commutation relations*

(2.18)
$$[\mathbf{C},\mathbf{B}] = f(\mathbf{A}),$$
$$\mathbf{C}\mathbf{A}_\mu = q^{m_\mu} \mathbf{A}_\mu \mathbf{C}, \quad \mathbf{A}_\mu \mathbf{B} = q^{m_\mu} \mathbf{B}\mathbf{A}_\mu, \quad \mu = 1,\ldots,k,$$

all irreducible Hermitian representations that possess a vacuum vector (1.2) *can be described as follows. Choose* $a \in \mathbb{R}^k$ *so that* $f(a) > 0$ *and* $\sum_{m\cdot\alpha=0} f_\alpha a^\alpha = 0$. *Then there is an integer* L *such that*

(2.19) $$\frac{(1-q)q^{nL}}{1-q^{n+1}} \sum_{j=0}^n \sum_{|\alpha|=0}^M f_\alpha (q^{m\cdot\alpha})^j a^\alpha = \Gamma_a(q^n), \quad \forall n \in \mathbb{Z}_+,$$

where Γ_a *is a polynomial such that* $\Gamma_a(0) \neq 0$. *If the polynomial* Γ_a *is positive on the entire lattice* $\{q^n \mid n \in \mathbb{Z}_+\}$, *then each factorization of* Γ_a *into two polynomial factors*

(2.20) $$\Gamma_a = \Xi_a \cdot \Upsilon_a$$

and each decomposition

(2.20a) $$L = \ell' + \ell'', \quad \ell', \ell' \in \mathbb{Z},$$

generate an irreducible Hermitian representation of the algebra (2.18) *in the space* $\mathcal{P}_{s(a)}$ *by the q-differential operators*

(2.21)
$$\overset{\circ}{B} = \bar{z}\, \Xi_a(I^{(q)}) \cdot (I^{(q)})^{-\ell'},$$
$$\overset{\circ}{C} = \Upsilon(I^{(q)}) \cdot (I^{(q)})^{-\ell''} \cdot \delta^{(q)},$$
$$\overset{\circ}{A} = a_\mu \cdot (I^{(q)})^{m_\mu}.$$

Here the q-dilation and q-derivative operators $I^{(q)}$ and $\delta^{(q)}$ are given by (2.13) and (2.13a), and the norm (1.8) in the space $\mathcal{P}_{s(a)}$ is determined by the sequence

$$s_0(a) = 1,$$
(2.22)
$$s_n(a) = \frac{(q;q)_n}{(1-q)^n} \cdot \frac{\overline{\Upsilon}_a(q^{n-1})\ldots\overline{\Upsilon}_a(1)}{\Xi_a(q^{n-1})\ldots\Xi_a(1)} \cdot q^{\frac{n(n-1)}{2}(\ell'-\ell'')}, \qquad n \geq 1.$$

We use the following notation for the shifted q-factorial (or the Pochhammer q-symbol):

(2.22a)
$$(\alpha;q)_n \stackrel{\text{def}}{=} \begin{cases} 1, & n = 0, \\ (1-\alpha)(1-\alpha q)\ldots(1-\alpha q^{n-1}). \end{cases}$$

If the polynomial Γ_a is positive on the subset $\{q^n \mid 0 \leq n < N\}$ and

(2.23)
$$\Gamma_a(q^N) = 0,$$

then each factorization (2.20) such that $\Xi(q^N) = 0$ and each decomposition (2.20a) generate an irreducible representation of the algebra (2.18) by q-differential operators (2.21) in the space of polynomials of degree N.

For a given a, the representations that correspond to different factorizations (2.20) are equivalent to each other.

2.4. Relations with basic q-hypergeometric series. According to the notation of [**GaR**], the basic hypergeometric series are defined as follows:

$$_p\Phi_r[b;c;q;z] = \sum_{n=0}^{\infty} \frac{(b;q)_n}{(c;q)_n} \cdot \left((-1)^n q^{\frac{n(n-1)}{2}}\right)^{1+r-p} \cdot \frac{z^n}{(q;q)_n},$$

where $(\cdot;q)_n$ is a shifted q-factorial (2.22a), $b = (b_1,\ldots,b_p)$, and $c = (c_1,\ldots,c_r)$.

Similarly to §2.2, we also need finite sums of this series which we call *basic hypergeometric jets*:

$$_p\Phi_r[b;c;q;z;N] = \sum_{n=0}^{N} \frac{(b;q)_n}{(c;q)_n} \cdot \left((-1)^n q^{\frac{n(n-1)}{2}}\right)^{1+r-p} \cdot \frac{z^n}{(q;q)_n}.$$

Now let us factor each polynomial Ξ and Υ in (2.20) as follows:

$$\Xi(x) = \lambda \prod_{j=1}^{p}(1-\xi_j x), \qquad \Upsilon(x) = \rho \prod_{j=1}^{r}(1-\upsilon^j x).$$

Thus, $1/\xi_j$ and $1/\upsilon^j$ are roots of the polynomial Γ_a (2.19), $p+r$ is the degree of this polynomial, and

(2.20b)
$$\lambda\rho = \frac{f(a)}{\prod_{j=1}^{p}(1-\xi_j)\prod_{j=1}^{r}(1-\upsilon^j)} = \Gamma_a(0).$$

THEOREM 2.4. *Suppose that the roots of the polynomial Γ_a (2.19) are divided into two groups $1/\xi_1, \ldots, 1/\xi_p$ and $1/\upsilon_1, \ldots, 1/\upsilon_r$ (in particular, $1/\xi_j = q^N$ is one of the roots in case (2.23)). We also choose integers ℓ' and ℓ'' from condition (2.20a) and complex constants λ and ρ from condition (2.20b).*

Then under the assumptions of Theorem 2.3, the norm in the Hilbert space $\mathcal{P}_{s(a)}$ of the irreducible representation is given by the sequence

$$s_n(a) = \left(\frac{\overline{\rho}}{\lambda}\right)^n \cdot \frac{(q;q)_n}{(1-q)^n} q^{\frac{n(n-1)}{2}(\ell'-\ell'')} \cdot \frac{(\overline{\upsilon};q)_n}{(\xi;q)_n},$$

the coherent states are given by the formula

$$(2.24) \qquad \mathfrak{P}_z = \sum_{n=0}^{N} \frac{1}{(\overline{\upsilon};q)_n} \cdot \frac{q^{\frac{n(n-1)}{2}\ell''}}{(q;q)_n} \left(\frac{(1-q)}{\overline{\rho}} z \mathbf{B}\right)^n \mathfrak{P}_0,$$

and the reproducing kernel is given by the formula

$$(2.25) \qquad \mathcal{K}(\overline{w},z) = \sum_{n=0}^{N} \frac{(\xi;q)_n}{(\overline{\upsilon};q)_n} \cdot \frac{q^{\frac{n(n-1)}{2}(\ell''-\ell')}}{(q;q)_n} \left(\frac{\lambda(1-q)}{\overline{\rho}} \overline{w} z\right)^n.$$

Here $1 \leq N \leq \infty$, and N is finite in the case of (2.23). In addition, if we choose $\ell'' = 1 + r$, then formula (2.24) reads

$$(2.26) \qquad \mathfrak{P}_z = {}_0\Phi_r\left[\overline{\upsilon}; q; \frac{(-1)^{1+r}(1-q)}{\overline{\rho}} z\mathbf{B}\right]\mathfrak{P}_0,$$

and if we choose $\ell'' - \ell' = 1 + r - p$, then formula (2.25) reads

$$(2.27) \qquad \mathcal{K}(\overline{w},z) = {}_p\Phi_r\left[\xi; \overline{\upsilon}; q; \frac{(-1)^{1+r-p}\lambda(1-q)}{\overline{\rho}} \overline{w}z\right].$$

Here ${}_p\Phi_r$ are basic hypergeometric series. In the case $N < \infty$, both series (2.26) and (2.27) have to be replaced by basic hypergeometric jets of degree N.

REMARK 2.1. In Theorems 2.3 and 2.4, we consider only the case of a diagonal mapping φ. In fact, one can generalize these theorems to the case of mappings φ with additional Jordan blocks (and with spectrum (2.12)). But the calculations related to this case are too cumbersome, so we do not present them here. Nevertheless, the diagonal case (2.12a) is itself very interesting and covers many important examples (see in the sequel). On the contrary, the presence of Jordan blocks is crucial in the situation of Theorems 2.1 and 2.2 and, in fact, controls the most exciting phenomena.

REMARK 2.2. The relationship between q-hypergeometric series and coherent states was studied in the recent interesting papers [**MOd**, **Od**$_2$] but in a different way, beyond the scope of the representation theory for algebras with permutation relations of type (2.18). The paper [**Od**$_2$] deals with the special case $\overset{\circ}{A} = \overline{z} d/d\overline{z}$ and with the relation $\overset{\circ}{B}\overset{\circ}{C} = \rho(\overset{\circ}{A})$ that holds in a particular irreducible component, while the basic q-polynomial $\Gamma_a(q^n)$ (2.19), as well as the ordinary hypergeometric series for $q = 1$, are not considered (that was the starting point of our considerations; see [**KaN**$_1$]).

2.5. Further q-deformation.
Now we consider the following permutation relations

$$\mathbf{CB} = q^\ell \mathbf{BC} + \sum_{|\alpha|=0}^{M} f_\alpha \mathbf{A}^\alpha,$$

(2.28)

$$\mathbf{CA}_\mu = q^{m_\mu} \mathbf{A}_\mu \mathbf{C}, \qquad \mathbf{A}_\mu \mathbf{B} = q^{m_\mu} \mathbf{BA}_\mu, \qquad \mu = 1, \ldots, k,$$

where q is a positive number, $\ell, m_\mu \in \mathbb{Z}$, and f_α are real constants.

The algebra (2.28) is a special case of the algebra (1.48), where

(2.29) $\quad \omega(A) \equiv q^\ell, \qquad f(A) = \sum_{|\alpha|=0}^{M} f_\alpha A^\alpha, \qquad \varphi(A)_\mu = q^{m_\mu} A_\mu, \qquad \mu = 1, \ldots, k.$

Previously (see §1.4), it was shown how to construct irreducible representations of the algebra (1.48), and hence of the algebra (2.28). The aim of this subsection is to state conditions on the structural constants ℓ, m_μ, and f_α under which the algebra (2.28) admits irreducible representations via q-differential operators (defined in §2.3), and to present these representations and the coherent transforms related to them.

By $f_*^{(\ell)}$ we denote the part of the polynomial (2.29) that satisfies the Floquet condition with respect to the mapping φ (2.29) with Floquet multiplier q^ℓ, that is,

$$f_*^{(\ell)}(A) \overset{\text{def}}{=} \sum_{m \cdot \alpha = \ell} f_\alpha A^\alpha, \qquad f_*^{(\ell)}(\varphi(A)) = q^\ell f_*^{(\ell)}(A).$$

By $\mathcal{R}_*^{(\ell)}$ we denote the zero set of the function $f_*^{(\ell)}$, that is,

(2.30) $\quad \mathcal{R}_*^{(\ell)} = \{A \in \mathbb{R}^k \mid f_*^{(\ell)}(A) = 0\}.$

LEMMA 2.3. *Let ℓ be an integer, and let $m = (m_1, \ldots, m_k)$ be an integer-valued vector from (2.28). Suppose that $a \in \mathcal{R}_*^{(\ell)}$. Then the function $\mathcal{F}_a^\omega$ (1.50) has the form*

(2.31) $\quad \mathcal{F}_a^\omega(n) = \dfrac{1 - q^{n+1}}{(n+1)(1-q)} \cdot \dfrac{\Gamma_a^{(\ell)}(q^n)}{q^{nL}},$

where $L \in \mathbb{Z}$, $\Gamma_a^{(\ell)}$ is a polynomial, $\Gamma_a^{(\ell)}(1) = f(a)$, and $\Gamma_a^{(\ell)}(0) \neq 0$ if $f(a) \neq 0$.

PROOF. By (1.50) and (2.29), we have

$$\mathcal{F}_a^\omega(n) = q^{\ell n} \sum_{m \cdot \alpha \neq \ell} f_\alpha a^\alpha \frac{1 - (q^{n+1})^{m \cdot \alpha - \ell}}{(1 - q^{m \cdot \alpha - \ell})(n+1)}.$$

So, (2.30) is satisfied with

(2.32) $\quad \Gamma_a^{(\ell)}(x) = x^{L+\ell} \sum_{m \cdot \alpha > \ell} \dfrac{1 + qx + (qx)^2 + \cdots + (qx)^{m \cdot \alpha - \ell - 1}}{1 + q + q^2 + \cdots + q^{m \cdot \alpha - \ell - 1}} f_\alpha a^\alpha$

$\qquad + \sum_{m \cdot \alpha < \ell} x^{L + m \cdot \alpha} \dfrac{1 + qx + (qx)^2 + \cdots + (qx)^{|m \cdot \alpha - \ell| - 1}}{1 + q + q^2 + \cdots + q^{|m \cdot \alpha - \ell| - 1}} f_\alpha a^\alpha.$

Here the number L is determined as follows: either $L \geq 1 - \ell$ is the maximal integer such that $\sum_{m \cdot \alpha = -L} f_\alpha a^\alpha \neq 0$, or if there are no such integers, then $L \leq -\ell$ is the maximal integer such that

$$\sum_{m \cdot \alpha \geq 1 - L} \frac{f_\alpha a^\alpha}{1 + q + \cdots + q^{m \cdot \alpha - \ell - 1}} \neq 0.$$

THEOREM 2.5. *The statements of Theorems* 2.3 *and* 2.4 *hold for the commutation relations* (2.28), *where the set* $\mathcal{R}_*$ *and the polynomial* Γ_a *are replaced by* $\mathcal{R}_*^{(\ell)}$ *and* $\Gamma_a^{(\ell)}$ *defined in* (2.30), (2.32).

§3. Quantum embeddings of surfaces

3.1. Surfaces of revolution and families of hypergeometric functions. Now we return to surfaces of revolution Ω embedded by the equations

(3.1) $$\Omega : \begin{cases} S_1^2 + S_2^2 - \rho(A) = \text{const}, \\ \varkappa_j(A) = \text{const}, \quad j = 1, \ldots, k - 1, \end{cases}$$

into the space $\mathbb{R}^{k+2}$ with coordinates $S_1, S_2, A_1, \ldots, A_k$.

This is a classical embedding $\Omega \subset \mathbb{R}^{k+2}$. The problem of quantum embedding can be formulated as follows.

We recall that a manifold $\mathcal{M}$ is called a *quantum manifold* if a linear space $\mathcal{F}_\hbar$ of functions over $\mathcal{M}$ is endowed with an associative algebra structure possessing the unit 1 (see in [**B**$_{2,4}$, **BFFLS**$_{1,2}$, **C**]). The elements of the center of this algebra are called *Casimir functions*.

DEFINITION 3.1. A *quantum embedding* of a surface Ω into $\mathbb{R}^{k+2}$ is an epimorphism of associative algebras

(3.2) $$\mathcal{F}_\hbar(\mathbb{R}^{k+2}) \to \mathcal{F}_\hbar(\Omega).$$

DEFINITION 3.2. The quantum embedding (3.2) is called a *quantum surface of revolution* if the Casimir functions in $\mathcal{F}_\hbar(\mathbb{R}^{k+2})$ are of the form $S_1^2 + S_2^2 - \rho(A)$, $\varkappa_1(A), \ldots, \varkappa_{k-1}(A)$.

The problem is to construct such a quantum surface starting with the classical surface (3.1).

We have introduced the index $\hbar$ that labels quantum algebras of functions. This means that sometimes it is interesting to consider not a single quantization but a family of quantizations, say, for $0 \leq \hbar \leq 1$, such that the quantum embedding is trivial for the limit value $\hbar = 0$ (that is, the products in all algebras $\mathcal{F}_0$ coincide with the usual commutative products).

By §1.3, we know that the existence of Casimir elements of type (3.1) follows from the quasiquadratic permutation relations (1.1) between generators (that is, quantum coordinates) $\mathbf{B} = \mathbf{S}_1 - i\mathbf{S}_2$, $\mathbf{C} = \mathbf{S}_1 + i\mathbf{S}_2$, and $\mathbf{A}$.

We slightly change these relations, introducing a parameter $\hbar$ and taking equation (1.28) into account:

(3.3) $$[\mathbf{C}, \mathbf{B}] = \rho(\varphi^\hbar(\mathbf{A})) - \rho(\mathbf{A}), \qquad \mathbf{CA} = \varphi^\hbar(\mathbf{A})\mathbf{C}, \qquad \mathbf{AB} = \mathbf{B}\varphi^\hbar(\mathbf{A}).$$

Here $\varphi^\hbar$ is the shift by time $\hbar$ along the trajectories of a vector field v (see Lemma 1.4).

The classical Poisson algebra structure corresponding to relations (3.3) is given by the following brackets between the classical coordinates on $\mathbb{R}^{k+2}$:

(3.3a)
$$\begin{aligned} \{C,B\} &= if^0(A), & \{S_1,S_2\} &= -(1/2)f^0(A), \\ \{C,A\} &= iv(A)C, \quad \text{or} \quad & \{S_1,A\} &= -v(A)S_2, \\ \{B,A\} &= -iv(A)B, & \{S_2,A\} &= v(A)S_1, \end{aligned}$$

where

$$f^0(A) \stackrel{\text{def}}{=} \langle v, d\rho \rangle = \sum_{\mu=1}^{k} v_\mu(A) \frac{\partial \rho(A)}{\partial A_\mu}.$$

Symplectic leaves of (3.3a) are determined by the classical Casimir functions

$$\Omega = \{BC - \rho(A) \equiv S_1^2 + S_2^2 - \rho(A) = \text{const}, \; \varkappa_\mu = \text{const}\},$$

where the $\varkappa_\mu$ ($\mu = 1, \ldots, k-1$) are functions on $\mathbb{R}^k$ constant along the trajectories of the field v.

Possible ways to quantize Ω are the following ones:
- Consider the classical symplectic form ω_0 on Ω, corresponding to the brackets (3.3a), and try to construct representations of some functional algebras, starting from a line bundle over Ω with the curvature form ω_0 (this is the basic geometric quantization way).
- Consider deformations of the classical Poisson brackets (3.3a) on Ω, as well as deformations of the commutative product of functions over Ω, by using formal power series in a parameter $\hbar$ (this is the deformation quantization way).
- Construct the sheaves of functions over Ω that oscillate as $\hbar \to 0$, and define a noncommutative product on such functions, as well as its irreducible representation (an analog of the Fourier integral calculus) modulo $O(\hbar^\infty)$ (this is the asymptotic quantization way).
- Employing the algebra (3.3) and its quantum groupoid, construct a complex structure, a quantum symplectic form, as well as transition amplitudes (in particular, the probability function) on the symplectic (complex) groupoid over Ω; then construct a noncommutative product and its irreducible representation, using these amplitudes and avoiding deformations, or asymptotic approximations in $\hbar$ (this is the groupoid geometric quantization way).

Here we follow this last way and show how to solve the quantum embedding problem for symplectic leaves

(3.1a) $\quad \Omega_a = \{BC = \rho(A) - \rho(a), \; \varkappa_\mu(A) = \varkappa_\mu(a), \; \mu = 1, \ldots, k-1\}, \quad a \in \mathcal{R}.$

Here $\mathcal{R} \subset \mathbb{R}^k$ is the subset labeling irreducible representations of the algebra (3.3) that are Hermitian and possess a vacuum vector.

To conclude this preliminary subsection, we briefly reformulate the results of §1 for relations (3.3).

By $\mathfrak{A}_A(t)$ we denote the trajectory of the field v that starts at $A \in \mathbb{R}^k$:

(3.4) $$\frac{d}{dt}\mathfrak{A}_A = v(\mathfrak{A}_A), \quad \mathfrak{A}_A\Big|_{t=0} = A.$$

We recall that the subset $\mathcal{R} \subset \mathbb{R}^k$ can be defined as follows: $a \in \mathcal{R}$ if and only if either

(3.5) $\qquad \rho\big(\mathfrak{A}_a(\hbar n)\big) > \rho(a) \qquad$ for all $\quad n = 1, 2, \ldots,$

or

(3.5a) $\qquad \begin{aligned} \rho\big(\mathfrak{A}_a(\hbar n)\big) &> \rho(a) \qquad \text{for} \quad n = 1, 2, \ldots, N, \\ \rho\big(\mathfrak{A}_a(\hbar(N+1))\big) &= \rho(a). \end{aligned}$

Let us factor the difference $\rho(A) - \rho(a)$ similarly to (1.36):

(3.5b) $\qquad \rho(A) - \rho(a) = \mathcal{D}_a(A)\,\mathcal{E}_a(A)$

so that $\mathcal{E}_a(a) = 0$, $\mathcal{D}_a(a) \neq 0$, and

(3.5c) $\qquad \begin{aligned} \mathcal{D}_a\big(\mathfrak{A}_a(\hbar(N+1))\big) &= 0, \\ \mathcal{E}_a\big(\mathfrak{A}_a(\hbar(N+1))\big) \neq 0, \qquad \frac{\partial \mathcal{D}_a}{\partial a'}\big(\mathfrak{A}_a(\hbar(N+1))\big) &\neq 0. \end{aligned}$

Compared to (1.37) and (1.38), here we have introduced some additional inequalities that provide useful simplifications in the sequel but, in fact, are not necessary.

COROLLARY 3.1. *If $a \in \mathcal{R}$, then the Hermitian irreducible representation of relations (3.3) is realized in the space $\mathcal{P}$ of antiholomorphic distributions $g(\bar{z}) = \sum_{n=0}^{N} g_n \bar{z}^n$ with the norm*

$$\|g\|_\mathcal{P}^2 = |g_0|^2 + \sum_{n=1}^{N} \mathcal{H}(\hbar n) \ldots \mathcal{H}(\hbar) |g_n|^2,$$

where $N = \infty$ in the case (3.5), and

(3.6) $\qquad \mathcal{H}(t) \stackrel{\mathrm{def}}{=} X_a(\mathfrak{A}_a(t)), \qquad X_a \stackrel{\mathrm{def}}{=} \overline{\mathcal{E}}_a / \mathcal{D}_a.$

This representation is given by the operators

(3.7) $\qquad \overset{\circ}{B} = \mathcal{D}_a(\overset{\circ}{A}) \cdot \bar{z}, \qquad \overset{\circ}{C} = \frac{1}{\bar{z}} \cdot \mathcal{E}_a(\overset{\circ}{A}), \qquad \overset{\circ}{A} = \mathfrak{A}_a\!\left(\hbar \bar{z} \frac{d}{d\bar{z}}\right).$

The corresponding coherent states in an abstract Hilbert space H_a and the reproducing kernel are

(3.8) $\qquad \mathfrak{P}_z = \mathfrak{P}_0 + \sum_{n=1}^{N} \frac{1}{\overline{\mathcal{E}}_a\big(\mathfrak{A}_a(\hbar n)\big) \ldots \overline{\mathcal{E}}_a\big(\mathfrak{A}_a(\hbar)\big)} (z\mathbf{B})^n \mathfrak{P}_0,$

(3.9) $\qquad \mathcal{K}(\overline{w}, z) = 1 + \sum_{n=1}^{N} \frac{(\overline{w} z)^n}{\mathcal{H}(\hbar n) \ldots \mathcal{H}(\hbar)}.$

In the polynomial case (that is, if the functions ρ and φ are polynomial), these reproducing kernels and coherent states can be represented by hypergeometric series of type (2.7) as in §2.2. The parameters of these hypergeometric series are $(-b; -\overline{c})$, where $\hbar(b_j + 1)$ and $\hbar(c_j + 1)$ are the roots of the polynomial $\big(\rho(\mathfrak{A}_a(t)) - \rho(a)\big)/t$ in the variable t. So, these parameters are controlled by the function ρ and the vector field v from (3.3); thus, *these parameters correspond to the surface of revolution* (3.1).

It is interesting to consider the opposite question. Suppose that we start with a family of hypergeometric series whose parameters b and c depend on a variable $A \in \mathbb{R}^k$. When and how is it possible to assign to this family some surface of revolution of type (3.1) and some quantum algebra with nonlinear permutation relations of type (3.3) or (1.1)?

We briefly describe the answer to this question.

Let v be a vector field on $\mathbb{R}^k$ with polynomial coefficients. Suppose that all equilibrium points of v are simple (nondegenerate) and the Jacobi matrix Dv is nilpotent at each point.

Let us consider a polynomial $\mathcal{F}^0(A, t)$ in variables $A \in \mathbb{R}^k$ and $t \in \mathbb{R}$. We say that $\mathcal{F}^0$ is *compatible* with the vector field v if the following conditions are satisfied:

(i) The degree of $\mathcal{F}^0(A, t)$ with respect to t does not depend on A.

(ii) The identity

$$(t + t') \mathcal{F}^0(A, t + t') = t \mathcal{F}^0(A, t) + t' \mathcal{F}^0(\mathfrak{A}_A(t), t')$$

holds for any $t, t' \in \mathbb{R}^k$ (here $\mathfrak{A}_A(t)$ is the trajectory of the dynamical system (3.4) generated by the vector field v).

(iii) There exists $A \in \mathbb{R}^k$ such that $\mathcal{F}^0(A, 0) > 0$.

(iv) The function $f^0(A) \stackrel{\text{def}}{=} \mathcal{F}^0(A, 0)$ vanishes on the set of equilibrium points of the field v.

For each polynomial $\mathcal{F}^0$ compatible with a vector field v, on $\mathbb{R}^k$ there exist polynomial functions $\varkappa_\mu$, $\mu = 1, \ldots, k-1$, and ρ such that

$$v(\varkappa_\mu) = 0, \quad v(\rho) = f^0.$$

Thus, *a surface of revolution of type* (3.1) *is assigned to any compatible pair* $\mathcal{F}^0, v$.

Now we can define the set

$[\mathcal{R}] = \{A \in \mathbb{R}^k \mid \mathcal{F}^0(A, t) > 0 \quad \text{for all } t \in [0, T),$

where either $t = \infty$ or $0 < T < \infty$;

in the latter case, $T = T(A)$ is a simple root of $\mathcal{F}^0(A, t)$ in the variable $t\}$.

This set is not empty.

Let us represent the polynomial $\mathcal{F}^0$ in the form

$$\mathcal{F}^0(A, t) = e(A) \prod_{j=1}^{r} (t - t_j(A)) \cdot \prod_{j=1}^{p} (t - \tau_j(A)),$$

where $e(A)$ does not vanish everywhere on $\mathbb{R}^k$ and $t_j(A)$, $\tau_j(A)$ are the roots of $\mathcal{F}^0$ with respect to the variable t. We divide these roots into two groups in an arbitrary way, requiring only that if $A \in [\mathcal{R}]$ and $T(A) < \infty$, then the root $T(A)$ should be among the roots $\tau_j(A)$, $j = 1, \ldots, p$.

We claim that the following statement holds. Let $\mathcal{F}^0$ be compatible with the field v, and let $a \in [\mathcal{R}]$. Then the operators

$$\overset{\circ}{B} = \prod_{j=1}^{p} \left(\hbar \bar{z} \frac{d}{d\bar{z}} - \tau_j(a)\right) \cdot \bar{z}, \quad \overset{\circ}{C} = e(a) \cdot \hbar \frac{d}{d\bar{z}} \cdot \prod_{j=1}^{r} \left(\hbar \bar{z} \frac{d}{d\bar{z}} - t_j(a)\right), \quad \overset{\circ}{A} = \mathfrak{A}_a\left(\hbar \bar{z} \frac{d}{d\bar{z}}\right)$$

generate an irreducible Hermitian representation of the relations (3.3) with Casimir elements $\varkappa_\mu(\mathbf{A})$ and $\mathbf{K} = \mathbf{BC} - \rho(\mathbf{A})$ satisfying (1.32), (1.33).

Here $\hbar > 0$ is arbitrary if $T(a) = \infty$, but

if $T(a) < \infty$, then $\dfrac{T(a)}{\hbar}$ should be a positive integer (*the quantization condition*).

In the latter case, the number $T(a)/\hbar$ is the dimension of the representation. The norm in the Hilbert space of the representation is defined in the same way as in Corollary 3.1, where

$$\mathcal{H}(t) \stackrel{\text{def}}{=} \overline{e(a)}\, t \prod_{j=1}^{r} \left(t - \overline{t_j(a)}\right) \cdot \prod_{j=1}^{p} \left(t - \tau_j(a)\right)^{-1}.$$

The coherent states corresponding to this representation are given by (3.8), where $\mathcal{E}_a\bigl(\mathfrak{A}_a(t)\bigr) \stackrel{\text{def}}{=} e(a)\, t \prod_{j=1}^{r}\left(t - t_j(a)\right)$. The reproducing kernel is given by (3.9), which reads as follows:

$$\mathcal{K}(\overline{w}, z) = {}_pF_r\left(1 - \dfrac{\tau(a)}{\hbar};\, 1 - \dfrac{\overline{t(a)}}{\hbar};\, \dfrac{\overline{w}z}{\overline{e(a)}\,\hbar^{r-p+1}}\right),$$

where ${}_pF_r$ is the hypergeometric series (2.7).

Thus, *if the parameters $1 - \tau_j/\hbar$, $1 - \overline{t_j}/\hbar$, and $1/\overline{e}\,\hbar^{r-p+1}$ of the hypergeometric series ${}_pF_r$ are related to the roots τ_j, t_j and to the main coefficient $e(a)$ of the polynomial $\mathcal{F}^0$ which is compatible with a vector field v over $\mathbb{R}^k$, then to this hypergeometric series one can assign a quantum algebra of type* (3.3) *or* (1.1) *and its irreducible representations, as well as the classical surface of revolution* (3.1).

As we shall see below (§§3.3, 3.6, and 4.6), in this situation, one can obtain much more:

– not only classical, but quantum surfaces of revolution related to a given hypergeometric function,

– the classical (as well as quantum) complex structure on the surfaces of revolution,

– the semiclassical approximation of hypergeometric functions as $\hbar \to 0$ expressed in geometric terms (via the Kähler potential related to the complex structure and the symplectic form on the surface of revolution), and

– analogs of these results for hypergeometric functions of several variables and corresponding hypersurfaces of multirevolution.

3.2. Antimeromorphic coherent states and the corresponding irreducible representations.

The coherent states $\mathfrak{P}_z$ considered above were holomorphic in z, and the basic Hilbert space $\mathcal{P}$ was made up of antiholomorphic distributions in the variable z.

Now we would like to introduce the dual picture, replacing holomorphic functions by meromorphic ones. By $\widetilde{\mathcal{P}}$ we denote the Hilbert space of meromorphic distributions over $\mathbb{C} \setminus \{0\}$ of the type

$$\widetilde{g}(z) = \sum_{n=0}^{N} \dfrac{\widetilde{g}_n}{z^{n+1}}$$

with the norm

$$(3.10) \qquad \|\widetilde{g}\|_{\widetilde{\mathcal{P}}}^2 \stackrel{\text{def}}{=} |\widetilde{g}_0|^2 + \sum_{n=1}^{N} \dfrac{|\widetilde{g}_n|^2}{\mathcal{H}(\hbar n)\ldots\mathcal{H}(\hbar)}.$$

Here $N \leq \infty$, and N is finite in the case of (3.5a).

THEOREM 3.1. (a) *The norm (3.10) reads*

$$\|\widetilde{g}\|_{\widetilde{\mathcal{P}}}^2 = \frac{1}{(2\pi)^2} \oint \oint \mathcal{K}(\overline{w}, z)\, \widetilde{g}(z)\, \overline{\widetilde{g}(w)}\, dz dw, \tag{3.11}$$

where the contour integrals are taken over cycles around the origin $z = 0$ (oriented anticlockwise) and $\mathcal{K}$ is the reproducing kernel corresponding to the coherent states $\mathfrak{P}_z$ (3.8).

(b) *The duality between the spaces $\widetilde{\mathcal{P}}$ and $\mathcal{P}$ is given by the mapping*

$$\mathcal{K} \colon \widetilde{\mathcal{P}} \to \mathcal{P}, \qquad (\mathcal{K}\widetilde{g})(\overline{w}) \stackrel{\text{def}}{=} \frac{1}{2\pi i} \oint \mathcal{K}(\overline{w}, z)\, \widetilde{g}(z)\, dz. \tag{3.12}$$

The inverse mapping is

$$(\mathcal{K}^{-1}g)(z) = -\frac{1}{2\pi i} \oint \widetilde{\mathcal{L}}(\overline{w}, z)\, g(\overline{w})\, d\overline{w}, \tag{3.13}$$

where

$$\widetilde{\mathcal{L}}(\overline{w}, z) \stackrel{\text{def}}{=} \frac{1}{\overline{w}z} + \sum_{n=1}^{N} \frac{\mathcal{H}(\hbar n) \ldots \mathcal{H}(\hbar)}{(\overline{w}z)^{n+1}}. \tag{3.14}$$

(c) *The function (distribution) $\widetilde{\mathcal{L}}$ is a "reproducing kernel" corresponding to the antimeromorphic coherent states*

$$\widetilde{\mathcal{L}}(\overline{w}, z) = (\widetilde{\mathfrak{P}}_{\overline{z}}, \widetilde{\mathfrak{P}}_{\overline{w}})_{H_a}, \tag{3.15}$$

$$\widetilde{\mathfrak{P}}_{\overline{w}} \stackrel{\text{def}}{=} \overline{\mathcal{K}^{-1}\mathfrak{P}_z} = \frac{1}{2\pi i} \oint \widetilde{\mathcal{L}}(\overline{w}, z)\mathfrak{P}_z\, dz$$

$$= \frac{1}{\overline{w}}\mathfrak{P}_0 + \sum_{n=1}^{N} \frac{1}{\mathcal{D}_a(\mathfrak{A}_a(\hbar n)) \ldots \mathcal{D}_a(\mathfrak{A}_a(\hbar))} \cdot \frac{\mathbf{B}^n}{\overline{w}^{n+1}}\mathfrak{P}_0. \tag{3.16}$$

(d) *The Hermitian irreducible representation of the algebra (3.3) in the space $\widetilde{\mathcal{P}}$ dual to the representation (3.7) is given by*

$$\overset{\circ}{B} = \mathcal{K}^{-1}\overset{\circ}{B}\mathcal{K} = \overline{\mathcal{E}}_a(\overset{\circ}{A}) \cdot \frac{1}{z},$$

$$\overset{\circ}{C} = \mathcal{K}^{-1}\overset{\circ}{C}\mathcal{K} = z \cdot \overline{\mathcal{D}}_a(\overset{\circ}{A}), \tag{3.17}$$

$$\overset{\circ}{A} = \mathcal{K}^{-1}\overset{\circ}{A}\mathcal{K} = \mathfrak{A}_a\!\left(-\hbar z\frac{d}{dz} - \hbar\right).$$

Note that the second formula for the operator $\overset{\circ}{C}$ in (3.17) should be considered modulo holomorphic functions, that is,

$$\overset{\circ}{C}\!\left(\frac{1}{z^n}\right) = z\overline{\mathcal{D}}_a(\overset{\circ}{A})\!\left(\frac{1}{z^n}\right) = \overline{\mathcal{D}}_a\!\left(\mathfrak{A}_a(\hbar(n-1))\right)\frac{1}{z^{n-1}}$$

for $n = 2, 3, \ldots$, but for $n = 1$, one has

$$\overset{\circ}{C}(1/z) = 0 \qquad \text{(modulo holomorphic functions).} \tag{3.17a}$$

Note that the norm of any holomorphic function calculated using (3.11) is automatically zero.

The proof of the theorem readily follows from Lemma 1.1 and the definitions (3.10), (3.14), and (3.16).

REMARK 3.1. For the case of Heisenberg commutation relations, a version of meromorphic coherent states, as well as a transform in a sense similar to (3.12), (3.13), was introduced by Paul Dirac in the framework of quantum field theory [**D**]; see applications and the discussion, for instance, in [**VoB**].[11]

REMARK 3.2. The reproducing property of $\widetilde{\mathcal{L}}$ reads

$$(3.18) \qquad \frac{1}{(2\pi)^2} \oint \oint \frac{\widetilde{\mathcal{L}}(\overline{v},w)\widetilde{\mathcal{L}}(\overline{w},z)}{\widetilde{\mathcal{L}}(\overline{w},w)} \, d\widetilde{m}(\overline{w},w) = \widetilde{\mathcal{L}}(\overline{v},z)$$

and the reproducing property of $\mathcal{K}$ reads

$$(3.19) \qquad \frac{1}{(2\pi)^2} \oint \oint \frac{\mathcal{K}(\overline{v},w)\mathcal{K}(\overline{w},z)}{\mathcal{K}(\overline{w},w)} \, d\widetilde{m}(\overline{w},w) = \mathcal{K}(\overline{v},z).$$

Here the "measure" $d\widetilde{m}$ is formally determined by

$$d\widetilde{m}(\overline{w},w) \stackrel{\text{def}}{=} \mathcal{K}(\overline{w},w)\widetilde{\mathcal{L}}(\overline{w},w) \, d\overline{w}dw.$$

Using the same contour integrals, we can represent the resolution of the identity in the Hilbert space H_a in two forms:

$$(3.20) \qquad \frac{1}{(2\pi)^2} \oint \oint \mathbf{\Pi}(\overline{z},z) d\widetilde{m}(\overline{z},z) = \mathbf{I},$$

$$(3.21) \qquad \frac{1}{(2\pi)^2} \oint \oint \widetilde{\mathbf{\Pi}}(\overline{z},z) d\widetilde{m}(\overline{z},z) = \mathbf{I},$$

where $\mathbf{\Pi}$ and $\widetilde{\mathbf{\Pi}}$ are one-dimensional projections onto coherent states $\mathfrak{P}_z$ and $\widetilde{\mathfrak{P}}_{\overline{z}}$, respectively.

All these resolutions of the identity are written over the torus, that is, over the product of two contours. The usual way is to use integrals over the phase space. This will be considered later.

3.3. Equations for the quantum Kähler potential and the density of the reproducing measure.
According to Theorems 1.1 and 3.1, the coherent states and the reproducing kernel are defined as distributions without any convergence conditions for the power series. Specific examples show that, in fact, the radius of convergence

$$(3.22) \qquad \underline{R} = \lim_{n\to\infty} \left(\mathcal{H}(\hbar n)\ldots\mathcal{H}(\hbar)\right)^{1/n}$$

of the series (3.9) can be equal to zero: $\underline{R} = 0$; and thus, the kernel $\mathcal{K}(\overline{w},z)$, as well as the coherent state $\mathfrak{P}_z$, does not exist in the usual sense (that is, as a function analytic in a neighborhood of the origin).

On the other hand, the freedom in choosing the factors $\mathcal{D}_a$ and $\mathcal{E}_a$ in (3.5b) and statement (c) of Theorem 1.1 show that the numbers $\mathcal{H}(\hbar n)$ in (3.9) can always be chosen to increase sufficiently rapidly so that $\underline{R} = \infty$. Thus, later in this section

---

[11]The authors are grateful to Dr. A. Vourdas for referring them to Dirac's paper and sending a copy.

we will not worry about this problem and *assume that $\mathcal{K}(\overline{w}, z)$ is analytic in z and $\overline{w}$ on the entire plane $\mathbb{C}$.*

REMARK 3.3. At the same time, the function $\widetilde{\mathcal{L}}(\overline{w}, z)$ determined by the series (3.14) is analytic in a neighborhood of infinity:

$$|z| > \sqrt{\overline{R}}, \qquad |w| > \sqrt{\overline{R}},$$

where

$$\overline{R} = \varlimsup_{n \to \infty} \left(\mathcal{H}(\hbar n) \ldots \mathcal{H}(\hbar) \right)^{1/n}.$$

If $\underline{R} = \infty$, then also $\overline{R} = \infty$, and this neighborhood is empty.

LEMMA 3.1. *The function $\mathcal{K}(\overline{z}, z) = k(|z|^2)$ determined by the series (3.9) in the case $N = \infty$ can be obtained as a solution of the Cauchy problem*

(3.23)
$$\mathcal{H}\left(\hbar x \frac{d}{dx}\right) k(x) = x k(x), \quad x > 0,$$
$$k(0) = 1.$$

In the case $N < \infty$, the differential equation (3.23) should be modified as follows:

$$\mathcal{H}\left(\hbar x \frac{d}{dx}\right) k(x) = x k(x) - \frac{x^{N+1}}{\mathcal{H}(\hbar N) \ldots \mathcal{H}(\hbar)}.$$

This statement readily follows from (3.9) if we note that $\mathcal{H}(0) = 0$.

Now let us consider the equation dual to (3.23).

LEMMA 3.2. *Let ℓ be a smooth solution of the equation*

(3.24)
$$\mathcal{H}\left(-\hbar x \frac{d}{dx}\right) \ell(x) = x \ell(x), \qquad x > 0,$$

such that $\int_0^\infty x^n |\ell(x)| \, dx < \infty$ for $0 \leq n \leq N$. Let us normalize ℓ as follows:

(3.25)
$$\frac{1}{\hbar} \int_0^\infty \ell(x) \, dx = 1.$$

Then

(3.26)
$$\frac{1}{\hbar} \int_0^\infty x^n \ell(x) \, dx = \mathcal{H}(\hbar n) \ldots \mathcal{H}(\hbar), \qquad 1 \leq n \leq N,$$

and, for any distribution $g \in \mathcal{P}$,

(3.27)
$$\|g\|_\mathcal{P}^2 = \frac{1}{2\pi\hbar} \int_\mathbb{C} |g(\overline{z})|^2 \, \ell(|z|^2) \, d\overline{z} dz.$$

Here $d\overline{z}dz = dx d\varphi$ and $z = \sqrt{x} \exp\{i\varphi\}$. So, in this case all elements from $\mathcal{P}$ are regular L^2-functions on the plane.

PROOF. It follows from (3.24) that

$$\int_0^\infty x^n \ell(x)\, dx = \int_0^\infty x^{n-1} \mathcal{H}\Big(-\hbar x \frac{d}{dx}\Big) \ell(x)\, dx$$
$$= \int_0^\infty \ell(x) \mathcal{H}\Big(\hbar x \frac{d}{dx} + \hbar\Big)(x^{n-1})\, dx = \mathcal{H}(\hbar n) \int_0^\infty x^{n-1} \ell(x)\, dx.$$

So, the normalization condition (3.25) implies (3.26). Then we have

$$\|g\|_{\mathcal{P}}^2 = |g_0|^2 + \sum_{n\geq 1} \mathcal{H}(\hbar n)\ldots\mathcal{H}(\hbar)\,|g_n|^2 = \frac{1}{\hbar}\int_0^\infty dx\,\ell(x) \sum_{n\geq 0} x^n g_n \bar{g}_n$$
$$= \frac{1}{2\pi\hbar}\int_0^\infty dx \int_0^{2\pi} d\varphi\, \ell(|z|^2) \sum_{n\geq 0}(z\bar{z})^n g_n \bar{g}_n$$
$$= \frac{1}{2\pi\hbar}\int_{\mathbb{C}} d\bar{z}dz\, \ell(|z|^2) \Big(\sum_{n\geq 0} \bar{z}^n g_n\Big) \Big(\sum_{m\geq 0} z^m \bar{g}_m\Big).$$

This is equality (3.27). The proof of the lemma is complete. □

It follows from (3.26) that the function

(3.28) $$\mathcal{L}(\bar{z}, z) \stackrel{\text{def}}{=} \ell(\bar{z}z)$$

satisfies the equality

$$\frac{1}{2\pi\hbar}\int_{\mathbb{C}} z^n \bar{z}^m \mathcal{L}(\bar{z}, z)\, d\bar{z}dz = \mathcal{H}(\hbar m)\ldots\mathcal{H}(\hbar)\,\delta_m^n$$

for any $0 \leq n, m \leq N$. Thus,

(3.29) $$\frac{1}{(2\pi)^2} \oint\oint z^n \bar{z}^m \widetilde{\mathcal{L}}(\bar{z}, z)\, d\bar{z}dz = \frac{1}{2\pi\hbar} \int_{\mathbb{C}} z^n \bar{z}^m \mathcal{L}(\bar{z}, z)\, d\bar{z}dz.$$

Relations (3.29) provide a possibility of replacing the contour integrals in (3.19) and (3.20) by an integral over the phase space Ω. In our case, Ω is covered by the closure of a single chart:

(3.30) $$\Omega \approx [\mathbb{C}].$$

COROLLARY 3.2. *Under the assumptions of Lemma 3.2, the resolutions of the identity* (3.19) *and* (3.20) *read*

(3.31) $$\frac{1}{2\pi\hbar}\int_\Omega p(\xi;\xi')\, dm(\xi') = 1, \qquad \forall\, \xi \in \Omega,$$

(3.32) $$\frac{1}{2\pi\hbar}\int_\Omega \mathbf{\Pi}(\xi)\, dm(\xi) = \mathbf{I}.$$

Here ξ is a point from the phase space (3.30) *with the complex coordinate $z = z(\xi)$. The function p in* (3.31) *is defined by*

(3.33) $$p(\xi;\xi') \stackrel{\text{def}}{=} \operatorname{tr}\big(\mathbf{\Pi}(\xi)\mathbf{\Pi}(\xi')\big) = \frac{\mathcal{K}(\bar{z},z')\mathcal{K}(\bar{z}',z)}{\mathcal{K}(\bar{z},z)\mathcal{K}(\bar{z}',z')},$$

and the measure dm over Ω is defined by

(3.34) $$dm(\xi) = M(|z|^2)\, d\bar{z}dz, \qquad M(x) \stackrel{\text{def}}{=} k(x)\ell(x).$$

The function p on $\Omega \times \Omega$ is called a *probability function* because it has the following properties:

$$0 \leq p(\xi; \xi') \leq 1, \qquad p(\xi; \xi') = 1 \iff \xi = \xi'$$

(together with (3.31)). The measure dm is said to be a *reproducing measure on Ω*.

As usual, one can represent the reproducing kernel in the exponential form

$$\mathcal{K} = e^{F/\hbar}, \tag{3.35}$$

and consider the function $F = F(z\bar{z})$ as a Kähler potential of the "quantum" symplectic form over Ω:

$$\omega = i\bar{\partial}\partial F. \tag{3.36}$$

COROLLARY 3.3. *The quantum Kähler potential F is a solution of the Cauchy problem*

$$\mathcal{H}\left(x\frac{dF(x)}{dx} + \hbar x \frac{d}{dx}\right)1 = x - \frac{x^{N+1}}{\mathcal{H}(\hbar N)\ldots\mathcal{H}(\hbar)}e^{-F(x)/\hbar}, \qquad F(0) = 0, \tag{3.37}$$

where the function $\mathcal{H}$ is defined by (3.6) and the last summand on the right of the differential equation (3.37) is, of course, zero if $N = \infty$. The density of the reproducing measure dm from the resolution of the identity (3.31), (3.32) satisfies the equations

$$\mathcal{H}\left(x\frac{dF(x)}{dx} - \hbar x \frac{d}{dx}\right)M(x) = xM(x), \qquad \frac{1}{\hbar}\int_0^\infty e^{-F(x)/\hbar}M(x)\,dx = 1. \tag{3.38}$$

We note that the form ω (3.36) can be analytically continued onto the complex groupoid $\Omega \times \Omega$ by the formula

$$\omega_{(\overline{w},z)} = i\frac{\partial^2 F(z\overline{w})}{\partial z \partial \overline{w}}d\overline{w} \wedge dz.$$

This form has singularities on the two-dimensional submanifold

$$\Omega^0 = \{(\overline{w}, z) \mid \mathcal{K}(\overline{w}, z) = 0\} \subset \Omega \times \Omega.$$

Nevertheless, integrals of the closed form ω over two-dimensional membranes (films) in $\Omega \times \Omega$ are well defined if the boundaries of the membranes are transversal to Ω^0. In particular, we have the following statement.

COROLLARY 3.4. *The probability function can be written as a membrane amplitude generated by the quantum Kähler form:*

$$p(\xi; \xi') = \exp\left\{\frac{i}{\hbar}\int_{\Sigma(\xi;\xi')} \omega\right\}. \tag{3.39}$$

Here $\Sigma(\xi; \xi')$ is a membrane in $\Omega \times \Omega$ with quadrangle boundary consisting of four vertical and horizontal paths:

$$\partial \Sigma = \{\xi' \equiv (z', \overline{z}') \to (z', \overline{z}) \to \xi \equiv (z, \overline{z}) \to (z, \overline{z}') \to (z', \overline{z}')\}.$$

Each path goes along a leaf of the Kähler polarization $\{z = \text{const}\}$ or of the conjugate polarization $\{\bar{z} = \text{const}\}$.

The representation (3.39) has numerous generalizations and applications, in particular, because of the use of membranes with different static configurations (triangle, pentagon, hexagon, etc.) and various dynamic configurations (see [**Ka**$_{11-16}$]). This representation relates the commutative Kähler geometry to the theory of quantum phase spaces. An important point is that, starting from the structural functions f and φ in relations (1.1), one can construct the quantum form ω (3.36), membrane amplitudes (that is, dynamical wave functions), and quantum products of functions (see [**Ka**$_{17}$], and also §6.3 below).

Of course, the main object is the reproducing kernel $\mathcal{K}$ defined by the series (3.9). We know that the freedom in choosing the facors $\mathcal{D}_a$ and $\mathcal{E}_a$ in (3.5b) shows that this kernel can be represented, for instance, by hypergeometric functions of general type (see §2). An important alternative way for obtaining the function $\mathcal{K}$ is to determine the function F in (3.35) directly. From this point of view, the nonlinear equation (3.37) for the quantum Kähler potential and the linear equation (3.38) for the density of the reproducing measure play the central role.

Note that the equations for the quantum potential are substantially simplified in the classical limit $\hbar = 0$.

THEOREM 3.2. *Suppose that $a \in \mathcal{R}$ and $\mathcal{D}_a \neq 0$ on $\Omega = \Omega_a$.*

(a) *If the classical potential $F_0 = F_0(z\bar{z})$ over Ω is determined as a solution of the problem*

$$(3.40) \qquad \mathcal{H}\big(xF_0'(x)\big) = x, \qquad F_0(0) = 0,$$

then the Kähler form

$$(3.41) \qquad \omega_0 = i\,\bar{\partial}\partial F_0 \equiv i\,g_0(|z|^2)\,d\bar{z} \wedge dz, \qquad g_0(x) = \big(xF_0'(x)\big)',$$

coincides with the symplectic form on the leaf Ω of the Poisson algebra (3.3a).

(b) *The classical complex structure on $\Omega \subset \mathbb{R}^{k+2}$ is given by the complex coordinate*

$$(3.42) \qquad z = \frac{S_1 + iS_2}{\overline{\mathcal{D}}_a(A)} \equiv \frac{C}{\overline{\mathcal{D}}_a(A)}, \qquad |z|^2 = X_a(A),$$

where (S_1, S_2, A) are classical (commuting) coordinates on $\mathbb{R}^{k+2}$ and $C = S_1 + iS_2$.

(c) *The quantum form on Ω can be represented as*

$$(3.43) \qquad \omega = \omega_0 + \hbar\lambda, \qquad \lambda = i\,\bar{\partial}\partial \ln\sqrt{g},$$

where g is a quantum correcting metric over Ω, and

$$(3.44) \qquad \mathcal{K} = e^{F_0/\hbar}\sqrt{g}\big(1 + O(\hbar)\big).$$

(d) *In the case (3.5a), for $N < \infty$, we have*

$$\frac{1}{2\pi\hbar}\int_\Omega \omega = N, \qquad \frac{1}{2\pi\hbar}\int_\Omega \omega_0 = N+1,$$

$$c_1(\Omega) = -\frac{1}{\pi}\int_\Omega \lambda = 2 \quad \text{is the Chern class of the leaf } \Omega.$$

In particular, the usual quantization rule is acquired on Ω:

$$\frac{1}{2\pi\hbar}[\omega_0] - \frac{1}{2}c_1 \in H^2(\Omega, \mathbb{Z}).$$

(e) *The classical limit of the quantum correcting metric g is proportional to the Kähler metric g_0* (3.41):

(3.45) $$g = \text{const} \cdot g_0 + O(\hbar),$$

where $\text{const} = f^0(a)/|\mathcal{D}_a(a)|^2 > 0$.

The classical limit of the quantum correcting form λ coincides with one-half of the classical Ricci form:

(3.46) $$\lambda = i\bar{\partial}\partial \ln\sqrt{g_0} + O(\hbar).$$

The classical limit of the reproducing measure coincides with Liouville measure:

(3.47) $$M = g_0 + O(\hbar), \qquad dm(z,\bar{z}) = \big(g_0(|z|^2) + O(\hbar)\big)d\bar{z}dz.$$

REMARK 3.4. The representation (3.44) shows the importance of the quantum correcting metric g, which controls a part of the transition amplitudes that is regular in $\hbar$. In fact, the formulas in (c) and (e) are general (see [**Ka**$_{11-16}$]). Of course, in the multidimensional case, one should replace g and g_0 by $\det g$ and $\det g_0$ in formulas (3.43), (3.44), (3.46), and (3.47).

In the general case, it follows from the resolution of the identity that

$$\frac{1}{(2\pi\hbar)^{\dim\Omega/2}} \int_\Omega dm \quad \text{is the dimension of the representation space.}$$

In our case $\dim\Omega = 2$, and the later formula reads

$$\frac{1}{2\pi\hbar} \int_\Omega dm = N + 1.$$

Comparing this with statement (d), one can see that the remainder $O(\hbar)$ in (3.47) gives a zero contribution to the total integral over Ω. So, we again see that the classical symplectic form ω_0 controls the dimension of the irreducible representation. This gives us an opportunity to start the basic geometric quantization process [**Ko, S$_1$**] from the classical form ω_0, and this actually works in the case of Lie algebras. But in general, the derivation of the quantum correcting metric g along these lines does not seem to be a simple problem.

PROOF OF THEOREM 3.2. Formulas (3.42) directly follow from (1.46). So, from (3.3a), we obtain $\{z, A\} = iv(A)z$. Thus,

$$\{z, X_a(A)\} = iz\langle v, dX_a\rangle(A),$$

or

$$\{z, z\bar{z}\} = iz\frac{d}{dt}X_a\big(\mathfrak{A}_a(t)\big)\bigg|_{t=t(x)},$$

where $x = |z|^2$ and $t = t(x)$ satisfies the equations

$$\mathcal{H}(t) = x, \qquad t\bigg|_{x=0} = 0$$

(in fact, $t(x) = xF_0'(x)$; see (3.40)).

Now we have
$$\{z,\bar{z}\} = i\mathcal{H}'(t)\Big|_{t=t(x)},$$
but (3.40) implies
(3.48) $$\mathcal{H}'(t)t'(x) = 1, \qquad t' = (xF_0')' \equiv g_0.$$

Thus, $\{z,\bar{z}\} = i\,g_0(|z|^2)^{-1}$. This means that the symplectic structure on the leaf Ω has the form (3.41).

For $N < \infty$, we have

(3.49)
$$\mathcal{K}(x) = 1 + \sum_{n=1}^{N} \frac{x^n}{\mathcal{H}(\hbar n)\ldots\mathcal{H}(\hbar)},$$
$$F(x) = \hbar\ln\mathcal{K}(x) = \hbar N \ln x - \hbar\ln\bigl(\mathcal{H}(\hbar N)\ldots\mathcal{H}(\hbar)\bigr)$$
$$+ \hbar\ln\left(1 + \frac{\mathcal{H}(\hbar N)}{x} + O\Bigl(\frac{1}{x^2}\Bigr)\right) \qquad \text{as} \quad x \to \infty,$$
$$xF'(x) = \hbar N - \hbar\left(\frac{\mathcal{H}(\hbar N)}{x} + O\Bigl(\frac{1}{x^2}\Bigr)\right),$$
$$(xF'(x))' = \frac{\hbar\mathcal{H}(\hbar N)}{x^2} + O\Bigl(\frac{1}{x^3}\Bigr).$$

Thus,

(3.50) $$\frac{1}{2\pi\hbar}\int_\Omega \omega = \frac{i}{2\pi\hbar}\int_{\mathbb{C}} \bar{\partial}\partial F = \frac{1}{2\pi\hbar}\int_0^{2\pi} d\varphi \int_0^\infty (xF'(x))'\,dx$$
$$= \frac{1}{\hbar} xF'(x)\Big|_{x=0}^{x=\infty} = N.$$

It follows from (3.40) that
$$xF_0'(x) = \mathcal{H}^{-1}(x) = \mathcal{H}^{-1}(\infty) + O(1/x),$$
but (3.5c) implies $\mathcal{H}^{-1}(\infty) = \hbar(N+1)$. Thus, we have
$$x(\ln\sqrt{g(x)})' \equiv \frac{xF'(x)}{\hbar} - \frac{xF_0'(x)}{\hbar}$$
$$= N - \left(\frac{\mathcal{H}(\hbar N)}{x} + O\Bigl(\frac{1}{x^2}\Bigr)\right) - \left(N+1+O\Bigl(\frac{1}{x}\Bigr)\right) = -1 + O\Bigl(\frac{1}{x}\Bigr),$$
and hence
$$\frac{1}{2\pi}\int_\Omega \lambda = \int_0^\infty \bigl(x(\ln\sqrt{g(x)})'\bigr)'\,dx = -1.$$

Together with (3.50), this proves statement (d) of the theorem.

Let us consider statement (e). First, if we set $F = F_0 + \hbar F_1 + O(\hbar^2)$ and introduce the new coordinate $y = \ln x$, then it follows from equations (3.37) and (3.40) that
$$\mathcal{H}\left(\frac{dF}{dy} + \hbar\frac{d}{dy}\right)1 + O(\hbar^\infty) = e^y = \mathcal{H}\left(\frac{dF_0}{dy}\right)$$
(here the remainder $O(\hbar^\infty)$ appears only in the case $N < \infty$), or
$$\mathcal{H}\left(\frac{dF}{dy}\right) + \frac{\hbar}{2}\frac{d^2 F}{dy^2}\mathcal{H}''\left(\frac{dF}{dy}\right) + O(\hbar^2) = \mathcal{H}\left(\frac{dF_0}{dy}\right),$$

or

$$(3.51) \qquad \frac{dF_1}{dy}\mathcal{H}'\left(\frac{dF_0}{dy}\right) + \frac{1}{2}\frac{d^2 F_0}{dy^2}\mathcal{H}''\left(\frac{dF_0}{dy}\right) = 0.$$

Hence,

$$\frac{dF_1}{dy} = -\frac{1}{2}\frac{\frac{d}{dy}\left(\mathcal{H}'\left(\frac{dF_0}{dy}\right)\right)}{\mathcal{H}'\left(\frac{dF_0}{dy}\right)} = -\frac{1}{2}\frac{d}{dy}\ln\mathcal{H}'\left(\frac{dF_0}{dy}\right),$$

Therefore,

$$F_1(x) = -\frac{1}{2}\ln\left(\frac{\mathcal{H}'\left(xF_0'(x)\right)}{\text{const}}\right) = \ln\left(\sqrt{g_0(x)\cdot\text{const}}\right),$$

Here const $= \mathcal{H}'(0)$, and hence

$$g(x) = \mathcal{H}'(0)g_0(x) + O(\hbar).$$

Then we derive

$$\mathcal{H}'(0) = \langle v(a), dX_a(a)\rangle = \frac{\langle v(a), d\rho(a)\rangle}{|\mathcal{D}_a(a)|^2}$$

(see (3.6), (3.5b)). This constant is positive because of conditions (3.5). So, we have proved (3.45) and (3.46).

As previously, from the differential equation (3.38), we obtain a first order differential equation for the classical density $M_0 = M\big|_{\hbar=0}$, namely,

$$\frac{d(\ln M_0)}{dy}\mathcal{H}'\left(\frac{dF_0}{dy}\right) + \frac{d^2 F_0}{dy^2}\mathcal{H}''\left(\frac{dF_0}{dy}\right) = 0,$$

which is similar to (3.51). Thus,

$$M_0(x) = c_0 g_0(x), \quad \text{where } c_0 \text{ is a constant.}$$

It follows from the normalization condition (3.38) that

$$1 = \frac{c_0}{\sqrt{\text{const}}}\int_0^\infty \sqrt{g_0(\hbar x')}\exp\left\{-\frac{1}{\hbar}F_0(\hbar x')\right\}dx' + O(\hbar),$$

where the constant is the same as before (see (3.45)).

This condition implies the explicit formula

$$c_0 = F_0'(0)\frac{\sqrt{\text{const}}}{\sqrt{g_0(0)}}.$$

But $g_0(0) = \text{const}^{-1}$ and $F_0'(x) = \mathcal{H}^{-1}(x)/x$, $F_0'(0) = \mathcal{H}'(0)^{-1} = \text{const}^{-1}$. Thus, $c_0 = 1$, and we have proved (3.47). The proof of Theorem 3.2 is complete. $\square$

3.4. Symbols of operators with respect to holomorphic and meromorphic coherent states.
Let $\mathbf{R}$ be a linear operator in the space H_a of the irreducible representation of the algebra (3.3). For simplicity, we suppose that $\mathbf{R}$ is bounded (or, at least, its domain contains all vectors $\mathbf{B}^n \mathfrak{P}_0$).

Following Berezin and Klauder [$\mathbf{B}_{0,1}$, $\mathbf{Kl}_2$] (see also [$\mathbf{Si}$]), one can introduce the *Wick symbol* R (also called the low symbol or the covariant symbol) of the operator $\mathbf{R}$ as follows:

$$(3.52) \qquad R = \operatorname{tr}(\mathbf{R}\Pi),$$

and also the *anti-Wick symbol* r (or the upper symbol, or the contravariant symbol)[12] by the formula

$$(3.53) \qquad \mathbf{R} = \frac{1}{2\pi\hbar} \int_\Omega r\Pi\, dm.$$

We shall use the notation

$$\mathbf{R} = \widehat{R} = \underset{\wedge}{r}$$

if the functions R and r relate to the operator $\mathbf{R}$ via (3.52) and (3.53).[13]

First, note that the notions of "Wick" and "anti-Wick" symbols change places if we replace holomorphic coherent states by meromorphic states, that is, if we replace Π by $\widetilde{\Pi}$.

LEMMA 3.3. *The following formula holds*:

$$(3.54) \qquad \mathbf{R} \equiv \widehat{R} = \frac{1}{(2\pi)^2} \oint\oint R\widetilde{\Pi}\, d\widetilde{m}.$$

PROOF. Using definition (3.52), we derive

$$\frac{1}{(2\pi)^2} \oint\oint R\widetilde{\Pi}\, d\widetilde{m} = \frac{1}{(2\pi)^2} \oint\oint R(\overline{w},z)\mathcal{K}(\overline{w},z)\widetilde{\mathfrak{P}}_{\overline{w}} \otimes \widetilde{\mathfrak{P}}_{\overline{z}}^*\, d\overline{w}dz$$

$$= \frac{1}{(2\pi)^2} \oint\oint (\mathbf{R}\mathfrak{P}_z, \mathfrak{P}_w)_H\, \widetilde{\mathfrak{P}}_{\overline{w}} \otimes \widetilde{\mathfrak{P}}_{\overline{z}}^*\, d\overline{w}dz$$

$$= \frac{1}{(2\pi)^2} \oint\oint \left(\frac{1}{2\pi i} \oint \widetilde{\mathcal{L}}(\overline{w},v)\mathfrak{P}_v \otimes \mathfrak{P}_w^*\, dv \right)$$

$$\times \mathbf{R}\left(-\frac{1}{2\pi i} \oint \widetilde{\mathcal{L}}(\overline{v},z)\mathfrak{P}_z \otimes \mathfrak{P}_v^*\, d\overline{v} \right) d\overline{w}dz$$

$$= \left(\frac{1}{(2\pi)^2} \oint\oint \Pi\, d\widetilde{m} \right) \mathbf{R} \left(\frac{1}{(2\pi)^2} \oint\oint \Pi\, d\widetilde{m} \right) = \mathbf{R}.$$

In this relation we took into account the resolution of the identity (3.20). The lemma is proved. □

---

[12]We use the names "Wick" and "anti-Wick" symbols in order to relate them with complex structure on Ω (normal and anti-normal ordering of creation-annihilation operators). Other possible names (low and upper, covariant and contravariant) are, in fact, much more general and could be employed for many different types of quantizers Π.

[13]In [$\mathbf{Ka}_{14,15}$] the notation $\mathbf{R} = \widehat{r}$ was used.

REMARK 3.5. Formula (3.54) is not absolutely similar to (3.53), but, under the additional assumption that $\widetilde{\boldsymbol{\Pi}}(z, \bar{z})$ is smooth in a neighborhood of the origin and in view of (3.29), formula (3.54) reads

$$(3.54a) \qquad \mathbf{R} \equiv \widehat{R} = \frac{1}{2\pi\hbar} \int_\Omega R\widetilde{\boldsymbol{\Pi}}\, dm.$$

COROLLARY 3.5. *In the finite-dimensional case $N < \infty$ (that is, in the case (3.5a)), formula (3.54a) holds, as well as the following analog of (3.52):*

$$(3.55) \qquad r = \operatorname{tr}(\mathbf{R}\widetilde{\boldsymbol{\Pi}}).$$

One also has the resolutions of the identity

$$(3.21a) \qquad \mathbf{I} = \frac{1}{2\pi\hbar} \int_\Omega \widetilde{\boldsymbol{\Pi}}\, dm,$$

$$(3.18a) \qquad 1 = \frac{1}{2\pi\hbar} \int_\Omega \widetilde{p}(\xi; \xi')\, dm(\xi'), \qquad \forall \xi \in \Omega,$$

where

$$(3.56) \qquad \widetilde{p}(\xi; \xi') \stackrel{\text{def}}{=} \operatorname{tr}\left(\widetilde{\boldsymbol{\Pi}}(\xi)\widetilde{\boldsymbol{\Pi}}(\xi')\right), \qquad 0 \leq \widetilde{p} \leq 1, \qquad \widetilde{p}(\xi; \xi') = 1 \iff \xi = \xi'.$$

For $N = \infty$, how to construct a well-defined function $\widetilde{p}$ is a problem, since in this case $\widetilde{\mathcal{L}}$ has an infinite Laurent expansion at zero (see (3.14) and Remark 3.3). In the finite-dimensional case, the *dual probability function* $\widetilde{p}$ is well defined and can be interpreted as a membrane amplitude (similar to (3.39)) via the *dual quantum form* $\widetilde{\omega} \stackrel{\text{def}}{=} i\overline{\partial}\partial \widetilde{F}(z\bar{z})$. Here the dual quantum Kähler potential $\widetilde{F}$ is determined by the relation

$$\widetilde{\mathcal{L}}(\bar{z}, z) \stackrel{\text{def}}{=} \frac{1}{|z|^{2(N+1)}} e^{\widetilde{F}(z\bar{z})/\hbar}.$$

From (3.52), (3.53), and the definition of the probability function (3.33), it follows that the symbols r and R are related by

$$(3.57) \qquad R = \mathbb{P}_\hbar(r),$$

where $\mathbb{P}_\hbar$ is the *probability operator*:

$$(3.58) \qquad \left(\mathbb{P}_\hbar(r)\right)(\xi) \stackrel{\text{def}}{=} \frac{1}{2\pi\hbar} \int_\Omega p(\xi; \xi')r(\xi')\, dm(\xi').$$

On the other hand, formulas (3.54a) and (3.55) imply the dual relation.

COROLLARY 3.6. *If $N < \infty$, then*

$$(3.59) \qquad r = \widetilde{\mathbb{P}}_\hbar(R),$$

where the dual probability operator $\widetilde{\mathbb{P}}_\hbar$ acts in the same way as in (3.58) but with the dual probability function $\widetilde{p}$ (3.56) instead of p.

For $N = \infty$, the singular (over Ω) distribution $\widetilde{\boldsymbol{\Pi}}$ can be understood as a linear functional on the space of functions, holomorphic in z and $\bar{z}$, with respect to the following pairing:

$$\langle \widetilde{\boldsymbol{\Pi}}, f \rangle = \frac{1}{(2\pi)^2} \oint \oint f \widetilde{\boldsymbol{\Pi}}\, d\widetilde{m}.$$

In the same way, we can generalize the definition of the operator $\widetilde{\mathbb{P}_\hbar}$ as follows:

$$\widetilde{\mathbb{P}_\hbar}(f)(\xi) = \frac{1}{(2\pi)^2} \oint \oint \widetilde{p}(\xi; \xi') f(\xi') \, d\widetilde{m}(\xi').$$

COROLLARY 3.5A. *Formulas (3.55) and (3.59) hold both for $N = \infty$ and $N < \infty$ if $\widetilde{\Pi}$ and $\widetilde{\mathbb{P}_\hbar}$ are understood in the sense of contour integrals.*

So, by using integration along double paths (in fact, along a torus) in the complex groupoid $\Omega \times \Omega$, we obtain more general formulas than we do by using integration along the original manifold Ω.

We also see a parallel correspondence between the formal measure $d\widetilde{m} = \mathcal{K}\widetilde{\mathcal{L}} \, d\bar{z} dz$ and the reproducing measure $dm = \mathcal{K}\mathcal{L} \, d\bar{z} dz$. To establish the duality between Wick and anti-Wick quantizations, we use the function $\widetilde{\mathcal{L}}$, not $\mathcal{L}$. Certainly, $\widetilde{\mathcal{L}}$ and $\mathcal{L}$ have somewhat different properties, but there is an interesting question: is it possible to interpret $\mathcal{L}$ similarly to $\widetilde{\mathcal{L}}$ as a reproducing kernel for something else?

Now we note that *the probability operator $\mathbb{P}_\hbar$ determines the product of Wick symbols*. Indeed (see [**B**$_1$]), if R and R' are the Wick symbols of the operators **R** and **R**$'$, and $R * R'$ is the Wick symbol of **RR**$'$, then in view of (3.52) and (3.32), we have

$$\begin{aligned}(R * R')(\xi) &= \operatorname{tr}\left(\mathbf{RR}'\Pi(\xi)\right) = \frac{1}{2\pi\hbar} \int_\Omega \operatorname{tr}\left(\mathbf{R}\Pi(\xi')\mathbf{R}'\Pi(\xi)\right) dm(\xi') \\ (3.60) \qquad &= \frac{1}{2\pi\hbar} \int_\Omega \operatorname{tr}\left(\mathbf{R}\Pi(\bar{z}, z')\right) \operatorname{tr}\left(\mathbf{R}'\Pi(\bar{z}', z)\right) p(\xi; \xi') \, dm(\xi') \\ &= \frac{1}{2\pi\hbar} \int_\Omega p(\xi; \xi') R(\bar{z}, z') R'(\bar{z}', z) \, dm(\xi').\end{aligned}$$

Here $z, z' \in \mathbb{C}$ are the complex coordinates of points $\xi, \xi' \in \mathbb{R}^2$. In this notation, we have $\widehat{RR'} = \widehat{R * R'}$.

In the special case where R is holomorphic and R' is antiholomorphic, formula (3.60) reads

$$(3.61) \qquad R * R' = \mathbb{P}_\hbar(RR').$$

At the same time, if R is antiholomorphic or if R' is holomorphic, then

$$(3.62) \qquad R * R' = RR'.$$

In this sense, the probability operator $\mathbb{P}_\hbar$ (3.58) completely determines the Wick product $*$. Below in §6.3 we present explicit formulas for the expansion of $\mathbb{P}_\hbar$, as well as the expansion of the reproducing measure dm, to power series in $\hbar$.

The definition of the anti-Wick product is more delicate, since the operator $\mathbb{P}_\hbar$ is not invertible (for details, see [**Ka**$_{12-15}$]). But in the finite-dimensional case $N < \infty$, we can use equalities (3.55) and (3.21a) and similarly derive the product of two anti-Wick symbols r and r':

$$(r \circledast r')(\xi) = \operatorname{tr}\left(\mathbf{R}\mathbf{R}'\widetilde{\mathbf{\Pi}}(\xi)\right) = \frac{1}{2\pi\hbar}\int_{\Omega}\operatorname{tr}\left(\mathbf{R}\widetilde{\mathbf{\Pi}}(\xi')\mathbf{R}'\widetilde{\mathbf{\Pi}}(\xi)\right)dm(\xi')$$

(3.60a)
$$= \frac{1}{2\pi\hbar}\int_{\Omega}\operatorname{tr}\left(\mathbf{R}\widetilde{\mathbf{\Pi}}(\bar{z}',z)\right)\operatorname{tr}\left(\mathbf{R}'\widetilde{\mathbf{\Pi}}(\bar{z},z')\right)\widetilde{p}(\xi;\xi')dm(\xi')$$

$$= \frac{1}{2\pi\hbar}\int_{\Omega}\widetilde{p}(\xi;\xi')r(\bar{z}',z)r'(\bar{z},z')dm(\xi').$$

In this notation, we have $r \underset{\frown}{\,} r' = r \circledast r'$.

If r is antiholomorphic and r' is holomorphic, then

(3.61a) $$r \circledast r' = \widetilde{\mathbb{P}_\hbar}(rr').$$

At the same time, if r is holomorphic or if r' is antiholomorphic, we have

(3.62a) $$r \circledast r' = rr'.$$

Note that the unit function 1 is the unit element of both multiplications (3.60) and (3.60a).

THEOREM 3.3. *Let $\overset{\circ}{R}$ and $\widetilde{R}$ be operators in $\mathcal{P}$ and $\widetilde{\mathcal{P}}$ that correspond to an operator $\mathbf{R}$ in H_a after coherent transforms via holomorphic and anti-holomorphic coherent states, respectively. Let R and r be the Wick and anti-Wick symbols of $\mathbf{R}$, and let $R*$ and $r\circledast$ be the left multiplication operators with respect to the Wick and anti-Wick products $*$ (3.60) and $\circledast$ (3.60a). Then*

(3.63)
$$R* = \mathcal{K}^{-1} \cdot \overset{\circ}{R} \cdot \mathcal{K} = e^{-F/\hbar} \cdot \overset{\circ}{R} \cdot e^{F/\hbar},$$
$$r\circledast = \widetilde{\mathcal{L}}^{-1} \cdot \underset{\circ}{R} \cdot \widetilde{\mathcal{L}} = e^{-\widetilde{F}/\hbar} \cdot |z|^{2(N+1)} \cdot \underset{\circ}{R} \cdot |z|^{-2(N+1)} \cdot e^{\widetilde{F}/\hbar}.$$

The last equality holds in the finite-dimensional case $N < \infty$.

In particular, the following formulas relate the Wick and anti-Wick symbols to the operators of antiholomorphic and holomorphic representations:

(3.64) $$R = e^{-F/\hbar}\overset{\circ}{R}(e^{F/\hbar}), \qquad r = e^{-\widetilde{F}/\hbar}|z|^{2(N+1)}\underset{\circ}{R}\left(|z|^{-2(N+1)}e^{\widetilde{F}/\hbar}\right)$$

(the last formula holds for $N < \infty$).

PROOF. Let us prove, say, the first formula in (3.63). By the definition (3.60), we have

(3.65) $$R * R' = \mathcal{K}^{-1}(\mathfrak{P}, \mathbf{R}\mathbf{R}'\mathfrak{P})_{H_a} = \mathcal{K}^{-1}(\mathbf{R}^*\mathfrak{P}, \mathbf{R}'\mathfrak{P})_{H_a}.$$

By the definition of the operator $\overset{\circ}{R}$, we have $\mathbf{R}^*\mathfrak{P}_z = \overline{\overset{\circ}{R}}\mathfrak{P}_z$, where $\overline{\overset{\circ}{R}}$ is the complex conjugate operator acting with respect to the variable z. Thus, (3.65) implies

$$R * R' = \mathcal{K}^{-1}(\overline{\overset{\circ}{R}}\mathfrak{P}, \mathbf{R}'\mathfrak{P})_{H_a} = \mathcal{K}^{-1}\overset{\circ}{R}(\mathfrak{P}, \mathbf{R}'\mathfrak{P})_{H_a} = \mathcal{K}^{-1}\overset{\circ}{R}(\mathcal{K}R').$$

Recalling that $\mathcal{K} = e^{F/\hbar}$, we obtain the first formula in (3.63). The second one can be obtained similarly. The theorem is proved. □

3.5. The product in the enveloping algebra (in the space of polynomials over $\mathbb{R}^{k+2}$).
The enveloping algebra corresponding to the relations (3.3) can be described as a set of elements

$$\widehat{g} \stackrel{\text{def}}{=} g(\overset{3}{\mathbf{B}}, \overset{2}{\mathbf{A}}, \overset{1}{\mathbf{C}}), \tag{3.66}$$

where the symbols g are polynomials over $\mathbb{R}^{k+2}$.

LEMMA 3.4 [M]. *The product of two elements of the enveloping algebra is given by*

$$\widehat{g}\widehat{g}' = \widehat{g \star g'}.$$

Here

$$(g \star g')(B, A, C) \stackrel{\text{def}}{=} g(\overset{3}{L_B}, \overset{2}{L_A}, \overset{1}{L_C})g'(B, A, C), \tag{3.67}$$

and $\mathcal{L}_B$, $\mathcal{L}_A$, *and* $\mathcal{L}_C$ *are operators of the left regular representation defined by the formulas*

$$\mathbf{B}\widehat{g} = \widehat{L_B g}, \qquad \mathbf{A}\widehat{g} = \widehat{L_A g}, \qquad \mathbf{C}\widehat{g} = \widehat{L_C g}.$$

Thus, it suffices to calculate the operators L_B, L_A, and L_C. Here we follow the methods developed in [**Ka**$_{2,5}$]; see also [**KaM**$_4$, Appendix II].

Obviously, $L_B = B$ is the operator of multiplication by the coordinate function B on $\mathbb{R}^{k+2}$.

LEMMA 3.5.

$$L_A = \mathfrak{A}_A\left(\hbar B \frac{\partial}{\partial B}\right). \tag{3.68}$$

PROOF. By (3.3) we have

$$\mathbf{A}g(\mathbf{B}) = \sum_n g_n \mathbf{A}\mathbf{B}^n = \sum_n g_n \mathbf{B}^n \underbrace{\varphi^\hbar \circ \cdots \circ \varphi^\hbar}_{n}(\mathbf{A}) = \sum_n g_n \mathbf{B}^n \varphi^{\hbar n}(\mathbf{A}),$$

and if we recall that $\mathfrak{A}_A(\hbar) = \varphi^\hbar(A) = e^{\hbar V}A$, $V \stackrel{\text{def}}{=} \langle v(A), \frac{\partial}{\partial A}\rangle$, then

$$L_A(g) = \sum_n g_n B^n \varphi^{\hbar n}(A) = \sum_n g_n (Be^{\hbar V})^n A$$

$$= g(Be^{\hbar V})A = \left(e^{\hbar V B \frac{\partial}{\partial B}} A\right) g(B) = \mathfrak{A}_A\left(\hbar B\frac{\partial}{\partial B}\right)g(B).$$

The lemma is proved. $\square$

LEMMA 3.6.

$$L_C = Ce^{\hbar V} + \hbar \rho_\hbar\left(\overset{2}{A}, \hbar B\overset{1}{\frac{\partial}{\partial B}}\right)\frac{\partial}{\partial B}, \tag{3.69}$$

where

$$V = \left\langle v(A), \frac{\partial}{\partial A}\right\rangle, \qquad \rho_\hbar(A, t) \stackrel{\text{def}}{=} \frac{e^{t\delta_\hbar} - 1}{t}\rho(A), \qquad \delta_\hbar \stackrel{\text{def}}{=} \frac{e^{\hbar V} - I}{\hbar}. \tag{3.70}$$

PROOF. Using the commutation formula (see [**KaM**$_4$, Appendix I]), we derive

$$(3.71) \qquad \mathbf{C}g(\overset{2}{\mathbf{B}}, \overset{1}{\mathbf{A}}) = g(\overset{2}{\mathbf{B}}, \varphi^\hbar(\overset{1}{\mathbf{A}}))\mathbf{C} + \hbar(\delta_\hbar\rho)(\overset{3}{\mathbf{A}}) \cdot \delta g(\overset{4}{\mathbf{B}}, \overset{2}{\mathbf{B}}; \overset{1}{\mathbf{A}}),$$

where $\delta g(B', B; A) = \int_0^1 \frac{\partial g}{\partial B}((1-\tau)B' + \tau B, A)\, d\tau$.

We can apply Lemma 3.5 to the first summand on the right-hand side of (3.71) and arrange the operators in normal order (3.66). Thus, we have

$$(3.72) \qquad L_C g = Cg(B, \varphi^\hbar(A)) + \hbar\left[(\delta_\hbar\rho)\left(\mathfrak{A}_A\left(\hbar B\frac{\partial}{\partial B}\right)\right)\delta g(B', B; A)\right]\bigg|_{B'=B}.$$

Now let us consider the following chain of identities:

$$e^{\mu\hbar B\frac{\partial}{\partial B}}\delta g(B', B; A)\bigg|_{B'=B} = \int_0^1 d\tau \left[e^{\mu\hbar B\frac{\partial}{\partial B}}e^{\tau B\frac{\partial}{\partial B''}}\frac{\partial g}{\partial B''}(B'', A)\right]_{B''=(1-\tau)B}$$

$$= \int_0^1 d\tau \left[e^{\tau B\frac{\partial}{\partial B''}}\left(e^{\mu\hbar B(\frac{\partial}{\partial B}+\tau\frac{\partial}{\partial B''})}1(B)\right)\frac{\partial g}{\partial B''}(B'', A)\right]_{B''=(1-\tau)B}$$

$$= \left[\int_0^1 d\tau \left(e^{\mu(\hbar B\frac{\partial}{\partial B}+\nu B)}1(B)\right)_{\nu=\hbar\tau\frac{\partial}{\partial B''}}\frac{\partial g}{\partial B''}(B'', A)\right]_{B''=B}.$$

Note that

$$e^{\mu(\hbar B\frac{\partial}{\partial B}+\nu B)}1(B) = \exp\left\{\nu B\frac{e^{\hbar\mu}-1}{\hbar}\right\},$$

and thus,

$$(3.73) \qquad e^{\mu\hbar B\frac{\partial}{\partial B}}\delta g(B', B; A)\bigg|_{B'=B} = \int_0^1 d\tau \exp\left\{\tau(e^{\hbar\mu}-1)B\overset{2}{\frac{\partial}{\partial B}}\right\}\overset{1}{\frac{\partial g}{\partial B}}(B, A).$$

Now we derive

$$(\delta_\hbar\rho)\left(\mathfrak{A}_A\left(\hbar B\frac{\partial}{\partial B}\right)\right) = e^{V\hbar B\frac{\partial}{\partial B}}\delta_\hbar\rho(A),$$

and using (3.73), we simplify (3.72) as follows:

$$L_C = Ce^{\hbar V} + \hbar\int_0^1 d\tau\left[\exp\left\{\tau(e^{\hbar V}-1)B\overset{2}{\frac{\partial}{\partial B}}\right\}(\delta_\hbar\rho)(A)\right]\overset{1}{\frac{\partial}{\partial B}}.$$

This is exactly (3.69). The lemma is proved. □

Of course, we have to point out that the statements of Lemmas 3.4, 3.5, and 3.6 hold if and only if the space of polynomials in the variables B, A, and C is invariant with respect to all operators that were used in the proofs. This assumption is guaranteed by the following conditions:
  (i) the functions ρ and v_μ are polynomial on $\mathbb{R}^k$, and
  (ii) the Jacobi matrix $\partial v(A)/\partial A$ is nilpotent for all $A \in \mathbb{R}^k$.

THEOREM 3.4. *Under conditions* (i) *and* (ii), *the space of polynomials over* $\mathbb{R}^{k+2}$ *is correctly endowed with the product* (3.67), *where* $L_B = B$ *and* L_A, L_C *are given by formulas* (3.68) *and* (3.69). *This product determines the structure of an associative algebra with the unit element* 1. *In the classical limit* $\hbar = 0$, *this algebra coincides with the usual commutative algebra of polynomials. The operators* L_B, L_A, *and* L_C *of the left regular representation in this algebra satisfy relations* (3.3). *For each representation of* (3.3), *the mapping*

$$g \to \widehat{g} = g(\overset{3}{\mathbf{B}}, \overset{2}{\mathbf{A}}, \overset{1}{\mathbf{C}})$$

is a homomorphism.

3.6. Quantum restriction to symplectic leaves.
Now to each polynomial $g = g(B, A, C)$ on $\mathbb{R}^{k+2}$, we can assign a function on the leaf

$$\Omega = \{BC = \rho(A) - \rho(a), \ \varkappa_j(A) = \varkappa_j(a)\}, \qquad a \in \mathcal{R},$$

by the following formula [**Ka**$_{17}$]:

(3.74) $$g\Big|_{\widehat{\Omega}} \stackrel{\text{def}}{=} e^{-F/\hbar} g(\overset{3}{\overset{\circ}{B}}, \overset{2}{\overset{\circ}{A}}, \overset{1}{\overset{\circ}{C}})(e^{F/\hbar}),$$

where $\overset{\circ}{B}$, $\overset{\circ}{A}$, *and* $\overset{\circ}{C}$ *are the operators* (3.7) *of the irreducible representation of* (3.3) *corresponding to the vector* $a \in \mathcal{R}$, *and* F *is the quantum Kähler potential.*

In particular, each coordinate function on $\mathbb{R}^{k+2}$ can be restricted to the leaf $\widehat{\Omega}$ in the quantum sense (3.74):

$$A\Big|_{\widehat{\Omega}} = e^{-F/\hbar}\overset{\circ}{A}(e^{F/\hbar}), \qquad B\Big|_{\widehat{\Omega}} = e^{-F/\hbar}\overset{\circ}{B}(e^{F/\hbar}), \qquad C\Big|_{\widehat{\Omega}} = e^{-F/\hbar}\overset{\circ}{C}(e^{F/\hbar}).$$

If we compare (3.74) with (3.63) and (3.64), we see that $g\big|_{\widehat{\Omega}}$ is just the polynomial g with $B\big|_{\widehat{\Omega}}$, $A\big|_{\widehat{\Omega}}$, and $C\big|_{\widehat{\Omega}}$ substituted, and multiplication understood in the noncommutative sense (in the sense of the Wick algebra).

THEOREM 3.5 [**Ka**$_{17}$]. *The mapping*

(3.75) $$g \to g\big|_{\widehat{\Omega}}$$

defined by (3.74) *is a homomorphism of the algebra of polynomials over* $\mathbb{R}^{k+2}$ *with product* $\star$ (3.67) *and the Wick algebra over the symplectic leaf* Ω *with product* $*$ (3.61), (3.62) *defined by the probability operator. The homomorphism* (3.75) *transforms Casimir functions into Casimir functions, hence determines a quantum embedding of the surface of revolution* Ω *into* $\mathbb{R}^{k+2}$ (*or a quantum restriction of functions from* $\mathbb{R}^{k+2}$ *onto* Ω). *The quantization commutes with the quantum restriction*

(3.76) $$\widehat{g}\Big|_{H_a} = \widehat{g\big|_{\widehat{\Omega}}}.$$

PROOF. The function (3.74) is the Wick symbol of the operator $\widehat{g} = g(\overset{3}{\mathbf{B}}, \overset{2}{\mathbf{A}}, \overset{1}{\mathbf{C}})$ that acts in the space H_a of irreducible representation. So, the product of two such functions $g\big|_{\widehat{\Omega}} * g'\big|_{\widehat{\Omega}}$ is the Wick symbol of the product of two operators

$$\widehat{g}\widehat{g'} = (g \star g')(\overset{3}{\mathbf{B}}, \overset{2}{\mathbf{A}}, \overset{1}{\mathbf{C}}).$$

The Wick symbol of this operator is $(g \star g')|_{\widehat{\Omega}}$; hence

$$(g \star g')|_{\widehat{\Omega}} = g|_{\widehat{\Omega}} \star g'|_{\widehat{\Omega}}.$$

The set of all Casimir functions on $\mathbb{R}^{k+2}$ with respect to the product (3.67) are generated by the elementary Casimir functions

$$K_0 = BC - \rho(A), \qquad K_j = \varkappa_j(A) \qquad (j = 1, \ldots, k-1).$$

However, in the irreducible representation, we have

$$\overset{\circ}{B}\overset{\circ}{C} - \rho(\overset{\circ}{A}) = -\rho(a) \cdot I, \qquad \varkappa_j(\overset{\circ}{A}) = \varkappa_j(a) \cdot I;$$

hence

$$K_0|_{\widehat{\Omega}} = -\rho(a) = \text{const}, \qquad K_j|_{\widehat{\Omega}} = \varkappa_j(a) = \text{const}.$$

Thus, the homomorphism (3.75) transforms K_0, K_j into constants, that is, into the Casimir functions of the Wick algebra over Ω. Theorem 3.5 is proved. $\square$

COROLLARY 3.7. *The quantum restriction of the complex coordinate function $\mathcal{C}_a/\overline{\mathcal{D}}_a$ from $\mathbb{R}^{k+2}$ onto the leaf Ω_a coincides with the complex coordinate function z on Ω_a.*

So, we see that quantum and classical complex coordinates on the leaf coincide. For general functions this is not true; in general, a function restricted to a leaf in the quantum way differs from the classical restriction.

But, of course, in the limit $\hbar = 0$, the quantum restriction (3.75) coincides with the classical one:

$$(3.77) \qquad g|_{\Omega} = e^{-F/\hbar} g(\overset{\circ}{B}, \overset{\circ}{A}, \overset{\circ}{C})(e^{F/\hbar})|_{\hbar=0} = g(B|_{\Omega}, A|_{\Omega}, C|_{\Omega}),$$

where

$$(3.78) \qquad B|_{\Omega} = \overline{C}|_{\Omega} = \bar{z} \mathcal{D}_a(A|_{\Omega}), \qquad A|_{\Omega} = \mathfrak{A}_a(|z|^2 F_0'(|z|^2)).$$

COROLLARY 3.8. *The mapping $g \to g|_{\Omega}$ defined by (3.77) and (3.78) is a homomorphism of the Poisson algebra (3.3a) onto the Poisson algebra of functions over the symplectic leaf Ω corresponding to the classical form (3.41).*

REMARK 3.6. By substituting the formula

$$F = F_0 + \frac{1}{\hbar} \ln g_0 + \frac{1}{\hbar} \ln \text{const}$$

(see Theorem 3.2) into (3.74), one can readily derive the first quantum correction ($\sim \hbar$), which distinguishes the quantum restriction $|_{\widehat{\Omega}}$ from the classical one $|_{\Omega}$.

REMARK 3.7. In fact, all statements of this section (and their proofs) are sufficiently general and can be applied to any Kähler manifold Ω and to any algebra with general permutation relations (for which the product $\star$ is known). See [**Ka**$_{17}$].

§4. Multidimensional generalizations. Surfaces of birevolution

4.1. General relations and Jacobi identities. Let us consider the following multidimensional generalization of the basic quasiquadratic relations (1.1), (1.48):

(4.1)
$$\mathbf{C}^q \mathbf{B}_p = \sum_{r,s} \mathbf{B}_r \omega_{sp}^{qr}(\mathbf{A}) \mathbf{C}^s + f_p^q(\mathbf{A}),$$
$$\mathbf{C}^p \mathbf{A} = \varphi_p(\mathbf{A}) \mathbf{C}^p, \qquad \mathbf{A} \mathbf{B}_p = \mathbf{B}_p \varphi_p(\mathbf{A}).$$

Here $\mathbf{A} = (\mathbf{A}_1, \ldots, \mathbf{A}_k)$, $\mathbf{B} = (\mathbf{B}_1, \ldots, \mathbf{B}_d)$, and $\mathbf{C} = (\mathbf{C}^1, \ldots, \mathbf{C}^d)$ are sets of mutually commuting elements:

(4.1a)
$$[\mathbf{A}_\mu, \mathbf{A}_\nu] = 0, \qquad \mu, \nu = 1, \ldots, k,$$
$$[\mathbf{B}_p, \mathbf{B}_q] = 0, \quad [\mathbf{C}^p, \mathbf{C}^q] = 0, \qquad p, q = 1, \ldots, d.$$

The functions φ_p in (4.1) are vector-valued. For simplicity, we assume that all ω_{sp}^{qr}, f_p^q, and φ_p are polynomial functions. We also assume that the Poincaré–Birkhoff–Witt property holds for relations (4.1). In this case, the enveloping algebra related to (4.1) can be described as a set of polynomial functions $\widehat{g} = g(\overset{3}{\mathbf{B}}, \overset{2}{\mathbf{A}}, \overset{1}{\mathbf{C}})$ with the condition that the zero element $\widehat{g} = 0$ can be obtained only from the zero symbol $g = 0$.

The last condition implies a number of (Jacobi) identities for the structural functions over $\mathbb{R}^k$:

(4.2)
- (a) $\varphi_p(\varphi_q(A)) = \varphi_q(\varphi_p(A))$,
- (b) $f_p^q(A)(\varphi_p(A) - \varphi_q(A)) = 0$,
- (c) $\omega_{sp}^{qr}(A)\big(\varphi_q(\varphi_r(A)) - \varphi_p(\varphi_s(A))\big) = 0$,
- (d) $\displaystyle\sum_{s=1}^d \omega_{sp}^{qr}(A) f_t^s(A) + \delta_t^r f_p^q(\varphi_t(A)) = \sum_{s=1}^d \omega_{st}^{qr}(A) f_p^s(A) + \delta_p^r f_t^q(\varphi_p(A)),$
- (e) $\displaystyle\sum_{s=1}^d \omega_{sp}^{qr}(\varphi_t(A)) \omega_{uv}^{st}(A) + \sum_{s=1}^d \omega_{sp}^{qt}(\varphi_r(A)) \omega_{uv}^{sr}(A)$
  $\displaystyle= \sum_{s=1}^d \omega_{sv}^{qr}(\varphi_t(A)) \omega_{up}^{st}(A) + \sum_{s=1}^d \omega_{sv}^{qt}(\varphi_r(A)) \omega_{up}^{sr}(A).$

On the set of the indices $\{1, 2, \ldots, d\}$ that enumerate the coordinates B or C, we define the equivalence relation

$$p \sim q \quad \Longleftrightarrow \quad \varphi_p(A) = \varphi_q(A) \qquad \forall A \in \mathbb{R}^k.$$

Let us enumerate the coordinates B and, simultaneously, the coordinates C so that the following condition will be satisfied:

$$p \sim r, \quad p \leq q \leq r \quad \Longrightarrow \quad p \sim q.$$

Then if the mapping φ has no fixed points, the identities (4.2b) and (4.2c) imply that the matrix $F = ((f_s^r))$ and the matrices $\Omega_p^q = ((\omega_{sp}^{qr}))$ are block-diagonal for

$p \sim q$, that is,
$$f_s^r = 0 \quad \text{for} \quad r \not\sim s,$$
$$\omega_{sp}^{qr} = 0 \quad \text{for} \quad p \sim q, \quad r \not\sim s,$$

and, on the contrary, for $p \not\sim q$, the matrices Ω_p^q have nonzero elements only outside the diagonal blocks corresponding to the equivalence classes
$$\omega_{sp}^{qr} = 0 \quad \text{for} \quad p \not\sim q, \quad r \sim s.$$

In what follows, we also assume that the algebra (4.1), (4.1a) is endowed with involution such that
$$\mathbf{B}_p^* = \mathbf{C}^p, \quad p = 1, \ldots, d, \qquad \mathbf{A}_\mu^* = \mathbf{A}_\mu, \quad \mu = 1, \ldots, k.$$

Then the additional conditions

(4.3) $$\overline{\varphi_p} = \varphi_p, \qquad \overline{f_p^q} = f_q^p, \qquad \overline{\omega_{sp}^{qr}} = \omega_{rq}^{ps}$$

are satisfied.

In conclusion, note that by analogy with §1.4, one can try to reduce relations (4.1) with general tensor ω_{sp}^{qr} to the special "Kronecker" case $\omega_{sp}^{qr} = \delta_s^q \delta_p^r$. Namely, if there exists a solution $\mathfrak{S} = ((\mathfrak{S}_p^q))$ of the equation
$$[\mathfrak{S}^*(\varphi_q(A)) \cdot \widetilde{\Omega}_p^q(A) \cdot \mathfrak{S}(\varphi_p(A))]_s^r = \mathfrak{S}_p^{*r}(A) \mathfrak{S}_s^q(A),$$

where $\widetilde{\Omega}_p^q \stackrel{\text{def}}{=} ((\omega_{ps}^{rq}))$, and if the condition
$$\mathfrak{S}_p^s(\varphi_r(A)) \mathfrak{S}_q^r(A) + \mathfrak{S}_p^r(\varphi_s(A)) \mathfrak{S}_q^s(A) = \mathfrak{S}_p^s(A) \mathfrak{S}_q^r(\varphi_s(A)) + \mathfrak{S}_p^r(A) \mathfrak{S}_q^s(\varphi_r(A))$$

is satisfied, then the change of variables
$$\widetilde{\mathbf{B}}_p \stackrel{\text{def}}{=} \sum_{s=1}^d \mathbf{B}_s \mathfrak{S}_p^s(\mathbf{A}), \qquad \widetilde{\mathbf{C}}^q \stackrel{\text{def}}{=} \sum_{r=1}^d \mathfrak{S}_r^{*q}(\mathbf{A}) \mathbf{C}^r$$

reduces relations (4.1), (4.1a) to the following commutation relations:
$$[\widetilde{\mathbf{C}}^q, \widetilde{\mathbf{B}}_p] = \widetilde{f}_p^q(\mathbf{A}), \qquad \widetilde{\mathbf{C}}^p \mathbf{A} = \varphi_p(\mathbf{A}) \widetilde{\mathbf{C}}^p, \qquad \mathbf{A} \widetilde{\mathbf{B}}_p = \widetilde{\mathbf{B}}_p \varphi_p(\mathbf{A}),$$
$$[\mathbf{A}_\mu, \mathbf{A}_\nu] = 0, \qquad [\widetilde{\mathbf{B}}_p, \widetilde{\mathbf{B}}_q] = 0, \qquad [\widetilde{\mathbf{C}}^p, \widetilde{\mathbf{C}}^q] = 0.$$

Here $\widetilde{f}_p^q(A) \stackrel{\text{def}}{=} (\mathfrak{S}^*(A) \cdot F(A) \cdot \mathfrak{S}(A))_p^q$.

4.2. Multidimensional Heisenberg block. Let us consider the particular case

(4.4) $$\omega_{sp}^{qr} = \delta_s^q \delta_p^r.$$

Then the first permutation relation in (4.1) turns into the commutation relation
$$[\mathbf{C}^q, \mathbf{B}_p] = f_p^q(\mathbf{A}).$$

In this case, according to (4.2b), the partition of the indices corresponds to the partition of the elements $\mathbf{B}$ and $\mathbf{C}$ into commuting blocks, that is,
$$[\mathbf{C}^q, \mathbf{B}_p] = 0 \quad \text{for} \quad p \not\sim q.$$

Conditions (4.2c) and (4.2e) are satisfied automatically, and condition (4.2d) implies

(4.5) $$f_p^q(\varphi_t(A)) = f_p^q(A) \quad \text{for} \quad p \sim q, \quad t \neq p.$$

This means that if the equivalence class (block) $[p] = [q]$ contains more than one element, then the element $f_p^q(\mathbf{A})$ commutes with all elements $\mathbf{B}$ and $\mathbf{C}$:

$$[f_p^q(\mathbf{A}), \mathbf{B}_t] = \mathbf{B}_t \big(f_p^q(\varphi_t(\mathbf{A})) - f_p^q(\mathbf{A})\big) = 0.$$

The last relation is obvious for $p \not\sim q$ or for $p \sim q$ and $t \neq p$. But if $p \sim q$ and $t = p$, then in the multidimensional equivalence class $[p]$ there is an r such that $r \neq p$; then (4.5) implies

$$f_p^q(\varphi_p(A)) - f_p^q(A) = f_p^q(\varphi_r(A)) - f_p^q(A) = 0.$$

Thus, in this case, to each multidimensional equivalence class (block) $[s]$ there corresponds a subalgebra with generators $\mathbf{B}_p, \mathbf{C}^q, f_p^q(\mathbf{A}), p, q \in [s]$, and relations

(4.6) $$[\mathbf{C}^q, \mathbf{B}_p] = f_p^q(\mathbf{A}), \quad [\mathbf{C}^r, f_p^q(\mathbf{A})] = [f_p^q(\mathbf{A}), \mathbf{B}_r] = 0, \quad p, q, r \in [s].$$

This subalgebra can be reduced to a Heisenberg algebra.

We point out that to one-dimensional equivalence classes there correspond much more complicated algebras of type (1.1).

The elements of subalgebras that correspond to different equivalence classes commute with each other.

First, we construct an antiholomorphic representation and coherent states for one multidimensional Heisenberg block (4.6); then (in §4.4) we show how to "sew" together several blocks.

By m we denote the dimension of the block $[s]$; $m > 1$. Let $F^{[s]}(A)$ be the corresponding diagonal $(m \times m)$-block of the matrix $F(A)$.

The relations

(4.7) $$f_p^q(\mathbf{A})\mathfrak{P}_0^{[s]} = f_p^q(a)\mathfrak{P}_0^{[s]}, \quad \mathbf{C}^p \mathfrak{P}_0^{[s]} = 0, \quad p, q \in [s]; \quad \|\mathfrak{P}_0^{[s]}\| = 1$$

define a vacuum vector from the Hilbert space of a representation of the algebra (4.6).

We define the Hilbert space $\mathcal{P}^{[s]}$ of functions that are antiholomorphic with respect to the variables $z_{[s]} \in \mathbb{C}^m$ ($z_{[s]}$ is a vector with components z^p, where $p \in [s]$); this space is equipped with the inner product

$$(g', g)_{\mathcal{P}^{[s]}} = \int_{\mathbb{C}^m} g(\overline{z}) \overline{g'(\overline{z})} \, d\mu^{[s]}(z_{[s]}, \overline{z_{[s]}}),$$

where

$$d\mu^{[s]}(z_{[s]}, \overline{z_{[s]}}) = \frac{\det F^{[s]}(a)}{(2\pi)^m} \exp\{-\langle F^{[s]}(a) z_{[s]}, \overline{z_{[s]}} \rangle\} \, d\overline{z_{[s]}} dz_{[s]}.$$

THEOREM 4.1. (a) *The set $\{a \in \mathbb{R}^k \mid \det F^{[s]}(a) \geq 0\}$ is in a one-to-one correspondence with the set of irreducible Hermitian representations of the algebra (4.6) that possess a vacuum vector. If $a \in \mathcal{R}^{[s]} = \{a \in \mathbb{R}^k \mid \det F^{[s]}(a) > 0\}$, then such a representation is infinite-dimensional.*

(b) *For each $a \in \mathcal{R}^{[s]}$, the operators*

(4.8) $$\overset{\circ}{B}_p = \sum_{r \in [s]} f_p^r(a) \overline{z^r}, \quad \overset{\circ}{C}^q = \frac{\partial}{\partial \overline{z^q}}, \quad \overset{\circ}{f_p^q}(A) = f_p^q(a), \quad p, q \in [s],$$

represent the algebra (4.6). This representation is irreducible Hermitian and possesses the vacuum vector 1 in the Hilbert space $\mathcal{P}^{[s]}$.

(c) *An abstract Hermitian representation of the algebra* (4.6) *in a Hilbert space H_a with the vacuum vector $\chi_0^{[s]} = \chi_0^{[s]}(a)$, satisfying* (4.7) *for some $a \in \mathcal{R}^{[s]}$, can be intertwined with the representation* (4.8) *by means of the coherent states*

$$(4.9) \qquad \mathfrak{P}_{z_{[s]}}^{[s]} = \exp\Big\{ \sum_{p \in [s]} z_{[s]}^p \mathbf{B}_p \Big\} \mathfrak{P}_0^{[s]}.$$

The reproducing kernel corresponding to the states (4.9) *is given by the formula*

$$\mathcal{K}(\overline{w_{[s]}}, z_{[s]}) \stackrel{\text{def}}{=} \big(\mathfrak{P}_{w_{[s]}}^{[s]}, \mathfrak{P}_{z_{[s]}}^{[s]}\big)_{H_a} = \exp\{\langle F^{[s]}(a) z_{[s]}, \overline{w_{[s]}} \rangle\}$$

and satisfies the reproducing property

$$\int_{\mathbb{C}^m} \mathcal{K}(\overline{w_{[s]}}, v_{[s]}) \mathcal{K}(\overline{v_{[s]}}, z_{[s]}) \, d\mu^{[s]}(v_{[s]}, \overline{v_{[s]}}) = \mathcal{K}(\overline{w_{[s]}}, z_{[s]}).$$

4.3. Two one-dimensional blocks. Hilbert space of bi-antiholomorphic distributions. Now let us consider the case of two one-dimensional blocks (that is, the case $d = 2$, $\varphi_1 \neq \varphi_2$).

LEMMA 4.1. *Let $d = 2$. Suppose that $\varphi_1(A) \neq \varphi_2(A)$ for each $A \in \mathbb{R}^k$. Then the following conditions are sufficient for identities* (4.2) *and* (4.3) *to be satisfied*:

(a) $\varphi_1(\varphi_2(A)) = \varphi_2(\varphi_1(A))$,
(b) $f_2^1 \equiv f_1^2 \equiv 0$,
(c) $\omega_{11}^{12} \equiv \omega_{21}^{11} \equiv \omega_{12}^{22} \equiv \omega_{22}^{21} \equiv \omega_{12}^{11} \equiv \omega_{11}^{21} \equiv \omega_{22}^{12} \equiv \omega_{21}^{22} \equiv \omega_{22}^{11} \equiv \omega_{11}^{22} \equiv 0$,
(d) $\omega_{12}^{12}(A) \equiv \omega_{21}^{21}(A) \stackrel{\text{def}}{=} \theta(A)$,

$f_1^1(\varphi_2(A)) = \theta(A) f_1^1(A) - \alpha_1^2(A) f_2^2(A), \quad$ where $\alpha_1^2 \stackrel{\text{def}}{=} \omega_{21}^{12}$,

$f_2^2(\varphi_1(A)) = \theta(A) f_2^2(A) - \alpha_2^1(A) f_1^1(A), \quad$ where $\alpha_2^1 \stackrel{\text{def}}{=} \omega_{12}^{21}$,

(e) $\alpha_1^1(\varphi_2(A))\theta(A) + \alpha_1^2(\varphi_1(A))\alpha_2^1(A) = \theta(\varphi_1(A))\alpha_1^1(A),$

(4.10) $\hspace{5cm}$ where $\alpha_1^1 \stackrel{\text{def}}{=} \omega_{11}^{11}$,

$\alpha_2^2(\varphi_1(A))\theta(A) + \alpha_2^1(\varphi_2(A))\alpha_1^2(A) = \theta(\varphi_2(A))\alpha_2^2(A),$

$\hspace{5cm}$ where $\alpha_2^2 \stackrel{\text{def}}{=} \omega_{22}^{22}$,

$\alpha_1^2(\varphi_2(A))\alpha_2^2(A) = \theta(\varphi_2(A))\alpha_1^2(A),$

$\alpha_2^1(\varphi_1(A))\alpha_1^1(A) = \theta(\varphi_1(A))\alpha_2^1(A),$

(f) *the functions* $\varphi_1, \varphi_2, f_1^1, f_2^2, \alpha_1^1, \alpha_2^2, \alpha_1^2, \alpha_2^1$, *and* θ *are real*.

REMARK 4.1. In fact, the relations (4.10) necessarily follow from (4.2), (4.3) and the condition $\varphi_1(A) \neq \varphi_2(A)$ on the spectrum of the operators **A** (but not on the entire $\mathbb{R}^k$). But, for simplicity, we shall require that the identities (4.10) are satisfied for all $A \in \mathbb{R}^k$.

Throughout §4.3, we assume that conditions (4.10) are satisfied.

Then the commutation relations (4.1), (4.1a) have the following form:

(4.11)
$$\begin{aligned}
\mathbf{C}^1\mathbf{B}_1 &= \mathbf{B}_1\alpha_1^1(\mathbf{A})\mathbf{C}^1 + \mathbf{B}_2\alpha_1^2(\mathbf{A})\mathbf{C}^2 + f_1^1(\mathbf{A}),\\
\mathbf{C}^2\mathbf{B}_2 &= \mathbf{B}_2\alpha_2^2(\mathbf{A})\mathbf{C}^2 + \mathbf{B}_1\alpha_2^1(\mathbf{A})\mathbf{C}^1 + f_2^2(\mathbf{A}),\\
\mathbf{C}^1\mathbf{B}_2 &= \mathbf{B}_2\theta(\mathbf{A})\mathbf{C}^1, \qquad \mathbf{C}^2\mathbf{B}_1 = \mathbf{B}_1\theta(\mathbf{A})\mathbf{C}^2,\\
\mathbf{C}^1\mathbf{A} &= \varphi_1(\mathbf{A})\mathbf{C}^1, \qquad \mathbf{A}\mathbf{B}_1 = \mathbf{B}_1\varphi_1(\mathbf{A}),\\
\mathbf{C}^2\mathbf{A} &= \varphi_2(\mathbf{A})\mathbf{C}^2, \qquad \mathbf{A}\mathbf{B}_2 = \mathbf{B}_2\varphi_2(\mathbf{A}),\\
[\mathbf{A}_\mu, \mathbf{A}_\nu] &= 0, \qquad [\mathbf{B}_1, \mathbf{B}_2] = 0, \qquad [\mathbf{C}^1, \mathbf{C}^2] = 0,
\end{aligned}$$

and

$$\mathbf{B}_1^* = \mathbf{C}^1, \qquad \mathbf{B}_2^* = \mathbf{C}^2, \qquad \mathbf{A}_\mu^* = \mathbf{A}_\mu, \qquad \mu = 1, \dots, k.$$

LEMMA 4.2. (a) *If a function $\varkappa$ on $\mathbb{R}^k$ is preserved by the mappings $\varphi_1\colon \mathbb{R}^k \to \mathbb{R}^k$ and $\varphi_2\colon \mathbb{R}^k \to \mathbb{R}^k$, then $\varkappa(\mathbf{A})$ is an element of the center of the algebra (4.11).*
(b) *If functions σ_1, σ_2, and ρ on $\mathbb{R}^k$ satisfy the system of equations*

(∗)
$$\begin{aligned}
\sigma_1\bigl(\varphi_1(A)\bigr)\alpha_1^1(A) &= \sigma_1(A),\\
\sigma_1\bigl(\varphi_2(A)\bigr)\alpha_1^2(A) + \sigma_2\bigl(\varphi_1(A)\bigr)\theta(A) &= \sigma_2(A),\\
\rho\bigl(\varphi_1(A)\bigr) - \rho(A) &= \sigma_1(A)f_1^1(A),
\end{aligned}$$

(∗∗)
$$\begin{aligned}
\sigma_2\bigl(\varphi_2(A)\bigr)\alpha_2^2(A) &= \sigma_2(A),\\
\sigma_2\bigl(\varphi_1(A)\bigr)\alpha_2^1(A) + \sigma_1\bigl(\varphi_2(A)\bigr)\theta(A) &= \sigma_1(A),\\
\rho\bigl(\varphi_2(A)\bigr) - \rho(A) &= \sigma_2(A)f_2^2(A),
\end{aligned}$$

then

$$\mathbf{K} = \mathbf{B}_1\sigma_1(\mathbf{A})\mathbf{C}^1 + \mathbf{B}_2\sigma_2(\mathbf{A})\mathbf{C}^2 - \rho(\mathbf{A})$$

is a Casimir element of the algebra (4.11).

PROOF. The proof is similar to that of Lemma 1.3. We only note that the relations $\mathbf{KB}_1 = \mathbf{B}_1\mathbf{K}_1$ and $\mathbf{C}_1\mathbf{K} = \mathbf{KC}_1$ follow from (∗), whereas $\mathbf{KB}_2 = \mathbf{B}_2\mathbf{K}_1$ and $\mathbf{C}_2\mathbf{K} = \mathbf{KC}_2$ follow from (∗∗). □

We consider only Hermitian representations of (4.11) in a Hilbert space H that possess a vacuum:

(4.12)
$$\begin{aligned}
\mathbf{A}\mathfrak{P}_0 &= a\mathfrak{P}_0, \qquad \mathbf{C}\mathfrak{P}_0 = 0,\\
a &= (a_1, \dots, a_k) \in \mathbb{R}^k, \qquad \|\mathfrak{P}_0\| = 1.
\end{aligned}$$

Now we are going to describe a subset $\mathcal{R} \subset \mathbb{R}^k$ of all possible values of the vector-parameter a that labels irreducible representations of the algebra (4.11).

Let us define the following functions over $\mathbb{Z}_+$:

(4.13)
$$\mathcal{F}_a^{\omega,1}(p) \stackrel{\text{def}}{=} \begin{cases} f_1^1(a), & p = 0, \\ \frac{1}{p+1}\Big(f_1^1\big(\mathcal{A}_a(p,0)\big) \\ \quad + \sum_{r=0}^{p-1} f_1^1\big(\mathcal{A}_a(r,0)\big) \prod_{u=r}^{p-1} \alpha_1^1\big(\mathcal{A}_a(u,0)\big)\Big), & p \geq 1, \end{cases}$$

$$\mathcal{F}_a^{\omega,2}(q) \stackrel{\text{def}}{=} \begin{cases} f_2^2(a), & q = 0, \\ \frac{1}{q+1}\Big(f_2^2\big(\mathcal{A}_a(0,q)\big) \\ \quad + \sum_{s=0}^{q-1} f_2^2\big(\mathcal{A}_a(0,s)\big) \prod_{v=s}^{q-1} \alpha_2^2\big(\mathcal{A}_a(0,v)\big)\Big), & q \geq 1, \end{cases}$$

(4.14)
$$\mathcal{G}_a^1(p,q) \stackrel{\text{def}}{=} \begin{cases} 1, & q = 0, \\ \prod_{s=0}^{q-1} \theta(\mathcal{A}_a(p,s)), & q \geq 1, \end{cases}$$

$$\mathcal{G}_a^2(p,q) \stackrel{\text{def}}{=} \begin{cases} 1, & p = 0, \\ \prod_{r=0}^{p-1} \theta(\mathcal{A}_a(r,q)), & p \geq 1, \end{cases}$$

(4.15)
$$\mathcal{G}_a(p,q) \stackrel{\text{def}}{=} \prod_{r=0}^{p} \prod_{s=0}^{q} \theta(\mathcal{A}_a(r,s)),$$

where

(4.16)
$$\mathcal{A}_a(p,q) \stackrel{\text{def}}{=} \underbrace{\varphi_1\big(\ldots\big(\varphi_1}_{p}\big(\underbrace{\varphi_2\big(\ldots\big(\varphi_2}_{q}(a)\big)\ldots\big)\big)\big)\ldots\big)$$

is the composition of p mappings φ_1 and q mappings φ_2.

Simultaneously, we note that the following identities hold:

(4.17)
$$\mathcal{G}_a(p,q) = \prod_{r=0}^{p} \mathcal{G}_a^1(r,q+1) = \prod_{s=0}^{q} \mathcal{G}_a^2(p+1,s).$$

Now let us define the subset $\mathcal{R} \subset \mathbb{R}^k$. First, we introduce a partial ordering in the set $\mathbb{Z}_+ \times \mathbb{Z}_+$; namely,

$$(p',q') \prec (p,q) \iff p' \leq p, \quad q' \leq q.$$

We shall say that $a \in \mathcal{R}$ if the following three conditions are satisfied:
  (i) if $\mathcal{F}_a^{\omega,1}(p) < 0$ for some $p \in \mathbb{Z}_+$, then there exists $p' \in \mathbb{Z}_+$, $p' \leq p$, such that $\mathcal{F}_a^{\omega,1}(p') = 0$;
  (ii) if $\mathcal{F}_a^{\omega,2}(q) < 0$ for some $q \in \mathbb{Z}_+$, then there exists $q' \in \mathbb{Z}_+$, $q' \leq q$, such that $\mathcal{F}_a^{\omega,2}(q') = 0$;
  (iii) if $\theta(\mathcal{A}_a(p,q)) < 0$ for some $(p,q) \in \mathbb{Z}_+ \times \mathbb{Z}_+$, then there exists $(p',q') \in \mathbb{Z}_+ \times \mathbb{Z}_+$, $(p',q') \prec (p,q)$, such that

(4.18)
$$\mathcal{F}_a^{\omega,1}(p')\, \mathcal{F}_a^{\omega,2}(q')\, \theta\big(\mathcal{A}_a(p',q')\big) = 0.$$

For each $a \in \mathcal{R}$, by N_j we denote the minimal root of the function $\mathcal{F}_a^{\omega,j}$ on $\mathbb{Z}_+$ (here $j = 1, 2$). If $\mathcal{F}_a^{\omega,j}$ does not have roots on $\mathbb{Z}_+$, then we set $N_j = \infty$.

We also introduce a subset $J_a \subset \mathbb{Z}_+ \times \mathbb{Z}_+$ as follows:

(4.18a)
$$J_a = \{(p'', q'') \mid \exists (p', q') \in \mathbb{Z}_+ \times \mathbb{Z}_+ : \\ (p'', q'') \prec (p', q') \text{ and equality (4.18) holds}\}.$$

In other words, J_a is the set of points from the integer lattice that are smaller (in the sense $\prec$) than some root of the function (4.18).

By $\widetilde{J}_a$, let us denote the set of points from $\mathbb{Z}_+ \times \mathbb{Z}_+$ that are greater (in the sense $\prec$) than some root of the function (4.18). The intersection $J_a \cap \widetilde{J}_a \stackrel{\text{def}}{=} \mathcal{L}_a$ is the "north-east" boundary of the set J_a (see Figure 1). The boundary $\mathcal{L}_a$ consists of three parts, $\mathcal{L}_a = \mathcal{L}_a^1 \cup \mathcal{L}_a^2 \cup \mathcal{L}_a^0$, where $\mathcal{L}_a^1$ and $\mathcal{L}_a^2$ are the "pure east" and "pure north" boundaries, that is,

$$\mathcal{L}_a^1 = \begin{cases} \varnothing & \text{if } N_1 = \infty, \\ \{(N_1, q) \mid (N_1, q) \in \mathcal{L}_a\} & \text{if } N_1 < \infty, \end{cases}$$
$$\mathcal{L}_a^2 = \begin{cases} \varnothing & \text{if } N_2 = \infty, \\ \{(p, N_2) \mid (p, N_2) \in \mathcal{L}_a\} & \text{if } N_2 < \infty, \end{cases}$$

and $\mathcal{L}_a^0$ is the mixed "north-east" boundary, that is,

$$\mathcal{L}_a^0 = \{(p, q) \mid (p, q) \in \mathcal{L}_a, \ p < N_1, \ q < N_2\}.$$

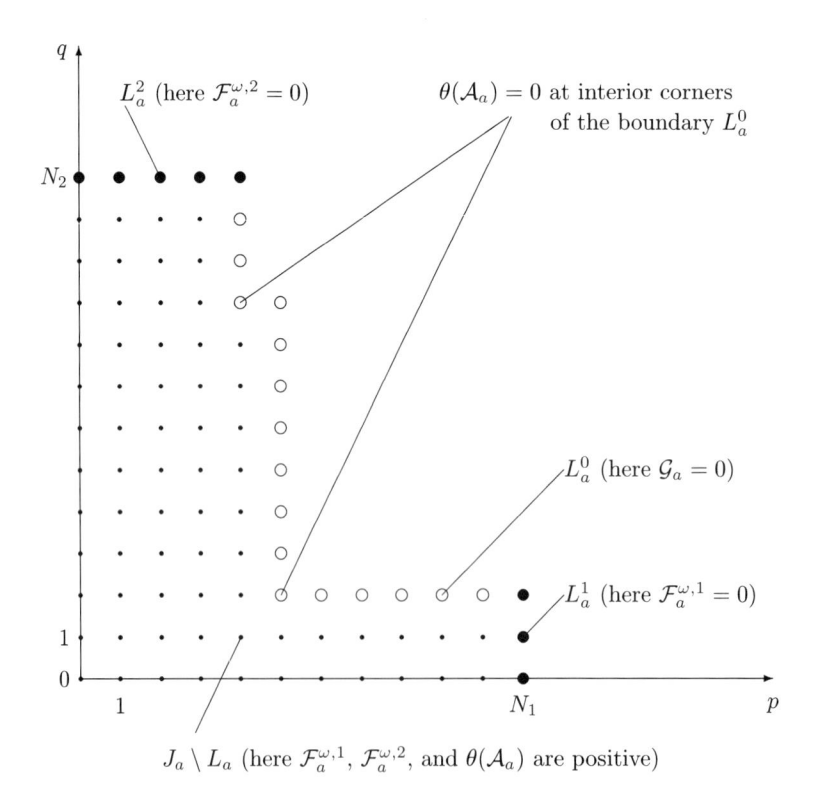

FIGURE 1. The set J_a with boundary $\mathcal{L}_a$

Note that for each $a \in \mathcal{R}$,
$$\mathcal{G}_a^1(p,q) > 0, \qquad \mathcal{G}_a^2(p,q) > 0, \qquad \mathcal{G}_a(p,q) > 0$$
for all $(p,q) \in J_a \setminus \mathcal{L}_a$, and
$$(4.19) \qquad \mathcal{G}_a^1(p,q)\mathcal{G}_a^2(p+1,q) = \mathcal{G}_a^1(p,q+1)\mathcal{G}_a^2(p,q) = 0$$
for all $(p,q) \in \mathcal{L}_a^0$.

For each $a \in \mathcal{R}$, we decompose each of the functions $\mathcal{F}_a^{\omega,1}\mathcal{G}_a^1$ and $\mathcal{F}_a^{\omega,2}\mathcal{G}_a^2$ into two factors (real or complex):
$$(4.20) \qquad \begin{aligned} \mathcal{F}_a^{\omega,1}(p)\mathcal{G}_a^1(p,q) &= \mathcal{L}_a^1(p,q)\mathcal{M}_a^1(p,q), \\ \mathcal{F}_a^{\omega,2}(q)\mathcal{G}_a^2(p,q) &= \mathcal{L}_a^2(p,q)\mathcal{M}_a^2(p,q), \qquad p \in \mathbb{Z}_+, \quad q \in \mathbb{Z}_+, \end{aligned}$$
with the "compatibility" conditions
$$(4.21) \qquad \begin{aligned} \mathcal{L}_a^1(p,q)\mathcal{L}_a^2(p+1,q) &= \mathcal{L}_a^1(p,q+1)\mathcal{L}_a^2(p,q), \\ \mathcal{M}_a^1(p,q)\mathcal{M}_a^2(p+1,q) &= \mathcal{M}_a^1(p,q+1)\mathcal{M}_a^2(p,q), \qquad p \in \mathbb{Z}_+, \quad q \in \mathbb{Z}_+, \end{aligned}$$
and the boundary conditions
$$(4.22) \qquad \begin{aligned} \mathcal{L}_a^1(N_1, q) &= 0 \quad \text{for all} \quad (N_1, q) \in \mathcal{L}_a^1, & & \text{if} \quad \mathcal{L}_a^1 \neq \varnothing, \\ \mathcal{L}_a^2(p, N_2) &= 0 \quad \text{for all} \quad (p, N_2) \in \mathcal{L}_a^2, & & \text{if} \quad \mathcal{L}_a^2 \neq \varnothing, \\ \mathcal{L}_a^1(p,q)\mathcal{L}_a^2(p+1,q) &= \left(\mathcal{L}_a^1(p,q+1)\mathcal{L}_a^2(p,q)\right) = 0 \\ &\text{for all} \quad (p,q) \in \mathcal{L}_a^0, & & \text{if} \quad \mathcal{L}_a^0 \neq \varnothing. \end{aligned}$$

Note that such factorizations exist for each $a \in \mathcal{R}$. For example, if we set
$$(4.23) \qquad \begin{aligned} \mathcal{L}_a^1(p,q) &= \mathcal{F}_a^{\omega,1}(p)\mathcal{G}_a^1(p,q), & \mathcal{M}_a^1(p,q) &\equiv 1, \\ \mathcal{L}_a^2(p,q) &= \mathcal{F}_a^{\omega,2}(q)\mathcal{G}_a^2(p,q), & \mathcal{M}_a^2(p,q) &\equiv 1, \end{aligned}$$
then conditions (4.21), (4.22) are satisfied. Actually, the compatibility condition for $\mathcal{L}_a^1$, $\mathcal{L}_a^2$ follows from (4.14):
$$\begin{aligned} \mathcal{L}_a^1(p,q)\mathcal{L}_a^2(p+1,q) &= \mathcal{F}_a^{\omega,1}(p)\,\mathcal{F}_a^{\omega,2}(q)\,\mathcal{G}_a^1(p,q)\,\mathcal{G}_a^2(p+1,q) \\ &= \mathcal{F}_a^{\omega,1}(p)\,\mathcal{F}_a^{\omega,2}(q)\,\mathcal{G}_a^1(p,q)\,\mathcal{G}_a^2(p,q)\,\theta\big(\mathcal{A}_a(p,q)\big) \\ &= \mathcal{F}_a^{\omega,1}(p)\,\mathcal{F}_a^{\omega,2}(q)\,\mathcal{G}_a^1(p,q+1)\,\mathcal{G}_a^2(p,q) = \mathcal{L}_a^1(p,q+1)\mathcal{L}_a^2(p,q); \end{aligned}$$
the boundary conditions on $\mathcal{L}_a^1$ and $\mathcal{L}_a^2$ follow from the equalities $\mathcal{F}_a^{\omega,1}(N_1) = 0$ and $\mathcal{F}_a^{\omega,2}(N_2) = 0$, and on $\mathcal{L}_a^0$ from (4.19).

Similarly, one can easily see that if we decompose the functions $\mathcal{F}_a^{\omega,1}$ and $\mathcal{F}_a^{\omega,2}$ into factors
$$\mathcal{F}_a^{\omega,1}(p) = \widetilde{\mathcal{L}}_a^1(p)\widetilde{\mathcal{M}}_a^1(p), \qquad \mathcal{F}_a^{\omega,2}(q) = \widetilde{\mathcal{L}}_a^2(q)\widetilde{\mathcal{M}}_a^2(q),$$
so that
$$\mathcal{F}_a^{\omega,1}(p) = 0 \implies \widetilde{\mathcal{L}}_a^1(p) = 0, \qquad \mathcal{F}_a^{\omega,2}(q) = 0 \implies \widetilde{\mathcal{L}}_a^2(q) = 0,$$
then
$$(4.24) \qquad \begin{aligned} \mathcal{L}_a^1(p,q) &= \widetilde{\mathcal{L}}_a^1(p)\mathcal{G}_a^1(p,q), & \mathcal{M}_a^1(p,q) &= \widetilde{\mathcal{M}}_a^1(p), \\ \mathcal{L}_a^2(p,q) &= \widetilde{\mathcal{L}}_a^2(q)\mathcal{G}_a^2(p,q), & \mathcal{M}_a^2(p,q) &= \widetilde{\mathcal{M}}_a^2(q), \end{aligned}$$
satisfy all conditions (4.21) and (4.22).

Now let us define the following positive numbers:

$$t_{(0,0)}(a) = 1,$$

$$t_{(p,0)}(a) = \frac{p! \prod_{r=0}^{p-1} \overline{\mathcal{M}_a^1}(r,0)}{\prod_{r=0}^{p-1} \mathcal{L}_a^1(r,0)}, \quad 1 \leq p \leq N_1,$$

(4.25) $\quad t_{(0,q)}(a) = \dfrac{q! \prod_{s=0}^{q-1} \overline{\mathcal{M}_a^2}(0,s)}{\prod_{s=0}^{q-1} \mathcal{L}_a^2(0,s)}, \quad 1 \leq q \leq N_2,$

$$t_{(p,q)}(a) = \frac{p!q! \prod_{r=0}^{p-1} \overline{\mathcal{M}_a^1}(r,0) \prod_{s=0}^{q-1} \overline{\mathcal{M}_a^2}(p,s)}{\prod_{r=0}^{p-1} \mathcal{L}_a^1(r,0) \prod_{s=0}^{q-1} \mathcal{L}_a^2(p,s)},$$
$$(p,q) \in J_a, \quad p \geq 1, \quad q \geq 1,$$

$$t_{(p,q)}(a) = \infty, \quad (p,q) \notin J_a.$$

Note that for $p \geq 1$ and $q \geq 1$, one can rewrite formula (4.25) in the form symmetric with respect to $(\mathcal{L}^1, \mathcal{M}^1)$ and $(\mathcal{L}^2, \mathcal{M}^2)$ by using the identities

(4.21a)
$$\prod_{r=0}^{p-1} \mathcal{L}_a^1(r,0) \prod_{s=0}^{q-1} \mathcal{L}_a^2(p,s) = \prod_{r=0}^{p-1} \mathcal{L}_a^1(r,q) \prod_{s=0}^{q-1} \mathcal{L}_a^2(0,s),$$

$$\prod_{r=0}^{p-1} \mathcal{M}_a^1(r,0) \prod_{s=0}^{q-1} \mathcal{M}_a^2(p,s) = \prod_{r=0}^{p-1} \mathcal{M}_a^1(r,q) \prod_{s=0}^{q-1} \mathcal{M}_a^2(0,s),$$

which follow from the compatibility conditions (4.21).

By using the numbers $t_{(p,q)}(a)$, we introduce a Hilbert space of distributions over $\mathbb{R}^4$ that are antiholomorphic, that is, annihilated by the operators $\partial/\partial z^1$ and $\partial/\partial z^2$, where z^1 and z^2 are the standard complex coordinates on $\mathbb{R}^4$. Namely, let us consider distributions g over $\mathbb{R}^4$ that can be represented by power series

$$g(\overline{z^1}, \overline{z^2}) = \sum_{p=0}^{\infty} \sum_{q=0}^{\infty} g_{(p,q)} (\overline{z^1})^p (\overline{z^2})^q.$$

By $\mathcal{P}_{t(a)}$ we denote the space of all series that satisfy the condition

$$\|g\|_{\mathcal{P}_{t(a)}} \stackrel{\text{def}}{=} \left(\sum_{p=0}^{\infty} \sum_{q=0}^{\infty} t_{(p,q)}(a) |g_{(p,q)}|^2 \right)^{1/2} < \infty.$$

In the case $N_1 < \infty$ and $N_2 < \infty$, the space $\mathcal{P}_{t(a)}$ is finite-dimensional and consists of all polynomials of the form

(4.26)
$$g(\overline{z^1}, \overline{z^2}) = \sum_{(p,q) \in J_a} g_{(p,q)} (\overline{z^1})^p (\overline{z^2})^q.$$

In any case, the set of polynomials is dense in $\mathcal{P}_{t(a)}$, the unit function 1 belongs to $\mathcal{P}_{t(a)}$, and the basis of monomials $\{(\overline{z^1})^p (\overline{z^2})^q\}$ is generated from 1 by powers of the following multiplication operator:

$$\overline{z^1}: \mathcal{P}_{t(a)} \to \mathcal{P}_{t(a)}, \quad \overline{z^1} \sum_{(p,q) \in J_a} g_{(p,q)} (\overline{z^1})^p (\overline{z^2})^q = \sum_{(p+1,q) \in J_a} g_{(p,q)} (\overline{z^1})^{p+1} (\overline{z^2})^q,$$

$$\overline{z^2}: \mathcal{P}_{t(a)} \to \mathcal{P}_{t(a)}, \quad \overline{z^2} \sum_{(p,q) \in J_a} g_{(p,q)} (\overline{z^1})^p (\overline{z^2})^q = \sum_{(p,q+1) \in J_a} g_{(p,q)} (\overline{z^1})^p (\overline{z^2})^{q+1}.$$

Obviously, different factorizations (4.20) generate isomorphic Hilbert spaces; if $\mathcal{P}_{t(a)}$ and $\mathcal{P}_{t'(a)}$ are such two spaces, then an isomorphism between them is given by the mapping

$$(4.27) \quad \sum_{(p,q)} g_{(p,q)} (\overline{z^1})^p (\overline{z^2})^q \to \sum_{(p,q)} g'_{(p,q)} (\overline{z^1})^p (\overline{z^2})^q, \qquad g'_{(p,q)} = \sqrt{\frac{t_{(p,q)}(a)}{t'_{(p,q)}(a)}} g_{(p,q)}.$$

4.4. Irreducible antiholomorphic representations with two complex variables. Now we can describe irreducible representations of the algebra (4.11) and objects related to them.

THEOREM 4.2. (a) *There is a one-to-one correspondence between the set $\mathcal{R}$ and the set of representations of the algebra (4.11) that are irreducible, Hermitian, and possess a vacuum vector. Such a representation corresponding to $a \in \mathcal{R}$ is finite-dimensional if and only if both functions $\mathcal{F}_a^{\omega,j}$ ($j = 1, 2$) defined in (4.13) have roots on $\mathbb{Z}_+$; in this case the number[14] $\dim J_a$ is the dimension of the representation. Here J_a is defined in (4.18a).*

(b) *For each $a \in \mathcal{R}$ and for each factorization (4.20), the operators*

$$(4.28) \quad \begin{aligned} \overset{\circ}{B}_j &= \overline{z^j} \mathcal{L}_a^j \left(\overline{z^1} \frac{\partial}{\partial \overline{z^1}}, \overline{z^2} \frac{\partial}{\partial \overline{z^2}} \right), \\ \overset{\circ}{C}^j &= \mathcal{M}_a^j \left(\overline{z^1} \frac{\partial}{\partial \overline{z^1}}, \overline{z^2} \frac{\partial}{\partial \overline{z^2}} \right) \frac{\partial}{\partial \overline{z^j}}, \qquad j = 1, 2, \\ \overset{\circ}{A}_j &= \mathcal{A}_a \left(\overline{z^1} \frac{\partial}{\partial \overline{z^1}}, \overline{z^2} \frac{\partial}{\partial \overline{z^2}} \right) \end{aligned}$$

represent the algebra (4.11), and moreover, this representation is irreducible Hermitian and possesses the vacuum 1 in the Hilbert space $\mathcal{P}_{t(a)}$.

(c) *Representations (4.28) assigned to different vectors $a \in \mathcal{R}$ are not equivalent, but for each chosen $a \in \mathcal{R}$, representations assigned to different factorizations (4.20) are equivalent under the isomorphism (4.27).*

(d) *An abstract Hermitian representation of the algebra (4.11) in a Hilbert space H_a with the vacuum vector $\mathfrak{P}_0 = \mathfrak{P}_0(a)$, satisfying (4.12) for some $a \in \mathbb{R}^k$, can be intertwined with the representation (4.28) by means of the following generalized coherent states:*[15]

$$(4.29) \quad \begin{aligned} \mathfrak{P}_{z^1,z^2} = \mathfrak{P}_0 &+ \sum_{1 \leq p \leq N_1} \frac{1}{p! \prod_{r=0}^{p-1} \overline{\mathcal{M}_a^1(r,0)}} (z^1 \mathbf{B}_1)^p \mathfrak{P}_0 \\ &+ \sum_{1 \leq q \leq N_2} \frac{1}{q! \prod_{s=0}^{q-1} \overline{\mathcal{M}_a^2(0,s)}} (z^2 \mathbf{B}_2)^q \mathfrak{P}_0 \\ &+ \sum_{\substack{(p,q) \in J_a \\ p \geq 1, q \geq 1}} \frac{1}{p! q! \prod_{r=0}^{p-1} \overline{\mathcal{M}_a^1(r,0)} \prod_{s=0}^{q-1} \overline{\mathcal{M}_a^2(p,s)}} (z^1 \mathbf{B}_1)^p (z^2 \mathbf{B}_2)^q \mathfrak{P}_0. \end{aligned}$$

---

[14] This number satisfies the inequality $\dim J_a \leq (N_1 + 1)(N_2 + 1)$.

[15] The coefficient in the last sum in (4.29) is written in a form that is not symmetric with respect to $\mathcal{M}_a^1$ and $\mathcal{M}_a^2$. However, one can rewrite (4.29) in a symmetric form by using the second identity from (4.21a).

This family of coherent states is an H_a-valued holomorphic distribution over $\mathbb{R}^4$, or a holomorphic polynomial in the case when $N_1 < \infty$ and $N_2 < \infty$. The generalized "reproducing kernel" corresponding to the states (4.29),
$$\mathcal{K}_{t(a)}(\overline{w^1}, \overline{w^2}, z^1, z^2) = (\mathfrak{P}_{w^1,w^2}, \mathfrak{P}_{z^1,z^2})_{H_a},$$
is the kernel of the identity operator in $\mathcal{P}_{t(a)}$ and is given by the following distribution over $\mathbb{R}^4 \times \mathbb{R}^4$:

$$(4.30) \qquad \mathcal{K}_{t(a)}(\overline{w^1}, \overline{w^2}, z^1, z^2) = \sum_{(p,q) \in J_a} \frac{(\overline{w^1} z^1)^p (\overline{w^2} z^2)^q}{t_{(p,q)}(a)}.$$

This distribution satisfies the equations

$$\mathcal{M}_a^j\left(\overline{w^1}\frac{\partial}{\partial \overline{w^1}}, \overline{w^2}\frac{\partial}{\partial \overline{w^2}}\right) \frac{\partial}{\partial \overline{w^j}} \mathcal{K}_{t(a)} = z^j \overline{\mathcal{L}_a^j}\left(z^1\frac{\partial}{\partial z^1}, z^2\frac{\partial}{\partial z^2}\right) \mathcal{K}_{t(a)}, \qquad j = 1, 2,$$

$$\mathcal{A}_a\left(\overline{w^1}\frac{\partial}{\partial \overline{w^1}}, \overline{w^2}\frac{\partial}{\partial \overline{w^2}}\right) \mathcal{K}_{t(a)} = \mathcal{A}_a\left(z^1\frac{\partial}{\partial z^1}, z^2\frac{\partial}{\partial z^2}\right) \mathcal{K}_{t(a)}.$$

To prove Theorem 4.2, we need the following two lemmas that are similar to Lemmas 1.1 and 1.2, respectively.

LEMMA 4.3. *Suppose that an Hermitian representation of the algebra* (4.11) *in a Hilbert space H_a possesses the vacuum vector $\mathfrak{P}_0 = \mathfrak{P}_0(a)$* (4.12). *Then the following equalities hold:*

(a) $\quad \mathbf{C}^1(\mathbf{B}_1)^p \mathfrak{P}_0 = p\mathcal{F}_a^{\omega,1}(p-1)(\mathbf{B}_1)^{p-1}\mathfrak{P}_0, \quad p \geq 1;$
$\quad \mathbf{C}^2(\mathbf{B}_2)^q \mathfrak{P}_0 = q\mathcal{F}_a^{\omega,2}(q-1)(\mathbf{B}_2)^{q-1}\mathfrak{P}_0, \quad q \geq 1;$
$\quad (\mathbf{C}^1)^p(\mathbf{B}_2)^q = (\mathbf{B}_2)^q \, \mathcal{G}_{\mathbf{A}}(p-1, q-1)(\mathbf{C}_1)^p, \quad p \geq 1, \quad q \geq 1;$
$\quad (\mathbf{C}^2)^q(\mathbf{B}_1)^p = (\mathbf{B}_1)^p \, \mathcal{G}_{\mathbf{A}}(p-1, q-1)(\mathbf{C}_2)^q, \quad p \geq 1, \quad q \geq 1;$

(b) $\quad \|(\mathbf{B}_1)^p \mathfrak{P}_0\|_{H_a}^2 = p! \prod_{r=0}^{p-1} \mathcal{F}_a^{\omega,1}(r), \quad p \geq 1,$

$\quad \|(\mathbf{B}_2)^q \mathfrak{P}_0\|_{H_a}^2 = q! \prod_{s=0}^{q-1} \mathcal{F}_a^{\omega,2}(s), \quad q \geq 1,$

$\quad \|(\mathbf{B}_1)^p(\mathbf{B}_2)^q \mathfrak{P}_0\|_{H_a}^2 = p!q! \prod_{r=0}^{p-1} \mathcal{F}_a^{\omega,1}(r) \prod_{s=0}^{q-1} \mathcal{F}_a^{\omega,2}(s) \, \mathcal{G}_a(p-1, q-1),$
$\hfill p \geq 1, \, q \geq 1;$

(c) $\quad \left((\mathbf{B}_1)^p(\mathbf{B}_2)^q \mathfrak{P}_0, (\mathbf{B}_1)^{p'}(\mathbf{B}_2)^{q'} \mathfrak{P}_0\right)_{H_a} = 0 \quad \textit{if } p \neq p' \textit{ or } q \neq q'.$

PROOF. According to (4.11) and the definition (4.14), we have

$$\mathbf{C}^1(\mathbf{B}_2)^q = \left(\mathbf{B}_2 \theta(\mathbf{A})\right)^q \mathbf{C}^1 = (\mathbf{B}_2)^q \prod_{s=0}^{q-1} \theta\big(\underbrace{\varphi_2(\ldots(\varphi_2}_{s}(\mathbf{A}))\ldots)\big)\mathbf{C}^1$$

$$= (\mathbf{B}_2)^q \, \mathcal{G}_{\mathbf{A}}^1(0, q)\, \mathbf{C}^1.$$

Therefore, from (4.17) we obtain the third formula in statement (a):

$$(\mathbf{C}^1)^p(\mathbf{B}_2)^q = (\mathbf{B}_2)^q \left(\mathcal{G}^1_\mathbf{A}(0,q)\mathbf{C}^1\right)^p = (\mathbf{B}_2)^q \prod_{r=0}^{p-1} \mathcal{G}^1_{\underbrace{\varphi_1(\ldots(\varphi_1(\mathbf{A}))\ldots)}_{r}}(0,q)\,(\mathbf{C}^1)^p$$

$$= (\mathbf{B}_2)^q \prod_{r=0}^{p-1} \mathcal{G}^1_\mathbf{A}(r,q)\,(\mathbf{C}^1)^p = (\mathbf{B}_2)^q\,\mathcal{G}_\mathbf{A}(p-1,q-1)\,(\mathbf{C}^1)^p.$$

The fourth formula in (a) follows from the third formula after the conjugation.

The other formulas are proved similarly to those in Lemma 1.1. $\square$

LEMMA 4.4. *Let $\mathcal{A}$, $\mathcal{L}^j$, and $\mathcal{M}^j$, $j = 1, 2$, be functions over $\mathbb{Z}_+ \times \mathbb{Z}_+$. Then the operators*

(4.31)
$$\begin{aligned}
\overset{\circ}{B}_j &= \overline{z^j}\mathcal{L}^j\left(\overline{z^1}\frac{\partial}{\partial\overline{z^1}}, \overline{z^2}\frac{\partial}{\partial\overline{z^2}}\right), \\
\overset{\circ}{C}^j &= \mathcal{M}^j\left(\overline{z^1}\frac{\partial}{\partial\overline{z^1}}, \overline{z^2}\frac{\partial}{\partial\overline{z^2}}\right)\frac{\partial}{\partial\overline{z^j}}, \\
\overset{\circ}{A} &= \mathcal{A}\left(\overline{z^1}\frac{\partial}{\partial\overline{z^1}}, \overline{z^2}\frac{\partial}{\partial\overline{z^2}}\right)
\end{aligned}$$

are well defined on all polynomials and, in particular, on the unit function 1. *These operators generate an irreducible Hermitian representation of the algebra (4.11) in a Hilbert space of distributions with the vacuum* 1 *if and only if the functions* $\mathcal{A}$, $\mathcal{L}^j$, *and* $\mathcal{M}^j$, $j = 1, 2$, *satisfy the following conditions:*
- *the vector $a = \mathcal{A}(0)$ belongs to the set $\mathcal{R}$, that is, $a \in \mathcal{R}$;*
- $\mathcal{A}$ *coincides with the function $\mathcal{A}_a$ given by (4.16);*
- *the products $\mathcal{L}^1\mathcal{M}^1$ and $\mathcal{L}^2\mathcal{M}^2$ coincide with the functions $\mathcal{F}^{\omega,1}_a\mathcal{G}^1_a$ and $\mathcal{F}^{\omega,2}_a\mathcal{G}^2_a$, respectively, given by (4.13) and (4.14);*
- *the functions $\mathcal{L}^1$, $\mathcal{L}^2$, and $\mathcal{M}^1$, $\mathcal{M}^2$ satisfy the compatibility conditions (4.21);*
- *the functions $\mathcal{L}^1$ and $\mathcal{L}^2$ satisfy the boundary conditions (4.22).*

In this case, the Hilbert space of the representation automatically coincides with $\mathcal{P}_{t(a)}$, where the sequence $t(a)$ is given by (4.25).

PROOF. The proof of this lemma is similar to that of Lemma 1.2. Therefore, we outline only the basic steps of this proof.

1. The permutation relations

$$\overset{\circ}{C}^j\overset{\circ}{A} = \varphi_j(\overset{\circ}{A})\overset{\circ}{C}^j, \qquad \overset{\circ}{A}\overset{\circ}{B}_j = \overset{\circ}{B}_j\varphi_j(\overset{\circ}{A}), \quad j = 1, 2,$$

for the operators (4.31) are equivalent to the condition that the function $\mathcal{A}$ coincide with $\mathcal{A}_a$ (4.16), where $a = \mathcal{A}(0)$.

The permutation relations

$$\overset{\circ}{C}^j\overset{\circ}{B}_j = \overset{\circ}{B}_j\,\alpha^j_j(\overset{\circ}{A})\,\overset{\circ}{C}^j + \overset{\circ}{B}_\ell\,\alpha^\ell_j(\overset{\circ}{A})\,\overset{\circ}{C}^\ell + f^j_j(\overset{\circ}{A}), \qquad j, \ell = 1, 2, \quad j \neq \ell,$$

are equivalent to the following system of equations with respect to the products $\mathcal{L}^1\mathcal{M}^1$ and $\mathcal{L}^2\mathcal{M}^2$:

$$
\begin{aligned}
(p+1)\,(\mathcal{L}^1\mathcal{M}^1)(p,q) &= p\,\alpha_1^1\big(\mathcal{A}(p-1,q)\big)\,(\mathcal{L}^1\mathcal{M}^1)(p-1,q)\\
&\quad + q\,\alpha_1^2\big(\mathcal{A}(p,q-1)\big)\,(\mathcal{L}^2\mathcal{M}^2)(p,q-1) + f_1^1\big(\mathcal{A}(p,q)\big),\\
(q+1)\,(\mathcal{L}^2\mathcal{M}^2)(p,q) &= q\,\alpha_2^2\big(\mathcal{A}(p,q-1)\big)\,(\mathcal{L}^2\mathcal{M}^2)(p,q-1)\\
&\quad + p\,\alpha_2^1\big(\mathcal{A}(p-1,q)\big)\,(\mathcal{L}^1\mathcal{M}^1)(p-1,q) + f_2^2\big(\mathcal{A}(p,q)\big).
\end{aligned}
\tag{4.32}
$$

The permutation relations

$$
\overset{\circ}{C}{}^1 \overset{\circ}{B}_2 = \overset{\circ}{B}_2\,\theta(\overset{\circ}{A})\,\overset{\circ}{C}{}^1, \qquad \overset{\circ}{C}{}^2 \overset{\circ}{B}_1 = \overset{\circ}{B}_1\,\theta(\overset{\circ}{A})\,\overset{\circ}{C}{}^2
$$

lead to the conditions

$$
\begin{aligned}
\mathcal{L}^2(p+1,q)\mathcal{M}^1(p,q+1) &= \theta\big(\mathcal{A}(p,q)\big)\mathcal{L}^2(p,q)\mathcal{M}^1(p,q),\\
\mathcal{L}^1(p,q+1)\mathcal{M}^2(p+1,q) &= \theta\big(\mathcal{A}(p,q)\big)\mathcal{L}^1(p,q)\mathcal{M}^2(p,q),
\end{aligned}
\tag{4.33}
$$

and the relations

$$
[\overset{\circ}{B}_1,\overset{\circ}{B}_2] = 0, \qquad [\overset{\circ}{C}{}^1,\overset{\circ}{C}{}^2] = 0
$$

lead to the compatibility conditions (4.21).

Taking the (Jacobi) identities (4.10) into account, one can verify that the unique solution

$$\mathcal{L}^j(p,q), \quad \mathcal{M}^j(p,q) \quad (j=1,2), \qquad p\in\mathbb{Z}_+,\quad q\in\mathbb{Z}_+,$$

of system (4.32), (4.33), (4.21) is described by formulas (4.20) and (4.21), where $\mathcal{F}_a^{\omega,1}$, $\mathcal{F}_a^{\omega,2}$, $\mathcal{G}_a^1$, $\mathcal{G}_a^2$ are given by (4.13) and (4.14).

2. The boundary conditions (4.22) provide that the space of functions of the form (4.24) is invariant with respect to the operators $\overset{\circ}{B}_j$ (4.31), $j=1,2$.

3. The condition $a\in\mathcal{R}$ is equivalent to the fact that in the space of functions of form (4.26) there is an inner product with respect to which the representation (4.31) (where $\mathcal{A}=\mathcal{A}_a$, $\mathcal{L}^j=\mathcal{L}_a^j$, and $\mathcal{M}^j=\mathcal{M}_a^j$) is Hermitian.

4. Finally, the condition $a\in\mathcal{R}$ implies that the functions $\mathcal{L}_a^j$ and $\mathcal{M}_a^j$ ($j=1,2$) do not vanish for $(p,q)\in J_a\setminus\mathcal{L}_a$. Therefore, in the space $\mathcal{P}_{t(a)}$ there does not exist an invariant (with respect to (4.31)) subspace whose dimension is less than that of $\mathcal{P}_{t(a)}$; that is, the representation (4.31) is irreducible. $\square$

PROOF OF THEOREM 4.2. Theorem 4.2 is proved using Lemmas 4.3 and 4.4 in the same way as Lemmas 1.1 and 1.2 were used in the proof of Theorem 1.1.

Here we present only the system of equations for the coefficients of coherent states,

$$\mathfrak{P}_{z^1,z^2} = \sum_{(p,q)\in J_a} \eta_{(p,q)}(z^1\mathbf{B}_1)^p(z^2\mathbf{B}_2)^q \mathfrak{P}_0, \tag{4.29a}$$

which is equivalent to the intertwining property

$$
\begin{aligned}
\eta_{(0,0)} &= 1,\\
\eta_{(p+1,q)}\,(p+1)\overline{\mathcal{M}_a^1}(p,q) &= \eta_{(p,q)},\\
\eta_{(p,q+1)}\,(q+1)\overline{\mathcal{M}_a^2}(p,q) &= \eta_{(p,q)}.
\end{aligned}
\tag{4.34}
$$
$\square$

COROLLARY 4.1. *The following orthonormal eigenbases of the operators* $\overset{\circ}{\mathbf{A}}_\mu$ ($\mu = 1, \ldots, k$) (4.28):

$$\left\{ \frac{(\overline{z^1})^p (\overline{z^2})^q}{\sqrt{t_{p,q}(a)}} \;\bigg|\; (p,q) \in J_a \right\} \quad \text{in} \quad \mathcal{P}_{t(a)}$$

and of the operators $\mathbf{A}_\mu$:

$$\left\{ \frac{1}{\sqrt{t_{p,q}(a)} \prod_{r=0}^{p-1} \mathcal{L}_a^1(r,0) \prod_{s=0}^{q-1} \mathcal{L}_a^2(p,s)} (\mathbf{B}_1)^p (\mathbf{B}_2)^q \mathfrak{P}_0 \;\bigg|\; (p,q) \in J_a \right\} \quad \text{in} \quad H_a$$

(*the products* $\prod_{r=0}^{p-1} \ldots$ *or* $\prod_{s=0}^{q-1} \ldots$ *in the denominator are replaced by* 1 *if* $p = 0$ *or* $q = 0$) *correspond to each other under the coherent transform*

$$\mathcal{P}_{t(a)} \ni g \mapsto \mathfrak{p} \in H_a, \qquad (\mathfrak{p}', \mathfrak{p})_{H_a} = \big((\mathfrak{P}, \mathfrak{p}')_{H_a}, g\big)_{\mathcal{P}_{t(a)}}, \quad \forall \mathfrak{p}' \in H_a,$$

$$H_a \ni \mathfrak{p} \mapsto g \in \mathcal{P}_{t(a)}, \qquad g = (\mathfrak{P}, \mathfrak{p})_{H_a}.$$

4.5. Quantum surfaces of birevolution. Irreducible representations.
Let us consider the following important particular case of permutation relations (4.11):

$$(4.35) \quad \begin{aligned} [\mathbf{C}^1, \mathbf{B}_1] &= f_1^1(\mathbf{A}), & [\mathbf{C}^2, \mathbf{B}_2] &= f_2^2(\mathbf{A}), \\ [\mathbf{C}^1, \mathbf{B}_2] &= 0, & [\mathbf{C}^2, \mathbf{B}_1] &= 0, \\ \mathbf{C}^1 \mathbf{A} &= \varphi_1(\mathbf{A}) \mathbf{C}^1, & \mathbf{A}\mathbf{B}_1 &= \mathbf{B}_1 \varphi_1(\mathbf{A}), \\ \mathbf{C}^2 \mathbf{A} &= \varphi_2(\mathbf{A}) \mathbf{C}^2, & \mathbf{A}\mathbf{B}_2 &= \mathbf{B}_2 \varphi_2(\mathbf{A}), \\ [\mathbf{A}_\mu, \mathbf{A}_\nu] = 0, \quad [\mathbf{B}_1, \mathbf{B}_2] &= 0, \quad [\mathbf{C}^1, \mathbf{C}^2] = 0, \end{aligned}$$

By Lemma 4.1, for the Jacobi identities (4.2) and conditions (4.3) to be satisfied in this case, it suffices to require that the functions f_1^1, f_2^2 and φ_1, φ_2 are real and satisfy the conditions

$$(4.36) \quad \begin{aligned} &\text{(a)} \quad \varphi_1\big(\varphi_2(A)\big) = \varphi_2\big(\varphi_1(A)\big), \\ &\text{(b)} \quad f_1^1\big(\varphi_2(A)\big) = f_1^1(A), \qquad f_2^2\big(\varphi_1(A)\big) = f_2^2(A). \end{aligned}$$

Throughout §4.5, we assume that these conditions are satisfied.

Note that conditions (4.36)(b) imply

$$[f_p^p(\mathbf{A}), \mathbf{B}_q] = [f_p^p(\mathbf{A}), \mathbf{C}^q] = 0, \qquad p \neq q,$$

and therefore each of the operators $\mathbf{B}_1$, $\mathbf{C}^1$, and $f_1^1(\mathbf{A})$ in the algebra (4.35) commutes with each of the operators $\mathbf{B}_2$, $\mathbf{C}^2$, and $f_2^2(\mathbf{A})$.

Since the algebra (4.35) is a particular case of (4.11), all statements of the previous §§4.3 and 4.4 are satisfied for this algebra. However, in this case, all formulas for the irreducible representations and objects related to these representations are drastically simplified.

The functions (4.13), (4.14), and (4.15) turn into the following functions:

(4.13a)
$$\mathcal{F}_a^{\omega,1}(p) = \mathcal{F}_a^1(p), \qquad \mathcal{F}_a^1(p) \stackrel{\text{def}}{=} \frac{1}{p+1}\sum_{r=0}^{p} f_1^1(\mathcal{A}_a(r,0)),$$
$$\mathcal{F}_a^{\omega,2}(q) = \mathcal{F}_a^2(q), \qquad \mathcal{F}_a^2(q) \stackrel{\text{def}}{=} \frac{1}{q+1}\sum_{s=0}^{q} f_2^2(\mathcal{A}_a(0,s)),$$

where $\mathcal{A}_a(p,q)$ is the composition (4.16);

(4.14a) $$\mathcal{G}_a^1(p,q) = \mathcal{G}_a^2(p,q) \equiv 1;$$
(4.15a) $$\mathcal{G}_a(p,q) \equiv 1.$$

In this case the set $\mathcal{R}$ that labels irreducible representations of the algebra (4.35) can be represented as the intersection of two subsets: $\mathcal{R} = \mathcal{R}_1 \cap \mathcal{R}_2$, where, by definition, $a \in \mathcal{R}_j$ if $\mathcal{F}_a^j(n) > 0$ for all $\in \mathbb{Z}_+$ or if there exists an integer $N_j \in \mathbb{Z}_+$ such that
$$\mathcal{F}_a^j(n) > 0 \text{ for } 0 \leq n < N_j \text{ and } \mathcal{F}_a^j(N_j) = 0.$$

If $\mathcal{F}_a^j(n) > 0$ everywhere on $\mathbb{Z}_+$, then $N_j \stackrel{\text{def}}{=} \infty$.

The set J_a of pairs of integer numbers and its boundary $\mathcal{L}_a$ take the form

$$J_a = \{(p,q) \in \mathbb{Z}_+ \times \mathbb{Z}_+ \mid p \leq N_1, q \leq N_2\},$$
$$\mathcal{L}_a = \mathcal{L}_a^1 \cup \mathcal{L}_a^2, \qquad \mathcal{L}_a^1 = \begin{cases} \varnothing & \text{if } N_1 = \infty, \\ \{(N_1, q) \mid 0 \leq q \leq N_2\} & \text{if } N_1 < \infty, \end{cases}$$
$$\mathcal{L}_a^2 = \begin{cases} \varnothing & \text{if } N_2 = \infty, \\ \{(p, N_2) \mid 0 \leq p \leq N_1\} & \text{if } N_2 < \infty. \end{cases}$$

Furthermore, since the function $\mathcal{F}_a^{\omega,1}\mathcal{G}_a^1 = \mathcal{F}_a^1$ is now independent of q and the function $\mathcal{F}_a^{\omega,2}\mathcal{G}_a^2 = \mathcal{F}_a^2$ is independent of p, the function $\mathcal{F}_a^{\omega,1}\mathcal{G}_a^1$ is decomposed into the factors $\mathcal{L}_a^1(p,q) = \mathcal{B}_a^1(p)$ and $\mathcal{M}_a^1(p,q) = \mathcal{C}_a^1(p)$ independent of q, and $\mathcal{F}_a^{\omega,2}\mathcal{G}_a^2$ into the factors $\mathcal{L}_a^2(p,q) = \mathcal{B}_a^2(q)$ and $\mathcal{M}_a^2(p,q) = \mathcal{C}_a^2(q)$ independent of p:

(4.20a)
$$\mathcal{F}_a^1(p) = \mathcal{B}_a^1(p)\mathcal{C}_a^1(p), \qquad p \in \mathbb{Z}_+,$$
$$\mathcal{F}_a^2(q) = \mathcal{B}_a^2(q)\mathcal{C}_a^2(q), \qquad q \in \mathbb{Z}_+.$$

Then the compatibility conditions (4.21) are automatically satisfied, and the boundary conditions (4.22) can be written in the form

(4.22a)
$$\mathcal{B}_a^1(N_1) = 0, \quad \text{if } N_1 < \infty,$$
$$\mathcal{B}_a^2(N_2) = 0, \quad \text{if } N_2 < \infty.$$

The formulas for the numbers $t_{(p,q)}(a)$ (that are used to construct the inner product in the space $\mathcal{P}_{t(a)}$ of antiholomorphic distributions) are simplified:

(4.25a)
$$t_{(p,q)}(a) = s_p^1(a)s_q^2(a) \stackrel{\text{def}}{=} s_{(p,q)}(a),$$
$$s_0^j(a) = 1,$$
$$s_n^j(a) = \frac{n!\prod_{m=0}^{n-1}\overline{\mathcal{C}_a^j(m)}}{\prod_{m=0}^{n-1}\mathcal{B}_a^j(m)} = \frac{n!\mathcal{F}_a^j(n-1)\dots\mathcal{F}_a^j(0)}{|\mathcal{B}_a^j(n-1)|^2\dots|\mathcal{B}_a^j(0)|^2} \qquad \text{for } 1 \leq n \leq N_j,$$
$$s_n^j(a) = \infty \qquad \text{for } n \geq N_j + 1, \qquad\qquad\qquad\qquad j = 1,2.$$

For $N_1, N_2 < \infty$, the space $\mathcal{P}_{t(a)} = \mathcal{P}_{s(a)}$ is of dimension $(N_1 + 1)(N_2 + 1)$ and consists of all polynomials in $\overline{z^1}$ and $\overline{z^2}$ whose degree with respect to $\overline{z^1}$ does not exceed N_1 and whose degree with respect to $\overline{z^2}$ does not exceed N_2.

The operators of irreducible representation of algebra (4.35) in the space $\mathcal{P}_{s(a)}$ are given by the formulas

(4.28a)
$$\overset{\circ}{B}_j = \overline{z^j}\,\mathcal{B}_a^j\!\left(\overline{z^j}\frac{\partial}{\partial \overline{z^j}}\right),$$
$$\overset{\circ}{C}^j = \mathcal{C}_a^j\!\left(\overline{z^j}\frac{\partial}{\partial \overline{z^j}},\right)\frac{\partial}{\partial \overline{z^j}}, \qquad j = 1, 2,$$
$$\overset{\circ}{A} = \mathcal{A}_a\!\left(\overline{z^1}\frac{\partial}{\partial \overline{z^1}}, \overline{z^2}\frac{\partial}{\partial \overline{z^2}}\right).$$

The coherent states that intertwine an abstract Hermitian representation of the algebra (4.35) in a Hilbert space H_a (with the vacuum vector $\mathfrak{P}_0$ (4.12)) and the irreducible representation (4.28a) have the form

$$\mathfrak{P}_{z^1,z^2} = \prod_{j=1}^{2}\left(\mathbf{I} + \sum_{n=1}^{N_j} \frac{1}{n!\,\overline{C_a^j}(n-1)\ldots\overline{C_a^j}(0)}(z^j\mathbf{B}_j)^n\right)\mathfrak{P}_0.$$

The reproducing kernel $\mathcal{K}_{t(a)} = \mathcal{K}_{s(a)}$ is given by

(4.30a) $\qquad \mathcal{K}_{s(a)}(\overline{w^1}, \overline{w^2}, z^1, z^2) = \mathcal{K}_{s^1(a)}^1(\overline{w^1}, z^1)\,\mathcal{K}_{s^2(a)}^2(\overline{w^2}, z^2),$

where

$$\mathcal{K}_{s^j(a)}^j(\overline{w^j}, z^j) \overset{\text{def}}{=} \sum_{n=0}^{N_j} \frac{(\overline{w^j}z^j)^n}{s_n^j(a)}, \qquad j = 1, 2.$$

Now let us consider the regular case of relations of type (4.35). First, we reformulate Lemma 4.2(b) for relations (4.35).

If a function $\rho_j(A)$ (where $j = 1$ or $j = 2$) satisfies the system of equations

(4.37) $\qquad \rho_j\big(\varphi_\ell(A)\big) - \rho_j(A) = \delta_{j\ell}f_j^j(A), \qquad \ell = 1, 2,$

then

(4.38) $\qquad \mathbf{K}_j = \mathbf{B}_j\mathbf{C}^j - \rho_j(\mathbf{A})$

is a Casimir element of the algebra (4.35).

Now let us also formulate an analog of Lemma 1.4.

LEMMA 4.5. Let $k > 1$.

(a) Assume that the mapping φ_j is the shift per unit time along the trajectories of the vector field $v_j = \sum_{\mu=1}^k v_{j\mu}(A)\partial/\partial A_\mu$ on $\mathbb{R}^k$, $j = 1, 2$. Moreover, suppose that $[v_1, v_2] = 0$ on $\mathbb{R}^k$, and $\mathbb{R}^k$ (or an appropriate domain in $\mathbb{R}^k$) is fibered by trajectories of the vector fields v_1 and v_2. Then there are independent smooth real functions $\varkappa_1, \ldots, \varkappa_{k-2}$ on $\mathbb{R}^k$ that are preserved by φ_1 and φ_2, and therefore, there are independent Casimir elements $\varkappa_1(\mathbf{A}), \ldots, \varkappa_{k-2}(\mathbf{A})$ of the algebra (4.35).

(b) In addition to (a), assume that f_1^1 and f_2^2 are smooth tempered functions on $\mathbb{R}^k$, the fields v_1 and v_2 are also tempered, and the Jacobi matrices Dv_j $(j = 1, 2)$

are nilpotent. Then there exists a smooth real solution ρ_j $(j = 1, 2)$ of equations (4.37), and therefore, there is a Casimir element $\mathbf{K}_j$ $(j = 1, 2)$ of type (4.38).

Lemma 4.4 is proved similarly to Lemma 1.4 if we take the identities (4.36) into account.

DEFINITION 4.1. *The permutation relations (4.35) are called regular if there are independent real smooth (φ_1, φ_2)-invariant functions $\varkappa_1, \ldots, \varkappa_{k-2}$ on $\mathbb{R}^k$, as well as a smooth real solution ρ_1, ρ_2 of the difference equations (4.37).*

By using the Hermitian generators

$$\mathbf{S}_1^j = \frac{1}{2}(\mathbf{B}_j + \mathbf{C}^j), \qquad \mathbf{S}_2^j = \frac{i}{2}(\mathbf{B}_j - \mathbf{C}^j), \quad j = 1, 2,$$

similarly to the one-dimensional case, one can write equations for symplectic leaves $\Omega \subset \mathbb{R}^{k+4}$ that correspond to the irreducible representation of (4.35) assigned to $a \in \mathcal{R}$:

(4.39)
$$(\mathbf{S}_1^j)^2 + (\mathbf{S}_2^j)^2 = \frac{1}{2}\Big(\rho_j(\mathbf{A}) + \rho_j(\varphi_j(\mathbf{A}))\Big) - \rho_j(a) \cdot \mathbf{I}, \quad j = 1, 2,$$
$$\varkappa_\ell(\mathbf{A}) = \varkappa_\ell(a) \cdot \mathbf{I}, \qquad \qquad \ell = 1, \ldots, k-2.$$

So, in this case, Ω are 4-dimensional *surfaces of birevolution*, that is, 4-submanifolds in $\mathbb{R}^{k+4}$ fibered by tori.

In this case, in view of (4.37), the basic functions $\mathcal{F}_a^1, \mathcal{F}_a^2$ (4.13a) can be derived as follows:

$$\mathcal{F}_a^1(p) = \frac{1}{p+1}\Big(\rho_1\big(\mathcal{A}_a(p+1, 0)\big) - \rho_1(a)\Big),$$
$$\mathcal{F}_a^2(q) = \frac{1}{q+1}\Big(\rho_2\big(\mathcal{A}_a(0, q+1)\big) - \rho_2(a)\Big).$$

Thus, one has the following statement.

LEMMA 4.6. *The inclusion $a \in \mathcal{R}$ is equivalent to the following two conditions:*
- *either*

$$\rho_1\big(\mathcal{A}_a(p, 0)\big) > \rho_1(a) \qquad \text{for any} \quad p \in \mathbb{Z}_+, \quad p \geq 1,$$

or

(4.40.1) $\qquad \begin{cases} \rho_1\big(\mathcal{A}_a(p, 0)\big) > \rho_1(a) & \text{for } 1 \leq p \leq N_1, \\ \rho_1\big(\mathcal{A}_a(N_1 + 1, 0)\big) = \rho_1(a), \end{cases}$

- *either*

$$\rho_2\big(\mathcal{A}_a(0, q)\big) > \rho_2(a) \qquad \text{for any} \quad q \in \mathbb{Z}_+, \quad q \geq 1,$$

or

(4.40.2) $\qquad \begin{cases} \rho_2\big(\mathcal{A}_a(0, q)\big) > \rho_2(a) & \text{for } 1 \leq q \leq N_2, \\ \rho_2\big(\mathcal{A}_a(0, N_2 + 1)\big) = \rho_2(a). \end{cases}$

Now, instead of the factorization (4.20a), one can consider the following procedure. For any $a, a' \in \mathbb{R}^k$, let us decompose the difference

(4.41) $\qquad \rho_j(a') - \rho_j(a) = \mathcal{D}_a^j(a') \mathcal{E}_a^j(a'), \qquad j = 1, 2,$

into two smooth factors and assume that

(4.42)
$$\mathcal{D}_a^j(\varphi_\ell(a')) = \mathcal{D}_a^j(a'), \quad j,\ell = 1,2, \quad j \neq \ell,$$
$$\mathcal{E}_a^j(a) = 0, \quad j = 1,2,$$

as well as

(4.43)
$$\mathcal{D}_a^1(\mathcal{A}_a(N_1+1,0)) = 0 \quad \text{in case (4.40.1)},$$
$$\mathcal{D}_a^2(\mathcal{A}_a(0,N_2+1)) = 0 \quad \text{in case (4.40.2)}.$$

Note that the conditions (4.42) are new as compared with the one-dimensional case (see §1.3); for $j \neq \ell$, (4.42) and (4.37) imply

(4.44) $\quad\quad\quad\quad \mathcal{E}_a^j(\varphi_\ell(a')) = \mathcal{E}_a^j(a'), \quad j,\ell = 1,2, \quad j \neq \ell.$

Let us choose

$$\mathcal{B}_a^1(p) = \mathcal{D}_a^1(\mathcal{A}_a(p+1,0)), \quad \mathcal{C}_a^1(p) = \frac{1}{p+1}\mathcal{E}_a^1(\mathcal{A}_a(p+1,0)),$$
$$\mathcal{B}_a^2(q) = \mathcal{D}_a^2(\mathcal{A}_a(0,q+1)), \quad \mathcal{C}_a^2(q) = \frac{1}{q+1}\mathcal{E}_a^2(\mathcal{A}_a(0,q+1)).$$

Then, in view of (4.41) and (4.43), conditions (4.20a) and (4.22a) hold, and we arrive at the following statement.

THEOREM 4.3. *Suppose that relations* (4.35) *are regular, the functions* ρ_1 *and* ρ_2 *are assigned to* f_1^1 *and* f_2^2 *by* (4.37), *and* $a \in \mathcal{R}$. *Then:*
* *Each decomposition* (4.41) *generates a Hilbert space of antiholomorphic distributions*

$$g(\overline{z^1},\overline{z^2}) = \sum_{p=0}^{N_1}\sum_{q=0}^{N_2} g_{(p,q)} (\overline{z^1})^p (\overline{z^2})^q$$

with the inner product

$$(g',g) \stackrel{\text{def}}{=} \sum_{p=0}^{N_1}\sum_{q=0}^{N_2} s_p^1(a) s_q^2(a)\, g_{(p,q)} \overline{g'_{(p,q)}},$$

where the numbers $s_n^j(a)$ (4.25a) *are expressed in terms of the functions* $\mathcal{D}_a^j$ *and* $\mathcal{E}_a^j$ *as follows:*

$$s_0^1(a) = s_0^2(a) = 1,$$
$$s_p^1(a) = X_a^1(\mathcal{A}_a^1(p,0))\ldots X_a^1(\mathcal{A}_a^1(1,0)), \quad 1 \leq p \leq N_1,$$
$$s_q^2(a) = X_a^2(\mathcal{A}_a^1(0,q))\ldots X_a^2(\mathcal{A}_a^1(0,1)), \quad 1 \leq q \leq N_2,$$
$$X_a^j(a) \stackrel{\text{def}}{=} 1,$$
$$X_a^j(a') \stackrel{\text{def}}{=} \mathcal{D}_a^j(a')^{-1} \cdot \overline{\mathcal{E}_a^j(a')} \quad \text{if} \quad \rho_j(a') \neq \rho_j(a),$$

and the numbers $N_j \in \mathbb{Z}_+ \cup \{\infty\}$ *are taken from* (4.40).
* *In this Hilbert space the irreducible Hermitian representation of the algebra* (4.35) *is given by the operators*

(4.45) $\quad \overset{\circ}{B}_j = \mathcal{D}_a^j(\overset{\circ}{A}) \cdot \overline{z^j}, \quad \overset{\circ}{C}^j = \frac{1}{\overline{z^j}} \cdot \mathcal{E}_a^j(\overset{\circ}{A}), \quad \overset{\circ}{A} = \mathcal{A}_a\left(\overline{z^1}\frac{\partial}{\partial \overline{z^1}}, \overline{z^2}\frac{\partial}{\partial \overline{z^2}}\right).$

- *The corresponding coherent states are given by*

(4.46)
$$\mathfrak{P}_{z^1,z^2} = \left(\mathbf{I} + \sum_{p=1}^{N_1} \frac{1}{\overline{\mathcal{E}_a^1}(\mathcal{A}_a(p,0))\dots\overline{\mathcal{E}_a^1}(\mathcal{A}_a(1,0))}(z^1\mathbf{B}_1)^p\right)$$
$$\times \left(\mathbf{I} + \sum_{q=1}^{N_2} \frac{1}{\overline{\mathcal{E}_a^2}(\mathcal{A}_a(0,q))\dots\overline{\mathcal{E}_a^2}(\mathcal{A}_a(0,1))}(z^2\mathbf{B}_2)^q\right)\mathfrak{P}_0$$

$$= \begin{cases} [\mathbf{I} - (z^1\mathbf{B}_1^+)^{N_1+1}][\mathbf{I} - (z^2\mathbf{B}_2^+)^{N_2+1}](\mathbf{I} - z^1\mathbf{B}_1^+)^{-1}(\mathbf{I} - z^2\mathbf{B}_2^+)^{-1}\mathfrak{P}_0, \\ \qquad\qquad \text{if }\ N_1 < \infty,\ \ N_2 < \infty, \\ [\mathbf{I} - (z^1\mathbf{B}_1^+)^{N_1+1}](\mathbf{I} - z^1\mathbf{B}_1^+)^{-1}(\mathbf{I} - z^2\mathbf{B}_2^+)^{-1}\mathfrak{P}_0, \\ \qquad\qquad \text{if }\ N_1 < \infty,\ \ N_2 = \infty, \\ [\mathbf{I} - (z^2\mathbf{B}_2^+)^{N_2+1}](\mathbf{I} - z^1\mathbf{B}_1^+)^{-1}(\mathbf{I} - z^2\mathbf{B}_2^+)^{-1}\mathfrak{P}_0, \\ \qquad\qquad \text{if }\ N_1 = \infty,\ \ N_2 < \infty, \\ (\mathbf{I} - z^1\mathbf{B}_1^+)^{-1}(\mathbf{I} - z^2\mathbf{B}_2^+)^{-1}\mathfrak{P}_0, \quad \text{if }\ N_1 = \infty,\ \ N_2 = \infty, \end{cases}$$

where $\mathfrak{P}_0$ is the vacuum (4.12); in this "resolvent form," the operators $\mathbf{B}_1^+$ and $\mathbf{B}_2^+$ are given by the formula

$$\mathbf{B}_j^+ \stackrel{\text{def}}{=} \overline{\mathcal{E}_a^j}(\mathbf{A})^{-1}\mathbf{B}_j, \qquad j = 1, 2,$$

and commute with each other,[16]

$$[\mathbf{B}_1^+, \mathbf{B}_2^+] = 0.$$

- *The reproducing kernel* $\mathcal{K}(\overline{w^1}, \overline{w^2}, z^1, z^2) = (\mathfrak{P}_{w^1,w^2}, \mathfrak{P}_{z^1,z^2})$ *can also be written in the "resolvent form"*

$$\mathcal{K}(\overline{w^1}, \overline{w^2}, z^1, z^2) = \left(1 + \sum_{p\geq 1} \frac{(\overline{w^1}z^1)^p}{X_a^1(\mathcal{A}_a(p,0))\dots X_a^1(\mathcal{A}_a(1,0))}\right)$$
$$\times \left(1 + \sum_{q\geq 1} \frac{(\overline{w^2}z^2)^q}{X_a^2(\mathcal{A}_a(0,q))\dots X_a^2(\mathcal{A}_a(0,1))}\right)$$
$$= \left(I - X_a^1(\overset{\circ}{A})^{-1}\cdot x^1\right)^{-1}\left(I - X_a^2(\overset{\circ}{A})^{-1}\cdot x^2\right)^{-1}1(x^1, x^2)\Big|_{\substack{x^1=\overline{w^1}z^1,\\x^2=\overline{w^2}z^2}};$$

here $\overset{\circ}{A} = \mathcal{A}_a\left(x^1\frac{\partial}{\partial x^1}, x^1\frac{\partial}{\partial x^1}\right)$, and all formulas are written under the assumption that $N_1 = \infty$ and $N_2 = \infty$ (for $N_1 < \infty$ or $N_2 < \infty$, the same version as in (4.46) should be used).

- *The operators of quantum complex structure are given by*

$$\widehat{\overline{z^j}} = \mathcal{D}_a^j(\mathbf{A})^{-1}\mathbf{B}_j,$$
$$\widehat{z^j} \equiv \widehat{\overline{z^j}}^* = \overline{\mathcal{D}_a^j}(\varphi_j(\mathbf{A}))^{-1}\mathbf{C}_j,$$
$$|\widehat{z^j}|^2 \equiv \widehat{\overline{z^j}}\,\widehat{z^j} = X_a^j(\mathbf{A}), \qquad j = 1, 2.$$

---

[16]This is a consequence of the identities (4.44).

The proof of Theorem 4.3 is similar to that of its one-dimensional analog, Theorem 1.2.

Theorem 4.3 readily allows us to apply all results about the quantum embedding problem discussed in §3 to the case of quantum surfaces of birevolution (4.39).

4.6. Surfaces of birevolution and bihypergeometric series. Let us begin with a simple case of permutation relations (4.35). We still assume that conditions (4.36) are satisfied. Then, similarly to the one-dimensional case, one can readily prove the following analog of Lemma 2.1.

LEMMA 4.7. *Suppose that the following conditions are satisfied:*
(a) *the functions* $f_1^1\colon \mathbb{R}^k \to \mathbb{R}$ *and* $f_2^2\colon \mathbb{R}^k \to \mathbb{R}$ *are polynomials;*
(b) *the components of the mappings* $\varphi_1\colon \mathbb{R}^k \to \mathbb{R}^k$ *and* $\varphi_2\colon \mathbb{R}^k \to \mathbb{R}^k$ *are polynomials;*
(c) *the Jacobi matrices* $D\varphi_1$ *and* $D\varphi_2$ *are upper triangular, and their diagonal elements are equal to 1.*

Then the relations (4.35) are regular, the basic functions $\mathcal{F}_a^1$ *and* $\mathcal{F}_a^2$ *(4.13a) are polynomials over* $\mathbb{Z}_+$, *and the functions* $\mathcal{A}_a$ *(4.16) are polynomials over* $\mathbb{Z}_+ \times \mathbb{Z}_+$.

THEOREM 4.4. *Under assumptions* (a), (b), (c) *of Lemma 4.7, there exist irreducible representations of the algebra (4.35) by antiholomorphic differential operators with polynomial coefficients (given by formula (4.28a) or (4.45)).*

Under assumptions (a), (b), (c) of Lemma 4.7, the functions $\mathcal{F}_a^1$ and $\mathcal{F}_a^2$ are polynomials; thus $\mathcal{B}_a^1, \mathcal{C}_a^1$ and $\mathcal{B}_a^2, \mathcal{C}_a^2$ in (4.20a) can be chosen as polynomials

$$\mathcal{B}_a^j(n) = \beta_j \prod_{\ell=1}^{p_j} \left(n - b_\ell^j(a) \right), \qquad \mathcal{C}_a^j(n) = \gamma_j \prod_{\ell=1}^{r_j} \left(n - c_\ell^j(a) \right), \qquad j=1,2.$$

Here by $b_\ell^j = b_\ell^j(a)$ we denote the (complex) roots of $\mathcal{B}_a^j$ and by $c_\ell^j = c_\ell^j(a)$ the roots of $\mathcal{C}_a^j$. The numbers $p_j = p_j(a)$ and $r_j = r_j(a)$ are the degrees of the polynomials $\mathcal{B}_a^j$ and $\mathcal{C}_a^j$. The sum $p_j + r_j$ is fixed and is the degree of the polynomial $\mathcal{F}_a^j$, which is uniquely defined by the structural polynomials f_j^j and φ_j from (4.35). The degree $p_j + r_j$ is independent of $a \in \mathcal{R}$ at generic points, but it can fall down on some exclusive surfaces.

THEOREM 4.5. *Under assumptions* (a), (b), (c) *of Lemma 4.7, the coherent states* $\mathfrak{P}_{z^1,z^2}$ *and the reproducing kernel* $\mathcal{K}_{s(a)}$, *corresponding to the polynomial antiholomorphic representations in Theorem 4.4, can be derived via products of hypergeometric series and hypergeometric jets as follows.*

• *In the case* $N_1 = N_2 = \infty$ *(that is, if the polynomials* $\mathcal{B}_a^j$ *and* $\mathcal{C}_a^j$ $(j=1,2)$ *have no roots in* $\mathbb{Z}_+$),

$$(4.47) \qquad \mathfrak{P}_z = \prod_{j=1}^{2} \sum_{n=0}^{\infty} \frac{1}{n!\,(-\overline{c^j})_n} \left(\frac{z^j \mathbf{B}_j}{\overline{\gamma_j}} \right)^n \mathfrak{P}_0$$

$$= {}_0F_{r_1}\left(-\overline{c^1}; \frac{z^1 \mathbf{B}_1}{\overline{\gamma_1}} \right) {}_0F_{r_2}\left(-\overline{c^2}; \frac{z^2 \mathbf{B}_2}{\overline{\gamma_2}} \right) \mathfrak{P}_0,$$

$$(4.48) \qquad \mathcal{K}_{s(a)}(\overline{w^1}, \overline{w^2}, z^1, z^2) = {}_{p_1}F_{r_1}\left(-b^1; -\overline{c^1}; \frac{\beta^1 \overline{w^1} z^1}{\overline{\gamma_1}} \right) {}_{p_2}F_{r_2}\left(-b^2; -\overline{c^2}; \frac{\beta^2 \overline{w^2} z^2}{\overline{\gamma_2}} \right),$$

where $_pF_r$ is the hypergeometric series (2.7).

• In the case $N_1 < \infty$ (or $N_2 < \infty$), that is, if the polynomial $\mathcal{B}_a^1$ (or $\mathcal{B}_a^2$) has a root in $\mathbb{Z}_+$, one should replace the series $_0F_{r_1}$ (or $_0F_{r_2}$) in (4.47) by the hypergeometric jet $_0f_{r_1}$ (or $_0f_{r_2}$) (2.7a) of degree N_1 (or N_2) and replace the series $_{p_1}F_{r_1}$ (or $_{p_2}F_{r_2}$) in (4.48) by the hypergeometric jet $_{p_1}f_{r_1}$ (or $_{p_2}f_{r_2}$) (2.7a) of degree N_1 (or N_2).

Now let us briefly consider the permutation relations (4.11) under the assumption that conditions (4.10) are satisfied. Assume that the structural functions in (4.11) are such that $\mathcal{F}_a^{\omega,1}\mathcal{G}_a^1$ and $\mathcal{F}_a^{\omega,2}\mathcal{G}_a^2$ are polynomials over $\mathbb{Z}_+ \times \mathbb{Z}_+$. Then $\mathcal{L}_a^1$, $\mathcal{M}_a^1$ and $\mathcal{L}_a^2$, $\mathcal{M}_a^2$ in (4.20) can be chosen as polynomials (for example, as $\mathcal{L}_a^1$, $\mathcal{M}_a^1$ and $\mathcal{L}_a^2$, $\mathcal{M}_a^2$, one can take (4.23) or (4.24)). In this case, the operators (4.28a) of the antiholomorphic representation are differential operators, and the corresponding coherent states and reproducing kernel can be expressed via hypergeometric series of two variables defined, according to J. Horn, as follows (see [**BE**]).

A power series in two variables

$$(4.49) \qquad \sum_{p=0}^{\infty}\sum_{q=0}^{\infty} T_{p,q}\,(x^1)^p (x^2)^q$$

is a *hypergeometric series* if the two ratios

$$\frac{T_{p+1,q}}{T_{p,q}} = u(p,q), \qquad \frac{T_{p,q+1}}{T_{p,q}} = v(p,q)$$

are rational functions of p and q. Obviously, u and v must satisfy the condition

$$(4.50) \qquad u(p,q)\,v(p+1,q) = u(p,q+1)\,v(p,q) \qquad \text{for all} \quad p,q \in \mathbb{Z}_+,$$

since both sides of this equality are equal to $T_{p+1,q+1}/T_{p,q}$. It is easy to see that any pair of rational functions of p and q that satisfy (4.50) generates a hypergeometric series.

Under our assumptions, the functions

$$u(p,q) = \frac{1}{(p+1)\overline{\mathcal{M}_a^1}(p,q)}, \qquad v(p,q) = \frac{1}{(q+1)\overline{\mathcal{M}_a^2}(p,q)}$$

are rational and, by (4.21), satisfy conditions (4.50). Thus, from (4.34) it follows that for $N_1 = \infty$ and $N_2 = \infty$, the series in the formula for coherent states (4.29a) (or, which is the same, in (4.29)) is a *hypergeometric series of two variables* of type (4.49), where $T_{p,q} \stackrel{\text{def}}{=} \eta_{(p,q)}$, $x^j \stackrel{\text{def}}{=} z^j \mathbf{B}_j$.

Similarly, from the fact that the functions

$$u(p,q) = \frac{\mathcal{L}_a^1(p,q)}{(p+1)\overline{\mathcal{M}_a^1}(p,q)}, \qquad v(p,q) = \frac{\mathcal{L}_a^2(p,q)}{(q+1)\overline{\mathcal{M}_a^2}(p,q)}$$

are rational and from the compatibility conditions (4.21) it follows that for $N_1 = \infty$ and $N_2 = \infty$ the reproducing kernel (4.30) is a *hypergeometric series of two variables* of type (4.49) with $T_{p,q} \stackrel{\text{def}}{=} 1/t_{(p,q)}(a)$, $x^j \stackrel{\text{def}}{=} \overline{w^j} z^j$.

If $N_1 < \infty$ (or $N_2 < \infty$), then one should replace the infinite sum $\sum_{p=0}^{\infty}$ (or $\sum_{q=0}^{\infty}$) in (4.49) by the series $\sum_{p=0}^{N_1}$ (or $\sum_{q=0}^{N_2}$).

§5. Hypersurfaces and twisted hypergeometric functions of several variables

5.1. Floquet pseudometric and twisted algebra.
Now we consider more complicated manifolds $\Omega = \Omega^{2d}$ embedded into $\mathbb{R}^{k+2d}$ as hypersurfaces:

(5.1) $$\Omega = \begin{cases} \langle S_1, \sigma(A)S_1 \rangle + \langle S_2, \sigma(A)S_2 \rangle - \rho(A) = \text{const}, \\ \varkappa_\mu(A) = \text{const} \quad (\mu = 1, \ldots, k-1). \end{cases}$$

Here $S_1, S_2 \in \mathbb{R}^d$, $A \in \mathbb{R}^k$; the functions ρ, $\varkappa_\mu$ are scalar; $\sigma(A) = ((\sigma_p^q(A)))$ is a nondegenerate pseudometric,[17]

$$\sigma^*(A) = \sigma(A), \qquad \det \sigma(A) \neq 0.$$

We assume that there is an invertible mapping $\varphi \colon \mathbb{R}^k \to \mathbb{R}^k$ and a function $\omega \colon \mathbb{R}^k \to \mathbb{R}$ such that

(5.2) $$\sigma(\varphi(A)) \cdot \omega(A) = \sigma(A),$$
$$\rho(\varphi(\varphi(A))) - 2\rho(\varphi(A)) + \rho(A) = 0, \qquad \varkappa_\mu(\varphi(A)) = \varkappa_\mu(A) \quad \text{on } \mathbb{R}^k.$$

So, we assume that the pseudometric σ possesses the *Floquet property* with respect to φ^{-1} with a multiplicator ω, all functions $\varkappa_\mu$ are φ-invariant, and ρ is a φ-harmonic function.

Note that hypersurfaces (5.1) represent a $2d$-dimensional Floquet generalization of surfaces of revolution (3.1). So, now we look at the case $d \geq 2$. The most interesting situation appears when the off-diagonal part of σ is not zero.

We would like to construct a quantum embedding of the hypersurface Ω^{2d} following the general scheme of §3. To this end, we need a noncommutative algebra whose symplectic leaves are the surfaces Ω. We also need coherent states and irreducible representations of this algebra.

Let us consider the algebra generated by the permutation relations

(5.3) $$\mathbf{C}^q \mathbf{B}_p = \mathbf{B}_p \omega(\mathbf{A}) \mathbf{C}^q + f_p^q(\mathbf{A}),$$
$$\mathbf{C}^p \mathbf{A} = \varphi(\mathbf{A}) \mathbf{C}^p, \qquad \mathbf{A}\mathbf{B}_p = \mathbf{B}_p \varphi(\mathbf{A}),$$
$$[\mathbf{A}_\mu, \mathbf{A}_\nu] = 0, \qquad [\mathbf{B}_p, \mathbf{B}_q] = 0, \qquad [\mathbf{C}^p, \mathbf{C}^q] = 0$$

and the Hermitian conditions

$$\mathbf{B}_p^* = \mathbf{C}^p, \qquad \mathbf{A}_\mu^* = \mathbf{A}_\mu, \qquad p, q = 1, \ldots, d, \quad \mu, \nu = 1, \ldots, k.$$

On one hand, relations (5.3) are a special case of relations (4.1) when $\varphi_1 = \cdots = \varphi_k = \varphi$ and $\omega_{sp}^{qr} = \omega \cdot \delta_s^q \cdot \delta_p^r$. On the other hand, the special case of relations (5.3) with $d = 1$ was studied in §1.4; this is the Floquet generalization of the basic algebra.

We call the algebra (5.3) a *twisted algebra*; the twisting is given by the off-diagonal part of the Hermitian matrix $F(A) = ((f_p^q(A))) = F^*(A)$.

---

[17]Here we use the following agreement about the notation for the action of a matrix on a vector:
$$(\sigma S)^q \stackrel{\text{def}}{=} \sum_{p=1}^d \sigma_p^q S^p.$$

The Jacobi condition for the algebra (5.3) in the case $d \geq 2$ is the following:[18]

(5.4) $$F(\varphi(A)) = \omega(A) \cdot F(A).$$

LEMMA 5.1. *If*

(5.5) $$F(A) = \bigl(\rho(\varphi(A)) - \rho(A)\bigr)\sigma(A)^{-1},$$

then condition (5.4) *holds and the twisted algebra* (5.3) *possesses the Casimir elements*

$$\widehat{\varkappa}_\mu = \varkappa_\mu(\mathbf{A}), \qquad \mathbf{K} = \langle \mathbf{B}, \sigma(\mathbf{A})\mathbf{C}\rangle - \rho(\mathbf{A})$$

(*about the correctness of the definition of* $\mathbf{K}$ *see Remark* 5.3, *below*).

So, under conditions (5.2), (5.5) in the generic case, the algebra (5.3) admits symplectic leaves Ω of type (5.1), where $C = S_1 + iS_2$ and $B = S_1 - iS_2$.

This algebra also admits the $*$-product operation defined on $\mathbb{R}^{2d+k}$ in the same way as in §3.5.

Thus, in order to determine a quantum embedding of Ω into $\mathbb{R}^{2d+k}$, it is sufficient to construct the Wick quantization on Ω (that is, to construct a probability function or a reproducing kernel), as well as irreducible representations in spaces of antiholomorphic distributions.

REMARK 5.1. If ω does not vanish anywhere, then we can define the ω-commutator $[\mathbf{X}, \mathbf{Y}]^{(\omega)} \stackrel{\text{def}}{=} \mathbf{XY} - \mathbf{Y}\omega(\mathbf{A})\mathbf{X}$ instead of the usual commutator $[\mathbf{X}, \mathbf{Y}] = \mathbf{XY} - \mathbf{YX}$. Then for $d \geq 2$, instead of the algebra (5.3) with generators $\mathbf{B}_p$, $\mathbf{C}^p$, $\mathbf{A}_\mu$, we obtain the following Heisenberg algebra with respect to the ω-commutator with generators $\mathbf{B}_p$, $\mathbf{C}^p$, $f_p^q(\mathbf{A})$:

$$[\mathbf{C}^q, \mathbf{B}_p]^{(\omega)} = f_p^q(\mathbf{A}),$$
$$[f_p^q(\mathbf{A}), \mathbf{B}_r]^{(\omega)} = 0, \qquad [f_p^q(\mathbf{A}), \mathbf{C}^r]^{(\omega)} = 0.$$

In particular, if $\omega(A) > 0$ and the matrix $F(A)$ is positive definite on $\mathbb{R}^k$, then the change $\mathbf{B} \to \widetilde{\mathbf{B}}$, $\mathbf{C} \to \widetilde{\mathbf{C}}$ described in §4.1 with matrix $\mathfrak{S}(A) = \bigl(F(A)\bigr)^{-1/2}$ leads to the Heisenberg algebra (with respect to the usual commutator) with generators $\mathbf{B}_p$, $\mathbf{C}^p$:

$$[\mathbf{C}^q, \mathbf{B}_p] = \delta_p^q.$$

A similar change for $\omega(A) < 0$ on $\mathbb{R}^k$ results in an analog of the Heisenberg algebra with respect to the anticommutator with the same generators.

Thus, we are most interested in the case when ω has zeros on $\mathbb{R}^k$, and therefore the algebra (5.3) cannot be reduced to the Heisenberg algebra by the change described above.

EXAMPLE 5.1. Suppose that in equations (5.1) we have

$$k = 1, \qquad \rho(A) = \ln A, \qquad \sigma(A) = \frac{1}{\sin A} \cdot \sigma',$$

where σ' is an arbitrary constant invertible Hermitian matrix. Then conditions (5.2) hold for

$$\varphi(A) = 2A, \qquad \omega(A) = 2\cos A.$$

---

[18]In the case $d = 1$ the Jacobi conditions for the algebra (5.3) are satisfied for arbitrary structural functions f, ω, and φ.

From (5.5) we derive the matrix $F(A) = \sin A \cdot F'$, where $F' = \ln 2 \cdot (\sigma')^{-1}$, satisfying condition (5.4). In this case, the algebra (5.3) has the form

$$\mathbf{C}^q \mathbf{B}_p = 2\mathbf{B}_p \cos \mathbf{A}\, \mathbf{C}^q + \sin \mathbf{A}\, f'^q_p,$$
$$\mathbf{C}^p \mathbf{A} = 2\mathbf{A}\mathbf{C}^p, \qquad \mathbf{A}\mathbf{B}_p = 2\mathbf{B}_p \mathbf{A}.$$

In this example, ω has infinitely many zeros on $\mathbb{R}$, namely, $A_j = \pi j/2$. Therefore, the above change does not exist.

EXAMPLE 5.2. Let

$$k = 1, \qquad \rho(A) = A, \qquad \sigma(A) = \frac{1}{\Gamma(A)}\sigma', \qquad \varphi(A) = A + 1, \qquad \omega(A) = A,$$
$$F(A) = \Gamma(A) \cdot F', \qquad \text{where } F' = (\sigma')^{-1},$$

where Γ is the Γ-function and σ' is an arbitrary constant invertible Hermitian matrix. Then relations (5.2), (5.4), and (5.5) are satisfied and the commutation relations (5.3) have the form

$$\mathbf{C}^q \mathbf{B}_p = 2\mathbf{B}_p \mathbf{A} \mathbf{C}^q + \Gamma(\mathbf{A}) f'^q_p,$$
$$\mathbf{C}^p \mathbf{A} = (\mathbf{A} + 1)\mathbf{C}^p, \qquad \mathbf{A}\mathbf{B}_p = \mathbf{B}_p(\mathbf{A} + 1).$$

5.2. Basic matrix and its factorization. If $d = 2$, then the dimension of Ω (5.1) is equal to four. Below we consider this case in detail.

Let $d = 2$.

We consider representations of algebra (5.3) with vacuum vector $\mathfrak{P}_0$ such that

(5.6) $$\mathbf{A}\mathfrak{P}_0 = a\mathfrak{P}_0, \qquad \mathbf{C}^1 \mathfrak{P}_0 = \mathbf{C}^2 \mathfrak{P}_0 = 0,$$
$$a = (a_1, \ldots, a_k), \qquad \|\mathfrak{P}_0\| = 1.$$

Let us define the set $\mathcal{R} \subset \mathbb{R}^k$ as follows: a vector a belongs to $\mathcal{R}$ if the following two conditions are satisfied:

1°. The matrix $F(a)$ is non-negative, that is, $F(a) \geq 0$.
2°. If $F(a) \neq 0$, then
 (1) either

(5.7a) $$\omega(\mathcal{A}_a(n)) > 0 \qquad \text{for all} \quad n \in \mathbb{Z}_+,$$

 (2) or there exists an integer $N \geq 1$ such that

(5.7b) $\quad \omega(\mathcal{A}_a(n)) > 0$ for $0 \leq n < N - 1$ $\quad$ and $\quad \omega(\mathcal{A}_a(N-1)) = 0.$

Here $\mathcal{A}_a(n)$ is the composition (1.4) of n mappings φ. In the case of (5.7a) we set $N = \infty$.

Let us introduce the Hermitian *basic matrix*

(5.8) $$\mathcal{F}_a^\omega(n) = \begin{cases} F(a), & n = 0, \\ F(a) \prod_{j=0}^{n-1} \omega(\mathcal{A}_a(j)), & n \geq 1. \end{cases}$$

By using the basic matrix $\mathcal{F}_a^\omega$, we can describe the set $\mathcal{R}$ in an equivalent way as follows: $a \in \mathcal{R}$ if and only if

– either $\mathcal{F}_a^\omega(n) \geq 0$ and $\mathcal{F}_a^\omega(n) \neq 0$ for all $n \in \mathbb{Z}_+$,

– or there exists an integer $N \in \mathbb{Z}_+$ such that

$$[\mathcal{F}_a^\omega(n) \geq 0 \quad \text{and} \quad \mathcal{F}_a^\omega(n) \neq 0 \quad \text{for} \quad 0 \leq n < N] \quad \text{and} \quad \mathcal{F}_a^\omega(N) = 0.$$

We write $\mathcal{R}_0 \stackrel{\text{def}}{=} \{a \in \mathcal{R} \mid \det F(a) = 0\}$.

First, let us assume that $a \in \mathcal{R}_0$. Then either $F(a) = 0$, or one eigenvalue of the matrix $F(a)$ is zero and the other is positive. In the first case, we have $\mathbf{A} = a \cdot \mathbf{I}$ and $\mathbf{C}^q = \mathbf{B}_p = 0$. In the second case, the vectors $\mathbf{B}_1 \mathfrak{P}_0$ and $\mathbf{B}_2 \mathfrak{P}_0$ are linearly dependent since the matrix of inner products $(\mathbf{B}_q \mathfrak{P}_0, \mathbf{B}_p \mathfrak{P}_0) = (\mathfrak{P}_0, \mathbf{C}^q \mathbf{B}_p \mathfrak{P}_0) = (\mathfrak{P}_0, f_p^q(\mathbf{A})\mathfrak{P}_0) = f_p^q(a)$ is degenerate. However, since this matrix is not identically zero, we have either $f_1^1(a) \neq 0$ or $f_2^2(a) \neq 0$. Then we arrive at the case $d = 1$, considered in §1.4, either by the change $\mathbf{B}_1 = \mathbf{B}$, $\mathbf{B}_2 = \frac{f_2^1(a)}{f_1^1(a)}\mathbf{B}$ or by the change $\mathbf{B}_1 = \frac{f_1^2(a)}{f_2^2(a)}\mathbf{B}$, $\mathbf{B}_2 = \mathbf{B}$, respectively.

Next we assume that $a \in \mathcal{R} \setminus \mathcal{R}_0$. Then the matrices $\mathcal{F}_a^\omega(n) > 0$ for $n < N$ and $\mathcal{F}_a^\omega(n) = 0$ for $n \geq N$.

EXAMPLE 5.1 (continuation). Let us return to Example 5.1, in which the structural functions from relations (5.3) are given by the formulas

$$\varphi(A) = 2A, \qquad \omega(A) = 2\cos A, \qquad F(A) = \sin A \cdot F',$$

where F' is an arbitrary constant invertible Hermitian matrix. We assume that $F' > 0$. Then the set $\mathcal{R}_0$ consists of roots of the function $\sin A$, i.e., $\mathcal{R}_0 = \{\pi n \mid n \in \mathbb{Z}\}$. Let us find the set $\mathcal{R} \setminus \mathcal{R}_0$ explicitly. Note that if $a \notin \mathcal{R}_0$, then there is $n \in \mathbb{Z}_+$ such that $\omega(\mathcal{A}_a(n)) = 2\cos(2^n a) < 0$. Therefore, condition (5.7a) cannot be satisfied for $a \notin \mathcal{R}_0$, and hence

$$a \in \mathcal{R} \setminus \mathcal{R}_0 \iff$$
$$[\sin a > 0; \exists N \geq 1\colon \cos(2^n a) > 0 \text{ for } 0 \leq n < N-1 \text{ and } \cos(2^{N-1}a) = 0].$$

This implies that in this example

$$\mathcal{R} \setminus \mathcal{R}_0 = \left\{\frac{\pi}{2^N} + 2\pi k \mid N \in \mathbb{N}, k \in \mathbb{Z}\right\}.$$

Let us return to the general algebra (5.3). For each $a \in \mathcal{R} \setminus \mathcal{R}_0$ and for any multi-index $\alpha = (\alpha_1, \alpha_2) \in \mathbb{Z}_+^2$, we decompose the matrix $\mathcal{F}_a^\omega(\alpha_1 + \alpha_2)$ into two factors,

(5.9) $$\mathcal{F}_a^\omega(|\alpha|) = \mathcal{M}_a(\alpha) \cdot \mathcal{L}_a(\alpha), \qquad |\alpha| \stackrel{\text{def}}{=} \alpha_1 + \alpha_2,$$

and assume that the following conditions hold:

(i) the nondegeneracy condition[19]

(5.10) $$\det \mathcal{M}_a(\alpha) \neq 0 \qquad \text{for all} \quad \alpha \in \mathbb{Z}_+^2;$$

---

[19] For $|\alpha| < N$ the nondegeneracy of the matrices $\mathcal{M}_a(\alpha)$ follows from relation (5.9) and the nondegeneracy of the matrices $\mathcal{F}_a^\omega(|\alpha|)$.

For $|\alpha| \geq N$ the condition $\det \mathcal{M}_a(\alpha) \neq 0$ is not necessary. However, since the matrices $\mathcal{F}_a^\omega(|\alpha|) = 0$ for $|\alpha| \geq N$, one can always choose nondegenerate matrices $\mathcal{M}_a(\alpha)$ for convenience, for example, by setting $\mathcal{L}_a(\alpha) = 0$ for $|\alpha| \geq N$. In the sequel we shall see that the choice of the matrices $\mathcal{M}_a(\alpha)$ for $|\alpha| \geq N$ does not affects either coherent states or antiholomorphic representation or any other objects related to them.

(ii) the compatibility condition

(5.11) $\quad \mathcal{M}_{a\,p}^{\ q}(\alpha + 1_r)\,\mathcal{M}_{a\,r}^{\ s}(\alpha) = \mathcal{M}_{a\,r}^{\ s}(\alpha + 1_p)\,\mathcal{M}_{a\,p}^{\ q}(\alpha) \qquad$ for all $\alpha \in \mathbb{Z}_+^2$,

where we denote $1_1 \stackrel{\text{def}}{=} (1,0)$, $1_2 \stackrel{\text{def}}{=} (0,1)$;
(iii) the boundary condition

(5.12) $\qquad\qquad \mathcal{L}_a(\alpha) = 0 \quad$ for all α such that $|\alpha| = N$.

Note that the nondegeneracy condition (5.10) and the compatibility condition (5.11) imply the identity

(5.11*) $\qquad \mathcal{M}_{a\,p}^{\ p}(\alpha + 1_q)\mathcal{M}_{a\,p}^{\ q}(\alpha) = \mathcal{M}_{a\,p}^{\ q}(\alpha + 1_q)\mathcal{M}_{a\,p}^{\ p}(\alpha).$

Indeed, if $\det \mathcal{M}_a(\alpha) \neq 0$, then for a given $q \in \{1,2\}$, at least one of the two numbers $\mathcal{M}_{a\,q}^{\ 1}(\alpha)$ and $\mathcal{M}_{a\,q}^{\ 2}(\alpha)$ is nonzero. For definiteness, let $\mathcal{M}_{a\,q}^{\ 1}(\alpha) \neq 0$. Then it follows from (5.11) that

$$\mathcal{M}_{a\,p}^{\ p}(\alpha + 1_q) = \frac{\mathcal{M}_{a\,q}^{\ 1}(\alpha + 1_p)\mathcal{M}_{a\,p}^{\ p}(\alpha)}{\mathcal{M}_{a\,q}^{\ 1}(\alpha)}, \qquad \mathcal{M}_{a\,p}^{\ q}(\alpha + 1_q) = \frac{\mathcal{M}_{a\,q}^{\ 1}(\alpha + 1_p)\mathcal{M}_{a\,p}^{\ q}(\alpha)}{\mathcal{M}_{a\,q}^{\ 1}(\alpha)},$$

which implies (5.11*).

LEMMA 5.2. *Let $a \in \mathcal{R} \setminus \mathcal{R}_0$. Then:*
(a) *The matrices $\mathcal{L}_a(\alpha)$ satisfy the following compatibility condition:*

(5.13) $\quad \mathcal{L}_{a\,p}^{\ q}(\alpha + 1_s)\,\mathcal{L}_{a\,r}^{\ s}(\alpha) = \mathcal{L}_{a\,r}^{\ s}(\alpha + 1_q)\,\mathcal{L}_{a\,p}^{\ q}(\alpha), \qquad$ *for all* $p,q,s,r.$

(b) *The following commutation relations are satisfied:*

(5.14)
$$\left[\sum_{r=1}^2 \mathcal{M}_{a\,r}^{\ 1}\!\left(x\overrightarrow{\frac{\partial}{\partial x}}\right)\frac{\partial}{\partial x^r},\ \sum_{s=1}^2 \mathcal{M}_{a\,s}^{\ 2}\!\left(x\overrightarrow{\frac{\partial}{\partial x}}\right)\frac{\partial}{\partial x^s}\right] = 0,$$
$$\left[\sum_{r=1}^2 x^r \mathcal{M}_a^{-1\,*\,1}{}_r\!\left(x\overrightarrow{\frac{\partial}{\partial x}}\right),\ \sum_{s=1}^2 x^s \mathcal{M}_a^{-1\,*\,2}{}_s\!\left(x\overrightarrow{\frac{\partial}{\partial x}}\right)\right] = 0,$$
$$\left[\sum_{r=1}^2 x^r \mathcal{L}_{a\,1}^{\ r}\!\left(x\overrightarrow{\frac{\partial}{\partial x}}\right),\ \sum_{s=1}^2 x^s \mathcal{L}_{a\,2}^{\ s}\!\left(x\overrightarrow{\frac{\partial}{\partial x}}\right)\right] = 0,$$

where $x\overrightarrow{\frac{\partial}{\partial x}} = \left(x^1 \frac{\partial}{\partial x^1}, x^2 \frac{\partial}{\partial x^2}\right).$

PROOF. (a) It follows from (5.9) and (5.10) that the matrix $\mathcal{L}_a(\alpha)$ satisfies the relation $\mathcal{L}_a(\alpha) = \mathcal{M}_a^{-1}(\alpha) \cdot \mathcal{F}_a^\omega(|\alpha|)$. Therefore, to prove (5.13), it suffices to apply the identity

(5.15) $\qquad \mathcal{F}_{a\ p}^{\omega\ q}(|\alpha|+1)\,\mathcal{F}_{a\ r}^{\omega\ s}(|\alpha|) = \mathcal{F}_{a\ r}^{\omega\ s}(|\alpha|+1)\,\mathcal{F}_{a\ p}^{\omega\ q}(|\alpha|),$

which is a consequence of definition (5.8), and to take into account the relations[20]

(5.11′) $\qquad \mathcal{M}_a^{-1\,q}{}_p(\alpha + 1_s)\,\mathcal{M}_a^{-1\,s}{}_r(\alpha) = \mathcal{M}_a^{-1\,s}{}_r(\alpha + 1_q)\,\mathcal{M}_a^{-1\,q}{}_p(\alpha)$

---

[20]Simultaneously, note that (5.11′) implies the identities

$$\mathcal{M}_a^{-1\,q}{}_q(\alpha + 1_p)\,\mathcal{M}_a^{-1\,q}{}_p(\alpha) = \mathcal{M}_a^{-1\,q}{}_p(\alpha + 1_p)\,\mathcal{M}_a^{-1\,q}{}_q(\alpha).$$

equivalent to (5.11) under the nondegeneracy condition (5.10).

(b) If the nondegeneracy condition (5.10) is satisfied, then identity (5.11) is equivalent to the pair of identities

(5.11a)
$$\mathcal{M}_{a\,p}^{\,\,\,q}(\alpha+1_r)\,\mathcal{M}_{a\,r}^{\,\,\,s}(\alpha) + \mathcal{M}_{a\,r}^{\,\,\,q}(\alpha+1_p)\,\mathcal{M}_{a\,p}^{\,\,\,s}(\alpha) \\ = \mathcal{M}_{a\,p}^{\,\,\,s}(\alpha+1_r)\,\mathcal{M}_{a\,r}^{\,\,\,q}(\alpha) + \mathcal{M}_{a\,r}^{\,\,\,s}(\alpha+1_p)\,\mathcal{M}_{a\,p}^{\,\,\,q}(\alpha),$$

(5.11b)
$$\mathcal{M}_{a}^{-1\,*\,q}{}_{p}(\alpha+1_r)\,\mathcal{M}_{a}^{-1\,*\,s}{}_{r}(\alpha) + \mathcal{M}_{a}^{-1\,*\,q}{}_{r}(\alpha+1_p)\,\mathcal{M}_{a}^{-1\,*\,s}{}_{p}(\alpha) \\ = \mathcal{M}_{a}^{-1\,*\,s}{}_{p}(\alpha+1_r)\,\mathcal{M}_{a}^{-1\,*\,q}{}_{r}(\alpha) + \mathcal{M}_{a}^{-1\,*\,s}{}_{r}(\alpha+1_p)\,\mathcal{M}_{a}^{-1\,*\,q}{}_{p}(\alpha).$$

The first identity is equivalent to the first commutation relation in (5.14), and the second identity to the second relation in (5.14). The third commutation relation is equivalent to the identity

(5.13a)
$$\mathcal{L}_{a\,p}^{\,\,\,r}(\alpha+1_s)\,\mathcal{L}_{a\,q}^{\,\,\,s}(\alpha) + \mathcal{L}_{a\,p}^{\,\,\,s}(\alpha+1_r)\,\mathcal{L}_{a\,q}^{\,\,\,r}(\alpha) \\ = \mathcal{L}_{a\,q}^{\,\,\,r}(\alpha+1_s)\,\mathcal{L}_{a\,p}^{\,\,\,s}(\alpha) + \mathcal{L}_{a\,q}^{\,\,\,s}(\alpha+1_r)\,\mathcal{L}_{a\,p}^{\,\,\,r}(\alpha)$$

that follows from (5.13). □

Note that the factorization (5.9) of the basic matrix exists for an arbitrary $a \in \mathcal{R} \setminus \mathcal{R}_0$. For example, if we set

$$\mathcal{L}_a(\alpha) = \mathcal{F}_a^\omega(|\alpha|), \qquad \mathcal{M}_a(\alpha) = I \quad \text{(the identity matrix)},$$

then, obviously, conditions (5.10), (5.11), (5.12) are satisfied.

Furthermore, one can readily see that if we decompose the matrix $F(a)$ and the function $\prod_{j=0}^{n-1} \omega(\mathcal{A}_a(j))$ into factors,

$$F(a) = \widetilde{\mathcal{M}}_a \cdot \widetilde{\mathcal{L}}_a, \qquad \prod_{j=0}^{n-1} \omega(\mathcal{A}_a(j)) = m_a(n) \cdot \ell_a(n),$$

so that $m_a(n) \neq 0$ for all $n \subset \mathbb{Z}_+$, then the matrices

$$\mathcal{L}_a(\alpha) = \ell_a(|\alpha|) \cdot \widetilde{\mathcal{L}}_a, \qquad \mathcal{M}_a(\alpha) = m_a(|\alpha|) \cdot \widetilde{\mathcal{M}}_a$$

satisfy all conditions (5.10), (5.11), (5.12).

Finally, for the factors $\mathcal{L}_a(\alpha)$ and $\mathcal{M}_a(\alpha)$ in (5.9), we can take the matrices

$$\mathcal{L}_a(\alpha) = \mathcal{M}_a^{-1}(\alpha) \cdot \mathcal{F}_a^\omega(|\alpha|), \qquad \mathcal{M}_a(\alpha) = \begin{pmatrix} m_a^1(\alpha) & 0 \\ 0 & m_a^2(\alpha) \end{pmatrix}.$$

Here m_a^1, m_a^2 are arbitrary functions on $\mathbb{Z}_+^2$ such that

$$m_a^1(\alpha) \cdot m_a^2(\alpha) \neq 0, \qquad m_a^1(\alpha+1_2)\,m_a^2(\alpha) = m_a^2(\alpha+1_1)\,m_a^1(\alpha) \quad \text{for all } \alpha \in \mathbb{Z}_+^2.$$

5.3. Matrix factorial.
Now we need a definition of one general object related to families of matrices. We shall call it the *matrix factorial*.[21] This is a natural generalization of the usual factorials

$$n \to n!,$$
$$n \to (\alpha)_n = \alpha(\alpha+1)\dots(\alpha+n-1),$$
$$n \to (\alpha;q)_n = (1-\alpha)(1-\alpha q)\dots(1-\alpha q^{n-1}),$$
$$n \to \frac{(b)_n}{(c)_n}, \qquad n \to \frac{(b;q)_n}{(c;q)_n},$$

which were intensively used in §§2.2 and 2.4 in connection with the representation of coherent states via hypergeometric functions for permutation relations of the type (5.3) with a *scalar* $F = f \cdot I$. Now we want to consider the general matrix $F = ((f_p^q))$ and the matrix factorization (5.9), but still want to preserve the "hypergeometric" representation for this matrix case.

For any $\alpha \in \mathbb{Z}_+^d$, by $S(\alpha)$ we denote the set of all paths in $\mathbb{Z}_+^d$ from 0 to α. By definition, a path can go stepwise in the positive direction only, so that the coordinates of each next vertex in the path are larger than or equal to (but never less than) the corresponding coordinates of the previous vertex. In the case $d = 2$, this means that the path can go upwards and to the right but cannot go downwards and to the left.

The number of elements in $S(\alpha)$ is equal to

$$\#(S(\alpha)) = \frac{|\alpha|!}{\alpha!}.$$

Each path $\vec{p} \in S(\alpha)$ can be represented by a sequence

$$\vec{p} = (p_1, \dots, p_n), \qquad n = |\alpha|,$$

where $1 \leq p_j \leq d$; namely, we assume that the jth link of the path $\vec{p}$ goes by unit step along the p_j-axis in $\mathbb{Z}_+^d$. Say, in the case $d = 2$, Figure 2 shows the path $\vec{p} = (1, 2, 1, 1, 1, 1, 2, 2, 1, 1, 2)$. Also note that in this notation we have the representation

$$\alpha = 1_{p_1} + \dots + 1_{p_n},$$

where $1_s \stackrel{\text{def}}{=} (0, \dots, 1, \dots, 0) \in \mathbb{Z}_+^d$ (the only element 1 is at the sth place).

Let $G = G(\alpha; \beta)$ be an arbitrary family of $d \times d$ matrices depending on a pair of multi-indices $\alpha, \beta \in \mathbb{Z}_+^d$, $|\alpha| = |\beta|$. For any $\vec{p} \in S(\alpha)$, $\vec{q} \in S(\beta)$ we define the multiplicative integral of G along the paths $\vec{p}$ and $\vec{q}$ as follows:

$$(5.16) \quad \prod G\{\vec{p}; \vec{q}\} = \prod G\{p_1, \dots, p_n; q_1, \dots, q_n\}$$
$$\stackrel{\text{def}}{=} \begin{cases} 1, & n = 0, \\ G_{p_n}^{q_n}(1_{p_{n-1}} + \dots + 1_{p_1}; 1_{q_{n-1}} + \dots + 1_{q_1}) \dots G_{p_2}^{q_2}(1_{p_1}; 1_{q_1}) G_{p_1}^{q_1}(0;0), & n \geq 1, \end{cases}$$

where n denotes the length of the paths $\vec{p}$ and $\vec{q}$.

---

[21] Perhaps, it would be better to use the term "*matrix Γ-function*."

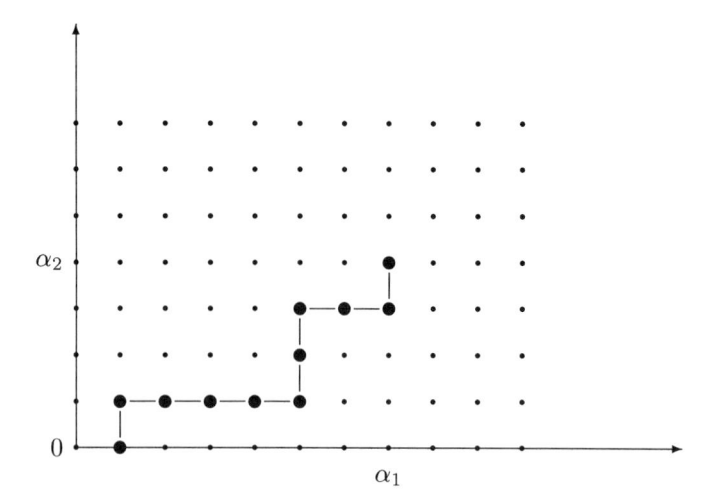

FIGURE 2. Path $\vec{p} = (1, 2, 1, 1, 1, 1, 2, 2, 1, 1, 2)$

DEFINITION 5.1. The *matrix factorial* $G[n]$ is a family of numbers $G(\cdot \mid \cdot)_\alpha^\beta$ (where $|\alpha| = |\beta| = n$), which are determined by the "path integral"

$$G(\cdot \mid \cdot)_\alpha^\beta = \begin{cases} 1, & \text{if } n = 0, \\ \dfrac{\alpha!\beta!}{|\alpha|!|\beta|!} \sum_{\vec{p} \in S(\alpha),\, \vec{q} \in S(\beta)} \prod G\{\vec{p};\vec{q}\}, & \text{if } n \geq 1. \end{cases}$$

If a family $G = G(\alpha)$ depends only on one multi-index $\alpha \in \mathbb{Z}_+^d$, then from this family we can construct a two-index family in two ways: $G(\alpha;\ ) = G(\alpha)$ or $G(\ ;\alpha) = G(\alpha)$. (Needless to say, as before, G efficiently depends only on one multi-index). If a family $G(\alpha;\beta)$ depends only on $|\alpha| = |\beta|$, we denote it by $G(|\alpha|)$. In these cases, we shall denote the matrix factorial by $G(\cdot \mid)_\alpha^\beta$, $G(\mid \cdot)_\alpha^\beta$, and $G(|\cdot|)_\alpha^\beta$, respectively. For a constant matrix G, the matrix factorial will be denoted by $G(\mid)_\alpha^\beta$.

THEOREM 5.1. *The matrix factorial satisfies the relation*

$$G(\cdot \mid \cdot)_\alpha^\beta = \sum_{s=1}^d \sum_{r=1}^d \frac{\alpha_r}{|\alpha|} \frac{\beta_s}{|\beta|} G_r^s(\alpha - 1_r; \beta - 1_s)\, G(\cdot \mid \cdot)_{\alpha - 1_r}^{\beta - 1_s}$$

for any $n \geq 1$.

Note that in this formula the sum is taken over all points $\alpha' = \alpha - 1_r \in \mathbb{Z}_+^d$ and $\beta' = \beta - 1_s \in \mathbb{Z}_+^d$ lying at the unit distance from α and β. Also note that the coefficients $\frac{\alpha_r}{|\alpha|}$ and $\frac{\beta_s}{|\beta|}$ generate partitions of unity, that is,

$$\sum_{r=1}^d \frac{\alpha_r}{|\alpha|} = \sum_{s=1}^d \frac{\beta_s}{|\beta|} = 1.$$

So, formula of Theorem 5.1 is a natural matrix generalization of the fundamental property of the Γ-function: $\Gamma(n) = (n-1)\Gamma(n-1)$.

PROOF OF THEOREM 5.1. Note that all paths $\vec{p} \in S(\alpha)$ from 0 to α can be divided into groups: the rth group consists of paths passing through the point $\alpha - 1_r$, that is, $\vec{p} = (r, p_{n-1}, \ldots, p_1) = (r, \vec{p}')$, where $\vec{p}' \in S(\alpha - 1_r)$. In a similar way, all paths $\vec{q} \in S(\beta)$ from 0 to β can be divided into groups, which we label by the index s, that is, $\vec{q} = (s, \vec{q}')$. Therefore, Definition 5.1 of the matrix factorial is equivalent to

$$G(\cdot \mid \cdot)^\beta_\alpha = \frac{\alpha!\beta!}{|\alpha|!|\beta|!} \sum_{r,s=1}^{d} \sum_{\substack{\vec{p}' \in S(\alpha - 1_r) \\ \vec{q}' \in S(\beta - 1_s)}} G^s_r(1_{p_{n-1}} + \cdots + 1_{p_1}; 1_{q_{n-1}} + \cdots + 1_{q_1})$$
$$\times G^{q_n-1}_{p_n-1}(1_{p_{n-2}} + \cdots + 1_{p_1}; 1_{1_{q_{n-2}}} + \cdots + 1_{q_1}) \ldots G^{q_2}_{p_2}(1_{p_1}; 1_{q_1}) G^{q_1}_{p_1}(0; 0).$$

Here $1_{p_{n-1}} + \cdots + 1_{p_1} = \alpha - 1_r$, $1_{q_{n-1}} + \cdots + 1_{q_1} = \beta - 1_s$, and

$$G^s_r(1_{p_{n-1}} + \cdots + 1_{p_1}; 1_{q_{n-1}} + \cdots + 1_{q_1}) = G^s_r(\alpha - 1_r; \beta - 1_s)$$

are independent of $\vec{p}'$, $\vec{q}'$. Hence, we have

$$G(\cdot \mid \cdot)^\beta_\alpha = \sum_{r,s=1}^{d} \frac{\alpha_r}{|\alpha|} \frac{\beta_s}{|\beta|} G^s_r(\alpha - 1_r; \beta - 1_s) \frac{(\alpha - 1_r)! (\beta - 1_s)!}{|\alpha - 1_r|! |\beta - 1_s|!} \sum_{\substack{\vec{p}' \in S(\alpha - 1_r) \\ \vec{q}' \in S(\beta - 1_s)}} G\{\vec{p}'; \vec{q}'\}$$

$$= \sum_{r,s=1}^{d} \frac{\alpha_r}{|\alpha|} \frac{\beta_s}{|\beta|} G^s_r(\alpha - 1_r; \beta - 1_s) G(\cdot \mid \cdot)^{\beta - 1_s}_{\alpha - 1_r}. \quad \square$$

In some cases, our definition of the matrix factorial is reduced to the usual definition. Say, let $G(\alpha; \beta)$ be scalar matrices such that

$$G(\alpha; \beta) = g(n) \cdot I, \qquad |\alpha| = |\beta| = n,$$

where g is an arbitrary function on $\mathbb{Z}_+$. Then for $n \geq 1$, we have

$$G(\cdot \mid \cdot)^\beta_\alpha = \begin{cases} 0, & \alpha \neq \beta, \\ \dfrac{\alpha!}{n!} g(n-1) \ldots g(1) g(0), & \alpha = \beta. \end{cases}$$

In the following we shall apply the construction of the matrix factorial to the families $\mathcal{M}_a$, $\mathcal{L}_a$, $\mathcal{F}^\omega_a$ defined in §5.2. These families satisfy certain additional compatibility conditions. Let us now look at the general matrix factorial under such additional conditions.

For simplicity, we consider only the case $d = 2$.

Let us fix an integer $M \geq 1$ and consider the set $\mathcal{G}_M$ of all 2×2 matrix families $G = G(\alpha; \beta)$ (where $\alpha, \beta \in \mathbb{Z}_+^2$) which satisfy the condition

(5.17) $\quad G^{q_2}_{p_2}(\alpha + 1_{p_1}; \beta + 1_{q_1}) G^{q_1}_{p_1}(\alpha; \beta) = G^{q_1}_{p_1}(\alpha + 1_{p_2}; \beta + 1_{q_2}) G^{q_2}_{p_2}(\alpha; \beta),$
$\qquad p_1, p_2, q_1, q_2 \in \{1, 2\}, \qquad |\alpha| = |\beta| < M - 1.$

If the identity (5.17) holds for all $|\alpha| = |\beta| \in \mathbb{Z}_+$, then we set $M = \infty$ and denote the corresponding matrix class by $\mathcal{G}_\infty$.

Let us point out some properties of the set $\mathcal{G}_M$.

LEMMA 5.3. (a) $G(\alpha;\beta) \in \mathcal{G}_M \Longrightarrow G^*(\beta;\alpha) \in \mathcal{G}_M$.
(b) $H(\ ;\beta), J(\alpha;\ ) \in \mathcal{G}_M \Longrightarrow H(\ ;\beta) \cdot J(\alpha;\ ) \in \mathcal{G}_M$.
(c) $H(\ ;\beta) \in \mathcal{G}_M$, $\det H(\ ;\beta) \neq 0$ for $|\beta| \leq M-1 \Longrightarrow H^{-1}(\alpha;\ ) \in \mathcal{G}_M$.

LEMMA 5.4 (graphic interpretation of condition (5.17)). Let $G \in \mathcal{G}_M$. Then for $|\alpha| = |\beta| \leq M$ the following statements are satisfied:
(a) The sum $\sum_{\vec{q} \in S(\beta)} \prod G\{\vec{p};\vec{q}\}$ is independent of the choice of the path $\vec{p}$ from 0 to α. That is, if $\vec{p}, \vec{p}' \in S(\alpha)$, then

$$\sum_{\vec{q} \in S(\beta)} \prod G\{\vec{p};\vec{q}\} = \sum_{\vec{q} \in S(\beta)} \prod G\{\vec{p}';\vec{q}\}. \tag{5.18a}$$

(b) The sum $\sum_{\vec{p} \in S(\alpha)} \prod G\{\vec{p};\vec{q}\}$ is independent of the choice of the path $\vec{q}$ from 0 to β. That is, if $\vec{q}, \vec{q}' \in S(\beta)$, then

$$\sum_{\vec{p} \in S(\alpha)} \prod G\{\vec{p};\vec{q}\} = \sum_{\vec{p} \in S(\alpha)} \prod G\{\vec{p};\vec{q}'\}. \tag{5.18b}$$

PROOF. (a) 1. First we assume that the path $\vec{p}' \in S(\alpha)$ can be obtained from the path $\vec{p}$ by replacing the part (p_j, p_{j+1}) by (p_{j+1}, p_j), that is,

$$\vec{p}' = (p_1, \ldots, p_{j-1}, p_{j+1}, p_j, p_{j+2}, \ldots, p_n).$$

Then to each path $\vec{q} \in S(\beta)$ we assign the path

$$\vec{q}' = (q_1, \ldots, q_{j-1}, q_{j+1}, q_j, q_{j+2}, \ldots, q_n)$$

and compare the products $\prod G\{\vec{p};\vec{q}\}$ and $\prod G\{\vec{p}';\vec{q}'\}$ (5.16). These products differ only by the jth and $(j+1)$st factors (if the factors in (5.16) are numbered from right to left). However, in view of (5.17), the product of the jth and $(j+1)$st factors

$$G_{p_{j+1}}^{q_{j+1}}(1_{p_j}+1_{p_{j-1}}+\cdots+1_{p_1}; 1_{q_j}+1_{q_{j-1}}+\cdots+1_{q_1}) G_{p_j}^{q_j}(1_{p_{j-1}}+\cdots+1_{p_1}; 1_{q_{j-1}}+\cdots+1_{q_1})$$

does not change if we simultaneously replace (p_j, p_{j+1}) by (p_{j+1}, p_j) and (q_j, q_{j+1}) by (q_{j+1}, q_j). Therefore,

$$\prod G\{\vec{p};\vec{q}\} = \prod G\{\vec{p}';\vec{q}'\}.$$

Summing over all $\vec{q} \in S(\beta)$ and replacing the summation index $\vec{q} \to \vec{q}'$ on the right-hand side of the relation obtained, we arrive at identity (5.18a) if the paths $\vec{p}$ and $\vec{p}'$ differ only by the jth and $(j+1)$st steps.

2. Now let $\vec{p}, \vec{p}'$ be arbitrary paths lying in the set $S(\alpha)$. Then, obviously, the path $\vec{p}'$ can be obtained from the path $\vec{p}$ by finitely many replacements of pairs of path steps: $(p_j, p_{j+1}) \to (p_{j+1}, p_j)$. We have already proved that the sum does not change after such a replacement. This means that identity (5.18a) holds for arbitrary paths $\vec{p}, \vec{p}' \in S(\alpha)$.

(b) Identity (5.18b) can be proved in a similar way. □

COROLLARY 5.1. *Let $G \in \mathcal{G}_M$ and let $|\alpha| = |\beta| = n \leq M$. Then there are the following formulas for the matrix factorial:*

(a) $$G(\cdot \mid \cdot)_\alpha^\beta = \frac{\beta!}{n!} \sum_{\vec{q}' \in S(\beta)} \prod G\{\vec{p}; \vec{q}'\},$$

where $\vec{p} \in S(\alpha)$ is an arbitrary path from 0 to α;

(b) $$G(\cdot \mid \cdot)_\alpha^\beta = \frac{\alpha!}{n!} \sum_{\vec{p}' \in S(\alpha)} \prod G\{\vec{p}'; \vec{q}\},$$

where $\vec{q} \in S(\beta)$ is an arbitrary path from 0 to β.

REMARK 5.3. Identity (5.17) is equivalent to the following pair of identities:

(5.17a)
$$G_{p_2}^{q_2}(\alpha + 1_{p_1}; \beta + 1_{q_1}) G_{p_1}^{q_1}(\alpha; \beta) + G_{p_1}^{q_1}(\alpha + 1_{p_1}; \beta + 1_{q_2}) G_{p_2}^{q_2}(\alpha; \beta)$$
$$= G_{p_1}^{q_2}(\alpha + 1_{p_2}; \beta + 1_{q_1}) G_{p_2}^{q_1}(\alpha; \beta) + G_{p_1}^{q_1}(\alpha + 1_{p_2}; \beta + 1_{q_2}) G_{p_2}^{q_2}(\alpha; \beta),$$

(5.17b)
$$G_{p_2}^{q_2}(\alpha + 1_{p_1}; \beta + 1_{q_1}) G_{p_1}^{q_1}(\alpha; \beta) + G_{p_1}^{q_2}(\alpha + 1_{p_2}; \beta + 1_{q_1}) G_{p_2}^{q_1}(\alpha; \beta)$$
$$= G_{p_2}^{q_1}(\alpha + 1_{p_1}; \beta + 1_{q_2}) G_{p_1}^{q_2}(\alpha; \beta) + G_{p_1}^{q_1}(\alpha + 1_{p_2}; \beta + 1_{q_2}) G_{p_2}^{q_2}(\alpha; \beta).$$

If the matrix G satisfies (5.17a) for $|\alpha| = |\beta| < M$, then Lemma 5.4(a) (and Corollary 5.1(a)) hold for G. Therefore, on the set of matrices satisfying (5.17a), one can define the matrix factorial by the formula

(5.19a) $$G(\cdot \mid \cdot)_\alpha^\beta \stackrel{\text{def}}{=} \begin{cases} 1, & n = 0, \\ \dfrac{\beta!}{n!} \displaystyle\sum_{\vec{q} \in S(\beta)} \prod G\{\vec{p}; \vec{q}\}, & n \geq 1, \end{cases}$$

where $\vec{p} \in S(\alpha)$ is an arbitrary path from 0 to α. This matrix factorial possesses the basic property

$$G(\cdot \mid \cdot)_\alpha^{\beta+1_r} = \sum_{s=1}^{2} \frac{\beta_s}{|\beta|} G_r^s(\alpha; \beta - 1_s) G(\cdot \mid \cdot)_\alpha^{\beta-1_s}, \qquad |\alpha| + 1 = |\beta| = n \leq M.$$

In a similar way, in view of Lemma 5.4(b), one can define the matrix factorial on the set of matrices satisfying (5.17b) as follows:

(5.19b) $$G(\cdot \mid \cdot)_\alpha^\beta \stackrel{\text{def}}{=} \begin{cases} 1, & n = 0, \\ \dfrac{\alpha!}{n!} \displaystyle\sum_{\vec{p} \in S(\alpha)} \prod G\{\vec{p}; \vec{q}\}, & n \geq 1, \end{cases}$$

where $\vec{q} \in S(\beta)$ is an arbitrary path from 0 to β. This factorial possesses the basic property

$$G(\cdot \mid \cdot)_{\alpha}^{\beta+1_s} = \sum_{r=1}^{2} \frac{\alpha_r}{|\alpha|} G_r^s(\alpha - 1_r; \beta) G(\cdot \mid \cdot)_{\alpha-1_r}^{\beta}, \qquad |\alpha| = |\beta| + 1 = n \leq M.$$

Now we shall prove that some matrices composed of matrix factorials can be multiplied.

THEOREM 5.2. *Let $H = H(\ ;\beta)$ and $J = J(\alpha;\ )$ be two families of matrices from the class $\mathcal{G}_M$. Then for any $|\alpha| = |\beta| = n \leq M$ we have:*
(a) *The following equality holds:*

$$\sum_{|\gamma|=n} \frac{n!}{\gamma!} H(\ |\ \cdot)^\beta_\gamma J(\cdot\ |\ )^\gamma_\alpha = G(\cdot\ |\ \cdot)^\beta_\alpha, \qquad \text{where } G(\alpha;\beta) \stackrel{\text{def}}{=} H(\ ;\beta) \cdot J(\alpha;\ ).$$

(b) *If, in addition, the product $J(\alpha;\ ) \cdot H(\ ;\alpha)$ depends only on $|\alpha|$, that is,*

$$J(\alpha;\ ) \cdot H(\ ;\alpha) = \widetilde{G}(|\alpha|), \qquad 0 \leq |\alpha| \leq n-1,$$

then the following equality holds:

$$\sum_{|\gamma|=n} \frac{n!}{\gamma!} J(\cdot\ |\ )^\beta_\gamma H(\ |\ \cdot)^\gamma_\alpha = \widetilde{G}(|\ \cdot\ |)^\beta_\alpha.$$

PROOF. (a) It follows from Lemma 5.3(b) that the matrix $G(\alpha;\beta) = H(\ ;\beta) \cdot J(\alpha;\ ) \in \mathcal{G}_M$. By Definition 5.1, we have

$$G(\cdot\ |\ \cdot)^\beta_\alpha = \frac{\alpha!\,\beta!}{n!\,n!} \sum_{\substack{\vec{p}\in S(\alpha)\\ \vec{q}\in S(\beta)}} \prod G\{\vec{p};\vec{q}\}$$

$$= \frac{\alpha!\,\beta!}{n!\,n!} \sum_{\substack{\vec{p}\in S(\alpha)\\ \vec{q}\in S(\beta)}} \sum_{r_1,\ldots,r_n=1}^{2} H^{q_n}_{r_n}(\ ;1_{q_{n-1}}+\cdots+1_{q_1}) J^{r_n}_{p_n}(1_{p_{n-1}}+\cdots+1_{p_1};\ )$$

$$\cdot\ldots\cdot H^{q_2}_{r_2}(\ ;1_{q_1}) J^{r_2}_{p_2}(1_{p_1};\ ) \cdot H^{q_1}_{r_1}(\ ;0) J^{r_1}_{p_1}(0;\ )$$

$$= \sum_{|\gamma|=n}\sum_{\vec{r}\in S(\gamma)} \left\{ \frac{\beta!}{n!} \sum_{\vec{q}\in S(\beta)} H^{q_n}_{r_n}(\ ;1_{q_{n-1}}+\cdots+1_{q_1})\ldots H^{q_2}_{r_2}(\ ;1_{q_1}) H^{q_1}_{r_1}(\ ;0) \right\}$$

$$\times \left\{ \frac{\alpha!}{n!} \sum_{\vec{p}\in S(\alpha)} J^{r_n}_{p_n}(1_{p_{n-1}}+\cdots+1_{p_1};\ )\ldots J^{r_2}_{p_2}(1_{p_1};\ ) J^{r_1}_{p_1}(0;\ ) \right\}.$$

Lemma 5.4 implies that the expressions in curly brackets are independent of the choice of the path $\vec{r} \in S(\gamma)$ from 0 to γ. Therefore, the sum over $\vec{r} \in S(\gamma)$ can be replaced by the product of $\frac{n!}{\gamma!}$ (this is the number of summands in this sum) by an arbitrary summand from this sum. Then it follows from the definition (5.19a) that

$$G(\cdot\ |\ \cdot)^\beta_\alpha = \sum_{|\gamma|=n} \frac{n!}{\gamma!} H(\ |\ \cdot)^\beta_\gamma J(\cdot\ |\ )^\gamma_\alpha.$$

(b) It follows from the definitions (5.19b) and (5.19a) that

$$\sum_{|\gamma|=n} \frac{n!}{\gamma!} J(\cdot\ |\ )^\beta_\gamma H(\ |\ \cdot)^\gamma_\alpha$$

$$= \sum_{|\gamma|=n} \frac{n!}{\gamma!} \left\{ \frac{\alpha!}{n!} \sum_{\vec{p}\in S(\alpha)} J^{q_n}_{r_n}(1_{r_{n-1}}+\cdots+1_{r_1};\ )\ldots J^{q_2}_{r_2}(1_{r_1};\ ) J^{q_1}_{r_1}(0;\ ) \right\}$$

$$\times \left\{ \frac{\beta!}{n!} \sum_{\vec{q}\in S(\beta)} H^{r_n}_{p_n}(\ ;1_{r_{n-1}}+\cdots+1_{r_1})\ldots H^{r_2}_{p_2}(\ ;1_{r_1}) H^{r_1}_{p_1}(\ ;0) \right\},$$

where $\vec{r} \in S(\gamma)$ is an arbitrary path from 0 to γ. Furthermore, by analogy with the proof of statement (a), we can rewrite this expression as follows:

$$\frac{\alpha!\,\beta!}{n!\,n!} \sum_{\substack{\vec{p}\in S(\alpha) \\ \vec{q}\in S(\beta)}} \sum_{r_1,\ldots,r_n=1}^{2} J_{r_n}^{q_n}(1_{r_{n-1}}+\cdots+1_{r_1};\,)H_{p_n}^{r_n}(\,;1_{r_{n-1}}+\cdots+1_{r_1})$$
$$\cdots \cdot J_{r_2}^{q_2}(1_{r_1};\,)H_{p_2}^{r_2}(\,;1_{r_1})\cdot J_{r_1}^{q_1}(0;\,)H_{p_1}^{r_1}(\,;0).$$

In our sum only the factors $J_{r_n}^{q_n}(1_{r_{n-1}}+\cdots+1_{r_1};\,)$ and $H_{p_n}^{r_n}(\,;1_{r_{n-1}}+\cdots+1_{r_1})$ depend on r_n. Moreover, it follows from the assumption of the theorem that the sum

$$\sum_{r_n=1}^{2} J_{r_n}^{q_n}(1_{r_{n-1}}+\cdots+1_{r_1};\,)\cdot H_{p_n}^{r_n}(\,;1_{r_{n-1}}+\cdots+1_{r_1}) = \widetilde{G}_{p_n}^{q_n}(n-1)$$

is independent of r_{n-1}. Therefore, taking the sum over r_n, we obtain the sum $\sum_{r_1,\ldots,r_{n-1}=1}^{2}$ in which again only the two factors $J_{r_{n-1}}^{q_{n-1}}(1_{r_{n-2}}+\cdots+1_{r_1};\,)$ and $H_{p_{n-1}}^{r_{n-1}}(\,;1_{r_{n-2}}+\cdots+1_{r_1})$ depend on r_{n-1}, and

$$\sum_{r_{n-1}=1}^{2} J_{r_{n-1}}^{q_{n-1}}(1_{r_{n-2}}+\cdots+1_{r_1};\,)\cdot H_{p_{n-1}}^{r_{n-1}}(\,;1_{r_{n-2}}+\cdots+1_{r_1}) = \widetilde{G}_{p_{n-1}}^{q_{n-1}}(n-2)$$

is independent of r_{n-2}. Thus, taking the sum successively over $r_n, r_{n-1},\ldots,r_1$, we obtain the identity

$$\sum_{|\gamma|=n} \frac{n!}{\gamma!} J(\cdot\,|\,)_\gamma^\beta H(\,|\,\cdot)_\alpha^\gamma = \frac{\alpha!\,\beta!}{n!\,n!} \sum_{\substack{\vec{p}\in S(\alpha) \\ \vec{q}\in S(\beta)}} \widetilde{G}_{p_n}^{q_n}(n-1)\cdots\cdot \widetilde{G}_{p_2}^{q_2}(1)\,\widetilde{G}_{p_1}^{q_1}(0),$$

which, by Definition 5.1, coincides with the desired identity. $\square$

COROLLARY 5.2. *Let* $H(\,;\beta) \in \mathcal{G}_M$, *and let* $\det H(\,;\beta) \neq 0$ *for* $|\beta| \leq M-1$. *Then*

$$\sum_{|\gamma|=n} \frac{n!}{\gamma!} H^{-1}(\cdot\,|\,)_\gamma^\beta H(\,|\,\cdot)_\alpha^\gamma = I(\,|\,)_\alpha^\beta = \begin{cases} 0, & \alpha \neq \beta, \\ \dfrac{\alpha!}{n!}, & \alpha = \beta, \end{cases}$$

where I *is the identity matrix.*

Corollary 5.2 follows from Lemma 5.3(c) and Theorem 5.2(b).

As an example, we consider the special case in which $G(\alpha;\beta)$ is a diagonal matrix, that is, $G(\alpha;\beta) = \begin{pmatrix} g^1(\alpha;\beta) & 0 \\ 0 & g^2(\alpha;\beta) \end{pmatrix}$. In this case, identity (5.17) is equivalent to the compatibility condition

$$g^1(\alpha+1_2;\beta+1_2)\,g^2(\alpha;\beta) = g^2(\alpha+1_1;\beta+1_1)\,g^1(\alpha;\beta).$$

Assume that this condition holds for $|\alpha|=|\beta| < M-1$. Then $G \in \mathcal{G}_M$. When we calculate $G(\cdot\,|\,\cdot)_\alpha^\beta$ according to formula (5.19a), to be definite, we choose a path

$\vec{p} \in S(\alpha)$ so that $(p_1, \ldots, p_n) = (\underbrace{1, \ldots, 1}_{\alpha_1}, \underbrace{2, \ldots, 2}_{\alpha_2})$. Then for $n \geq 1$ we obtain

$$G(\cdot \mid \cdot)_\alpha^\beta = \begin{cases} 0, & \alpha \neq \beta, \\ \dfrac{\beta!}{n!} G_{p_n}^{p_n}(1_{p_{n-1}} + \cdots + 1_{p_1}; 1_{p_{n-1}} + \cdots + 1_{p_1}) \\ \qquad \cdot \ldots \cdot G_{p_2}^{p_2}(1_{p_1}; 1_{p_1}) G_{p_1}^{p_1}(0; 0), & \alpha = \beta, \end{cases}$$

$$= \begin{cases} 0, & \alpha \neq \beta, \\ \prod_{r=0}^{\alpha_1-1} g^1(r, 0; r, 0), & \alpha = \beta = (\alpha_1, 0), \\ \prod_{s=0}^{\alpha_2-1} g^2(0, s; 0, s), & \alpha = \beta = (0, \alpha_2), \\ \dfrac{\alpha!}{n!} \prod_{r=0}^{\alpha_1-1} g^1(r, 0; r, 0) \cdot \prod_{s=0}^{\alpha_2-1} g^2(\alpha_1, s; \alpha_1, s), & \alpha = \beta = (\alpha_1, \alpha_2), \\ & \alpha_1, \alpha_2 \geq 1. \end{cases}$$

One can see such products in many formulas in §4.3 (the case $d = 2$, two one-dimensional blocks); see, for example, formulas (4.25) and (4.29). These products contain basic pairs of functions $\bigl(\mathcal{F}_a^{\omega,1}(\alpha_1) \mathcal{G}_a^1(\alpha), \mathcal{F}_a^{\omega,2}(\alpha_2) \mathcal{G}_a^2(\alpha)\bigr)$, $\bigl(\mathcal{L}_a^1(\alpha), \mathcal{L}_a^2(\alpha)\bigr)$, and $\bigl(\mathcal{M}_a^1(\alpha), \mathcal{M}_a^2(\alpha)\bigr)$ that are defined only for $\beta = \alpha$. The compatibility condition for diagonal elements $\widetilde{g}^j(\alpha) \stackrel{\text{def}}{=} g^j(\alpha; \alpha)$ has the form

$$\widetilde{g}^1(\alpha + 1_2) \widetilde{g}^2(\alpha) = \widetilde{g}^2(\alpha + 1_1) \widetilde{g}^1(\alpha).$$

For the diagonal matrix $G(\alpha, \beta)$, the basic property of the matrix factorial has the form

$$G(\cdot \mid \cdot)_{\alpha+1_r}^{\beta+1_r} = \begin{cases} 0, & \alpha \neq \beta, \\ \dfrac{\alpha_r}{|\alpha|} g^r(\alpha; \alpha) \, G[n-1]_\alpha^\alpha, & \alpha = \beta. \end{cases}$$

The matrix factorial from Definition 5.1 is a generalization of products of the form $g(n-1) \cdot \ldots \cdot g(1) g(0)$ or $\prod_{r=0}^{\alpha_1-1} g^1(r, 0; r, 0) \cdot \prod_{s=0}^{\alpha_2-1} g^2(\alpha_1, s; \alpha_1, s)$ to the case of nondiagonal matrices $G \in \mathcal{G}_M$. Such nondiagonal matrices appear in §5.2, namely, they are the basic matrices $\mathcal{F}_a^\omega$, $\mathcal{L}_a$, and $\mathcal{M}_a$. Indeed, for $a \in \mathcal{R} \setminus \mathcal{R}_0$, it follows from properties (5.15), (5.13), and (5.11) that the matrices $\mathcal{F}_a^\omega(|\alpha|)$, $\mathcal{L}_a(\ ; \beta) = \mathcal{L}_a(\beta)$, and $\mathcal{M}_a(\alpha;\ ) = \mathcal{M}_a(\alpha)$ also satisfy the identity (5.17), and hence

(5.20a) $$\mathcal{F}_a^\omega(|\alpha|), \ \mathcal{L}_a(\ ; \beta), \ \mathcal{M}_a(\alpha;\ ) \in \mathcal{G}_\infty.$$

Since the matrices $\mathcal{L}_a(\beta)$ are invertible for $|\beta| \leq N - 1$ (where N is the positive integer from (5.7b)) and since the matrices $\mathcal{M}_a$ are invertible, Lemma 5.3 implies the inclusions

(5.20b)
$$\mathcal{L}_a^*(\alpha;\ ), \ \mathcal{L}_a^{-1}(\alpha;\ ), \ \mathcal{L}_a^{-1*}(\ ;\beta) \in \mathcal{G}_N,$$
$$\mathcal{M}_a^*(\ ;\beta), \ \mathcal{M}_a^{-1}(\ ;\beta), \ \mathcal{M}_a^{-1*}(\alpha;\ ) \in \mathcal{G}_N,$$
$$\mathcal{L}_a(\ ;\beta) \cdot \mathcal{M}_a^{-1*}(\alpha;\ ), \ \mathcal{M}_a^*(\ ;\beta) \cdot \mathcal{L}_a^{-1}(\alpha;\ ) \in \mathcal{G}_N.$$

5.4. Hilbert spaces of antiholomorphic distributions.
Using the notion of the matrix factorial, we define the following complex numbers:

$$(5.21) \quad t_{(\alpha;\beta)}(a) = \begin{cases} 0, & |\alpha| \neq |\beta|, \\ 1, & |\alpha| = |\beta| = 0, \\ n!\left(\mathcal{M}_a^*(\,|\,\cdot\,)\mathcal{L}_a^{-1}(\,\cdot\,|\,)\right)_\alpha^\beta, & 1 \leq |\alpha| = |\beta| = n \leq N, \\ \infty, & |\alpha| = |\beta| > N. \end{cases}$$

In terms of the numbers $t_{(\alpha;\beta)}(a)$, we introduce a Hilbert space of antiholomorphic distributions over $\mathbb{R}^4$; that is, these distributions are annihilated by the operators $\partial/\partial z^1$ and $\partial/\partial z^2$, where z^1 and z^2 are the standard complex coordinates on $\mathbb{R}^4$. Namely, let us consider distributions g over $\mathbb{R}^4$ that can be represented by power series

$$g(\overline{z^1}, \overline{z^2}) = \sum_{|\alpha| \geq 0} g_\alpha \overline{z}^\alpha \qquad \left(\alpha = (\alpha_1, \alpha_2) \in \mathbb{Z}_+^2\right).$$

By $\mathcal{P}_{t(a)}$ we denote the space of such series with the inner product

$$(5.22) \qquad (g'', g')_{\mathcal{P}_{t(a)}} \stackrel{\text{def}}{=} \sum_{|\alpha| \geq 0} \sum_{|\beta| \geq 0} t_{(\alpha;\beta)}(a)\, g'_\alpha \cdot \overline{g''_\beta}.$$

LEMMA 5.5. *Let $a \in \mathcal{R} \setminus \mathcal{R}_0$. Then (5.22) satisfy all axioms of the inner product.*

PROOF. Since $t_{(\alpha;\beta)}(a) = 0$ for $|\alpha| \neq |\beta|$, it suffices to prove that for each $n\colon 0 \leq n \leq N$, the matrix of inner products $T_a(n) \stackrel{\text{def}}{=} \left(\!\left(t_{(\alpha;\beta)}(a)\right)\!\right) = \left(\!\left((\overline{z}^\beta, \overline{z}^\alpha)_{\mathcal{P}_{t(a)}}\right)\!\right)$, $|\alpha| = |\beta| = n$, is:
(a) Hermitian, that is, $t_{(\alpha;\beta)}(a) = \overline{t_{(\beta;\alpha)}(a)}$,
(b) positive definite, that is, $\sum_{|\alpha|=|\beta|=n} t_{(\alpha;\beta)}(a)\, x^\alpha \cdot \overline{x^\beta} > 0$ for $x = (x^\alpha) \neq 0$.
Let us prove these statements.
(a) Since the matrices $\mathcal{F}_a^\omega(|\alpha|)$ are Hermitian, it follows from (5.9) that

$$\mathcal{L}_a^*(\beta)\mathcal{M}_a^*(\beta) = \mathcal{F}_a^{\omega *}(|\beta|) = \mathcal{F}_a^\omega(|\alpha|) = \mathcal{M}_a(\alpha)\mathcal{L}_a(\alpha)$$

for $|\alpha| = |\beta|$. If $|\beta| \leq N-1$ and $|\alpha| \leq N-1$, then the matrices $\mathcal{L}_a^*(\beta)$ and $\mathcal{L}_a(\alpha)$ are invertible. Therefore, for $|\alpha| = |\beta| \leq N-1$, we can rewrite the last identity in the form

$$G(\alpha; \beta) = \mathcal{M}_a^*(\beta) \cdot \mathcal{L}_a^{-1}(\alpha) = \mathcal{L}_a^{*-1}(\beta) \cdot \mathcal{M}_a(\alpha) = G^*(\beta; \alpha).$$

It remains to note that if $G(\alpha;\beta) = G^*(\beta;\alpha) \in \mathcal{G}_M$, then it follows from Definition 5.1 that $G(\,\cdot\,|\,\cdot\,)_\alpha^\beta = \overline{G(\,\cdot\,|\,\cdot\,)_\beta^\alpha}$. Hence,

$$t_{(\alpha;\beta)}(a) = n!\left(\mathcal{M}_a^*(\,|\,\cdot\,)\mathcal{L}_a^{-1}(\,\cdot\,|\,)\right)_\alpha^\beta = n!\,\overline{\left(\mathcal{M}_a^*(\,|\,\cdot\,)\mathcal{L}_a^{-1}(\,\cdot\,|\,)\right)_\beta^\alpha} = \overline{t_{(\beta;\alpha)}(a)}.$$

(b) For each n, $0 \leq n \leq N$, it suffices to prove that the matrix

$$R_a(n) = \left(\!\left(r_{(\alpha;\beta)}(a)\right)\!\right) \stackrel{\text{def}}{=} \left(\!\left((p_\beta(\overline{z}), p_\alpha(\overline{z}))_{\mathcal{P}_{t(a)}}\right)\!\right), \qquad |\alpha| = |\beta| = n,$$

of pairwise inner products of homogeneous (of degree n) polynomials

$$p_\alpha(\overline{z}) \stackrel{\text{def}}{=} \sum_{|\gamma|=n} \frac{n!}{\gamma!}\, \mathcal{L}_a(\,|\,\cdot\,)_\alpha^\gamma \cdot \overline{z}^\gamma, \qquad |\alpha| = n,$$

is positive definite. The elements $r_{(\alpha;\beta)}(a)$ of the matrix $R_a(n)$ can be expressed via the elements $t_{(\alpha;\beta)}(a)$ of the matrix $T_a(n)$ by the formula

$$r_{(\alpha;\beta)}(a) = \sum_{|\gamma|=|\delta|=n} \frac{n!\, n!}{\gamma!\, \delta!} \mathcal{L}_a(\,|\cdot)_\alpha^\gamma \, \overline{\mathcal{L}_a(\,|\cdot)_\beta^\delta} \, t_{(\gamma;\delta)}(a).$$

By Definition 5.1, we can rewrite the matrix factorial $\overline{\mathcal{L}_a(\,|\cdot)_\beta^\delta}$ in this formula as follows:

$$\overline{\mathcal{L}_a(\,|\cdot)_\beta^\delta} = \mathcal{L}_a^*(\cdot\,|\,)_\delta^\beta.$$

Furthermore, using the relation

$$\mathcal{M}_a^*(\beta) \cdot \mathcal{L}_a^{-1}(\alpha) = \mathcal{L}_a^{-1*}(\beta) \cdot \mathcal{F}_a^\omega(n) \cdot \mathcal{L}_a^{-1}(\alpha), \qquad |\alpha|=|\beta|=n,$$

and Theorem 5.2(a), we represent the numbers $t_{(\gamma;\delta)}$ in the form
(5.21a)
$$t_{(\gamma;\delta)}(a) = n! \sum_{|\zeta|=|\xi|=n} \frac{n!\, n!}{\zeta!\, \xi!} \mathcal{L}_a^{-1*}(\,|\,)_\zeta^\delta \cdot \mathcal{F}_a^\omega(|\cdot|)_\xi^\zeta \cdot \mathcal{L}_a^{-1}(\cdot\,|\,)_\gamma^\xi, \qquad |\gamma|=|\delta|=n.$$

Then from Corollary 5.2 and Theorem 5.1(b), we obtain

$$r_{(\alpha;\beta)}(a) = n! \sum_{|\zeta|=|\xi|=n} \frac{n!\, n!}{\zeta!\, \xi!} \left(\sum_{|\delta|=n} \frac{n!}{\delta!} \mathcal{L}_a^*(\cdot\,|\,)_\delta^\beta \cdot \mathcal{L}_a^{-1*}(\,|\cdot)_\zeta^\delta \right)$$

$$\cdot \mathcal{F}_a^\omega(|\cdot|)_\xi^\zeta \left(\sum_{|\gamma|=n} \frac{n!}{\gamma!} \mathcal{L}_a^{-1}(\cdot\,|\,)_\gamma^\xi \cdot \mathcal{L}_a(\,|\cdot)_\alpha^\gamma \right)$$

$$= n! \sum_{|\zeta|=|\xi|=n} \frac{n!\, n!}{\zeta!\, \xi!} I(\,|\,)_\zeta^\beta \cdot \mathcal{F}_a^\omega(|\cdot|)_\xi^\zeta \cdot I(\,|\,)_\alpha^\xi = n! \mathcal{F}_a^\omega(|\cdot|)_\alpha^\beta,$$

where I is the unit matrix. In the following (see Corollary 5.3), we show that the matrix $R_a(n) = \big((n! \mathcal{F}_a^\omega(|\cdot|)_\alpha^\beta) \big)$ is positive definite for $a \in \mathcal{R} \setminus \mathcal{R}_0$. $\square$

In the case $N < \infty$, the space $\mathcal{P}_{t(a)}$ is finite-dimensional (with $\dim \mathcal{P}_{t(a)} = (N+1)(N+2)/2$) and consists of all polynomials of the form

(5.23) $$g(\overline{z^1}, \overline{z^2}) = \sum_{0 \le |\alpha| \le N} g_\alpha \overline{z}^\alpha.$$

In any case, the set of polynomials is dense in $\mathcal{P}_{t(a)}$ and the unit function 1 belongs to $\mathcal{P}_{t(a)}$.

LEMMA 5.6. *Let $a \in \mathcal{R} \setminus \mathcal{R}_0$. Then different factorizations (5.9) generate isomorphic Hilbert spaces. If $\mathcal{P}_{t(a)}$ and $\mathcal{P}_{l'(a)}$ are two such spaces, then an isomorphism between them is given by the mapping*

(5.24) $$\sum_{0 \le |\alpha| \le N} g_\alpha \overline{z}^\alpha \to \sum_{0 \le |\alpha| \le N} g'_\alpha \overline{z}^\alpha, \qquad g'_\alpha = \frac{n!}{\alpha!} \sum_{|\beta|=n} (\mathcal{L}'(\,|\cdot) \mathcal{L}(\cdot\,|\,))_\beta^\alpha \cdot g_\beta.$$

PROOF. Identity (5.9) and the fact that the matrix $\mathcal{F}_a^\omega(n)$ is nondegenerate for $a \in \mathcal{R} \setminus \mathcal{R}_0$ imply that the matrix $\mathcal{L}_a(\alpha)$, where $|\alpha| = n$, is invertible for each n, $0 \leq n \leq N - 1$. It follows from (5.20a) and (5.20b) that the matrix $\mathcal{L}'_a(\ ;\alpha)\mathcal{L}(\beta;\ ) \in \mathcal{G}_N$, due to Lemma 5.3(b), (c). Therefore, Corollary 5.1 and Theorem 5.2 work for the matrix factorial $\bigl(\mathcal{L}'(\ |\ \cdot)\mathcal{L}(\cdot\ |\ )\bigr)_\beta^\alpha$.

The equality $\|g\|_{\mathcal{P}_{t(a)}} = \|g'\|_{\mathcal{P}_{t'(a)}}$ follows from the system of equalities

$$\sum_{|\alpha|=|\beta|=n} t_{(\alpha;\beta)}(a)\, g_\alpha \cdot \overline{g_\beta} = \sum_{|\alpha|=|\beta|=n} t'_{(\alpha;\beta)}(a)\, g'_\alpha \cdot \overline{g'_\beta}, \qquad 0 \leq n \leq N,$$

each of which can be verified by using (5.21a) (see similar calculations in the proof of Lemma 5.5(b)). $\square$

5.5. Irreducible antiholomorphic representations of a twisted algebra. Now we can describe irreducible representations of the algebra (5.3) and objects related to them.

THEOREM 5.3. (a) *There is a one-to-one correspondence between the set $\mathcal{R}$ and the set of representations of the algebra (5.3) that are irreducible Hermitian and possess a vacuum vector. Such a representation corresponding to*[22] *$a \in \mathcal{R} \setminus \mathcal{R}_0$ is finite-dimensional if and only if the function $\chi_a(n) = \prod_{j=0}^{n-1} \omega(\mathcal{A}_a(j))$ has roots on $\mathbb{N}$; in this case, the number $(N+1)(N+2)/2$, where $N \in \mathbb{N}$ is the smallest possible root of the function χ_a, is the dimension of the representation.*

(b) *For each $a \in \mathcal{R} \setminus \mathcal{R}_0$ and for each factorization (5.9), the operators*

(5.25)
$$\overset{\circ}{B}_p = \sum_{q=1}^{2} \overline{z^q} \mathcal{L}_{a\,p}^{\ q}\!\left(\overrightarrow{\overline{z}\frac{\partial}{\partial \overline{z}}}\right),$$
$$\overset{\circ}{C}{}^p = \sum_{q=1}^{2} \mathcal{M}_{a\,q}^{\ p}\!\left(\overrightarrow{\overline{z}\frac{\partial}{\partial \overline{z}}}\right) \frac{\partial}{\partial \overline{z^q}},$$
$$\overset{\circ}{A} = \mathcal{A}_a\!\left(\overline{z}\cdot\frac{\partial}{\partial\overline{z}}\right),$$

where

$$\overrightarrow{\overline{z}\frac{\partial}{\partial\overline{z}}} = \left(\overline{z^1}\frac{\partial}{\partial\overline{z^1}}, \overline{z^2}\frac{\partial}{\partial\overline{z^2}}\right), \qquad \overline{z}\cdot\frac{\partial}{\partial\overline{z}} = \overline{z^1}\frac{\partial}{\partial\overline{z^1}} + \overline{z^2}\frac{\partial}{\partial\overline{z^2}},$$

represent the algebra (5.3), and moreover, this representation is irreducible Hermitian and possesses the vacuum vector 1 in the Hilbert space $\mathcal{P}_{t(a)}$.

(c) *Representations (5.25) assigned to different vectors $a \in \mathcal{R} \setminus \mathcal{R}_0$ are not equivalent, but for each chosen $a \in \mathcal{R}$, representations assigned to different factorizations (5.9) are equivalent under the isomorphism (5.24).*

(d) *An abstract Hermitian representation of the algebra (5.3) in a Hilbert space H_a with the vacuum vector $\mathfrak{P}_0 = \mathfrak{P}_0(a)$, satisfying (5.6) for some $a \in \mathcal{R}$, can be intertwined with the representation (5.25) by means of the generalized coherent*

---

[22]Recall that the case $a \in \mathcal{R}_0$ can be reduced to the one-dimensional case (see §1).

states

(5.26) $$\mathfrak{P}_{z^1,z^2} = \sum_{\substack{\alpha,\beta; \\ 0\leq|\alpha|=|\beta|=n\leq N}} \frac{n!}{\alpha!\beta!} \mathcal{M}_a^{-1*}(\cdot\,|\,)_\alpha^\beta \, z^\alpha \, \mathbf{B}^\beta \, \mathfrak{P}_0.$$

This family of coherent states is an H_a-valued holomorphic distribution over $\mathbb{R}^4$, or a holomorphic polynomial in the case $N < \infty$.

The generalized reproducing kernel corresponding to the states (5.26),

$$\mathcal{K}_{t(a)}(\overline{w^1},\overline{w^2};z^1,z^2) = (\mathfrak{P}_{w^1,w^2},\mathfrak{P}_{z^1,z^2})_{H_a}$$

is the kernel of the identity operator in $\mathcal{P}_{t(a)}$ and is given by the following distribution over $\mathbb{R}^4 \times \mathbb{R}^4$:

(5.27) $$\mathcal{K}_{t(a)}(\overline{w^1},\overline{w^2};z^1,z^2) = \sum_{\substack{\alpha,\beta; \\ 0\leq|\alpha|=|\beta|=n\leq N}} \frac{n!}{\alpha!\beta!} \left(\mathcal{L}_a(\,|\,\cdot)\mathcal{M}_a^{-1*}(\cdot\,|\,)\right)_\alpha^\beta z^\alpha \overline{w}^\beta.$$

This distribution satisfies the equations

$$\sum_{q=1}^2 \mathcal{M}_a{}_q^p\left(\overline{w}\frac{\overrightarrow{\partial}}{\partial\overline{w}}\right)\frac{\partial}{\partial\overline{w}^q}\mathcal{K}_{t(a)} = \sum_{q=1}^2 z^q \overline{\mathcal{L}_a}{}_q^p\left(z\frac{\overrightarrow{\partial}}{\partial z}\right)\mathcal{K}_{t(a)},$$

$$\mathcal{A}_a\left(\overline{w}\cdot\frac{\partial}{\partial\overline{w}}\right)\mathcal{K}_{t(a)} = \mathcal{A}_a\left(z\cdot\frac{\partial}{\partial z}\right)\mathcal{K}_{t(a)}.$$

It is natural to call the function on the right-hand side of (5.27) a *twisted hypergeometric* function. For a comparison with the usual hypergeometric function, see §2.2.

To prove Theorem 5.3, we need the following three statements.

LEMMA 5.7. *Suppose that an Hermitian representation of the algebra* (5.3) *in a Hilbert space H_a possesses the vacuum vector $\mathfrak{P}_0 = \mathfrak{P}_0(a)$* (5.6). *Then the following relations hold*:

(a) $\mathbf{C}^q \mathbf{B}^\alpha \mathfrak{P}_0 = \sum_{p=1}^2 \alpha_p \mathcal{F}_a{}_p^{\omega q}(n-1) \mathbf{B}^{\alpha-1_p}\mathfrak{P}_0, \qquad |\alpha| = n \geq 1,$

(b) $(\mathbf{B}^\alpha\mathfrak{P}_0, \mathbf{B}^\beta\mathfrak{P}_0)_{H_a} = \begin{cases} 0, & |\alpha| \neq |\beta|, \\ n!\,\mathcal{F}_a^\omega(|\cdot|)_\alpha^\beta, & |\alpha| = |\beta| = n. \end{cases}$

PROOF. (a) According to (5.3), (5.4), and (5.8), we have

$$\mathbf{C}^q \mathbf{B}_{p_1}\mathfrak{P}_0 = \big(\mathbf{B}_{p_1}\omega(\mathbf{A})\mathbf{C}^q + f_{p_1}^q(\mathbf{A})\big)\mathfrak{P}_0 = f_{p_1}^q(a)\mathfrak{P}_0,$$

$$\mathbf{C}^q \mathbf{B}_{p_2}\mathbf{B}_{p_1}\mathfrak{P}_0 = \big(\mathbf{B}_{p_2}\omega(\mathbf{A})\mathbf{C}^q + f_{p_2}^q(\mathbf{A})\big)\mathbf{B}_{p_1}\mathfrak{P}_0$$
$$= \omega(a)\big(f_{p_1}^q(a)\mathbf{B}_{p_2} + f_{p_2}^q(a)\mathbf{B}_{p_1}\big)\mathfrak{P}_0,$$

$$\mathbf{C}^q \mathbf{B}_{p_3}\mathbf{B}_{p_2}\mathbf{B}_{p_1}\mathfrak{P}_0 = \big(\mathbf{B}_{p_3}\omega(\mathbf{A})\mathbf{C}^q + f_{p_3}^q(\mathbf{A})\big)\mathbf{B}_{p_2}\mathbf{B}_{p_1}\mathfrak{P}_0$$
$$= \omega(a)\omega(\varphi(a))\big(f_{p_1}^q(a)\mathbf{B}_{p_2}\mathbf{B}_{p_3} + f_{p_2}^q(a)\mathbf{B}_{p_1}\mathbf{B}_{p_3}$$
$$+ f_{p_3}^q(a)\mathbf{B}_{p_1}\mathbf{B}_{p_2}\big)\mathfrak{P}_0,$$

etc., that is,

$$\mathbf{C}^q \mathbf{B}^\alpha \mathfrak{P}_0 = \mathbf{C}^q \mathbf{B}_{p_n} \mathbf{B}_{p_{n-1}} \ldots \mathbf{B}_{p_1} \mathfrak{P}_0 = \prod_{k=0}^{n-2} \omega(\varphi^k(a)) \sum_{p=1}^{n} f_p^q(a) \cdot \alpha_p \mathbf{B}^{\alpha-1_p} \mathfrak{P}_0$$

$$= \sum_{p=1}^{n} \alpha_p \, \mathcal{F}_{a\ p}^{\omega\ q}(n-1) \, \mathbf{B}^{\alpha-1_p} \mathfrak{P}_0, \qquad n \geq 2.$$

(b) Using equality (a), we obtain

$$\mathbf{C}^{q_n} \mathbf{B}^\alpha \mathfrak{P}_0 = \sum_{p_n=1}^{n} \alpha_{p_n} \mathcal{F}_{a\ p_n}^{\omega\ q_n}(n-1) \, \mathbf{B}^{\alpha-1_{p_n}} \mathfrak{P}_0,$$

$$\mathbf{C}^{q_{n-1}} \mathbf{C}^{q_n} \mathbf{B}^\alpha \mathfrak{P}_0 = \sum_{p_{n-1},p_n=1}^{n} \alpha_{p_n}(\alpha - 1_{p_n})_{p_{n-1}} \mathcal{F}_{a\ p_n}^{\omega\ q_n}(n-1) \mathcal{F}_{a\ p_{n-1}}^{\omega\ q_{n-1}}(n-2)$$

$$\times \mathbf{B}^{\alpha-1_{p_n}-1_{p_{n-1}}} \mathfrak{P}_0,$$

etc., that is,

$$\mathbf{C}^\beta \mathbf{B}^\alpha \mathfrak{P}_0 = \sum_{p_1,\ldots,p_n=1}^{n} \alpha_{p_n}(\alpha - 1_{p_n})_{p_{n-1}} \cdot \ldots \cdot (\alpha - 1_{p_n} - \cdots - 1_{p_2})_{p_1}$$

$$\times \mathcal{F}_{a\ p_n}^{\omega\ q_n}(n-1) \mathcal{F}_{a\ p_{n-1}}^{\omega\ q_{n-1}}(n-2) \cdot \ldots \cdot \mathcal{F}_{a\ p_1}^{\omega\ q_1}(0) \mathfrak{P}_0$$

$$= \alpha! \sum_{\vec{p} \in S(\alpha)} \mathcal{F}_{a\ p_n}^{\omega\ q_n}(n-1) \mathcal{F}_{a\ p_{n-1}}^{\omega\ q_{n-1}}(n-2) \cdot \ldots \cdot \mathcal{F}_{a\ p_1}^{\omega\ q_1}(0) \mathfrak{P}_0,$$

$$|\alpha| = |\beta| = n \geq 2.$$

In view of (5.6) and (5.19b), this implies that

$$(\mathbf{B}^\alpha \mathfrak{P}_0, \mathbf{B}^\beta \mathfrak{P}_0)_{H_a} = (\mathbf{C}^\beta \mathbf{B}^\alpha \mathfrak{P}_0, \mathfrak{P}_0,)_{H_a} = n! \, \mathcal{F}_a^\omega(|\cdot|)_\alpha^\beta,$$

$$(\mathbf{B}^\alpha \mathfrak{P}_0, \mathbf{B}^{\beta+\gamma} \mathfrak{P}_0)_{H_a} = n! \, \mathcal{F}_a^\omega(|\cdot|)_\alpha^\beta (\mathbf{C}^\gamma \mathfrak{P}_0, \mathfrak{P}_0,)_{H_a} = 0,$$

$$|\alpha| = |\beta|, \qquad |\gamma| \geq 1. \quad \square$$

COROLLARY 5.3. *The following two statements are equivalent:*
1°. $a \in \mathcal{R} \setminus \mathcal{R}_0$.
2°. *The matrix* $R_a(n) = \left((n! \, \mathcal{F}_a^\omega(|\cdot|)_\alpha^\beta)\right)$, *where* $|\alpha| = |\beta| = n$, *is*
   (1) *positive definite for* $0 \leq n \leq N$,
   (2) *zero for* $n \geq N + 1$.

PROOF. Since the matrix $F(a)$ is Hermitian, there exist a unitary matrix $U(a)$ such that $U^{-1}(a)F(a)U(a) = \widetilde{F}(a)$ is a diagonal matrix. Let us introduce new "creation" and "annihilation" operators as follows:

$$\widetilde{\mathbf{B}}_p = \sum_{r=1}^{2} \mathbf{B}_r U_p^r(a), \qquad \widetilde{\mathbf{C}}^q = \widetilde{\mathbf{B}}_q^* = \sum_{s=1}^{2} U^{*q}{}_s(a) \mathbf{C}^s.$$

Since $\widetilde{\mathbf{B}}_p$, $\widetilde{\mathbf{C}}^p$, $\mathbf{A}$ satisfy the commutation relations

(5.3)
$$\widetilde{\mathbf{C}}^q \widetilde{\mathbf{B}}_p = \widetilde{\mathbf{B}}_p \omega(\mathbf{A}) \widetilde{\mathbf{C}}^q + \left(U^{-1}(a) \cdot F(\mathbf{A}) \cdot U(a)\right)_p^q,$$
$$\widetilde{\mathbf{C}}^p \mathbf{A} = \varphi(\mathbf{A}) \widetilde{\mathbf{C}}^p, \qquad \mathbf{A} \widetilde{\mathbf{B}}_p = \widetilde{\mathbf{B}}_p \varphi(\mathbf{A}),$$

of the type (5.3) and since $\mathbf{A}\mathfrak{P}_0 = a\mathfrak{P}_0$, $\widetilde{\mathbf{C}}^1 \mathfrak{P}_0 = \widetilde{\mathbf{C}}^2 \mathfrak{P}_0 = 0$, Lemma 5.7 implies the identities

(a) $\quad \widetilde{\mathbf{C}}^q \widetilde{\mathbf{B}}^\alpha \mathfrak{P}_0 = \sum_{p=1}^{2} \alpha_p \widetilde{\mathcal{F}}_{a\,p}^{\omega\,q}(n-1) \widetilde{\mathbf{B}}^{\alpha - 1_p} \mathfrak{P}_0, \qquad |\alpha| = n \geq 1,$

(b) $\quad (\widetilde{\mathbf{B}}^\alpha \mathfrak{P}_0, \widetilde{\mathbf{B}}^\beta \mathfrak{P}_0)_{H_a} = \begin{cases} 0, & |\alpha| \neq |\beta|, \\ n!\, \widetilde{\mathcal{F}}_a^\omega(|\cdot|)_\alpha^\beta, & |\alpha| = |\beta| = n, \end{cases}$

where the matrix $\widetilde{\mathcal{F}}_a^\omega(n)$ is determined by (5.8) in which the structural matrix $F(A)\big|_{A=a} = F(a)$ is replaced by the structural matrix $U^{-1}(a) F(A) U(a)\big|_{A=a} = \widetilde{F}(a)$ from $\widetilde{(5.3)}$, that is,

$$\widetilde{\mathcal{F}}_a^\omega(n) = \begin{cases} \widetilde{F}(a), & n = 0, \\ \widetilde{F}(a) \prod_{j=0}^{n-1} \omega(\mathcal{A}_a(j)), & n \geq 0. \end{cases}$$

The matrices $\widetilde{\mathcal{F}}_a^\omega(n)$ are diagonal for all $n \in \mathbb{Z}_+$. Therefore, the matrices $\widetilde{R}_a(n) = \big((\widetilde{r}_\alpha^{\,\beta}(n))\big) = \big((n!\, \widetilde{\mathcal{F}}_a^\omega(|\cdot|)_\alpha^\beta)\big)$, where $|\alpha| = |\beta| = n$, are also diagonal, that is,

$$\widetilde{r}_\alpha^{\,\beta}(n) = \begin{cases} 0, & \alpha \neq \beta, \\ n!\, \prod_{r=0}^{\alpha_1 - 1} \widetilde{\mathcal{F}}_{a\,1}^{\omega\,1}(r), & \alpha = \beta = (\alpha_1, 0), \\ n!\, \prod_{s=0}^{\alpha_2 - 1} \widetilde{\mathcal{F}}_{a\,2}^{\omega\,2}(s), & \alpha = \beta = (0, \alpha_2), \\ \alpha!\, \prod_{r=0}^{\alpha_1 - 1} \widetilde{\mathcal{F}}_{a\,1}^{\omega\,1}(r) \cdot \prod_{s=0}^{\alpha_2 - 1} \widetilde{\mathcal{F}}_{a\,2}^{\omega\,2}(\alpha_1 + s), & \alpha = \beta = (\alpha_1, \alpha_2), \\ & \alpha_1, \alpha_2 \geq 1. \end{cases}$$

Obviously, statement 2° about the matrices $R_a(n)$ is equivalent to a similar statement about the matrices $\widetilde{R}_a(n)$, that is, to the conditions

$$\widetilde{r}_\alpha^{\,\alpha}(n) > 0 \quad \text{for} \quad 0 \leq n \leq N, \qquad \widetilde{r}_\alpha^{\,\alpha}(n) = 0 \quad \text{for} \quad n \geq N+1 \quad (|\alpha| = n)$$

or, as one can see from the formulas for $\widetilde{r}_\alpha^{\,\beta}(n)$, to the conditions

$$\widetilde{\mathcal{F}}_{a\,1}^{\omega\,1}(n) > 0, \quad \widetilde{\mathcal{F}}_{a\,2}^{\omega\,2}(n) > 0 \quad \text{for} \quad 0 \leq n \leq N,$$
$$\widetilde{\mathcal{F}}_{a\,1}^{\omega\,1}(n) = \widetilde{\mathcal{F}}_{a\,2}^{\omega\,2}(n) = 0 \quad \text{for} \quad n \geq N+1.$$

The last conditions are equivalent to the following statement: the matrix $\mathcal{F}_a^\omega(n) = U(a) \widetilde{\mathcal{F}}_a^\omega(n) U^{-1}(a)$ is positive definite for $0 \leq n \leq N$ and zero for $n \geq N+1$; that is, the last conditions are equivalent to the statement that $a \in \mathcal{R} \setminus \mathcal{R}_0$. $\square$

LEMMA 5.8. *Let* $\mathcal{A}_\mu$ $(\mu = 1, \ldots, k)$ $\mathcal{L}_p^q$, *and* $\mathcal{M}_p^q$ $(p, q = 1, 2)$ *be functions over* $\mathbb{Z}_+^2$. *Then the operators*

(5.28)
$$\overset{\circ}{B}_p = \sum_{q=1}^2 \overline{z^q} \mathcal{L}_p^q\left(\overline{z} \frac{\overrightarrow{\partial}}{\partial \overline{z}}\right),$$
$$\overset{\circ}{C}^p = \sum_{q=1}^2 \mathcal{M}_q^p\left(\overline{z} \frac{\overrightarrow{\partial}}{\partial \overline{z}}\right) \frac{\partial}{\partial \overline{z^q}},$$
$$\overset{\circ}{A} = \mathcal{A}_a\left(\overline{z} \cdot \frac{\partial}{\partial \overline{z}}\right)$$

are well defined on all polynomials and, in particular, on the unit function 1. *The following conditions are sufficient for these operators to generate an irreducible Hermitian representation of the algebra* (5.3) *in a Hilbert space of distributions with the vacuum vector* 1:

- *the vector* $a = \mathcal{A}(0)$ *belongs to the set* $\mathcal{R} \setminus \mathcal{R}_0$, *that is,* $a \in \mathcal{R} \setminus \mathcal{R}_0$;
- *the functions* $\mathcal{A}$ *coincide with the functions* $\mathcal{A}_a$ *given by* (1.4);
- *the product* $\mathcal{M}_a(\alpha)\mathcal{L}_a(\alpha)$ *coincides with the matrix* $\mathcal{F}_a^\omega(|\alpha|)$ *given by* (5.8);
- *the functions* $\mathcal{M}_p^q$ *satisfy the compatibility conditions* (5.11);
- *the functions* $\mathcal{L}_p^q$ *satisfy the boundary conditions* (5.12).

In this case, the Hilbert space of the representations automatically coincides with $\mathcal{P}_{t(a)}$, *where the sequence* $t(a)$ *is given by* (5.21).

PROOF. By analogy with the proof of Lemma 4.4, we outline only the basic steps of this proof.

1. The permutation relations

$$\overset{\circ}{C}^p \overset{\circ}{A} = \varphi(\overset{\circ}{A}) \overset{\circ}{C}^p, \qquad \overset{\circ}{A} \overset{\circ}{B}_p = \overset{\circ}{B}_p \varphi(\overset{\circ}{A}), \qquad p = 1, 2,$$

for operators (5.28) are equivalent to the condition that the functions $\mathcal{A}$ coincide with $\mathcal{A}_a$ (1.4), where $a = \mathcal{A}(0)$.

The permutation relations

$$\overset{\circ}{C}^q \overset{\circ}{B}_p = \overset{\circ}{B}_p \omega(\overset{\circ}{A}) \overset{\circ}{C}^q + f_p^q(\overset{\circ}{A}), \qquad p = 1, 2,$$

are equivalent to the system of equations

$$\mathcal{M}_r^q(\alpha + 1_s) \mathcal{L}_p^s(\alpha + 1_r) = \omega(\mathcal{A}(|\alpha|)) \mathcal{M}_r^q(\alpha) \mathcal{L}_p^s(\alpha),$$
$$s \neq r, \quad 0 \leq |\alpha| \leq N - 1,$$

(5.29)
$$\sum_{r=1}^2 (\alpha_r + 1) \mathcal{M}_r^q(\alpha) \mathcal{L}_p^s(\alpha)$$
$$= \omega(\mathcal{A}(|\alpha| - 1)) \sum_{r=1}^2 \alpha_r \mathcal{M}_r^q(\alpha - 1_r) \mathcal{L}_p^s(\alpha - 1_r) + f_p^q(\mathcal{A}(|\alpha|)),$$
$$0 \leq |\alpha| \leq N,$$

and the relations

$$[\overset{\circ}{B}_1, \overset{\circ}{B}_2] = 0, \qquad [\overset{\circ}{C}^1, \overset{\circ}{C}^2] = 0$$

lead to the compatibility conditions (5.11.a) and (5.13.a).

Taking into account the (Jacobi) identity (5.4), one can verify that the functions
$$\mathcal{L}_p^q(\alpha), \quad \mathcal{M}_p^q(\alpha), \qquad p,q = 1,2, \quad 0 \le |\alpha| \le N-1,$$
described by formulas (5.9) and (5.11), satisfy the system of equations (5.29) and conditions (5.11a) and (5.13a).

2. The boundary conditions (5.12) imply that the space of functions of the form (5.23) is invariant with respect to the operators $\overset{\circ}{B}_p$, $p = 1, 2$.

3. The condition $a \in \mathcal{R} \setminus \mathcal{R}_0$ is sufficient for the existence in the space of functions of the form (5.23) of an inner product with respect to which the representation (5.28) is Hermitian.

4. Finally, the condition $a \in \mathcal{R}\setminus\mathcal{R}_0$ implies that the matrix $T_a(n) = ((t_{(\alpha;\beta)}(a)))$ (5.21) is nondegenerate for all n, $0 \le n \le N-1$. Therefore, in the space $\mathcal{P}_{t(a)}$ there does not exist an invariant (with respect to (5.28)) subspace whose dimension is less than that of $\mathcal{P}_{t(a)}$; that is, the representation (5.28) is irreducible. $\square$

PROOF OF THEOREM 5.3. Theorem 5.3 can be proved using Lemma 5.7, Corollary 5.3, and Lemma 5.8 similarly to the proof of Theorem 1.1.

Here we present only the system of equations for the coefficients of coherent states
$$\mathfrak{P}_{z^1,z^2} = \sum_{\substack{\alpha,\beta \\ 0\le|\alpha|=|\beta|=n\le N}} \eta_{(\alpha;\beta)}\, z^\alpha \mathbf{B}^\beta \mathfrak{P}_0,$$
which is equivalent to the intertwining property

$$\eta_{(0;0)} = 1,$$

(5.30a) $\qquad (\alpha_r + 1)\eta_{(\alpha+1_r;\beta)} = \sum_{p:\ 1\le p\le 2;\ \beta_p \ge 1} \mathcal{M}^{-1*\,p}_{\ \ r}(\alpha)\, \eta_{(\alpha;\beta-1_p)},$

(5.30b) $\qquad \sum_{p=1}^{2}(\beta_p+1)\mathcal{F}^{\omega\,q}_{a\ p}(\beta)\,\eta_{(\alpha;\beta+1_p)} = \sum_{r=1}^{2}\mathcal{L}^{*q}_r(\alpha - 1_r)\,\eta_{(\alpha-1_r;\beta)}.$

The numbers $\eta_{(\alpha;\beta)} = \frac{n!}{\alpha!\beta!}\mathcal{M}_a^{-1*}(\cdot\,|\,)_\alpha^\beta$ satisfy equations (5.30a) and (5.30b) due to Theorem 5.1 (see also Remark 5.2). $\square$

REMARK 5.3 (on the definition of the Casimir operator). Note that if $a \in \mathcal{R} \setminus \mathcal{R}_0$ and the function $\chi_a(n) = \prod_{j=0}^{n-1}\omega(\varphi^j(a))$ has roots in $\mathbb{N}$, then the condition (5.2) implies that the pseudometric σ has singularities. Nevertheless, the Casimir operator $\mathbf{K} = \langle \mathbf{B}, \sigma(\mathbf{A})\mathbf{C}\rangle - \rho(\mathbf{A})$ is well defined on the irreducible component H_a generated by the vacuum vector $\mathfrak{P}_0 = \mathfrak{P}_0(a)$ satisfying (5.6). Indeed, the space H_a is the linear span of vectors of the form $\mathbf{B}^\alpha \mathfrak{P}_0$, where $0 \le |\alpha| \le N$ and $N \in \mathbb{N}$ is the least root of the function χ_a. Since the operators $\mathbf{C}^q$ map H_a to the subspace $\mathrm{span}\{\mathbf{B}^\alpha \mathfrak{P}_0 \mid 0 \le |\alpha| \le N-1\} \subset H_a$ and since the operators $\sigma_p^q(\mathbf{A})$ are well defined on this subspace:
$$\sigma_p^q(\mathbf{A})\mathbf{B}^\alpha \mathfrak{P}_0 = \sigma_p^q(\varphi^n(a))\mathbf{B}^\alpha \mathfrak{P}_0 = \frac{\sigma_p^q(a)}{\chi_a(n)}\mathbf{B}^\alpha \mathfrak{P}_0, \qquad 0 \le |\alpha| = n \le N-1,$$
the operator $\mathbf{K}$ is also well defined on the entire space H_a.

COROLLARY 5.4. *The following orthonormal eigenbases of the operators $\overset{\circ}{A}_\mu$ ($\mu = 1, \ldots, k$) (5.28):*

$$\left\{ \frac{1}{\sqrt{\alpha!}} \sum_{|\beta|=n} \frac{n!}{\beta!} \left(\mathcal{L}_a(|\cdot|)\mathcal{F}_a^{\omega-1/2}(|\cdot|)\right)_\alpha^\beta \overline{z}^\beta \mid 0 \leq |\alpha| = n \leq N \right\} \in \mathcal{P}_{t(a)}$$

and of the operators $\mathbf{A}_\mu$

$$\left\{ \frac{1}{\sqrt{\alpha!}} \sum_{|\beta|=n} \frac{n!}{\beta!} \mathcal{F}_a^{\omega-1/2}(|\cdot|)_\alpha^\beta \mathbf{B}^\beta \mathfrak{P}_0 \mid 0 \leq |\alpha| = n \leq N \right\} \in H_a$$

correspond to each other under the coherent transform

$$\mathcal{P}_{t(a)} \ni g \to \mathfrak{p} \in H_a, \qquad (\mathfrak{p}', \mathfrak{p})_{H_a} = ((\mathfrak{P}, \mathfrak{p}')_{H_a}, g)_{\mathcal{P}_{t(a)}} \qquad \forall p' \in H_a,$$

$$H_a \ni \mathfrak{p} \to g \in \mathcal{P}_{t(a)}, \qquad g = (\mathfrak{P}, \mathfrak{p})_{H_a}.$$

5.6. Exponential form of coherent states. There is a remarkable representation of all coherent states obtained above for different classes of algebras. Let us first demonstrate this form and list the above algebras. Later we shall use this form of coherent states for the investigation of more general permutation relations.

THEOREM 5.4. (a) *The generalized coherent states for the algebra* (1.48)

$$\mathfrak{P}_z = \mathfrak{P}_0 + \sum_{n \geq 1} \frac{1}{n! \overline{\mathcal{M}_a}(n-1) \ldots \overline{\mathcal{M}_a}(0)} (z\mathbf{B})^n \mathfrak{P}_0$$

(described in §1.4) can be written in the form

$$\mathfrak{P}_z = \exp\left\{ z\overline{\mathcal{M}}_a^{-1}\left(z\frac{d}{dz}\right)\mathbf{B} \right\} \mathfrak{P}_0.$$

(b) *Formula* (4.29) *for generalized coherent states of the algebra* (4.11) *is equivalent to the formula*

$$\mathfrak{P}_{z^1,z^2} = \exp\left\{ \sum_{p=1}^{2} z^p (\overline{\mathcal{M}_a^p})^{-1}\left(z_p \frac{\partial}{\partial z_p}\right) \mathbf{B}_p \right\} \mathfrak{P}_0.$$

(c) *Formula* (5.26) *for generalized coherent states of the algebra* (5.3) *is equivalent to the formula*

(5.31) $$\mathfrak{P}_{z^1,z^2} = \exp\left\{ \sum_{p,q=1}^{2} z^p (\mathcal{M}_a^{*-1})_p^q\left(z\frac{\overrightarrow{\partial}}{\partial z}\right) \mathbf{B}_q \right\} \mathfrak{P}_0.$$

Note that in all these formulas the creation operators $\mathbf{B}$ act on the vacuum vector $\mathfrak{P}_0$ and the differential operators in z-variables act on the unit function $1 = 1(z)$.

PROOF. All statements (a), (b), (c) can be proved by expanding the operator exponential in a power series. Next, for example, in case (c), we apply the operator

$$\left(\sum_{p,q=1}^{2} z^p (\mathcal{M}_a^{*-1})_p^q \left(z\overrightarrow{\frac{\partial}{\partial z}}\right) \mathbf{B}_q\right)^n$$

to the unit functioin $1(z_1, z_2)$ and obtain

$$\left(\sum_{p,q=1}^{2} z^p (\mathcal{M}_a^{*-1})_p^q \left(z\overrightarrow{\frac{\partial}{\partial z}}\right) \mathbf{B}_q\right)^n (1(z_1, z_2))$$

$$= \sum_{p_1,\ldots,p_n=1}^{2} \sum_{q_1,\ldots,q_n=1}^{2} z^{p_n} \ldots z^{p_1} \mathbf{B}_{q_n} \ldots \mathbf{B}_{q_n}$$
$$\cdot \mathcal{M}_a^{-1*}{}_{p_n}^{q_n}(1_{p_{n-1}} + \cdots + 1_{p_1}) \cdot \ldots \cdot \mathcal{M}_a^{-1*}{}_{p_2}^{q_2}(1_{p_1}) \mathcal{M}_a^{-1*}{}_{p_1}^{q_1}(0)$$

$$= \sum_{|\alpha|=|\beta|=n} z^\alpha \mathbf{B}^\beta \sum_{\vec{p}\in S(\alpha), \vec{q}\in S(\beta)} \mathcal{M}_a^{-1*}{}_{p_n}^{q_n}(1_{p_{n-1}} + \cdots + 1_{p_1})$$
$$\cdot \ldots \cdot \mathcal{M}_a^{-1*}{}_{p_2}^{q_2}(1_{p_1}) \mathcal{M}_a^{-1*}{}_{p_1}^{q_1}(0)$$

$$= \sum_{|\alpha|=|\beta|=n} \frac{(n!)^2}{\alpha!\beta!} \mathcal{M}_a^{-1*}(\cdot\,|\,)_\alpha^\beta z^\alpha \mathbf{B}^\beta.$$

Hence, formulas (5.26) and (5.31) are equivalent.

In the case of a finite-dimensional representation, we have $\mathbf{B}^\alpha \mathfrak{P}_0 = 0$ for $|\alpha| > N$. Then we apply the series for the exponential to the vacuum vector $\mathfrak{P}_0$ and obtain a finite sum over n, $0 \leq n \leq N$. So, in this case, formulas (5.26) and (5.31) are also equivalent. □

Thus, there is a general exponential form for all coherent states constructed above. It is important that this "exponential" form also works in the case of more general commutation relations. In the following, we consider an example of such relations that admit coherent states of type (5.31).

5.7. Nonscalar generalization of twisted algebra. Let us consider the following special case of commutation relations (4.1) (for simplicity, we consider the dimension $d = 2$):

$$\varphi_p(A) = \varphi(A), \qquad \omega_{sp}^{qr}(A) = \delta_p^r \cdot \delta_s^q \cdot \omega_p^q(A), \qquad p, q, r, s = 1, 2.$$

In this case, relations (4.1) have the form

(5.32)
$$\mathbf{C}^q \mathbf{B}_p = \mathbf{B}_p \omega_p^q(\mathbf{A}) \mathbf{C}^q + f_p^q(\mathbf{A}),$$
$$\mathbf{C}^p \mathbf{A} = \varphi(\mathbf{A}) \mathbf{C}^p, \qquad \mathbf{A}\mathbf{B}_p = \mathbf{B}_p \varphi_p(\mathbf{A}).$$

Now the matrix ω_p^q is not scalar. The scalar case $\omega_p^q = \omega$ was considered above in §§5.1–5.5.

The Jacobi identities (4.2) and conditions (4.3) imply the following conditions on the structural functions:

(5.33)
$$\overline{\varphi} = \varphi, \qquad \overline{\omega_p^q} = \omega_p^q = \omega_q^p, \qquad \overline{f_p^q} = f_q^p,$$
$$\omega_p^q(\varphi(A))\omega_t^q(A) = \omega_t^q(\varphi(A))\omega_p^q(A),$$
$$f_p^q(\varphi(A)) = \omega_t^q(A) \cdot f_p^q(A), \qquad t \neq p,$$
$$\omega_1^1 \neq \omega_2^2 \implies f_2^1 = f_1^2 = 0.$$

We consider only Hermitian representations of (5.32) in a Hilbert space H that possess a vacuum (4.12).

Let us define the following matrix over $\mathbb{Z}_+^2$:

(5.34)
$$\mathcal{F}_a^\omega(\alpha) = ((\mathcal{F}_{a\ p}^{\omega\ q}(\alpha))), \qquad p, q = 1, 2,$$
$$\mathcal{F}_{a\ p}^{\omega\ q}(\alpha) \stackrel{\text{def}}{=} \frac{1}{\alpha_p + 1} \sum_{j=0}^{\alpha_p} f_p^q(\mathcal{A}_a(|\alpha| - j)) \cdot \prod_{m=|\alpha|-j}^{|\alpha|-1} \omega_p^q(\mathcal{A}_a(m)),$$

where $\mathcal{A}_a(n)$ is the composition (1.4).

In formula (5.34) and everywhere in the sequel, products of the type $\prod_{\ell=\ell_1}^{\ell_2} \cdots$ should be replaced by 1 if $\ell_1 > \ell_2$.

LEMMA 5.10. *The functions $\mathcal{F}_{a\ p}^{\omega\ q}$ possess the following properties*:

(5.35) $\qquad \mathcal{F}_{a\ r}^{\omega\ q}(\alpha + 1_p) = \omega_p^q(\mathcal{A}_a(|\alpha|))\mathcal{F}_{a\ r}^{\omega\ q}(\alpha), \qquad r \neq p;$

(5.36) $\quad (\alpha_p + 1)\mathcal{F}_{a\ p}^{\omega\ q}(\alpha) = \alpha_p \omega_p^q(\mathcal{A}_a(\alpha - 1))\mathcal{F}_{a\ p}^{\omega\ q}(\alpha - 1_p) + f_p^q(\mathcal{A}_a(|\alpha|));$

(5.37) $\qquad \mathcal{F}_{a\ p_2}^{\omega\ q_2}(\alpha + 1_{p_1})\mathcal{F}_{a\ p_1}^{\omega\ q_1}(\alpha) = \mathcal{F}_{a\ p_1}^{\omega\ q_1}(\alpha + 1_{p_2})\mathcal{F}_{a\ p_2}^{\omega\ q_2}(\alpha).$

It follows from Lemma 5.10 that the matrix $\mathcal{F}_a^\omega(\alpha, \ ) \stackrel{\text{def}}{=} \mathcal{F}_a^\omega(\alpha) \in \mathcal{G}_\infty$.

PROOF OF LEMMA 5.10. By definition (5.34), for $r \neq p$ we have

$$\mathcal{F}_{a\ r}^{\omega\ q}(\alpha + 1_p) = \frac{1}{\alpha_r + 1} \sum_{j=0}^{\alpha_r} f_r^q(\mathcal{A}_a(|\alpha| + 1 - j)) \prod_{m=|\alpha|+1-j}^{|\alpha|} \omega_r^q(\mathcal{A}_a(m)).$$

Using the identities

$$f_r^q(\mathcal{A}_a(|\alpha| + 1 - j)) = f_r^q(\mathcal{A}_a(|\alpha| - j))\, \omega_p^q(\mathcal{A}_a(|\alpha| - j)),$$
$$\omega_p^q(\mathcal{A}_a(|\alpha| - j)) \prod_{m=|\alpha|+1-j}^{|\alpha|} \omega_r^q(\mathcal{A}_a(m)) = \omega_p^q(\mathcal{A}_a(|\alpha|)) \prod_{m=|\alpha|-j}^{|\alpha|-1} \omega_r^q(\mathcal{A}_a(m)),$$

that hold for $r \neq p$ and are consequences of (5.33) (see also (1.4)), we can rewrite the formula for $\mathcal{F}_{a\ r}^{\omega\ q}(\alpha + 1_p)$ as follows:

$$\mathcal{F}_{a\ r}^{\omega\ q}(\alpha + 1_p) = \omega_p^q(\mathcal{A}_a(|\alpha|)) \frac{1}{\alpha_r + 1} \sum_{j=0}^{\alpha_r} f_r^q(\mathcal{A}_a(|\alpha| - j)) \prod_{m=|\alpha|-j}^{|\alpha|-1} \omega_r^q(\mathcal{A}_a(m))$$
$$= \omega_p^q(\mathcal{A}_a(|\alpha|))\, \mathcal{F}_{a\ r}^{\omega\ q}(\alpha), \qquad r \neq p.$$

Thus, we have proved (5.35). (5.36) can be proved in a similar way.

Relation (5.37) follows from (5.35), (5.36), and the Jacobi identities (5.33). For example, let us prove (5.37) for $p_1 \neq p_2$. In this case, according to (5.35) and (5.36), identity (5.37) is equivalent to

$$\omega_{p_1}^{q_2}\bigl(\mathcal{A}_a(|\alpha|)\bigr)\, \mathcal{F}_{a\ p_2}^{\omega\ q_2}(\alpha)\, \mathcal{F}_{a\ p_1}^{\omega\ q_1}(\alpha) = \omega_{p_2}^{q_1}\bigl(\mathcal{A}_a(|\alpha|)\bigr)\, \mathcal{F}_{a\ p_1}^{\omega\ q_1}(\alpha)\, \mathcal{F}_{a\ p_2}^{\omega\ q_2}(\alpha).$$

If $q_1 = p_1 \neq p_2 = q_2$, then (5.33) implies $\omega_{p_1}^{q_2}(A) = \omega_{p_2}^{q_1}(A)$ and the last relation holds. But if $q_2 = p_1 \neq p_2 = q_1$, then either we again have $\omega_{p_1}^{q_2}(A) = \omega_{p_2}^{q_1}(A)$, or we have $\omega_{p_1}^{q_2}(A) \neq \omega_{p_2}^{q_1}(A)$ together with $\mathcal{F}_{a\ p_1}^{\omega\ q_1}(\alpha) = \mathcal{F}_{a\ p_2}^{\omega\ q_2}(\alpha) = 0$. Thus, the desired relation is also satisfied. $\square$

LEMMA 5.11. *Suppose that an Hermitian representation of the algebra (5.32) in a Hilbert space H_a possesses the vacuum vector $\mathfrak{P}_0 = \mathfrak{P}_0(a)$ (4.12). Then the following relations hold:*

(a) $\mathbf{C}^q \mathbf{B}^\alpha \mathfrak{P}_0 = \displaystyle\sum_{p=1}^{2} \alpha_p\, \mathcal{F}_{a\ p}^{\omega q}(\alpha - 1_p)\, \mathbf{B}^{\alpha - 1_p} \mathfrak{P}_0,$

(b) $(\mathbf{B}^\beta \mathfrak{P}_0, \mathbf{B}^\alpha \mathfrak{P}_0)_{H_a} = n!\, \mathcal{F}_a^\omega(\,\cdot\,|\,)_\alpha^\beta.$

PROOF. (a) According to (5.32), we have

$$\mathbf{C}^q (\mathbf{B}_p)^n = (\mathbf{B}_p)^n \prod_{j=0}^{n-1} \omega_p^q\bigl(\mathcal{A}_{\mathbf{A}}(j)\bigr) \mathbf{C}^q + (\mathbf{B}_p)^{n-1} \sum_{j=0}^{n-1} f_p^q\bigl(\mathcal{A}_{\mathbf{A}}(j)\bigr) \prod_{\ell=j}^{n-2} \omega_p^q\bigl(\mathcal{A}_{\mathbf{A}}(\ell)\bigr).$$

To calculate $\mathbf{C}^q \mathbf{B}^\alpha = \mathbf{C}^q (\mathbf{B}_1)^{\alpha_1}(\mathbf{B}_2)^{\alpha_2}$, we must apply this formula two times, first for $p = 1$ and $n = \alpha_1$, then for $p = 2$ and $n = \alpha_2$ (or vice versa). To derive (a), it remains to apply the obtained operator relation to the vector $\mathfrak{P}_0$ (4.12) and to take into account the Jacobi identities (5.33).

(b) Relation (b) follows from (a). Indeed, let $\vec{q} \in S(\beta)$ be an arbitrary path from 0 to β: $\vec{q} = (q_1, \ldots, q_n)$, $n = |\beta|$. Then, in view of (a), we have

$$(\mathbf{B}^\beta \mathfrak{P}_0, \mathbf{B}^\alpha \mathfrak{P}_0)_{H_a} = (\mathfrak{P}_0, \mathbf{C}^{q_1} \ldots \mathbf{C}^{q_n} \mathbf{B}^{q_n} \mathfrak{P}_0)_{H_a}$$

$$= \sum_{p_1, \ldots, p_n = 1}^{2} \alpha_{p_n}(\alpha - 1_{p_n})_{p_{n-1}} \ldots (\alpha - 1_{p_n} - \cdots - 1_{p_2})_{p_1}$$

$$\times \mathcal{F}_{a\ p_n}^{\omega\ q_n}(\alpha - 1_{p_n}) \ldots \mathcal{F}_{a\ p_1}^{\omega\ q_1}(\alpha - 1_{p_n} - \cdots - 1_{p_1})$$

$$= \alpha! \sum_{\vec{p} \in S(\alpha)} \mathcal{F}_{a\ p_n}^{\omega\ q_n}(1_{p_{n-1}} + \cdots + 1_{p_1}) \cdot \ldots \cdot \mathcal{F}_{a\ p_2}^{\omega\ q_2}(1_{p_1})\, \mathcal{F}_{a\ p_1}^{\omega\ q_1}(0)$$

$$= n!\, \mathcal{F}_a^\omega(\,\cdot\,|\,)_\alpha^\beta. \quad \square$$

5.8. Application of the left regular representation. For the algebra (5.32) it is possible to construct operators of the left regular representation in the same way as in §3.5. Restricting this representation to the subspace $C = 0$, $A = a$, one obtains a representation in variables B only. The general scheme of such a restriction procedure and of its application to calculating irreducible antiholomorphic representations is given in [**Ka**[17]]. In §§5.8 and 5.9, we apply this scheme to relations (5.32).

We define the following pseudodifferential operators with respect to the variables $B = (B_1, B_2) \in \mathbb{C}^2$ as follows:

$$(5.38) \quad \overset{\square}{B}_p = B_p, \quad \overset{\square}{C}{}^p = \sum_{q=1}^{2} \mathcal{F}_{a\,q}^{\omega p}\left(B\dfrac{\overrightarrow{\partial}}{\partial B}\right)\dfrac{\partial}{\partial B_q}, \quad \overset{\square}{A} = \mathcal{A}_a\left(B \cdot \dfrac{\partial}{\partial B}\right).$$

PROPOSITION 5.1. (a) *The operators $\overset{\square}{B}_p$, $\overset{\square}{C}{}^p$, $\overset{\square}{A}$ satisfy the commutation relations* (5.32).

(b) *If the representation B, C, A of algebra* (5.32) *possesses the vacuum vector $\mathfrak{P}_0 = \mathfrak{P}_0(a)$* (4.12), *then the following equalities hold*:

$$(5.39) \quad \begin{aligned} \mathbf{B}_p \mathbf{B}^\alpha \mathfrak{P}_0 &= (\overset{\square}{B}_p B^\alpha)\Big|_{B=\mathbf{B}} \mathfrak{P}_0, \\ \mathbf{C}^p \mathbf{B}^\alpha \mathfrak{P}_0 &= (\overset{\square}{C}_p B^\alpha)\Big|_{B=\mathbf{B}} \mathfrak{P}_0, \\ \mathbf{A} \mathbf{B}^\alpha \mathfrak{P}_0 &= (\overset{\square}{A} B^\alpha)\Big|_{B=\mathbf{B}} \mathfrak{P}_0. \end{aligned}$$

PROOF. The relations $[\overset{\square}{B}_p, \overset{\square}{B}_q] = 0$ and $[\overset{\square}{A}_\mu, \overset{\square}{A}_\nu] = 0$ are obvious. The commutation relation $[\overset{\square}{C}{}^p, \overset{\square}{C}{}^q] = 0$ is equivalent to the identity

$$\mathcal{F}_{a\,p_2}^{\omega q_2}(\alpha + 1_{p_1})\,\mathcal{F}_{a\,p_1}^{\omega q_1}(\alpha) + \mathcal{F}_{a\,p_1}^{\omega q_2}(\alpha + 1_{p_2})\,\mathcal{F}_{a\,p_2}^{\omega q_1}(\alpha)$$
$$= \mathcal{F}_{a\,p_2}^{\omega q_1}(\alpha + 1_{p_1})\,\mathcal{F}_{a\,p_1}^{\omega q_2}(\alpha) + \mathcal{F}_{a\,p_1}^{\omega q_1}(\alpha + 1_{p_2})\,\mathcal{F}_{a\,p_2}^{\omega q_2}(\alpha).$$

which holds due to (5.37).

Next, it follows from the definition of the functions $\mathcal{A}_a(n)$ that

$$\overset{\square}{A}\overset{\square}{B}_p = \mathcal{A}_a\left(B \cdot \dfrac{\partial}{\partial B}\right) \cdot B_p = B_p \mathcal{A}_a\left(B \cdot \dfrac{\partial}{\partial B} + 1\right) = B_p \varphi\left(\mathcal{A}_a\left(B \cdot \dfrac{\partial}{\partial B}\right)\right) = \overset{\square}{B}_p \varphi(\overset{\square}{A}),$$

and similarly,

$$\overset{\square}{C}{}^p \overset{\square}{A} = \sum_{q=1}^{2} \mathcal{F}_{a\,q}^{\omega p}\left(B\dfrac{\overrightarrow{\partial}}{\partial B}\right)\dfrac{\partial}{\partial B_q}\mathcal{A}_a\left(B \cdot \dfrac{\partial}{\partial B}\right) = \sum_{q=1}^{2} \mathcal{F}_{a\,q}^{\omega p}\left(B\dfrac{\overrightarrow{\partial}}{\partial B}\right)\mathcal{A}_a\left(B \cdot \dfrac{\partial}{\partial B} + 1\right)\dfrac{\partial}{\partial B_q}$$
$$= \varphi\left(\mathcal{A}_a\left(B \cdot \dfrac{\partial}{\partial B}\right)\right)\sum_{q=1}^{2} \mathcal{F}_{a\,q}^{\omega p}\left(B\dfrac{\overrightarrow{\partial}}{\partial B}\right)\dfrac{\partial}{\partial B_q} = \varphi(\overset{\square}{A})\overset{\square}{C}{}^p.$$

Finally, we prove the commutation relation

$$\overset{\square}{C}{}^q \overset{\square}{B}_p = \overset{\square}{B}_p \omega_p^q(\overset{\square}{A})\overset{\square}{C}{}^q + f_p^q(\overset{\square}{A}).$$

Writing the left- and right-hand parts of this relation in the form

$$\overset{\square}{C}{}^q \overset{\square}{B}_p = \sum_{r=1}^{2} \mathcal{F}_{a\,r}^{\omega q}\left(B\dfrac{\overrightarrow{\partial}}{\partial B}\right)\dfrac{\partial}{\partial B_r} B_p$$
$$= B_p \sum_{\substack{r=1 \\ r \neq p}}^{2} \mathcal{F}_{a\,r}^{\omega q}\left(B\dfrac{\overrightarrow{\partial}}{\partial B} + 1_p\right)\dfrac{\partial}{\partial B_r} + \left(B_p \dfrac{\partial}{\partial B_p} + 1\right)\mathcal{F}_{a\,p}^{\omega q}\left(B\dfrac{\overrightarrow{\partial}}{\partial B}\right),$$

$$\overset{\square}{B}_p\,\omega_p^q(\overset{\square}{A})\overset{\square}{C}{}^q + f_p^q(\overset{\square}{A})$$

$$= B_p\,\omega_p^q\Big(\mathcal{A}_a\Big(B\cdot\frac{\partial}{\partial B}\Big)\Big)\sum_{r=1}^{2}\mathcal{F}_{a\;r}^{\omega\,q}\Big(B\overrightarrow{\frac{\partial}{\partial B}}\Big)\frac{\partial}{\partial B_r} + f_p^q\Big(\mathcal{A}_a\Big(B\cdot\frac{\partial}{\partial B}\Big)\Big)$$

$$= B_p\,\omega_p^q\Big(\mathcal{A}_a\Big(B\cdot\frac{\partial}{\partial B}\Big)\Big)\sum_{\substack{r=1\\r\neq p}}^{2}\mathcal{F}_{a\;r}^{\omega\,q}\Big(B\overrightarrow{\frac{\partial}{\partial B}}\Big)\frac{\partial}{\partial B_r}$$

$$+ \Big(B_p\frac{\partial}{\partial B_p}\Big)\omega_p^q\Big(\mathcal{A}_a\Big(B\cdot\frac{\partial}{\partial B}-1\Big)\Big)\mathcal{F}_{a\;p}^{\omega\,q}\Big(B\overrightarrow{\frac{\partial}{\partial B}}-1_p\Big) + f_p^q\Big(\mathcal{A}_a\Big(B\cdot\frac{\partial}{\partial B}\Big)\Big),$$

we see that the desired commutation relation follows from the system of identities (5.35), (5.36).

(b) The first identity (5.39) is obvious. The second identity is a consequence of Lemma 5.11(a):

$$\mathbf{C}^p\mathbf{B}^\alpha\mathfrak{P}_0 = \sum_{q=1}^{2}\alpha_q\mathcal{F}_{a\;q}^{\omega\,p}(\alpha-1_q)\,\mathbf{B}^{\alpha-1_q}\mathfrak{P}_0$$

$$= \Big(\sum_{q=1}^{2}\mathcal{F}_{a\;q}^{\omega\,p}\Big(B\overrightarrow{\frac{\partial}{\partial B}}\Big)\frac{\partial}{\partial B_q}(B^\alpha)\Big)\Big|_{B=\mathbf{B}}\mathfrak{P}_0 = (\overset{\square}{C}B^\alpha)\Big|_{B=\mathbf{B}}\mathfrak{P}_0.$$

The third identity in (5.39) can be derived from the commutation relation $\mathbf{AB}^\alpha = \varphi^\alpha(\mathbf{A})\mathbf{B}^\alpha = \mathcal{A}_\mathbf{A}(|\alpha|)\mathbf{B}^\alpha$. Indeed,

$$\mathbf{AB}^\alpha\mathfrak{P}_0 = \mathcal{A}_a(|\alpha|)\mathbf{B}^\alpha\mathfrak{P}_0 = \Big(\mathcal{A}_a\Big(B\cdot\frac{\partial}{\partial B}\Big)(B^\alpha)\Big)\Big|_{B=\mathbf{B}}\mathfrak{P}_0 = (\overset{\square}{A}B^\alpha)\Big|_{B=\mathbf{B}}\mathfrak{P}_0. \quad\square$$

5.9. Antiholomorphic representation of mixed type.
Now we use the operators (5.38) to calculate the antiholomorphic representation of the algebra (5.32).

Let $\mathcal{M}(\alpha)$ be a family of nondegenerate matrices. We introduce the notation

$$p(z,B) \overset{\text{def}}{=} \exp\Big\{\sum_{p,q=1}^{2}z^p\mathcal{M}^{*-1}{}_p^{\;q}\Big(z\overrightarrow{\frac{\partial}{\partial z}}\Big)B_q\Big\},$$

$$\overset{\vee}{B}_p = \sum_{q=1}^{2}\overline{\mathcal{M}_q^p}\Big(z\overrightarrow{\frac{\partial}{\partial z}}\Big)\frac{\partial}{\partial z^q},$$

$$\overset{\vee}{\partial}{}^p = \sum_{q=1}^{2}z^q\overline{\mathcal{M}^{-1}{}_p^{\;q}}\Big(z\overrightarrow{\frac{\partial}{\partial z}}\Big),\qquad p=1,2.$$

LEMMA 5.12. *Let $\mathcal{M}(\alpha)$ satisfy the compatibility condition (5.11).*

(a) *The operators $\overset{\vee}{B}_p$ and $\overset{\vee}{\partial}{}^p$ satisfy the Heisenberg relations*

(5.40) $\quad [\overset{\vee}{B}_p,\overset{\vee}{B}_q]=0,\qquad [\overset{\vee}{\partial}{}^p,\overset{\vee}{\partial}{}^q]=0,\qquad [\overset{\vee}{B}_p,\overset{\vee}{\partial}{}^q]=\delta_p^q,\qquad p,q=1,2.$

(b) *The following identities hold:*

$$B_p p(z,B) = \overset{\vee}{B}_p p(z,B),\qquad \frac{\partial}{\partial B_p}p(z,B) = \overset{\vee}{\partial}{}^p p(z,B),\qquad p=1,2.$$

PROOF. (a) The first two commutation relations (5.40) hold due to Lemma 5.2.(b). Let us prove the third relation. We write the commutator $[\check{B}_p, \check{\partial}^q]$ in the form

$$[\check{B}_p, \check{\partial}^q] = \left[\sum_{r=1}^{2} \overline{\mathcal{M}^p_r}\left(z\frac{\vec{\partial}}{\partial z}\right)\frac{\partial}{\partial z^r}, \sum_{s=1}^{2} z^s \overline{\mathcal{M}^{-1}{}^s_q}\left(z\frac{\vec{\partial}}{\partial z}\right)\right]$$

$$= \sum_{r,s=1}^{2} z^s \left(\overline{\mathcal{M}^p_r}\left(z\frac{\vec{\partial}}{\partial z}+1_s\right)\overline{\mathcal{M}^{-1}{}^s_q}\left(z\frac{\vec{\partial}}{\partial z}+1_r\right)\right.$$

$$\left.- \overline{\mathcal{M}^p_r}\left(z\frac{\vec{\partial}}{\partial z}\right)\overline{\mathcal{M}^{-1}{}^s_q}\left(z\frac{\vec{\partial}}{\partial z}\right)\right)\frac{\partial}{\partial z^r} + \delta^q_p.$$

Then we see that the commutation relation $[\check{B}_p, \check{\partial}^q] = \delta^q_p$ follows from the identities

$$\mathcal{M}^p_r(\alpha + 1_s)\mathcal{M}^{-1}{}^s_q(\alpha + 1_r) = \mathcal{M}^p_r(\alpha)\mathcal{M}^{-1}{}^s_q(\alpha),$$

which can be readily verified using (5.11) and (5.11*). $\square$

It follows from Lemma 5.12 that the operators

$$\overset{\circ}{d}_p \overset{\text{def}}{=} \overline{\check{\partial}^p \cdot \check{B}_p} = \sum_{r,s} \bar{z}^r \mathcal{M}^{-1}{}^r_p\left(\bar{z}\frac{\vec{\partial}}{\partial \bar{z}}\right) \cdot \mathcal{M}^p_s\left(\bar{z}\frac{\vec{\partial}}{\partial \bar{z}}\right)\frac{\partial}{\partial \bar{z}^s}, \qquad p=1,2,$$

commute with each other.

Now we can define the following antiholomorphic operators:

(5.41)
$$\overset{\circ}{B}_p \overset{\text{def}}{=} \sum_{q=1}^{2} \bar{z}^q \left(\mathcal{M}^{-1}\left(\bar{z}\frac{\vec{\partial}}{\partial \bar{z}}\right) \cdot \mathcal{F}^{\omega*}_a(\overset{\circ}{d})\right)^q_p,$$

$$\overset{\circ}{C}^p \overset{\text{def}}{=} \sum_{q=1}^{2} \mathcal{M}^p_q\left(\bar{z}\frac{\vec{\partial}}{\partial \bar{z}}\right)\frac{\partial}{\partial \bar{z}^q},$$

$$\overset{\circ}{A} \overset{\text{def}}{=} \mathcal{A}_a\left(\bar{z}\cdot\frac{\partial}{\partial \bar{z}}\right).$$

Recall that the operators $\overset{\circ}{d}_p$ involve not only combinations $\bar{z}^q \cdot \partial/\partial \bar{z}^q$ but also general combinations $\bar{z}^r \cdot \partial/\partial \bar{z}^s$. We call them *operators of mixed type*.

THEOREM 5.5. *Suppose that an Hermitian representation* $\mathbf{B}_p$, $\mathbf{C}^p$, $\mathbf{A}$ *of the algebra* (5.32) *in a Hilbert space* H_a *possesses the vacuum vector* $\mathfrak{P}_0 = \mathfrak{P}_0(a)$ (4.12). *Let the matrices* $\mathcal{M}(\alpha) = \mathcal{M}_a(\alpha)$ *be nondegenerate and satisfy the compatibility conditions* (5.11). *Then*:

(a) *The operators* (5.41) *of mixed type satisfy the permutation relations* (5.32).

(b) *The coherent states of the algebra* (5.32), *intertwining a given Hermitian representation* $\mathbf{B}$, $\mathbf{C}$, $\mathbf{A}$ *with the operators* (5.41), *are given by formula* (5.31); *this means that*

$$\mathbf{B}_p \mathfrak{P}_z = \overline{\overset{\circ}{C}{}^p} \mathfrak{P}_z, \qquad \mathbf{C}^p \mathfrak{P}_z = \overline{\overset{\circ}{B}_p} \mathfrak{P}_z, \qquad \mathbf{A}\mathfrak{P}_z = \overline{\overset{\circ}{A}}\mathfrak{P}_z.$$

PROOF. Both statements (a) and (b) follow from the construction of the operators (5.41). Namely, the operators $\overset{\circ}{\overline{C^p}}$, $\overset{\circ}{\overline{B_p}}$, and $\overset{\circ}{\overline{A}}$ are constructed from the operators $\overset{\square}{B_p}$, $\overset{\square}{C^p}$, and $\overset{\square}{A}$ (5.38), respectively, according to the rule:

1) the operators B_p and $\partial/\partial B_p$ are replaced by the operators $\overset{\vee}{B}_p$ and $\overset{\vee}{\partial}{}^p$, respectively, and
2) the order of action of the operators B_p and $\overset{\vee}{\partial}{}^p$ in (5.41) is opposite to that of the operators B_p and $\partial/\partial B_p$ in (5.38).

So, for example, $B_p \frac{\partial}{\partial B_p} \to \overset{\vee}{\partial}{}^p \cdot \overset{\vee}{B}_p = \overset{\circ}{\overline{d_p}}$, and therefore, we have

$$\overset{\circ}{\overline{B}_p} = \sum_{q=1}^{2} \mathcal{F}^{\omega p}_{a\,q}(\overset{\overset{1}{\rightarrow}}{\overset{\circ}{d}}) \overset{\overset{2}{\vee}}{\partial}{}^q = \sum_{q=1}^{2} \overset{\overset{2}{\vee}}{\partial}{}^q \mathcal{F}^{\omega p}_{a\,q}(\overset{\overset{\rightarrow}{\circ}}{d}).$$

Constructing the operator $\overset{\circ}{\overline{A}}$, we also took into account the obvious relation

$$\sum_{p=1}^{2} \overset{\circ}{d}_p = \bar{z} \cdot \frac{\partial}{\partial \bar{z}}.$$

Statement (a) follows from Proposition 5.1(a) and Lemma 5.12(a); statement (b) follows from Proposition 5.1(b) and Lemma 5.12(b). □

§6. Quantum tensors and explicit asymptotic expansions for Weyl and Wick products

Having in mind a series of multidimensional quantum algebras and their quantum leaves described in previous sections, we would like to demonstrate some general procedures for calculating $\hbar$-expansions of star-products over general Poisson and Kähler manifolds.

We briefly describe some of results of [**Ka**$_{17}$] related to the deformation quantization of general nonlinear Poisson brackets. We start with an explicit construction of the Weyl star-product and then, in a similar way, consider the procedure for calculating the $\hbar$-expansion of the Wick product on Kähler manifolds (exploited in §3).

6.1. Weyl symmetrization and quantum tensors. Let $\mathcal{M}^n$ be a manifold with a formal $\star$-product given by a power series in $\hbar$. So, the space $C^\infty(\mathcal{M}, [\hbar])$ of formal power series in $\hbar$ with smooth coefficients is an associative algebra with unit 1 [**BFFLS**]. The product $\star$ is called *local* [**OMY**$_2$] if at each point the function $f \star g$ depends only on the germs of f and g at this point.

In the following, we assume, for simplicity, that $\mathcal{M}$ is an analytic manifold, all functions are from the class C^ω, and the $\star$-product is of C^ω-type.

The local $\star$-product on $\mathcal{M}$ is called the *Weyl product* if the following identities hold in some local coordinates $\xi = (\xi^1, \ldots, \xi^n)$ in each chart (the Weyl symmetrization rule):

$$\xi^\alpha = \langle \xi \rangle^\alpha, \qquad \forall \alpha = (\alpha_1, \ldots, \alpha_n) \in \mathbb{Z}_+^n.$$

Here, on the left we denote $\xi^\alpha = (\xi^1)^{\alpha_1} \ldots (\xi^n)^{\alpha_n}$ and on the right

(6.1) $$\langle\xi\rangle^\alpha \stackrel{\text{def}}{=} \frac{\alpha!}{|\alpha|!} \sum_{j \in S(\alpha)} \xi^{j_1} \star \cdots \star \xi^{j_{|\alpha|}},$$

where the sum is taken over the set $S(\alpha)$ of all mappings

$$j \colon (1, \ldots, |\alpha|) \to (1, \ldots, n)$$

that take each value k exactly at α_k points (or, in other words, α_k times).

The quantum manifold $\mathcal{M}$ with such a $\star$-product is called a *Weyl manifold* (compare with [**OMY**$_{1,2}$, **OMMY**] in the nondegenerate (that is, symplectic) case).

Now let us introduce the notation for the quantum brackets of functions as follows:

$$[f, g]_\hbar \stackrel{\text{def}}{=} \frac{i}{\hbar}(f \star g - g \star f).$$

In each local chart with Weyl local coordinates $\xi^1, \ldots, \xi^n$ we can define a set of functions

(6.2) $$K^{\ell s}(\xi) = [\xi^\ell, \xi^s]_\hbar, \qquad \ell, s = 1, \ldots, n,$$

and call this set a $\star$-*tensor* on $\mathcal{M}$. We stress that it is not a tensor in the usual sense.

Of course, the classical limit of the $\star$-tensor $P^{\ell s} = \lim_{\hbar \to 0} K^{\ell s}$ is a classical Poisson tensor on $\mathcal{M}$. In the expansion

(6.3) $$K^{\ell s} = P^{\ell s} + \hbar^2 P^{\ell s}_{(1)} + \hbar^4 P^{\ell s}_{(2)} + \ldots,$$

all higher order coefficients $P^{\ell s}_{(1)}, P^{\ell s}_{(2)}, \ldots$ are called [**KaM**$_4$] *quantum corrections* to the Poisson tensor. These corrections carry some information about the associativity of the $\star$-product. The question is: *which conditions for $K^{\ell s}$ are equivalent to the associativity of the $\star$-product?*

Let $a = (a^1, \ldots, a^n)$ be a vector-valued function in a local chart. For each polynomial f, we can define a Weyl-symmetrized function in the noncommuting variables $\xi^1 + \varepsilon a^1(\xi), \ldots, \xi^n + \varepsilon a^n(\xi)$ using (6.1). Denoting this function by $f\langle\xi + \varepsilon a(\xi)\rangle$ and expanding in ε, we obtain

(6.4) $$f\langle\xi + \varepsilon a(\xi)\rangle = f(\xi) + \varepsilon \hat{a}(f)(\xi) + O(\varepsilon^2).$$

By $\hat{a}(f)(\xi)$ we denote the coefficient at the first order term with respect to ε. Thus one obtains a linear operator $\hat{a}$ on the space of functions, which will be called a *quantum vector field*.

The functions a^j will be called *components* of the quantum vector field over a given local chart. The rule for transformation of components from one chart to another follows from (6.4).

Now let us derive quantum vector fields explicitly via the $\star$-product.

Denote by $\mathrm{ad}_\hbar = (\mathrm{ad}^1_\hbar, \ldots, \mathrm{ad}^n_\hbar)$ the vector-operator whose components act as the brackets $\mathrm{ad}^j_\hbar f \stackrel{\text{def}}{=} [\xi^j, f]_\hbar$. Also consider the vector-operator $d = (d_1, \ldots, d_n)$ with $d_j = d/d\xi^j$.

Let us introduce the "lunar" product of functions as follows (see [**Ka**$_{17}$]):

(6.5) $$g \odot f \stackrel{\text{def}}{=} g \, \mathbf{p}\bigl(-i\hbar \, \underset{\leftarrow}{\mathrm{ad}_\hbar} \star \underset{\rightarrow}{d}\bigr) f.$$

Here $\mathbf{p}(x) = (e^x - 1)/x$; the left and right arrows indicate on which factor, g or f, the given operator acts.

So, in (6.5) the operators $\text{ad}_\hbar$ act on g, the operators d act on f, and then the $\star$-product is evaluated. Expanding the functions $\mathbf{p}$ in (6.5) in power series, we obtain the following formal power series in $\hbar$ for the lunar product:

$$(6.5a) \qquad g \odot f = \sum_{m=0}^\infty \frac{(-i\hbar)^m}{m+1} \sum_{|\alpha|=m} \frac{1}{\alpha!} \langle \text{ ad}_\hbar \rangle^\alpha g \star d^\alpha f.$$

This product is not associative, but it is very useful.

Let us also introduce the following multi-component generalization of the lunar product:

$$(6.5b) \qquad ((g_1 \vee \cdots \vee g_m) \odot f)(\xi) \stackrel{\text{def}}{=} \langle g_1 \ldots g_m f(\xi^1, \ldots, \xi^n) \rangle.$$

Here on the right-hand side we take the symmetrization (6.1) of noncommuting elements $g_1(\xi) \ldots g_m(\xi), \xi^1, \ldots, \xi^n$ in the $\star$-product algebra.

THEOREM 6.1. (i) *The multi-component lunar product defined by (6.5b) in the case $m = 1$ coincides with the lunar product defined by (6.5).*

(ii) *Quantum vector fields can be expressed via the lunar product as follows:*

$$(6.6) \qquad \hat{a} = a \odot d.$$

(iii) *The quantum vector fields act on the lunar product as follows:*

$$(6.7) \qquad \hat{a}(g \odot f) = \hat{a}(g) \odot f + (g \vee a^s) \odot d_s f.$$

Here and everywhere in the following the usual rule of summation over twice repeated (up and down) indices is applied.

(iv) *The space of quantum vector fields is a Lie algebra with respect to the commutator*

$$(6.8) \qquad [\hat{a}, \hat{b}] = \widehat{[a,b]_\hbar}.$$

On the right the brackets $[\cdot, \cdot]_\hbar$ are defined by

$$(6.9) \qquad [a,b]_\hbar^s \stackrel{\text{def}}{=} \hat{a}(b^s) - \hat{b}(a^s), \qquad s = 1, \ldots, n.$$

For the proof of this theorem, see [**Ka**$_{17}$].

Now let us note that, over each quantum manifold, there are operators of special type, namely, inner derivations (for which the standard Leibniz rule holds) [**D-V, D-VM**].

Let us fix a function f. The *inner derivation* is given by the mapping $g \to [f, g]_\hbar$.

COROLLARY 6.1. *Inner derivations of the $\star$-product are quantum vector fields. Namely,*

$$(6.10) \qquad [f,g]_\hbar = \text{ad}_\hbar(f)\hat{\ }g = -\text{ad}_\hbar(g)\hat{\ }f,$$

where the components of the quantum field $\text{ad}_\hbar(f)$ are defined by

$$(6.11) \qquad \text{ad}_\hbar(f)^j \stackrel{\text{def}}{=} -\text{ad}_\hbar^j(f) = K^{si} \odot d_s f, \qquad j = 1, \ldots, n.$$

In particular,

(6.12) $$\mathrm{ad}_\hbar(\xi^j)\hat{\,} = \mathrm{ad}_\hbar^j, \qquad \mathrm{ad}_\hbar(\xi^j)^s = K^{js}.$$

Now denote by $V_\hbar^0(\mathcal{M})$ the space of quantum tensors of rank 0, that is, of functions. Denote by $V_\hbar^1(\mathcal{M})$ the space of quantum tensors of rank 1, that is, of quantum vector fields. The quantum (skew symmetric) tensor of rank r is an element of the exterior product of r copies $V_\hbar^r(\mathcal{M}) \stackrel{\mathrm{def}}{=} V_\hbar^1(\mathcal{M}) \wedge \cdots \wedge V_\hbar^1(\mathcal{M})$. Elements of $V_\hbar^r(\mathcal{M})$ can be represented in the form

$$A = \frac{1}{r!} A^{j_r \ldots j_1} \odot d_{j_1} \wedge \cdots \wedge d_{j_r},$$

where the set of functions $A^{j_1 \ldots j_r}$ is skew-symmetric in the multi-index j. Such tensors are identified with quantum r-vector fields as follows:

$$A(f_1, \ldots, f_r) \stackrel{\mathrm{def}}{=} \frac{1}{r!} \big((A^{j_1 \ldots j_1} \odot d_{j_1} f_1) \odot \ldots \big) \odot d_{j_r} f_r.$$

The quantum Schouten brackets

$$[\![\cdot, \cdot]\!]_\hbar \colon V_\hbar^r \times V_\hbar^m \to V_\hbar^{r+m-1}, \qquad r+m \geq 1,$$

can be defined by analogy with the commutative case [**Ksz**]:

$$[\![u_1 \wedge \cdots \wedge u_k, v_1 \wedge \cdots \wedge v_l]\!]_\hbar$$
$$= \sum_{i,j} (-1)^{i+j} [\![u_j, v_i]\!]_\hbar \wedge v_1 \wedge \cdots \wedge \breve{v}_i \wedge \cdots \wedge v_m \wedge \cdots \wedge u_1 \wedge \cdots \wedge \breve{u}_j \wedge \cdots \wedge u_r,$$

where u_j and v_i are quantum vector fields (elements from $V_\hbar^1(\mathcal{M})$).

The explicit formula for the *quantum Schouten brackets of two quantum tensors* A and B is

(6.13)
$$[\![A, B]\!]_\hbar^{j_{k+\ell-1} \ldots j_1} = \sum_{i \sim j} \Big(\frac{1}{\ell! (k-1)!} A^{i_{\ell+k-1} \ldots i_{\ell+1} s} \odot d_s B^{i_\ell \ldots i_1}$$
$$+ \frac{(-1)^{\ell k + \ell + k}}{(\ell-1)! k!} B^{i_{k+\ell-1} \ldots i_{k+1} s} \odot d_s A^{i_k \ldots i_1} \Big) (-1)^{\sigma(i)}.$$

Here $k = \mathrm{rank}\, A$, $\ell = \mathrm{rank}\, B$, the sum $\sum_{i \sim j}$ is taken over all permutations i of the multi-index $j = (j_{k+\ell-1}, \ldots, j_1)$, and $\sigma(i)$ denotes the sign (the parity) of the permutation i.

Note that quantum tensors over $\mathcal{M}$ form a graded Lie algebra (a superalgebra) with respect to brackets (6.13).

Now let us look at the definition (6.2). From the Jacobi identities for commutators, we obtain the following result [**Ka**$_{17}$].

THEOREM 6.2. *The $\star$-tensor K (6.2) is an element of $V_\hbar^2(\mathcal{M})$ such that $[\![K, K]\!]_\hbar = 0$, or in other words,*

(6.14) $$\underset{(j_1, j_2, j_3)}{\mathfrak{S}} K^{j_3 s} \odot d_s K^{j_2 j_1} = 0.$$

The operation $A \to [\![K, A]\!]_\hbar$ can be interpreted as the differential in the quantum tensor complex $\{V_\hbar^r(\mathcal{M}) \mid r = 0, 1, 2, \ldots\}$.

EXAMPLE 6.1. The quantum Euler–Poisson vector fields $\mathrm{ad}_\hbar(f)$ are coboundaries in this complex, since $\mathrm{ad}_\hbar(f) = -[\![K, f]\!]_\hbar$.

Theorem 6.2 represents the quantum analog of the Lichnerowicz description of Poisson manifolds via the classical Schouten brackets [**L**] (see also [**Ksm**, **Mck**]).

6.2. Quantum Jacobi identity and the Weyl product.
Now we derive the expansion for the quantum brackets $[\cdot, \cdot]_\hbar$ in power series in $\hbar$, and hence, represent the equality (6.14) as a system of equations for the quantum corrections (6.3).

There is the following key multiplication formula in the Weyl calculus [**Ka₁**]:

$$(6.15) \qquad T^s f\langle T\rangle = \sum_{m=0}^\infty \frac{b_m}{m!} \langle [T^{j_1}, \ldots [T^{j_m}, T^s]\ldots] d_{j_1} \ldots d_{j_m} f(T)\rangle.$$

Here $T = (T^1, \ldots, T^n)$ is a set of noncommuting elements; $s \in (1, \ldots, n)$; the b_m are Bernoulli numbers; the angular brackets denote the symmetrization (6.1). The infinite sum in (6.15), in fact, is finite if f is a polynomial.

Applying (6.15) to our Weyl $\star$-product, we obtain

$$(6.16) \quad \begin{aligned} \xi^s \star f(\xi) &= \sum_{m=0}^\infty \frac{(-i\hbar)^m b_m}{m!} \big[\xi^{j_1}, \ldots [\xi^{j_m}, \xi^s]_\hbar, \ldots\big]_\hbar \odot d_{j_1} \ldots d_{j_m} f(\xi) \\ &= \sum_{|\alpha|=0}^\infty \frac{(-i\hbar)^{|\alpha|} b_{|\alpha|}}{\alpha!} \langle \mathrm{ad}_\hbar\rangle^\alpha (\xi^s) \odot d^\alpha f(\xi) \\ &= \sum_{m=0}^\infty \frac{(-i\hbar)^m b_m}{m!} \xi^s \big(\underset{\leftarrow}{\mathrm{ad}_\hbar} \odot \underset{\rightarrow}{d}\big)^m f(\xi). \end{aligned}$$

So, we arrive at the following statement.

LEMMA 6.1. *The operators of the left regular representation $(L^s f)(\xi) \stackrel{\mathrm{def}}{=} \xi^s \star f(\xi)$ for the Weyl $\star$-product are given by the formula*

$$(6.17) \qquad L^s = \xi^s \mathbf{q}\big(-i\hbar\,\underset{\leftarrow}{\mathrm{ad}_\hbar} \odot \underset{\rightarrow}{d}\big),$$

where

$$(6.18) \qquad \mathbf{q}(x) \stackrel{\mathrm{def}}{=} \frac{x}{e^x - 1} = \sum_{m=0}^\infty \frac{b_m}{m!} x^m.$$

Formula (6.17) represents the star product $\star$ via the lunar product $\odot$. On the other hand, there is another formula (6.5), which also connects these two products. Combining (6.17) and (6.5), we find a procedure for evaluating the $\star$ and $\odot$ products separately as power series in $\hbar$.

First, note that, in view of (6.12), $\mathrm{ad}_\hbar^j = K^{js} \odot d_s$, and from (6.5a) we obtain

$$(6.19) \qquad \mathrm{ad}_\hbar^j = \sum_{m=0}^\infty \frac{(-i\hbar)^m}{m+1} \sum_{|\alpha|=m} \frac{1}{\alpha!} \langle \mathrm{ad}_\hbar\rangle^\alpha K^{js} \star d^\alpha d_s.$$

Recurrently, step by step, from (6.19) we derive the following expansion of $\mathrm{ad}_\hbar^j$ with respect to the deformation parameter $\hbar$:

$$\mathrm{ad}_\hbar^j = K^{js} \star d_s - \frac{i\hbar}{2} K^{\ell r} \star d_r K^{js} \star d_\ell d_s$$
(6.20)
$$- \hbar^2 \Big(\frac{1}{4} K^{qr} \star d_r K^{\ell s} \star d_q d_s K^{jp} \star d_\ell d_p$$
$$+ \frac{1}{6} K^{\ell q} \star d_q (K^{rp} \star d_p K^{js}) \star d_\ell d_r d_s \Big) + \dots .$$

Substituting this expansion into (6.5a), we express the lunar product $\odot$ via the star product $\star$ as follows:

(6.21)
$$f \odot = f \star - \frac{i\hbar}{2} K^{js} \star d_s f \star d_j$$
$$- \hbar^2 \Big(\frac{1}{4} K^{\ell r} \star d_r K^{js} \star d_\ell d_s f \star d_j + \frac{1}{6} K^{is} \star d_s (K^{\ell r} \star d_r f) \star d_i d_\ell \Big) + \dots .$$

On the other hand, formula (6.17) (or (6.16)) provides the following expansion of operators L^s via the lunar product:

(6.22)
$$L^s = \xi^s + \frac{i\hbar}{2} \xi^s \underset{\leftarrow}{\mathrm{ad}_\hbar} \odot \underset{\rightarrow}{d} - \frac{\hbar^2}{12} \xi^s \big(\underset{\leftarrow}{\mathrm{ad}_\hbar} \odot \underset{\rightarrow}{d} \big)^2 + \dots$$
$$= \xi^s - \frac{i\hbar}{2} K^{s\ell} \odot d_\ell + \frac{\hbar^2}{12} (K^{rp} \odot d_p K^{s\ell}) \odot d_\ell d_r + \dots .$$

Applying (6.21) to (6.22), we obtain the following formula for L^s via the star product:

(6.23) $$L^s = \xi^s - \frac{i\hbar}{2} K^{s\ell} \star d_\ell - \frac{\hbar^2}{6} K^{jr} \star d_r K^{s\ell} \star d_j d_\ell + \dots .$$

Note that the Weyl $\star$-product can be expressed by using the set of operators of the left regular representation $L = (L^1, \dots, L^n)$ as follows:

(6.24) $$f \star g = f \langle L \rangle g.$$

So, formulas (6.23), in fact, are recurrent equations for the set of operators L. By solving these equations, we obtain the final expansion of L^s in the power series in $\hbar$:

(6.25) $$L^s = \xi^s - \frac{i\hbar}{2} K^{s\ell}(\xi) d_\ell + \frac{\hbar^2}{12} K^{jr} d_r K^{s\ell} d_j d_\ell + \dots .$$

The substitution of (6.25) into the right-hand side of (6.24) leads to the final expansion of the $\star$-product:

(6.26) $$f \star g = fg - \frac{i\hbar}{2} K^{sj} d_s f d_j g$$
$$+ \hbar^2 \Big(\frac{1}{12} K^{\ell r} d_r K^{sj} (d_s f \cdot d_\ell d_j g - d_\ell d_s f \cdot d_j g) - \frac{1}{8} K^{\ell r} K^{sj} d_\ell d_s f \cdot d_r d_j g \Big) + \dots .$$

Next, the expansion for the lunar product can readily be obtained from (6.21) and (6.26) as follows:

(6.27) $$f \odot g = fg - \hbar^2 \Big(\frac{1}{24} K^{\ell r} K^{sj} d_\ell d_s f \cdot d_r d_j g + \frac{1}{12} K^{\ell r} d_r K^{sj} d_\ell d_s f \cdot d_j g \Big) + \dots .$$

The substitution of this expansion into the quantum Jacobi conditions (6.14) leads to closed (highly nonlinear) equations for the quantum tensor K. The first terms in the $\hbar$-expansion for these equations are the following:

$$(6.28) \quad \mathop{\mathfrak{S}}_{(j_1,j_2,j_3)} K^{j_3\ell} \cdot d_\ell K^{j_2 j_1} - \hbar^2 \mathop{\mathfrak{S}}_{(j_1,j_2,j_3)} \Big(\frac{1}{24} K^{pr} K^{sq} \cdot d_p d_s K^{j_3\ell} \cdot d_r d_q d_\ell K^{j_2 j_1} + \frac{1}{12} K^{pr} \cdot d_r K^{sq} \cdot d_p d_s K^{j_3\ell} \cdot d_q d_\ell K^{j_2 j_1} \Big) + \cdots = 0$$

(all odd powers of $\hbar$ are absent).

Thus, we have proved the following statement [**Ka**$_{17}$].

THEOREM 6.3. (i) *Formulas (6.5) and (6.17) determine an explicit expansion of the Weyl star product $\star$ in a power series in $\hbar$ via the quantum $\star$-tensor K. The first terms of this expansion are given by (6.26).*

(ii) *Formulas (6.5) and (6.17) determine the expansion of the lunar product $\odot$ as well. The first terms are given in (6.27) (all odd powers of $\hbar$ are absent in (6.27)).*

(iii) *Any skew-symmetric quantum tensor $K \in V_\hbar^2(\mathcal{M})$ satisfying equations (6.28) determines via (6.26) a unique associative $\star$-product, which makes $\mathcal{M}$ a quantum Weyl-manifold with $\star$-tensor K.*

REMARK 6.1. If we seek the solution of (6.28) in the form (6.3), then for the quantum corrections $P_{(1)}, P_{(2)}, \ldots$ to the Poisson tensor P we obtain a chain of linear equations of the following type:

$$[\![P, P_{(m)}]\!] = \Gamma_{(m)}, \qquad m = 1, 2, \ldots.$$

Here $[\![\cdot, \cdot]\!]$ are the classical Schouten brackets, and the right-hand sides $\Gamma_{(m)}$ for each fixed m are given explicitly via $P, P_{(1)}, \ldots, P_{(m-1)}$. For the first time, the formula for $\Gamma_{(1)}$ was derived in [**KaM**$_4$]. Now we have represented the explicit procedure for calculation of all $\Gamma_{(m)}$, $m \geq 1$.

6.3. Explicit power series for the Wick product and the reproducing measure. Now we make some remarks and clarify the geometrical meaning of the Wick–Klauder–Berezin procedure of quantization over Kähler manifolds. Here we use the formal power series approach as in [**BFFLS**$_{1,2}$].

Let F be a potential of the Kähler form over the local chart:

$$(6.29) \qquad \omega = i\overline{\partial}_j \partial_k F \, d\bar{z}^j \wedge dz^k \equiv i\omega_{jk} \, d\bar{z}^j \wedge dz^k.$$

We denote

$$(6.30) \qquad \mathcal{K}(\bar{z}, z) = \exp\{F(\bar{z}, z)/\hbar\}$$

and introduce the Hilbert space $\mathcal{P}$ of antiholomorphic distributions for which $\mathcal{K}$ is used as the integral kernel of the identity operator in $\mathcal{P}$:

$$(6.31) \qquad g(\bar{z}) = \big(\mathcal{K}(\cdot, z), g\big)_\mathcal{P}, \qquad \forall\, g \in \mathcal{P}.$$

Now on the space of functions $\varphi(\bar{z}, z)$ holomorphic in both $\bar{z}$ and z, we can define the natural "matrix" multiplication

$$(R \circ R')(\bar{z}, z) \stackrel{\text{def}}{=} \big(\overline{R(\bar{z}, \cdot)}, R'(\cdot, z)\big)_\mathcal{P}.$$

Here the scalar product in $\mathcal{P}$ is considered with respect to the arguments marked by dots. The unit element for this "matrix" multiplication is the function $\mathcal{K}$.

Now, one can introduce a new multiplication

$$(6.32) \qquad R*R' \stackrel{\text{def}}{=} \frac{1}{\mathcal{K}}\Big((\mathcal{K}R) \circ (\mathcal{K}R')\Big).$$

The unit for this multiplication is 1. The algebra of functions with this product will be denoted by $\mathcal{F}_\hbar$. The product (6.32) is exactly the same as that considered in §3 (see (3.60)).

The key properties of the product (6.32) are

$$(6.33) \qquad \begin{aligned} \overline{R*R'} &= \overline{R'}*\overline{R}, \\ R*R' - R'*R &= -i\hbar\{R, R'\} + O(\hbar^2), \\ R*R' &= RR' \qquad \text{if } R' \text{ is holomorphic}, \end{aligned}$$

where $\{\cdot, \cdot\}$ are the Poisson brackets generated by the symplectic form

$$(6.29\text{a}) \qquad \omega_0 \stackrel{\text{def}}{=} i\bar{\partial}\partial F^{(0)}\, d\bar{z} \wedge dz, \qquad \omega_0 = \omega\Big|_{\hbar=0}.$$

The first and the last properties in (6.33) can readily be verified by using the definition (6.32). But the second one is not simple. We present a proof of the second property using and developing the Fock–Dirac idea of creation-annihilation operators. This provides us both with (6.33) and with explicit formulas for all terms of power $\hbar$-expansions of the product $*$ in terms of the Kähler form (6.29) exclusively.

In the Hilbert space $\mathcal{P}$, let us consider the operators of multiplication by the coordinate functions $\bar{z}^1, \ldots, \bar{z}^d$ and introduce the adjoint operators

$$(6.34) \qquad \widehat{z}^s \stackrel{\text{def}}{=} (\bar{z}^s)^*, \qquad s = 1, \ldots, d.$$

Since $\mathcal{K}(\bar{z}, z)$ is the integral kernel of the identity operator (6.31), for each fixed z, $\mathcal{K}(\cdot, z)$ is an eigenfunction for all operators $\widehat{z}^s$, that is,

$$(6.35) \qquad \begin{aligned} \widehat{z}^s \pi(z) &= z^s \cdot \pi(z), \qquad s = 1, \ldots, d, \\ \pi(z) &\stackrel{\text{def}}{=} \mathcal{K}(\cdot, z). \end{aligned}$$

In particular,

$$(6.35\text{a}) \qquad \widehat{z}^s 1 = 0, \qquad s = 1, \ldots, d.$$

In view of (6.34) and (6.35a), the operators $\widehat{z}^s$ could be considered as *annihilation operators* in $\mathcal{P}$ with the vacuum element $1 \in \mathcal{P}$.

To each function $R = R(\bar{z}, z) \in \mathcal{F}_\hbar$, we assign an operator $\widehat{R}$ in the Hilbert space $\mathcal{P}$ by the formula

$$(6.36) \qquad \widehat{R} \stackrel{\text{def}}{=} R(\overset{2}{\bar{z}}, \overset{1}{\widehat{z}}).$$

In view of (6.35), this formula explicitly reads

$$(6.36\text{a}) \qquad (\widehat{R}g)(\bar{z}) = \Big(\overline{R(\bar{z}, \cdot)}\pi(z), g\Big)_{\mathcal{P}}, \qquad g \in \mathcal{P}.$$

LEMMA 6.2. *The following equalities hold:*

$$\widehat{R}_1 \widehat{R}_2 = \widehat{R_1 * R_2}, \tag{6.37}$$

$$R* = \frac{1}{\mathcal{K}} \circ \widehat{R} \circ \mathcal{K}. \tag{6.38}$$

In particular, if $R = \overline{\partial} F$, then (6.36a) implies

$$\widehat{\overline{\partial} F} = \hbar \overline{\partial}, \tag{6.39}$$

where $\overline{\partial} \equiv \partial/\partial \overline{z}$.

An explicit form of (6.39) is

$$\overline{\partial}_s F(\overset{2}{\overline{z}}, \overset{1}{\widehat{z}}) = \hbar \overline{\partial}_s, \qquad s = 1, \ldots, d. \tag{6.39a}$$

Note that *from this equation one can derive the annihilation operator $\widehat{z}$ explicitly as a function in $\overline{z}$ and $\overline{\partial}$* as long as the matrix

$$\omega_{\ell s} \overset{\text{def}}{=} \partial_\ell \overline{\partial}_s F \tag{6.40}$$

is not degenerate.

Let us now derive an explicit formula for the quantum $*$-tensor from (6.39a).

Following the general definition (6.2), in order to determine the quantum $*$-tensor in our case, we have to express the commutators $[\widehat{z}^r, \overline{z}^\ell]$ via functions of type (6.36), that is,

$$[\widehat{z}^r, \overline{z}^\ell] = \hbar \psi^{\ell r}(\overset{2}{\overline{z}}, \overset{1}{\widehat{z}}). \tag{6.41}$$

The matrix $\psi = ((\psi^{\ell r}))$ represents the nonzero part of the quantum $*$-tensor.

At the same time we introduce analogs of the operators $\mathrm{ad}_\hbar^j$ from §6.1; let us denote

$$[\widehat{z}^r, \widehat{R}] \overset{\text{def}}{=} \hbar \widehat{\overline{D}^r R}, \qquad [\widehat{R}, \overline{z}^\ell] \overset{\text{def}}{=} \hbar \widehat{D^\ell R}. \tag{6.42}$$

The operators $\overline{D}^r$ and D^ℓ are Euler–Poisson quantum vector fields with respect to the product $*$. They can be derived via the quantum $*$-tensor ψ using the general commutation formulas (see, for example, [**KaM**$_4$], Appendix I). Say, the formula for D^ℓ is the following:

$$\widehat{D^\ell R} = \int_0^1 \overset{2}{\widehat{\psi^{\ell r}}} \partial_r R\left(\overset{4}{\overline{z}}, \tau \overset{3}{\widehat{z}} + (1-\tau)\overset{2}{\widehat{z}}\right) d\tau$$

$$= \sum_{|\alpha| \geq 0} \frac{1}{(|\alpha|+1)\alpha!} [\underbrace{\overset{2}{\widehat{z}}, \ldots, [\overset{2}{\widehat{z}}, \widehat{\psi^{\ell r}}]}_{\alpha} \ldots] \partial^\alpha \partial_r R(\overset{3}{\overline{z}}, \overset{1}{\widehat{z}}).$$

The multi-commutator case can be determined from the definition (6.42) as follows:

$$[\underbrace{\widehat{z}, \ldots, [\widehat{z}, \widehat{\psi^{\ell r}}]}_{\alpha} \ldots] = \hbar^{|\alpha|} \widehat{\overline{D}^\alpha \psi^{\ell r}}.$$

Thus we obtain

$$D^\ell = \sum_{|\alpha| \geq 0} \frac{\hbar^{|\alpha|}}{(|\alpha|+1)\alpha!} \overline{D}^\alpha(\psi^{\ell r}) \partial^\alpha \partial_r = \psi^{\ell r} \mathbf{p}(\hbar \overset{\leftarrow}{\overline{D}} \cdot \overset{\rightarrow}{\partial}) \partial_r, \tag{6.43}$$

where $\mathbf{p}(x) = (e^x - 1)/x$ (compare with (6.5), (6.11), and (6.35)).

For the conjugate operators $\overline{D}^\ell$, an analog of (6.43) looks as follows:

$$\overline{D}^\ell = \sum_{|\alpha|\geq 0} \frac{\hbar^{|\alpha|}}{(|\alpha|+1)\alpha!} D^\alpha(\psi^{r\ell}) \overline{\partial}^\alpha \overline{\partial}_r = \psi^{r\ell} \mathbf{p}(\hbar \overleftarrow{D} \cdot \overrightarrow{\overline{\partial}}) \overline{\partial}_r. \tag{6.44}$$

From the pair of equations (6.43), (6.44) we readily calculate the $\hbar$-expansion for both D and $\overline{D}$. The first three terms are the following:

$$D^\ell = \psi^{\ell s}\partial_s + \frac{\hbar}{2}(\psi^{pq}\overline{\partial}_p\psi^{\ell s})\partial_q\partial_s \tag{6.45}$$
$$+ \hbar^2\Big(\frac{1}{4}(\psi^{rm}\partial_m\psi^{pq}\overline{\partial}_r\overline{\partial}_p\psi^{\ell s})\partial_q\partial_s + \frac{1}{6}(\psi^{rm}\overline{\partial}_r(\psi^{pq}\overline{\partial}_p\psi^{\ell s}))\partial_m\partial_q\partial_s\Big) + \ldots.$$

The formula for $\overline{D}^\ell$ differs from (6.45) just by the changes $\partial \to \overline{\partial}$ and $\psi^{ij} \to \psi^{ji}$.

Now let us consider the operators of the left and right regular representation

$$\widehat{z}\widehat{R} \stackrel{\text{def}}{=} \widehat{LR}.$$

Then $L^r = z^r + \hbar\overline{D}^r$, $r = 1,\ldots,d$.

Since $[L^r, z^s] = 0$, we have $[\overline{D}^r, z^s] = 0$ for all s, r. Thus, the product $*$, which is the product of normal symbols of operators (6.36), can be derived as follows:

$$R * R' = R(\overset{2}{\overline{z}}, \overset{1}{L})R' = R(\overset{2}{\overline{z}}, z + \hbar\overline{D})R' \tag{6.46}$$
$$= \sum_{|\alpha|\geq 0} \frac{\hbar^{|\alpha|}}{\alpha!} \partial^\alpha R \cdot \overline{D}^\alpha R' = R \exp\{\hbar\overleftarrow{\partial} \cdot \overrightarrow{\overline{D}}\} R',$$

or, simultaneously, as follows:

$$R * R' = R'(\overline{z} + \hbar D, \overset{2}{z})R \tag{6.46a}$$
$$= \sum_{|\alpha|\geq 0} \frac{\hbar^{|\alpha|}}{\alpha!} \overline{\partial}^\alpha R' \cdot D^\alpha R = R \exp\{\hbar\overleftarrow{D} \cdot \overrightarrow{\overline{\partial}}\} R'.$$

The leading terms of the $\hbar$-expansion for the $*$-product are

$$R * R' = R \cdot R' + \hbar \psi^{rs}\partial_s R \cdot \overline{\partial}_r R' \tag{6.46b}$$
$$+ \frac{\hbar^2}{2}\Big(\psi^{pq}\overline{\partial}_p\psi^{rs}\partial_q\partial_s R \cdot \overline{\partial}_r R' + \psi^{rs}\partial_s(\psi^{pq}\partial_q R) \cdot \overline{\partial}_p\overline{\partial}_r R'\Big) + \ldots.$$

If the quantum tensor ψ in (6.41) is given *a priori*, then formulas for the star product (6.46), (6.46a,b) are final. But, on the other hand, the quantum tensor ψ itself could be determined, say, via the form (6.29).

To this end, let us calculate the commutator with $\overline{z}^\ell$ on both sides of (6.39a):

$$D^\ell(\overline{\partial}_s F) = \delta_s^\ell, \qquad s, \ell = 1, \ldots, d.$$

On the right, δ_s^ℓ denotes the Kronecker symbol. After the substitution of (6.43) and (6.40), we obtain

$$\psi^{\ell r} \mathbf{p}(\hbar\overleftarrow{\overline{D}} \cdot \overrightarrow{\partial})\omega_{rs} = \delta_s^\ell \tag{6.47}$$

or, if we use the notation of §6.1,

(6.47a) $$\psi^{\ell r} \odot \omega_{rs} = \delta^{\ell}_{s}.$$

From this relation we can easily derive the quantum tensor ψ via the Kähler form ω. Let us denote the elements of the inverse matrix $((\omega_{\ell s}))^{-1}$ by $g^{\ell s} = (\omega^{-1})^{\ell s}$. By (6.47), the leading terms of the $\hbar$-expansion for the quantum tensor ψ are the following:

(6.48)
$$\psi^{\ell r} = g^{\ell r} + \frac{\hbar}{2}\partial_m g^{pr}\overline{\partial}_p g^{\ell m} - \hbar^2 \Big[\frac{1}{4}g^{sr}g^{jk}(\partial_k g^{pi})(\overline{\partial}_p\overline{\partial}_j g^{\ell m})\partial_i \omega_{ms}$$
$$+ \frac{1}{6}g^{sr}g^{jk}\overline{\partial}_j(g^{pq}\overline{\partial}_p g^{\ell m})\partial_q\partial_k \omega_{ms} + \frac{1}{4}g^{sr}(\partial_q g^{ji})(\overline{\partial}_j g^{pq})(\overline{\partial}_p g^{\ell m})\partial_i \omega_{ms}$$
$$- \frac{1}{4}(\partial_i g^{pr})\overline{\partial}_p\big((\partial_m g^{ji})(\overline{\partial}_j g^{\ell m})\big)\Big] + O(\hbar^3).$$

After the substitution of this expansion into (6.46b) we obtain an explicit $\hbar$-expansion for the $*$-product:

(6.49)
$$R * R' = R \cdot R' + \hbar g^{rs}\partial_s R \cdot \overline{\partial}_r R'$$
$$+ \frac{\hbar^2}{2}\Big[(g^{pq}\overline{\partial}_p g^{rs})\partial_q\partial_s R \cdot \overline{\partial}_r R' + g^{rs}\partial_s(g^{pq}\partial_q R) \cdot \overline{\partial}_p\overline{\partial}_r R'$$
$$+ (\partial_m g^{ps})(\overline{\partial}_p g^{rm})\partial_s R \cdot \overline{\partial}_r R'\Big] + O(\hbar^3).$$

As a simple consequence of this expansion we prove the second relation in (6.33) with the Poisson brackets given by

(6.50) $$\{R, R'\} = ig_0^{rs}\big(\partial_s R\overline{\partial}_r R' - \partial_s R'\overline{\partial}_r R\big),$$

where $g_0 = g|_{\hbar=0}$ is the metric corresponding to the classical symplectic form (6.29a).

The operators D^{ℓ} can also be represented via the metric from expansions (6.45) and (6.48):

(6.51) $$D^{\ell} = g^{\ell s}\partial_s + \frac{\hbar}{2}\Big[g^{pq}(\overline{\partial}_p g^{\ell s})\partial_q\partial_s + (\partial_m g^{ps})(\overline{\partial}_p g^{\ell m})\partial_s\Big] + O(\hbar^2).$$

Note that although all above calculations were performed in a local chart, the resulting expansions (6.49), (6.50) are invariant with respect to a change of local coordinates $\overline{z}$, z. Thus, we obtain the following statement [**Ka**$_{17}$].

THEOREM 6.4. *Any (pseudo-) Kähler 2-form ω over a manifold Ω generates the associative star product on Ω with properties (6.33), (6.32) (the function $\mathcal{K}$ in (6.32) is given by (6.30), and F is the local potential of the form ω (6.29)). This product can be calculated by formula (6.46) or (6.46a), where the operators D, $\overline{D}$ are derived from (6.43), (6.44) (and the quantum tensor ψ is taken from (6.47a)). Several terms of $\hbar$-expansions for the star product, for D, and for ψ are given in (6.49), (6.51) and (6.48). All terms of (6.49) are differential invariants on Ω.*

Thus we have obtained remarkable differential invariants of orders $2, 3, \ldots$ on an arbitrary (pseudo-) Kähler manifold. They are just coefficients at powers of $\hbar$

in the expansion

$$R(z) * R'(\overline{z}) = R(z)R'(\overline{z}) + \sum_{m=1}^{\infty} \frac{\hbar^m}{m!} I_m(R(z)R'(\overline{z})).$$

As it follows from (6.49), the first invariant I_1 is the half of the Laplace–Beltrami operator:

$$I_1 = g^{rs}\partial_s\overline{\partial}_r = \frac{1}{2}\Delta.$$

The next invariant I_2 one can see in (6.49) as the coefficient at $\hbar^2$:

$$I_2 = \frac{1}{4}\Delta^2 + \left[(\overline{\partial}_r g^{ps})(\partial_s g^{rq}) - g^{rs}\partial_s\overline{\partial}_r g^{pq}\right]\partial_q\overline{\partial}_p.$$

Here the fourth order summand is just the square of the Laplace–Beltrami operator, but we also obtain an additional second order differential invariant determined by the following tensor of $(1,1)$-type:

(6.52) $$\widetilde{g}^{pq} = (\overline{\partial}_r g^{ps})(\partial_s g^{rq}) - g^{rs}\partial_s\overline{\partial}_r g^{pq}.$$

There are many other invariants in the next terms of (6.49). The above procedure guarantees explicit formulas for all invariants I_m (see also [**Ka**$_6$]; in particular, Remarks 1.1 and 1.2 therein).

REMARK 6.2. The anti-Wick quantization $r \to \underset{\wedge}{r}$ is defined by (3.53) as follows:

(6.53) $$\underset{\wedge}{r} = \frac{1}{(2\pi\hbar)^d} \int_\Omega r(\overline{z},z)\mathbf{\Pi}(\overline{z},z)\,dm(\overline{z},z).$$

Here $\mathbf{\Pi}(\overline{z},z)$ is the orthogonal projection in $\mathcal{P}$ onto the linear subspace generated by the function $\pi(z)$ (6.35), and dm is the *reproducing measure* taking part in the resolutions of the identity (see (3.21a)):
(6.53a)
$$\frac{1}{(2\pi\hbar)^d}\int_\Omega \mathbf{\Pi}(\overline{z},z)\,dm(\overline{z},z) = I, \quad \frac{1}{(2\pi\hbar)^d}\int_\Omega p(\overline{w},w;\overline{z},z)\,dm(\overline{z},z) = 1, \quad \forall \overline{w},w,$$

where the probability function p is given by (3.33), and $2d = \dim\Omega$.

Note that in terms of the differential invariants I_m obtained above, the transformation from anti-Wick to Wick symbols is given by the formula

(6.54) $$\underset{\wedge}{r} = \widehat{R}, \qquad R = \left(I + \sum_{m=1}^{\infty}\frac{\hbar^m}{m!}I_m\right)r.$$

In the special case of symplectic leaves in semisimple Lie algebras, this transformation was studied in [**B**$_3$], and the invariants I_m were calculated in terms of Casimir elements of the Lie algebra.

Thus we have derived the probability operator (3.58) in terms of the invariants I_m or the operators D and $\overline{D}$ as follows:

(6.55) $$\mathbb{P} = I + \sum_{m=1}^{\infty}\frac{\hbar^m}{m!}I_m = \exp\{\hbar\overset{2}{D}\cdot\overset{1}{\overline{\partial}}\} = \exp\{\hbar\overset{2}{\overline{D}}\cdot\overset{1}{\partial}\}.$$

Formula (6.55) implies that the probability operator $\mathbb{P}$ is formally invertible, and the inverse operator can be readily calculated as the $\hbar$-series

$$(6.55a) \quad \mathbb{P}^{-1} = \left(\exp\{\hbar\overset{2}{\overline{D}} \cdot \overset{1}{\partial}\}\right)^{-1} = I - \frac{\hbar}{2}\Delta - \frac{\hbar^2}{2}\left(\widetilde{g}^{pq}\overline{\partial}_p\partial_q - \frac{1}{4}\Delta^2\right) + O(\hbar^3).$$

Here the tensor $\widetilde{g}$ is given by (6.52). Now we denote

$$(6.56) \quad \mathbb{Q} \overset{\text{def}}{=} \frac{1}{\hbar^2}\left(I - \frac{\hbar}{2}\Delta - \mathbb{P}^{-1}\right) = \widetilde{g}^{pq}\overline{\partial}_p\partial_q - \frac{1}{4}\Delta^2 + O(\hbar).$$

We also introduce operators $\mathbb{Q}^{pq}$ by using the definition

$$(6.56a) \quad \mathbb{Q} \overset{\text{def}}{=} \mathbb{Q}^{pq}\overline{\partial}_p\partial_q$$

and expand these operators into the $\hbar$-power series

$$(6.57) \quad \mathbb{Q}^{pq} = \mathbb{Q}^{pq}_0 + \hbar\mathbb{Q}^{pq}_1 + \hbar^2\mathbb{Q}^{pq}_2 + \ldots.$$

Here the leading term is the second order differential operator

$$(6.58) \quad \mathbb{Q}^{pq}_0 = (\overline{\partial}_r g^{ps})(\partial_s g^{rq}) - \Delta(g^{pq}) - g^{rs}(\partial_s g^{pq})\overline{\partial}_r - g^{rs}(\overline{\partial}_r g^{pq})\partial_s - \frac{1}{2}g^{pq}\Delta.$$

All other terms in (6.57) can also be calculated explicitly via the quantum metric g by using formulas (6.56), (6.55a), and (6.51).

On Ω we introduce the Liouville measure generated by the form ω (6.29), namely,

$$dm^\omega \overset{\text{def}}{=} \det\omega \cdot d\bar{z}dz.$$

Now we can express the reproducing measure dm via the Liouville measure as follows:

$$(6.59) \quad dm = e^{\hbar f}\, dm^\omega.$$

Let us show how to calculate the "reproducing function" f. Here we follow [**Ka**$_{17}$].

THEOREM 6.5. *The reproducing function f in (6.59) has to be a solution of the differential equation*

$$(6.60) \quad df = \sum_{p,q=1}^{d} \mathbb{Q}^{pq}\, d\omega_{pq}.$$

Here $\omega_{r\ell}$ are coefficients of the Kähler form and $\mathbb{Q}^{pq}$ are (pseudo)differential operators defined by (6.56).

The operators in the expansion (6.57) determine closed 1-forms over the Kähler manifold Ω:

$$\theta_m \overset{\text{def}}{=} \sum_{p,q=1}^{d} \mathbb{Q}^{pq}_m\, d\omega_{pq}, \qquad m = 0, 1, 2, \ldots,$$

so that equation (6.60) reads

$$(6.60a) \quad df = \sum_{m=0}^{\infty} \hbar^m \theta_m.$$

If Ω is simply connected, then (6.60a) is globally solvable. In general, the cohomology classes $[\theta_m] \in H^1(\Omega, \mathbb{R})$ are the obstructions to the existence of the reproducing measure on Ω (smoothly depending on $\hbar$). The leading class $[\theta_0]$ is given by the 1-form on Ω:

$$\theta_0 = \sum_{p,q} \mathbb{Q}_0^{pq} \, d\omega_{pq},$$

where the second order differential operators $\mathbb{Q}_0^{pq}$ are determined by (6.58).

PROOF. The anti-Wick operators of type (6.53) act on functions $g \in \mathcal{P}$ by the formula

$$(6.61) \qquad (\underset{\wedge}{r}g)(\overline{w}) = \frac{1}{(2\pi\hbar)^d} \int_\Omega e^{(F(\overline{w},z)-F(\overline{z},z))/\hbar} \, r(\overline{z},z) g(\overline{z}) \, dm(\overline{z},z).$$

Let us denote by M the density of the measure dm, that is, $dm = M(\overline{z},z) \, d\overline{z}dz$. Then from (6.61) we obtain

$$(\underline{\overline{\partial}_s \ln M})(g) = \frac{1}{(2\pi\hbar)^d} \int_\Omega e^{(F(\overline{w},z)-F(\overline{z},z))/\hbar} g(\overline{z}) \, \overline{\partial}_s M \, d\overline{z}dz$$

$$= \frac{1}{(2\pi\hbar)^d} \int_\Omega e^{(F(\overline{w},z)-F(\overline{z},z))/\hbar} \Big(-\overline{\partial}_s g(\overline{z}) + \frac{1}{\hbar} \overline{\partial}_s F(\overline{z},z) \Big) M \, d\overline{z}dz$$

$$= \Big(-\overline{\partial}_s + \frac{1}{\hbar} \underline{\overline{\partial}_s F} \Big) g(\overline{w}).$$

Thus, using (6.36) and (6.54), we find that

$$\mathbb{P}\big(\overline{\partial}_s(\ln M)\big) = \frac{1}{\hbar}\big(\mathbb{P}(\overline{\partial}_s F) - \overline{\partial}_s F\big), \qquad s = 1, \ldots, d.$$

Applying the formal inverse operator $\mathbb{P}^{-1}$ and taking into account the notation (6.56), (6.56a), we derive

$$\overline{\partial}_s f = \mathbb{Q}(\overline{\partial}_s F) \quad \text{or} \quad \overline{\partial}_s f = \sum_{r,\ell} \mathbb{Q}^{r\ell} \overline{\partial}_s \omega_{r\ell}.$$

Since $\mathbb{Q}$ is a real operator, we finally obtain equation (6.60). □

REMARK 6.3. We stress that all formal expansions (6.48), (6.49), (6.55), (6.57), (6.60a) are not purely $\hbar$-power series; their coefficients implicitly depend on $\hbar$, since the quantum metric g and the quantum symplectic form ω in general can depend on $\hbar$ in a very complicated way. For instance, if the Kähler manifold Ω is compact or just possesses a nontrivial 2-cycle, then the form ω has to satisfy the quantization condition $\frac{1}{2\pi\hbar}[\omega] \in H^2(\Omega, \mathbb{Z})$, which explicitly includes $\hbar$.

REMARK 6.4. Equation (6.60a) determines the reproducing function f up to an arbitrary additive constant only. This constant can readily be obtained from the "normalization" condition (6.53a). Note that in the compact case the first relation (6.53a) implies a formula for the dimension of the quantum function space $\mathcal{P}$:

$$\frac{1}{(2\pi\hbar)^d} \int_\Omega e^{\hbar f} \, dm^\omega = \dim \mathcal{P}.$$

Part II. Examples of Quantum Embeddings and Irreducible Representations

§1. Simplest Lie algebras

In this section we show how the general formulas obtained in Part I work in simple examples of Lie algebras. Examples of non-Lie algebras will be considered in the next sections.

1.1. The Lie algebra su(2). Let us consider the relations

(1.1) $\qquad [\mathbf{S}_1, \mathbf{S}_2] = i\hbar \mathbf{S}_3, \qquad [\mathbf{S}_2, \mathbf{S}_3] = i\hbar \mathbf{S}_1, \qquad [\mathbf{S}_3, \mathbf{S}_1] = i\hbar \mathbf{S}_2,$

where $\hbar > 0$ and $\mathbf{S}_j^* = \mathbf{S}_j$.

We define

$$\mathbf{B} = \mathbf{S}_1 - i\mathbf{S}_2, \qquad \mathbf{C} = \mathbf{S}_1 + i\mathbf{S}_2, \qquad \mathbf{A} = \mathbf{S}_3.$$

Then we obtain

(1.1a) $\qquad [\mathbf{C}, \mathbf{B}] = 2\hbar \mathbf{A}, \qquad \mathbf{AB} = \mathbf{B}(\mathbf{A} - \hbar), \qquad \mathbf{CA} = (\mathbf{A} - \hbar)\mathbf{C}.$

So, in this example, as compared with the general notation in § I.1, we have

$$k = 1, \qquad f(A) = 2\hbar A, \qquad \varphi(A) = A - \hbar,$$

and so, by (I.1.3) and (I.1.4),

$$\mathcal{F}_a(n) = \hbar^2(2a/\hbar - n), \qquad \mathcal{A}_a(n) = a - n\hbar.$$

The function $\mathcal{F}_a$ can take nonpositive values on $\mathbb{Z}_+$ only if $a \geq 0$. In this case, conditions (I.1.6) and (I.1.6a) allow only integer values of $2a/\hbar$. So, we obtain the following quantization rule:

(1.2) $\qquad \mathcal{R} = \{a = N\hbar/2 \mid N \in \mathbb{Z}_+\}.$

Under the assumption that $a \in \mathcal{R}$, we have $\mathcal{F}_a(n) > 0$ for $0 \leq n < N - 1$ and $\mathcal{F}_a(N) = 0$; thus, the case (I.1.5) is realized here.

In this example, the factorization (I.1.6) looks like

(1.3) $\qquad \mathcal{F}_a = \mathcal{B}_a \mathcal{C}_a, \qquad \mathcal{B}_a(n) = 2a - n\hbar, \qquad \mathcal{C}_a(n) = \hbar.$

All other possible factorizations either differ from (1.3) only by constant factors in $\mathcal{B}_a$ and $\mathcal{C}_a$, or generate pseudodifferential (not differential) operators of antiholomorphic representation. For (1.3), one obtains the following irreducible representation of the algebra (1.1a):

(1.4) $\qquad \overset{\circ}{B} = \hbar \bar{z}\left(N - \bar{z}\frac{d}{d\bar{z}}\right), \qquad \overset{\circ}{C} = \hbar \frac{d}{d\bar{z}}, \qquad \overset{\circ}{A} = \hbar\left(\frac{N}{2} - \bar{z}\frac{d}{d\bar{z}}\right),$

whose Hilbert space $\mathcal{P}$ consists of polynomials $g(\bar{z}) = \sum_{n=0}^{N} g_n \bar{z}^n$ of degree less than or equal to N and is endowed with the norm

$$\|g\|_{\mathcal{P}}^2 = \sum_{n=0}^{N} \frac{n!\,(N-n)!}{N!}|g_n|^2.$$

The coherent states (I.1.12) corresponding to the representation (1.4) are

$$\mathfrak{P}_z = \sum_{n=0}^{N} \frac{1}{n!} \left(\frac{z\mathbf{B}}{\hbar}\right)^n \mathfrak{P}_0 = e^{z\mathbf{B}/\hbar} \mathfrak{P}_0, \tag{1.5}$$

where $\mathfrak{P}_0$ is the vacuum vector in an abstract Hilbert space:

$$\mathbf{A}\mathfrak{P}_0 = \frac{N\hbar}{2}\mathfrak{P}_0, \qquad \mathbf{C}\mathfrak{P}_0 = 0, \qquad \|\mathfrak{P}_0\| = 1.$$

The reproducing kernel (I.1.13) generated by the states (1.5) is

$$\mathcal{K}(\bar{z}, z) = \sum_{n=0}^{N} \frac{N!}{n!(N-n)!} (\bar{z}z)^n = (1+\bar{z}z)^N.$$

The representation (1.4), as well as the coherent states and the reproducing kernel, is well known [$\mathbf{P}_{1,2}$].

In our example, the antimeromorphic coherent states (I.3.16) and the corresponding meromorphic "reproducing kernel" (I.3.15) take the form

$$\widetilde{\mathfrak{P}}_{\bar{z}} = \frac{1}{N!} \sum_{n=0}^{N} \frac{(N-n)!}{\bar{z}^{n+1}} \left(\frac{\mathbf{B}}{\hbar}\right)^n \mathfrak{P}_0, \qquad \widetilde{L}(\bar{z}, z) = \sum_{n=0}^{N} \frac{n!(N-n)!}{N!(\bar{z}z)^{n+1}}.$$

The corresponding irreducible representation of the algebra (1.1a) dual to the representation (1.4) is derived by (I.3.17):

$$\overset{\circ}{A} = \hbar\left(\frac{N+2}{2} + z\frac{d}{dz}\right), \qquad \overset{\circ}{B} = -\hbar\frac{d}{dz}, \qquad \overset{\circ}{C} = \hbar z\left(N+2+z\frac{d}{dz}\right). \tag{1.6}$$

The Casimir element for the algebra (1.1), (1.1a) is

$$\mathbf{K} = \mathbf{BC} + \mathbf{A}(\mathbf{A}+\hbar) = \mathbf{S}_1^2 + \mathbf{S}_2^2 + \mathbf{S}_3^2,$$

and the function ρ from (I.1.28) in this example takes the form $\rho(A) = -A(A+\hbar)$.

In the irreducible representation (1.4) or (1.6), the Casimir element is, of course, scalar:

$$\overset{\circ}{B}\overset{\circ}{C} + \overset{\circ}{A}(\overset{\circ}{A}+\hbar) = \underset{\circ\circ}{BC} + \underset{\circ\circ}{A(A+\hbar)} = \hbar^2 \frac{N}{2}\left(\frac{N}{2}+1\right) \equiv -\rho(a).$$

From (I.3.5b) and (I.3.6) we obtain

$$\mathcal{E}_a(A) = a - A, \qquad \mathcal{D}_a(A) = a + A + \hbar,$$
$$X_a(A) = \frac{a-A}{a+A+\hbar}, \qquad \mathcal{H}(t) = \frac{t}{\hbar(N+1) - t}.$$

In this case, equations (I.3.23), (I.3.24) for k and ℓ look as follows:

$$(1+x)\frac{dk}{dx} = Nk(x), \qquad k(0) = 1 \qquad \Longrightarrow \qquad k(x) = (1+x)^N,$$
$$(1+x)\frac{d\ell}{dx} + (N+2)\ell(x) = 0, \qquad \frac{1}{\hbar}\int_0^\infty \ell(x)\,dx = 1 \qquad \Longrightarrow \qquad \ell(x) = \frac{\hbar(N+1)}{(1+x)^{N+2}}.$$

The density of the measure (I.3.34) and the quantum Kähler potential (I.3.35) are the following:

$$M(x) = \frac{\hbar(N+1)}{(1+x)^2}, \qquad F(x) = \hbar N \ln(1+x) = 2a \ln(1+x).$$

The quantum form (I.3.36) is

$$\omega = \frac{i\hbar N \, d\bar{z} \wedge dz}{(1+|z|^2)^2} = \frac{2i \, d\bar{z} \wedge dz}{(1+|z|^2)^2} \cdot a. \tag{1.7}$$

At the same time, the classical Kähler potential (I.3.40), the classical symplectic form (I.3.41), the quantum correcting metric g, and the form λ (I.3.43), (I.3.44) look as follows:

$$F_0(x) = \hbar(N+1) \ln(1+x), \qquad \omega_0 = \frac{i\hbar (N+1) \, d\bar{z} \wedge dz}{(1+|z|^2)^2},$$

$$g(x) = \frac{1}{(1+x)^2}, \qquad \lambda = -\frac{i\bar{z} \wedge dz}{(1+|z|^2)^2}. \tag{1.8}$$

In this example, the general asymptotic formulas (I.3.45)–(I.3.47) are, in fact, exact (already in the first term).

All differential forms in (1.7), (1.8) are globally determined over symplectic leaves (I.3.1a)

$$\Omega_a = \left\{ S_1^2 + S_2^2 + S_3^2 + \hbar S_3 = \frac{\hbar^2 N(N+2)}{4} \right\}, \qquad a = N\frac{\hbar}{2} \in \mathcal{R}. \tag{1.9}$$

The complex coordinate z on these leaves is given by (I.3.42):

$$z = \frac{S_1 + iS_2}{\hbar(\frac{N}{2}+1) + S_3} \equiv \frac{C}{\hbar(\frac{N}{2}+1) + A}. \tag{1.10}$$

The symplectic form corresponding to the Poisson brackets (I.3.3a) (with $f^0(A) = 2A + \hbar$ and $v(A) = -1$) restricted to the leaf Ω_a coincides with the form ω_0 (1.8).

The quantum restriction of the coordinate functions S_1, S_2, and S_3 from $\mathbb{R}^3$ onto Ω_a is derived by using the general formula (I.3.74):

$$\begin{aligned} S_1 \Big|_{\widehat{\Omega}_a} &= e^{-F/\hbar} \overset{\circ}{S}_1(e^{F/\hbar}) = \frac{z + \bar{z}}{1+|z|^2} \cdot a \overset{\text{def}}{=} s_1, \\ S_2 \Big|_{\widehat{\Omega}_a} &= e^{-F/\hbar} \overset{\circ}{S}_2(e^{F/\hbar}) = \frac{i(\bar{z}-z)}{1+|z|^2} \cdot a \overset{\text{def}}{=} s_2, \\ S_3 \Big|_{\widehat{\Omega}_a} &= e^{-F/\hbar} \overset{\circ}{S}_3(e^{F/\hbar}) = \frac{1-|z|^2}{1+|z|^2} \cdot a \overset{\text{def}}{=} s_3. \end{aligned} \tag{1.11}$$

These three functions s_1, s_2, and s_3 determine the quantum embedding

$$s : \Omega_a \to \mathbb{R}^3. \tag{1.12}$$

The image of the mapping s is the sphere of radius $a = \hbar N/2$ centered at $(0,0,0)$:

$$\sum_{\ell=1}^{3} s_\ell^2 = a^2.$$

Note that the classical leaf $\Omega_a \subset \mathbb{R}^3$ is the sphere of radius $\hbar(N+1)/2$ centered at $(0,0,-\hbar/2)$.

Three functions s_1, s_2, and s_3 represent the algebra (1.1) via the Wick $*$-product on Ω_a. The Wick product is generated by the general probability function (I.3.33); in our example it reads

$$p(\xi;\xi') = \frac{|1+z\overline{z'}|^{2N}}{(1+|z|^2)^N(1+|z'|^2)^N}.$$

Here $z = z(\xi)$ and $z' = z(\xi')$ are the complex coordinates (1.10) of points $\xi,\xi' \in \Omega_a$.

If we transport the function p to the sphere $\mathbb{S}_a^2 = \{s \in \mathbb{R}^3 \mid |s| = a\}$ by means of the quantum embedding (1.12), we obtain the probability function

$$(1.13) \qquad p(s;s') = \left(\frac{1+\cos\alpha_{s,s'}}{2}\right)^N$$

over $\mathbb{S}_a^2 \times \mathbb{S}_a^2$. Here $N = 2a/\hbar \in \mathbb{Z}_+$, and $\alpha_{s,s'}$ denotes the angle between the vectors s and s' in $\mathbb{R}^3$. So, the coordinate functions s_1, s_2, and s_3 on $\mathbb{S}_a^2$ represent the algebra (1.1) with respect to the Wick $*$-product generated by the probability function (1.13). Thus we can avoid the inconsistency of quantizations on $\mathbb{R}^3$ and $\mathbb{S}^2$ mentioned, for instance, in [**Ka**$_{12}$].

1.2. The Lie algebra su$(1,1)$. Let us consider the relations

$$(1.14) \qquad [\mathbf{S}_1,\mathbf{S}_2] = i\hbar\mathbf{S}_3, \qquad [\mathbf{S}_2,\mathbf{S}_3] = -i\hbar\mathbf{S}_1, \qquad [\mathbf{S}_3,\mathbf{S}_1] = -i\hbar\mathbf{S}_2,$$

where $\hbar > 0$ and $\mathbf{S}_j^* = \mathbf{S}_j$.

We define

$$\mathbf{B} = \mathbf{S}_1 - i\mathbf{S}_2, \qquad \mathbf{C} = \mathbf{S}_1 + i\mathbf{S}_2, \qquad \mathbf{A} = \mathbf{S}_3;$$

then

$$(1.14\text{a}) \qquad [\mathbf{C},\mathbf{B}] = 2\hbar\mathbf{A}, \qquad \mathbf{A}\mathbf{B} = \mathbf{B}(\mathbf{A}+\hbar), \qquad \mathbf{C}\mathbf{A} = (\mathbf{A}+\hbar)\mathbf{C}.$$

The Casimir element has the form $\mathbf{K} = \mathbf{B}\mathbf{C} - \mathbf{A}^2 + \hbar\mathbf{A}$, and therefore in this case we have

$$k = 1, \qquad f(A) = 2\hbar A, \qquad \varphi(A) = A + \hbar,$$
$$\rho(A) = A^2 - \hbar A, \qquad f^0(A) = 2A - \hbar, \qquad v(A) = 1,$$
$$\mathcal{F}_a(n) = \hbar^2(2a/\hbar + n), \qquad \mathcal{A}_a(n) = a + n\hbar.$$

Thus $\mathcal{R} = \{a \geq 0\}$. For each $a > 0$, the function $\mathcal{F}_a$ is positive everywhere on $\mathbb{Z}_+$. The corresponding irreducible representation is trivial (one-dimensional) for $a = 0$ and is infinite-dimensional for $a > 0$.

If $a > 0$, the factorization of $\mathcal{F}_a$ related to the representation by differential (not pseudodifferential) operators can be chosen in two ways:

$$(1.15) \qquad \begin{array}{lll} \text{I.} & \mathcal{B}_a = 2a + n\hbar, & \mathcal{C}_a = \hbar; \\ \text{II.} & \mathcal{B}_a = 1, & \mathcal{C}_a = \hbar(2a + n\hbar). \end{array}$$

Let us consider both these variants.

Variant I. The irreducible representation corresponding to the factorization (1.15) (I) reads

(1.16)
$$\overset{\circ}{S}_1 = \frac{\hbar}{2}\left(2b\bar{z} + \bar{z}^2 \frac{d}{d\bar{z}} + \frac{d}{d\bar{z}}\right), \qquad \overset{\circ}{B} = \hbar\bar{z}\left(2b + \bar{z}\frac{d}{d\bar{z}}\right),$$
$$\overset{\circ}{S}_2 = \frac{i\hbar}{2}\left(2b\bar{z} + \bar{z}^2 \frac{d}{d\bar{z}} - \frac{d}{d\bar{z}}\right), \quad \text{or} \quad \overset{\circ}{C} = \hbar\frac{d}{d\bar{z}},$$
$$\overset{\circ}{S}_3 = \hbar\left(b + \bar{z}\frac{d}{d\bar{z}}\right), \qquad \overset{\circ}{A} = \hbar\left(b + \bar{z}\frac{d}{d\bar{z}}\right).$$

Here we denote $a = b\hbar$, $b \geq 0$. The Hilbert space $\mathcal{P}$ of this representation consists of the antiholomorphic distributions

$$g(\bar{z}) = \sum_{n=0}^{\infty} g_n \bar{z}^n, \qquad \|g\|_{\mathcal{P}}^2 = \sum_{n=0}^{\infty} \frac{n!}{(2b)_n} |g_n|^2,$$

where $(2b)_n \equiv 2b(2b+1)\ldots(2b+n-1) = \Gamma(2b+n)/\Gamma(2b)$. The coherent states and the reproducing kernel are

(1.17)
$$\mathfrak{P}_z = e^{z\mathbf{B}/\hbar}\mathfrak{P}_0, \qquad \text{where } \mathbf{A}\mathfrak{P}_0 = b\hbar\,\mathfrak{P}_0, \quad \mathbf{C}\mathfrak{P}_0 = 0,$$
$$\mathcal{K}(\bar{z}, z) = \sum_{n=0}^{\infty} \frac{(2b)_n}{n!}(\bar{z}z)^n \equiv k(|z|^2).$$

The last series converges in the disc $z\bar{z} < 1$, so, in this case, the radius $\underline{R}$ (I.3.22) is equal to 1.

The equations for k and ℓ are

(1.18)
$$(1-x)\frac{dk}{dx} = 2b\,k(x), \qquad k(0) = 1$$
$$\implies k(x) = \frac{1}{(1-x)^{2b}},$$

(1.19)
$$(1-x)\frac{d\ell}{dx} = (2-2b)\,\ell(x), \qquad \frac{1}{\hbar}\int_0^1 \ell(x)\,dx = 1$$
$$\implies \ell(x) = \hbar(2b-1)(1-x)^{2b-2}.$$

The last formula works only for $b > 1/2$. If $0 < b \leq 1/2$, then there does not exist a function ℓ with suitable properties, and hence, a measure dm. If $b > 1/2$, then the measure $dm = M\,d\bar{z}dz$ has the following density:

(1.20)
$$M(x) = \frac{(2b-1)\hbar}{(1-x)^2}, \qquad x = |z|^2.$$

The representation (1.16), the coherent states (1.17), the reproducing kernel (1.18), and the measure (1.20) are well known [$\mathbf{P}_{1,2}$].

In this example, the symplectic leaf Ω_a (I.3.1a) is given by the equations

(1.21)
$$\Omega_a = \{S_1^2 + S_2^2 - S_3^2 + \hbar S_3 = -\hbar^2 b\,(b-1)\}, \qquad a = b\hbar.$$

If $b \neq 1/2$, then Ω_a is a hyperboloid of two sheets; the down sheet (where $S_3 < 0$ as $S_1^2 + S_2^2 \to \infty$) is assumed to be cut off. If $b = 1/2$, then this is a cone whose lower part (where $S_3 < \hbar/2$) is also assumed to be cut off.

The quantum symplectic form and the quantum Kähler potential are determined by

(1.22) $$\omega = \frac{i 2\hbar b \, d\bar{z} \wedge dz}{(1 - |z|^2)^2}, \qquad F(x) = 2\hbar b \ln \frac{1}{1-x}.$$

One can transport them to the leaf (1.21) by means of the complex coordinate z defined by (I.3.42):

(1.23) $$z = \frac{S_1 + iS_2}{\hbar(b-1) + S_3}.$$

Here we have to assume that $b > 1/2$, and hence, the denominator in (1.23) is positive on the considered part of the hyperboloid (1.21).

From (I.3.5b) and (I.3.6) we obtain for $b > 1/2$:

$$\mathcal{E}_a(A) = A - a, \qquad \mathcal{D}_a(A) = A + a - \hbar, \qquad X_a(A) = \frac{A-a}{A+a-\hbar},$$

$$\mathcal{H}(t) = \frac{t}{t + (2b-1)\hbar}, \qquad F_0(x) = \hbar(2b-1) \ln \frac{1}{1-x} \quad \text{(where } x < 1\text{)}.$$

Thus,

(1.24) $$\omega_0 = \frac{i\hbar(2b-1) \, d\bar{z} \wedge dz}{(1-|z|^2)^2}, \qquad \lambda = \frac{i \, d\bar{z} \wedge dz}{(1-|z|^2)^2}, \qquad g(x) = \frac{1}{(1-x)^2}.$$

We can see that the classical symplectic form is Kähler only for $b > 1/2$. In this example, asymptotics (I.3.45)–(I.3.47) are exact in the main term.

The quantum restriction of coordinate functions S_1, S_2, and S_3 from $\mathbb{R}^3$ onto Ω_a is derived as

(1.25) $$\begin{aligned} S_1|_{\widehat{\Omega}_a} &= \frac{z + \bar{z}}{1 - |z|^2} \cdot a \stackrel{\text{def}}{=} s_1, \\ S_2|_{\widehat{\Omega}_a} &= \frac{i(\bar{z} - z)}{1 - |z|^2} \cdot a \stackrel{\text{def}}{=} s_2, \\ S_3|_{\widehat{\Omega}_a} &= \frac{1 + |z|^2}{1 - |z|^2} \cdot a \stackrel{\text{def}}{=} s_3. \end{aligned}$$

Since

(1.26) $$s_1^2 + s_2^2 = s_3^2 - a^2, \qquad s_3 \geq a,$$

the functions s_1, s_2, and s_3 determine a mapping from Ω_a to the standard hyperboloid (1.26) embedded into $\mathbb{R}^3$ (this is the quantum embedding of Ω_a into $\mathbb{R}^3$). *The functions s_1, s_2, and s_3 represent the algebra (1.14) via the Wick $*$-product on Ω_a.*

Variant II. The irreducible Hermitian representation corresponding to the factorization (1.15) (II) reads

$$
(1.27) \quad \begin{aligned}
\overset{\circ}{S}_1 &= \frac{1}{2}\left(\overline{w} + \hbar^2\left(2b + \overline{w}\frac{d}{d\overline{w}}\right)\frac{d}{d\overline{w}}\right), & \overset{\circ}{B} &= \overline{w}, \\
\overset{\circ}{S}_2 &= \frac{i}{2}\left(\overline{w} - \hbar^2\left(2b + \overline{w}\frac{d}{d\overline{w}}\right)\frac{d}{d\overline{w}}\right), \quad \text{or} \quad & \overset{\circ}{C} &= \hbar^2\left(2b + \overline{w}\frac{d}{d\overline{w}}\right)\frac{d}{d\overline{w}}, \\
\overset{\circ}{S}_3 &= \hbar\left(b + \overline{w}\frac{d}{d\overline{w}}\right), & \overset{\circ}{A} &= \hbar\left(b + \overline{w}\frac{d}{d\overline{w}}\right).
\end{aligned}
$$

Here we denote the complex coordinate on $\mathbb{C}$ by w (instead of z) to distinguish this case from Variant I.

The Hilbert space $\mathcal{P}$ of the representation (1.27) consists of the antiholomorphic distributions

$$
(1.28) \quad g(\overline{w}) = \sum_{n=0}^{\infty} g_n \overline{w}^n, \qquad \|g\|_{\mathcal{P}}^2 = \sum_{n=0}^{\infty} n!\, (2b)_n\, \hbar^{2n}\, |g_n|^2.
$$

The coherent states and the reproducing kernel look as follows:

$$
(1.29) \quad \begin{aligned}
\mathfrak{P}_w &= \sum_{n=0}^{\infty} \frac{1}{n!(2b)_n}\left(\frac{w\mathbf{B}}{\hbar^2}\right)^n \mathfrak{P}_0 = \widetilde{I}_{2b-1}\left(\frac{2}{\hbar}\sqrt{w\mathbf{B}}\right)\mathfrak{P}_0, \\
\mathcal{K}(\overline{w}, w) &= \widetilde{I}_{2b-1}\left(\frac{2}{\hbar}\sqrt{\overline{w}w}\right).
\end{aligned}
$$

Here the vacuum $\mathfrak{P}_0$ is the same as in (1.17), and $\widetilde{I}_\alpha$ is a modified *Bessel function* of imaginary argument (it differs from the function I_α in [**BE**] by a factor):

$$
(1.30) \quad \widetilde{I}_\alpha \overset{\text{def}}{=} \sum_{n=0}^{\infty} \left(\frac{y}{2}\right)^{2n} \frac{\Gamma(\alpha+1)}{n!\,\Gamma(\alpha+n+1)}.
$$

The equations for the function ℓ (I.3.24) in this example read

$$
\hbar^2\left(x\frac{d}{dx} + 2 - 2b\right)\frac{d\ell}{dx} = \ell(x), \qquad \frac{1}{\hbar}\int_0^\infty \ell(x)\,dx = 1,
$$

whose solution is the *modified MacDonald function*

$$
(1.31) \quad \ell(x) = \frac{(\sqrt{x})^{2b-1}}{\hbar^{2b}\,\Gamma(2b)} \int_{-\infty}^{\infty} \exp\left\{-\frac{2\sqrt{x}}{\hbar}\cosh t - (2b-1)t\right\} dt.
$$

In this case, the parameter $b > 0$ is arbitrary; so, in contrast with the previous case, the assumption $b > 1/2$ does not appear.

The factors $\mathcal{E}_a$ and $\mathcal{D}_a$ in (I.3.5b) are the following:

$$
\mathcal{E}_a(A) = (A-a)(A+a-\hbar), \qquad \mathcal{D}_a(A) \equiv 1,
$$

and hence, $\mathcal{H}(t) = t(t+2q)$, where $q \overset{\text{def}}{=} \hbar(b-1/2)$. The complex structure on Ω_a is determined by the complex coordinate $w = S_1 + iS_2$. The classical symplectic form and its potential are

$$\omega_0 = \frac{i\operatorname{sgn}(q)\,d\overline{w} \wedge dw}{2\sqrt{|w|^2 + q^2}} = i\overline{\partial}\partial F_0,$$

$$F_0(x) = 2\operatorname{sgn}(q)\sqrt{x+q^2} - 2q - 2q\ln\left(\frac{\sqrt{x+q^2}+|q|}{2|q|}\right), \qquad x = |w|^2.$$

If $b = 1/2$ (or $q = 0$, or $a = \hbar/2$), then the leaf Ω_a (1.21) is a cone, and the form ω_0 is singular at the vertex $w = 0$.

1.3. Transition from one holomorphic representation to another. We have already mentioned in Theorem I.1.1 that representations of the same algebra corresponding to different factorizations (I.1.6) are equivalent. Suppose we have two such equivalent representations realized in the spaces of antiholomorphic distributions $\mathcal{P}_{s(a)}$ and $\mathcal{P}_{s'(a)}$. Then the isomorphism $\#\colon \mathcal{P}_{s(a)} \to \mathcal{P}_{s'(a)}$ (see (I.1.10)) can be realized by an integral kernel. This kernel $\mathcal{K}^\#$ is a holomorphic distribution with values in the Hilbert space $\mathcal{P}_{s'(a)}$, that is,

$$\mathcal{K}^\# = \mathcal{K}^\#(\overline{w}, z) = \sum_{n\geq 0} \mathcal{K}_n^\#(\overline{w}) z^n, \qquad \mathcal{K}_n^\# \in \mathcal{P}_{s'(a)}.$$

More exactly, the isomorphism $\#$ is defined by the kernel $\mathcal{K}^\#$ as follows:

$$g(\overline{z}) \to g^\#(\overline{w}) = \sum_n \mathcal{K}_n^\#(\overline{w})(g, \overline{z}^n)_{\mathcal{P}_{s(a)}} = \sum_n \mathcal{K}_n^\#(\overline{w}) g_n s_n(a), \qquad \forall\, g \in \mathcal{P}_{s(a)}.$$

LEMMA 1.1. *The kernel $\mathcal{K}^\#$ can be considered as coherent states in the space $\mathcal{P}_{s'(a)}$; the corresponding coherent transform is the isomorphism $\#\colon \mathcal{P}_{s(a)} \to \mathcal{P}_{s'(a)}$.*

This lemma allows us to calculate the kernel $\mathcal{K}^\#$ directly by using the technique for constructing coherent states. Let us show how this idea works in the case of two representations of the algebra $\mathbf{su}(1,1)$ obtained in § 1.2.

In this example, the space $\mathcal{P}_{s(a)}$ corresponds to Variant I (see (1.16), (1.17)), and the space $\mathcal{P}_{s'(a)}$ corresponds to Variant II (see (1.27), (1.28)). In this case, the intertwining kernel $\mathcal{K}^\#$ gives coherent states of type I realized in the Hilbert space (1.28). In view of (1.17), coherent states of type I have the form

$$\exp\left\{\frac{z}{\hbar}\cdot \text{creation operator}\right\} \text{vacuum}.$$

The creation operator in the space (1.28) is $\overset{\circ}{B} = \overline{w}$ (see (1.27)), and the vacuum vector is the function 1. Thus,

(1.32) $$\mathcal{K}^\#(\overline{w}, z) = e^{z\overline{w}/\hbar}.$$

If we label the coherent states (in an abstract space H_a), as well as the reproducing kernels (1.17) and (1.29), by I and II, respectively, then we immediately arrive at the following identities:

(1.33)
$$\mathfrak{P}_z^I = \frac{1}{2\pi\hbar}\int_{\mathbb{C}} \mathfrak{P}_w^{II}\, e^{z\overline{w}/\hbar}\, \ell^{II}(|w|^2)\, d\overline{w}\, dw,$$

$$\mathfrak{P}_w^{II} = \frac{1}{2\pi\hbar}\int_{|z|<1} \mathfrak{P}_z^I\, e^{\overline{z}w/\hbar}\, \ell^I(|z|^2)\, d\overline{z}\, dz.$$

Here ℓ^I and ℓ^{II} are the functions (1.19) and (1.31), respectively; the last identity in (1.33) holds only for $b > 1/2$ in the notation of §1.2.

Similar identities hold for the reproducing kernels $\mathcal{K}^I$ and $\mathcal{K}^{II}$ from (1.17), (1.18), and (1.29). Thus, we obtain the following statement.

COROLLARY 1.1.
$$(1 - \overline{v}z)^{-2b} = \frac{1}{2\pi\hbar} \int_{\mathbb{C}} e^{(w\overline{v} + \overline{w}z)/\hbar} \ell^{II}(|w|^2) \, d\overline{w} dw,$$
$$\widetilde{I}_{2b-1}\left(\frac{2}{\hbar}\sqrt{\overline{u}w}\right) = \frac{2b-1}{2\pi} \int_{|z|<1} e^{(z\overline{u} + \overline{z}w)/\hbar} (1 - |z|^2)^{2b-2} \, d\overline{z} dz,$$

where $b > 1/2$ in the second formula, and the function ℓ^{II} in the first formula is given by (1.31).

§2. Quadratic algebra related to the Zeeman effect

In this section we consider in detail irreducible representations of an algebra with non-Lie commutation relations, which appeared in [**KaN**$_1$] while a well-known physical model, the hydrogen atom in a uniform magnetic field, was considered.

Let us first recall this construction. We use the "regularized" Hamiltonian of the hydrogen atom in the form
$$S_0 = |q|\left(\frac{1}{4} + p^2\right), \qquad q, p \in \mathbb{R}^3.$$

Functions that are in involution with S_0 are generated by the angular momentum $M = q \times p$ and by the modified Laplace–Runge–Lentz vector
$$L = q\left(\frac{1}{4} + p^2\right) - 2(p \times M).$$

With respect to the symplectic form $dp \wedge dq$ we have
$$\{M_j, M_k\} = -\varepsilon_{jkl} M_l, \qquad \{L_j, L_k\} = -\varepsilon_{jkl} M_l, \qquad \{M_j, L_k\} = -\varepsilon_{jkl} L_l.$$

The Casimir identities are
$$M \cdot L = 0, \qquad M^2 + L^2 = S_0^2.$$

Now let us introduce the vectors
$$J = (M + L)/2, \qquad K = (M - L)/2$$

and define
$$T_0 = J_3 - K_3, \quad T_1 = J_1 K_2 - J_2 K_1, \quad T_2 = J_1 K_1 + J_2 K_2, \quad T_3 = J_3 K_3 + J^2.$$

In the Cartesian coordinates q, p these functions look as follows:
$$T_0 = \left(\frac{1}{4} - p^2\right) q_3 + 2(q \cdot p) p_3,$$
$$T_1 = \frac{1}{2}\left((q \cdot p)\left(\frac{1}{4} + p^2\right) q_3 - \left(q^2\left(\frac{1}{4} - p^2\right) + 2(q \cdot p)^2\right) p_3\right),$$
$$T_2 = \frac{1}{4}\left(q^2 p^2 - (q \cdot p)^2 - q^2\left(\frac{1}{4} + p^2\right)^2 + q^2 p_3^2 + \left(\left(\frac{1}{4} + p^2\right) q_3 - 2(q \cdot p) p_3\right)^2\right),$$
$$T_3 = \frac{1}{4}\left(q^2 p^2 - (q \cdot p)^2 + q^2\left(\frac{1}{4} + p^2\right)^2 - q^2 p_3^2 - \left(\left(\frac{1}{4} + p^2\right) q_3 - 2(q \cdot p) p_3\right)^2\right).$$

The algebra of functions over $\mathbb{R}^6$ which are in involution with S_0 and M_3 is generated by T_0, T_1, T_2, T_3; the pairwise relations between them are [**KaN**$_1$]

$$\{T_1, T_2\} = -T_0 T_3, \qquad \{T_0, T_1\} = -2T_2,$$
$$\{T_2, T_3\} = T_0 T_1, \qquad \{T_0, T_2\} = 2T_1,$$
$$\{T_3, T_1\} = T_0 T_2, \qquad \{T_0, T_3\} = 0.$$

The quantum version of this algebra is constructed in the same way. Let us determine

$$\mathbf{S}_0 = |\mathbf{q}|\left(\frac{1}{4} + \mathbf{p}^2\right), \qquad \mathbf{p} = -i\hbar\frac{\partial}{\partial q}, \qquad \mathbf{q} = q.$$

The set of operators commuting with $\mathbf{S}_0$ is generated by

$$\mathbf{M} = \mathbf{q} \times \mathbf{p}, \qquad \mathbf{L} = \mathbf{q}\left(\frac{1}{4} + \mathbf{p}^2\right) - \mathbf{p} \times \mathbf{M} - \mathbf{M} \times \mathbf{p}.$$

Then we define $\mathbf{J} = (\mathbf{M} + \mathbf{L})/2$, $\mathbf{K} = (\mathbf{M} - \mathbf{L})/2$ and note that the set of operators commuting with both $\mathbf{S}_0$ and $\mathbf{M}_3$ is generated by

$$\mathbf{T}_0 = \mathbf{J}_3 - \mathbf{K}_3, \quad \mathbf{T}_1 = \mathbf{J}_1 \mathbf{K}_2 - \mathbf{J}_2 \mathbf{K}_1, \quad \mathbf{T}_2 = \mathbf{J}_1 \mathbf{K}_1 + \mathbf{J}_2 \mathbf{K}_2, \quad \mathbf{T}_3 = \mathbf{J}_3 \mathbf{K}_3 + \mathbf{J}^2.$$

These operators can be represented in the form

$$\mathbf{T}_j = \mathcal{T}_j(\overset{2}{\mathbf{q}}, \overset{1}{\mathbf{p}})$$

with the following polynomial symbols $\mathcal{T}_j$:

$$\mathcal{T}_0 = B_0 - 2i\hbar p_3,$$
$$\mathcal{T}_1 = B_1 - i\hbar\left(q_3\left(\frac{1}{8} + \frac{3}{2}p^2\right) - 3(q \cdot p)\, p_3\right) + \hbar^2 p_3,$$
$$\mathcal{T}_2 = B_2 + i\hbar\left((q \cdot p)\left(\frac{1}{8} + p^2\right) + 2q_3 p_3\, , p^2 + \frac{5}{8} q \cdot p - 4(q \cdot p)\, p_3^2\right)$$
$$\quad + \hbar^2\left(p^2 - 2p_3^2 + \frac{1}{4}\right),$$
$$\mathcal{T}_3 = B_3 - i\hbar\left((q \cdot p)\left(\frac{1}{8} + p^2\right) + 2q_3 p_3\, p^2 - \frac{5}{8} q \cdot p - 4(q \cdot p)\, p_3^2\right)$$
$$\quad - \hbar^2\left(p^2 - 2p_3^2 + \frac{1}{4}\right),$$

The operator $\mathbf{T}_j$ satisfy the commutation relations [**KaN**$_1$]

$$[\mathbf{T}_1, \mathbf{T}_2] = i\hbar\,\mathbf{T}_0 \mathbf{T}_3, \qquad\qquad [\mathbf{T}_0, \mathbf{T}_1] = 2i\hbar\,\mathbf{T}_2,$$
$$[\mathbf{T}_2, \mathbf{T}_3] = -\frac{i\hbar}{2}(\mathbf{T}_0 \mathbf{T}_1 + \mathbf{T}_1 \mathbf{T}_0), \qquad [\mathbf{T}_0, \mathbf{T}_2] = -2i\hbar\,\mathbf{T}_1,$$
$$[\mathbf{T}_3, \mathbf{T}_1] = -\frac{i\hbar}{2}(\mathbf{T}_0 \mathbf{T}_2 + \mathbf{T}_2 \mathbf{T}_0), \qquad [\mathbf{T}_0, \mathbf{T}_3] = 0.$$

All the above operators are self-adjoint in the space $L^2_-(\mathbb{R}^3)$ with the norm

$$\|\varphi\|_- = \left(\frac{\pi}{4}\int_{\mathbb{R}^3} |\varphi(q)|^2 \frac{dq}{|q|}\right)^{1/2}$$

(about this space and the natural interpretation of the basic operator $\mathbf{S}_0$, see below in §6; all details and applications can be found in [**KaN**$_{2-4}$]).

2.1. The set of irreducible representations.
Let us consider abstract Hermitian generators $\mathbf{S}_0$, $\mathbf{S}_1$, $\mathbf{S}_2$, and $\mathbf{S}_3$ with relations

(2.1)
$$[\mathbf{S}_1, \mathbf{S}_2] = -\frac{i\hbar}{2}(\mathbf{S}_0\mathbf{S}_3 + \mathbf{S}_3\mathbf{S}_0), \qquad [\mathbf{S}_0, \mathbf{S}_1] = -2i\hbar\,\mathbf{S}_2,$$
$$[\mathbf{S}_2, \mathbf{S}_3] = \frac{i\hbar}{2}(\mathbf{S}_0\mathbf{S}_1 + \mathbf{S}_1\mathbf{S}_0), \qquad [\mathbf{S}_0, \mathbf{S}_2] = 2i\hbar\,\mathbf{S}_1,$$
$$[\mathbf{S}_3, \mathbf{S}_1] = \frac{i\hbar}{2}(\mathbf{S}_0\mathbf{S}_2 + \mathbf{S}_2\mathbf{S}_0), \qquad [\mathbf{S}_0, \mathbf{S}_3] = 0,$$

where $\hbar > 0$.

Let us denote
$$\mathbf{B} = \mathbf{S}_1 - i\mathbf{S}_2, \quad \mathbf{C} = \mathbf{S}_1 + i\mathbf{S}_2, \quad \mathbf{A}_1 = \mathbf{S}_0, \quad \mathbf{A}_2 = \mathbf{S}_3.$$

Then relations (2.1) acquire the form

(2.1a)
$$[\mathbf{C}, \mathbf{B}] = -2\hbar\,\mathbf{A}_1\mathbf{A}_2,$$
$$\mathbf{A}_1\mathbf{B} = \mathbf{B}(\mathbf{A}_1 + 2\hbar), \qquad \mathbf{C}\mathbf{A}_1 = (\mathbf{A}_1 + 2\hbar)\mathbf{C},$$
$$\mathbf{A}_2\mathbf{B} = \mathbf{B}(\mathbf{A}_2 - \hbar\mathbf{A}_1 - \hbar^2), \qquad \mathbf{C}\mathbf{A}_2 = (\mathbf{A}_2 - \hbar\mathbf{A}_1 - \hbar^2)\mathbf{C},$$
$$[\mathbf{A}_1, \mathbf{A}_2] = 0,$$

plus the conditions of being Hermitian, $\mathbf{B}^* = \mathbf{C}$, $\mathbf{A}_\mu^* = \mathbf{A}_\mu$. This is a special case of our general algebra (I.1.1) with

$$k = 2, \quad f(A) = -2\hbar A_1 A_2, \quad \varphi_1(A) = A_1 + 2\hbar, \quad \varphi_2(A) = A_2 - \hbar A_1 - \hbar^2.$$

There are two Casimir elements (see § I.1.3)

(2.2)
$$\mathbf{A}_1^2 + 4\mathbf{A}_2, \qquad \mathbf{K} = \mathbf{B}\mathbf{C} - \left(\mathbf{A}_2 + \frac{\hbar}{2}\mathbf{A}_1\right)^2;$$

thus, in our case,

(2.3)
$$\varkappa(A) = A_1^2 + 4A_2, \qquad \rho(A) = \left(A_2 + \frac{\hbar}{2}A_1\right)^2$$

(one can readily check that $\varkappa$ is a φ-invariant function on $\mathbb{R}^2$ and ρ satisfies the difference equation (I.1.28)).

Iterations of the mapping $\varphi\colon \mathbb{R}^2 \to \mathbb{R}^2$ are derived as follows:

(2.4)
$$\mathcal{A}_a(n) = \varphi^n(a) = (a_1 + 2n\hbar,\ a_2 - a_1 n\hbar - n^2\hbar^2), \qquad n \in \mathbb{Z}_+,$$

and so, the basic function $\mathcal{F}_a$ (I.1.3) in this example reads

(2.5)
$$\mathcal{F}_a(n) = \frac{1}{n+1}\sum_{j=0}^{n}(-2\hbar)(a_1 + 2j\hbar)(a_2 - a_1 j\hbar - j^2\hbar^2)$$
$$= \frac{1}{n+1}\left(4\hbar^4\sum_{j=0}^{n}j^3 + 6\hbar^3 a_1 \sum_{j=0}^{n}j^2 + 2\hbar^2(a_1^2 - 2a_2)\sum_{j=0}^{n}j - 2\hbar\,a_1 a_2 \sum_{j=0}^{n}1\right)$$
$$= \hbar\Big(\hbar^3 n^3 + \hbar^2(2a_1 + \hbar)n^2 + \hbar(a_1^2 - 2a_2 + \hbar a_1)n - 2a_1 a_2\Big)$$
$$= \hbar^4(n + a_1/\hbar)\big(n^2 + (1 + a_1/\hbar)n - 2a_2/\hbar^2\big)$$
$$= \hbar^4(n - f_1)(n - f_2)(n - f_3),$$

where

(2.5a) $$f_1 = -a_1/\hbar, \qquad f_2 = (\sqrt{(a_1+\hbar)^2 + 8a_2} - a_1 - \hbar)/2\hbar,$$
$$f_3 = -(\sqrt{(a_1+\hbar)^2 + 8a_2} + a_1 + \hbar)/2\hbar.$$

Here a_1 and a_2 are the values of the operators $\mathbf{A}_1$ and $\mathbf{A}_2$ on the vacuum vector

$$\mathbf{A}_j \mathfrak{P}_0 = a_j \mathfrak{P}_0 \quad (j = 1, 2), \qquad \mathbf{C} \mathfrak{P}_0 = 0.$$

If $a_1 a_2 > 0$, then the point $a = (a_1, a_2) \in \mathbb{R}^2$ does not belong to the subset $\mathcal{R}$, since $f(a) < 0$ (see § I.1.1). If $a_1 a_2 = 0$, then $a \in \mathcal{R}$, but the corresponding irreducible representation of the algebra (2.1a) is trivial. Thus, we consider only the case $a_1 a_2 < 0$.

If $a_1 > 0$, then $a_2 < 0$, and so the polynomial $\mathcal{F}_a$ (2.5) is positive on $\mathbb{Z}_+$. In this situation, the representations are infinite-dimensional ($N = \infty$): see §I.1.1.

If $a_1 < 0$ and $a_2 > 0$, then the roots (2.5a) have the following signs:

$$f_1 > 0, \qquad f_2 > 0, \qquad f_3 < 0.$$

Therefore, the following three cases are possible:
 • If $a_2 > -\hbar a_1/2$, then $f_1 < f_2$, and condition (I.1.6a) allows us to fix an integer value of f_1, say, $f_1 = k_1$; thus, we obtain the quantization condition

(2.6) $$a_1 = -k_1 \hbar, \qquad k_1 = 1, 2, \ldots;$$

the dimension of the corresponding representations is $N + 1 = k_1 + 1$.
 • If $a_2 < -\hbar a_1/2$, then $f_1 > f_2$, and we have to choose an integer value of f_2, say, $f_2 = k_2$; thus, we obtain the quantization condition

(2.7) $$a_2 = \frac{\hbar^2}{2} k_2 \left(k_2 + \frac{a_1}{\hbar} + 1 \right), \qquad k_2 = 1, 2, \ldots;$$

the dimension of the corresponding representations is $N + 1 = k_2 + 1$.
 • If $a_2 = -\hbar a_1/2$, then $f_1 = f_2$ and $\mathcal{F}_a(n)$ is positive for $n \in \mathbb{Z}_+$, $n \neq f_1$; thus there are two possible situations:
   (i) if $f_1 \notin \mathbb{Z}_+$, then $\mathcal{F}_a > 0$ everywhere on $\mathbb{Z}_+$, so that the corresponding representations are infinite-dimensional ($N = \infty$);
   (ii) if $f_1 \in \mathbb{Z}_+$, that is,

(2.8) $$a_1 = -\hbar k_3, \qquad a_2 = \hbar^2 k_3/2, \qquad k_3 = 1, 2, \ldots,$$

then the dimension of the corresponding representations is $N + 1 = k_3 + 1$. We present this information in Table 1.

2.2. Coherent states and representations by differential operators of order ≤ 4. In each case (a)–(e), we can factorize the function $\mathcal{F}_a$ in several ways as in (I.1.6) and (I.2.8). Such a factorization is not arbitrary: condition (I.1.6a) must be satisfied. In Table 2 we show all possible factorizations (up to the multiplication by a constant). All together we have eight variants of factorization and each variant can be realized in several cases listed in Table 1.

Note that Variant IV, (b) and (e), was first studied in [**KaN**[1]].

TABLE 1. THE SET $\mathcal{R}$ OF IRREDUCIBLE REPRESENTATIONS

Case	Subsets in $\mathcal{R}$	Quantization conditions	Positive roots of $\mathcal{F}_a$	dim $N+1$
(a)	$a_1 > 0$, $a_2 < 0$	no	no	∞
(b)	$a_1 < 0$, $a_2 > -\frac{\hbar a_1}{2}$	$a_1 = -k_1\hbar$, $k_1 \in \mathbb{N}$	$f_1 = k_1$, $f_2 = \frac{k_1-1}{2} + \frac{1}{2\hbar}\sqrt{(k_1-1)^2\hbar^2 + 8a_2}$	$k_1 + 1$
(c)	$0 < a_2 < -\frac{\hbar a_1}{2}$	$a_2 = \frac{\hbar^2}{2}k_2(k_2 + \frac{a_1}{\hbar} + 1)$, $k_2 \in \mathbb{N}$	$f_1 = -\frac{a_1}{\hbar}$, $f_2 = k_2$	$k_2 + 1$
(d)	$0 < a_2 = -\frac{\hbar a_1}{2}$, $\frac{a_1}{\hbar} \notin \mathbb{Z}$	no	$f_1 = f_2 = -\frac{a_1}{\hbar}$	∞
(e)	$0 < a_2 = -\frac{\hbar a_1}{2}$, $\frac{a_1}{\hbar} \in \mathbb{Z}$	$a_1 = -\hbar k_3$, $a_2 = \frac{\hbar^2 k_3}{2}$, $k_3 \in \mathbb{N}$	$f_1 = f_2 = k_3$	$k_3 + 1$
(f)	$a_1 a_2 = 0$	no		1

TABLE 2. VARIANTS OF FACTORIZATION OF THE FUNCTION $\mathcal{F}_a$

Number of the variant	Factor $\mathcal{B}_a(n)$	Factor $\mathcal{C}_a(n)$	Cases in which there exists an irreducible representation
I	$\hbar^3(n-f_1)(n-f_2)(n-f_3)$	$\hbar$	(a)–(e)
II	$\hbar^2(n-f_2)(n-f_3)$	$\hbar^2(n-f_1)$	(a), (c)–(e)
III	$\hbar^2(n-f_3)(n-f_1)$	$\hbar^2(n-f_2)$	(a), (b), (d), (e)
IV	$\hbar^2(n-f_1)(n-f_2)$	$\hbar^2(n-f_3)$	(a)–(e)
V	$\hbar(n-f_1)$	$\hbar^3(n-f_2)(n-f_3)$	(a), (b), (d), (e)
VI	$\hbar(n-f_2)$	$\hbar^3(n-f_3)(n-f_1)$	(a), (c)–(e)
VII	$\hbar(n-f_3)$	$\hbar^3(n-f_1)(n-f_2)$	(a), (d)
VIII	1	$\hbar^4(n-f_1)(n-f_2)(n-f_3)$	(a), (d)

Variant I, cases (a)–(e). The antiholomorphic representation of the algebra (2.1a) is given by the differential operators of order ≤ 3

$$(2.9) \quad \overset{\circ}{B} = \hbar^3 \bar{z}\left(\bar{z}\frac{d}{d\bar{z}} - f_1\right)\left(\bar{z}\frac{d}{d\bar{z}} - f_2\right)\left(\bar{z}\frac{d}{d\bar{z}} - f_3\right), \quad \overset{\circ}{C} = \hbar\frac{d}{d\bar{z}},$$
$$\overset{\circ}{A}_1 = a_1 + 2\hbar \bar{z}\frac{d}{d\bar{z}}, \quad \overset{\circ}{A}_2 = a_2 - \hbar a_1 \frac{d}{d\bar{z}} - \hbar^2\left(\bar{z}\frac{d}{d\bar{z}}\right)^2,$$

where f_1, f_2, and f_3 were defined in (2.5a).

The inner product in the space $\mathcal{P}$ of antiholomorphic distributions (or polynomials) is

$$(2.10) \qquad (g_1, g_2)_{\mathcal{P}} = \sum_{n=0}^{N} \frac{n!}{(-f_1)_n(-f_2)_n(-f_3)_n \hbar^{2n}} \overline{g_{1n}}\, g_{2n},$$

where

$$g_j(\overline{z}) = \sum_{n=0}^{N} g_{jn}\, \overline{z}^n ;$$

the dimension of the representation is $N+1$ (see Table 1).

Recall the notation $(f_j)_0 = 1$, $(f_j)_n \equiv f_j(f_j+1)\ldots(f_j+n-1)$, $n \in \mathbb{N}$.

In cases (d) and (e), formula (2.10) becomes simpler:

$$(g_1, g_2)_{\mathcal{P}} = \sum_{n=0}^{N} \frac{1}{\left((-f_1)_n \hbar^n\right)^2} \overline{g_{1n}}\, g_{2n};$$

in case (e), the same formula can be written as follows:

$$(g_1, g_2)_{\mathcal{P}} = \sum_{n=0}^{N} \left(\frac{(k_3-n)!}{k_3!\,\hbar^n}\right)^2 \overline{g_{1n}}\, g_{2n}.$$

The coherent states are

$$(2.11) \qquad \mathfrak{P}_z = \exp\{z\mathbf{B}/\hbar\}\mathfrak{P}_0.$$

The reproducing kernel is

$$(2.12) \quad \mathcal{K}(\overline{z}, z) = \sum_{n \geq 0} \frac{(-f_1)_n(-f_2)_n(-f_3)_n}{n!} (\hbar^2 \overline{z} z)^n = {}_3F_0(-f_1, -f_2, -f_3;\, \hbar^2|z|^2),$$

where ${}_3F_0$ is the *hypergeometric series* (I.2.7).

In fact, in cases (b), (c), and (e), the series (2.12) is a polynomial, since $(-f_1)_n = 0$ for $n \geq k_1 + 1$ in case (b), $(-f_2)_n = 0$ for $n \geq k_2 + 1$ in case (c), and $(-f_1)_n = (-f_2)_n = 0$ for $n \geq k_3 + 1$ in case (e). So, in these cases, using the notation (I.2.7a), we can write

$$\mathcal{K}(\overline{z}, z) = {}_3f_0(-f_1, -f_2, -f_3;\, \hbar^2|z|^2;\, N),$$

where ${}_3f_0$ is a *hypergeometric jet* (I.2.7a), and the number N is given in Table 1.

In cases (a) and (d), the series (2.12) is infinite, and moreover, it diverges: the radius of convergence (I.3.22) $\underline{R} = 0$.

In cases (b), (c), and (e), the "reproducing" measure $dm = k(|z|^2)\ell(|z|^2)\, d\overline{z}dz$ for the resolution of the identity and for the reproducing property can be obtained by solving problem (I.3.24), (I.3.25) for the function ℓ.

In our example, this problem reads

$$(2.13) \qquad \hbar^2 \frac{d}{dx}\left(x^3 \frac{d^2\ell}{dx^2} + \alpha x^2 \frac{d\ell}{dx} + \beta x \ell(x)\right) + \hbar^2 \gamma\, \ell(x) - \frac{d\ell}{dx} = 0,$$

$$\frac{1}{\hbar}\int_0^\infty \ell(x)\, dx = 1,$$

where
$$\alpha = f_1 + f_2 + f_3 + 6,$$
$$\beta = f_1 f_2 + f_2 f_3 + f_3 f_1 + 3(f_1 + f_2 + f_3) + 7,$$
$$\gamma = (f_1 + 1)(f_2 + 1)(f_3 + 1).$$

Probably, in cases (b) and (c), there does not exist a solution of (2.13).

In case (e), $f_3 = -1$, and therefore, since $\gamma = 0$, equation (2.13) can be integrated. We obtain
$$x^3 \frac{d^2\ell}{dx^2} + (2k_3 + 5)x^2 \frac{d\ell}{dx} + (k_3 + 2)^2 x\ell(x) = \frac{1}{\hbar^2}\ell(x).$$

The solution must decrease at infinity, so,
$$\ell(x) = \frac{c}{x^{k_3+2}} K_0\left(\frac{2}{\hbar\sqrt{x}}\right),$$

where K_0 is the *MacDonald function* [**BE**]
$$K_0(y) \sim \sqrt{\frac{\pi}{2y}} e^{-y} \quad \text{as} \quad y \to \infty, \qquad K_0(y) \sim \ln\frac{1}{y} \quad \text{as} \quad y \to 0.$$

The normalization constant c can be derived from the normalization condition in (2.13); therefore,
$$(2.14) \qquad \ell(x) = \frac{2}{(k_3!)^2 \, \hbar^{2k_3+1}} \frac{1}{x^{k_3+2}} K_0\left(\frac{2}{\hbar\sqrt{x}}\right).$$

In case (e), (2.12) implies
$$(2.15) \qquad k(x) = \sum_{n=0}^{k_3} \left(\frac{k_3!}{(k_3-n)!}\right)^2 (\hbar^2 x)^n.$$

Thus, in case I (e), we obtain the density $M = k\ell$ of the measure $dm = M(|z|^2)d\bar{z}dz$, where k and ℓ are given by (2.15) and (2.14).

In each variant discussed below the operators $\overset{\circ}{A}_1$ and $\overset{\circ}{A}_2$ are given by the same formulas as in (2.9); only the expressions for $\overset{\circ}{B}$ and $\overset{\circ}{C}$ must be changed. That is why we shall write only formulas for $\overset{\circ}{B}$ and $\overset{\circ}{C}$.

Variant II, cases (a), (c)–(e). The antiholomorphic representation is
$$(2.16) \qquad \overset{\circ}{B} = \hbar^2 \bar{z}\left(\bar{z}\frac{d}{d\bar{z}} - f_2\right)\left(\bar{z}\frac{d}{d\bar{z}} - f_3\right), \qquad \overset{\circ}{C} = \hbar^2\left(\bar{z}\frac{d}{d\bar{z}} - f_1\right)\frac{d}{d\bar{z}}.$$

The inner product in the space $\mathcal{P}$ of antiholomorphic distributions is
$$(g_1, g_2)_\mathcal{P} = \sum_{n=0}^N \frac{n!\,(-f_1)_n}{(-f_2)_n(-f_3)_n} \overline{g_{1n}} g_{2n},$$

where N should be taken from Table 1.

The coherent states in cases (a), (c), and (d) are
$$\mathfrak{P}_z = \sum_{n=0}^\infty \frac{\Gamma(-f_1)}{n!\,\Gamma(n-f_1)}\left(\frac{z\mathbf{B}}{\hbar^2}\right)^n \mathfrak{P}_0 = \widetilde{I}_{-f_1-1}\left(\frac{2}{\hbar}\sqrt{z\mathbf{B}}\right)\mathfrak{P}_0,$$

where $\widetilde{I}_\alpha$ is the *modified Bessel function* of imaginary argument (1.30).

In case (e), the coherent states are

$$\mathfrak{P}_z = \sum_{n=0}^{k_3} \frac{(-1)^n (k_3 - n)!}{n!\, k_3!} \left(\frac{z\mathbf{B}}{\hbar^2}\right)^n \mathfrak{P}_0.$$

The reproducing kernel is

(2.17) $\quad \mathcal{K}(\bar{z}, z) = k(|z|^2), \quad k(x) = \sum_{n=0}^{N} \frac{(-f_2)_n (-f_3)_n}{n!\, (-f_1)_n} x^n = {}_2F_1(-f_2, -f_3; -f_1; x),$

where ${}_2F_1$ is the *Gauss hypergeometric function*.

The radius of convergence is $\underline{R} = 1$ in cases (a) and (d).

In cases (d) and (e), since $f_1 = f_2$ and $f_3 = -1$, we obtain $k(x) = \sum_{n=0}^N x^n$, and thus,

$$k(x) = \frac{1}{1-x} \quad \text{in case (d)}, \qquad k(x) = \frac{1 - x^{k_3+1}}{1-x} \quad \text{in case (e)}.$$

In case (c), the function k (2.17) is the *Jacobi polynomial*

$$k(x) = {}_2F_1(-k_2, k_2 + \nu + 1; \nu; x) = \frac{x^{1-\nu}}{(\nu)_{k_2}(1-x)} \frac{d^{k_2}}{dx^{k_2}}[x^{\nu + k_2 - 1}(1-x)^{k_2+1}],$$

where $\nu = a_1/\hbar$.

Now let us consider the equation for the function ℓ (which defines the density $M = \ell k$ of the reproducing measure):

$$x(1-x)\frac{d^2\ell}{dx^2} + \big(\gamma - (\alpha + \beta + 1)x\big)\frac{d\ell}{dx} - \alpha\beta\,\ell(x) = 0.$$

Here

$$\alpha = 2 + f_2, \qquad \beta = 2 + f_3, \qquad \gamma = 2 + f_1$$

are parameters of the hypergeometric equation.

In cases (d) and (e), this equation can be solved explicitly, but the normalization condition $\frac{1}{\hbar}\int_0^1 \ell(x)\,dx = 1$ cannot be satisfied.

In case (c), the desired solution exists:

$$\ell(x) = \frac{(k_2 + 1)(k_2 + \nu)\,\hbar}{1 - \nu}\, {}_2F_1(k_2 + 2, 1 - k_2 - \nu; 2 - \nu; x),$$

where $\nu = a_1/\hbar$.

Case (a) can also be investigated; the solution ℓ exists but looks cumbersome. Namely, if $f_1, f_2, f_3 \notin \mathbb{Z}_+$, then

$$\ell(x) = \frac{\hbar}{c}\bigg(\frac{\Gamma(2+f_2)\,\Gamma(2+f_3)}{\Gamma(2+f_1)}\, {}_2F_1(2+f_2, 2+f_3; 2+f_1; x)$$
$$- \frac{\Gamma(-f_2)\,\Gamma(-f_3)}{\Gamma(-f_1)\,x^{1+f_1}}\, {}_2F_1(f_2 - f_1 + 1, f_3 - f_1 + 1; -f_1; x)\bigg),$$

where

$$c = \pi\Big(\cot\big(\pi(1+f_2)\big) + \cot\big(\pi(1+f_3)\big)\Big) - \frac{\Gamma(1+f_2)\,\Gamma(1+f_3)}{\Gamma(1+f_1)} - \frac{1}{1+f_2} - \frac{1}{1+f_3}.$$

If one of the f_j is an integer, this formula simplifies. Say, if $f_1 \in \mathbb{Z}$ but $f_2, f_3 \notin \mathbb{Z}$, then
$$\ell(x) = \frac{(1+f_2)(1+f_3)\hbar}{c_1} \, {}_2F_1(2+f_2, 2+f_3; 2; 1-x),$$
where
$$c_1 = \frac{\Gamma(-f_1)}{\Gamma(f_2-f_1+1)\,\Gamma(f_3-f_1+1)} - 1.$$

Variant III, cases (a), (b), (d), and (e). In cases (d) and (e), Variant III is identical to Variant II, since $f_1 = f_2$. Thus, here we consider only cases (a) and (b).

The antiholomorphic representation is
$$\overset{\circ}{B} = \hbar^3 \bar{z}\Big(\bar{z}\frac{d}{d\bar{z}} - f_1\Big)\Big(\bar{z}\frac{d}{d\bar{z}} - f_3\Big), \qquad \overset{\circ}{C} = \hbar^2\Big(\bar{z}\frac{d}{d\bar{z}} - f_2\Big)\frac{d}{d\bar{z}}.$$

The inner product is
$$(g_1, g_2)_{\mathcal{P}} = \sum_{n=0}^{N} \frac{n!\,(-\overline{f_2})_n}{(-f_1)_n(-f_3)_n} \overline{g_{1n}}\, g_{2n},$$
where $N = \infty$ in case (a) and $N = k_1$ in case (b).

The coherent states in cases (a) and (b) with $f_2 \notin \mathbb{Z}_+$ are given by
$$\mathfrak{P}_z = \sum_{n=0}^{\infty} \frac{\Gamma(-\overline{f_2})}{n!\,\Gamma(n-\overline{f_2})}\Big(\frac{z\mathbf{B}}{\hbar^2}\Big)^n \mathfrak{P}_0 = \widetilde{I}_{-\overline{f_2}-1}\Big(\frac{2}{\hbar}\sqrt{z\mathbf{B}}\Big)\mathfrak{P}_0,$$
where $\widetilde{I}_\alpha$ is a *modified Bessel function* of imaginary argument.

In case (b), if $f_2 \in \mathbb{Z}_+$, then we have
$$\mathfrak{P}_z = \sum_{n=0}^{f_2} \frac{(-1)^n(f_2-n)!}{n!\, f_2!}\Big(\frac{z\mathbf{B}}{\hbar^2}\Big)^n \mathfrak{P}_0.$$

The reproducing kernel is
$$\mathcal{K}(\bar{z}, z) = k(|z|^2), \qquad k(x) = {}_2F_1(-f_1, -f_3; -\overline{f_2}; x).$$

In case (b), it is a *Jacobi polynomial*. In case (a), for $a_2 < -(a_1+\hbar)^2/8$, since $f_3 = \overline{f_2}$, we have $k(x) = {}_1F_0(-f_1; x)$. In case (a), the radius of convergence $\underline{R} = 1$.

Variant IV, cases (a)–(e). The antiholomorphic representation is
$$\overset{\circ}{B} = \hbar^2 \bar{z}\Big(\bar{z}\frac{d}{d\bar{z}} - f_1\Big)\Big(\bar{z}\frac{d}{d\bar{z}} - f_2\Big), \qquad \overset{\circ}{C} = \hbar^2\Big(\bar{z}\frac{d}{d\bar{z}} - f_3\Big)\frac{d}{d\bar{z}}.$$

The inner product is
$$(g_1, g_2)_{\mathcal{P}} = \sum_{n=0}^{N} \frac{n!\,(-\overline{f_3})_n}{(-f_1)_n(-f_2)_n} \overline{g_{1n}}\, g_{2n},$$
where N can be taken from Table 1.

The coherent states are
$$\mathfrak{P}_z = \sum_{n=0}^{\infty} \frac{\Gamma(-\overline{f_3})}{n!\,\Gamma(n-\overline{f_3})}\Big(\frac{z\mathbf{B}}{\hbar^2}\Big)^n \mathfrak{P}_0 = \widetilde{I}_{-\overline{f_3}-1}\Big(\frac{2}{\hbar}\sqrt{z\mathbf{B}}\Big)\mathfrak{P}_0,$$
where $\widetilde{I}_\alpha$ is a *modified Bessel function* of imaginary argument.

The reproducing kernel is
$$k(x) = {}_2F_1(-f_1, -f_2; -\overline{f_3}; x).$$

In cases (b), (c), and (e), this is a Jacobi polynomial. For the function $\ell(x)$ in cases (b) and (e), see [**KaN**$_{1,3}$]. In case (a), for $a_2 < -(a_1 + \hbar)^2/8$, we have
$$k(x) = {}_1F_0(-f_1; x).$$

The radius of convergence $\underline{R} = 1$ in cases (a) and (d).

Variant V, cases (a), (b), (d), (e). The antiholomorphic representation is
$$\overset{\circ}{B} = \hbar \overline{z}\left(\overline{z}\frac{d}{d\overline{z}} - f_1\right), \qquad \overset{\circ}{C} = \hbar^3\left(\overline{z}\frac{d}{d\overline{z}} - f_2\right)\left(\overline{z}\frac{d}{d\overline{z}} - f_3\right)\frac{d}{d\overline{z}}.$$

The inner product is
$$(g_1, g_2)_{\mathcal{P}} = \sum_{n=0}^{N} \frac{n!\,(-\overline{f_2})_n(-\overline{f_3})_n \hbar^{2n}}{(-f_1)_n} \overline{g_{1n}}\, g_{2n},$$

where N is taken from Table 1. The coherent states are
$$\mathfrak{P}_z = \sum_{n=0}^{N} \frac{1}{n!\,(-\overline{f_2})_n(-\overline{f_3})_n} \left(\frac{z\mathbf{B}}{\hbar^3}\right)^n \mathfrak{P}_0.$$

In cases (a) and (d), as well as in (b), if $f_2 \notin \mathbb{Z}_+$, these coherent states can be written via the hypergeometric series:
$$\mathfrak{P}_z = {}_0F_2\left(-\overline{f_2}, -\overline{f_3}; \frac{z\mathbf{B}}{\hbar^3}\right)\mathfrak{P}_0.$$

In case (b), if $f_2 \in \mathbb{Z}_+$, then we have
$$\mathfrak{P}_z = \frac{(f_2 - k_1)!}{(f_2)!} \sum_{n=0}^{k} \frac{(f_2 - n)!}{n!\,(f_2 - k_1 + n)!} \left(-\frac{z\mathbf{B}}{\hbar^3}\right)^n \mathfrak{P}_0,$$

where k is any integer from the interval $k_1 \leq k \leq f_2$.

In case (e) we have
$$\mathfrak{P}_z = \frac{1}{k_3!} \sum_{n=0}^{k_3} \frac{(k_3 - n)!}{(n!)^2} \left(-\frac{z\mathbf{B}}{\hbar^3}\right)^n \mathfrak{P}_0.$$

In cases (a), (b), and (d), the reproducing kernel is
$$k(x) = \sum_{n=0}^{\infty} \frac{(-f_1)_n}{n!\,(-\overline{f_2})_n(-\overline{f_3})_n} \left(\frac{x}{\hbar^2}\right)^n = {}_1F_2(-f_1; -\overline{f_2}, -\overline{f_3}; x/\hbar^2).$$

In cases (a) and (d), the radius of convergence $\underline{R} = \infty$. In case (b), k is a polynomial.

In case (d), we have
$$k(x) = \sum_{n=0}^{\infty} \frac{1}{(n!)^2} \left(\frac{x}{\hbar^2}\right)^n = I_0(2\sqrt{x}/\hbar) \equiv \widetilde{I}_0(2\sqrt{x}/\hbar),$$

where I_0 is the *Bessel function* of imaginary argument.

In case (e),
$$k(x) = \sum_{n=0}^{k_3} \frac{1}{(n!)^2} \left(\frac{x}{\hbar^2}\right)^n.$$

Variant VI, cases (a), (c)–(e). In cases (d) and (e), Variant VI is identical to Variant V, since $f_1 = f_2$. Thus, here we consider only cases (a) and (c).

The antiholomorphic representation is
$$\overset{\circ}{B} = \hbar \bar{z}\left(z\frac{d}{dz} - f_2\right), \qquad \overset{\circ}{C} = \hbar^3 \left(z\frac{d}{dz} - f_1\right)\left(z\frac{d}{dz} - f_3\right)\frac{d}{dz}.$$

The inner product is
$$(g_1, g_2)_{\mathcal{P}} = \sum_{n=0}^{N} \frac{n!\,(-f_1)_n(-\overline{f_3})_n \hbar^{2n}}{(-f_2)_n} \overline{g_{1n}}\, g_{2n},$$

where $N = \infty$ in case (a) and $N = k_2$ in case (c). Hypergeometric series also appear in this variant.

The coherent states are
$$\mathfrak{P}_z = \sum_{n=0}^{\infty} \frac{1}{n!\,(-f_1)_n(-\overline{f_3})_n} \left(\frac{z\mathbf{B}}{\hbar^3}\right)^n \mathfrak{P}_0 = {}_0F_2\left(-f_1, -\overline{f_3}; \frac{z\mathbf{B}}{\hbar^3}\right)\mathfrak{P}_0.$$

The reproducing kernel is
$$k(x) = \sum_{n=0}^{\infty} \frac{(-f_2)_n}{n!\,(-f_1)_n(-\overline{f_3})_n} \left(\frac{x}{\hbar^2}\right)^n = {}_1F_2(-f_2; -f_1, -\overline{f_3}; x/\hbar^2).$$

The hypergeometric series ${}_pF_r$ are again used here. In case (c), this is a polynomial.

In case (a), for $a_2 < -(a_1+\hbar)^2/8$, we have $k(x) = {}_0F_1(-f_1; x/\hbar^2)$. The radius of convergence $\underline{R} = \infty$ in case (a).

Variant VII, cases (a) and (d). The antiholomorphic representation is
$$\overset{\circ}{B} = \hbar \bar{z}\left(z\frac{d}{dz} - f_3\right), \qquad \overset{\circ}{C} = \hbar^3 \left(z\frac{d}{dz} - f_1\right)\left(z\frac{d}{dz} - f_2\right)\frac{d}{dz}.$$

The inner product is
$$(g_1, g_2)_{\mathcal{P}} = \sum_{n=0}^{\infty} \frac{n!\,(-f_1)_n(-\overline{f_2})_n \hbar^{2n}}{(-f_3)_n} \overline{g_{1n}}\, g_{2n},$$

In case (a), if $a_2 < -(a_1+\hbar^2)/8$, this formula becomes simpler:
$$(g_1, g_2)_{\mathcal{P}} = \sum_{n=0}^{\infty} n!\,(-f_1)_n\, \hbar^{2n}\, \overline{g_{1n}}\, g_{2n}.$$

In case (d), we have
$$(g_1, g_2)_{\mathcal{P}} = \sum_{n=0}^{\infty} \left((-f_1)_n \hbar^n\right)^2 \overline{g_{1n}}\, g_{2n}.$$

The coherent states and the reproducing kernel in Variant VII can also be expressed in terms of hypergeometric series.

The coherent states are
$$\mathfrak{P}_z = \sum_{n=0}^{\infty} \frac{1}{n!\,(-f_1)_n(-\overline{f_2})_n} \left(\frac{z\mathbf{B}}{\hbar^3}\right)^n \mathfrak{P}_0 = {}_0F_2\left(-f_1, -\overline{f_2}; \frac{z\mathbf{B}}{\hbar^3}\right)\mathfrak{P}_0.$$

The reproducing kernel is
$$k(x) = \sum_{n=0}^{\infty} \frac{(-f_3)_n}{n!\,(-f_1)_n(-\overline{f_2})_n} \left(\frac{x}{\hbar^2}\right)^n = {}_1F_2(-f_3; -f_1, -\overline{f_2}; x/\hbar^2).$$

In case (a), for $a_2 < -(a_1+\hbar)^2/8$, we have
$$k(x) = {}_0F_1(-f_1; x/\hbar^2).$$

In case (d), we have
$$k(x) = \sum_{n=0}^{\infty} \left(\frac{\Gamma(f_1+1-n)}{\Gamma(f_1+1)}\right)^2 \left(\frac{x}{\hbar^2}\right)^n.$$

The radius of convergence $\underline{R} = \infty$.

Variant VIII, cases (a) and (d). The antiholomorphic representation is
$$\overset{\circ}{B} = \overline{z}, \qquad \overset{\circ}{C} = \hbar^4 \left(\overline{z}\frac{d}{d\overline{z}} - f_1\right)\left(\overline{z}\frac{d}{d\overline{z}} - f_2\right)\left(\overline{z}\frac{d}{d\overline{z}} - f_3\right)\frac{d}{d\overline{z}}.$$

The inner product is
$$(g_1, g_2)_\mathcal{P} = \sum_{n=0}^{\infty} n!\,(-f_1)_n(-f_2)_n(-f_3)_n\,\hbar^{4n}\,\overline{g_{1n}}\,g_{2n}.$$

In case (d), this formula becomes simpler:
$$(g_1, g_2)_\mathcal{P} = \sum_{n=0}^{\infty} \left(n!\,(-f_1)_n\,\hbar^{2n}\right)^2 \overline{g_{1n}}\,g_{2n}.$$

The coherent states are
$$\mathfrak{P}_z = \sum_{n=0}^{\infty} \frac{1}{n!\,(-f_1)_n(-f_2)_n(-f_3)_n} \left(\frac{z\mathbf{B}}{\hbar^4}\right)^n \mathfrak{P}_0 = {}_0F_3\left(-f_1, -f_2, -f_3; \frac{z\mathbf{B}}{\hbar^4}\right)\mathfrak{P}_0,$$

where ${}_0F_3$ is a hypergeometric series.

In case (d), we have
$$\mathfrak{P}_z = \sum_{n=0}^{\infty} \frac{1}{(n!\,(-f_1)_n)^2} \left(\frac{z\mathbf{B}}{\hbar^4}\right)^n \mathfrak{P}_0.$$

The reproducing kernel is
$$k(x) = \sum_{n=0}^{\infty} \frac{1}{n!\,(-f_1)_n(-f_2)_n(-f_3)_n} \left(\frac{x}{\hbar^4}\right)^n = {}_0F_3(-f_1, -f_2, -f_3; x/\hbar^4).$$

2.3. Symplectic leaves and complex structures. The classical version of the algebra (2.1), including the structure of the symplectic leaves, was investigated in [**KaN**$_3$]. Here we discuss the geometry of the symplectic leaves and related objects using general notation from § I.3.

First of all, let us look at the functions $\varkappa$ and ρ (2.3), which determine the symplectic leaves
$$\Omega_a = \{BC = \rho(A) - \rho(a),\ \varkappa(A) = \varkappa(a)\}.$$
It is convenient to introduce the new variables
$$\widetilde{A}_1 = A_1 - \hbar, \qquad \widetilde{A}_2 = A_2 + \frac{\hbar}{2}A_1,$$
and the corresponding combinations of parameters

(2.18) $$\widetilde{a}_1 = a_1 - \hbar, \qquad \widetilde{a}_2 = a_2 + \frac{\hbar}{2}a_1.$$

Then $\rho = \widetilde{A}_2^2$, $\varkappa = 4\widetilde{A}_2 + \widetilde{A}_1^2 - \hbar^2$, and so the leaves Ω_a are given by the equations

(2.19) $$S_1^2 + S_2^2 = \widetilde{A}_2^2 - \widetilde{a}_2^2, \qquad \widetilde{A}_2 + \frac{1}{4}\widetilde{A}_1^2 = \widetilde{a}_2 + \frac{1}{4}\widetilde{a}_1^2.$$

Note that, using the notation (I.3.3a), in our example, we have
$$v = (v_1, v_2), \qquad v_1(A) = 2, \qquad v_2(A) = -A_1,$$
$$f^0 = \langle v, d\rho \rangle = 2\frac{\partial \rho}{\partial A_1} - A_1\frac{\partial \rho}{\partial A_2} = -2\widetilde{A}_1\widetilde{A}_2.$$

Thus, the Poisson algebra corresponding to the quantum relations (2.1a) has the form

(2.20)
$$\begin{array}{ll}
\{C, B\} = -2i\widetilde{A}_1\widetilde{A}_2, & \{S_1, S_2\} = \widetilde{A}_1\widetilde{A}_2, \\
\{C, \widetilde{A}_1\} = 2iC, & \{S_1, \widetilde{A}_1\} = -2S_2, \\
\{C, \widetilde{A}_2\} = -i\widetilde{A}_1 C, \quad \text{or} & \{S_1, \widetilde{A}_2\} = \widetilde{A}_1 S_2, \\
\{B, \widetilde{A}_1\} = -2iB, & \{S_2, \widetilde{A}_1\} = 2S_1, \\
\{B, \widetilde{A}_2\} = i\widetilde{A}_1 B, & \{S_2, \widetilde{A}_2\} = -\widetilde{A}_1 S_1,
\end{array}$$

and, of course,
$$\{\widetilde{A}_1, \widetilde{A}_2\} = 0.$$

It follows from (2.19) that the sign of $\widetilde{a}_1$ is not felt by the geometry of the symplectic leaf Ω_a.

Note that (2.19) implies the inequalities
$$\widetilde{A}_2^2 \geq \widetilde{a}_2^2, \qquad \widetilde{A}_2 \leq \widetilde{a}_2 + \frac{1}{4}\widetilde{a}_1^2 \quad \text{over each leaf} \quad \Omega_a.$$

We obtain the following list of different configurations of leaves.
(i) Let $\widetilde{a}_2 > 0$ and $\widetilde{a}_1 \neq 0$. Then the set (2.19) consists of three leaves:
• the leaf with
$$-|\widetilde{a}_1| \leq \widetilde{A}_1 \leq |\widetilde{a}_1|, \qquad \widetilde{a}_2 \leq \widetilde{A}_2 \leq \frac{1}{4}\widetilde{a}_1^2 + \widetilde{a}_2$$
is diffeomorphic to the sphere $\mathbb{S}^2$;
• the leaf with
$$\widetilde{A}_1 \leq -(\widetilde{a}_1^2 + 8\widetilde{a}_2)^{1/2}, \qquad \widetilde{A}_2 \leq -\widetilde{a}_2, \quad \widetilde{a}_1 < 0$$
is the down sheet of a hyperboloid of two sheets;

- the leaf with
$$\widetilde{A}_1 \geq (\widetilde{a}_1^2 + 8\widetilde{a}_2)^{1/2}, \qquad \widetilde{A}_2 \leq -\widetilde{a}_2, \quad \widetilde{a}_1 > 0$$
is the down sheet of a hyperboloid of two sheets.

(ii) Let $\widetilde{a}_2 > 0$ and $\widetilde{a}_1 = 0$. Then the set (2.19) is almost the same as in the previous case (i), but the "sphere" contracts to a single point $\{S_1 = S_2 = \widetilde{A}_1 = 0, \widetilde{A}_2 = \widetilde{a}_2\}$ on which the brackets (2.20) degenerate.

(iii) Let $\widetilde{a}_2 = 0$ and $\widetilde{a}_1 \neq 0$. Then, as compared with case (i), we have to cut out two points $\{S_1 = S_2 = \widetilde{A}_2 = 0, \widetilde{A}_1 = \pm\widetilde{a}_1\}$ from the "sphere" and to cut out two points $\{S_1 = S_2 = \widetilde{A}_2 = 0, \widetilde{A}_1 = \pm\widetilde{a}_1\}$ from the "hyperboloids" (in fact, from the cones).

In this case, all two-dimensional leaves are diffeomorphic to cylinders. The closure of the first cylinder (obtained from the sphere) is compact; the closures of the two other cylinders are not compact.

The four cut-out points are zero-dimensional leaves on which the brackets (2.20) degenerate.

(iv) Let $\widetilde{a}_1 = \widetilde{a}_2 = 0$. Then there are two leaves ($\widetilde{A}_1 > 0$ or $\widetilde{A}_1 < 0$) that are cones without their vertex and a zero-dimensional leaf that is this vertex $\{S_1 = S_2 = \widetilde{A}_1 = \widetilde{A}_2 = 0\}$ on which the brackets (2.20) degenerate.

(v) Let $-\widetilde{a}_1^2/8 < \widetilde{a}_2 < 0$ and $\widetilde{a}_1 \neq 0$. Then the set (2.19) consists of three leaves:

- the leaf with
$$-(\widetilde{a}_1^2 + 8\widetilde{a}_2)^{1/2} \leq \widetilde{A}_1 \leq (\widetilde{a}_1^2 + 8\widetilde{a}_2)^{1/2}, \qquad -\widetilde{a}_2 \leq \widetilde{A}_2 \leq \frac{1}{4}\widetilde{a}_1^2 + \widetilde{a}_2$$
is diffeomorphic to the sphere $\mathbb{S}^2$;

- the leaf with
$$\widetilde{A}_1 \leq \widetilde{a}_1, \qquad \widetilde{A}_2 \leq \widetilde{a}_2, \quad \widetilde{a}_1 < 0$$
is the down sheet of a hyperboloid of two sheets;

- the leaf with
$$\widetilde{A}_1 \geq \widetilde{a}_1, \qquad \widetilde{A}_2 \leq \widetilde{a}_2, \quad \widetilde{a}_1 > 0$$
is the down sheet of a hyperboloid of two sheets.

(vi) Let $\widetilde{a}_2 = -\widetilde{a}_1^2/8$ and $\widetilde{a}_1 \neq 0$. Then the set (2.19) is almost the same as in the previous case (v), but the "sphere" contracts to a single point $\{S_1 = S_2 = \widetilde{A}_1 = 0, \widetilde{A}_2 = -\widetilde{a}_2\}$ on which the brackets (2.20) degenerate.

(vii) Let $\widetilde{a}_2 < -\widetilde{a}_1/8$. Then we have only noncompact leaves (the down sheets of the "hyperboloids") from (v).

In this list of leaves, only those entries that contain the point $\{\widetilde{A} = \widetilde{a}\}$ are assigned to irreducible representations from §2.2; in (i) this is the first leaf only; in (v) these are the second and the third leaves.

Now let us describe the complex structures on Ω_a. Following (I.1.36), we factor the difference:
$$\rho(A) - \rho(a) = \widetilde{A}_2^2 - \widetilde{a}_2^2 = (\widetilde{A}_2 - \widetilde{a}_2)(\widetilde{A}_2 + \widetilde{a}_2).$$

So, we have a natural way for choosing the factors $\mathcal{D}_a$ and $\mathcal{E}_a$ in (I.1.36):

(2.21) $$\mathcal{D}_a(A) = -(\widetilde{A}_2 + \widetilde{a}_2), \qquad \mathcal{E}_a(A) = -(\widetilde{A}_2 - \widetilde{a}_2),$$

where $\widetilde{a}$ is related to $a = (a_1, a_2)$ by (2.18) (for convenience, we choose the sign "minus" in both factors (2.21)).

The factorization (2.21) generates functions $\mathcal{B}_a$ and $\mathcal{C}_a$ by (I.1.39). This way of choosing $\mathcal{B}_a$ and $\mathcal{C}_a$ is exactly Variant II from Table 2.

In our particular situation, we can obtain the function X_a and the complex structure on Ω_a from (I.3.6) and (I.3.42):

$$X_a(A) = \frac{\widetilde{A}_2 - \widetilde{a}_2}{\widetilde{A}_2 + \widetilde{a}_2}, \qquad \frac{\widetilde{A}_2 - \widetilde{a}_2}{\widetilde{A}_2 + \widetilde{a}_2} = |z|^2,$$

$$z = -\frac{C}{\widetilde{A}_2 + \widetilde{a}_2} = -\frac{S_1 + iS_2}{A_2 + \hbar A_1/2 + a_2 + \hbar a_1/2},$$

or

$$C = S_1 + iS_2 = -\frac{2(a_2 + \hbar a_1/2)}{1 - |z|^2} z, \qquad |z| < 1.$$

Note that this complex structure degenerates in cases (d) and (e) in which $a_2 = -\hbar a_1/2$.

We conclude this section with the list of complex structures on the sphere, the hyperboloid, or the cylinder that refer to other Variants I, III–VIII from Table 2.

$$\text{Variant I:} \qquad z = -\frac{2(S_1 + iS_2)}{(\widetilde{A}_1 + \widetilde{a}_1)(\widetilde{A}_2 + \widetilde{a}_2)},$$

$$\text{Variant III:} \qquad z = \frac{4(S_1 + iS_2)}{(\widetilde{A}_1 + \widetilde{a}_1)(\widetilde{A}_2 + \sqrt{\widetilde{a}_1^2 + 8\widetilde{a}_2})},$$

$$\text{Variant IV:} \qquad z = \frac{4(S_1 + iS_2)}{(\widetilde{A}_1 + \widetilde{a}_1)(\widetilde{A}_1 - \sqrt{\widetilde{a}_1^2 + 8\widetilde{a}_2})},$$

$$\text{Variant V:} \qquad z = \frac{2(S_1 + iS_2)}{\widetilde{A}_1 + \widetilde{a}_1},$$

$$\text{Variant VI:} \qquad z = \frac{2(S_1 + iS_2)}{\widetilde{A}_1 - \sqrt{\widetilde{a}_1^2 + 8\widetilde{a}_2}},$$

$$\text{Variant VII:} \qquad z = \frac{2(S_1 + iS_2)}{\widetilde{A}_1 + \sqrt{\widetilde{a}_1^2 + 8\widetilde{a}_2}},$$

$$\text{Variant VIII:} \qquad z = S_1 + iS_2.$$

§3. Quadratic Faddeev–Sklyanin algebra

3.1. The set of irreducible representations. There is a deep theory of algebras with quadratic commutation relations generated by solutions of the quantum Yang–Baxter equation; see [**FRT, J**]. Here we consider an important example known as the Faddeev–Sklyanin algebra (see, for instance, [**KaM**[4], Appendix II]), which is a degenerate case of the elliptic Sklyanin algebra [**Sk**]. We apply our general results to this particular example, construct irreducible representations and coherent states, and connect this algebra with the theory of basic q-hypergeometric functions.

Let us consider four Hermitian generators $\mathbf{S}_0$, $\mathbf{S}_1$, $\mathbf{S}_2$, and $\mathbf{S}_3$ with quadratic commutation relations

(3.1)
$$[\mathbf{S}_0, \mathbf{S}_1] = i\mu(\mathbf{S}_2\mathbf{S}_3 + \mathbf{S}_3\mathbf{S}_2), \qquad [\mathbf{S}_1, \mathbf{S}_2] = i\hbar(\mathbf{S}_0\mathbf{S}_3 + \mathbf{S}_3\mathbf{S}_0),$$
$$[\mathbf{S}_0, \mathbf{S}_2] = -i\mu(\mathbf{S}_1\mathbf{S}_3 + \mathbf{S}_3\mathbf{S}_1), \qquad [\mathbf{S}_2, \mathbf{S}_3] = i\mu(\mathbf{S}_0\mathbf{S}_1 + \mathbf{S}_1\mathbf{S}_0),$$
$$[\mathbf{S}_0, \mathbf{S}_3] = 0, \qquad [\mathbf{S}_3, \mathbf{S}_1] = i\mu(\mathbf{S}_0\mathbf{S}_2 + \mathbf{S}_2\mathbf{S}_0),$$
$$\mathbf{S}_j^* = \mathbf{S}_j, \qquad j = 0, 1, 2, 3.$$

Here μ and $\hbar$ are two parameters, $1 > \mu > 0$, $\hbar > 0$.

First of all, note that *relations* (3.1) *cannot be represented by differential operators*. Indeed, if there is such a representation and p_j are the orders of differential operators that represent the generators $\mathbf{S}_j$, then the first relation for $[\mathbf{S}_0, \mathbf{S}_1]$ in (3.1) implies $p_2 + p_3 \leq p_0 + p_1 - 1$; the relation for $[\mathbf{S}_2, \mathbf{S}_3]$ implies $p_0 + p_1 \leq p_2 + p_3 - 1$; and hence $p_0 + p_1 + 1 \leq p_2 + p_3 \leq p_0 + p_1 - 1$, which is impossible for finite p_j.

LEMMA 3.1. *Suppose that the common spectrum of the commuting operators* $\mathbf{S}_0$ *and* $\mathbf{S}_3$ *has a discrete component in a certain Hermitian representation. Then this representation of the algebra* (3.1) *has a vacuum vector.*

PROOF. The algebra (3.1) has the following Casimir elements:

(3.2) $$\widehat{\varkappa} = \mathbf{S}_3^2 - \mathbf{S}_0^2, \qquad \mathbf{K} = \mathbf{S}_1^2 + \mathbf{S}_2^2 + \hbar \frac{1+\mu^2}{2\mu}(\mathbf{S}_3^2 + \mathbf{S}_0^2).$$

Thus, the discrete spectrum of $\mathbf{S}_0$ and $\mathbf{S}_3$ is bounded in any irreducible representation. The same holds for the operators

$$\mathbf{A}_1 \stackrel{\text{def}}{=} \mathbf{S}_3 + \mathbf{S}_0, \qquad \mathbf{A}_2 \stackrel{\text{def}}{=} \mathbf{S}_3 - \mathbf{S}_0.$$

We denote $\mathbf{S}_\pm = \mathbf{S}_1 \pm i\mathbf{S}_2$. Then it follows from (3.1) that

$$\mathbf{A}_1 \mathbf{S}_- = q \mathbf{S}_- \mathbf{A}_1, \qquad \mathbf{A}_1 \mathbf{S}_+ = \frac{1}{q} \mathbf{S}_+ \mathbf{A}_1,$$
$$\mathbf{A}_2 \mathbf{S}_- = \frac{1}{q} \mathbf{S}_- \mathbf{A}_2, \qquad \mathbf{A}_2 \mathbf{S}_+ = q \mathbf{S}_+ \mathbf{A}_2,$$

where

$$q \stackrel{\text{def}}{=} \frac{1-\mu}{1+\mu}, \qquad 0 < q < 1.$$

Thus, if the maximal eigenvalue of the operator $\mathbf{A}_1$ in a certain irreducible representation is positive ($a_1 > 0$), and if $\mathfrak{P}_0$ is the corresponding eigenvector, then $\mathbf{S}_+ \mathfrak{P}_0 = 0$ (since $\mathbf{A}_1 \mathbf{S}_+ \mathfrak{P}_0 = (a_1/q) \mathbf{S}_+ \mathfrak{P}_0$ but $a_1/q > a_1$, and so a_1/q cannot be an eigenvalue of $\mathbf{A}_1$). So, the spectrum of $\mathbf{A}_1$ consists of positive numbers $q^n a_1$, $n = 0, 1, 2, \ldots$, and the space of the irreducible representation is the linear envelope of the eigenvectors $\mathbf{S}_-^n \mathfrak{P}_0$. In this case, the spectrum of $\mathbf{A}_2$ consists of the numbers $q^{-n} a_2$, $n = 0, 1, 2, \ldots$, where a_2 is a value of $\mathbf{A}_2$ on $\mathfrak{P}_0$ (the sign of a_2 is arbitrary).

The same is true if the minimal eigenvalue a_1 of the operator $\mathbf{A}_1$ is negative ($a_1 < 0$) and $\mathfrak{P}_0$ is the corresponding eigenvector. In this case, the spectrum of $\mathbf{A}_1$ consists of negative numbers $q^n a_1$, $n = 0, 1, 2, \ldots$.

If the spectrum of $\mathbf{A}_1$ consists of a single zero point, $a_1 = 0$, then we must study the operator $\mathbf{A}_2$. If the maximal eigenvalue a_2 of $\mathbf{A}_2$ is positive ($a_2 > 0$), and $\mathfrak{P}_0$ is the corresponding eigenvector, then $\mathbf{S}_- \mathfrak{P}_0 = 0$, the space of the representation is the span of vectors $\mathbf{S}_+^n \mathfrak{P}_0$, and the spectrum of $\mathbf{A}_2$ consists of positive numbers $q^n a_2$, $n = 0, 1, 2, \ldots$.

Finally, if $\mathbf{A}_1 = 0$ and the minimal eigenvalue a_2 of $\mathbf{A}_2$ is negative ($a_2 < 0$), then for $\mathfrak{P}_0$ we choose the corresponding eigenvector, and $\mathbf{S}_- \mathfrak{P}_0 = 0$.

In the trivial case $\mathbf{A}_1 = \mathbf{A}_2 = 0$, the representation is one-dimensional, $\mathbf{S}_1$ and $\mathbf{S}_2$ commute, and $\mathbf{S}_- = \mathbf{S}_+ = 0$.

In all cases, the representation possesses the vacuum vector $\mathfrak{P}_0$ with properties (I.1.2); for the corresponding annihilation operator, we have $\mathbf{C} = \mathbf{S}_+$ in the first two cases and $\mathbf{C} = \mathbf{S}_-$ in the second two cases. The lemma is proved. $\square$

So, if one studies only the representations with discrete spectrum of $\mathbf{S}_0$ and $\mathbf{S}_3$, then without loss of generality it is possible to choose

$$\mathbf{B} = \mathbf{S}_1 - i\mathbf{S}_2, \qquad \mathbf{C} = \mathbf{S}_1 + i\mathbf{S}_2, \qquad \mathbf{A}_1 = \mathbf{S}_3 + \mathbf{S}_0, \qquad \mathbf{A}_2 = \mathbf{S}_3 - \mathbf{S}_0,$$

to reduce the algebra (3.1) to the following one:

(3.3)
$$[\mathbf{C}, \mathbf{B}] = \hbar\,(\mathbf{A}_1^2 - \mathbf{A}_2^2),$$
$$\mathbf{A}_1 \mathbf{B} = q\,\mathbf{B}\mathbf{A}_1, \qquad \mathbf{C}\mathbf{A}_1 = q\,\mathbf{A}_1 \mathbf{C},$$
$$\mathbf{A}_2 \mathbf{B} = \frac{1}{q}\,\mathbf{B}\mathbf{A}_2, \qquad \mathbf{C}\mathbf{A}_2 = \frac{1}{q}\,\mathbf{A}_2 \mathbf{C},$$
$$\mathbf{B}^* = \mathbf{C}, \qquad \mathbf{A}_\mu^* = \mathbf{A}_\mu,$$

and to consider representations with the vacuum vector

(3.3a) $\qquad \mathbf{C}\mathfrak{P}_0 = 0, \qquad \mathbf{A}_1 \mathfrak{P}_0 = a_1 \mathfrak{P}_0, \qquad \mathbf{A}_2 \mathfrak{P}_0 = a_2 \mathfrak{P}_0.$

The Casimir elements in the algebra (3.3) are derived from (3.2):

(3.3b) $\qquad \varkappa(\mathbf{A}) = \mathbf{A}_1 \mathbf{A}_2, \qquad \mathbf{K} = \mathbf{B}\mathbf{C} + \lambda(\mathbf{A}_1^2 + q^2 \mathbf{A}_2^2),$

where

(3.3c) $\qquad \lambda = \hbar/(1 - q^2).$

The mapping $\varphi \colon \mathbb{R}^2 \to \mathbb{R}^2$ and the function $f \colon \mathbb{R}^2 \to \mathbb{R}$ (from the general notation (I.1.1)) in this example read

$$f(A) = \hbar(A_1^2 - A_2^2), \qquad \varphi(A) = \left(qA_1, \frac{1}{q}A_2\right).$$

Thus, the function (I.1.4) has the form

(3.4)
$$\mathcal{F}_a(n) \equiv \frac{1}{n+1}\sum_{j=0}^{n} f(\varphi^j(a)) = \frac{\hbar}{n+1}\sum_{j=0}^{n}\left(q^{2j} a_1^2 - \frac{a_2^2}{q^{2j}}\right)$$
$$= \frac{\lambda(1 - q^{2n+2})(q^{2n} a_1^2 - a_2^2)}{q^{2n}(n+1)}, \qquad a = (a_1, a_2) \in \mathbb{R}^2.$$

Now we can describe the set $\mathcal{R}$ of irreducible Hermitian representations of (3.3) and (3.3a).

If $a_1^2 < a_2^2$, then $a \notin \mathcal{R}$, since $\mathcal{F}_a(0) < 0$.

If $a_1^2 = a_2^2$, then $a \in \mathcal{R}$, but in this case the representation is trivial: $N = 0$. So, now we assume that $a_1^2 > a_2^2$.

Case (a): $a_2 = 0$, $a_1 \neq 0$. In this case, $\mathcal{F}_a > 0$ everywhere on $\mathbb{Z}_+$; hence $a \in \mathcal{R}$ and the corresponding representations of the algebra (3.3) are infinite-dimensional, $N = \infty$.

Case (b): $a_2 \neq 0$, $a_1^2 > a_2^2$. In this case, $\mathcal{F}_a(n)$ is negative for sufficiently large $n \in \mathbb{Z}_+$. Thus, the only possibility for the inclusion $a \in \mathcal{R}$ is given by the quantization condition

$$\left(\frac{a_2}{a_1}\right)^2 = q^{2N}, \quad N \in \mathbb{N}. \tag{3.5}$$

Then $\mathcal{F}_a(n) > 0$ for $0 \leq n \leq N - 1$ and $\mathcal{F}_a(N) = 0$; so, the case (I.1.5) is realized and, in this case, the irreducible representation is of dimension $N + 1$.

3.2. Formulas for representations and coherent states.
The next step in our general scheme (§ I.1) is to choose a factorization (I.1.6). There are infinitely many ways for obtaining a factorization of the function (3.4). Here we consider only three variants. In the first variant, the coherent states and the reproducing kernel cannot be realized by basic hypergeometric series (in the second and third variants, they can be realized). The third (remarkable!) variant will be considered in most detail.

Variant I, cases (a) and (b). Let us take

$$\mathcal{B}_a(n) = \frac{(1 - q^{2n+2})(a_1^2 q^{2n} - a_2^2)}{q^{2n}(n+1)}, \quad \mathcal{C}_a(n) = \lambda.$$

Then the operators of the irreducible representation of the algebra (3.3) are

$$\overset{\circ}{B} = a_1^2 \mathcal{S}_{(q^2)}^{(1)} - a_2^2 q^2 \mathcal{S}_{(1)}^{(1/q^2)}, \quad \overset{\circ}{C} = \lambda \frac{d}{d\bar{z}}, \quad \overset{\circ}{A}_1 = a_1 I^{(q)}, \quad \overset{\circ}{A}_2 = a_2 I^{(1/q)},$$

where $I^{(q)}$ is the q-dilation operator (I.2.13) and $\mathcal{S}_{(\alpha)}^{(\beta)}$ is the integral operator defined by

$$(\mathcal{S}_{(\alpha)}^{(\beta)} g)(\bar{z}) = \int_{\alpha\bar{z}}^{\beta\bar{z}} g(\bar{z}')\, d\bar{z}'.$$

The inner product in the space $\mathcal{P}$ of antiholomorphic distributions with the vacuum 1, where the operators of representation act, is as follows:
- case (a)

$$(g_1, g_2)_{\mathcal{P}} = \sum_{n=0}^{\infty} \frac{(n!)^2}{(q^2; q^2)_n} \left(\frac{\lambda}{a_1^2}\right)^n \overline{g_{1n}}\, g_{2n};$$

- case (b)

$$(g_1, g_2)_{\mathcal{P}} = \sum_{n=0}^{N} \frac{(q^2; q^2)_{N-n}(n!)^2}{(q^2; q^2)_N\,(q^2; q^2)_n} \left(\frac{\lambda}{a_1^2}\right)^n \overline{g_{1n}}\, g_{2n}.$$

Here the shifted q-factorial $(\cdot; q)$ was defined in (I.2.22a).

The coherent states are

$$\mathfrak{P}_z = \exp\left\{\frac{z\mathbf{B}}{\lambda}\right\} \mathfrak{P}_0.$$

The reproducing kernel is $\mathcal{K}(\bar{z}, z) = k(|z|^2)$, where

- case (a)
$$k(x) = \sum_{n=0}^{\infty} \frac{(q^2;q^2)_n}{(n!)^2}\left(\frac{a_1^2 x}{\lambda}\right)^n, \qquad 0 \le x < \infty;$$

- case (b)
$$k(x) = \sum_{n=0}^{N} \frac{(q^2;q^2)_N (q^2;q^2)_n}{(q^2;q^2)_{N-n}(n!)^2}\left(\frac{a_1^2 x}{\lambda}\right)^n.$$

Variant II, cases (a) and (b). Let us take
$$\mathcal{B}_a(n) = a_1^2 q^{2n} - a_2^2, \qquad \mathcal{C}_a(n) = \frac{\lambda(1-q^{2n+2})}{q^{2n}(n+1)}.$$

In this variant, the antiholomorphic representation of the algebra (3.3) in both cases (a) and (b) is
$$\overset{\circ}{B} = \bar{z}\left(a_1^2 I^{(q^2)} - a_2^2\right), \qquad \overset{\circ}{C} = \hbar\, \delta^{(1/q^2)};$$
and $\overset{\circ}{A}_1$ and $\overset{\circ}{A}_2$ are the same as in the previous variant. Here the difference operator $\delta^{(\alpha)}$ is defined in (I.2.13a).

The inner product is
- case (a)
$$(g_1,g_2)_\mathcal{P} = \sum_{n=0}^{\infty} \frac{(q^2;q^2)_n}{q^{2n(n-1)}}\left(\frac{\lambda}{a_1^2}\right)^n \overline{g_{1n}}\, g_{2n};$$

- case (b)
$$(g_1,g_2)_\mathcal{P} = \sum_{n=0}^{N} \frac{(q^2;q^2)_n (q^2;q^2)_{N-n}}{(q^2;q^2)_N\, q^{2n(n-1)}}\left(\frac{\lambda}{a_1^2}\right)^n \overline{g_{1n}}\, g_{2n}.$$

The coherent states are
$$\mathfrak{P}_z = \sum_{n=0}^{\infty} \frac{q^{n(n-1)}}{(q^2;q^2)_n}\left(\frac{z\mathbf{B}}{\lambda}\right)^n \mathfrak{P}_0 = {}_0\Phi_0\left(q^2; -\frac{z\mathbf{B}}{\lambda}\right)\mathfrak{P}_0 = \prod_{n=0}^{\infty}\left(1 + q^{2n}\cdot\frac{z\mathbf{B}}{\lambda}\right)^n \mathfrak{P}_0.$$

In the last expression we used the famous Euler summation formula; ${}_r\Phi_t$ is the basic hypergeometric series (see [**GaR**] and § I.2.4).

The reproducing kernel $\mathcal{K}(\bar{z},z) = k(|z|^2)$ satisfies the Cauchy problem for the difference equation
$$\frac{1}{x}\left(k(x/q^2) - k(x)\right) = \frac{1}{\lambda q^2}\left(a_1^2 k(q^2 x) - a_2^2 k(x)\right), \qquad k(0) = 1.$$

The solution is
- case (a)
$$k(x) = {}_0\Phi_1\left(0; q^2; \frac{a_1^2 x}{\lambda}\right) = \sum_{n=0}^{\infty} \frac{q^{2n(n-1)}}{(q^2;q^2)_n}\left(\frac{a_1^2 x}{\lambda}\right)^n, \qquad 0 \le x < \infty;$$

- case (b)
$$k(x) = \sum_{n=0}^{N} \frac{(q^2;q^2)_N\, q^{2n(n-1)}}{(q^2;q^2)_n (q^2;q^2)_{N-n}}\left(\frac{a_1^2 x}{\lambda}\right)^n.$$

Variant III, cases (a) and (b). Let us take

$$\mathcal{B}_a(n) = a_1^2 q^n - a_2^2 q^{-n}, \qquad \mathcal{C}_a(n) = \frac{\lambda(1 - q^{2n+2})}{q^n(n+1)}.$$

In this variant, the antiholomorphic representation of the algebra (3.3) in both cases (a) and (b) is given by

(3.6)
$$\overset{\circ}{B} = \overline{z}\left(a_1^2 I^{(q)} - a_2^2 I^{(1/q)}\right), \qquad \overset{\circ}{C} = \frac{\lambda q}{\overline{z}}\left(I^{(1/q)} - I^{(q)}\right),$$
$$\overset{\circ}{A}_1 = a_1 I^{(q)}, \qquad \overset{\circ}{A}_2 = a_2 I^{(1/q)}.$$

The inner product is the following:
- case (a)
$$(g_1, g_2)_{\mathcal{P}} = \sum_{n=0}^{\infty} \frac{(q^2; q^2)_n}{q^{n(n-1)}} \left(\frac{\lambda}{a_1^2}\right)^n \overline{g_{1n}}\, g_{2n};$$

- case (b)
$$(g_1, g_2)_{\mathcal{P}} = \sum_{n=0}^{N} \frac{(q^2; q^2)_n (q^2; q^2)_{N-n}}{(q^2; q^2)_N\, q^{n(n-1)}} \left(\frac{\lambda}{a_1^2}\right)^n \overline{g_{1n}}\, g_{2n}.$$

The coherent states are

$$\mathfrak{P}_z = \sum_{n=0}^{\infty} \frac{q^{n(n-1)/2}}{(q^2;q^2)_n} \left(\frac{z\mathbf{B}}{\lambda}\right)^n \mathfrak{P}_0 = {}_1\Phi_1\!\left(0; -q; q; -\frac{z\mathbf{B}}{\lambda}\right)\mathfrak{P}_0.$$

The reproducing kernel $\mathcal{K}(\overline{z}, z) = k(|z|^2)$ can be calculated as follows:
- case (a)

(3.7)
$$k(x) = \sum_{n=0}^{\infty} \frac{q^{n(n-1)}}{(q^2;q^2)_n} \left(\frac{a_1^2 x}{\lambda}\right)^n = {}_0\Phi_0\!\left(q^2; -\frac{a_1^2 x}{\lambda}\right)$$
$$= \prod_{n=0}^{\infty}\left(1 + q^{2n}\cdot \frac{a_1^2 x}{\lambda}\right), \qquad 0 \le x < \infty;$$

- case (b)

(3.8) $$k(x) = \sum_{n=0}^{N} \frac{(q^2;q^2)_N\, q^{n(n-1)}}{(q^2;q^2)_n (q^2;q^2)_{N-n}} \left(\frac{a_1^2 x}{\lambda}\right)^n = \prod_{n=0}^{N-1}\left(1 + q^{2n}\cdot\frac{a_1^2 x}{\lambda}\right).$$

We can also derive the quantum complex structure in this variant. Indeed, we have $\mathcal{A}_a(n) = (q^n a_1; q^{-n} a_2)$, and the formula for $\mathcal{B}_a(n)$ implies

$$\mathcal{B}_a(n) = \mathcal{D}_a\bigl(\mathcal{A}_a(n+1)\bigr), \qquad \mathcal{D}_a(A) = \frac{a_1 A_1}{q} - q a_2 A_2.$$

Simultaneously, we note that the function $\mathcal{E}_a$ from (I.1.39) is given by the formula

$$\mathcal{E}_a(A) = \lambda q\left(\frac{A_2}{a_2} - \frac{A_1}{a_1}\right).$$

These functions $\mathcal{D}_a$ and $\mathcal{E}_a$ satisfy the identity (I.1.36) provided that $\varkappa(A) = A_1 A_2 = a_1 a_2 = \varkappa(a)$; see the last remark at the end of § I.1.3.

Thus, we have found the function $\mathcal{D}_a$ from (I.1.39), and so, it follows from (I.1.46) that
$$\widehat{\overline{z}} = q\,(a_1\mathbf{A}_1 - q^2 a_2 \mathbf{A}_2)^{-1}\mathbf{B}.$$
Taking into account the values of Casimir elements (3.3b)
$$\mathbf{A}_1\mathbf{A}_2 = a_1 a_2 \cdot \mathbf{I}, \qquad \mathbf{BC} = \lambda(a_1^2 + q^2 a_2^2 - \mathbf{A}_1^2 - q^2 \mathbf{A}_2^2),$$
we obtain the following formula for the operators of complex structure:

(3.9)
$$\widehat{\overline{z}} = \frac{q}{a_1} \cdot \frac{\mathbf{A}_1}{\mathbf{A}_1^2 - q^2 a_2^2} \cdot \mathbf{B}, \qquad \widehat{z} = \widehat{\overline{z}}^* = \frac{q}{a_1}\cdot \mathbf{C}\cdot \frac{\mathbf{A}_1}{\mathbf{A}_1^2 - q^2 a_2^2},$$
$$|\widehat{z}|^2 = \frac{\lambda q^2}{a_1^2} \cdot \frac{a_1^2 - \mathbf{A}_1^2}{\mathbf{A}_1^2 - q^2 a_2^2}.$$

Note that the operator $\mathbf{A}_1$ in the irreducible representation with vacuum (3.3a) has the following spectrum:
$$\operatorname{spec}(\mathbf{A}_1) = \{a_1 q^n \mid 0 \le n \le N\},$$
where $N = \infty$ in case (a). Note that $a_1 \neq 0$, $q < 1$. Therefore, the spectrum of the operator $\mathbf{A}_1^2 - q^2 a_2^2$ is positive and the spectrum of the operator $a_1^2 - \mathbf{A}_1^2$ is nonnegative; so, all operators in (3.9) are well defined and the operator $|\widehat{z}|^2$ is positive.

3.3. Reproducing measure and the Wick product. It is remarkable that in Variant III, the reproducing measure dm in the resolution of the identity and in the reproducing property (see (I.3.31), (I.3.32)) can be calculated explicitly. Namely, $dm = M(|z|^2)\,d\bar{z}dz$, where $M = k\ell$, and the function ℓ satisfies problem (I.3.24), which in our example reads
$$\frac{1 - q^{2D}}{a_1^2 q^{-2} - a_2^2 q^{2D}}\ell(x) + \frac{x}{\lambda}\ell(x) = 0, \qquad \frac{1}{\hbar}\int_0^\infty \ell(x)\,dx = 1,$$
where $D = x\frac{d}{dx}$. Note that $q^D = I^{(q)}$ is a q-dilation operator. Therefore, this problem for the function ℓ can be solved by using a simple iteration process. We obtain the following result:

- case (a)

(3.10)
$$\ell(x) = \frac{\hbar\, a_1^2}{\lambda q^2 \ln(1/q^2)} \sum_{n=0}^\infty \frac{1}{(q^2;q^2)_n}\left(-\frac{a_1^2 x}{q^2 \lambda}\right)^n$$
$$= \frac{\hbar\, a_1^2}{\lambda q^2 \ln(1/q^2)} \cdot \prod_{n=-1}^\infty \frac{1}{1 + q^{2n} a_1^2 x/\lambda};$$

- case (b)

(3.11)
$$\ell(x) = \frac{\hbar\, a_1^2 (1 - q^{2N+2})}{\lambda q^2 \ln(1/q^2)} \cdot \prod_{n=-1}^N \frac{1}{1 + q^{2n} a_1^2 x/\lambda}.$$

Note that the function $\ell(x)$ decays exponentially as $x \to \infty$ in case (a) and decays as x^{-N-2} in case (b).

Now we can derive the product $M = k\ell$.

In case (a), using (3.7) and (3.10), we obtain

$$M(x) = \frac{\hbar a_1^2}{\lambda q^2 \ln(1/q^2)(1 + a_1^2 x/(\lambda q^2))}. \tag{3.12}$$

In case (b), using (3.8) and (3.11), we obtain

$$M(x) = \frac{\hbar a_1^2(1 - q^{2N+2})}{\lambda q^2 \ln(1/q^2)(1 + q^{2N} a_1^2 x/\lambda)(1 + a_1^2 x/\lambda q^2)}. \tag{3.13}$$

Thus, in this variant the reproducing measure dm is calculated explicitly. Moreover, we obtain a simple formula for the quantum potential:

$$F = \hbar \ln k = \hbar \sum_n \ln(1 + q^{2n} a_1^2 x/\lambda), \qquad x = |z|^2. \tag{3.14}$$

Here the summation is taken over $0 \leq n < \infty$ in case (a) and over $0 \leq n \leq N - 1$ in case (b).

Thus, in this example the quantum form is

$$\omega = i\bar{\partial}\partial F = \frac{\hbar a_1^2}{\lambda} \sum_n \frac{q^{2n}}{(1 + q^{2n} a_1^2 |z|^2/\lambda)^2} \, i\, d\bar{z} \wedge dz \tag{3.15}$$

with the same agreement about the limits of summation in cases (a) and (b).

Formulas (3.7) and (3.8) also yield the probability function (I.3.33)

$$p(\xi; \xi') = \prod_n \frac{|1 + q^{2n} a_1^2 z\bar{z}'/\lambda|^2}{(1 + q^{2n} a_1^2 |z|^2/\lambda)(1 + q^{2n} a_1^2 |z'|^2/\lambda)}.$$

Here $\lambda = \hbar(1-q^2)^{-1}$, $q = (1-\mu)(1+\mu)^{-1}$, and the product is taken over $0 \leq n < \infty$ in case (a) or over $0 \leq n \leq N-1$ in case (b). Thus, we obtain a very interesting formula for the probability function:

$$p(\xi; \xi') = \prod_n p_a^{(n)}(\xi; \xi'), \tag{3.16}$$

where the $p_a^{(n)}$ are "partial" probability functions

$$p_a^{(n)}(\xi; \xi') \stackrel{\text{def}}{=} \frac{1 + \langle e_a^n(\xi), e_a^n(\xi') \rangle}{2}.$$

The last expression coincides with the **su**(2)-probability function (1.13) corresponding to the dilated complex structure; namely, by $e_a^n(\xi)$ we denote a vector from the unit sphere, related by the stereographic projection (from the north pole) $\mathbb{S}^2 \to \mathbb{C}$ to the complex number $zq^n a_1/\sqrt{\lambda}$, where $z = z(\xi)$ is the complex coordinate of the point $\xi \in \Omega_a$.

In case (a), for large n the vectors $e_a^n(\xi)$ are close to the south pole of the sphere $\mathbb{S}^2$. Thus, the angles between $e_a^n(\xi)$ and $e_a^n(\xi')$ tend to zero and the cosines of the angles tend to 1, $\langle e_a^n(\xi), e_a^n(\xi') \rangle \to 1$, so that the product (3.16) converges in case (a).

Using the probability function, the reproducing measure, and the quantum potential, one can derive the Wick and anti-Wick products, as well as the quantum restriction onto the symplectic leaves Ω_a (or the quantum embedding of Ω_a into $\mathbb{R}^4$) following the general formulas from § I.3.

For instance, the Wick product on the leaf Ω_a corresponding to the irreducible representation (3.6) of the algebra (3.3) (or (3.1)) is given by the formula

$$(R_1 * R_2)(\overline{z}, z) = \frac{1}{2\pi\hbar} \int_{\mathbb{C}} \prod_n p_a^{(n)}(\overline{z}, w; \overline{w}, z) \, R_1(\overline{z}, w) \, R_2(\overline{w}, z) \, M(|w|^2) \, d\overline{w} \, dw,$$

where $p_a^{(n)}$ is the partial probability function and the density M is given by (3.12) and (3.13).

3.4. Symplectic leaves and complex structures. From (3.3b) we obtain the function ρ (see the general notation in § I.3):

$$\rho(A) = -\lambda(A_1^2 + q^2 A_2^2).$$

Thus, the symplectic leaf $\Omega_a \subset \mathbb{R}^4$ that corresponds to the vector $a \in \mathcal{R}$ and labels the irreducible representation of the algebra (3.3) is determined by the equations

$$S_1^2 + S_2^2 + \lambda A_+^2 = \lambda a_+^2, \qquad A_+^2 - A_-^2 = a_+^2 - a_-^2.$$

Here we denote $A_\pm = A_1 \pm qA_2$, $a_\pm = a_1 \pm qa_2$. On any leaf, we have

$$S_1^2 + S_2^2 \leq \lambda \cdot \min\{a_-^2, a_+^2\}, \qquad a_+^2 - a_-^2 \leq A_+^2 \leq a_+^2, \qquad a_-^2 - a_+^2 \leq A_-^2 \leq a_-^2.$$

Thus, the following possibilities occur.

(i) The symplectic leaf is zero-dimensional, that is, it consists of a single point; this happens either if $a_- a_+ = 0$ (in this case, each point at which $S_1 = S_2 = 0$ and $|A_\pm| = |a_\pm|$ is a separate leaf) or if $a_-^2 = a_+^2$ (in this case, all points of the circle $S_1^2 + S_2^2 = \lambda a_+^2$, $A_+ = A_- = 0$, are zero-dimensional leaves).

(ii) Let $a_+^2 = a_-^2 \neq 0$. Then the symplectic leaf Ω_a is a half of the ellipsoid given by $S_1^2 + S_2^2 + \lambda A_+^2 = \lambda a_+^2$ and by one of the following relations: $A_+ = A_- > 0$, or $A_+ = A_- < 0$, or $A_+ = -A_- > 0$, or $A_+ = -A_- < 0$. These four cases naturally correspond to the four possible combinations of signs of the numbers a_+ and a_-.

(iii) Let $a_-^2 > a_+^2 > 0$. Then the symplectic leaf Ω_a is the ellipsoid

$$(3.17) \qquad \Omega_a = \left\{ S_1^2 + S_2^2 + \lambda A_+^2 = \lambda a_+^2, \; A_- = \operatorname{sgn}(a_-)\sqrt{A_+^2 + a_-^2 - a_+^2} \right\}.$$

(iv) Let $a_+^2 > a_-^2 > 0$. Then the symplectic leaf Ω_a is the ellipsoid

$$(3.18) \qquad \Omega_a = \left\{ S_1^2 + S_2^2 + \lambda A_-^2 = \lambda a_-^2, \; A_+ = \operatorname{sgn}(a_+)\sqrt{A_-^2 + a_+^2 - a_-^2} \right\}.$$

Note that the irreducible representations constructed in §§ 3.1–3.3 in case (a) correspond to the symplectic leaves (ii), while those constructed in case (b) correspond to (iii) and (iv). Namely,

• in case (a), that is, for $a_2 = 0$ and $a_1 \neq 0$, the representation corresponds to the leaf

$$\Omega_a = \{ S_1^2 + S_2^2 + \lambda A_1^2 = \lambda a_1^2, \; A_2 = 0, \; \operatorname{sgn}(A_1) = \operatorname{sgn}(a_1)\};$$

• in case (b), that is, for $a_1^2 > a_2^2 > 0$, the representation corresponds to the leaf Ω_a (3.17) if $a_1 a_2 < 0$ and to the leaf Ω_a (3.18) if $a_1 a_2 > 0$.

In Variant III, the quantum complex structure is given by (3.9), and so, the classical complex structure on the leaf Ω_a is

$$z = C \frac{qA_1}{a_1(A_1^2 - q^2 a_2^2)}, \qquad |z|^2 = \frac{\lambda q^2}{a_1^2} \cdot \frac{(a_1^2 - A_1^2)}{(A_1^2 - q^2 a_2^2)}.$$

Here the complex coordinate and its absolute value on Ω_a are expressed via the Euclidean coordinates $A_1, C = S_1 + iS_2$ on $\mathbb{R}^4$. The inverse formulas are

$$A_1 = a_1 q \left(\frac{\lambda + a_2^2 |z|^2}{\lambda q^2 + a_1^2 |z|^2} \right)^{1/2}, \qquad A_2 = \frac{a_2}{q} \left(\frac{\lambda q^2 + a_1^2 |z|^2}{\lambda + a_2^2 |z|^2} \right)^{1/2},$$

$$\overline{B} = C = z \cdot \frac{(a_1^2 - q^2 a_2^2) \lambda}{(\lambda q^2 + a_1^2 |z|^2)^{1/2} (\lambda + a_2^2 |z|^2)^{1/2}}.$$

The function $\mathcal{H}$ (I.3.6) is given by

$$\mathcal{H}(t) = \lambda q^2 \frac{1 - q^{2t/\hbar}}{a_1^2 q^{2t/\hbar} - a_2^2 q^2}.$$

From the equation $\mathcal{H} = x$, we obtain

$$x \frac{dF_0}{dx} = \hbar n = \frac{\hbar}{2} \ln \left(\frac{1 + x a_1^2 / \lambda q^2}{1 + x a_2^2 / \lambda} \right) \bigg/ \ln \frac{1}{q}, \qquad F_0(0) = 0.$$

These equations determine the classical potential $F_0 = F_0(x)$ on Ω_a. The corresponding Kähler metric on Ω is

$$g_0 = \frac{d}{dx} \left(x \frac{dF_0}{dx} \right) = \frac{\hbar (a_1^2 - q^2 a_2^2) \lambda}{2 \ln(1/q)(\lambda q^2 + x a_1^2)(\lambda + x a_2^2)}.$$

The classical symplectic form on Ω_a is

$$\omega_0 = \frac{\hbar (a_1^2 - q^2 a_2^2) \lambda}{2 \ln(1/q)} \cdot \frac{i \, d\bar{z} \wedge dz}{(\lambda q^2 + a_1^2 |z|^2)(\lambda + a_2^2 |z|^2)}$$

(the formula for the quantum form ω was given in (3.15)).

§4. Weakly nonlinear algebras and representations with two vacuum vectors

4.1. Hermitian irreducible representations. Let us consider the following almost linear commutation relations:

(4.1)
$$[\mathbf{C}^q, \mathbf{B}_p] = \sum_{\alpha=1}^{\ell} f^{\alpha q}_{\ p} \mathbf{R}_\alpha + f^q_p(\mathbf{A}),$$

$$[\mathbf{R}_\alpha, \mathbf{B}_p] = \sum_{r=1}^{d} \mathbf{B}_r \psi^{\ r}_{\alpha p}, \qquad [\mathbf{C}^q, \mathbf{R}_\alpha] = \sum_{r=1}^{d} \psi^{\ p}_{\alpha r} \mathbf{C}^r,$$

$$\mathbf{A} \mathbf{B}_p = \mathbf{B}_p \varphi(\mathbf{A}), \qquad \mathbf{C}^p \mathbf{A} = \varphi(\mathbf{A}) \mathbf{C}^p,$$

$$[\mathbf{R}_\alpha, \mathbf{R}_\beta] = \sum_{\gamma=1}^{\ell} \chi^\gamma_{\alpha\beta} \mathbf{R}_\gamma, \qquad [\mathbf{A}_\mu, \mathbf{R}_\alpha] = 0,$$

$$[\mathbf{A}_\mu, \mathbf{A}_\nu] = 0, \qquad [\mathbf{B}_p, \mathbf{B}_q] = 0, \qquad [\mathbf{C}^p, \mathbf{C}^q] = 0,$$

$$p, q = 1, \ldots, d, \quad \mu, \nu = 1, \ldots, k, \quad \alpha, \beta = 1, \ldots, \ell.$$

Here $F^\alpha = ((f^{\alpha q}_{\ p}))$, $\Psi_\alpha = ((\psi^{\ q}_{\alpha p}))$, and $X^\alpha = ((\chi^\alpha_{\beta\gamma}))$ are matrices whose elements are either constants or φ-invariant functions of A, that is,

$$f^{\alpha q}_{\ p}(\varphi(A)) = f^{\alpha q}_{\ p}(A), \qquad \psi^{\ q}_{\alpha p}(\varphi(A)) = \psi^{\ q}_{\alpha p}(A), \qquad \chi^\alpha_{\beta\gamma}(\varphi(A)) = \chi^\alpha_{\beta\gamma}(A).$$

We are interested in the Hermitian representations

$$\mathbf{B}_p^* = \mathbf{C}^p, \qquad \mathbf{R}_\alpha^* = \mathbf{R}_\alpha, \qquad \mathbf{A}_\mu^* = \mathbf{A}_\mu$$

possessing the vacuum vector $\mathfrak{P}_0$, namely,

(4.2)
$$\mathbf{A}\mathfrak{P}_0 = a\mathfrak{P}_0, \quad \mathbf{R}_\alpha \mathfrak{P}_0 = 0 \quad (\alpha = 1, \ldots, \ell), \quad \mathbf{C}^q \mathfrak{P}_0 = 0 \quad (q = 1, \ldots, d),$$
$$a = (a_1, \ldots, a_k) \in \mathbb{R}^k, \qquad \|\mathfrak{P}_0\| = 1.$$

Of course, the structural constants and the structural functions in (4.1) must be compatible with the condition of being Hermitian:

(4.3) $\quad \varphi = \overline{\varphi}, \quad F = F^*, \quad F^\alpha = (F^\alpha)^*, \quad \Psi_\alpha = \Psi_\alpha^*, \quad X^\alpha = -\overline{X^\alpha} = (X^\alpha)^*,$

where $F = ((f_p^q))$.

For (4.1), the Jacobi identities have the form

(4.4)
(a) $\displaystyle\sum_{\varepsilon=1}^{\ell}(\chi_{\alpha\beta}^\varepsilon \chi_{\varepsilon\gamma}^\delta + \chi_{\beta\gamma}^\varepsilon \chi_{\varepsilon\alpha}^\delta + \chi_{\gamma\alpha}^\varepsilon \chi_{\varepsilon\beta}^\delta) = 0, \qquad \alpha, \beta, \gamma, \delta = 1, \ldots, \ell;$

(b) $[\Psi_\alpha, \Psi_\beta] = \displaystyle\sum_\gamma \chi_{\alpha\beta}^\gamma \Psi_\gamma, \qquad \alpha, \beta = 1, \ldots, \ell;$

(c) $\displaystyle\sum_{\alpha=1}^{\ell}(f^{\alpha q}{}_r \psi_{\alpha p}{}^s - f^{\alpha q}{}_p \psi_{\alpha s}{}^r) = \delta_r^s \bigl(f_p^q(\varphi(A)) - f_p^q(A)\bigr)$
$\qquad\qquad - \delta_p^s \bigl(f_r^q(\varphi(A)) - f_r^q(A)\bigr), \qquad p, q, r, s = 1, \ldots, d;$

(d) $[F^\alpha, \Psi_\beta] = \displaystyle\sum_{\gamma=1}^{\ell} \chi_{\beta\gamma}^\alpha F^\gamma, \qquad \alpha, \beta = 1, \ldots, \ell;$

(e) $[F(A), \Psi_\alpha] = 0, \qquad \alpha = 1, \ldots, \ell.$

We assume that (4.3) and (4.4) are satisfied. We also assume that the dimension $d > 1$. In this case, identity (4.4c) implies

$$f_p^q(\varphi(A)) - f_p^q(A) \equiv \sum_{\alpha=1}^{\ell}(f^{\alpha q}{}_r \psi_{\alpha p}{}^r - f^{\alpha q}{}_p \psi_{\alpha r}{}^r) \stackrel{\text{def}}{=} \lambda_p^q, \qquad r \neq p,$$

and therefore,

(4.5) $$F(\mathcal{A}_a(n)) = F(a) + n\Lambda,$$

where $\Lambda = ((\lambda_p^q))$ and $\mathcal{A}_a(n)$ is the former composition (I.1.4).

LEMMA 4.1. *Let* $x \in \mathbb{C}^d$, $n \in \mathbb{Z}_+$. *Then the following commutation relations hold*:[23]

(a) $[\mathbf{R}_\alpha, \langle x, \mathbf{B}\rangle^n] = n\langle x, \mathbf{B}\rangle^{n-1}\langle \Psi_\alpha x, \mathbf{B}\rangle$,

(b) $[\mathbf{C}^q, \langle x, \mathbf{B}\rangle^n] = n\langle x, \mathbf{B}\rangle^{n-2}\bigg(\langle x, \mathbf{B}\rangle \sum_{\alpha=1}^\ell (F^\alpha x)^q \mathbf{R}_\alpha + \dfrac{n-1}{2}\sum_{\alpha=1}^\ell (F^\alpha x)^q \langle \Psi_\alpha x, \mathbf{B}\rangle$
$+ \langle x, \mathbf{B}\rangle (F(A)x)^q + \dfrac{n-1}{2}\langle x, \mathbf{B}\rangle (\Lambda x)^q \bigg).$

PROOF. We derive
- commutator (a):

$$[\mathbf{R}_\alpha, \langle x, \mathbf{B}\rangle] = \langle \Psi_\alpha x, \mathbf{B}\rangle,$$
$$[\mathbf{R}_\alpha, \langle x, \mathbf{B}\rangle^n] = \sum_{j=0}^{n-1} \langle x, \mathbf{B}\rangle^{n-1-j}\langle \Psi_\alpha x, \mathbf{B}\rangle \langle x, \mathbf{B}\rangle^j = n\langle x, \mathbf{B}\rangle^{n-1}\langle \Psi_\alpha x, \mathbf{B}\rangle;$$

- commutator (b):

$$[\mathbf{C}^q, \langle x, \mathbf{B}\rangle] = \sum_{\alpha=1}^\ell (F^\alpha x)^q \mathbf{R}_\alpha + (F(\mathbf{A})x)^q,$$

$$[\mathbf{C}^q, \langle x, \mathbf{B}\rangle^n] = \sum_{j=0}^{n-1}\langle x, \mathbf{B}\rangle^{n-1-j}\bigg(\sum_{\alpha=1}^\ell (F^\alpha x)^q \mathbf{R}_\alpha + (F(\mathbf{A})x)^q\bigg)\langle x, \mathbf{B}\rangle^j$$

$$= \sum_{j=0}^{n-1}\langle x, \mathbf{B}\rangle^{n-1-j}\bigg(j\sum_{\alpha=1}^\ell (F^\alpha x)^q \langle x, \mathbf{B}\rangle^{j-1}\langle \Psi_\alpha x, \mathbf{B}\rangle$$
$$+\sum_{\alpha=1}^\ell (F^\alpha x)^q \langle x, \mathbf{B}\rangle^j \mathbf{R}_\alpha + \langle x, \mathbf{B}\rangle^j \Big(F\big(\underbrace{\varphi(\ldots(\varphi(\mathbf{A}))\ldots)}_j\big)x\Big)^q\bigg)$$

$$= \dfrac{n(n-1)}{2}\langle x, \mathbf{B}\rangle^{n-2}\sum_{\alpha=1}^\ell (F^\alpha x)^q\langle \Psi_\alpha x, \mathbf{B}\rangle$$
$$+ n\langle x, \mathbf{B}\rangle^{n-1}\sum_{\alpha=1}^\ell (F^\alpha x)^q \mathbf{R}_\alpha + n\langle x, \mathbf{B}\rangle^{n-1}(F(\mathbf{A})x)^q$$
$$+ \dfrac{n(n-1)}{2}\langle x, \mathbf{B}\rangle^{n-1}(\Lambda x)^q$$

by (I.1.4) and (4.5). □

THEOREM 4.1. *Let* $\mathcal{M}_a(n)$ *be a locally bounded function on* $\mathbb{Z}_+$ *and let* $\mathcal{M}_a(n) \neq 0$ *for all* $n \in \mathbb{Z}_+$. *Then*:

---

[23] Here and in the following we use the notation $\langle x, B\rangle \stackrel{\text{def}}{=} \sum_{p=1}^d x^p B_p$ and $(Gx)^q \stackrel{\text{def}}{=} \sum_{p=1}^d g_p^q x^p$ for $G = ((g_p^q))$.

(i) *The operators*

(4.6)
$$\overset{\circ}{B}_p = \Big(\frac{1}{2}\sum_{\alpha=1}^{\ell}(\overline{F^\alpha z})_p\langle \bar z, \Psi_\alpha \frac{\partial}{\partial \bar z}\rangle + (\overline{F(a)\bar z})_p + \frac{1}{2}(\overline{\Lambda \bar z})_p\langle \bar z, \frac{\partial}{\partial \bar z}\rangle\Big)\Big(\mathcal{M}_a\big(\langle \bar z, \frac{\partial}{\partial \bar z}\rangle\big)\Big)^{-1},$$

$$\overset{\circ}{C}{}^p = \mathcal{M}_a\big(\langle \bar z, \frac{\partial}{\partial \bar z}\rangle\big)\frac{\partial}{\partial \bar z^p}, \qquad p = 1,\ldots,d,$$

$$\overset{\circ}{R}_\alpha = \langle \bar z, \Psi_\alpha \frac{\partial}{\partial \bar z}\rangle, \qquad \alpha = 1,\ldots,\ell,$$

$$\overset{\circ}{A}_\mu = \mathcal{A}_a\big(\langle \bar z, \frac{\partial}{\partial \bar z}\rangle\big)_\mu, \qquad \mu = 1,\ldots,k,$$

determine the representation of the algebra (4.1).

(ii) *The coherent states have the form*

$$\mathfrak{P}_z = \mathfrak{P}_0 + \sum_{n\geq 1}\frac{1}{n!\,\overline{\mathcal{M}}_a(n-1)\ldots\overline{\mathcal{M}}_a(0)}\langle z,\mathbf{B}\rangle^n\mathfrak{P}_0$$

and possess the intertwining property

(4.7) $\quad \mathbf{B}_p\mathfrak{P}_z = \overline{\overset{\circ}{C}{}^p}\mathfrak{P}_z, \quad \mathbf{C}^p\mathfrak{P}_z = \overline{\overset{\circ}{B}_p}\mathfrak{P}_z, \quad \mathbf{R}_\alpha\mathfrak{P}_z = \overline{\overset{\circ}{R}_\alpha}\mathfrak{P}_z, \quad \mathbf{A}_\mu\mathfrak{P}_z = \overline{\overset{\circ}{A}_\mu}\mathfrak{P}_z.$

Here the operators $\overline{\overset{\circ}{B}_p}$, $\overline{\overset{\circ}{C}{}^p}$, $\overline{\overset{\circ}{R}_\alpha}$, *and* $\overline{\overset{\circ}{A}_\mu}$ *are obtained by complex conjugation from* (4.6); *they act with respect to the parameter* z *of the coherent states* $\mathfrak{P}_z$.

PROOF. Let us prove statement (i). To verify the commutation relations (4.1) for the operators (4.6), we use the commutation formulas

$$\langle \bar z, \frac{\partial}{\partial \bar z}\rangle\frac{\partial}{\partial \bar z^p} = \frac{\partial}{\partial \bar z^p}\Big(\langle \bar z, \frac{\partial}{\partial \bar z}\rangle - 1\Big), \qquad \langle \bar z, \frac{\partial}{\partial \bar z}\rangle \overline{z^p} = \overline{z^p}\Big(\langle \bar z, \frac{\partial}{\partial \bar z}\rangle + 1\Big),$$

$$\Big[\langle \bar z, \frac{\partial}{\partial \bar z}\rangle, \overline{z^p}\frac{\partial}{\partial \bar z^q}\Big] = 0,$$

the relation

$$f_p^q(\overset{\circ}{A}) = f(a) + \lambda_p^q \langle \bar z, \frac{\partial}{\partial \bar z}\rangle$$

that follows from (4.5), as well as conditions (4.3) and the Jacobi identities (4.4). More exactly, to prove the relation for $[\overset{\circ}{C}{}^q, \overset{\circ}{B}_p]$, one should take into account the identity (4.4c). The relation for $[\overset{\circ}{R}_\alpha, \overset{\circ}{R}_\beta]$ can be proved by using (4.4b), and that for $[\overset{\circ}{R}_\alpha, \overset{\circ}{B}_p]$ by taking into account the identities (4.4b), (4.4d), and (4.4e). For example, let us prove the formula for the last commutator:

$$\overline{[\overset{\circ}{R}_\alpha, \overset{\circ}{B}_p]} = \Big[\langle \Psi_\alpha z, \frac{\partial}{\partial z}\rangle, \frac{1}{2}\sum_{\beta=1}^{\ell}(F^\beta z)^p \langle \Psi_\beta z, \frac{\partial}{\partial z}\rangle$$

$$+ (F(a)z)^p + \frac{1}{2}(\Lambda z)^p\langle z, \frac{\partial}{\partial z}\rangle\Big]\cdot \Big(\mathcal{M}_a\big(\langle z, \frac{\partial}{\partial z}\rangle\big)\Big)^{-1}$$

$$= \Big(\frac{1}{2}\sum_{\beta=1}^{\ell}(\Psi_\alpha F^\beta z)^p \langle \Psi_\beta z, \frac{\partial}{\partial z}\rangle + \frac{1}{2}\sum_{\beta=1}^{\ell}(F^\beta z)^p \langle [\Psi_\alpha, \Psi_\beta]z, \frac{\partial}{\partial z}\rangle$$

$$+ (\Psi_\alpha F(a)z)^p + \frac{1}{2}(\Psi_\alpha \Lambda z)^p \langle z, \frac{\partial}{\partial z}\rangle \Big) \cdot \Big(\mathcal{M}_a\Big(\langle z, \frac{\partial}{\partial z}\rangle\Big)\Big)^{-1}$$

$$= \Big(\frac{1}{2}\sum_{\beta=1}^\ell (\Psi_\alpha F^\beta z)^p \langle \Psi_\beta z, \frac{\partial}{\partial z}\rangle + \frac{1}{2}\sum_{\beta,\gamma=1}^\ell \chi_{\alpha\beta}^\gamma (F^\beta z)^p \langle \Psi_\gamma z, \frac{\partial}{\partial z}\rangle$$

$$+ (F(a)\Psi_\alpha z)^p + \frac{1}{2}(\Lambda\Psi_\alpha z)^p \langle z, \frac{\partial}{\partial z}\rangle\Big) \cdot \Big(\mathcal{M}_a\Big(\langle z, \frac{\partial}{\partial z}\rangle\Big)\Big)^{-1}$$

by (4.4b), (4.4e)

$$= \Big(\frac{1}{2}\sum_{\beta=1}^\ell (\Psi_\alpha F^\beta z)^p \langle \Psi_\beta z, \frac{\partial}{\partial z}\rangle + \frac{1}{2}\sum_{\gamma=1}^\ell ([F^\gamma, \Psi_\alpha]z)^p \langle \Psi_\gamma z, \frac{\partial}{\partial z}\rangle$$

$$+ (F(a)\Psi_\alpha z)^p + \frac{1}{2}(\Lambda\Psi_\alpha z)^p \langle z, \frac{\partial}{\partial z}\rangle\Big) \cdot \Big(\mathcal{M}_a\Big(\langle z, \frac{\partial}{\partial z}\rangle\Big)\Big)^{-1} \quad \text{by (4.4d)}$$

$$= \Big(\frac{1}{2}\sum_{\gamma=1}^\ell (F^\gamma \Psi_\alpha z)^p \langle \Psi_\gamma z, \frac{\partial}{\partial z}\rangle + (F(a)\Psi_\alpha z)^p + \frac{1}{2}(\Lambda\Psi_\alpha z)^p \langle z, \frac{\partial}{\partial z}\rangle\Big)$$

$$\times \Big(\mathcal{M}_a\Big(\langle z, \frac{\partial}{\partial z}\rangle\Big)\Big)^{-1} = \sum_{r=1}^d \overline{\overset{\circ}{B}_r \psi_{\alpha p}^r}.$$

Now let us prove statement (ii). The last equality in (4.7) can be proved as follows:

$$\mathbf{A}\mathfrak{P}_z = \mathbf{A}\mathfrak{P}_0 + \sum_{n\geq 1} \frac{1}{n!\,\overline{\mathcal{M}}_a(n-1)\dots\overline{\mathcal{M}}_a(0)} \langle z, \mathbf{B}\rangle^n \underbrace{\varphi(\dots(\varphi(\mathbf{A}))\dots)}_n \mathfrak{P}_0$$

$$= a\mathfrak{P}_0 + \sum_{n\geq 1} \frac{\mathcal{A}_a(n)}{n!\,\overline{\mathcal{M}}_a(n-1)\dots\overline{\mathcal{M}}_a(0)} \langle z, \mathbf{B}\rangle^n \mathfrak{P}_0 = \overline{\overset{\circ}{A}}\mathfrak{P}_z.$$

The second equality in (4.7) is proved as follows:

$$\mathbf{B}_p\mathfrak{P}_z = \mathbf{B}_p\mathfrak{P}_0 + \sum_{n\geq 1} \frac{\overline{\mathcal{M}_a(n)}}{n!\,\overline{\mathcal{M}}_a(n)\dots\overline{\mathcal{M}}_a(0)} \langle z, \mathbf{B}\rangle^n \mathbf{B}_p\mathfrak{P}_0$$

$$= \overline{\mathcal{M}}_a\Big(\langle z, \frac{\partial}{\partial z}\rangle\Big)\Big(\frac{1}{\overline{\mathcal{M}}_a(0)}\mathbf{B}_p\mathfrak{P}_0 + \sum_{n\geq 1} \frac{1}{n!\,\overline{\mathcal{M}}_a(n)\dots\overline{\mathcal{M}}_a(0)} \langle z, \mathbf{B}\rangle^n \mathbf{B}_p\mathfrak{P}_0\Big)$$

$$= \overline{\mathcal{M}}_a\Big(\langle z, \frac{\partial}{\partial z}\rangle\Big)\frac{\partial}{\partial z^p}\Big(\mathfrak{P}_0 + \frac{1}{\overline{\mathcal{M}}_a(0)}\langle z, \mathbf{B}\rangle\mathfrak{P}_0$$

$$+ \sum_{n\geq 1} \frac{1}{(n+1)!\,\overline{\mathcal{M}}_a(n)\dots\overline{\mathcal{M}}_a(0)} \langle z, \mathbf{B}\rangle^{n+1}\mathfrak{P}_0\Big)$$

$$= \overline{\mathcal{M}}_a\Big(\langle z, \frac{\partial}{\partial z}\rangle\Big)\frac{\partial}{\partial z^p}\Big(\mathfrak{P}_0 + \sum_{n\geq 1} \frac{1}{n!\,\overline{\mathcal{M}}_a(n-1)\dots\overline{\mathcal{M}}_a(0)} \langle z, \mathbf{B}\rangle^n \mathfrak{P}_0\Big)$$

$$= \overline{\overset{\circ}{C}^p}\mathfrak{P}_z.$$

The third one follows from (4.2) and Lemma 4.1(a):

$$\mathbf{R}_\alpha \mathfrak{P}_z = \sum_{n\geq 1} \frac{1}{n!\,\overline{\mathcal{M}_a}(n-1)\ldots\overline{\mathcal{M}_a}(0)} n\langle \Psi_\alpha z, \mathbf{B}\rangle \langle z, \mathbf{B}\rangle^{n-1} \mathfrak{P}_0$$

$$= \sum_{n\geq 1} \frac{1}{n!\,\overline{\mathcal{M}_a}(n-1)\ldots\overline{\mathcal{M}_a}(0)} \langle \Psi_\alpha z, \frac{\partial}{\partial z}\rangle \langle z, \mathbf{B}\rangle^{n} \mathfrak{P}_0 = \overset{\circ}{R}_\alpha \mathfrak{P}_z.$$

Similarly, the first relation in (4.7) can be obtained from (4.2) and Lemma 4.1(b):

$$\mathbf{C}^p \mathfrak{P}_z = \sum_{n\geq 1} \frac{n}{n!\,\overline{\mathcal{M}_a}(n-1)\ldots\overline{\mathcal{M}_a}(0)} \left(\frac{n-1}{2} \sum_{\alpha=1}^{\ell} (F^\alpha z)^p \langle \Psi_\alpha z, \mathbf{B}\rangle \right.$$

$$\left. + (F(a)z)^p \langle z, \mathbf{B}\rangle + \frac{n-1}{2}(\Lambda z)^p \langle z, \mathbf{B}\rangle \right) \langle z, \mathbf{B}\rangle^{n-2} \mathfrak{P}_0$$

$$= \sum_{n\geq 0} \frac{1}{n!\,\overline{\mathcal{M}_a}(n)\ldots\overline{\mathcal{M}_a}(0)} \left(\frac{n}{2} \sum_{\alpha=1}^{\ell} (F^\alpha z)^p \langle \Psi_\alpha z, \mathbf{B}\rangle + (F(a)z)^p \langle z, \mathbf{B}\rangle \right.$$

$$\left. + \frac{n}{2}(\Lambda z)^p \langle z, \mathbf{B}\rangle \right) \langle z, \mathbf{B}\rangle^{n-1} \mathfrak{P}_0$$

$$= \sum_{n\geq 0} \frac{1}{n!\,\overline{\mathcal{M}_a}(n)\ldots\overline{\mathcal{M}_a}(0)} \left(\frac{1}{2} \sum_{\alpha=1}^{\ell} (F^\alpha z)^p \langle \Psi_\alpha z, \frac{\partial}{\partial z}\rangle + (F(a)z)^p \right.$$

$$\left. + \frac{1}{2}(\Lambda z)^p \langle z, \frac{\partial}{\partial z}\rangle \right) \langle z, \mathbf{B}\rangle^{n} \mathfrak{P}_0 = \overset{\circ}{B}_p \mathfrak{P}_z. \quad \square$$

4.2. The second "vacuum" and a new type of coherent states. For the algebra (4.1) we now consider an interesting method that allows us to construct coherent states of a completely different type than those we constructed previously.

Assume that there exists a symmetric matrix $\Gamma = \Gamma^T$ satisfying the system of equations

(4.8) $$\Gamma \Psi_\alpha + \overline{\Psi_\alpha} \Gamma = 0, \qquad \alpha = 1, \ldots, \ell.$$

Here the Ψ_α are matrices from the commutation relations (4.1).

By θ and θ_1 we denote the following functions of d variables:

$$\theta(z) \overset{\text{def}}{=} \sqrt{\langle \Gamma z, z\rangle}, \qquad \theta_1(\overline{z}) \overset{\text{def}}{=} \sqrt{\langle \Gamma^{-1}\overline{z}, \overline{z}\rangle},$$

where $\langle \Gamma z, z\rangle \overset{\text{def}}{=} \sum_{p,q=1}^{d} \Gamma_{pq} z^p z^q$, $\langle \Gamma^{-1}\overline{z}, \overline{z}\rangle \overset{\text{def}}{=} \sum_{p,q=1}^{d} (\Gamma^{-1})^{pq} \overline{z^p}\, \overline{z^q}$.

We shall consider only Hermitian representations (4.3) of the algebra (4.1) that satisfy the following conditions:

1°. The space of the representation is invariant under the action of the operator $\theta_1(\mathbf{B})$.

2°. In the space of the representation, there is a vacuum vector $\mathfrak{P}_0$ (4.2) that satisfies the following additional condition:

(4.2a) $$\mathbf{L}^p \mathfrak{P}_0 = 0, \qquad p = 1, \ldots, d,$$

where

(4.9) $$\mathbf{L}^p \stackrel{\text{def}}{=} [\mathbf{C}^p, \theta_1(\mathbf{B})].$$

LEMMA 4.2. *The operators $\theta_1(\mathbf{B})$ and $\mathbf{L}^p$ satisfy the following permutation relations*:

(a) $\mathbf{A}\theta_1(\mathbf{B}) = \theta_1(\mathbf{B})\varphi(\mathbf{A})$,
(b) $[\mathbf{R}_\alpha, \theta_1(\mathbf{B})] = 0$, $\quad \alpha = 1, \ldots, \ell$,
(c) $[\mathbf{L}^q, \mathbf{B}_p] = \lambda_p^q \theta_1(\mathbf{B})$, $\quad p, q = 1, \ldots, d$,
(d) $[\mathbf{L}^q, \theta_1(\mathbf{B})] = (\Lambda\Gamma^{-1}\mathbf{B})^q$, $\quad q = 1, \ldots, d$,

where $(\Lambda\Gamma^{-1}\mathbf{B})^q \stackrel{\text{def}}{=} \sum_{p,r=1}^d \lambda_p^q (\Gamma^{-1})^{pr} \mathbf{B}_r$.

PROOF. (a) Formula (a) is a consequence of the permutation relation $\mathbf{A}\mathbf{B}_p = \mathbf{B}_p \varphi(\mathbf{A})$ and the fact that the function θ_1 is homogeneous in the first degree, that is, $\theta_1(cB) = c\theta_1(B)$.

(b) Formula (b) follows from $[\mathbf{R}_\alpha, \mathbf{B}_p] = \sum_{q=1}^d \mathbf{B}_q \psi_{\alpha p}^q$ and equation (4.8):

$$\left[\mathbf{R}_\alpha, \left(\theta_1(\mathbf{B})\right)^2\right] = [\mathbf{R}_\alpha, \langle \Gamma^{-1}\mathbf{B}, \mathbf{B}\rangle] = \langle (\Psi_\alpha \Gamma^{-1} + \Gamma^{-1}\overline{\Psi_\alpha})\mathbf{B}, \mathbf{B}\rangle$$
$$= \langle \Gamma^{-1}(\Gamma\Psi_\alpha + \overline{\Psi_\alpha}\Gamma)\Gamma^{-1}\mathbf{B}, \mathbf{B}\rangle = 0.$$

(c) Relation (c) is derived by using the Jacobi identity for the triple of the operators $\mathbf{C}^q$, $\theta_1(\mathbf{B})$, and $\mathbf{B}_p$:

$$[\mathbf{L}^q, \mathbf{B}_p] = [[\mathbf{C}^q, \theta_1(\mathbf{B})], \mathbf{B}_p] = -[[\mathbf{B}_p, \mathbf{C}^q], \theta_1(\mathbf{B})] = \left[\sum_{\alpha=1}^\ell f^{\alpha q}_p \mathbf{R}_\alpha + f^q_p(\mathbf{A}), \theta_1(\mathbf{B})\right]$$
$$= \theta_1(\mathbf{B})\left(f^q_p(\varphi(\mathbf{A})) - f^q_p(\mathbf{A})\right) = \lambda^q_p \theta_1(\mathbf{B})$$

(see (4.5)).

(d) Since $[\mathbf{L}^q, \mathbf{B}_p]$ commutes with the operators $\mathbf{B}_r$, we have

$$[\mathbf{L}^q, \theta_1(\mathbf{B})] = \sum_{p=1}^d [\mathbf{L}^q, \mathbf{B}_p]\frac{\partial \theta_1}{\partial B_p}(\mathbf{B}) = \sum_{p,r} \lambda^q_p (\Gamma^{-1})^{pr}\mathbf{B}_r = (\Lambda\Gamma^{-1}\mathbf{B})^q. \quad \square$$

In the sequel we shall assume that the condition

(4.10) $$\text{tr}\left(\Gamma^{-1*} F(a)\left(2F(a) + \Lambda + \sum_{\alpha=1}^\ell \Psi_\alpha F^\alpha\right)\Gamma^{-1}\right) \neq 0$$

is satisfied, which implies that

(4.11) $$[\overline{\theta_1}(\mathbf{C}), \theta_1(\mathbf{B})]\mathfrak{P}_0 = c\mathfrak{P}_0, \quad c \neq 0,$$

and $\mathfrak{P}'_0 \stackrel{\text{def}}{=} \theta_1(\mathbf{B})\mathfrak{P}_0/\sqrt{c} \neq 0$.

This and Lemma 4.2 show that the vector $\mathfrak{P}'_0$ satisfies the following conditions of type (4.2):

(4.12) $$\mathbf{A}\mathfrak{P}'_0 = \varphi(a)\mathfrak{P}'_0, \quad \mathbf{R}_\alpha \mathfrak{P}'_0 = 0, \quad \mathbf{C}^q \mathfrak{P}'_0 = 0, \quad \|\mathfrak{P}'_0\| = 1$$

(but does not satisfy condition (4.2a)). Thus, $\mathfrak{P}'_0$ is the second vacuum. The space of the representation of the algebra (4.1) includes two irreducible components that correspond to nonequivalent representations in the case $\varphi(a) \neq a$. In other words, the space of the representation can be described as $H + \theta_1(\mathbf{B})H$, where H is the closure of the linear span of vectors of the form $\mathbf{B}^\alpha \mathfrak{P}_0$ (α is a multi-index) and $\theta_1(\mathbf{B})H$ is the image of the subspace H with respect to $\theta_1(\mathbf{B})$.

THEOREM 4.2. (a) *In the space of functions antiholomorphic with respect to* $z^0 \stackrel{\text{def}}{=} \theta(z), z^1, \ldots, z^d$, *the following operators determine a representation of the commutation relations* (4.1):

$$\overset{\circ}{B}_p = \left(\left(\overline{F(a)}\bar{z} \right)_p + \frac{1}{2}(\overline{\Lambda z})_p \langle \bar{z}, \frac{\partial}{\partial \bar{z}} \rangle + \frac{1}{2}\sum_{\alpha=1}^{\ell}(\overline{F^\alpha z})_p \langle \bar{z}, \Psi_\alpha \frac{\partial}{\partial \bar{z}} \rangle \right.$$
$$\left. + \frac{1}{2} \langle \overline{\Gamma z}, \bar{z} \rangle \left(\overline{\Lambda \Gamma^{-1}} \frac{\partial}{\partial \bar{z}} \right)_p \right) \cdot \left(\mathcal{C}_a \left(\langle \bar{z}, \frac{\partial}{\partial \bar{z}} \rangle \right) \right)^{-1},$$

(4.13) $\quad \overset{\circ}{C}^p = \mathcal{C}_a\left(\langle \bar{z}, \frac{\partial}{\partial \bar{z}}\rangle\right) \cdot \left(\frac{\partial}{\partial \bar{z}^p} - \left(2 \langle \bar{z}, \frac{\partial}{\partial \bar{z}} \rangle + d - 1\right)^{-1}(\overline{\Gamma z})^p \Delta_\Gamma \right),$

$$p = 1, \ldots, d,$$

$$\overset{\circ}{R}_\alpha = \langle \bar{z}, \Psi_\alpha \frac{\partial}{\partial \bar{z}} \rangle, \qquad \alpha = 1, \ldots, \ell,$$

$$\overset{\circ}{A}_\mu = \mathcal{A}_a\left(\langle \bar{z}, \frac{\partial}{\partial \bar{z}}\rangle\right)_\mu, \qquad \mu = 1, \ldots, k.$$

Here

$$\Delta_\Gamma \stackrel{\text{def}}{=} \sum_{p,q=1}^{d} \overline{(\Gamma^{-1})^{pq}} \frac{\partial}{\partial \bar{z}^p} \frac{\partial}{\partial \bar{z}^q},$$

$\mathcal{A}_a(n)$ *is the composition* (I.1.4), *and* $\mathcal{C}_a(n)$ *is an arbitrary locally bounded function,* $\mathcal{C}_a(n) \neq 0$ *for all* $n \in \mathbb{Z}_+$.

The coherent states that intertwine the abstract Hermitian representation of the algebra (4.1) *with the vacuum* (4.2), (4.2a) *and the antiholomorphic representation* (4.13) *are given by the formula*

(4.14) $\quad \mathfrak{P}_{z^0,z} = \mathfrak{P}_0 + \sum_{n \geq 1} \frac{1}{n! \, \overline{\mathcal{C}}_a(n-1) \ldots \overline{\mathcal{C}}_a(0)} \left(\langle z, \mathbf{B} \rangle + z^0 \theta_1(\mathbf{B}) \right)^n \mathfrak{P}_0.$

(b) *The space of antiholomorphic functions described in* (a) *can be decomposed into the direct sum* $\mathcal{P} + \mathcal{P}'$ *of two subspaces invariant with respect to* (4.13). *Here* $\mathcal{P}$ *is the space of functions antiholomorphic with respect to* $z^1, \ldots, z^d$ *with the vacuum vector* $g_0(\bar{z}) \equiv 1$ *such that* $\overset{\circ}{A} g_0 = a \, g_0$; *and* $\mathcal{P}'$ *is the space of functions of the form* $\bar{z}^0 g(\bar{z})$, *where* $g \in \mathcal{P}$, *with the vacuum vector* $g'_0(\bar{z}^0, \bar{z}) \equiv \bar{z}^0/\sqrt{c}$ *such that* $\overset{\circ}{A} g'_0 = \varphi(a) g'_0$.

The coherent states can be decomposed into the direct sum

(4.15) $\quad \mathfrak{P}_{z^0,z} = \mathfrak{P}_z + \sqrt{c} z^0 \cdot \mathfrak{P}'_z.$

Here $\mathfrak{P}_z \in H \stackrel{\text{def}}{=} \overline{\text{span}}\{\mathbf{B}^\alpha \mathfrak{P}_0 \mid \alpha \in \mathbb{Z}^d\}$ (the bar denotes the closure) and $\mathfrak{P}'_z \in H' = \theta_1(\mathbf{B})H$; the components of (4.15) are given by the formulas

(4.15a)
$$\mathfrak{P}_z = \mathfrak{P}_0 + \sum_{n \geq 1} \frac{1}{\overline{\mathcal{C}}_a(n-1)\ldots\overline{\mathcal{C}}_a(0)}$$
$$\times \sum_{j=0}^{[n/2]} \frac{\langle \Gamma z, z \rangle^j}{(2j)!(n-2j)!} \langle z, \mathbf{B} \rangle^{n-2j} \langle \Gamma^{-1}\mathbf{B}, \mathbf{B} \rangle^j \mathfrak{P}_0,$$

(4.15b)
$$\mathfrak{P}'_z = \mathfrak{P}'_0 + \sum_{n \geq 1} \frac{1}{\overline{\mathcal{C}}_a(n-1)\ldots\overline{\mathcal{C}}_a(0)}$$
$$\times \sum_{j=0}^{[(n-1)/2]} \frac{\langle \Gamma z, z \rangle^j}{(2j+1)!(n-2j-1)!} \langle z, \mathbf{B} \rangle^{n-2j-1} \langle \Gamma^{-1}\mathbf{B}, \mathbf{B} \rangle^j \mathfrak{P}'_0,$$

where $\mathfrak{P}'_0 = \theta_1(\mathbf{B})\mathfrak{P}_0/\sqrt{c}$.

(c) The coherent states $\mathfrak{P}_z$ (4.15a) intertwine the representation of the algebra (4.1) in the subspace H with the representation (4.13) in the subspace $\mathcal{P}$.

The coherent states $\mathfrak{P}'_z$ (4.15b) intertwine the representation of the algebra (4.1) in the subspace H' with the following representation in the subspace $\mathcal{P}'$:

$$\overset{\circ}{B}'_p = \left(\left(\overline{(F(a)\overline{z})}\right)_p + (\overline{\Lambda}\overline{z})_p + \frac{1}{2}(\overline{\Lambda}\overline{z})_p \langle \overline{z}, \frac{\partial}{\partial \overline{z}} \rangle + \frac{1}{2} \sum_{\alpha=1}^{\ell} (\overline{F^\alpha}\overline{z})_p \langle \overline{z}, \Psi_\alpha \frac{\partial}{\partial \overline{z}} \rangle \right.$$
$$\left. + \frac{1}{2} \langle \overline{\Gamma}\overline{z}, \overline{z} \rangle \left(\overline{\Lambda \Gamma^{-1}} \frac{\partial}{\partial \overline{z}}\right)_p \right) \cdot \left(\mathcal{C}_a\left(\langle \overline{z}, \frac{\partial}{\partial \overline{z}} \rangle + 1\right) \right)^{-1},$$

(4.16) $\overset{\circ}{C}'^p = \mathcal{C}_a\left(\langle \overline{z}, \frac{\partial}{\partial \overline{z}} \rangle + 1\right) \cdot \left(\frac{\partial}{\partial \overline{z}^p} - \left(2\langle \overline{z}, \frac{\partial}{\partial \overline{z}} \rangle + d + 1\right)^{-1} (\overline{\Gamma}\overline{z})^p \Delta_\Gamma \right),$

$$p = 1, \ldots, d,$$

$$\overset{\circ}{R}'_\alpha = \langle \overline{z}, \Psi_\alpha \frac{\partial}{\partial \overline{z}} \rangle, \qquad \alpha = 1, \ldots, \ell,$$

$$\overset{\circ}{A}'_\mu = \mathcal{A}_a\left(\langle \overline{z}, \frac{\partial}{\partial \overline{z}} \rangle + 1\right)_\mu, \qquad \mu = 1, \ldots, k.$$

PROOF. (a) The fact that the operators (4.13) satisfy (4.1) can be proved as previously (see the proof of Theorem 4.1).

Now we show that the intertwining property (4.7) is satisfied for the states (4.14). We shall write $\langle z, \mathbf{B} \rangle + \theta(z)\theta_1(\mathbf{B}) \equiv \mathfrak{b}(z, \mathbf{B})$. It follows from Lemma 4.2(a) that
$$\mathbf{A}\,\mathfrak{b}(z, \mathbf{B})^n = \mathfrak{b}(z, \mathbf{B})^n \underbrace{\varphi(\ldots(\varphi(\mathbf{A}))\ldots)}_{n}.$$

On the other hand, it follows from the definition of θ that

$$\langle z, \frac{\partial}{\partial z} \rangle \mathfrak{b}(z, \mathbf{B})^n = n\,\mathfrak{b}(z, \mathbf{B})^n.$$

Comparing these two identities, we obtain the relation

$$\mathbf{A}\,\mathfrak{b}(z,\mathbf{B})^n \mathfrak{P}_0 = \mathcal{A}_a\Big(\big\langle z, \frac{\partial}{\partial z}\big\rangle\Big)\mathfrak{b}(z,\mathbf{B})^n\mathfrak{P}_0,$$

which immediately yields the intertwining property of $\mathfrak{P}_z$ for the operators $\mathbf{A}$ and $\overset{\circ}{A}$.

Similarly, using Lemma 4.2(b), the condition $\mathbf{R}_\alpha \mathfrak{P}_0 = 0$, and equation (4.8), we obtain the identity

$$\mathbf{R}_\alpha\,\mathfrak{b}(z,\mathbf{B})^n\mathfrak{P}_0 = \sum_{j=0}^{n} \mathfrak{b}(z,\mathbf{B})^{n-j-1}\langle \Psi_\alpha z, \mathbf{B}\rangle\,\mathfrak{b}(z,\mathbf{B})^j \mathfrak{P}_0$$

$$= n\langle \Psi_\alpha z, \mathbf{B}\rangle\,\mathfrak{b}(z,\mathbf{B})^{n-1}\mathfrak{P}_0 = \big\langle \Psi_\alpha z, \frac{\partial}{\partial z}\big\rangle\,\mathfrak{b}(z,\mathbf{B})^n\mathfrak{P}_0,$$

which implies the intertwining property for the operators $\mathbf{R}_\alpha$ and $\overset{\circ}{R}_\alpha$.

Furthermore, taking into account the definition of the function θ, we obtain the identity[24]

$$\frac{\theta(z)}{(n+1)(2n+d-1)}\overline{\Delta_\Gamma}\,\mathfrak{b}(z,\mathbf{B})^{n+1} = \theta_1(\mathbf{B})\,\mathfrak{b}(z,\mathbf{B})^n,$$

and, as a consequence, the relation

$$\frac{1}{n+1}\Big(\frac{\partial}{\partial z^p} - \frac{1}{2n+d-1}\theta(z)\frac{\partial\theta}{\partial z^p}\overline{\Delta_\Gamma}\Big)\mathfrak{b}(z,\mathbf{B})^{n+1} = \mathbf{B}_p\,\mathfrak{b}(z,\mathbf{B})^n.$$

By using the last relation, we obtain the intertwining property for the operators $\mathbf{B}_p$ and $\overset{\circ}{C}{}^p$:

$$\mathbf{B}_p \mathfrak{P}_{z^0,z} = \Big(\frac{\partial}{\partial z^p} - \frac{1}{d-1}(\Gamma z)_p\overline{\Delta_\Gamma}\Big)\mathfrak{b}(z,\mathbf{B})\mathfrak{P}_0$$

$$+ \sum_{n\geq 2}\frac{1}{n!\,\overline{C}_a(n-2)\dots\overline{C}_a(0)}\Big(\frac{\partial}{\partial z^p} - \frac{1}{2n+d-3}(\Gamma z)_p\overline{\Delta_\Gamma}\Big)\mathfrak{b}(z,\mathbf{B})^n\mathfrak{P}_0$$

$$= \overset{\circ}{C}{}^p\mathfrak{P}_{z^0,z}.$$

Finally, Lemma 4.2(c), (d) implies

$$[\mathbf{L}^p, \mathfrak{b}(z,\mathbf{B})^j] = j\Big((\Lambda z)^p\theta_1(\mathbf{B}) + \theta(z)(\Lambda\Gamma^{-1}\mathbf{B})^p\Big)\mathfrak{b}(z,\mathbf{B})^{j-1},$$

---

[24]This identity implies the relation

$$\theta_1(\mathbf{B})\mathfrak{P}_{z^0,z} = \frac{\theta(z)}{d-1}\overline{\Delta_\Gamma}\,\mathfrak{b}(z,\mathbf{B})\mathfrak{P}_0 + \sum_{n\geq 2}\frac{\theta(z)}{n!\,\overline{C}_a(n-2)\dots\overline{C}_a(0)(2n+d-3)}\overline{\Delta_\Gamma}\,\mathfrak{b}(z,\mathbf{B})^n\mathfrak{P}_0$$

$$= \overline{C}_a\Big(\big\langle z,\frac{\partial}{\partial z}\big\rangle\Big)\Big(2\big\langle z,\frac{\partial}{\partial z}\big\rangle + d - 1\Big)^{-1}\theta(z)\overline{\Delta_\Gamma}\,\mathfrak{P}_{z^0,z}.$$

and, taking into account conditions (4.2) and (4.2a), we obtain

$$\mathbf{C}^p \mathfrak{b}(z,\mathbf{B})^n \mathfrak{P}_0 = n\Big((F(a)z)^p + \frac{n-1}{2}(\Lambda z)^p\Big)\mathfrak{b}(z,\mathbf{B})^{n-1}\mathfrak{P}_0$$
$$+ \frac{n(n-1)}{2}\Big(\sum_{\alpha=1}^{\ell}(F^\alpha z)^p\langle\Psi_\alpha z,\mathbf{B}\rangle + \theta(z)(\Lambda z)^p\theta_1(\mathbf{B})$$
$$+ \big(\theta(z)\big)^2(\Lambda\Gamma^{-1}\mathbf{B})^p\Big)\mathfrak{b}(z,\mathbf{B})^{n-2}\mathfrak{P}_0.$$

The last relation leads to the intertwining property for the operators $\mathbf{C}^p$ and $\overline{\overset{\circ}{B}}_p$:

$$\mathbf{C}^p \mathfrak{P}_{z^0,z} = \sum_{n\geq 1}\frac{1}{(n-1)!\overline{C}_a(n-1)\ldots\overline{C}_a(0)}\Big((F(a)z)^p + \frac{1}{2}(\Lambda z)^p\langle z,\frac{\partial}{\partial z}\rangle$$
$$+ \frac{1}{2}\sum_{\alpha=1}^{\ell}(F^\alpha z)^p\langle\Psi_\alpha z,\frac{\partial}{\partial z}\rangle + \frac{1}{2}(\theta(z))^2\Big(\Lambda\Gamma^{-1}\frac{\partial}{\partial z}\Big)^p\Big)\mathfrak{b}(z,\mathbf{B})^{n-1}\mathfrak{P}_0$$
$$= \overline{\overset{\circ}{B}}_p\mathfrak{P}_{z^0,z}.$$

(b) Obviously, the spaces $\mathcal{P}$ and $\mathcal{P}'$ are invariant with respect to the action of the operators (4.13) (see the explicit formulas (4.13)).

(c) This statement follows from (b). Indeed, since the operators $\mathbf{A}$, as well as the operators $\overset{\circ}{A}$, preserve the spaces H and H', the intertwining property

$$\mathbf{A}\mathfrak{P}_z + \sqrt{c}z^0\mathbf{A}\mathfrak{P}'_z = \mathbf{A}\mathfrak{P}_{z^0,z} = \overline{\overset{\circ}{A}}\mathfrak{P}_{z^0,z} = \overline{\overset{\circ}{A}}\mathfrak{P}_z + \sqrt{c}\overline{\overset{\circ}{A}}(z^0\mathfrak{P}'_z)$$

implies the identities

$$\mathbf{A}\mathfrak{P}_z = \overline{\overset{\circ}{A}}\mathfrak{P}_z, \qquad \mathbf{A}\mathfrak{P}'_z = \Big(\frac{1}{z^0}\cdot\overline{\overset{\circ}{A}}\cdot z^0\Big)\mathfrak{P}'_z.$$

Similar identities hold for the other generators of the algebra (4.1). Hence, the antiholomorphic representation corresponding to the coherent state $\mathfrak{P}_z$ coincides with (4.13), and the antiholomorphic representation corresponding to $\mathfrak{P}'_z$ can be obtained from (4.13) as follows:

$$\overset{\circ}{B}'_p = \frac{1}{\overline{\theta(z)}}\cdot\overset{\circ}{B}_p\cdot\overline{\theta(z)}, \qquad \overset{\circ}{C}'^p = \frac{1}{\overline{\theta(z)}}\cdot\overset{\circ}{C}^p\cdot\overline{\theta(z)},$$
$$\overset{\circ}{R}'_\alpha = \frac{1}{\overline{\theta(z)}}\cdot\overset{\circ}{R}_\alpha\cdot\overline{\theta(z)}, \qquad \overset{\circ}{A}'_\mu = \frac{1}{\overline{\theta(z)}}\cdot\overset{\circ}{A}_\mu\cdot\overline{\theta(z)}.$$

Explicit calculations lead to (4.16). □

§5. Eight-dimensional quadratic algebra
(quantum embedding of $T^*\mathbb{S}^3$) and Bessel coherent states

5.1. Transformation to a weakly nonlinear algebra. Let us consider the following quadratic commutation relations:

(5.1)
$$[\boldsymbol{\rho}_p, \boldsymbol{\rho}_q] = 0,$$
$$[\boldsymbol{\rho}_p, \boldsymbol{\sigma}_q] = -i\hbar\left(\delta_{pq}\boldsymbol{\rho}^2 - \boldsymbol{\rho}_p\boldsymbol{\rho}_q\right),$$
$$[\boldsymbol{\sigma}_p, \boldsymbol{\sigma}_q] = -i\hbar\left(\boldsymbol{\sigma}_p\boldsymbol{\rho}_q - \boldsymbol{\sigma}_q\boldsymbol{\rho}_p\right),$$

together with the conditions

$$\boldsymbol{\rho}_p^* = \boldsymbol{\rho}_p, \qquad \boldsymbol{\sigma}_p^* = \boldsymbol{\sigma}_p, \qquad p,q = 0,1,2,3.$$

Here $\boldsymbol{\rho}^2 \stackrel{\text{def}}{=} \sum_{p=0}^{3} \boldsymbol{\rho}_p^2$, $\hbar > 0$. The Jacobi identities are automatically satisfied for these relations.

LEMMA 5.1. *In the algebra* (5.1), *there are two Casimir elements*

$$\mathbf{K}_1 = \boldsymbol{\rho}^2, \qquad \mathbf{K}_2 = \frac{1}{2}\left(\langle \boldsymbol{\rho}, \boldsymbol{\sigma}\rangle + \langle \boldsymbol{\sigma}, \boldsymbol{\rho}\rangle\right),$$

where $\langle \boldsymbol{\rho}, \boldsymbol{\sigma}\rangle \stackrel{\text{def}}{=} \sum_{p=0}^{3} \boldsymbol{\rho}_p \boldsymbol{\sigma}_p$.

By H we denote the subspace on which

(5.2)
$$\mathbf{K}_1 = \mathbf{I}, \qquad \mathbf{K}_2 = 0.$$

These equations read as equations for a symplectic leaf corresponding to the irreducible representation of (5.1). This leaf is the surface $\{\rho^2 = 1, \langle \rho, \sigma\rangle = 0\}$ embedded into the space $\mathbb{R}^8$ with classical coordinates ρ, σ. This surface is diffeomorphic to $T^*\mathbb{S}^3$.

Our aim is to construct the coherent states, the reproducing kernel, and the quantum embedding of $T^*\mathbb{S}^3$ into $\mathbb{R}^8$ (see the general scheme in § I.3).

LEMMA 5.2. *On the subspace H, the eigenvalues of the operator $\boldsymbol{\sigma}^2$ satisfy the inequality*

$$\boldsymbol{\sigma}^2 \geq (9/4)\hbar^2.$$

PROOF. Let $x \in \mathbb{R}$. Then, obviously, the spectrum of the operator

$$\mathbf{X}(x) \stackrel{\text{def}}{=} \sum_{p=0}^{3} (\boldsymbol{\sigma}_p + ix\boldsymbol{\rho}_p)(\boldsymbol{\sigma}_p - ix\boldsymbol{\rho}_p) = \boldsymbol{\sigma}^2 + x^2\boldsymbol{\rho}^2 + ix\sum_{p=0}^{3}[\boldsymbol{\rho}_p, \boldsymbol{\sigma}_p] = \boldsymbol{\sigma}^2 + x^2\boldsymbol{\rho}^2 + 3\hbar x \boldsymbol{\rho}^2$$

is nonnegative. Hence, on the subspace H, the estimate

$$\boldsymbol{\sigma}^2 = \mathbf{X}(x) - x^2 - 3\hbar x \geq -x^2 - 3\hbar x$$

holds for all $x \in \mathbb{R}$. In particular, this inequality holds for $x = -3\hbar/2$ (the function $-x^2 - 3\hbar x$ attains its maximum at this point), that is,

$$\boldsymbol{\sigma}^2 \geq -(-3\hbar/2)^2 - 3\hbar(-3\hbar/2) = 9\hbar^2/4. \quad \square$$

REMARK 5.1. Note that for arbitrary values of Casimir elements $\mathbf{K}_1 \neq 0$ and $\mathbf{K}_2$ we have the following analog of Lemma 5.2: $\boldsymbol{\sigma}^2 \geq \frac{9}{4}\hbar^2 \mathbf{K}_1 + (\mathbf{K}_1)^{-1}(\mathbf{K}_2)^2$. For $\mathbf{K}_1 = 0$, we automatically have $\mathbf{K}_2 = 0$, and then $\boldsymbol{\sigma}^2 \geq 0$.

By Lemma 5.2, the operator

$$\mathbf{A} \stackrel{\text{def}}{=} \left(\boldsymbol{\sigma}^2 - \frac{\hbar^2}{4}\boldsymbol{\rho}^2 - \hbar^2\right)^{1/2} \tag{5.3}$$

is well defined on the subspace H, and its eigenvalues satisfy the inequality

$$\mathbf{A} \geq \hbar. \tag{5.4}$$

On the subspace H, we also define the following self-adjoint operators ($p = 0, 1, 2, 3$):

$$\begin{aligned}
\boldsymbol{\zeta}_p &\stackrel{\text{def}}{=} \mathbf{A}^{1/2}\boldsymbol{\rho}_p\mathbf{A}^{1/2}, \\
\boldsymbol{\eta}_p &\stackrel{\text{def}}{=} \frac{1}{2}\left(\mathbf{A}^{1/2}\boldsymbol{\sigma}_p\mathbf{A}^{-1/2} + \mathbf{A}^{-1/2}\boldsymbol{\sigma}_p\mathbf{A}^{1/2}\right) \\
&\quad + \frac{i\hbar}{4}\left(\mathbf{A}^{1/2}\boldsymbol{\rho}_p\mathbf{A}^{-1/2} - \mathbf{A}^{-1/2}\boldsymbol{\rho}_p\mathbf{A}^{1/2}\right).
\end{aligned} \tag{5.5}$$

THEOREM 5.1. (i) *On the subspace H defined by (5.2), the operators $\mathbf{A}$, $\boldsymbol{\zeta}_p$, and $\boldsymbol{\eta}_p$ satisfy the commutation relations*

$$\begin{aligned}
&[\mathbf{A}, \boldsymbol{\zeta}_p] = i\hbar\boldsymbol{\eta}_p, && [\mathbf{A}, \boldsymbol{\eta}_p] = -i\hbar\boldsymbol{\zeta}_p, \\
&[\boldsymbol{\zeta}_p, \boldsymbol{\zeta}_q] = -i\hbar(\boldsymbol{\eta}_p\boldsymbol{\zeta}_q - \boldsymbol{\eta}_q\boldsymbol{\zeta}_p)\mathbf{A}^{-1}, && [\boldsymbol{\eta}_p, \boldsymbol{\eta}_q] = -i\hbar(\boldsymbol{\eta}_p\boldsymbol{\zeta}_q - \boldsymbol{\eta}_q\boldsymbol{\zeta}_p)\mathbf{A}^{-1}, \\
&[\boldsymbol{\zeta}_p, \boldsymbol{\eta}_q] = -i\hbar\delta_{pq}\mathbf{A}, && p, q = 0, 1, 2, 3,
\end{aligned} \tag{5.6}$$

and the identities

$$\boldsymbol{\zeta}^2 = \boldsymbol{\eta}^2 = \mathbf{A}^2 + \hbar^2, \qquad \langle \boldsymbol{\zeta}, \boldsymbol{\eta} \rangle = -\langle \boldsymbol{\eta}, \boldsymbol{\zeta} \rangle = -2i\hbar\mathbf{A}. \tag{5.7}$$

If $\boldsymbol{\rho}_p$, $\boldsymbol{\sigma}_p$ are Hermitian, then $\boldsymbol{\zeta}_p$, $\boldsymbol{\eta}_p$ are Hermitian too.

(ii) *On the space H, the generators $\boldsymbol{\rho}_p$ and $\boldsymbol{\sigma}_p$ can be expressed via the operators $\mathbf{A}$, $\boldsymbol{\zeta}_p$, and $\boldsymbol{\eta}_p$ ($p = 0, 1, 2, 3$) as follows:*

$$\boldsymbol{\rho}_p = \mathbf{A}^{-1/2}\boldsymbol{\zeta}_p\mathbf{A}^{-1/2}, \qquad \boldsymbol{\sigma}_p = \frac{1}{2}\left(\mathbf{A}^{1/2}\boldsymbol{\eta}_p\mathbf{A}^{-1/2} + \mathbf{A}^{-1/2}\boldsymbol{\eta}_p\mathbf{A}^{1/2}\right). \tag{5.8}$$

(iii) *If the operators $\mathbf{A}$, $\boldsymbol{\zeta}$, $\boldsymbol{\eta}$ satisfy relations (5.6), (5.7) and inequality (5.4) holds, then the operators $\boldsymbol{\rho}$, $\boldsymbol{\sigma}$ defined by (5.8) satisfy relations (5.1) and (5.2).*

Prior to proving this theorem, we reformulate relations (5.6) in a "polarized" form. Let us define creation operators

$$\mathbf{B}_0 \stackrel{\text{def}}{=} -\boldsymbol{\zeta}_0 - i\boldsymbol{\eta}_0, \qquad \mathbf{B}_p \stackrel{\text{def}}{=} \boldsymbol{\eta}_p - i\boldsymbol{\zeta}_p \quad (p = 1, 2, 3),$$

and annihilation operators

$$\mathbf{C}^0 \stackrel{\text{def}}{=} -\boldsymbol{\zeta}_0 + i\boldsymbol{\eta}_0, \qquad \mathbf{C}^q \stackrel{\text{def}}{=} \boldsymbol{\eta}_q + i\boldsymbol{\zeta}_q \quad (q = 1, 2, 3).$$

Then

$$\begin{aligned}
\boldsymbol{\zeta}_0 &= -\frac{1}{2}(\mathbf{B}_0 + \mathbf{C}^0), & \boldsymbol{\zeta}_p &= \frac{i}{2}(\mathbf{B}_p - \mathbf{C}^p), \\
\boldsymbol{\eta}_0 &= \frac{i}{2}(\mathbf{B}_0 - \mathbf{C}^0), & \boldsymbol{\eta}_p &= \frac{1}{2}(\mathbf{B}_p + \mathbf{C}^p), & p &= 1, 2, 3.
\end{aligned} \tag{5.9}$$

The following lemma readily follows from (5.6) and (5.7).

LEMMA 5.3. *The operators $\mathbf{A}$, $\mathbf{B}_p$, and $\mathbf{C}^p$ satisfy the permutation relations*

(5.10)
$$\mathbf{C}^0 \mathbf{B}_0 = \mathbf{B}_0 \mathbf{C}^0 + 2\hbar \mathbf{A}, \qquad \mathbf{C}^q \mathbf{B}_p = \mathbf{B}_p(1+\hbar \mathbf{A}^{-1})\mathbf{C}^q - \hbar \mathbf{B}_q \mathbf{A}^{-1}\mathbf{C}^p + 2\hbar \delta_p^q \mathbf{A},$$
$$\mathbf{C}^p \mathbf{B}_0 = \mathbf{B}_0(1+\hbar \mathbf{A}^{-1})\mathbf{C}^p + \hbar \mathbf{B}_p \mathbf{A}^{-1}\mathbf{C}^0,$$
$$\mathbf{C}^0 \mathbf{B}_p = \mathbf{B}_p(1+\hbar \mathbf{A}^{-1})\mathbf{C}^0 + \hbar \mathbf{B}_0 \mathbf{A}^{-1}\mathbf{C}^p,$$
$$\mathbf{A}\mathbf{B}_0 = \mathbf{B}_0(\mathbf{A}+\hbar), \qquad \mathbf{C}^0 \mathbf{A} = (\mathbf{A}+\hbar)\mathbf{C}^0,$$
$$\mathbf{A}\mathbf{B}_p = \mathbf{B}_p(\mathbf{A}+\hbar), \qquad \mathbf{C}^p \mathbf{A} = (\mathbf{A}+\hbar)\mathbf{C}^p,$$
$$[\mathbf{B}_0, \mathbf{B}_p] = [\mathbf{B}_p, \mathbf{B}_q] = 0, \qquad [\mathbf{C}^0, \mathbf{C}^p] = [\mathbf{C}^p, \mathbf{C}^q] = 0,$$
$$p, q, = 1, 2, 3,$$

and the identities

(5.11)
$$\mathbf{B}_1^2 + \mathbf{B}_2^2 + \mathbf{B}_3^2 = \mathbf{B}_0^2, \qquad (\mathbf{C}^1)^2 + (\mathbf{C}^2)^2 + (\mathbf{C}^3)^2 = (\mathbf{C}^0)^2,$$
$$\frac{1}{4}(\langle \mathbf{C}, \mathbf{B}\rangle + \langle \mathbf{B}, \mathbf{C}\rangle) - \hbar^2 = \mathbf{A}^2.$$

We start proving Theorem 5.1 with the following preliminary identities.

LEMMA 5.4. *The following equalities hold for the operators $\boldsymbol{\rho}$, $\boldsymbol{\sigma}$ satisfying (5.1), (5.2) and for the operator $\mathbf{A}$ (5.3):*

(5.12)
(a) $[\boldsymbol{\rho}_p, \mathbf{A}^2] = -2i\hbar \boldsymbol{\sigma}_p, \qquad [\boldsymbol{\sigma}_p, \mathbf{A}^2] = i\hbar\left(\mathbf{A}^2 \boldsymbol{\rho}_p + \boldsymbol{\rho}_p \mathbf{A}^2 - \frac{\hbar^2}{2}\boldsymbol{\rho}_p\right);$

(b) $[\mathbf{A}, [\mathbf{A}, \boldsymbol{\rho}_p]] = \hbar^2 \boldsymbol{\rho}_p, \qquad [\mathbf{A}, [\mathbf{A}, \boldsymbol{\sigma}_p]] = \hbar^2 \boldsymbol{\sigma}_p;$

(c) $\mathbf{A}\boldsymbol{\rho}_p \mathbf{A} = \frac{1}{2}(\mathbf{A}^2 \boldsymbol{\rho}_p + \boldsymbol{\rho}_p \mathbf{A}^2) - \frac{\hbar^2}{2}\boldsymbol{\rho}_p;$

(d) $[\mathbf{A}\boldsymbol{\rho}_p \mathbf{A}, \mathbf{A}] = \frac{i\hbar}{2}(\mathbf{A}\boldsymbol{\sigma}_p + \boldsymbol{\sigma}_p \mathbf{A}) + \frac{\hbar^2}{4}[\mathbf{A}, \boldsymbol{\rho}_p];$

(e) $[\mathbf{A}, \boldsymbol{\rho}_p]\mathbf{A} = i\hbar \boldsymbol{\sigma}_p - \frac{\hbar^2}{2}\boldsymbol{\rho}_p.$

PROOF. Equalities (a) readily follow from (5.1), (5.2), (5.3) and the identity $\langle \boldsymbol{\sigma}, \boldsymbol{\rho}\rangle = \langle \boldsymbol{\rho}, \boldsymbol{\sigma}\rangle + 3i\hbar$. To prove (b), we calculate

$$[\mathbf{A}^2, [\mathbf{A}^2, \boldsymbol{\rho}_p]] = 2i\hbar[\mathbf{A}^2, \boldsymbol{\sigma}_p] = 2\hbar^2(\boldsymbol{\rho}\mathbf{A}^2 + \mathbf{A}^2 \boldsymbol{\rho}) - \hbar^4 \boldsymbol{\rho}.$$

On the other hand, we have

$$[\mathbf{A}^2, [\mathbf{A}^2, \boldsymbol{\rho}_p]] = [\mathbf{A}, [\mathbf{A}, [\mathbf{A}, [\mathbf{A}, \boldsymbol{\rho}_p]]]_+]_+,$$

where the brackets $[\cdot, \cdot]_+$ denote the anticommutator $[\mathbf{A}, \mathbf{R}]_+ \equiv \mathbf{A}\mathbf{R} + \mathbf{R}\mathbf{A}$. Thus we obtain

(5.13) $[\mathbf{A}, [\mathbf{A}, [\mathbf{A}, [\mathbf{A}, \boldsymbol{\rho}_p]]]_+]_+ = 2\hbar^2[\mathbf{A}, [\mathbf{A}, \boldsymbol{\rho}_p]_+]_+ - \hbar^4 \boldsymbol{\rho}_p - 4\hbar^2 \mathbf{A}\boldsymbol{\rho}_p \mathbf{A},$

since

$$\boldsymbol{\rho}\mathbf{A}^2 + \mathbf{A}^2 \boldsymbol{\rho} \equiv [\mathbf{A}, [\mathbf{A}, \boldsymbol{\rho}_p]_+]_+ - 2\mathbf{A}\boldsymbol{\rho}\mathbf{A}.$$

Now we substitute the expression

$$2\mathbf{A}\boldsymbol{\rho}_p \mathbf{A} = \frac{1}{2}[\mathbf{A}, [\mathbf{A}, \boldsymbol{\rho}_p]_+]_+ - \frac{1}{2}[\mathbf{A}, [\mathbf{A}, \boldsymbol{\rho}_p]]$$

into (5.13). Then (5.13) reads

(5.14) $$[\mathbf{A},[\mathbf{A},[\mathbf{A},[\mathbf{A},\boldsymbol{\rho}_p]]-\hbar^2\boldsymbol{\rho}_p]_+]_+ = \hbar^2\left([\mathbf{A},[\mathbf{A},\boldsymbol{\rho}_p]]-\hbar^2\boldsymbol{\rho}_p\right).$$

But inequality (5.4) implies the following embedding of spectra:

$$\operatorname{spec}([\mathbf{A},[\mathbf{A},\cdot]_+]_+) \subset \{(\lambda'+\lambda'')^2 \mid \lambda',\lambda'' \in \operatorname{spec}(\mathbf{A})\} \subset [4\hbar^2,\infty).$$

Thus the operator $[\mathbf{A},[\mathbf{A},\cdot]_+]_+ - \hbar^2\mathbf{I}$ is invertible, and (5.14) implies $[\mathbf{A},[\mathbf{A},\boldsymbol{\rho}_p]] - \hbar^2\boldsymbol{\rho}_p = 0$.

The second identity (b) can be proved in a similar way. Identity (c) readily follows from (b).

Let us prove (d). From (c) we have

$$\begin{aligned}[\mathbf{A},\mathbf{A}\boldsymbol{\rho}_p\mathbf{A}] &= \frac{1}{2}[\mathbf{A},\mathbf{A}^2\boldsymbol{\rho}_p + \boldsymbol{\rho}_p\mathbf{A}^2 - \hbar^2\boldsymbol{\rho}_p] \\ &= \frac{1}{2}\left(\mathbf{A}^2[\mathbf{A},\boldsymbol{\rho}_p] + [\mathbf{A},\boldsymbol{\rho}_p]\mathbf{A}^2\right) - \frac{\hbar^2}{2}[\mathbf{A},\boldsymbol{\rho}_p] \\ &= \frac{1}{2}\left(\mathbf{A}[\mathbf{A}^2,\boldsymbol{\rho}_p] - [\boldsymbol{\rho}_p,\mathbf{A}^2]\mathbf{A}\right) - \mathbf{A}^2\boldsymbol{\rho}_p\mathbf{A} + \mathbf{A}\boldsymbol{\rho}_p\mathbf{A}^2 - \frac{\hbar^2}{2}[\mathbf{A},\boldsymbol{\rho}_p].\end{aligned}$$

Now we apply (a) and obtain the desired identity (d).

Finally, let us consider the equality

(5.15) $$[\mathbf{A},\boldsymbol{\rho}_p]\mathbf{A} + \mathbf{A}[\mathbf{A},\boldsymbol{\rho}_p] = [\mathbf{A}^2,\boldsymbol{\rho}_p].$$

The second summand on the left-hand side is equal to

$$\mathbf{A}[\mathbf{A},\boldsymbol{\rho}_p] = [\mathbf{A},\boldsymbol{\rho}_p]\mathbf{A} + [\mathbf{A},[\mathbf{A},\boldsymbol{\rho}_p]] = [\mathbf{A},\boldsymbol{\rho}_p]\mathbf{A} + \hbar^2\boldsymbol{\rho}_p,$$

where we used (b). Thus from (5.15) we obtain

$$2[\mathbf{A},\boldsymbol{\rho}_p]\mathbf{A} + \hbar^2\boldsymbol{\rho}_p = [\mathbf{A}^2,\boldsymbol{\rho}_p].$$

Applying (a) to the right-hand side, we prove (e). □

PROOF OF THEOREM 5.1. (i) First, from the definition of the commutator and in view of (5.5), we derive

$$[\boldsymbol{\eta}_p,\mathbf{A}] = i\hbar\mathbf{A}^{-1/2}\left(\frac{1}{2}\mathbf{A}\boldsymbol{\rho}_p\mathbf{A} + \frac{1}{4}(\boldsymbol{\rho}_p\mathbf{A}^2 + \mathbf{A}^2\boldsymbol{\rho}_p) - \frac{\hbar^2}{4}\boldsymbol{\rho}_p\right)\mathbf{A}^{-1/2}.$$

Using (5.12) (c), we see that

$$[\boldsymbol{\eta}_p,\mathbf{A}] = i\hbar\mathbf{A}^{-1/2}\left(\frac{1}{2}\mathbf{A}\boldsymbol{\rho}_p\mathbf{A} + \frac{1}{2}\mathbf{A}\boldsymbol{\rho}_p\mathbf{A}\right)\mathbf{A}^{-1/2} = i\hbar\mathbf{A}^{1/2}\boldsymbol{\rho}_p\mathbf{A}^{1/2} = i\hbar\boldsymbol{\zeta}_p.$$

Thus *we have proved the second relation on the first line in* (5.6).

In a similar way, we have

$$\begin{aligned}[\mathbf{A},\boldsymbol{\zeta}_p] &= \mathbf{A}^{-1/2}(\mathbf{A}^2\boldsymbol{\rho}_p\mathbf{A} - \mathbf{A}\boldsymbol{\rho}_p\mathbf{A}^2)\mathbf{A}^{-1/2} = \mathbf{A}^{-1/2}[\mathbf{A},\mathbf{A}\boldsymbol{\rho}_p\mathbf{A}]\mathbf{A}^{-1/2} \\ &= i\hbar\mathbf{A}^{-1/2}\left(\frac{1}{2}(\mathbf{A}\boldsymbol{\sigma}_p + \boldsymbol{\sigma}_p\mathbf{A}) + \frac{i\hbar}{4}[\mathbf{A},\boldsymbol{\rho}_p]\right)\mathbf{A}^{-1/2},\end{aligned}$$

where we used (5.12) (d). Thus we have

$$[\mathbf{A}, \zeta_p] = i\hbar\Big(\frac{1}{2}(\mathbf{A}^{1/2}\boldsymbol{\sigma}_p\mathbf{A}^{-1/2} + \mathbf{A}^{-1/2}\boldsymbol{\sigma}_p\mathbf{A}^{1/2}) \\ + \frac{i\hbar}{4}(\mathbf{A}^{1/2}\boldsymbol{\rho}_p\mathbf{A}^{-1/2} - \mathbf{A}^{-1/2}\boldsymbol{\rho}_p\mathbf{A}^{1/2})\Big).$$

By using definition (5.4), *we obtain the first relation in* (5.6), namely, $[\mathbf{A}, \zeta_p] = i\hbar\boldsymbol{\eta}_p$.

Now let us prove the last relation in (5.6). We have

$$\boldsymbol{\eta}_q\zeta_p = \mathbf{A}^{-1/2}\Big(\frac{1}{2}(\mathbf{A}\boldsymbol{\sigma}_q\boldsymbol{\rho}_p\mathbf{A} + \boldsymbol{\sigma}_q\mathbf{A}\boldsymbol{\rho}_p\mathbf{A}) + \frac{i\hbar}{4}(\mathbf{A}\boldsymbol{\rho}_p\boldsymbol{\rho}_q\mathbf{A} - \boldsymbol{\rho}_q\mathbf{A}\boldsymbol{\rho}_p\mathbf{A})\Big)\mathbf{A}^{-1/2},$$

$$\zeta_p\boldsymbol{\eta}_q = \mathbf{A}^{-1/2}\Big(\frac{1}{2}(\mathbf{A}\boldsymbol{\rho}_p\mathbf{A}\boldsymbol{\sigma}_q + \mathbf{A}\boldsymbol{\rho}_p\boldsymbol{\sigma}_q\mathbf{A}) + \frac{i\hbar}{4}(\mathbf{A}\boldsymbol{\rho}_p\mathbf{A}\boldsymbol{\rho}_q - \mathbf{A}\boldsymbol{\rho}_q\boldsymbol{\rho}_p\mathbf{A})\Big)\mathbf{A}^{-1/2},$$

and therefore,

(5.16)
$$[\boldsymbol{\eta}_q, \zeta_p] = \mathbf{A}^{-1/2}\Big(\frac{1}{2}\mathbf{A}[\boldsymbol{\sigma}_q, \boldsymbol{\sigma}_p]\mathbf{A} + \frac{i\hbar}{2}\mathbf{A}\boldsymbol{\rho}_q\boldsymbol{\rho}_p\mathbf{A} \\ + \frac{1}{2}(\boldsymbol{\sigma}_q\mathbf{A}\boldsymbol{\rho}_p\mathbf{A} - \mathbf{A}\boldsymbol{\rho}_p\mathbf{A}\boldsymbol{\sigma}_q) \\ - \frac{i\hbar}{4}(\boldsymbol{\rho}_q\mathbf{A}\boldsymbol{\rho}_p\mathbf{A} + \mathbf{A}\boldsymbol{\rho}_p\mathbf{A}\boldsymbol{\rho}_q)\Big)\mathbf{A}^{-1/2}.$$

We apply (5.1) to the expression on the first line of the right-hand side of (5.16) and change these summands by

(5.17)
$$\frac{1}{2}\mathbf{A}[\boldsymbol{\sigma}_q, \boldsymbol{\sigma}_p]\mathbf{A} + \frac{i\hbar}{2}\mathbf{A}\boldsymbol{\rho}_q\boldsymbol{\rho}_p\mathbf{A} = \frac{i\hbar}{2}\delta_{qp}\mathbf{A}^2.$$

It follows from (5.12)(c) that the third term on the right-hand side of (5.16) is equal to

$$\frac{1}{2}[\boldsymbol{\sigma}_q, \mathbf{A}\boldsymbol{\rho}_p\mathbf{A}] = \frac{1}{4}[\boldsymbol{\sigma}_q, \mathbf{A}^2\boldsymbol{\rho}_p + \boldsymbol{\rho}_p\mathbf{A}^2 - \hbar^2\boldsymbol{\rho}_p].$$

Now we use commutators (5.1) and (5.12a) as follows:

$$\frac{1}{2}[\boldsymbol{\sigma}_q, \mathbf{A}\boldsymbol{\rho}_p\mathbf{A}] = \frac{i\hbar}{4}\Big(\mathbf{A}^2\boldsymbol{\rho}_p + \boldsymbol{\rho}_p\mathbf{A}^2 - \frac{\hbar^2}{2}\boldsymbol{\rho}_p\Big)\boldsymbol{\rho}_q + \frac{i\hbar}{4}\boldsymbol{\rho}_q\Big(\mathbf{A}^2\boldsymbol{\rho}_p + \boldsymbol{\rho}_p\mathbf{A}^2 - \frac{\hbar^2}{2}\boldsymbol{\rho}_p\Big) \\ + \frac{i\hbar}{4}\mathbf{A}^2(\delta_{pq} - \boldsymbol{\rho}_q\boldsymbol{\rho}_p) + \frac{i\hbar}{4}(\delta_{pq} - \boldsymbol{\rho}_q\boldsymbol{\rho}_p)\mathbf{A}^2 - \frac{i\hbar^3}{4}(\delta_{pq} - \boldsymbol{\rho}_q\boldsymbol{\rho}_p).$$

If we change the places of $\mathbf{A}^2$ and $\boldsymbol{\rho}_q$ to create the combinations $\mathbf{A}^2\boldsymbol{\rho}_q\boldsymbol{\rho}_p$ and $\boldsymbol{\rho}_q\boldsymbol{\rho}_p\mathbf{A}^2$, then this relation reads

$$\frac{1}{2}[\boldsymbol{\sigma}_q, \mathbf{A}\boldsymbol{\rho}_p\mathbf{A}] = \frac{i\hbar}{4}(\boldsymbol{\rho}_q\boldsymbol{\rho}_p\mathbf{A}^2 + \mathbf{A}^2\boldsymbol{\rho}_q\boldsymbol{\rho}_p) + \frac{i\hbar}{2}\delta_{pq}\mathbf{A}^2 - \frac{i\hbar^3}{4}\delta_{pq} \\ + \frac{i\hbar}{4}\boldsymbol{\rho}_p[\mathbf{A}^2, \boldsymbol{\rho}_q] + \frac{i\hbar}{4}[\boldsymbol{\rho}_q, \mathbf{A}^2]\boldsymbol{\rho}_p.$$

By using (5.12) (a) and (5.1), we transform the last two summands as follows:

$$\frac{i\hbar}{4}[\boldsymbol{\rho}_p, [\mathbf{A}^2, \boldsymbol{\rho}_q]] = -\frac{\hbar^2}{2}[\boldsymbol{\rho}_p, \boldsymbol{\sigma}_q] = \frac{i\hbar^3}{2}(\delta_{pq} - \boldsymbol{\rho}_q\boldsymbol{\rho}_p).$$

Thus we obtain

(5.18) $\quad \dfrac{1}{2}[\sigma_q, \mathbf{A}\rho_p\mathbf{A}] = \dfrac{i\hbar}{4}(\rho_q\rho_p\mathbf{A}^2 + \mathbf{A}^2\rho_q\rho_p) + \dfrac{i\hbar}{2}\delta_{pq}\mathbf{A}^2 + \dfrac{i\hbar^3}{4}\delta_{pq} - \dfrac{i\hbar^3}{2}\rho_q\rho_p.$

In the last summand of (5.16), we have

$$-\dfrac{i\hbar}{4}(\rho_q\mathbf{A}\rho_p\mathbf{A} + \mathbf{A}\rho_p\mathbf{A}\rho_q)$$
$$= -\dfrac{i\hbar}{4}(\rho_q\rho_p\mathbf{A}^2 + \mathbf{A}^2\rho_q\rho_p) + \dfrac{i\hbar}{4}(\mathbf{A}[\mathbf{A},\rho_p]\rho_q - \rho_q[\mathbf{A},\rho_p]\mathbf{A})$$
$$= -\dfrac{i\hbar}{4}(\rho_q\rho_p\mathbf{A}^2 + \mathbf{A}^2\rho_q\rho_p) + \dfrac{i\hbar}{4}[\mathbf{A},[\mathbf{A},\rho_p]]\rho_q + \dfrac{i\hbar}{4}[[\mathbf{A},\rho_p]\mathbf{A},\rho_q].$$

Applying (5.12) (b), (e) and (5.1), we now derive

(5.19) $\quad -\dfrac{i\hbar}{4}(\rho_q\mathbf{A}\rho_p\mathbf{A} + \mathbf{A}\rho_p\mathbf{A}\rho_q) = -\dfrac{i\hbar}{4}(\rho_q\rho_p\mathbf{A}^2 + \mathbf{A}^2\rho_q\rho_p) + \dfrac{i\hbar^3}{2}\rho_q\rho_p - \dfrac{i\hbar^3}{4}\delta_{qp}.$

Combining (5.17), (5.18), and (5.19) on the right-hand side of (5.16), we see that
$$[\eta_q, \zeta_p] = \mathbf{A}^{-1/2}(i\hbar\delta_{pq}\mathbf{A}^2)\mathbf{A}^{-1/2} = i\hbar\delta_{pq}\mathbf{A}.$$
So, *we have proved the last relations in* (5.6).

Now it follows from (5.5) and (5.1) that
$$0 = [\rho_p, \rho_q] = \mathbf{A}^{-1/2}\zeta_p\mathbf{A}^{-1}\zeta_q\mathbf{A}^{-1/2} - \mathbf{A}^{-1/2}\zeta_q\mathbf{A}^{-1}\zeta_p\mathbf{A}^{-1/2},$$
that is,

(5.20) $\quad\quad\quad\quad\quad\quad\quad\quad \zeta_p\mathbf{A}^{-1}\zeta_q = \zeta_q\mathbf{A}^{-1}\zeta_p.$

The last relation in (5.6) also implies

(5.21) $\quad \eta_q = \mathbf{A}^{-1}\eta_q\mathbf{A} - i\hbar\mathbf{A}^{-1}\zeta_q, \quad\quad \zeta_p = \mathbf{A}^{-1}\zeta_p\mathbf{A} + i\hbar\mathbf{A}^{-1}\eta_p,$
$$[\eta_p, \zeta_q] + [\zeta_p, \eta_q] = 0.$$

Combining (5.21) and (5.20), we obtain

(5.22) $\quad -i\hbar(\eta_p\zeta_q - \eta_q\zeta_p) - i\hbar(\zeta_p\eta_q - \zeta_q\eta_p) + \hbar^2(\zeta_p\mathbf{A}^{-1}\zeta_q - \zeta_q\mathbf{A}^{-1}\zeta_p) = 0.$

Now replacing η_q and η_p in the second bracket by (5.21), we arrive at
$$-i\hbar(\zeta_p\eta_q - \zeta_q\eta_p) = -i\hbar(\zeta_p\mathbf{A}^{-1}\eta_q - \zeta_q\mathbf{A}^{-1}\eta_p)\mathbf{A} - \hbar^2(\zeta_p\mathbf{A}^{-1}\zeta_q - \zeta_q\mathbf{A}^{-1}\zeta_p).$$
The substitution in (5.22) implies
$$-i\hbar(\eta_p\zeta_q - \eta_q\zeta_p) + \big(\zeta_p(-i\hbar\eta_q) - \zeta_q(-i\hbar\eta_p)\big)\mathbf{A} = 0.$$
In the second bracket, we replace $(-i\hbar\eta_q)$ and $(-i\hbar\eta_p)$ by (5.21), and obtain
$$-i\hbar(\eta_p\zeta_q - \eta_q\zeta_p) + \zeta_p(\mathbf{A}^{-1}\zeta_q\mathbf{A} - \zeta_q)\mathbf{A} - \zeta_q(\mathbf{A}^{-1}\zeta_p\mathbf{A} - \zeta_p)\mathbf{A} = 0.$$
Applying (5.20), we derive
$$-i\hbar(\eta_p\zeta_q - \eta_q\zeta_p) + [\zeta_q, \zeta_p]\mathbf{A} = 0.$$
Thus *we have proved the relation for* $[\zeta_p, \zeta_q]$ *in* (5.6).

Now from the relation for $[\mathbf{A}, \zeta_p]$ in (5.6), we get

$$[\eta_p, \eta_q] = \left(\dfrac{i}{\hbar}\right)^2[[\zeta_p, \mathbf{A}], [\zeta_q, \mathbf{A}]] = \left(\dfrac{i}{\hbar}\right)^2[\zeta_q, [\mathbf{A}, [\mathbf{A}, \zeta_p]]] + \left(\dfrac{i}{\hbar}\right)^2[\mathbf{A}, [[\mathbf{A}, \zeta_p], \zeta_q]].$$

Here the Jacobi identity for commutators was used. From (5.12) (b) and (5.6), we derive
$$[\eta_p, \eta_q] = -[\zeta_q, \zeta_p] - \frac{i}{\hbar}[\mathbf{A}, [\eta_p, \zeta_q]].$$
The last summand on the right-hand side is equal to zero in view of the relation for $[\zeta_p, \eta_q]$ in (5.6). So, we have $[\eta_p, \eta_q] = -[\zeta_q, \zeta_p]$, and thus *we have proved the relation for $[\eta_p, \eta_q]$ in (5.6)*.

Now it follows from (5.5) that $\mathbf{I} = \rho^2 = \sum_{p=0}^{3} \mathbf{A}^{-1/2}\zeta_p \mathbf{A}^{-1}\zeta_p \mathbf{A}^{-1/2}$, and thus

(5.23) $$\mathbf{A} = \sum_{p=0}^{3} \zeta_p \mathbf{A}^{-1}\zeta_p.$$

This equality and (5.21) imply

(5.24) $$\mathbf{A}^2 = \sum_{p=0}^{3} \zeta_p \mathbf{A}^{-1}\zeta_p \mathbf{A} = \zeta^2 - i\hbar \sum_{p=0}^{3} \zeta_p \mathbf{A}^{-1}\eta_p$$

and

(5.24a) $$\mathbf{A}^2 = \sum_{p=0}^{3} \mathbf{A}\zeta_p \mathbf{A}^{-1}\zeta_p = \zeta^2 + i\hbar \sum_{p=0}^{3} \eta_p \mathbf{A}^{-1}\zeta_p.$$

Again from (5.21) we have
$$\langle \zeta, \eta \rangle = \sum_{p=0}^{3} \zeta_p \mathbf{A}^{-1}\eta_p \mathbf{A} - i\hbar \sum_{p=0}^{3} \zeta_p \mathbf{A}^{-1}\zeta_p$$

and
$$\mathbf{A} \sum_{p=0}^{3} \eta_p \mathbf{A}^{-1}\zeta_p = \langle \eta, \zeta \rangle - i\hbar \sum_{p=0}^{3} \zeta_p \mathbf{A}^{-1}\zeta_p.$$

Applying formula (5.23) to both relations, we obtain
$$\sum_{p=0}^{3} \zeta_p \mathbf{A}^{-1}\eta_p = \langle \zeta, \eta \rangle \mathbf{A}^{-1} + i\hbar, \qquad \sum_{p=0}^{3} \eta_p \mathbf{A}^{-1}\zeta_p = \mathbf{A}^{-1}\langle \eta, \zeta \rangle - i\hbar.$$

Thus it follows from (5.24) and (5.24a) that

(5.25) $$\mathbf{A}^2 = \zeta^2 + \hbar^2 - i\hbar \langle \zeta, \eta \rangle \mathbf{A}^{-1}, \qquad \mathbf{A}^2 = \zeta^2 + \hbar^2 + i\hbar \mathbf{A}^{-1}\langle \eta, \zeta \rangle.$$

The last relation in (5.6) implies

(5.26) $$\langle \zeta, \eta \rangle = \langle \eta, \zeta \rangle - 4i\hbar \mathbf{A}.$$

Hence, the second relation in (5.25) is transformed to
$$\mathbf{A}^3 = \mathbf{A}\zeta^2 + i\hbar \langle \zeta, \eta \rangle - 3\hbar^2 \mathbf{A}.$$

The first relation in (5.25) reads

(5.27) $$\mathbf{A}^3 = \zeta^2 \mathbf{A} - i\hbar \langle \zeta, \eta \rangle + \hbar^2 \mathbf{A}.$$

From these two relations we derive
$$-i\hbar \langle \zeta, \eta \rangle = \frac{1}{2}[\mathbf{A}, \zeta^2] - 2\hbar^2 \mathbf{A}.$$

Substituting this relation into (5.27), we see that
$$2(\mathbf{A}^3 + \hbar^2 \mathbf{A}) = \mathbf{A}\boldsymbol{\zeta}^2 + \boldsymbol{\zeta}^2 \mathbf{A}.$$

Solving this equation, we obtain

(5.28)
$$\boldsymbol{\zeta}^2 = \mathbf{A}^2 + \hbar^2,$$

and hence

(5.29)
$$\langle \boldsymbol{\zeta}, \boldsymbol{\eta} \rangle = -2i\hbar \mathbf{A}.$$

On the other hand, (5.21) implies
$$\langle \boldsymbol{\zeta}, \boldsymbol{\eta} \rangle = \mathbf{A}^{-1}\sum_{p=0}^{3} \zeta_p \mathbf{A}\eta_p + i\hbar \mathbf{A}^{-1}\boldsymbol{\eta}^2, \qquad \sum_{p=0}^{3} \zeta_p \mathbf{A}\eta_p = \langle \boldsymbol{\zeta}, \boldsymbol{\eta} \rangle \mathbf{A} - i\hbar \boldsymbol{\zeta}^2,$$

and so,
$$\langle \boldsymbol{\zeta}, \boldsymbol{\eta} \rangle = \mathbf{A}^{-1}\langle \boldsymbol{\zeta}, \boldsymbol{\eta} \rangle \mathbf{A} - i\hbar \mathbf{A}^{-1}\boldsymbol{\zeta}^2 + i\hbar \mathbf{A}^{-1}\boldsymbol{\eta}^2.$$

From here, applying (5.26), (5.28), and (5.29), we derive $\boldsymbol{\eta}^2 = \boldsymbol{\zeta}^2$. Thus, *we have proved all relations in (5.7)*.

(ii) Only the second formula in (5.8) is not obvious.

Let us prove this formula. We denote
$$\mathbf{X}_p \stackrel{\text{def}}{=} \frac{i\hbar}{4}(\mathbf{A}^{1/2}\rho_p \mathbf{A}^{-1/2} - \mathbf{A}^{-1/2}\rho_p \mathbf{A}^{1/2}) = \frac{i\hbar}{4}(\zeta_p \mathbf{A}^{-1} - \mathbf{A}^{-1}\zeta_p).$$

Then we have the identity
$$\eta_p = \frac{1}{2}\big((\eta_p \mathbf{A} - i\hbar \zeta_p)\mathbf{A}^{-1} + \mathbf{A}^{-1}(\mathbf{A}\eta_p + i\hbar \zeta_p)\big) + 2\mathbf{X}_p.$$

Applying (5.21), we obtain
$$\eta_p = \frac{1}{2}(\mathbf{A}\eta_p \mathbf{A}^{-1} + \mathbf{A}^{-1}\eta_p \mathbf{A}) + 2\mathbf{X}_p$$

or
$$\eta_p = \frac{1}{4}(\mathbf{A}\eta_p \mathbf{A}^{-1} + \mathbf{A}^{-1}\eta_p \mathbf{A}) + \frac{1}{2}\eta_p + \mathbf{X}_p$$
$$= \frac{1}{2}(\mathbf{A}^{1/2}\mathbf{Y}_p \mathbf{A}^{-1/2} + \mathbf{A}^{-1/2}\mathbf{Y}_p \mathbf{A}^{1/2}) + \mathbf{X}_p,$$

where
$$\mathbf{Y}_p = \frac{1}{2}(\mathbf{A}^{1/2}\eta_p \mathbf{A}^{-1/2} + \mathbf{A}^{-1/2}\eta_p \mathbf{A}^{1/2}).$$

Taking into account the second relation in (5.5), we derive
$$\mathbf{A}^{1/2}\sigma_p \mathbf{A}^{-1/2} + \mathbf{A}^{-1/2}\sigma_p \mathbf{A}^{1/2} = \mathbf{A}^{1/2}\mathbf{Y}_p \mathbf{A}^{-1/2} + \mathbf{A}^{-1/2}\mathbf{Y}_p \mathbf{A}^{1/2}$$

or
$$\sigma_p + \mathbf{A}^{-1}\sigma_p \mathbf{A} = \mathbf{Y}_p + \mathbf{A}^{-1}\mathbf{Y}_p \mathbf{A}.$$

Solving this equation, we see that $\mathbf{Y}_p = \boldsymbol{\sigma}_p$, and this is exactly *the second formula in (5.8)*.

The proof of the last statement (iii) of Theorem 5.1 needs the following preliminary lemma.

LEMMA 5.5. *If $\mathfrak{P}_0$ is the eigenvector of $\mathbf{A}$, that is,*

$$\mathbf{A}\mathfrak{P}_0 = \hbar\mathfrak{P}_0, \tag{5.30}$$

then

$$\mathbf{C}^p\mathfrak{P}_0 = 0, \qquad \mathbf{C}^q\mathbf{B}_p\mathfrak{P}_0 = 2\hbar^2\delta_p^q\,\mathfrak{P}_0, \tag{5.31}$$

where $p, q = 0, 1, 2, 3$ and the operators $\mathbf{A}$, $\mathbf{C}$, $\mathbf{B}$ satisfy relations (5.10), and inequality (5.4) is assumed to be satisfied.

PROOF. It follows from (5.10) that

$$\hbar\mathbf{C}^p\mathfrak{P}_0 = \mathbf{C}^p\mathbf{A}\mathfrak{P}_0 = (\mathbf{A} + \hbar)\mathbf{C}^p\mathfrak{P}_0;$$

hence $\mathbf{A}\mathbf{C}^p\mathfrak{P}_0 = 0$. From (5.4) we obtain $\mathbf{C}^p\mathfrak{P}_0 = 0$. Then the second formula in (5.31) follows from the first relation in (5.10). □

Let us continue the proof of Theorem 5.1.

(iii) Using the commutation relations (5.6), we can write

$$[\mathbf{A}^{-1}, \boldsymbol{\zeta}_p] = -i\hbar\mathbf{A}^{-1}\boldsymbol{\eta}_p\mathbf{A}^{-1}, \qquad [\mathbf{A}^{-1}, \boldsymbol{\eta}_p] = i\hbar\mathbf{A}^{-1}\boldsymbol{\zeta}_p\mathbf{A}^{-1}. \tag{5.32}$$

If the $\boldsymbol{\rho}_p$ are determined by (5.8), then we obtain

$$\begin{aligned}[\boldsymbol{\rho}_p, \boldsymbol{\rho}_q] &= \mathbf{A}^{-1/2}(\boldsymbol{\zeta}_p\mathbf{A}^{-1}\boldsymbol{\zeta}_q - \boldsymbol{\zeta}_q\mathbf{A}^{-1}\boldsymbol{\zeta}_p)\mathbf{A}^{-1/2} \\
&= \mathbf{A}^{-1/2}\big([\boldsymbol{\zeta}_p, \boldsymbol{\eta}_q] - i\hbar(\boldsymbol{\zeta}_p\mathbf{A}^{-1}\boldsymbol{\eta}_q - \boldsymbol{\zeta}_q\mathbf{A}^{-1}\boldsymbol{\eta}_p)\big)\mathbf{A}^{-3/2} \\
&= -i\hbar\mathbf{A}^{-1/2}\big(\boldsymbol{\eta}_p\boldsymbol{\zeta}_q - \boldsymbol{\eta}_q\boldsymbol{\zeta}_p + \boldsymbol{\zeta}_p\boldsymbol{\eta}_q - \boldsymbol{\zeta}_q\boldsymbol{\eta}_p \\
&\qquad + i\hbar(\boldsymbol{\zeta}_p\mathbf{A}^{-1}\boldsymbol{\zeta}_q - \boldsymbol{\zeta}_q\mathbf{A}^{-1}\boldsymbol{\zeta}_p)\big)\mathbf{A}^{-5/2} \\
&= -i\hbar\mathbf{A}^{-1/2}\big([\boldsymbol{\eta}_p, \boldsymbol{\zeta}_q] + [\boldsymbol{\zeta}_p, \boldsymbol{\eta}_q]\big)\mathbf{A}^{-5/2} \\
&\qquad + \hbar^2\mathbf{A}^{-1/2}(\boldsymbol{\zeta}_p\mathbf{A}^{-1}\boldsymbol{\zeta}_q - \boldsymbol{\zeta}_q\mathbf{A}^{-1}\boldsymbol{\zeta}_p)\mathbf{A}^{-1/2}\cdot\mathbf{A}^{-2} \\
&= \hbar^2[\boldsymbol{\rho}_p, \boldsymbol{\rho}_q]\cdot\mathbf{A}^{-2}\end{aligned}$$

or

$$[\boldsymbol{\rho}_p, \boldsymbol{\rho}_q]\cdot(\mathbf{I} - \hbar^2\mathbf{A}^{-2}) = 0. \tag{5.33}$$

This implies that the first commutation relation $[\boldsymbol{\rho}_p, \boldsymbol{\rho}_q] = 0$ in (5.1) holds on the subspace $\mathcal{P}_0^\perp$ orthogonal to the eigenvector $\mathfrak{P}_0$ (5.30).

From (5.31), (5.9), and the permutation relation $\mathbf{A}^{-1}\mathbf{B}_q = \mathbf{B}_q(\mathbf{A}+\hbar)^{-1}$ (which is a consequence of (5.10)), we derive the chain of equalities

$$\begin{aligned}\boldsymbol{\rho}_p\boldsymbol{\rho}_q\mathfrak{P}_0 &= \mathbf{A}^{-1/2}\boldsymbol{\zeta}_p\mathbf{A}^{-1}\boldsymbol{\zeta}_q\mathbf{A}^{-1/2}\mathfrak{P}_0 - \frac{i}{2}\hbar^{-1/2}\mathbf{A}^{-1/2}\boldsymbol{\zeta}_p\mathbf{A}^{-1}(\mathbf{B}_q - \mathbf{C}^q)\mathfrak{P}_0 \\
&= \frac{i}{2}\hbar^{-1/2}\mathbf{A}^{-1/2}\boldsymbol{\zeta}_p\mathbf{B}_q(\mathbf{A}+\hbar)^{-1}\mathfrak{P}_0 \\
&= -\frac{1}{8}\hbar^{-3/2}\mathbf{A}^{-1/2}(\mathbf{B}_p - \mathbf{C}^p)\mathbf{B}_q\mathfrak{P}_0 = -\frac{1}{8}\hbar^{-3/2}\mathbf{A}^{-1/2}(\mathbf{B}_p\mathbf{B}_q - 2\hbar^2\delta_q^p)\mathfrak{P}_0,\end{aligned}$$

$$p, q = 1, 2, 3.$$

Since the operators $\mathbf{B}_p$ and $\mathbf{B}_q$ commute, the change $p \longleftrightarrow q$ does not change the right-hand side of the last relation. Hence, we have

$$\rho_p\rho_q\mathfrak{P}_0 = \rho_q\rho_p\mathfrak{P}_0, \qquad p,q = 1,2,3.$$

In a similar way, we consider the case $p=0$ or $q=0$.

So, we have completely proved the first commutation relation (5.1).

Now let us prove the first relation in (5.2). Using (5.32) and (5.7), we obtain

$$\rho^2 = \mathbf{A}^{-1/2}\sum_{p=0}^{3}\zeta_p\mathbf{A}^{-1}\zeta_p\mathbf{A}^{-1/2} = \mathbf{A}^{-1/2}\Big(\zeta^2 - i\hbar\sum_{p=0}^{3}\zeta_p\mathbf{A}^{-1}\eta_p\Big)\mathbf{A}^{-3/2}$$

$$= \mathbf{A}^{-1/2}\Big(\mathbf{A}^2 + \hbar^2 - i\hbar\langle\zeta,\eta\rangle\mathbf{A}^{-1} + \hbar^2\sum_{p=0}^{3}\zeta_p\mathbf{A}^{-1}\zeta_p\mathbf{A}^{-1}\Big)\mathbf{A}^{-3/2}$$

$$= \mathbf{I} - \hbar^2\mathbf{A}^{-2} + \hbar^2\rho^2\mathbf{A}^{-2},$$

or

$$\rho^2(\mathbf{I} - \hbar^2\mathbf{A}^{-2}) = \mathbf{I} - \hbar^2\mathbf{A}^{-2}.$$

On the subspace $\mathcal{P}_0^\perp$, where the operator $\mathbf{I} - \hbar^2\mathbf{A}^{-2}$ is invertible, this relation implies the desired first relation in (5.2), that is, $\rho^2 = \mathbf{I}$.

In view of (5.34), we have

$$\sum_{p=1}^{3}\rho_p^2\mathfrak{P}_0 = -\frac{1}{8}\hbar^{-3/2}\mathbf{A}^{-1/2}\Big(\sum_{p=1}^{3}\mathbf{B}_p^2 - 6\hbar^2\Big)\mathfrak{P}_0.$$

In a similar way, we can derive the relation

$$\rho_0^2\mathfrak{P}_0 = \frac{1}{8}\hbar^{-3/2}\mathbf{A}^{-1/2}(\mathbf{B}_0^2 + 2\hbar^2)\mathfrak{P}_0.$$

Summing the last two relations and taking into account the first relation in (5.11), we obtain

$$\rho^2\mathfrak{P}_0 = \hbar^{1/2}\mathbf{A}^{-1/2}\mathfrak{P}_0 = \mathfrak{P}_0.$$

Thus, the first relation in (5.2) is proved on the entire representation space.

Next, using the second relation in (5.32), as well as the last relation in (5.7), we calculate the commutator

$$[\rho_p,\sigma_q] = \frac{1}{2}(\mathbf{A}^{-1/2}\zeta_p\eta_q\mathbf{A}^{-1/2} - \mathbf{A}^{1/2}\eta_q\mathbf{A}^{-1}\zeta_p\mathbf{A}^{-1/2}$$

$$+ \mathbf{A}^{-1/2}\zeta_p\mathbf{A}^{-1}\eta_q\mathbf{A}^{1/2} - \mathbf{A}^{-1/2}\eta_q\zeta_p\mathbf{A}^{-1/2})$$

$$= \mathbf{A}^{-1/2}\Big([\zeta_p,\eta_q] + \frac{i\hbar}{2}(\zeta_q\mathbf{A}^{-1}\zeta_p + \zeta_p\mathbf{A}^{-1}\zeta_q)\Big)\mathbf{A}^{-1/2}$$

$$= -i\hbar\Big(\delta_{pq}\cdot\mathbf{I} - \frac{1}{2}(\rho_p\rho_q + \rho_q\rho_p)\Big).$$

From the relations $\rho_p\rho_q = \rho_q\rho_p$ and $\rho^2 = \mathbf{I}$ that we have already proved, we obtain the second commutation relation in (5.1), that is, $[\rho_p,\sigma_q] = -i\hbar(\delta_{pq}\rho^2 - \rho_p\rho_q)$.

Let us calculate the last commutator in (5.1). We again use the relations (5.6) and their consequences, that is, the second relation in (5.32) and the commutation relation $[\eta_p\zeta_q - \eta_q\zeta_p, \mathbf{A}^{-1}] = 0$. Then we obtain the chain of relations

$$\begin{aligned}[\sigma_p, \sigma_q] &= \frac{1}{4}\big(\mathbf{A}^{1/2}\eta_p\eta_q\mathbf{A}^{-1/2} - \mathbf{A}^{1/2}\eta_q\eta_p\mathbf{A}^{-1/2} + \mathbf{A}^{1/2}\eta_p\mathbf{A}^{-1}\eta_q\mathbf{A}^{1/2} \\
&\quad - \mathbf{A}^{-1/2}\eta_q\mathbf{A}\eta_p\mathbf{A}^{-1/2} + \mathbf{A}^{-1/2}\eta_p\mathbf{A}\eta_q\mathbf{A}^{-1/2} - \mathbf{A}^{1/2}\eta_q\mathbf{A}^{-1}\eta_p\mathbf{A}^{1/2} \\
&\quad + \mathbf{A}^{-1/2}\eta_p\eta_q\mathbf{A}^{1/2} - \mathbf{A}^{-1/2}\eta_q\eta_p\mathbf{A}^{1/2}\big) \\
&= \frac{1}{4}\big(2\mathbf{A}^{1/2}[\eta_p, \eta_q]\mathbf{A}^{-1/2} + 2\mathbf{A}^{-1/2}[\eta_p, \eta_q]\mathbf{A}^{1/2} \\
&\quad - i\hbar(\mathbf{A}^{-1/2}\zeta_p\mathbf{A}^{-1}\eta_q\mathbf{A}^{1/2} - \mathbf{A}^{-1/2}\zeta_q\mathbf{A}^{-1}\eta_p\mathbf{A}^{1/2} \\
&\quad - \mathbf{A}^{-1/2}\zeta_p\eta_q\mathbf{A}^{-1/2} + \mathbf{A}^{-1/2}\zeta_q\eta_p\mathbf{A}^{-1/2})\big) \\
&= -\frac{i\hbar}{4}\big(4\mathbf{A}^{-1/2}(\eta_p\zeta_q - \eta_q\zeta_p)\mathbf{A}^{-1/2} \\
&\quad + i\hbar(\mathbf{A}^{-1/2}\zeta_p\mathbf{A}^{-1}\zeta_q\mathbf{A}^{-1/2} - \mathbf{A}^{-1/2}\zeta_q\mathbf{A}^{-1}\zeta_p\mathbf{A}^{-1/2})\big) \\
&= -i\hbar\mathbf{A}^{-1/2}(\eta_p\zeta_q - \eta_q\zeta_p)\mathbf{A}^{-1/2} + \frac{\hbar^2}{4}[\rho_p, \rho_q].\end{aligned}$$

Since $\rho_q\rho_p = \rho_p\rho_q$, we arrive at

$$[\sigma_p, \sigma_q] = -i\hbar\mathbf{A}^{-1/2}(\eta_p\zeta_q - \eta_q\zeta_p)\mathbf{A}^{-1/2}. \tag{5.34}$$

On the other hand, we have

$$\begin{aligned}\sigma_p\rho_q - \sigma_q\rho_p &= \frac{1}{2}\big(\mathbf{A}^{1/2}\eta_p\mathbf{A}^{-1}\zeta_q\mathbf{A}^{-1/2} + \mathbf{A}^{-1/2}\eta_p\zeta_q\mathbf{A}^{-1/2} \\
&\quad - \mathbf{A}^{1/2}\eta_q\mathbf{A}^{-1}\zeta_p\mathbf{A}^{-1/2} - \mathbf{A}^{-1/2}\eta_q\zeta_p\mathbf{A}^{-1/2}\big) \\
&= \frac{1}{2}\big(2\mathbf{A}^{-1/2}(\eta_p\zeta_q - \eta_q\zeta_p)\mathbf{A}^{-1/2} \\
&\quad - i\hbar(\mathbf{A}^{-1/2}\zeta_p\mathbf{A}^{-1}\zeta_q\mathbf{A}^{-1/2} - \mathbf{A}^{-1/2}\zeta_q\mathbf{A}^{-1}\zeta_p\mathbf{A}^{-1/2})\big) \\
&= \mathbf{A}^{-1/2}(\eta_p\zeta_q - \eta_q\zeta_p)\mathbf{A}^{-1/2} - i\hbar[\rho_p, \rho_q].\end{aligned}$$

Since $\rho_q\rho_p = \rho_p\rho_q$, we obtain

$$\sigma_p\rho_q - \sigma_q\rho_p = \mathbf{A}^{-1/2}(\eta_p\zeta_q - \eta_q\zeta_p)\mathbf{A}^{-1/2}. \tag{5.35}$$

Comparing (5.34) and (5.35), we obtain the last commutator in (5.1), that is, $[\sigma_p, \sigma_q] = -i\hbar(\sigma_p\rho_q - \sigma_q\rho_p)$.

Finally, using the second relation in (5.7) and the first relation in (5.2), which we have already proved, as well as (5.1), we obtain

$$\begin{aligned}\langle\rho, \sigma\rangle &= \frac{1}{2}\Big(\mathbf{A}^{-1/2}\langle\zeta, \eta\rangle\mathbf{A}^{-1/2} + \mathbf{A}^{-1/2}\sum_{p=0}^{3}\zeta_p\mathbf{A}^{-1}\eta_p\mathbf{A}^{1/2}\Big) \\
&= \frac{1}{2}\Big(-2i\hbar + \mathbf{A}^{-1/2}\langle\zeta, \eta\rangle\mathbf{A}^{-1/2} + i\hbar\mathbf{A}^{-1/2}\sum_{p=0}^{3}\zeta_p\mathbf{A}^{-1}\zeta_p\mathbf{A}^{-1/2}\Big) \\
&= -2i\hbar + \frac{i\hbar}{2}\rho^2 = -\frac{3}{2}i\hbar\end{aligned}$$

and

$$\langle\sigma, \rho\rangle = \langle\rho, \sigma\rangle + \sum_{p=0}^{3}[\sigma_p, \rho_p] = -\frac{3}{2}i\hbar + 3i\hbar\rho^2 = \frac{3}{2}i\hbar.$$

This yields $\langle \boldsymbol{\rho}, \boldsymbol{\sigma} \rangle + \langle \boldsymbol{\sigma}, \boldsymbol{\rho} \rangle = 0$, that is, the second relation in (5.2) holds. Theorem 5.1 is proved. $\square$

It follows from (5.10) that the operators **A**, **B**, and **C** satisfy permutation relations of type (I.4.1), where

$$d = 4, \qquad k = 1, \qquad \omega_{sp}^{qr}(A) = \delta_s^q \delta_p^r (1 + \hbar A^{-1}) - \hbar \delta_{sp} \delta^{qr} A^{-1},$$
$$f_p^q(A) = 2\hbar \delta_p^q A, \qquad \varphi_p(A) \equiv \varphi(A) = A + \hbar.$$

However, it follows from (5.11) that the operators **A**, $\mathbf{B}_p$, and $\mathbf{C}^p$ in this example are not independent of each other and *the Jacobi identities do not hold for* (5.10).

On the other hand, using (5.11), one can prove that the operators **A**, $\mathbf{B}_p$, and $\mathbf{C}^p$ generate an algebra of type (4.1) considered in the previous section. Namely, let us define

$$(5.36) \qquad \mathbf{R}_p \stackrel{\text{def}}{=} \frac{i}{4} \varepsilon_{pqr} (\mathbf{C}^q \mathbf{B}_r + \mathbf{B}_r \mathbf{C}^q - \mathbf{C}^r \mathbf{B}_q - \mathbf{B}_q \mathbf{C}^r) \mathbf{A}^{-1}, \qquad p, q, r = 1, 2, 3,$$

where ε_{pqr} is a fundamental antisymmetric tensor. Then, from Lemma 5.3, we obtain the following statement.

PROPOSITION 5.1. *The operators* **A**, $\mathbf{B}_p$, $\mathbf{C}^p$, *and* $\mathbf{R}_p$ $(p = 1, 2, 3)$ *satisfy the commutation relations*

$$[\mathbf{C}^q, \mathbf{B}_p] = -2i\hbar \varepsilon_{qpr} \mathbf{R}_r + 2\hbar \delta_p^q \mathbf{A},$$
$$[\mathbf{R}_p, \mathbf{B}_q] = i\hbar \varepsilon_{pqr} \mathbf{B}_r, \qquad [\mathbf{C}^q, \mathbf{R}_p] = -i\hbar \varepsilon_{pqr} \mathbf{C}^r,$$
$$(5.37) \qquad \mathbf{A}\mathbf{B}_p = \mathbf{B}_p (\mathbf{A} + \hbar), \qquad \mathbf{C}^p \mathbf{A} = (\mathbf{A} + \hbar) \mathbf{C}^p,$$
$$[\mathbf{R}_p, \mathbf{R}_q] = i\hbar \varepsilon_{pqr} \mathbf{R}_r, \qquad [\mathbf{A}, \mathbf{R}_p] = 0,$$
$$[\mathbf{B}_p, \mathbf{B}_q] = 0, \qquad [\mathbf{C}^p, \mathbf{C}^q] = 0.$$

We have obtained an algebra of type (4.1) with

$$d = 3, \qquad k = 1, \qquad f^{rq}{}_p = -2i\hbar \varepsilon_{pqr}, \qquad f_p^q(A) = 2\hbar \delta_p^q A,$$
$$\psi_{pq}{}^r = i\hbar \varepsilon_{pqr}, \qquad \chi_{pq}{}^r = i\hbar \varepsilon_{pqr}, \qquad \varphi(A) = A + \hbar.$$

In this example, all Jacobi identities (4.4) are satisfied; the matrix Λ from (4.5) has the form $\lambda_p^q = 2\hbar^2 \delta_p^q$. Therefore, by using Theorem 4.1, we can construct the antiholomorphic representation and the coherent states of the algebra (5.37).

Let an Hermitian representation of the algebra (5.6) be given in some Hilbert space H, and let this representation possess a vacuum vector such that

$$(5.38) \qquad \mathbf{A}\mathfrak{P}_0 = \hbar \mathfrak{P}_0, \qquad \mathbf{C}^p \mathfrak{P}_0 = 0 \quad (p = 1, 2, 3), \qquad \|\mathfrak{P}_0\| = 1.$$

Hence, it follows from definition (5.36) and from the second permutation relation in (5.10) that

$$(5.39) \qquad \mathbf{R}_p \mathfrak{P}_0 = 0 \quad (p = 1, 2, 3).$$

Thus, $\mathfrak{P}_0$ satisfies all conditions (4.2).

Then the antiholomorphic *irreducible representation of the algebra* (5.37) is given by the following formulas (see (4.6)):

(5.40)
$$\overset{\circ}{B}_p = \hbar\Big(2\overline{z^p}\langle \overline{z}, \frac{\partial}{\partial \overline{z}}\rangle - (\overline{z})^2 \frac{\partial}{\partial \overline{z^p}} + 2\overline{z^p}\Big), \qquad \overset{\circ}{C}^p = \hbar \frac{\partial}{\partial \overline{z^p}},$$
$$\overset{\circ}{R}_p = -i\hbar\Big(\overline{z} \times \frac{\partial}{\partial \overline{z}}\Big)_p, \qquad p = 1,2,3,$$
$$\overset{\circ}{A} = \hbar\Big(\langle \overline{z}, \frac{\partial}{\partial \overline{z}}\rangle + 1\Big).$$

The coherent states have the form
$$\mathfrak{P}_z = \exp\Big\{\frac{1}{\hbar}\langle z, \mathbf{B}\rangle\Big\}\mathfrak{P}_0, \qquad z \in \mathbb{C}^3,$$

and the corresponding reproducing kernel $\mathcal{K}(\overline{w}, z) \overset{\text{def}}{=} (\mathfrak{P}_w, \mathfrak{P}_z)$ is
$$\mathcal{K}(\overline{w}, z) = \frac{1}{\langle \overline{w}, \overline{w}\rangle\langle z, z\rangle - 2\langle \overline{w}, z\rangle + 1}.$$

It follows from (5.40) that $\overset{\circ}{B}_p = \overline{z^p}(\overset{\circ}{A} + \hbar) + i(\overline{z} \times \overset{\circ}{R})_p$. Therefore,
$$\mathbf{C}^p \mathfrak{P}_z = \overline{\overset{\circ}{B}_p} \mathfrak{P}_z = \big((\mathbf{A} + \hbar)z^p + i(\mathbf{R} \times z)^p\big)\mathfrak{P}_z.$$

Hence, the operators $\widehat{z^1}$, $\widehat{z^2}$, and $\widehat{z^3}$ satisfy the relation
$$\begin{pmatrix} \mathbf{A}+\hbar & -i\mathbf{R}_3 & i\mathbf{R}_2 \\ i\mathbf{R}_3 & \mathbf{A}+\hbar & -i\mathbf{R}_1 \\ -i\mathbf{R}_2 & i\mathbf{R}_1 & \mathbf{A}+\hbar \end{pmatrix} \begin{pmatrix} \widehat{z^1} \\ \widehat{z^2} \\ \widehat{z^3} \end{pmatrix} = \begin{pmatrix} \mathbf{C}^1 \\ \mathbf{C}^2 \\ \mathbf{C}^3 \end{pmatrix}.$$

Inverting the operator matrix on the left-hand side, we obtain the following formula for *operators of complex structure*:

$$\begin{pmatrix} \widehat{z^1} \\ \widehat{z^2} \\ \widehat{z^3} \end{pmatrix} = \mathbf{A}^{-1}\big((\mathbf{A}+\hbar)\mathbf{A} - \mathbf{R}^2\big)^{-1}$$
$$\times \begin{pmatrix} \mathbf{A}^2 - \mathbf{R}_1^2 & -\mathbf{R}_1\mathbf{R}_2 + i\mathbf{A}\mathbf{R}_3 & -\mathbf{R}_1\mathbf{R}_3 - i\mathbf{A}\mathbf{R}_2 \\ -\mathbf{R}_2\mathbf{R}_1 - i\mathbf{A}\mathbf{R}_3 & \mathbf{A}^2 - \mathbf{R}_2^2 & -\mathbf{R}_2\mathbf{R}_3 + i\mathbf{A}\mathbf{R}_1 \\ -\mathbf{R}_3\mathbf{R}_1 + i\mathbf{A}\mathbf{R}_2 & -\mathbf{R}_3\mathbf{R}_2 - i\mathbf{A}\mathbf{R}_1 & \mathbf{A}^2 - \mathbf{R}_3^2 \end{pmatrix} \begin{pmatrix} \mathbf{C}^1 \\ \mathbf{C}^2 \\ \mathbf{C}^3 \end{pmatrix}.$$

5.2. The use of the second "vacuum". Our aim is to construct an antiholomorphic representation and coherent states for the initial relations (5.1) or (5.10); that is, it is necessary to remember the operators $\mathbf{B}_0$ and $\mathbf{C}^0$.

Note that, by (5.11), the operators $\mathbf{B}_0$ and $\mathbf{C}^0$ can be considered as the functions $\theta_1(\mathbf{B})$ and $\overline{\theta_1}(\mathbf{C})$ (see §4), where Γ is the unit matrix, namely,

$$\mathbf{B}_0 = \theta_1(\mathbf{B}), \qquad \mathbf{C}^0 = \overline{\theta_1}(\mathbf{C}), \qquad \theta_1(\overline{z}) = \sqrt{\langle \Gamma^{-1}\overline{z}, \overline{z}\rangle} = \sqrt{\overline{z}^2}, \qquad \Gamma = I.$$

Indeed, since in our example the matrices Ψ_p are pure imaginary, equations (4.8) become the commutation relations $[\Gamma, \Psi_p] = 0$, obviously satisfied by the matrix $\Gamma = I$.

In our example, the operators $\mathbf{L}_p$ (4.9) are given by the formula
$$\mathbf{L}_p = \hbar(\mathbf{C}^p \mathbf{B}_0 + \mathbf{B}_p \mathbf{C}^0)\mathbf{A}^{-1}, \qquad p = 1,2,3,$$

which, together with (5.10), implies that the vector $\mathfrak{P}_0$ (5.38) satisfies the additional conditions

(5.41) $$\mathbf{L}_p\mathfrak{P}_0 = 0, \qquad p = 1,2,3.$$

According to (5.3), we have

$$\|\theta_1(\mathbf{B})\mathfrak{P}_0\|^2 = \|\mathbf{B}_0\mathfrak{P}_0\|^2 = (\mathbf{C}^0\mathbf{B}_0\mathfrak{P}_0, \mathfrak{P}_0) = 2\hbar^2;$$

that is, the constant c in (4.11) is equal to $c = 2\hbar^2 \neq 0$. Thus, the assumptions of Theorem 4.2 are satisfied, and we can write, for the algebra (5.37), an anti-holomorphic representation of type (4.13) and the corresponding coherent states (4.14).

Since the operators $\mathbf{R}_r$ in (5.37) can be expressed via $\mathbf{C}$, $\mathbf{B}$, $\mathbf{A}$ by formula (5.36), we obtain a representation of the algebra in fact generated by $\mathbf{C}$, $\mathbf{B}$, $\mathbf{A}$. If we include the operators $\mathbf{C}_0$, $\mathbf{B}_0$ in this algebra, then the representation turns out to be irreducible.

Let us denote $z^0 \stackrel{\text{def}}{=} \sqrt{z^2}$ and apply Theorem 4.2 (see (4.13) with $\mathcal{C}_a(n) = 2\hbar^2(n+1)$).

THEOREM 5.2. *The operators*

(5.42)
$$\overset{\circ}{B}_0 = \overline{z^0}, \qquad \overset{\circ}{B}_p = \overline{z^p}, \qquad \overset{\circ}{C}^0 = \hbar^2 \overline{z^0}\Big(\frac{\partial}{\partial \overline{z}}\Big)^2,$$
$$\overset{\circ}{C}^p = \hbar^2\Big(2\langle \overline{z}, \frac{\partial}{\partial \overline{z}}\rangle \frac{\partial}{\partial \overline{z^p}} - \overline{z^p}\Big(\frac{\partial}{\partial \overline{z}}\Big)^2 + 2\frac{\partial}{\partial \overline{z^p}}\Big), \qquad p = 1,2,3,$$
$$\overset{\circ}{A} = \hbar\Big(\langle \overline{z}, \frac{\partial}{\partial \overline{z}}\rangle + 1\Big),$$

define an irreducible Hermitian representation of relations (5.10), (5.11) *in the space of functions antiholomorphic with respect to z^0 and z, endowed with the inner product*

$$(g_1, g_2) = \frac{1}{2\pi\hbar}\int_{\mathbb{C}^3} \overline{g_1(\overline{z^0}, \overline{z})}\, g_2(\overline{z^0}, \overline{z})\, \ell(|z^0|^2, |z|^2)\, dz\, d\overline{z}.$$

Here ℓ is given by the formula

(5.43) $$\ell(x^0, x) = \frac{1}{8\pi^2\hbar^3 x^0}K_0\Big(\frac{1}{\hbar}\sqrt{2(x^0+x)}\Big), \qquad x^0, x \in \mathbb{R}_+,$$

where K_0 is the MacDonald function [BE]. *The coherent states are*

(5.44) $$\mathfrak{P}_{z^0, z} = \widetilde{I}_0\Big(\frac{1}{\hbar}\sqrt{2(z^0\mathbf{B}_0 + \langle z, \mathbf{B}\rangle)}\Big)\mathfrak{P}_0,$$

where $\widetilde{I}_\alpha$ is the Bessel function of imaginary argument (1.30). *The reproducing kernel $\mathcal{K}(\overline{w^0}, \overline{w}, z^0, z) \stackrel{\text{def}}{=} (\mathfrak{P}_{w^0, w}, \mathfrak{P}_{z^0, z})$ is*

(5.45) $$\mathcal{K}(\overline{w^0}, \overline{w}, z^0, z) = I_0\Big(\frac{1}{\hbar}\sqrt{2(\overline{w^0}z^0 + \langle \overline{w}, z\rangle)}\Big).$$

The operators of complex structure are

$$\widehat{z^p} = \mathbf{C}^p, \qquad p = 1,2,3.$$

If we consider the algebra (5.37) and its representations, then the coherent state (5.11) can be decomposed into the sum (4.15). In the case considered, we obtain the following explicit formulas for this decomposition.

PROPOSITION 5.2. (a) *The coherent states* (4.15a) *in the space* H *look as follows*:
$$\mathfrak{P}_z = \sum_{n=0}^{\infty} \frac{1}{((2n)!)^2} \left(\frac{(z)^2 \mathbf{B}^2}{4\hbar^4}\right)^n \widetilde{I}_{2n}\left(\frac{1}{\hbar}\sqrt{2\langle z, \mathbf{B}\rangle}\right) \mathfrak{P}_0.$$

These states correspond to the following antiholomorphic irreducible representation of the algebra (5.37):

$$\overset{\circ}{B}_p = \overline{z}^p, \qquad \overset{\circ}{C}{}^p = \hbar^2\left(2\langle\overline{z}, \frac{\partial}{\partial \overline{z}}\rangle \frac{\partial}{\partial \overline{z}^p} - \overline{z}^p\left(\frac{\partial}{\partial \overline{z}}\right)^2 + 2\frac{\partial}{\partial \overline{z}^p}\right),$$
$$\overset{\circ}{R}_p = -i\hbar\left(\overline{z} \times \frac{\partial}{\partial \overline{z}}\right)_p, \qquad p = 1, 2, 3,$$
$$\overset{\circ}{A} = \hbar\left(\langle\overline{z}, \frac{\partial}{\partial \overline{z}}\rangle + 1\right).$$

This representation is Hermitian in the space $\mathcal{P}$ *of functions antiholomorphic with respect to* z, *endowed with the inner product*

$$(g_1, g_2)_{\mathcal{P}} = \frac{1}{2\pi\hbar} \int_{\mathbb{C}^3} \overline{g_1(\overline{z})}\, g_2(\overline{z})\, \ell(|z^2|, |z|^2)\, dz\, d\overline{z},$$

where ℓ *is still defined by* (5.43). *The corresponding reproducing kernel* $\mathcal{K}(\overline{w}, z) \overset{\text{def}}{=} (\mathfrak{P}_w, \mathfrak{P}_z)_H$ *has the form*

$$\mathcal{K}(\overline{w}, z) = \sum_{n=0}^{\infty} \frac{1}{((2n)!)^2} \left(\frac{\langle \overline{w}, \overline{w}\rangle\langle z, z\rangle}{4\hbar^4}\right)^n \widetilde{I}_{2n}\left(\frac{1}{\hbar}\sqrt{2\langle \overline{w}, z\rangle}\right).$$

(b) *The coherent states* (4.15b) *in the space* H' *are given by*

$$\mathfrak{P}'_z = \frac{1}{2\hbar^2} \sum_{n=0}^{\infty} \frac{1}{((2n+1)!)^2} \left(\frac{\langle z, z\rangle \mathbf{B}^2}{4\hbar^4}\right)^n \widetilde{I}_{2n+1}\left(\frac{1}{\hbar}\sqrt{2\langle z, \mathbf{B}\rangle}\right) \mathfrak{P}'_0,$$

where $\mathfrak{P}'_0 = \mathbf{B}_0 \mathfrak{P}_0 / \sqrt{2}\hbar$. *These states correspond to the antiholomorphic irreducible representation of the algebra* (5.6) *by the operators*

$$\overset{\circ}{B}{}'_p = \overline{z}^p, \qquad \overset{\circ}{C}{}'^p = \hbar^2\left(2\langle\overline{z}, \frac{\partial}{\partial \overline{z}}\rangle \frac{\partial}{\partial \overline{z}^p} - \overline{z}^p\left(\frac{\partial}{\partial \overline{z}}\right)^2 + 4\frac{\partial}{\partial \overline{z}^p}\right),$$
$$\overset{\circ}{R}{}'_p = -i\hbar\left(\overline{z} \times \frac{\partial}{\partial \overline{z}}\right)_p, \qquad p = 1, 2, 3,$$
$$\overset{\circ}{A}{}' = \hbar\left(\langle\overline{z}, \frac{\partial}{\partial \overline{z}}\rangle + 2\right)$$

in the space $\mathcal{P}'$ *of functions of the type* $g'(z) = \sqrt{z^2}\, g(z)$, *where* $g \in \mathcal{P}$. *This space is equipped with the inner product*

$$(g_1, g_2)_{\mathcal{P}'} = \frac{1}{2\pi\hbar} \int_{\mathbb{C}^3} \overline{g_1(\overline{z})}\, g_2(\overline{z})\, \ell'(|z^2|, |z|^2)\, dz\, d\overline{z},$$

where

$$\ell'(x^0, x) = 2\hbar^2 x^0 \ell(x^0, x).$$

The corresponding reproducing kernel $\mathcal{K}'(\overline{w}, z) \stackrel{\text{def}}{=} (\mathfrak{P}'_z, \mathfrak{P}'_w)_H$ has the form

$$\mathcal{K}'(\overline{w}, z) = \frac{1}{4\hbar^4} \sum_{n=0}^{\infty} \frac{1}{((2n+1)!)^2} \widetilde{I}_{2n+1}\left(\frac{1}{\hbar}\sqrt{2\langle \overline{w}, z\rangle}\right).$$

§6. Reduction of coherent states

In this section we derive irreducible representations and coherent states for algebras already considered previously in §1.4, §2, and §5.1, but now we shall use the unified procedure of reduction by symmetry groups. This is exactly the way in which the representations of quadratic algebras (2.1) and (5.1), as well as their coherent states, were obtained in [**Ka**$_{10}$, **KaN**$_{1-3}$].

We start with the ordinary Heisenberg algebra and the standard Gaussian coherent states over $\mathbb{R}^n$, and make several steps of reduction by its symmetries. At each stage, we apply one of the following three types of reduction.

The *reduction of the first kind* is performed in the space of an irreducible representation of the operator algebra. By this reduction, the coherent states that correspond to this algebra are projected on the eigenspace of an element (called the "*generator of reduction*") of this algebra. As the result, we obtain some new coherent state that corresponds to a subalgebra of the operators that commute with the generator of the reduction.

The *reduction of the second and third kinds* is performed in the space of parameters of coherent states: the second kind of reduction is intended for holomorphic (with respect to parameters) coherent states and is performed in the space of the holomorphic representation of the operator algebra. The third kind of reduction is used for coherent states whose parameters vary along Lagrangian submanifolds in symplectic leaves of the corresponding Poisson algebras [**Ka**$_{10}$, **KaN**$_{2-4}$]. In the case of the second and third kind of reduction, we consider both the generator and its symbol, which is an element of the corresponding Poisson algebra. The "new" coherent state is obtained from the "old" coherent state by averaging (with respect to the parameters) along the trajectories of the Hamilton field of this symbol.

In all examples considered here, as the generators of the reduction we use operators of "action" type. These operators divided by $\hbar$ have an integer-valued spectrum, and their symbols produce 2π-periodic Hamilton flows. Under these conditions, the (first kind) reduction of coherent states in the space of an irreducible representation is equivalent to the (second and third kind) reduction with respect to parameters. It is expedient to have in mind all these types of reduction and examine some properties by using the first kind of reduction and the other by the second and third kind of reduction.

6.1. Gaussian coherent states. Let operators $\widehat{c} = (\widehat{c}_1, \ldots, \widehat{c}_n)$, $\widehat{b} = (\widehat{b}_1, \ldots, \widehat{b}_n)$ acting in a certain Hilbert space satisfy the relations

(6.1) $\qquad [\widehat{c}_j, \widehat{c}_k] = [\widehat{b}_j, \widehat{b}_k] = 0, \qquad [\widehat{c}_j, \widehat{b}_k] = 2\hbar \delta_{jk},$

(6.2) $\qquad \widehat{b}_k = \widehat{c}_k^*,$

and let a vector $\mathfrak{G}$ be the vacuum vector for $\widehat{c}$, that is,

(6.3) $\qquad \widehat{c}_j \mathfrak{G}_0 = 0 \qquad (j = 1, \ldots, n),$

(6.4) $\qquad \|\mathfrak{G}_0\| = 1.$

The family of *Gaussian coherent states* is defined by

$$\mathfrak{G}_c \stackrel{\text{def}}{=} \exp\left\{\frac{1}{2\hbar}\langle c, \widehat{b}\rangle\right\}\mathfrak{G}_0, \qquad c \in \mathbb{C}^n, \quad \hbar > 0. \tag{6.5}$$

PROPOSITION 6.1. (a) *The following identities hold*:

$$\widehat{c}_j \mathfrak{G}_c = c_j \mathfrak{G}_c, \qquad \widehat{b}_j \mathfrak{G}_c = 2\hbar \frac{\partial}{\partial c_j} \mathfrak{G}_c. \tag{6.6}$$

(b) *The inner product of two Gaussian states has the form*

$$\widetilde{\mathcal{K}}(\overline{c''}; c') \stackrel{\text{def}}{=} (\mathfrak{G}_{c''}, \mathfrak{G}_{c'}) = \exp\left\{\frac{1}{2\hbar}\langle c', \overline{c''}\rangle\right\}. \tag{6.7}$$

(c) *The function $\widetilde{\mathcal{K}}$ satisfies the reproducing property*:

$$\frac{1}{2\pi\hbar}\int_{\mathbb{C}^n} \widetilde{\mathcal{K}}(\overline{c''}; c)\widetilde{\mathcal{K}}(\overline{c}; c')\, d\widetilde{\mu}(c) = \widetilde{\mathcal{K}}(\overline{c''}; c'), \tag{6.8}$$

where

$$d\widetilde{\mu}(c) = \frac{2\pi\hbar}{(4\pi\hbar)^n}\exp\left\{-\frac{1}{2\hbar}|c|^2\right\}d\overline{c}\,dc. \tag{6.9}$$

(d) *There is the resolution of the identity*

$$\frac{1}{2\pi\hbar}\int_{\mathbb{C}^n} \mathfrak{G}_c^* \otimes \mathfrak{G}_c\, d\widetilde{\mu}(c) = \widehat{I},$$

where $\widehat{I}$ is the identity operator and $\mathfrak{G}_c^* \otimes \mathfrak{G}_c$ is the projector on the one-dimensional subspace generated by the vector $\mathfrak{G}_c$.

In view of (c), the function $\widetilde{\mathcal{K}}$ can be considered as an example of the Gaussian coherent state $\mathfrak{G}_c = \widetilde{\mathcal{K}}(\cdot; c)$ in the space $\widetilde{\mathcal{P}}$ of antiholomorphic functions endowed with the inner product

$$(\mathfrak{p}'', \mathfrak{p}')_{\mathcal{P}} = \frac{1}{2\pi\hbar}\int_{\mathbb{C}^n} \mathfrak{p}'(\overline{c})\,\overline{\mathfrak{p}''(\overline{c})}\, d\widetilde{\mu}(c). \tag{6.10}$$

In this case, the vacuum vector in formula (6.3) is equal to $\mathfrak{G}_0 \equiv 1$, and the annihilation and creation operators have the form

$$\widehat{c}_j = 2\hbar\frac{\partial}{\partial \overline{c}_j}, \qquad \widehat{b}_j = \overline{c}_j, \qquad j = 1, \ldots, n.$$

These operators are adjoint to each other on the function space $\widetilde{\mathcal{P}}$ with inner product (6.10) and generate an irreducible *antiholomorphic representation* of relations (6.1). This is the well-known Fock–Bargmann representation.

The coherent states corresponding to the irreducible representation of relations (6.1) in the space $L^2(\mathbb{R}^n)$ provide another example, first considered by Schrödinger [**Sch**] and Heisenberg [**H**].

This representation is given by

$$\mathbf{c}_j = \mathbf{u}_j + i\mathbf{w}_j, \qquad \mathbf{b}_j = \mathbf{c}_j^* = \mathbf{u}_j - i\mathbf{w}_j, \qquad j = 1, \ldots, n,$$

where

$$\mathbf{u} \equiv u, \qquad \mathbf{w} \equiv -i\hbar\frac{\partial}{\partial u}.$$

PROPOSITION 6.2. *Let $\widehat{c}_j = \mathbf{c}_j$ and $\widehat{b}_j = \mathbf{b}_j$ in the Hilbert space $L^2(\mathbb{R}^n)$. Then the Gaussian exponential*

$$\mathfrak{G}_0(u) = \frac{1}{(\pi\hbar)^{n/4}} \exp\left\{-\frac{1}{2\hbar}u^2\right\}$$

satisfies conditions (6.3), (6.4) *for the vacuum vector.*

The holomorphic Gaussian states (6.5) *in the space $L^2(\mathbb{R}^n)$ are given by the formula*

(6.11) $$\mathfrak{G}_c(u) = \exp\left\{\frac{1}{4\hbar}(4\langle c, u\rangle - c^2)\right\} \cdot \mathfrak{G}_0(u).$$

6.2. Kustaanheimo transformation and representation of the eight-dimensional quadratic algebra.

Let us consider, for simplicity, the space $\mathbb{R}^4$ only (that is, $n = 4$). The Cartesian coordinates on $\mathbb{R}^4$ are denoted by $u = (u_1, \ldots, u_4)$ and Cartesian coordinates on $\mathbb{R}^3$ are denoted by $q = (q_1, q_2, q_3)$. The Kustaanheimo transformation is given by the formula

(6.12)
$$\sigma : \mathbb{R}^4 \to \mathbb{R}^3, \qquad \sigma : u \mapsto q = {}^\sigma u,$$
$$q_1 = 2(u_1 u_2 + u_2 u_4), \quad q_2 = 2(u_1 u_4 - u_2 u_3), \quad q_3 = u_1^2 + u_2^2 - u_3^2 - u_4^2,$$

Introducing the notation $\psi_1 = u_1 + iu_2$, $\psi_2 = u_3 + iu_4$ and $\psi = (\psi_1, \psi_2)$ and using the Pauli matrices

$$\sigma_1 = \begin{pmatrix} 0 & 1 \\ 1 & 0 \end{pmatrix}, \quad \sigma_2 = \begin{pmatrix} 0 & -i \\ i & 0 \end{pmatrix}, \quad \sigma_3 = \begin{pmatrix} 1 & 0 \\ 0 & -1 \end{pmatrix},$$

one can write the Kustaanheimo transformation (6.12) as follows:

$$q_j = \psi^* \sigma_j \psi, \qquad j = 1, 2, 3.$$

This formula explains why this transformation is often called the spinor regularization. The word "regularization" reflects the following property of the mapping σ: the function $|q| = \sqrt{q_1^2 + q_2^2 + q_3^2}$ is transformed to a smooth function in the variables u, namely $|q| = |{}^\sigma u| = u^2$. Here we collect some properties of transformation (6.12).

PROPOSITION 6.3. (a) *Any function $\varphi \in C_0^\infty(\mathbb{R}^3)$ satisfies the relation*

$$\int_{\mathbb{R}^4} \varphi({}^\sigma u)\, du = \frac{\pi}{4} \int_{\mathbb{R}^3} \varphi(q) \frac{dq}{|q|},$$

and so, with the standard inner product in $L^2(\mathbb{R}^4)$, the mapping σ associates the following inner product over $\mathbb{R}^3$:

(6.13) $$(\varphi', \varphi'')_{-} \stackrel{\text{def}}{=} \frac{\pi}{4}\int_{\mathbb{R}^3} \overline{\varphi'(q)}\, \varphi''(q) \frac{dq}{|q|} = \int_{\mathbb{R}^4} \overline{\varphi'({}^\sigma u)}\, \varphi''({}^\sigma u)\, du.$$

(b) *For any differentiable function $\varphi(q)$ on $\mathbb{R}^3$ one has*

$$\left(u_1 \frac{\partial}{\partial u_2} - u_2 \frac{\partial}{\partial u_1} + u_3 \frac{\partial}{\partial u_4} - u_4 \frac{\partial}{\partial u_3}\right)\varphi({}^\sigma u) \equiv 0, \qquad \frac{\partial \varphi}{\partial q}({}^\sigma u) \equiv \frac{1}{2u^2} D(u) \frac{\partial}{\partial u}(\varphi({}^\sigma u)),$$

where the matrix $D(u)$ is defined as follows:

$$D(u) = \begin{pmatrix} u_3 & u_4 & u_1 & u_2 \\ u_4 & -u_3 & -u_2 & u_1 \\ u_1 & u_2 & -u_3 & -u_4 \end{pmatrix}.$$

We consider $\hbar$-differential operators with respect to the variables $q \in \mathbb{R}^3$, that is, functions in the operators

$$\mathbf{q} \equiv q, \qquad \mathbf{p} \equiv -i\hbar \frac{\partial}{\partial q},$$

and $\hbar$-differential operators with respect to $u \in \mathbb{R}^4$, that is, functions in the variables

$$\mathbf{u} \equiv u, \qquad \mathbf{w} \equiv -i\hbar \frac{\partial}{\partial u}.$$

By Proposition 6.3(a), with each $\hbar$-differential operator $\mathbf{f} = f(\overset{2}{\mathbf{q}}, \overset{1}{\mathbf{p}})$, acting with respect to the variables q, we can associate some $\hbar$-differential operator $\mathbf{F}$ acting with respect to the variables u such that

(6.14) $$\forall \varphi \in C^\infty(\mathbb{R}^3): \qquad (\mathbf{f}\varphi)(^\sigma u) = \mathbf{F}(\varphi(^\sigma u)).$$

The set of operators $\mathbf{F}$ of this type will be denoted by $[\mathbf{F}]_\sigma$. For example, the operator $\mathbf{F}$ representing the class $[\mathbf{F}]_\sigma$ can be chosen as

(6.15) $$\mathbf{F} = f(\overset{2}{\mathbf{Q}}, \overset{1}{\mathbf{P}}),$$

where

(6.16) $$\mathbf{Q} = {}^\sigma \mathbf{u}, \qquad \mathbf{P} = \frac{1}{2\mathbf{u}^2} D(\mathbf{u})\mathbf{w}.$$

In the sequel, for simplicity, we use the notation $\mathbf{f} = [\mathbf{F}]_\sigma$, having in mind that $\mathbf{f}$ and $\mathbf{F}$ are related to each other by (6.14).

REMARK 6.1. The class $[\mathbf{F}]_\sigma$ contains not only one operator (6.15), since (by Proposition 6.3(b)) any differentiable function of the form $\varphi(^\sigma u)$ is annihilated by the operator

(6.17) $$\mathbf{D}_0 \overset{\mathrm{def}}{=} \mathbf{u}_1\mathbf{w}_2 - \mathbf{u}_2\mathbf{w}_1 + \mathbf{u}_3\mathbf{w}_4 - \mathbf{u}_4\mathbf{w}_3 = -\frac{i}{2}(\mathbf{b}_1\mathbf{c}_2 - \mathbf{b}_2\mathbf{c}_1 + \mathbf{b}_3\mathbf{c}_4 - \mathbf{b}_4\mathbf{c}_3),$$

that is, $\mathbf{D}_0 \varphi(^\sigma u) \equiv 0$.

PROPOSITION 6.4. (a) *The only operators commuting with $\mathbf{D}_0$ are operators of the type $[\mathbf{F}]_\sigma$.*
(b) *A function $\Phi(u)$ on $\mathbb{R}^4$ satisfies the equation $\mathbf{D}_0\Phi(u) = 0$ if and only if there is a function $\varphi(q)$ on $\mathbb{R}^3$ such that $\Phi(u) = \varphi(^\sigma u)$.*
(c) *The mapping $\varphi \to \Phi(u) \overset{\mathrm{def}}{=} \varphi(^\sigma u)$ is unitary, that is, $\|\Phi\|_{L^2(\mathbb{R}^4)} = \|\varphi\|_-$.*

Let us introduce the following operators over $\mathbb{R}^3$:

$$\mathbf{S}_0 \overset{\mathrm{def}}{=} |\mathbf{q}|\left(\frac{1}{4} + \mathbf{p}^2\right), \qquad \mathbf{M} \overset{\mathrm{def}}{=} \mathbf{q} \times \mathbf{p}, \qquad \mathbf{L} \overset{\mathrm{def}}{=} \mathbf{q}\left(\frac{1}{4} + \mathbf{p}^2\right) - (\mathbf{p} \times \mathbf{M}) + (\mathbf{M} \times \mathbf{p})$$

(see also the introduction to §2). Note that the vector-operator $\mathbf{L}$ can be called a *modified quantum Laplace–Runge–Lentz vector*, since one has the following commutation relations [**KaN**$_{3,4}$]:

$$[\mathbf{S}_0, \mathbf{M}] = [\mathbf{S}_0, \mathbf{L}] = 0,$$

$$[\mathbf{M}_j, \mathbf{M}_k] = i\hbar\varepsilon_{jk\ell}\mathbf{M}_\ell, \qquad [\mathbf{L}_j, \mathbf{L}_k] = i\hbar\varepsilon_{jk\ell}\mathbf{M}_\ell, \qquad [\mathbf{M}_j, \mathbf{L}_k] = i\hbar\,\varepsilon_{jkl}\mathbf{L}_l,$$

$$\langle \mathbf{M}, \mathbf{L} \rangle = 0, \qquad \mathbf{M}^2 + \mathbf{L}^2 = \mathbf{S}_0^2 - \hbar^2.$$

Note that the usual Laplace–Runge–Lentz vector-operator

$$\frac{x}{|x|} + \frac{\partial}{\partial x} \times \left(x \times \frac{\partial}{\partial x}\right) - \left(x \times \frac{\partial}{\partial x}\right) \times \frac{\partial}{\partial x}$$

corresponds to the hydrogen Hamiltonian $-\Delta - 1/|x|$ rather than to $\mathbf{S}_0$.

PROPOSITION 6.5. *The operators introduced above have the following form under the Kustaanheimo transformation:*

$$|\mathbf{q}| = [\mathbf{u}^2]_\sigma, \qquad \mathbf{p}^2 = \left[\frac{1}{4\mathbf{u}^2}\mathbf{w}^2\right]_\sigma, \qquad \langle \mathbf{q}, \mathbf{p} \rangle = \left[\frac{1}{2}\langle \mathbf{u}, \mathbf{w}\rangle\right]_\sigma,$$

$$\mathbf{S}_0 = \left[\frac{1}{4}(\mathbf{u}^2 + \mathbf{w}^2)\right]_\sigma, \qquad \mathbf{M} = \left[\frac{1}{2}C(\mathbf{u})\mathbf{w}\right]_\sigma, \qquad \mathbf{L} = \left[\frac{1}{4}({}^\sigma\mathbf{u} + {}^\sigma\mathbf{w})\right]_\sigma,$$

where

$$C(u) = \begin{pmatrix} u_4 & -u_3 & u_2 & -u_1 \\ -u_3 & -u_4 & u_1 & u_2 \\ u_2 & -u_1 & -u_4 & u_3 \end{pmatrix}.$$

Now note that on passing to the symbols q, p of the operators $\mathbf{q}$, $\mathbf{p}$ and to the symbols u, w of the operators $\mathbf{u}$, $\mathbf{w}$, we obtain the mapping Σ of the phase space $\mathbb{R}^8_{u,w}$ to the phase space $\mathbb{R}^6_{q,p}$:

(6.18) $$\Sigma \colon (u, w) \to (q, p) = (Q, P),$$

where

$$Q(u, w) \equiv {}^\sigma u, \qquad P(u, w) \equiv \frac{1}{2u^2}D(u)w.$$

By Proposition 6.4(a), the functions $Q(u, w)$ and $P(u, w)$ are the first integrals of the Hamilton system with the Hamiltonian

(6.19) $$D_0(u, w) = u_1w_2 - u_2w_1 + u_3w_4 - u_4w_3.$$

Therefore, the mapping Σ is constant along the trajectories of the Hamilton field $\mathrm{ad}(D_0)$.

According to the property (6.17) of the operator $\mathbf{D}_0$, it is natural to consider the level surface of its symbol D_0 in the phase space $\mathbb{R}^8_{u,w}$:

(6.20) $$\{(u, w) \in \mathbb{R}^8 \mid D_0(u, w) = 0\}.$$

PROPOSITION 6.6. *The pairwise Poisson brackets of the functions Q_j and P_k in $\mathbb{R}^8_{u,w}$ have the form*

$$\{Q_j, Q_k\} = 0, \qquad \{P_j, P_k\} = -\varepsilon_{jkl} \frac{({}^\sigma u)_l}{2(u^2)^3} D_0(u,w), \qquad \{P_j, Q_k\} = \delta_{j,k},$$

where ε_{jkl} is the basic antisymmetric tensor. On the surface $\{D_0 = 0\}$, these relations coincide with the canonical relations between coordinate functions on $\mathbb{R}^6_{q,p}$.

Note that the mapping Σ (6.18) fibers the seven-dimensional surface (6.20) by trajectories of the Hamilton field $\mathrm{ad}(D_0)$. The base of this fibration is a six-dimensional surface Ξ, and the natural coordinates on this base are the functions q_j and p_j ($j = 1, 2, 3$) related to the coordinates u_k and w_k ($k = 1, 2, 3, 4$) by the formulas

$$q \equiv {}^\sigma u, \qquad p = \frac{1}{2u^2} D(u) w.$$

The symplectic structure $dw \wedge du$ on the phase space $\mathbb{R}^8_{u,w}$ is reduced to the symplectic structure $dp \wedge dq$ on Ξ.

The mapping Σ associates the "action" $S_0 = |q|(\frac{1}{4} + p^2)$ on the symplectic manifold Ξ with the Hamilton function of the harmonic oscillator $\frac{1}{4}(u^2 + w^2)$ on the phase space $\mathbb{R}^8$.

Let us consider the symplectic manifold Ξ in more detail. First, we introduce a complex structure on this manifold. In the initial space $\mathbb{R}^8_{u,w}$ with complex coordinates $c_j = u_j + iw_j$ ($j = 1, 2, 3, 4$), we find holomorphic functions of a that are constant along the fibers of Σ. The scalar c^2 and the components of the vector ${}^\sigma c$ are such functions, since they are in involution with the function $D_0(u,w)$. Thus, for complex coordinates on the phase space Ξ one can take the functions

$$(6.21) \qquad v = \frac{1}{4} c^2 \bigg|_{\{D_0 = 0\}}, \qquad V_j = \frac{1}{4} ({}^\sigma c)_j \bigg|_{\{D_0 = 0\}} \qquad (j = 1, 2, 3),$$

satisfying the relation

$$(6.22) \qquad V_1^2 + V_2^2 + V_3^2 = v^2.$$

The complex functions v, V_j can be expressed in terms of the real coordinates q_j, p_j as follows:

$$(6.23) \qquad \begin{aligned} v &= |q|\left(\frac{1}{4} - p^2\right) + i\langle q, p\rangle, \\ V_j &= q_j\left(\frac{1}{4} - p^2\right) + 2(p \times M)_j + i|q|p_j, \qquad j = 1, 2, 3. \end{aligned}$$

This is a modification of the well-known Souriau complex structure [S₂, R₃].

The Poisson brackets between coordinates (6.23) are

$$\begin{aligned} &\{V_j, V_k\} = 0, \qquad \{V_j, v\} = 0, \\ &\{V_j, \overline{V_k}\} = 2i\, \delta_{jk} S_0 + i(V_j \overline{V_k} - V_k \overline{V_j}) S_0^{-1}, \\ &\{V_j, \overline{v}\} = i(V_j \overline{v} + \overline{V_j} v) S_0^{-1}, \\ &\{v, \overline{v}\} = 2i\, S_0, \qquad \{v, S_0\} = i\,v, \qquad \{V_j, S_0\} = i\,V_j, \end{aligned}$$

where

$$S_0 = \sqrt{(|v|^2 + |V|^2)/2}.$$

These relations imply the expression for the symplectic (Kähler) form on Ξ in terms of the complex coordinates:

$$(6.24) \qquad dp \wedge dq = d\varpi, \qquad \varpi = i\frac{\bar{v}\,dv + \langle V, dV\rangle}{\sqrt{2(|v|^2 + |V|^2)}} = i\partial\left(\sqrt{2(|v|^2 + |V|^2)}\right).$$

(Here we consider only one chart which covers the entire manifold Ξ except for the points at which $q_1 = q_2 = q_3 = 0$).

The quantum version of (6.23) is

$$(6.25) \qquad \begin{aligned} \mathbf{v} &\stackrel{\text{def}}{=} |\mathbf{q}|\left(\frac{1}{4} - \mathbf{p}^2\right) + \frac{i}{3}\Big(2\langle\mathbf{q},\mathbf{p}\rangle + \langle\mathbf{p},\mathbf{q}\rangle\Big), \\ \mathbf{V} &\stackrel{\text{def}}{=} \mathbf{q}\left(\frac{1}{4} - \mathbf{p}^2\right) + (\mathbf{p}\times\mathbf{M}) - (\mathbf{M}\times\mathbf{p}) + i|\mathbf{q}|\,\mathbf{p}. \end{aligned}$$

Under the Kustaanheimo transformation these operators look as follows:

$$\mathbf{v} = \left[\frac{1}{4}(\mathbf{u}+i\mathbf{w})^2\right]_\sigma, \qquad \mathbf{V} = \left[\frac{1}{4}\sigma(\mathbf{u}+i\mathbf{w})\right]_\sigma.$$

There is a list of very useful quantum relations between the operators introduced above. There are the following commutation relations:

$$\begin{aligned} [\mathbf{v}, \mathbf{S}_0] &= \hbar\,\mathbf{v}, & [\mathbf{V}_j, \mathbf{S}_0] &= \hbar\,\mathbf{V}_j, \\ [\mathbf{v}, \mathbf{M}_k] &= 0, & [\mathbf{V}_j, \mathbf{M}_k] &= i\hbar\,\varepsilon_{jkl}\mathbf{V}_l, \\ [\mathbf{v}, \mathbf{L}_k] &= \hbar\,\mathbf{V}_k, & [\mathbf{V}_j, \mathbf{L}_k] &= \hbar\,\delta_{jk}\mathbf{v}, \end{aligned}$$

$$\begin{aligned} [\mathbf{v}, \mathbf{V}_k] &= 0, & [\mathbf{v}, \mathbf{v}^*] &= 2\hbar\,\mathbf{S}_0, & [\mathbf{v}, \mathbf{V}_k^*] &= 2\hbar\,\mathbf{L}_k, \\ [\mathbf{V}_j, \mathbf{V}_k] &= 0, & [\mathbf{V}_j, \mathbf{V}_k^*] &= 2\hbar\big(\delta_{jk}\mathbf{S}_0 - i\varepsilon_{jkl}\mathbf{M}_l\big). \end{aligned}$$

There are the following identities:

$$\mathbf{V}^2 = \mathbf{v}^2,$$
$$\mathbf{v}\mathbf{v}^* + \mathbf{v}^*\mathbf{v} + \langle\mathbf{V},\mathbf{V}^*\rangle + \langle\mathbf{V}^*,\mathbf{V}\rangle = 4(\mathbf{S}_0^2 + \hbar^2),$$
$$\mathbf{v}\mathbf{v}^* = \mathbf{S}_0^2 - \mathbf{M}^2 + \hbar\mathbf{S}_0,$$
$$\mathbf{V}_j\mathbf{V}_j^* = \mathbf{S}_0^2 - \mathbf{L}^2 + \mathbf{L}_j^2 - \mathbf{M}_j^2 + \hbar\mathbf{S}_0,$$
$$\mathbf{V}\mathbf{v}^* = \mathbf{S}_0\,\mathbf{L} + i\,\mathbf{M}\times\mathbf{L} + 2\hbar\,\mathbf{L},$$
$$\mathbf{V}\times\mathbf{V}^* - \mathbf{V}^*\times\mathbf{V} = -4i\,\mathbf{S}_0\,\mathbf{M},$$
$$\mathbf{V}\mathbf{v}^* + \mathbf{V}^*\mathbf{v} = 2\,\mathbf{S}_0\,\mathbf{L},$$
$$\langle\mathbf{M},\mathbf{V}\rangle = 0,$$
$$\langle\mathbf{L},\mathbf{V}\rangle = \mathbf{S}_0\,\mathbf{v} - \hbar\,\mathbf{v},$$
$$i\,\mathbf{M}\times\mathbf{V} = \mathbf{S}_0\,\mathbf{V} - \mathbf{L}\,\mathbf{v} - \hbar\,\mathbf{V}.$$

As a corollary we have the following statement.

THEOREM 6.1. *The operators* (6.25) *satisfy the commutation relations*

$$(6.26) \qquad \begin{aligned} [\mathbf{V}_j, \mathbf{V}_k] &= 0, \qquad [\mathbf{V}_j, \mathbf{v}] = 0, \\ [\mathbf{V}_j, \mathbf{V}_k^*] &= 2\hbar\,\delta_{jk}\mathbf{S}_0 + \frac{\hbar}{2}(\mathbf{V}_j\mathbf{V}_k^* + \mathbf{V}_k^*\mathbf{V}_j - \mathbf{V}_k\mathbf{V}_j^* - \mathbf{V}_j^*\mathbf{V}_k)\,\mathbf{S}_0^{-1} \end{aligned}$$

and the identities

(6.27) $\quad \mathbf{V}_1^2 + \mathbf{V}_2^2 + \mathbf{V}_3^2 = \mathbf{v}^2, \qquad \dfrac{1}{4}\big(\mathbf{vv}^* + \mathbf{v}^*\mathbf{v} + \langle \mathbf{V}, \mathbf{V}^*\rangle + \langle \mathbf{V}^*, \mathbf{V}\rangle\big) - \hbar^2 = \mathbf{S}_0^2.$

If one identifies

$$\mathbf{V} \equiv \mathbf{C}, \qquad \mathbf{V}^* \equiv \mathbf{B}, \qquad \mathbf{v} = \mathbf{C}_0, \qquad \mathbf{v}^* = \mathbf{B}_0,$$
$$\mathbf{M} \equiv \mathbf{R}, \qquad \mathbf{S}_0 \equiv \mathbf{A},$$

then relations (6.26) *coincide with* (5.10) *(see also* (5.37)*), relations* (6.27) *coincide with* (5.11)*; hence, formulas* (5.8) *generate an irreducible representation of the quadratic algebra* (5.1) *in the Hilbert space* $L^2(\mathbb{R}^3)$ *of functions over* $\mathbb{R}^3$ *with the inner product* (6.13).

6.3. Reduction from Gaussian to Bessel coherent states. Now we show how the Bessel coherent states (5.44) and the antiholomorphic representation (5.42) (obtained above by a general technique) could be derived from the usual Gaussian states (6.5) and (6.11), exploiting the idea of reduction by a symmetry group.

In the space $L^2(\mathbb{R}^4)$ we have the *action operator* $\frac{1}{\hbar}\mathbf{D}_0$ (6.17), which can be written as

(6.28) $$\dfrac{1}{\hbar}\mathbf{D}_0 = -i\dfrac{\partial}{\partial \iota}.$$

Here $(^\sigma u, \iota)$ are coordinates in $\mathbb{R}^4$, $0 \leq \iota < 2\pi$,

(6.28a) $$\dfrac{\partial}{\partial \iota} \stackrel{\text{def}}{=} u_1 \dfrac{\partial}{\partial u_2} - u_2 \dfrac{\partial}{\partial u_1} + u_3 \dfrac{\partial}{\partial u_4} - u_4 \dfrac{\partial}{\partial u_3}, \qquad \dfrac{\partial}{\partial \iota}(^\sigma u) = 0.$$

The spectrum of operators (6.28) is integer, and the projector onto its null-subspace is given by the integral

(6.29) $\quad \boldsymbol{\sigma}[\Phi] \stackrel{\text{def}}{=} \dfrac{1}{2\pi} \displaystyle\int_0^{2\pi} e^{i\mathbf{D}_0 \tau/\hbar}(\Phi(u))\, d\tau = \dfrac{1}{2\pi} \displaystyle\int_0^{2\pi} \Phi(^\sigma u, \iota)\, d\iota, \qquad \Phi \in L^2(\mathbb{R}^4).$

Thus
$$\boldsymbol{\sigma}^2 = \boldsymbol{\sigma}, \qquad \mathbf{D}_0 \boldsymbol{\sigma}[\Phi] = \boldsymbol{\sigma}[\mathbf{D}_0 \Phi] = 0.$$

LEMMA 6.1. *Let* $\mathfrak{G}_c$ *be Gaussian coherent states* (6.11) *in the space* $L^2(\mathbb{R}^4)$, *and let* $\boldsymbol{\sigma}$ *be the reduction operator* (6.29). *Then*
$$\dfrac{\partial}{\partial \iota(c)}\boldsymbol{\sigma}[\mathfrak{G}_c] = 0,$$
and so, the function $\boldsymbol{\sigma}[\mathfrak{G}_c]$ *depends only on combinations* $^\sigma c$ *of the coordinates* $c = (c_1, \ldots, c_4)$.

PROOF. By (6.6) we have

$$i\hbar \dfrac{\partial}{\partial \iota}\boldsymbol{\sigma}[\mathfrak{G}_c(u)] = \dfrac{i\hbar}{2\pi}\int_0^{2\pi} \exp\Big\{\dfrac{i\tau}{\hbar}\mathbf{D}_0\Big\}\Big(c_1 \dfrac{\partial}{\partial c_2} - c_2 \dfrac{\partial}{\partial c_1} + c_3 \dfrac{\partial}{\partial c_4} - c_4 \dfrac{\partial}{\partial c_3}\Big)\mathfrak{G}_c(u)\,d\tau$$
$$= \dfrac{1}{2\pi}\int_0^{2\pi} \exp\Big\{\dfrac{i\tau}{\hbar}\mathbf{D}_0\Big\}(\mathbf{u}_1\mathbf{w}_2 - \mathbf{u}_2\mathbf{w}_1 + \mathbf{u}_3\mathbf{w}_4 - \mathbf{u}_4\mathbf{w}_3)\mathfrak{G}_c(u)\,d\tau$$
$$= \mathbf{D}_0 \boldsymbol{\sigma}[\mathfrak{G}_c(u)] = 0. \quad \square$$

DEFINITION 6.1. The holomorphic Bessel states (of order 0) are defined by the reduction operator as follows:

(6.30) $$\mathfrak{B}_{v,V}(q) = \sigma[\mathfrak{G}_c]\Big|_{\sigma_u=q,\ \sigma_c=4V,\ c^2=4v}$$

Lemma 6.1 proves the correctness of this definition. This method of reduction of coherent states was suggested in [**Ka**$_{10}$].

PROPOSITION 6.7. *The holomorphic Bessel states in the space $L^2_-(\mathbb{R}^3)$ can be represented by the formula*

(6.31) $$\mathfrak{B}_{v,V}(q) = \frac{1}{\pi\hbar}e^{-(2v+|q|)/2\hbar}\, I_0\!\left(\frac{1}{\hbar}\sqrt{2(|q|v+\langle q,V\rangle)}\right),$$

where
$$I_0(y) = \frac{1}{2\pi}\int_0^{2\pi} e^{y\cos\varphi}\,d\varphi$$

is the Bessel function of imaginary argument [**BE**].

PROOF. On the trajectories of the field $\partial/\partial\iota$, the holomorphic Gaussian coherent state (6.11) has the form

$$\mathfrak{G}_c({}^\sigma u,\iota) = \frac{1}{\pi\hbar}\exp\{-(c^2+2u^2)/4\hbar\}\exp\{(\langle u,c\rangle\cos\iota + D_0(u,c)\sin\iota))/\hbar\},$$

where D_0 is the function (6.19). Hence,

$$\mathfrak{B}_{c^2/4,\,\sigma_{c/4}}({}^\sigma u) = \frac{1}{2\pi}\int_0^{2\pi}\mathfrak{G}_c({}^\sigma u,\tau)\,d\tau$$
$$= \frac{1}{2\pi^2\hbar}e^{-(c^2+2u^2)/4\hbar}\int_0^{2\pi} e^{\sqrt{\langle u,c\rangle^2+D_0^2(u,c)}\cos(\tau-\tau^0)/\hbar}\,d\tau,$$

where τ^0 is determined by the relation $e^{i\tau^0} = \frac{\langle u,c\rangle + iD_0(u,c)}{\sqrt{\langle u,c\rangle^2+D_0^2(u,c)}}$. Applying the identity

$$\langle u,c\rangle^2 + D_0^2(u,c) \equiv \frac{1}{2}\bigl(u^2c^2 + \langle{}^\sigma u,{}^\sigma c\rangle\bigr)$$

and the integral representation of the Bessel function I_0, we obtain the following formula for reduced states:

$$\mathfrak{B}_{c^2/4,\,\sigma_{c/4}}({}^\sigma u) = \frac{1}{\pi\hbar}e^{-(c^2+2u^2)/4\hbar}\,I_0\!\left(\frac{1}{\hbar}\sqrt{\frac{1}{2}(u^2c^2 + \langle{}^\sigma u,{}^\sigma c\rangle)}\right).$$

The substitution ${}^\sigma u = q$, ${}^\sigma c = 4V$ leads to the desired equality (6.31). □

We note that the states (6.31) differ from those obtained in the now classical paper [**R**$_1$], as well as in [**HoOd**].

One can readily generalize the reduction (6.30) of a Gaussian coherent state $\mathfrak{G}_c(u)$ in the space $L^2(\mathbb{R}^4)$ to the case of a Gaussian coherent state $\mathfrak{G}_c$ (6.5) that corresponds to an abstract irreducible representation of the Heisenberg algebra in some Hilbert space $\widetilde{H}$. To this end, in the reduction operator $\boldsymbol{\sigma}[\cdot]$ (6.29) it suffices to take the projector on the null-eigensubspace of the operator

(6.32) $$\widehat{D}_0 = -\frac{i}{2}\bigl(\widehat{b}_1\widehat{c}_2 - \widehat{b}_2\widehat{c}_1 + \widehat{b}_3\widehat{c}_4 - \widehat{b}_4\widehat{c}_3\bigr).$$

It follows from the commutation relations (6.1) that the spectrum of the operator $\widehat{D}_0/\hbar$ is integer. Moreover, to the eigenvalue $k \in \mathbb{Z}$, there correspond the eigenvectors

$$(6.33) \qquad \begin{gathered} (\widehat{b_1} + i\widehat{b_2})^{l_1}(\widehat{b_1} - i\widehat{b_2})^{l_2}(\widehat{b_3} + i\widehat{b_4})^{r_1}(\widehat{b_3} - i\widehat{b_4})^{r_2}\mathfrak{G}_0, \\ l_1 - l_2 + r_1 - r_2 = k, \qquad l_j, r_j \in \mathbb{Z}_+, \end{gathered}$$

where $\mathfrak{G}_0$ is the vacuum vector given by equations (6.3). Hence, the exponential $\exp\{\frac{i\tau}{\hbar}\widehat{D}_0\}$ satisfies the periodicity condition $\exp\{\frac{2\pi i}{\hbar}\widehat{D}_0\} = \widehat{I}$, and therefore, the projector on the eigensubspace $H \subset \widetilde{H}$ of the operator $\widehat{D}_0$ corresponding to the zero eigenvalue has the form

$$\widehat{\sigma}[\Phi] = \frac{1}{2\pi} \int_0^{2\pi} \exp\left\{\frac{i\tau}{\hbar}\widehat{D}_0\right\} \Phi \, d\tau, \qquad \Phi \in \widetilde{H}.$$

DEFINITION 6.2. The formula

$$(6.34) \qquad \mathfrak{B}_{v,V} \stackrel{\text{def}}{=} \widehat{\sigma}[\mathfrak{G}_c]\Big|_{c^2=4v, \, \sigma c=4V},$$

determines the *family of holomorphic* (with respect to v, V) *Bessel states* in the subspace $H \subset \widetilde{H}$. Here $\mathfrak{G}_c$ is a holomorphic Gaussian coherent state (6.5) in the space $\widetilde{H}$, and H is the zero eigenspace of the operator $\widehat{D}_0$ (6.32).

Note that formula (6.34), where the projection to the eigenspace is used, is closely related to the construction of the constrained coherent states considered in [**Kl**₃].

By analogy with the special case $\widetilde{H} = L(\mathbb{R}^4)$, we can verify that $\mathfrak{B}_{v,V}$ in Definition 6.2 is well defined.

THEOREM 6.2. (a) *The holomorphic Bessel states defined by the reduction mapping* (6.34) *are represented in the form*

$$(6.35) \qquad \mathfrak{B}_{v,V} = I_0\left(\frac{1}{\hbar}\sqrt{2(v\widehat{v^*} + \langle V, \widehat{V^*}\rangle)}\right) \mathfrak{B}_{0,0}.$$

Here I_0 is the Bessel function of imaginary argument, and the "creation" operators $\widehat{v^*}$, $\widehat{V^*}$ are given by the formulas

$$(6.36) \qquad \widehat{v^*} \stackrel{\text{def}}{=} \frac{1}{4}\widehat{b}^2, \qquad \widehat{V^*} \stackrel{\text{def}}{=} \frac{1}{4}\sigma\widehat{b}.$$

The "vacuum" vector $\mathfrak{B}_{0,0}$ has the unit norm $\|\mathfrak{B}_{0,0}\|_H = 1$ and is given by the formula $\mathfrak{B}_{0,0} = \mathfrak{G}_0$, where $\mathfrak{G}_0$ is the vacuum (6.3).

(b) *The vacuum vector* $\mathfrak{B}_{0,0} \in H$ *is annihilated by the "annihilation" operators*

$$(6.37) \qquad \widehat{v} = \frac{1}{4}\widehat{c}^2, \qquad \widehat{V} = \frac{1}{4}\sigma\widehat{c};$$

that is, the following relations are satisfied:

$$(6.38) \qquad \widehat{v}\,\mathfrak{B}_{0,0} = 0, \qquad \widehat{V}\,\mathfrak{B}_{0,0} = 0.$$

(c) *On the subspace* $H \subset \widetilde{H}$, *the operators* $\widehat{v}$, $\widehat{V}$, $\widehat{v^*}$, $\widehat{V^*}$, *and* $\widehat{S}_0 = \frac{1}{8}(\langle\widehat{b},\widehat{c}\rangle + \langle\widehat{c},\widehat{b}\rangle)$ *determine the irreducible representation of the algebra* (6.26).

PROOF. (a) Calculating the commutator $[\langle c,\widehat{b}\rangle, \exp\{\frac{i\tau}{\hbar}\widehat{D}_0\}]$ and using the commutation relations

$$[\widehat{D}_0, \langle c,\widehat{b}\rangle] = i\hbar\, D_0(c,\widehat{b}), \qquad [\widehat{D}_0, D_0(c,\widehat{b})] = -i\hbar\,\langle c,\widehat{b}\rangle$$

(where $D_0(u,w)$ is the function (6.19)), we obtain the identity

$$\exp\left\{\frac{i\tau}{\hbar}\widehat{D}_0\right\}\langle c,\widehat{b}\rangle = \left(\cos\tau\,\langle c,\widehat{b}\rangle - \sin\tau\, D_0(c,\widehat{b})\right)\exp\left\{\frac{i}{\hbar}\widehat{D}_0\tau\right\}.$$

This identity implies

$$\exp\left\{\frac{i\tau}{\hbar}\widehat{D}_0\right\}\cdot\exp\left\{\frac{1}{2\hbar}\langle c,\widehat{b}\rangle\right\}$$
$$= \exp\left\{\frac{1}{2\hbar}\left(\cos\tau\,\langle c,\widehat{b}\rangle - \sin\tau\, D_0(c,\widehat{b})\right)\right\}\cdot\exp\left\{\frac{i\tau}{\hbar}\widehat{D}_0\right\}.$$

Thus the projection of the Gaussian coherent state $\mathfrak{G}_c$ on the subspace H is equal to

$$\widehat{\sigma}[\mathfrak{G}_c] = \frac{1}{2\pi}\int_0^{2\pi}\exp\left\{\frac{1}{2\hbar}\left(\cos\tau\,\langle c,\widehat{b}\rangle - \sin\tau\, D_0(c,\widehat{b})\right)\right\}\cdot\exp\left\{\frac{i\tau}{\hbar}\widehat{D}_0\right\}\mathfrak{G}_0\, d\tau.$$

Here

(6.39) $$\exp\left\{\frac{i\tau}{\hbar}\widehat{D}_0\right\}\mathfrak{G}_0 \equiv \mathfrak{G}_0 = \mathfrak{B}_{0,0},$$

since it follows from (6.3) that $\widehat{D}_0\mathfrak{G}_0 = 0$. Hence,

(6.40)
$$\widehat{\sigma}[\mathfrak{G}_c] = \frac{1}{2\pi}\int_0^{2\pi}\exp\left\{\frac{1}{2\hbar}\left(\cos\tau\,\langle c,\widehat{b}\rangle - \sin\tau\, D_0(c,\widehat{b})\right)\right\}\mathfrak{G}_0\, d\tau$$
$$= I_0\left(\frac{1}{2\hbar}\sqrt{\langle c,\widehat{b}\rangle^2 + (D_0(c,\widehat{b}))^2}\right)\mathfrak{G}_0.$$

One can prove the last relation (that is, the passage to the Bessel function I_0), say, by expanding the exponential $\exp\{\frac{1}{2\hbar}(\cos\tau\,\langle c,\widehat{b}\rangle - \sin\tau\, D_0(c,\widehat{b}))\}$ and the function $I_0(\frac{1}{2\hbar}\sqrt{\langle c,\widehat{b}\rangle^2 + (D_0(c,\widehat{b}))^2})$ in power series. In this case, we use the following representation of the Bessel function I_0 of imaginary argument:

$$I_0(y) = \sum_{k=0}^{\infty}\frac{1}{(k!)^2}\left(\frac{y}{2}\right)^{2k}.$$

By substituting

$$\langle c,\widehat{b}\rangle^2 + (D_0(c,\widehat{b}))^2 = \frac{1}{2}\left(c^2\widehat{b}^2 + \langle{}^\sigma c,{}^\sigma\widehat{b}\rangle\right)$$

into (6.40), we obtain the desired formula (6.35) for the projection $\widehat{\sigma}[\mathfrak{G}_c]$.

(b) This statement follows from (6.39) and properties (6.3), (6.4) of the vacuum vector $\mathfrak{G}_0$.

(c) The commutation relations (6.26) on the subspace $H \subset \widetilde{H}$, on which the operator $\widehat{D}_0$ is annihilated, can be verified by analogy with the special case $\widetilde{H} = L^2(\mathbb{R}^4)$, $\widehat{c} = \mathbf{c}$, and $\widehat{b} = \mathbf{b}$. The representation is irreducible in the space H, since this subspace is generated from the vacuum vector $\mathfrak{B}_{0,0}$ by powers of creation operators (see (6.33) for $k=0$):

$$\operatorname{const}(\widehat{v^*}+\widehat{V_3^*})^l\,(\widehat{v^*}-\widehat{V_3^*})^r\,(\widehat{V_1^*}-i\operatorname{sgn}(s)\widehat{V_2^*})^{|s|}\mathfrak{B}_{0,0}\qquad l,r\in\mathbb{Z}_+,\ s\in\mathbb{Z}$$

(here $l=l_2$, $r=r_1$, $s=l_1-l_2$, and $\operatorname{const}=2^{l+r+|s|}$). $\square$

Since the Bessel states $\mathfrak{B}_{v,V}$ are constructed by the reduction (6.34) from Gaussian coherent states $\mathfrak{G}_c$, it is natural to expect that the properties of Bessel states can be obtained from the properties of Gaussian states. But it is more convenient to study some of these properties by using another reduction, namely, the reduction with respect to parameters $c\in\mathbb{C}^4$ of the Gaussian coherent states $\mathfrak{G}_c$. Let us describe this reduction of the second kind.

LEMMA 6.2. *The reduction (6.34) in the space $\widetilde{H}$ of the irreducible representation of relations (6.1) is equivalent to the following reduction with respect to parameters $c\in\mathbb{C}^4$:*

$$(6.41)\qquad \mathfrak{B}_{v,V}\equiv\frac{1}{2\pi}\int_0^{2\pi}\mathfrak{G}_{c(c_0,\iota)}\Big|_{c_0^2=4v,\ \sigma c_0=4V}\,d\iota,$$

where $c(c_0,\iota)$ is the trajectory of the field $\partial/\partial\iota$ (6.28a) in the space $\mathbb{R}_c^4$.

PROOF. The Gaussian coherent state on the trajectories $c=c(c_0,\iota)$ can be written as follows:

$$\mathfrak{G}_{c(c_0,\iota)}=\exp\left\{\frac{1}{2\hbar}\Big(\cos\iota\,\langle c,\widehat{b}\rangle+\sin\iota\,D_0(c,\widehat{b})\Big)\right\}\mathfrak{G}_0,$$

where $D_0(u,w)$ is the function (6.29). Hence, the integral $\frac{1}{2\pi}\int_0^{2\pi}\mathfrak{G}_{c(4v,4V,\iota)}\,d\iota$ coincides with (6.40) and is equal to the Bessel state $\widehat{\sigma}[\mathfrak{G}_c]\big|_{c^2=4v,\ \sigma c=4V}=\mathfrak{B}_{v,V}$, as follows from (6.34). $\square$

PROPOSITION 6.8. *The inner product of two Bessel states can be calculated by the following formula:*

$$(6.42)\qquad \mathcal{K}(\overline{v},\overline{V};v,V)\stackrel{\mathrm{def}}{=}(\mathfrak{B}_{v,V},\mathfrak{B}_{v,V})_H\equiv I_0\Big(\frac{1}{\hbar}\sqrt{2(|v|^2+|V|^2)}\Big).$$

PROOF. It follows from Lemma 6.2 that the inner product of vectors (6.34), (6.41) can be calculated by the formula

$$(\mathfrak{B}_{v,V},\mathfrak{B}_{v,V})_H=\frac{1}{(2\pi)^2}\int_0^{2\pi}d\iota'\int_0^{2\pi}d\iota''\,\widetilde{\mathcal{K}}\big(\overline{c(4v,4V,\iota')};c(4v,4V,\iota'')\big),$$

where $\widetilde{\mathcal{K}}(\overline{c'};c'')$ is the inner product (6.7) of Gaussian states (6.5). To calculate the integral with respect to ι'' on the right-hand side, we use the formula

$$\frac{1}{2\pi}\int_0^{2\pi}\widetilde{\mathcal{K}}\big(2u;c(4v,4V,\iota)\big)\,d\iota=\frac{1}{2\pi}\int_0^{2\pi}\exp\left\{\frac{1}{\hbar}\langle u,c(4v,4V,\iota)\rangle\right\}d\iota$$

$$=I_0\Big(\frac{1}{\hbar}\sqrt{2(u^2v+\langle\sigma u,V\rangle)}\Big),$$

which was already derived in the proof of Proposition 6.7. Instead of ι, v, and V, we substitute ι'', $\overline{v}$, and $\overline{V}$, respectively, and replace the vector $2u$ by the vector $c(4v, 4V, \iota')$. Then we see that the integral with respect to ι'' is equal to

$$\text{(6.43)} \qquad \frac{1}{2\pi} \int_0^{2\pi} \widetilde{\mathcal{K}}\big(c(4v, 4V, \iota'); \overline{c(4v, 4V, \iota'')}\big) \, d\iota'' = I_0\left(\frac{1}{\hbar}\sqrt{2(|v|^2 + |V|^2)}\right)$$

and independent of ι'. Integrating with respect to ι' implies (6.42). □

6.4. Reduction of the reproducing property. We now prove that the reproducing property of Gaussian states implies a reproducing property of Bessel states.

LEMMA 6.3. *The function* (6.42) *satisfies the property of reproducing kernels:*

$$\text{(6.44)} \qquad \frac{1}{2\pi\hbar} \int_{\mathbb{C}^3} \mathcal{K}(\overline{v''}, \overline{V''}; v, V) \, \mathcal{K}(\overline{v}, \overline{V}; v', V') \, d\mu(V) = \mathcal{K}(\overline{v''}, \overline{V''}; v', V'),$$

where

(6.45)
$$d\mu(V) = \frac{1}{8\pi^2 \hbar^3 |v|^2} K_0\left(\frac{1}{\hbar}\sqrt{2(|v|^2 + |V|^2)}\right) d\overline{V} \, dV,$$

$$\text{(6.46)} \qquad K_0(r) = \int_0^\infty e^{-r \cosh t} \, dt \quad \text{is the McDonald function } [\mathbf{BE}].$$

PROOF. We average the identity (6.8) along the trajectories $c' = c(4v', 4V', \tau')$ and $c'' = c(4v'', 4V'', \tau'')$, that is, integrate it with respect to τ' and τ'':

$$\frac{1}{2\pi\hbar} \frac{1}{(2\pi)^2} \int_0^{2\pi} d\tau' \int_0^{2\pi} d\tau'' \int_{\mathbb{C}^4} \widetilde{\mathcal{K}}\big(\overline{c(4v'', 4V'', \tau'')}; c\big) \widetilde{\mathcal{K}}\big(\overline{c}; c(4v', 4V', \tau')\big) \, d\widetilde{\mu}(c)$$

$$= \frac{1}{(2\pi)^2} \int_0^{2\pi} d\tau' \int_0^{2\pi} d\tau'' \widetilde{\mathcal{K}}\big(\overline{c(4v'', 4V'', \tau'')}; c(4v', 4V', \tau')\big).$$

Then, by (6.43) and (6.42), we obtain the new identity

$$\text{(6.47)} \qquad \frac{1}{2\pi\hbar} \int_{\mathbb{C}^4} \mathcal{K}(\overline{v''}, \overline{V''}; v, V) \, \mathcal{K}(\overline{v}, \overline{V}; v', V') \, d\widetilde{\mu}(c) = \mathcal{K}(\overline{v''}, \overline{V''}; v', V'),$$

where we have $v = c^2/4$ and $V = {}^\sigma c/4$ under the integral. We can simplify the integral on the left-hand side by passing from the variables $(c, \overline{c})$ to the variables $(V, \overline{V}, D_0, \iota)$, where $D_0 = \frac{i}{2}(c_1\overline{c}_2 - c_2\overline{c}_1 + c_3\overline{c}_4 - c_4\overline{c}_3)$, and ι is the time on the trajectories of the Hamilton field $\frac{\partial}{\partial \iota} = \mathrm{ad}(D_0)$, that is, the coordinate ι is conjugate to D_0. The density of the measure $d\widetilde{\mu}$ (6.9) can be written in terms of the new coordinates as follows:

$$\frac{1}{(4\pi\hbar)^4} \exp\left\{-\frac{1}{2\hbar}|c|^2\right\} = \frac{1}{(4\pi\hbar)^4} \exp\left\{-\frac{1}{\hbar}\sqrt{2(|v|^2 + |V|^2) + D_0^2}\right\}.$$

For the Liouville measure on $\mathbb{C}^4$ we have

$$\frac{|dc \wedge d\overline{c}|}{16} = \frac{|dV \wedge d\overline{V}|}{|\det \Psi|} \cdot |dD_0 \wedge d\tau|.$$

Here Ψ is a 3×3 matrix with elements $\Psi_{jk} = \{\frac{1}{4}({}^\sigma c)_j, \frac{1}{4}({}^\sigma c)_k\}$;

$$|\det \Psi| = \frac{1}{8}|c|^2\,|c^2|^2 = 4|v|^2\sqrt{2(|v|^2 + |V|^2) + D_0^2}.$$

Therefore, the identity (6.47) is equivalent to

$$\frac{1}{2\pi\hbar}\int_{\mathbb{C}^3} \mathcal{K}(\overline{v''}, \overline{V''}; v, V)\,\mathcal{K}(\overline{v}, \overline{V}; v', V')\,\ell(|v|^2, |V|^2)\,|dV \wedge d\overline{V}| = \mathcal{K}(\overline{v''}, \overline{V''}; v', V'),$$

where

$$\ell(|v|^2, |V|^2) = \frac{1}{32(\pi\hbar)^3|v|^2}\int_{-\infty}^{\infty} dD_0 \int_0^{2\pi} d\tau\, \frac{\exp\left\{-\frac{1}{\hbar}\sqrt{2(|v|^2 + |V|^2) + D_0^2}\right\}}{\sqrt{2(|v|^2 + |V|^2) + D_0^2}}$$

is the density of the measure. In this formula, the integrand is independent of τ and is an even function of D_0. Therefore, after the change $D_0 = \sqrt{2(|v|^2 + |V|^2)}\cdot \sinh t$, we obtain

(6.48)
$$\rho(\overline{v}, \overline{V}; v, V) = \frac{1}{8\pi^2 \hbar^3 |v|^2}\int_0^{\infty} \exp\left\{-\frac{1}{\hbar}\sqrt{2(|v|^2 + |V|^2)}\cdot \cosh t\right\} dt$$
$$= \frac{1}{8\pi^2 \hbar^3 |v|^2} K_0\left(\frac{1}{\hbar}\sqrt{2(|v|^2 + |V|^2)}\right). \qquad \square$$

THEOREM 6.3. *Suppose that the Gaussian coherent state $\mathfrak{G}_c$ (6.5) corresponds to an irreducible representation of relations (6.1) in a Hilbert space $\widetilde{H}$. Then the family of vectors $\{\mathfrak{B}_{v,V} \mid v \in \mathbb{C}, V \in \mathbb{C}^3\}$ (6.34) is a family of coherent states in the null-subspace $H \subset \widetilde{H}$ of the operator $\widehat{D}_0$ (6.32). More precisely,*

$$\frac{1}{2\pi\hbar}\int_{\mathbb{C}^3} \overline{\mathfrak{B}_{v,V}} \otimes \mathfrak{B}_{v,V}\, d\mu(V) = \widehat{I},$$

where $\widehat{I}$ is the identity operator in H and $d\mu(V)$ is given by formula (6.45).

PROOF. By Proposition 6.1(d), the identity

$$\frac{1}{2\pi\hbar}\int_{\mathbb{C}^4} (\mathfrak{G}_c, \Phi)_{\widetilde{H}}\, \mathfrak{G}_c\, d\widetilde{\mu}(c) = \Phi$$

is satisfied for any vector $\Phi \in H \subset \widetilde{H}$. By (6.34) the projection of this identity on the subspace H has the form

$$\frac{1}{2\pi\hbar}\int_{\mathbb{C}^4} (\mathfrak{G}_c, \Phi)_{\widetilde{H}}\, \mathfrak{B}_{c^2/4,\, {}^\sigma c/4}\, d\widetilde{\mu}(c) = \Phi.$$

Here

$$(\mathfrak{G}_c, \Phi)_{\widetilde{H}} = (\mathfrak{G}_c, \widehat{\sigma}[\Phi])_{\widetilde{H}} = \frac{1}{2\pi}\int_0^{2\pi} \left(\mathfrak{G}_c, \exp\left\{\frac{i\tau}{\hbar}\widehat{D}_0\right\}\Phi\right)_{\widetilde{H}} d\tau$$
$$= \frac{1}{2\pi}\int_0^{2\pi}\left(\exp\left\{-\frac{i\tau}{\hbar}\widehat{D}_0\right\}\mathfrak{G}_c, \Phi\right)_{\widetilde{H}} d\tau = (\widehat{\sigma}[\mathfrak{G}_c], \Phi)_H = (\mathfrak{B}_{c^2/4,\, {}^\sigma c/4}, \Phi)_H,$$

since the operator $\widehat{D}_0$ (6.32) is self-adjoint in $\widetilde{H}$. Hence,

$$\frac{1}{2\pi\hbar}\int_{\mathbb{C}^4}(\mathfrak{B}_{c^2/4,\,{}^\sigma c/4}, \Phi)_H\, \mathfrak{B}_{c^2/4,\,{}^\sigma c/4}\, d\widetilde{\mu}(c) = \Phi.$$

If we transform the measure $d\widetilde{\mu}$ by analogy with the proof of Lemma 6.3, we obtain the desired representation of the vector $\Phi \in H$:

$$\frac{1}{2\pi\hbar} \int_{\mathbb{C}^4} (\mathfrak{B}_{v,V}, \Phi)_H \mathfrak{B}_{v,V} \, d\widetilde{\mu}(c) = \Phi. \quad \square$$

In particular, Theorem 6.3 implies that the family of Bessel states $\mathfrak{B}_{v,V}(q)$ given by (6.31) is complete in the space $L^2_-(\mathbb{R}^3)$ with the inner product (6.13).

6.5. Reduction of the intertwining property and of the antiholomorphic representation. By Proposition 6.1(a), the family of Gaussian coherent states $\mathfrak{G}_c$ has the following property: for any operator $\widehat{F} = F(\overset{1}{\widehat{c}}; \overset{2}{\widehat{b}})$, acting in the space $\widetilde{H}$, there exists an operator $\overset{\triangle}{F}$, acting by parameters $c \in \mathbb{C}^4$, such that $\widehat{F} \mathfrak{G}_c = \overset{\triangle}{F} \mathfrak{G}_c$. Moreover, we have the explicit formula

$$\overset{\triangle}{F} = F(\overset{2}{c}, \overset{1}{2\hbar \partial/\partial c}).$$

The reduced Bessel coherent states $\mathfrak{B}_{v,V}$ (6.34) also have a similar property. Namely, for each operator $\widehat{f} = f(\overset{1}{\widehat{v}}, \overset{1}{\widehat{V}}; \overset{2}{\widehat{v}^*}, \overset{2}{\widehat{V}^*})$, acting in the space H, there exists an operator $\overset{\triangle}{f}$, acting by parameters $V \in \mathbb{C}^3$, such that

$$\widehat{f} \mathfrak{B}_{v,V} = \overset{\triangle}{f} \mathfrak{B}_{v,V}. \tag{6.49}$$

To prove the property (6.49) in general, it suffices to write the formulas for the operators $\overset{\triangle}{\widehat{v}}, \overset{\triangle}{\widehat{V}_j}, \overset{\triangle}{\widehat{v}^*}$, and $\overset{\triangle}{\widehat{V}_j^*}$ associated by (6.49) with the operators $\widehat{v}, \widehat{V}_j$ (6.37) and $\widehat{v}^*, \widehat{V}^*j$ (6.36) (on the subspace H).

PROPOSITION 6.9. *The following relations are satisfied*:

$$\tag{6.50} \begin{array}{ll} \overset{\triangle}{v} = v, & \overset{\triangle}{v}^* = \hbar^2 v \left(\dfrac{\partial}{\partial V}\right)^2, \\[1em] \overset{\triangle}{V}_j = V_j, & \overset{\triangle}{V}_j^* = \hbar^2 \left(2 \langle V, \dfrac{\partial}{\partial V} \rangle \dfrac{\partial}{\partial V_j} - V_j \left(\dfrac{\partial}{\partial V}\right)^2 + 2 \dfrac{\partial}{\partial V_j}\right). \end{array}$$

PROOF. Using definitions (6.37) and (6.36) of the operators $\widehat{v}, \widehat{V}, \widehat{v}^*$, and $\widehat{V}^*$, formula (6.34), the commutation relations

$$[\widehat{v}, \widehat{D}_0] = 0, \quad [\widehat{v}^*, \widehat{D}_0] = 0, \quad [\widehat{V}, \widehat{D}_0] = 0, \quad [\widehat{V}^*, \widehat{D}_0] = 0$$

and Proposition 6.1(a), we obtain

$$\overset{\triangle}{v}\mathfrak{B}_{v,V} = \widehat{v}\,\mathfrak{B}_{v,V} = \frac{1}{4}(\widehat{c})^2\,\widehat{\sigma}[\mathfrak{G}_c]\Big|_{c^2=4v,\,\sigma_c=4V}$$

$$= \frac{1}{8\pi}\int_0^{2\pi} \exp\left\{\frac{i\tau}{\hbar}\widehat{D}_0\right\}(\widehat{c})^2\,\mathfrak{G}_c\,d\tau\Big|_{c^2=4v,\,\sigma_c=4V}$$

$$= \frac{1}{8\pi}\int_0^{2\pi} \exp\left\{\frac{i\tau}{\hbar}\widehat{D}_0\right\}c^2\,\mathfrak{G}_c\,d\tau\Big|_{c^2=4v,\,\sigma_c=4V}$$

$$= \frac{1}{4}c^2\,\widehat{\sigma}[\mathfrak{G}_c]\Big|_{c^2=4v,\,\sigma_c=4V} = v\,\mathfrak{B}_{v,V}.$$

Similarly, we have

$$\overset{\triangle}{V}\mathfrak{B}_{v,V} = V\mathfrak{B}_{v,V},$$

$$\overset{\triangle}{v^*}\mathfrak{B}_{v,V} = \hbar^2\left(\frac{\partial}{\partial c}\right)^2\mathfrak{B}\Big|_{c^2=4v,\,\sigma_c=4V},$$

$$\overset{\triangle}{V^*}\mathfrak{B}_{v,V} = \hbar^2\overset{\sigma}{\left(\frac{\partial}{\partial c}\right)}\mathfrak{B}\Big|_{c^2=4v,\,\sigma_c=4V}.$$

The first two formulas coincide with the desired ones. In the last two formulas, we replace the variables $c \in \mathbb{C}^4$ by the new independent variables $V = \frac{1}{4}\sigma c \in \mathbb{C}^3$, ι and obtain the operator equalities

$$\left(\frac{\partial}{\partial c}\right)^2 = v\left(\frac{\partial}{\partial V}\right)^2 + \left(\text{first order differential operator in } V,\,\iota\right)\cdot\frac{\partial}{\partial \iota},$$

$$\overset{\sigma}{\left(\frac{\partial}{\partial c}\right)} = 2\langle V,\frac{\partial}{\partial V}\rangle\frac{\partial}{\partial V} - V\left(\frac{\partial}{\partial V}\right)^2 + 2\frac{\partial}{\partial V}$$

$$+ \left(\text{first order differential operator in } V,\,\iota\right)\cdot\frac{\partial}{\partial \iota}.$$

From these relations and the identity $\frac{\partial}{\partial \iota}\widehat{\sigma}[\mathfrak{G}_c(u)] = 0$ we obtain

$$\overset{\triangle}{v^*}\mathfrak{B}_{v,V} = \hbar^2 v\left(\frac{\partial}{\partial V}\right)^2\mathfrak{B}_{v,V},$$

$$\overset{\triangle}{V^*}\mathfrak{B}_{v,V} = \hbar^2\left(2\langle V,\frac{\partial}{\partial V}\rangle\frac{\partial}{\partial V} - V\left(\frac{\partial}{\partial V}\right)^2 + 2\frac{\partial}{\partial V}\right)\mathfrak{B}_{v,V}. \quad\square$$

In addition to (6.50), it is also convenient to have explicit formulas for the operators $\overset{\triangle}{S}_0$, $\overset{\triangle}{M}$, $\overset{\triangle}{L}$, associated by (6.49) with the operators

(6.51) $\qquad \widehat{S}_0 = \frac{1}{8}(\langle \widehat{c},\widehat{b}\rangle + \langle \widehat{b},\widehat{c}\rangle), \qquad \widehat{M} = \frac{i}{4}C(\widehat{c})\widehat{b}, \qquad \widehat{L} = \frac{1}{8}D(\widehat{c})\widehat{b},$

where the matrices $C(u)$ and $D(u)$ are determined in Propositions 6.3(b) and 6.5. The operators (6.51) serve as a generalization of the above cited operators (see § 6.2)

$$\mathbf{S}_0 = \left[\frac{1}{8}(\langle \mathbf{c},\mathbf{b}\rangle + \langle \mathbf{b},\mathbf{c}\rangle)\right]_\sigma, \qquad \mathbf{M} = \left[\frac{i}{4}C(\mathbf{c})\mathbf{b}\right]_\sigma, \qquad \mathbf{L} = \left[\frac{1}{8}D(\mathbf{c})\mathbf{b}\right]_\sigma,$$

to the case of an abstract irreducible representation of relations (6.1).

PROPOSITION 6.10. *The following relations are satisfied:*

(6.52) $\quad \overset{\triangle}{S}_0 = \hbar\left(\langle V, \dfrac{\partial}{\partial V}\rangle + 1\right), \qquad \overset{\triangle}{M}_j = i\hbar\left(V \times \dfrac{\partial}{\partial V}\right)_j, \qquad \overset{\triangle}{L}_j = \hbar v \dfrac{\partial}{\partial V_j}.$

Note that properties (6.49) of the Bessel coherent states $\mathfrak{B}_{v,V}$ imply the following intertwining property of their inner product $\mathcal{K}(\overline{v''}, \overline{V''}; v', V')$ (6.42):
(6.53)
$$\left(\overset{\triangle}{f}\right)_{V'} \mathcal{K}(\overline{v''}, \overline{V''}; v', V') = (\mathfrak{B}_{v'',V''}, \widehat{f}\,\mathfrak{B}_{v',V'})_H$$
$$= (\widehat{f}^* \mathfrak{B}_{v'',V''}, \mathfrak{B}_{v',V'})_H = \overline{\left(\overset{\triangle}{f^*}\right)_{V''} \mathcal{K}(\overline{v''}, \overline{V''}; v', V')}.$$

(The subscripts assigned to the operators $\overset{\triangle}{f}$ and $\overset{\triangle}{f^*}$ show the variables with respect to which these operators act.)

THEOREM 6.4. *The operators*

(6.54) $\quad \overset{\circ}{v} = \hbar^2 \overset{\circ}{v}\left(\dfrac{\partial}{\partial \overline{V}}\right)^2, \qquad \overset{\circ}{V}_j = \hbar^2\left(2\left(\langle \overline{V}, \dfrac{\partial}{\partial \overline{V}}\rangle\right)\dfrac{\partial}{\partial \overline{V}_j} - \overline{V}_j\left(\dfrac{\partial}{\partial \overline{V}}\right)^2 + 2\dfrac{\partial}{\partial \overline{V}_j}\right),$

$\qquad \overset{\circ}{v}^* = \overline{v}, \qquad \overset{\circ}{V}_j^* = \overline{V}_j, \qquad \overset{\circ}{S}_0 = \hbar\left(\langle \overline{V}, \dfrac{\partial}{\partial \overline{V}}\rangle + 1\right),$

determine an irreducible representation of the algebra (6.26) *in the space* $\mathcal{P}$ *of antiholomorphic functions (with respect to v and V) with the inner product*

$$\dfrac{1}{2\pi\hbar}(\Phi', \Phi'')_{\mathcal{P}} = \dfrac{1}{2\pi\hbar}\int_{\mathbb{C}^3} \overline{\Phi'(\overline{v}, \overline{V})}\, \Phi''(\overline{v}, \overline{V})\, d\mu(V),$$

where $d\mu$ is given by formula (6.45).

Note that the operators $\overset{\circ}{f}$ are related to the operators $\overset{\triangle}{f}$ (see (6.49)) by transposition with respect to $d\mu$, that is, by conjugation in the space $\mathcal{P}$ in combination with the complex conjugation

(6.55) $\qquad\qquad\qquad\qquad \overset{\circ}{f} = \overline{\left(\overset{\triangle}{f}\right)^*}.$

It is also useful to compare formulas (6.54) with (5.42), (6.45) with (5.43), (6.35) with (5.44), and (6.42) with (5.45).

PROOF OF THEOREM 6.4. Since the MacDonald function $K_0(r)$ (6.46) satisfies the Bessel equation (for functions of imaginary argument)

$$ry''(r) + y'(r) - ry(r) = 0,$$

then the density $\rho(\overline{v}, \overline{V}; v, V)$ (1.38) of the measure $d\mu(V)$ satisfies the equations

$$\overset{\circ}{v}\rho(\overline{v}, \overline{V}; v, V) = \overline{\left(\overset{\circ}{v}^*\right)^T} \rho(\overline{v}, \overline{V}; v, V),$$

$$\overset{\circ}{V}_j \rho(\overline{v}, \overline{V}; v, V) = \overline{\left(\overset{\circ}{V}_j^*\right)^T} \rho(\overline{v}, \overline{V}; v, V;),$$

$$\overset{\circ}{S}_0 \rho(\overline{v}, \overline{V}; v, V) = \overline{\left(\overset{\circ}{S}_0\right)^T} \rho(\overline{v}, \overline{V}; v, V),$$

where $\ldots^T$ denotes transposition with respect to the measure $|dV \wedge d\overline{V}|$. Hence, the operator $\overset{\circ}{S}_0$ is self-adjoint in the space $\mathcal{P}$, and the operators $\overset{\circ}{v}$ and $\overset{\circ}{v}{}^*$, as well as $\overset{\circ}{V}_j$ and $\overset{\circ}{V}_j^*$, are adjoint to each other in $\mathcal{P}$.

According to the construction, the operators $\overset{\triangle}{v}$, $\overset{\triangle}{V}$, $\overset{\triangle}{v}{}^*$, $\overset{\triangle}{V}{}^*$, and $\overset{\triangle}{S}_0$ satisfy the commutation relations adjoint to (6.26), that is, relations with opposite signs on the right. Hence, the operators $\overset{\circ}{v}$, $\overset{\circ}{V}$, $\overset{\circ}{v}{}^*$, $\overset{\circ}{V}{}^*$, and $\overset{\circ}{S}_0$, obtained from $\overset{\triangle}{v}$, $\overset{\triangle}{V}$, $\overset{\triangle}{v}{}^*$, $\overset{\triangle}{V}{}^*$, and $\overset{\triangle}{S}_0$ by (6.55), determine the representation of the algebra (6.26).

To prove that this representation is irreducible in the space $\mathcal{P}$, we need the following lemma.

LEMMA 6.4. *The system of equations*
(6.56)
$$\left(\overset{\circ}{v}\right)\mathcal{K}(\overline{v},\overline{V};v,V;) = \overline{\left(\overset{\circ}{v}{}^*\right)}\mathcal{K}(\overline{v},\overline{V};v,V), \quad \left(\overset{\circ}{V}\right)\mathcal{K}(\overline{v},\overline{V};v,V;) = \overline{\left(\overset{\circ}{V}{}^*\right)}\mathcal{K}(\overline{v},\overline{V};v,V),$$
$$\left(\overset{\circ}{v}{}^*\right)\mathcal{K}(\overline{v},\overline{V};v,V) = \overline{\left(\overset{\circ}{v}\right)}\mathcal{K}(\overline{v},\overline{V};v,V), \quad \left(\overset{\circ}{V}{}^*\right)\mathcal{K}(\overline{v},\overline{V};v,V) = \overline{\left(\overset{\circ}{V}\right)}\mathcal{K}(\overline{v},\overline{V};v,V),$$
$$\left(\overset{\circ}{S}_0\right)\mathcal{K}(\overline{v},\overline{V};v,V) = \overline{\left(\overset{\circ}{S}_0\right)}\mathcal{K}(\overline{v},\overline{V};v,V)$$

has a unique solution that satisfies the condition
(6.57)
$$\mathcal{K}(0,0;0,0) = 1.$$

PROOF OF THEOREM 1.4 (Continuation). Assume that representation (6.54) is reducible. Then the space $\mathcal{P}$ can be split into the direct sum of the invariant subspaces
(6.58)
$$\mathcal{P} = \mathcal{P}_1 \oplus \mathcal{P}_2, \qquad \dim \mathcal{P}_k \neq 0, \qquad k = 1, 2.$$

By $\{\mathfrak{b}_j^k(\overline{v},\overline{V}) \mid j = 0, 1, 2, \ldots\}$ we denote a complete orthonormal system in $\mathcal{P}_k$ ($k = 1, 2$). Then the union $\bigcup_{k=1,2}\{\mathfrak{b}_j^k(\overline{v},\overline{V})\}$ is a complete orthonormalized system in $\mathcal{P}$. Since in the space $\mathcal{P}$ there is at least one function nonvanishing at the point $v = V = 0$ (this is the function $\mathcal{K}(\overline{v},\overline{V};0,0)$), the inequality $\mathfrak{b}_{j_0}^{k_0}(0,0) \neq 0$ is satisfied for at least one basis vector $\mathfrak{b}_{j_0}^{k_0}(\overline{v},\overline{V})$. Therefore, the following two functions are well defined:
$$\mathcal{K}_{\mathcal{P}_{k_0}}(\overline{v},\overline{V};v,V) = \frac{\sum_{j\geq 0}\overline{\mathfrak{b}_j^{k_0}(\overline{v},\overline{V})}\,\mathfrak{b}_j^{k_0}(\overline{v},\overline{V})}{\sum_{j\geq 0}|\mathfrak{b}_j^{k_0}(0,0)|^2};$$
$$\mathcal{K}_{\mathcal{P}}(\overline{v},\overline{V};v,V) = \frac{\sum_{k=1,2}\sum_{j\geq 0}\overline{\mathfrak{b}_j^k(\overline{v},\overline{V})}\,\mathfrak{b}_j^k(\overline{v},\overline{V})}{\sum_{k=1,2}\sum_{j\geq 0}|\mathfrak{b}_j^k(0,0)|^2}.$$

Both these functions satisfy system (6.56), (6.57). However, by (6.48) we have
$$\mathcal{K}_{\mathcal{P}_{k_0}}(\overline{v},\overline{V};v,V) \neq \mathcal{K}_{\mathcal{P}}(\overline{v},\overline{V};v,V).$$

This contradicts Lemma 6.4, and hence proves that the representation (6.54) is irreducible. $\square$

COROLLARY 6.1. *The Bessel coherent transform*

$$\mathfrak{B}: \mathcal{P} \to H, \qquad \mathfrak{B}(\mathfrak{p}) \stackrel{\text{def}}{=} \frac{1}{2\pi\hbar} \int_{\mathbb{C}^3} \mathfrak{p}(\overline{V}) \mathfrak{B}_{v,V} \, d\mu(V)$$

determines a unitary isomorphism between the spaces $\mathcal{P}$ and H. This transform intertwines the holomorphic representation of the algebra (6.26) by the operators $\overset{\circ}{v}$, $\overset{\circ}{V}_j$, $\overset{\circ}{v}{}^*$, $\overset{\circ}{V}_j^*$, $\overset{\circ}{S}_0$ (6.54) on the space $\mathcal{P}$ and the representation of the same algebra via the operators $\widehat{v}$, $\widehat{V}_j$, $\widehat{v^*}$, $\widehat{V_j^*}$, $\widehat{S}_0$ on the space H.

References

[AAn] S. T. Ali and J.-P. Antoine, *Quantum frames, quantization, and dequantization*, Quantization and Infinite-Dimensional Systems, Plenum, New York, 1994, pp. 133–145.

[AAG] S. T. Ali, J.-P. Antoine, and J.-P. Gazeau, *Coherent States, Wavelets, and their Generalizations*, Springer-Verlag, Berlin and New York, to appear.

[AEm] S. T. Ali and G. G. Emch, *Geometric quantization. Modular reduction theory and coherent states*, J. Math. Phys. **27** (1986), 2936–2943.

[ArC] M. Arik and D. Coon, *Hilbert spaces of analytic functions and generalized coherent states*, J. Math. Phys. **17** (1976), 524–527.

[Aro] N. Aronszajn, *Theory of reproducing kernels*, Trans. Amer. Math. Soc. **68** (1950), 337–401.

[AzE] J. A. de Azcarraga and D. Ellinas, *Complex analytic realizations for quantum algebras*, J. Math. Phys. **35** (1994), 1322–1333.

[Brg$_1$] V. Bargmann, *On a Hilbert space of analytic functions and an associated integral transform*, Comm. Pure Appl. Math. **14** (1961), 187–214.

[Brg$_2$] _____, *Remarks on a Hilbert space of analytic functions*, Proc. Nat. Acad. Sci. USA **48** (1962), 199–204.

[BE] H. Bateman and A. Erdelyi, *Higher Transcendental Functions*. Vols. I, II, McGraw-Hill, New York, 1953.

[BFFLS$_1$] F. Bayen, M. Flato, C. Fronsdal, A. Lichnerowicz, and D. Sternheimer, *Quantum mechanics as a deformation of classical mechanics*, Lett. Math. Phys. **1** (1975/77), 521–530.

[BFFLS$_2$] _____, *Deformation theory and quantization*, Ann. of Physics **111** (1978), 61–151.

[B$_0$] F. A. Berezin, *Wick and anti-Wick symbols of operators*, Mat. Sb. **86** (1971), 578–610; English transl. in Math. USSR-Sb. **15** (1971).

[B$_1$] _____, *Covariant and contravariant symbols of operators*, Izv. Akad. Nauk SSSR Ser. Mat. **36** (1972), 1134–1167; English transl., Math. USSR-Izv. **6** (1972), 1117–1151.

[B$_2$] _____, *Quantization*, Izv. Akad. Nauk SSSR Ser. Mat. **38** (1974), 1116–1175; English transl., Math. USSR Izv. **8** (1974), 1109–1165.

[B$_3$] _____, *Quantization of complex symmetric spaces*, Izv. Akad. Nauk SSSR Ser. Mat. **39** (1975), 363–402; English transl., Math. USSR Izv. **9** (1975), 341–379.

[B$_4$] _____, *General concept of quantization*, Comm. Math. Phys. **40** (1975), 153–174.

[BC] C. A. Berger and L. A. Coburn, *Toeplitz operators on the Segal–Bargmann space*, Trans. Amer. Math. Soc. **301** (1987), 813–829.

[Bgm] S. Bergmann, *The kernel functions and conformal mapping*, Math. Surveys, vol. 5, Amer. Math. Soc., Providence, RI, 1950.

[Bi] L. C. Biedenharn, *The quantum group $SU_q(2)$ and a q-analogue of the boson operators*, J. Phys. A **22** (1989), L873–L878.

[Bl$_1$] R. Blattner, *The metalinear geometry and nonlinear polarizations*, Lecture Notes in Math., vol. 570, Springer-Verlag, Berlin, 1977, pp. 11-45.

[Bl$_2$] _____, *On geometric quantization*, Lecture Notes in Math., vol. 1037, Springer-Verlag, Berlin, 1983, pp. 209–241.

[BD] D. Bonatsos and C. Daskaloyannis, *General deformation schemes and $N = 2$ supersymmetric quantum mechanics*, Phys. Lett. B **307** (1993), 100–105.

[BMS] M. Bordemann, E. Meinrenken, and M. Schlichenmaier, *Toeplitz quantization of Kähler manifolds and $gl(N)$, $N \to \infty$ limit*, Comm. Math. Phys. **165** (1994), 281–296.

[BKo1] R. Brylinski and B. Kostant, *Nilpotent orbits, normality, and Hamiltonian group actions*, J. Amer. Math. Soc. **7** (1994), 269–298.

[BKo2] _____, *Differential operators on conical Lagrangian manifolds*, Lie Theory and Geometry, Birkhäuser, Basel–Boston, 1994, pp. 65–96.

[CaGR] M. Cahen, S. Gutt, and J. Rawnsley, *Quantization of Kähler manifolds*. I, J. Geom. Phys. **7** (1990), 45–62; II, Trans. Amer. Math. Soc. **337** (1993), 73–98; III, Lett. Math. Phys. **30** (1994), 291–305; IV, Lett. Math. Phys. **30** (1995), 159–168.

[ChEK] M. Chaichian, D. Ellinas, and P. Kulish, *Quantum algebra as the dynamical symmetry of the deformed Jaynes–Cummings model*, Phys. Rev. Lett. **65** (1990), 980–983.

[Coo] J. M. Cook, *The mathematics of second quantization*, Trans. Amer. Math. Soc. **74** (1953), 222–245.

[C] A. Connes, *Noncommutative Geometry*, Academic Press, London, 1994.

[CFS] A. Connes, M. Flato, and D. Sternheimer, *Closed star products and cyclic cohomology*, Lett. Math. Phys. **24** (1992), 1–12.

[Das] S. Daskaloyannis, *Generalized deformed oscillator and nonlinear algebras*, J. Phys. A **24** (1991), L789–L794.

[DaG] I. Daubechies and A. Grossmann, *Frames in the Bargmann space of entire functions*, Comm. Pure Appl. Math. **41** (1988), 151–164.

[DaGM] I. Daubechies, A. Grossmann, and Y. Meyer, *Painless nonorthogonal expansions*, J. Math. Phys. **27** (1986), 1271–1283.

[DeB] S. De Bièvre, *Coherent states over symplectic homogeneous spaces*, J. Math. Phys. **30** (1989), 1401–1407.

[DQ1] C. Delbecq and C. Quesne, *Representation theory and q-boson realizations of Witten's* $su(2)$ *and* $su(1,1)$ *deformations*, Phys. Lett. B **300** (1993), 227–233.

[DQ2] _____, *Nonlinear deformations of* $su(2)$ *and* $su(1,1)$ *generalizing Witten's algebra*, J. Phys. A **26** (1993), L127–L134.

[D] P. A. M. Dirac, *Quantum electrodynamics*, Comm. Dublin Inst. Adv. Stud. Ser. A **1** (1943), 1–36.

[Dr] V. G. Drinfeld, *Quantum groups*, Proc. Intern. Congress Math. (Berkeley, 1986), Amer. Math. Soc., Providence, RI, 1987, pp. 789–820.

[D-V] M. Dubois-Violette, *Dérivations et calcul différentiel non commutatif*, C. R. Acad. Sci. Paris Sér. I Math. **307** (1988), 403–408.

[D-VM] M. Dubois-Violette, J. Madore, T. Masson, and J. Mourad, *Linear connections on the quantum plane*, Lett. Math. Phys. **35** (1995), 351–358.

[FRT] L. D. Faddeev, N. Yu. Reshetikhin, and L. A. Takhtajan, *Quantization of Lie groups and Lie algebras*, Algebra i Analiz **1** (1989), 178–206; English transl. in Leningrad Math. J. **1** (1990).

[FLS] M. Flato, A. Lichnerowicz, and D. Sternheimer, *Deformations of Poisson brackets, Dirac brackets, and applications*, J. Math. Phys **17** (1976), 1754–1762.

[FS] M. Flato and D. Sternheimer, *Closedness of star products and cohomologies*, Lie Theory and Geometry, Birkhäuser, Basel–Boston, 1994, pp. 241–259.

[Fo1] V. A. Fock, *Verallgemeinerung und Lösung der Diracschen statistischen Gleichung*, Z. Phys. **49** (1928), 339–357.

[Fo2] _____, *Konfigurationsraum und zweite Quantelung*, Z. Phys. **75** (1932), 622–647.

[GaR] G. Gasper and M. Rahman, *Basic Hypergeometric Series*, Cambridge Univ. Press, Cambridge and New York, 1990.

[GFa] I. M. Gelfand and D. B. Fairlie, *The algebra of Weyl symmetrized polynomials and its quantum extension*, Comm. Math. Phys. **136** (1991), 487–499.

[GGP-S] I. M. Gelfand, M. I. Graev, and I. I. Pyatetskii-Shapiro, *Representation Theory and Automorphic Functions*, Nauka, Moscow, 1966; English transl., Academic Press, Boston, MA, 1969; reprint, 1990.

[Gi1] R. Gilmore, *Geometry of symmetrized states*, Ann. of Physics **74** (1972), 391–463.

[Gi2] _____, *On properties of coherent states*, Rev. Mexicana Fís. **23** (1974), 143–187.

[Gl1] R. J. Glauber, *The quantum theory of optical coherence*, Phys. Rev. (2) **130** (1963), 2529–2539.

[Gl2] _____, *Coherent and incoherent states of radiation field*, Phys. Rev. (2) **131** (1963), 2766–2788.

[Go] Gong Ren-Shan, *A completeness relation for the coherent states of the (p,q)-oscillator by (p,q)-integration*, J. Phys. A **27** (1994), L375–L379.

[GLZh] Ya. I. Granovskii, I. M. Lutzenko, and A. S. Zhedanov, *Mutual integrability, quadratic algebras, and dynamical symmetry*, Ann. of Physics **217** (1992), 1–20.

[GZhG] Ya. I. Granovckii, A. S. Zhedanov, and O. B. Grakhovskaya, *Addition rule for nonlinear algebras*, Phys. Lett. B **278** (1992), 85–88.

[GZhL] Ya. I. Granovckii, A. S. Zhedanov, and I. M. Lutzenko, *Quadratic algebra as a "hidden" symmetry of the Hartmann potential*, J. Phys. A **24** (1991), 3887–3894.

[GrN] R. W. Gray and C. A. Nelson, *A completeness relation for the q-analogue coherent states by q-integration*, J. Phys. A **23** (1990), L945–L950.

[H] W. Heisenberg, *Über den anschaulichen Inhalt der quantentheoretischen Kinematik und Mechanik*, Z. Phys. **43** (1927), 172–198.

[HNØ1] J. Hilgert, K.-H. Neeb, and B. Ørsted, *The geometry of nilpotent coadjoint orbits of convex type in Hermitian Lie algebras*, J. of Lie Theory **4** (1994), 185–235.

[HNØ2] _____, *Conal Heisenberg algebras and associated Hilbert spaces*, J. Reine Angew. Math. **474** (1996), 67–112.

[HoOd] M. Horowski and A. Odzijewicz, *Geometry of the Kepler system in coherent states approach*, Ann. Inst. H. Poincaré Phys. Théor. **59** (1993), 69–89.

[Hu1] J. Huebschmann, *Poisson cohomology and quantization*, J. Reine Angew. Math. **408** (1990), 57–113.

[Hu2] _____, *On the quantization of Poisson algebras*, Symplectic Geometry and Mathematical Physics, Actes du colloque en l'honneur de J.-M. Souriau (P. Donato et al., eds.), Birkhäuser, Basel–Boston, 1991, pp. 204–233.

[Ja] A. Jannussis, *New deformed Heisenberg oscillator*, J. Phys. A **26** (1993), L233–L237.

[J] M. Jimbo, *A q-difference analog of Ug and the Yang–Baxter equation*, Lett. Math. Phys. **10** (1985), 63–69.

[Kb1] A. V. Karabegov, *Deformation quantization with separation of variables on a Kähler manifold*, Comm. Math. Phys. **180** (1996), 745–755.

[Kb2] _____, *Construction of the canonical trace density for deformation quantization with separation of variables*, Funktsional. Anal. i Prilozhen. (to appear); English transl., Functional Anal. Appl. (to appear).

[Kam] E. Kamke, *Gewöhnliche Differentialgleichungen*, Akademische Verlag. Geest & Portig, Leipzig, 1959.

[Ka1] M. Karasev, *Weyl calculus and the ordered calculus of noncommuting operators*, Mat. Zametki **26** (1979), 885–907; English transl. in Math. Notes **26** (1979).

[Ka2] _____, *Asymptotic spectrum and oscillation front for operators with nonlinear commutation relations*, Dokl. Akad. Nauk SSSR **243** (1978), 15–18; English transl., Soviet Math. Dokl. **19** (1978), 1300–1304.

[Ka3] _____, *Operators of regular representation for a class of non-Lie permutation relations*, Funktsional. Anal. i Prilozhen. **13** (1979), no. 3, 89–90; English transl. in Functional Anal. Appl. **13** (1979).

[Ka4] _____, *Problems in Operator Methods*, Moscow Inst. of Electronics & Math. (MIEM) Publ., Moscow, 1979. (Russian)

[Ka5] _____, *Quantization of nonlinear Lie–Poisson brackets in semiclassical approximation*, Inst. Theor. Phys., Kiev, Preprint ITP-85-72 (1985). (Russian)

[Ka6] _____, *Poisson algebras of symmetries and asymptotics of spectral series*, Funktsional. Anal. i Prilozhen. **20** (1986), no. 1, 21–32; English transl. in Functional Anal. Appl. **20** (1986).

[Ka7] _____, *Quantum reduction to orbits of symmetry algebras and the Ehrenfest problem*, Inst. Theor. Phys., Kiev, Preprint ITP-87-157 (1987). (Russian)

[Ka8] _____, *Connections on Lagrangian submanifolds and certain problems of the semiclassical approximation theory*, Zap. Nauchn. Sem. LOMI **172** (1989), 41–54; English transl., J. Soviet Math. **10** (1992), no. 5, 1053–1062.

[Ka9] _____, *New global asymptotics and anomalies in the problem of quantization of the adiabatic invariant*, Funktsional. Anal. i Prilozhen. **24** (1990), no. 2, 24–36; English transl., Functional Anal. Appl. **24** (1990), 104–114.

[Ka10] _____, *Simple quantization formula*, Symplectic Geometry and Mathematical Physics, Actes du colloque en l'honneur de J.-M. Souriau (P. Donato et al., eds.), Birkhäuser, Basel–Boston, 1991, pp. 234–243.

[Ka11] _____, *Integrals over membranes, transition amplitudes and quantization*, Russian J. Math. Phys. **1** (1993), 523–526.

[Ka12] _____, *Formulas for noncommutative products of functions in terms of membranes and strings*, Russian J. Math. Phys. **2** (1994), 445–462.

[Ka13] _____, *Quantization by means of two-dimensional surfaces (membranes). Geometrical formulas for wave-functions*, Contemp. Math. **179** (1994), 83–113.

[Ka14] _____, *Quantization and coherent states over Lagrangian submanifolds*, Russian J. Math. Phys. **3** (1995), 393–400.

[Ka15] _____, *Geometric coherent states, membranes, and star products*, Quantization, Coherent States, Complex Structures (J.-P. Antoine et al., eds.), Plenum, New York, 1995, pp. 185–199.

[Ka16] _____, *Representation of evolution operator via membrane amplitudes*, Mat. Zametki **60** (1996), 930–934; English transl. in Math. Notes **60** (1996).

[Ka17] _____, *Advances in quantization: quantum tensors, explicit ∗-products, and restriction to irreducible leaves*, Diff. Geom. Appl., to appear.

[KaKz1] M. V. Karasev and M. V. Kozlov, *Exact and semiclassical representation over Lagrangian submanifolds in $su(2)^*$, $so(4)^*$, and $su(1,1)^*$*, J. Math. Phys. **34** (1993), 4986–5006.

[KaKz2] _____, *Representations of compact semisimple Lie algebras over Lagrangian submanifolds*, Funktsional. Anal. i Prilozhen. **28** (1994), no. 4, 16–27; English transl., Functional Anal. Appl. **28** (1994), 238–246.

[KaM1] M. V. Karasev and V. P. Maslov, *Algebras with general permutation relations and their applications*. II, Itogi Nauki i Tekhniki: Sovremennye Problemy Mat., vol. 13, VINITI, Moscow, 1979, pp. 145–267; English transl., J. Soviet Math. **15** (1981), no. 3, 273–368.

[KaM2] _____, *Asymptotic and geometric quantization*, Uspekhi Mat. Nauk **39** (1984), no. 6, 115–173; English transl., Russian Math. Surveys **39** (1984), no. 6, 133–205.

[KaM3] _____, *Non-Lie permutation relations*, Uspekhi Mat. Nauk **45** (1990), no. 5, 41–79; English transl. in Russian Math. Surveys **45** (1990), no. 5.

[KaM4] _____, *Nonlinear Poisson Brackets. Geometry and Quantization*, Nauka, Moscow, 1991; English transl., Transl. Math. Monographs, vol. 119, Amer. Math. Soc., Providence, RI, 1993.

[KaN1] M. V. Karasev and E. M. Novikova, *Quadratic Poisson brackets in Zeeman effect. Irreducible representations and coherent states*, Uspekhi Mat. Nauk **49** (1994), no. 5, 169–170; English transl. in Russian Math. Surveys **49** (1994), no. 5.

[KaN2] _____, *Integral representation of eigenfunctions and coherent states for the Zeeman effect*, Quantization, Coherent States, Complex Structures (J.-P. Antoine et al., eds.), Plenum, New York, 1995, pp. 201–208.

[KaN3] _____, *Representation of exact and semiclassical eigenfunctions via coherent states. The hydrogen atom in a magnetic field*, Teoret. Mat. Fiz. **108** (1996), 339–387; English transl. in Theoret. and Math. Phys. **108** (1996).

[KaN4] _____, *Coherent transform of spectral problem and algebras with nonlinear commutation relations*, J. Math. Sci., to appear.

[KaV1] M. V. Karasev and Yu. M. Vorobjev, *Hermitian bundles over isotropic submanifolds and correction to Kostant–Souriau quantization rule*, Preprint Inst. Theor. Phys. ITP–90–85E (1991), Kiev.

[KaV2] _____, *Integral representations over isotropic submanifolds and equations of zero curvature*, Moscow Inst. of Electronics & Math., Preprint N AMath-QDS-92-01 (1992); Adv. Math **134** (1998).

[KQ] J. Katriel and C. Quesne, *Recursively minimal-deformed oscillators*, J. Math. Phys. **37** (1996), 1650–1661.

[K1] A. A. Kirillov, *Elements of the Theory of Representation*, Nauka, Moscow, 1972; English transl., Springer-Verlag, Berlin and New York, 1976.

[K2] _____, *Geometric quantization*, Itogi Nauki i Tekhniki: Sovremennye Problemy Mat.: Fundamental'nye Napravleniya, vol. 4, VINITI, Moscow, 1985, pp. 141–178; English transl. in Encyclopedia of Math. Sci., vol. 4 (Dynamical Systems, IV), Springer-Verlag, Berlin and New York, 1990.

[Kl$_1$] J. R. Klauder, *The action option and a Feynman quantization of spinor fields in terms of ordinary c-numbers*, Ann. of Physics **11** (1960), 123–168.

[Kl$_2$] _____, *Continuous representation theory*, J. Math. Phys. **4** (1963), 1055–1073.

[Kl$_3$] _____, *Coherent state quantization of constraint systems*, Ann. of Physics **254** (1997), 419–453.

[KlS] J. R. Klauder and B. S. Skagerstam, *Coherent States. Applications in Physics and Mathematics*, World Scientific, Singapore, 1985.

[Ksm] Y. Kosmann–Schwarzbach, *Exact Gerstenhaber algebras and Lie bialgebroids*, Acta Appl. Math. **41** (1995), 153–165.

[Ko] B. Kostant, *Quantization and unitary representations. I: Prequantization*, Lectures in Modern Analysis and Applications. III, Lecture Notes in Math., vol. 170, Springer-Verlag, Berlin, 1970, pp. 87–208.

[Ksz] S. L. Koszul, *Crochet de Schouten-Nijenhuis et cohomologie,*, Élie Cartan et les mathématiques d'adjourd'hui, Astérisque, Numéro Hors Série, Soc. Math. France, Paris, 1985, pp. 257–271.

[Ku] P. P. Kulish, *Contraction of quantum algebras and q-oscillators*, Teoret. Mat. Fiz. **86** (1991), 157–160; English transl. in Theoret. and Math. Phys. **86** (1991).

[KuR] P. P. Kulish and N. Yu. Reshetikhin, *Universal R-matrix of the quantum superalgebra osp(2|1)*, Lett. Math. Phys. **18** (1989), 143–149.

[L] A. Lichnerowicz, *Les variétés de Poisson et leurs algèbres de Lie associées*, J. Differential Geom. **12** (1977), 253–299.

[Le] B. M. Levitan, *Theory of Generalized Shift Operators*, Nauka, Moscow, 1973; English transl. of 1st ed., *Generalized translation operators and some of their applications*, Israel Program Sci. Transls., Jerusalem, and Davey, New York, 1964.

[Mac] A. F. Macfarlane, *On q-analogues of the quantum harmonic oscillator and the quantum group $SU(2)_q$*, J. Phys. A **22** (1989), 4581–4588.

[Mck] K. Mackenzie, *Lie algebroids and Lie pseudoalgebras*, Bull. London Math. Soc. **27** (1995), 97–147.

[M] V. P. Maslov, *Application of ordered operators method for obtaining exact solutions*, Teoret. Mat. Fiz. **33** (1977), 185–209; English transl. in Theoret. and Math. Phys. **33** (1977).

[MOd] V. Maximov and A. Odzijewicz, *The q-deformation of quantum mechanics of one degree of freedom*, J. Math. Phys. **36** (1995), 1681–1690.

[MMP] S. Meljanac, M. Milekovic, and S. Pallua, *Unified view of deformed single-mode oscillator algebras*, Phys. Lett. B **328** (1994), 55–59.

[Mo] C. Moreno, *∗-product on some Kähler manifolds*, Lett. Math. Phys. **11** (1986), 361–372.

[OKK] K. Odaka, T. Kishi, and S. Kamefuchi, *On quantization of simple harmonic oscillators*, J. Phys. A: Math. Geom. **24** (1991), L591–L596.

[Od$_1$] A. Odzijewicz, *On reproducing kernels and quantization of states*, Comm. Math. Phys. **114** (1988), 577–597.

[Od$_2$] _____, *Quantum algebras and q-special functions related to coherent states maps of the disk*, Comm. Math. Phys. **192** (1998), 183–215.

[OMMY] H. Omori, Y. Maeda, N. Miyazaki, and A. Yoshioka, *Poincaré–Cartan class and deformation quantization of Kähler manifolds*, Preprint (1997).

[OMY$_1$] H. Omori, Y. Maeda, and A. Yoshioka, *Weyl manifolds and deformation quantization*, Adv. Math. **85** (1991), 224–255.

[OMY$_2$] _____, *Deformation quantization of Poisson algebras*, Contemp. Math., vol. 179, Amer. Math. Soc., Providence, RI, 1994, pp. 213–240.

[O] E. Onofri, *A note on coherent state representation of Lie group*, J. Math. Phys. **16** (1975), 1087–1089.

[OPr] E. Onofri and M. Pauri, *Analyticity and quantization*, Lett Nuovo Cimento (2) **3** (1972), 35–42.

[P$_1$] A. M. Perelomov, *Coherent states for arbitrary Lie group*, Comm. Math. Phys. **26** (1972), 222–236.

[P2] _____, *Generalized Coherent States and Their Applications*, Springer-Verlag, Berlin and New York, 1986.

[Q] C. Quesne, *Coherent states, K-matrix theory, and q-boson realizations of the quantum algebra* $su_q(2)$, Phys. Lett. A **153** (1991), 303–307.

[R1] J. Rawnsley, *Coherent states and Kähler manifolds*, Quart. J. Math. Oxford Ser. (2) **28** (1977), 403–415.

[R2] _____, *On the pairing of polarizations*, Comm. Math. Phys. **58** (1978), 1–8.

[R3] _____, *A nonunitary pairing of polarizations for the Kepler problem*, Trans. Amer. Math. Soc. **250** (1979), 167–180.

[R4] _____, *Deformation quantization of Kähler manifolds*, Symplectic Geometry and Mathematical Physics, Actes du colloque en l'honneur de J.-M.Souriau (P. Donato et al., eds.), Birkhäuser, Basel–Boston, 1991, pp. 366–373.

[Ri1] M. A. Rieffel, *Deformation quantization of Heisenberg manifold*, Comm. Math. Phys. **122** (1989), 531–562.

[Ri2] _____, *Lie group convolution algebras as deformation quantization of linear Poisson structures*, Amer. J. Math. **112** (1990), 657–686.

[Ro] M. Roček, *Representation theory of the nonlinear* $su(2)$ *algebra*, Phys. Lett. B **255** (1991), 554–557.

[Sl] M. Schlichenmaier, *Berezin–Toeplitz quantization of compact Kähler manifolds*, Preprint q-ala/9601016 (1996).

[Sch] E. Schrödinger, *Der stetige Übergang von der Mikro- zur Makromechanik*, Naturwiss. **14** (1926), 664–666.

[Si] B. Simon, *The classical limit of quantum partition functions*, Comm. Math. Phys. **71** (1980), 247–276.

[Sk] E. K. Sklyanin, *Some algebraic structures connected with the Yang–Baxter equation*, Funktsional. Anal. i Prilozhen. **16** (1982), 27–34; English transl. in Functional Anal. Appl. **16** (1982).

[S1] J.-M. Souriau, *Structure des systemes dynamiques*, Dunod, Paris, 1970.

[S2] _____, *Sur la variété de Kepler*, Symposia Math. XIV, Academic Press, London and New York, 1974, pp. 343–360.

[Sp] V. Spiridonov, *Coherent states of the q-Weyl algebra*, Lett. Math. Phys. **35** (1995), 179–185.

[Tu] G. M. Tuynman, *Generalized Bergman kernels and geometric quantization*, J. Math. Phys. **28** (1987), 573–583.

[UU] A. Unterberger and H. Upmeier, *The Berezin transform and invariant differential operators*, Comm. Math. Phys. **164** (1994), 563–597.

[Ve] J. Vey, *Déformation du crochet de Poisson sur une variété symplectique*, Comment. Math. Helv. **50** (1975), 421–454.

[VVZ] V. S. Vladimirov, I. V. Volovich, and E. I. Zelenov, *p-Adic Analysis and Mathematical Physics*, Nauka, Moscow, 1994; English transl., World Scientific, Singapore, 1994.

[VoB] A. Vourdas and R. F. Bishop, *Dirac's contour representation in thermofield dynamics*, Phys. Rev. A **53** (1996), R1205–R1209.

[We] A. Weinstein, *Deformation quantization*, Sém. Bourbaki 1993/94, Astérisque, vol. 227, Soc. Math. France, Paris, 1995, pp. 389–409.

[W] E. Witten, *Gauge theories, vertex models, and quantum groups*, Nuclear Phys. B **30** (1990), 285–346.

[ZhFG] W.-M. Zhang, D. H. Feng, and R. Gilmore, *Coherent states. Theory and some applications*, Rev. Modern Phys. **26** (1990), 867–927.

[Zh] A. Zhedanov, *Nonlinear shift of q-Bose operators and q-coherent states*, J. Phys. A **24** (1991), L1129–L1131.

DEPARTMENT OF APPLIED MATHEMATICS, MOSCOW INSTITUTE OF ELECTRONICS AND MATHEMATICS, B.TREKHSVYAT. PER., 3/12, MOSCOW 109028, RUSSIA,

E-mail address: `karasev@qds.amath.msk.ru`

Translated by the authors

Adapted Connections, Hamilton Dynamics, Geometric Phases, and Quantization over Isotropic Submanifolds

Mikhail Karasev and Yurii Vorobjev

ABSTRACT. We investigate classical and quantum dynamics over an invariant isotropic submanifold using symplectic linear connections adapted to linearized Hamilton flows. Three types of results are presented: (i) stability, reducibility, integrability, topological and dynamical invariants of linearized Hamilton systems in terms of adapted connections; (ii) partial integrability of nonlinear Hamilton systems near a given invariant isotropic submanifold; the problem of "nice" isotropic embedding that admits a flat coisotropic extension; (iii) quantization of observables in a model phase space over an isotropic submanifold; applications of this quantization (employing deformed coherent states) to the derivation of semiclassical spectral data for Weyl pseudodifferential operators via basic geometric and topological characteristics of an isotropic submanifold (or invariants of the related Hamilton systems).

Contents

INTRODUCTION	204
PART 1. LINEARIZED HAMILTON DYNAMICS OVER ISOTROPIC SUBMANIFOLDS	209
§1.1. First variation operators and their flows	209
§1.2. Adapted symplectic connections and reducibility	214
§1.3. Flat symplectic connections in the normal bundle	221
§1.4. Invariant symplectic normal subbundles	229
§1.5. Symplectic vector bundles as model phase spaces	235
§1.6. Poisson brackets on vector bundles via Lie algebroids	242
§1.7. Geometric stability and integrability	247
§1.8. Variations of holonomy and characteristic multipliers	254
§1.9. Action-angle variables and the quasiperiodic motion	257
§1.10. Invariants of stable symplectic connections. Generalized rotation numbers and Gelfand–Lidskii indices	260
PART 2. SYMPLECTIC GEOMETRY AND NONLINEAR DYNAMICS NEAR ISOTROPIC SUBMANIFOLDS	269
§2.1. Symplectic and coisotropic extensions. General properties	269

1991 *Mathematics Subject Classification*. Primary 58F05, 70H05; Secondary 81Sxx, 53C05.

This research was partially supported by the Russian Foundation for Basic Research under grant No. 96-01-01426 and the International Science Foundation under grant MAL300.

§2.2. Poincaré extension theorem for tori 272
§2.3. Flat coisotropic extensions 277
§2.4. Inflations and torus actions 286
§2.5. Liouville tori 293

PART 3. GEOMETRIC QUANTIZATION OVER ISOTROPIC SUB-
MANIFOLDS 295
§3.1. Hermitian operators satisfying Dirac axioms 295
§3.2. Model spectral problem 302
§3.3. Geometric phases and integer cohomology classes 307
§3.4. Deformed coherent states 312
§3.5. Quantization conditions 314
REFERENCES 321

Introduction

Isotropic and Lagrangian submanifolds of symplectic manifolds are important objects in both classical and quantum mechanics [**AG, AKN, GuSt**$_{2,3}$**, Mas, MiFm, Nov, We**$_{1,2}$]. For completely integrable systems, Lagrangian submanifolds occur as regular level sets of integrals of motion. In this paper we are interested in a class of Hamilton systems that are not necessarily integrable. More precisely, we assume that a Hamilton system (M, Ω, H) admits at least one invariant isotropic submanifold $\Lambda \subset M$,

$$\dim \Lambda = k < n, \qquad \dim M = 2n.$$

For example, Λ may be a periodic trajectory, a low dimensional quasiperiodic torus, or a more complicated object. Our goal is to develop a geometric approach to investigation of the Hamilton dynamics near Λ via symplectic linear connections of Bott type.

It follows from Bott's general theory [**Bo**$_{1,2}$] that if $\mathcal{F}$ is a foliation of a manifold M, then there is a partial linear connection in the normal bundle $\nu(T\mathcal{F})$ of $\mathcal{F}$ (the corresponding covariant derivative is defined only along the leaves of $\mathcal{F}$). The restriction of this partial connection to each leaf L of $\mathcal{F}$ defines a flat linear connection (Bott's connection). Applications of Bott's connection to various problems can be found in [**Dz**$_{2,3}$**, L**$_1$**, Tab, We**$_1$**, Wo**]. In particular, if $\mathcal{F}$ is a Lagrangian foliation (that is, a real polarization) of a symplectic manifold (M, Ω), then each leaf L of $\mathcal{F}$ carries a flat affine connection without torsion [**We**$_1$] (see also [**Li**$_1$]). The appearance of flat torsion-free connections on leaves of isotropic foliations was investigated in [**Dz**$_{2,3}$**, L**$_1$].

In this paper we are interested in symplectic connections in the normal bundle over a given isotropic submanifold. Using such a connection, we develop a unique approach to investigation of both classical and quantum dynamics over an invariant isotropic submanifold.

Part 1 is devoted to investigation of the linearized dynamics. Let (M, Ω, H, Λ) be a Hamilton system and let Λ be an isotropic submanifold that is invariant with respect to the flow of the Hamilton vector field $\text{ad}(H)$ (in short, Λ is $\text{ad}(H)$-invariant). Let E be a symplectic normal bundle over Λ (we use the terminology introduced in [**We**₃]). In §1.1, we show that the linearized Hamilton flow of ad(H) on Λ induces a *first variation operator* $D_H \colon \Gamma(E) \to \Gamma(E)$, which is considered as a covariant differential operator acting on sections of E. We say that a symplectic linear connection ∇ in E is *adapted* [**KaVo**₁,₂] to the linearized Hamilton flow of ad(H) on Λ if

$$D_H = \nabla_v, \qquad v = \mathrm{ad}(H)\big|_\Lambda.$$

Geometrically, this means that the flow generated by the first variation operator on the symplectic normal bundle over Λ coincides with the parallel transport by the connection ∇ along trajectories in Λ.

The main problem is the existence of a *flat adapted* symplectic connection ∇ in E. If such a connection exists, then we can resolve the following basic problems of the linearized Hamilton dynamics:
- criteria of reducibility;
- criteria of stability and strong (parametric) stability;
- definition of monodromy operators and characteristic multipliers.

In the periodic case (dim $\Lambda = 1$, $\Lambda \approx \mathbb{S}^1$), the Floquet–Lyapunov theory and the Gelfand–Krein–Lidskii theory give complete answers to these questions. In the multidimensional quasiperiodic case (Λ is a quasiperiodic torus), there is no satisfactory analog of these theories, but some deep analytic results have been obtained in this area in [**E, JM, Jo, JoSe**].

In the present paper, using flat adapted connections, we try to extract and study geometric difficulties and obstructions arising in the multidimensional case.

Note that in the context of time-dependent linear Hamilton systems, the connection theory was used, for instance, by Buslaev, Nalimova [**BuNa**] and by Marsden, Ratiu, Raugel [**MRR**]. The paper [**MRR**] stimulated our approach most of all.

The main results concerning the existence of flat adapted connections and the reducibility of a linearized Hamilton flow over Λ are presented in §1.2 and §1.3.

Among others we study the problem of definition of the Hamilton structure for the first variation equations along a given invariant isotropic submanifold Λ. A similar problem for the first variation equation along a trajectory of a Hamilton vector field was considered in [**MRR**]. Our approach is based on the ideas of Sternberg's minimal coupling procedure [**St**₁] (see also [**GuSt**₂, **We**₄]). In §1.5, for a given isotropic submanifold Λ and a symplectic connection ∇ in E, we introduce a *model phase space* $(\mathcal{X} = T^*\Lambda \oplus E, \Omega^\nabla)$. Under the assumption that Λ is ad(H)-invariant, we consider the pull-back $D_H^\#$ of the first variation operator D_H to E. We say that D_H admits a *Hamiltonization* on a given model phase space $(\mathcal{X}, \Omega^\nabla)$ if there exists a Hamiltonian vector field on $(\mathcal{X}, \Omega^\nabla)$ whose restriction to $E \subset \mathcal{X}$ coincides with $D_H^\#$. The conditions under which D_H admits a Hamiltonization on $(\mathcal{X}, \Omega^\nabla)$ are formulated in Theorem 1.5.1. Geometrically, these conditions mean that the flow of $D_0^\#$ on E preserves the horizontal subbundle of TE associated with the connection ∇. In particular, if ∇ is adapted to the linearized Hamilton flow of ad(H) at Λ, then $D_H^\#$ admits a Hamiltonization on $\mathcal{X}$ under the following condition of compatibility with the curvature 2-form $\mathcal{C}$ of ∇:

$$v \lrcorner\, \mathcal{C} = 0, \qquad v = \mathrm{ad}(H)\big|_\Lambda.$$

In particular, this statement is always true if ∇ is flat.

Using these results in §1.7 and §1.9, we investigate the integrability of the *model Hamilton system* $(\mathcal{X}, \Omega^\nabla)$ associated with the linearized Hamilton flow of ad(H) at Λ. Under a certain stability condition, we give a geometric description of fiberwise quadratic integrals of motion of $D_H^\#$ which are analogs of the Lewis invariant for time dependent linear 2-dimensional Hamilton systems [**Lew**]. Furthermore, if $\Lambda \approx \mathbb{T}^k$ and the flow of $v = \mathrm{ad}(H)\big|_\Lambda$ is quasiperiodic, then we show that all trajectories of $D_H^\#$ are also quasiperiodic and the additional "normal" frequencies can be interpreted as a generalization of rotation numbers [**JM, JoNer**].

Later (in §1.10) we discuss phase space invariants of the linearized Hamilton dynamics over isotropic submanifolds, namely, some analogs of the Johnson–Moser rotation number and of the Gelfand–Lidskii index.

In §1.6, using Lie algebroids, we perform a parallel investigation of the Hamiltonization problem for covariant differential operators. We use fiberwise linear Poisson brackets on duals of Lie algebroids (see [**A-S, MRR, Cou**$_{1,2}$]). The main result about the Hamilton structures of covariant differential operators is formulated in Theorem 1.6.1. This approach may be useful if *a priori* we have no symplectic connections. We illustrate this thesis in Example 1.6.1.

Part 2 is devoted to investigation of the "nonlinear" Hamilton dynamics near a given invariant isotropic submanifold.

A typical situation in which a Hamilton system admits an invariant isotropic submanifold is the following. Assume that there are smooth mappings

$$f = (f^1, \ldots, f^k)\colon M \to \mathbb{R}^k \qquad (k < \dim M),$$
$$\Phi = (\Phi^1, \ldots, \Phi^{2r})\colon M \to \mathbb{R}^{2r} \qquad (2r = \dim M - 2k),$$

such that:

(i) the zero set $\Lambda = (f \times \Phi)^{-1}(0)$ is regular;

(ii) the pairwise Poisson brackets of the components f^i and Φ^i possess the properties

$$\{f^i, f^j\}\big|_\Lambda = 0, \qquad \{f^i, \Phi^\sigma\}\big|_\Lambda = 0, \qquad \det((\{\Phi^\sigma, \Phi^{\sigma'}\}\big|_\Lambda)) \neq 0;$$

(iii) a given Hamilton function H Poisson commutes with f^i and Φ^σ on Λ,

$$\{f^i, H\}\big|_\Lambda = \{\Phi^\sigma, H\}\big|_\Lambda = 0.$$

Then Λ is a k-dimensional isotropic submanifold invariant with respect to the flow of ad(H). In this situation, we are interested in the Hamilton dynamics of ad(H) near Λ.

Note that if Poisson brackets in (ii) and (iii) vanish everywhere on M, then we find themselves in the case of noncommutative integrability [**MiFm, Ne$_1$, Nov**].

To understand the situation in general, we point out that under conditions (i)–(iii) there is a symplectic connection ∇ in the normal bundle E over Λ (Theorem 1.3.1). This connection ∇ is determined by the zero-level set N of the mapping f, $N = f^{-1}(0)$, such that $T_\Lambda N = (T\Lambda)^\perp$, where $T\Lambda^\perp$ is the skew-orthogonal complement to $T\Lambda$. A submanifold $\mathcal{N} \subset M$ containing Λ and satisfying this property is called *coisotropic* on Λ [**KaVo$_2$**]. The next observation is that the connection ∇ is flat if the pairwise Poisson brackets of functions f^i takes the following form near Λ:

(iv) $$\{f^i, f^j\}\big|_\mathcal{N} = O(d_\Lambda^3), \quad \text{as} \quad d_\Lambda \to 0,$$

where d_Λ is the distance from a point of $\mathcal{N}$ to Λ (i.e., $d_\Lambda \approx |\Phi|$). In general, the right-hand side of (iv) is of order $O(d_\Lambda^2)$.

Let us assume that the Hamilton function H satisfies the condition

(v) $$H\big|_\mathcal{N} = \text{const} + O(d_\Lambda^3).$$

Then the flat symplectic connection ∇ (associated with $\mathcal{N}$) is adapted to the linearized Hamilton flow of $\text{ad}(H)$ on Λ (see [**KaVo**$_2$] and Theorem 1.3.1 below).

So, under conditions (i)–(v), the linearized Hamilton flow of $\text{ad}(H)$ on Λ possesses "good" properties, and we can consider the model Hamilton system

$$\big(M, \Omega, \text{ad}(H)\big) \Longrightarrow (\mathcal{X}, \Omega^\nabla, X_F).$$

Here X_F is the Hamiltonization of D_H on $(\mathcal{X}, \Omega^\nabla)$. A natural question arises: when is the original nonlinear Hamilton system equivalent to the linear model system? We consider this question in the following setup. Instead of (v), we assume that

(v′) $$H\big|_\mathcal{N} = \text{const}$$

and

(vi) $\qquad\qquad\qquad \mathcal{N}$ is coisotropic.

The submanifold $\mathcal{N}$ can be called a *coisotropic extension* of Λ in M. Since $\mathcal{N}$ is $\text{ad}(H)$-invariant, we can naturally formulate the equivalence problem for dynamical systems generated by the vector fields $\text{ad}(H)$ and $D_H^\#$ in some neighborhoods of Λ in $\mathcal{N}$ and in E. In the periodic case ($\Lambda \approx \mathbb{S}^1$), this question is usually transformed to the analysis of a Hamilton system on a given energy surface near a periodic orbit via the first variation system and the Birkhoff normal forms.

Our goal is to find geometric properties of the "collective energy surface" $\mathcal{N}$ under which the dynamical system of $\text{ad}(H)\big|_\mathcal{N}$ is equivalent to the model system. For this, we introduce the notion of a *flat coisotropic extension* of a given isotropic submanifold Λ. In §2.3 we give two equivalent definitions: in terms of flat Darboux atlases over Λ and in terms of flat torsionless symplectic Bott-type connections on $\mathcal{N}$ [**KaVo**$_3$]. The main result about the equivalence of flat coisotropic extensions is given in Theorem 2.3.2.

We also introduce the notion of *stable coisotropic extensions*. In the periodic case this notion coincides with linear stability of periodic orbits [**AbMa**].

In §2.4, we prove that if there exists a stable flat coisotropic extension $\mathcal{N}$ of Λ, then a neighborhood of Λ in $\mathcal{N}$ is fibered by invariant Lagrangian (at generic points) submanifolds. We call these submanifolds "*inflations*" of Λ. Each inflation itself is fibered over Λ by a torus of dimension $r = \frac{1}{2}\dim M - \dim \Lambda$. In fact, in this case we have an r-torus action in a neighborhood of Λ in $\mathcal{N}$ (Theorem 2.4.1). The complete classification of such torus actions is given in Theorem 2.4.2. In §2.5, the problem of integrability induced by torus actions is investigated for the special case $\Lambda \approx \mathbb{T}^k$.

Part 3 deals with quantization over an isotropic submanifold.

In §3.1 we construct a quantization mapping satisfying the Dirac axioms on the model phase space over an isotropic submanifold Λ. This is an irreducible representation on the level of Lie algebras only, but this mapping preserves the unity element (and so, is ready to be extended to the level of associative algebras).

We transform and adapt the van Hove, Segal, and fundamental Kostant–Souriau quantization procedures to our situation.

In our realization, quantum generators of the representation are first-order differential operators acting on fiberwise holomorphic functions (not sections, not half-densities!) over a certain Hermitian vector bundle related to the isotropic Λ. The quantization mapping which we define here, is strikingly similar to its Lagrangian version constructed in [**Ka**$_7$, **KaKz**$_1$]. But in the isotropic case, specifically, the Lie algebra of classical observables is controlled, in addition, by a symplectic connection in the normal bundle over Λ. We introduce an inner product of the Fock–Bargmann–Segal type over Λ and prove that our quantum generators (corresponding to real classical observables) are essentially self-adjoint operators.

In §3.2 we solve the spectral problem for these operators and express the spectral data in terms of certain fundamental geometric data related to Λ, as well the topology of Λ. In particular, we show how the spectral data depend on a choice of the symplectic connection.

In §3.3 and §3.4, we introduce *deformed coherent states* over an isotropic submanifold Λ that are associated with the same fundamental geometric data. Our definition follows the Lagrangian version of coherent states introduced in [**Ka**$_{1-5}$]; here we follow [**KaVo**$_{1,4,6,8}$]

In the isotropic case, we need a number of additional geometric objects over Λ. The most important difference from the Lagrangian case is the appearance of separation of the *geometric phases* over Λ into two parts: "tangential" and "normal."

The *tangential phases* along closed cycles determine an integer cohomology class, which can be viewed as an isotropic analog of the Maslov class (the latter is defined only on Lagrangian submanifolds). Indeed, the tangential class coincides with the Maslov class in the Lagrangian case; in the general case, this class turns out to be identically zero if the "set of singularities" on Λ (where a direction of the vertical polarization coincides with a direction of the tangent space $T\Lambda$) is empty.

The *normal phase* along closed curves determines an integer cohomology class. It is presented by an Arnold type form depending on the choice of an almost polarization in the symplectic normal bundle over an isotropic submanifold Λ.

Using all these objects, we define deformed coherent states as smooth L^2-valued functions over Λ under the additional condition that a certain quantization rule of the Bohr–Sommerfeld type holds.

The structure of the quantization rule is not simple. This structure is, of course, based on the fundamental and general theory [**A**$_{1,4}$, **Kel**, **Mas**, **MasF**]. But in the isotropic case, there still was an intriguing problem in the quantization rule, namely, to identify different contributions from different phase factors with some cohomology classes of Λ (as well as to define these classes). Here we summarize our results obtained during the last years [**KaVo**$_{1-8}$].

Besides of the usual Poincaré–Cartan contribution, the quantization rule in the isotropic case includes a cohomology class generalizing Floquet exponents, as well as the sum of tangential and normal integer-valued cohomology classes.

In §3.5, using deformed coherent states, we introduce a *very simple formula* for quasimodes (in Arnold's terminology [**A**$_2$]) which give asymptotic eigenvalues of Weyl operators associated with a compact isotropic submanifold of stable type.[1] Here the main point of interest is nontrivial geometric and topological conditions under which we can construct such quasimodes, hence derive the asymptotic expansion of spectral series of those Weyl operators. The conditions that we discuss in §3.4 and §3.5 allow us to generalize many well-known spectral results obtained within the framework of the semiclassical approximation associated with isotropic submanifolds [**BaLa, BeDo**$_{1,2}$**, DMasN, DuGu, Gu**$_2$**, GuWe, Kra**$_{1,2}$**, Mas, PU, Ral**].

We formulate our conditions in terms of adapted symplectic connection in the symplectic normal bundle over an isotropic submanifold Λ and in terms of topological properties of Λ, following [**KaVo**$_{1,4,6,8}$].

The final formula for spectral series (Theorem 3.5.2) includes the following phase space invariants of a Hamilton system with an invariant isotropic submanifold: the average of the Liouville form, the average of the "tangential" Arnold form on Λ and the rotation number of the first variation operator over Λ.

Part 1. Linearized Hamilton Dynamics over Isotropic Submanifolds

§1.1. First variation operators and their flows

We start with the definition of *first variation operators* of Hamilton vector fields. Here the most convenient viewpoint is the concept of *covariant differential operators* appearing in the theory of Lie algebroids [**Mck, Pr**]. We use the notions and terminology introduced in [**Mck**]. Throughout the paper we consider only manifolds and vector bundles of class C^∞.

Let $\pi\colon E \to B$ be a real vector bundle over a manifold B. By $\mathrm{Vect}(B)$ we denote the Lie algebra of smooth vector fields on B, and by $\Gamma(E)$ the $C^\infty(B)$-module of global sections of the vector bundle E.

DEFINITION 1.1.1. An $\mathbb{R}$-linear operator $D\colon \Gamma(E) \to \Gamma(E)$ is called a *covariant differential operator* in the vector bundle E over B if

(i) there is a (necessarily unique) vector field $v \in \mathrm{Vect}(B)$ such that

(1.1.1) $$D(fs) = fD(s) + v(f)s$$

for all $f \in C^\infty(B)$, $s \in \Gamma(E)$. The vector field v from (1.1.1) will be called *associated* with the operator D.

It follows from (1.1.1) that the set of all covariant differential operators in E is a *Lie algebra* with respect to the bracket

(1.1.2) $$[D, D'] = D \circ D' - D' \circ D.$$

We denote this Lie algebra by $\mathfrak{D}(E)$.

---

[1] The construction of quasimodes in this case has a long history: the general and fundamental theory of "complex germ" due to Maslov [**Mas**], the semiclassical theory of Gaussian packets due to Babich, Heller, and Voros [**Ba, Hel, Vor**], the Duistermaat–Guillemin–Ralston–Weinstein construction based on "symplectic spinors" [**DuGu, Gu**$_2$**, GuWe, Ral**]. Interesting results around this were obtained, for instance, in [**BaLa, BeDo**$_{1,2}$**, Bu, CdV, DMasN, Gut, Kra**$_{1,2}$**, Lj, MiSSh**].

There is the following interpretation of $\mathfrak{D}(E)$ in terms of Lie *algebroids* [**Mck**]. The set $\mathfrak{D}(E)$ is a finitely generated projective module over $C^\infty(B)$. Following [**Mck**], let us consider the vector bundle $\mathrm{CDO}(E)$ over B corresponding to $\mathfrak{D}(E)$, that is, $\mathfrak{D}(E) = \Gamma\big(\mathrm{CDO}(E)\big)$. Then $\mathrm{CDO}(E)$ is a Lie algebroid on B with bracket (1.1.2). The property (i) implies that the *anchor mapping* [**Mck**]

(1.1.3) $$D \mapsto v, \qquad D \in \mathfrak{D}(E),$$

defines a homomorphism between the Lie algebras $\mathfrak{D}(E)$ and $\mathrm{Vect}(B)$. We say that the vector field v on B from (1.1.3) is the *anchor* of an operator D. If the mapping (1.1.3) is a fiberwise submersion, then the Lie algebroid is called *transitive*.

If vector bundle E is equipped with an additional additive structure (a fiber metric, a symplectic structure, etc.), we define $\mathfrak{D}(E)$ as the Lie algebra of covariant differential operators D in E preserving the given structure. This means that the corresponding tensor field is covariantly constant along each $D \in \mathfrak{D}(E)$. Our focus is a symplectic vector bundle.

Let $(E, \mathcal{J})$ be a symplectic vector bundle. Here $\mathcal{J}\colon E \to E^*$ is a skew-symmetric isomorphism of vector bundles

$$\langle \mathcal{J} X, Y \rangle = -\langle \mathcal{J} Y, X \rangle \qquad (X, Y \in E_\alpha, \alpha \in B)$$

that gives the symplectic structure on each fiber E_α (here $\langle \, , \, \rangle \colon E^* \times E \to R$ denotes the pairing of dual spaces). In this case, we require that each $D \in \mathfrak{D}(E)$ satisfies the condition:

(ii) D is compatible with the symplectic structure of E, that is,

(1.1.4) $$v(\langle \mathcal{J} s', s'' \rangle) = \langle \mathcal{J}(Ds'), s'' \rangle + \langle \mathcal{J} s', Ds'' \rangle$$

for all $s', s'' \in \Gamma(E)$. Here $v \in \mathrm{Vect}(B)$ is the vector field on B from (1.1.1) associated with the operator D.

Note that each $D \in \mathfrak{D}(E)$ is an operator of local type, that is, for each $\alpha \in B$ and a section $s \in \Gamma(E)$, the value $D(s)(\alpha)$ depends only on the restriction of s to a neighborhood of α in B. So, for each submanifold $U \subset B$, the restriction $D_U \colon \Gamma(E_U) \to \Gamma(E_U)$ is a well-defined covariant differential operator.

Let $X_1(\alpha), \ldots, X_{2r}(\alpha)$ be a basis of local sections of the symplectic vector bundle E. Locally, each operator $D \in \mathfrak{D}(E)$ can be represented by the matrix-valued function $V(\alpha) = ((V_\sigma^{\sigma'}(\alpha)))$ $(\alpha \in B)$ whose elements are defined by the relations

(1.1.5) $$D(X_\sigma) = -\sum_{\sigma'=1}^{2r} V_\sigma^{\sigma'} X_{\sigma'} \qquad (\sigma = 1, \ldots, 2r).$$

In terms of V, the compatibility condition (1.1.4) takes the form

(1.1.6) $$v(J) + JV + V^T J = 0.$$

Here the vector field v is taken from (1.1.4) and

(1.1.7) $$J = ((J_{\sigma\sigma'}(\alpha))) \stackrel{\text{def}}{=} ((\langle \mathcal{J} X_{\sigma'}(\alpha), X_\sigma(\alpha) \rangle)).$$

It follows from (1.1.4) that a change of local bases $\widetilde{X}_\sigma(\alpha) = \sum_{\sigma'} T_\sigma^{\sigma'}(\alpha) X_{\sigma'}(\alpha)$ implies the following transformation rule for V:

(1.1.8) $$\widetilde{V} = -T^{-1} v(T) + T^{-1} V T,$$

where $T = ((T^{\sigma'}_\sigma(\alpha)))$ is the *transition matrix-valued function*.

Using the above properties, to each operator $D \in \mathfrak{D}(E)$ we can assign the vector field $D^\#\colon C^\infty(E) \to C^\infty(E)$ defined by the formula

$$(1.1.9) \qquad D^\# = \sum_i v_i(\alpha)\frac{\partial}{\partial \alpha_i} + \sum_{\sigma,\sigma'} V^{\sigma'}_\sigma(\alpha)x^\sigma \frac{\partial}{\partial x^{\sigma'}}.$$

Here the first term is the local representation of the vector field v associated with the operator D, $\alpha = (\alpha^1, \ldots, \alpha^k)$ are local coordinates on B, $x = (x^1, \ldots, x^{2k})$ are local coordinates along the fibers of E, and $V = ((V^{\sigma'}_\sigma(\alpha)))$ is the matrix-valued function from (1.1.5). In global terms, the vector field $D^\#$ can be defined in the following way.

By $\operatorname{Lin}(E) \subset C^\infty(E)$ we denote the subspace consisting of all smooth *fiberwise linear* functions on E. Then there is a natural isomorphism $l\colon \Gamma(E^*) \to \operatorname{Lin}(E)$, $l(\eta)(\alpha,x) = \langle \eta, x \rangle$, $x \in E_\alpha$, $\alpha \in B$, $\eta \in \Gamma(E^*)$. The vector field $D^\#$ is defined by the formula $D^\# = l \circ D^* \circ l^{-1}$, where $D^*\colon \Gamma(E^*) \to \Gamma(E^*)$ is the dual operator of D,

$$\langle D^*\eta, s \rangle + \langle \eta, Ds \rangle = v(\langle \eta, s \rangle)$$

for all $\eta \in \Gamma(E^*)$, $s \in \Gamma(E)$.

PROPOSITION 1.1.1. *If the base B of E is compact, then each covariant differential operator $D \in \mathfrak{D}(E)$ generates a one-parameter group of vector bundle symplectomorphisms $\mathbb{S}^t\colon E \to E$ ($t \in \mathbb{R}$),*

$$(\mathbb{S}^t)^* \circ \mathcal{J} \circ \mathbb{S}^t = \mathcal{J},$$

such that

$$\left.\frac{d}{dt}\right|_{t=0} \mathbb{S}^t(\xi) = D^\#(\xi), \qquad \xi \in E,$$

and the diagram

$$\begin{array}{ccc} E & \xrightarrow{\mathbb{S}^t} & E \\ \downarrow & & \downarrow \\ B & \xrightarrow[g^t]{} & B \end{array}$$

is commutative. Here $g^t\colon B \to B$ is the flow of the vector field $v \in \operatorname{Vect}(B)$ associated with the operator D.

So, for each $\alpha^0 \in B$, $t \in \mathbb{R}$, the restriction $\mathbb{S}^t\big|_{E_{\alpha^0}}\colon E_{\alpha^0} \to E_{g^t(\alpha^0)}$ is an isomorphism of symplectic vector spaces, which is locally defined by a solution $(\alpha(t), x(t))$ of the dynamical system

$$(1.1.10) \qquad \begin{aligned} \dot\alpha &= v(\alpha), & \alpha\big|_{t=0} &= \alpha^0, \\ \dot x &= V(\alpha)x, & x\big|_{t=0} &= x^0. \end{aligned}$$

We say that $\mathbb{S}^t$ is a *flow* generated by $D \in \mathfrak{D}(E)$.

Assume that the base B of E is compact. Choose a *norm* $\|\ \|$ on E, that is, a family $\|\ \|_\alpha\colon E_\alpha \to \mathbb{R}$ ($\alpha \in B$) of norms of the fibers E_α such that if $s \in \Gamma(U)$ is a local section of E over an open subset $U \subseteq B$, then $\|s(\alpha)\| = \|s(\alpha)\|_\alpha$ is a C^∞-function on U.

DEFINITION 1.1.2. A covariant differential operator $D \in \mathfrak{D}(E)$ is *metrically stable* (or stable in the sense of Lyapunov) if its flow $\mathbb{S}^t$ is uniformly bounded,

$$\|\mathbb{S}^t(\xi)\|_{g^t(\alpha)} \leq c\|\xi\|_\alpha \qquad (c = \text{const} > 0)$$

for all $t \in \mathbb{R}$, $\alpha \in B$, $\xi \in E_\alpha$.

Since B is compact, any two norms on E are equivalent, and hence, the definition of stability does not depend on the choice of a norm on E.

REMARK 1.1.1. Here we use the term of metric stability to emphasize the distinction between this notion and the notion of geometric stability, which appears in §1.7 in the context of integrability of linear Hamilton systems. In the periodic case, these notions are equivalent. For $\dim \Lambda > 1$, the condition of geometric stability is stronger than that of metric stability. For the general stability theory of linear skew-product flows, see [**Jo, JoSe**].

Now let us study the linearized Hamilton dynamics over isotropic submanifolds.

Let (M, Ω) be a symplectic manifold. Recall that the symplectic structure Ω of M, the Poisson bracket $\{\,,\,\}\colon C^\infty(M) \times C^\infty(M) \to C^\infty(M)$, and the Hamilton vector fields on M are related by the formula

$$\Omega\bigl(\mathrm{ad}(H_1), \mathrm{ad}(H_2)\bigr) = \{H_1, H_2\} = \mathrm{ad}(H_1) \lrcorner\, d(H_2).$$

Here $\mathrm{ad}(H)$ is the Hamilton vector field of a function $H \in C^\infty(M)$. The mapping $H \mapsto \mathrm{ad}(H)$ gives a homomorphism from the Poisson algebra $C^\infty(M)$ to the Lie algebra of Hamilton vector fields $\mathrm{Ham}(M)$.

A vector field a on M is said to be an *infinitesimal symplectic automorphism* of Ω if the Lie derivative of Ω along a vanishes, $L_a \Omega = 0$,

By $\mathcal{A} = \mathcal{A}(M)$ we denote the space of all infinitesimal symplectic automorphisms on M. As is well known [**LiMr**], $\mathcal{A}$ is a *Lie algebra*, and

$$\mathrm{Ham}(M) \subset \mathcal{A}, \qquad [\mathcal{A}, \mathcal{A}] \subseteq \mathrm{Ham}(M).$$

Locally, each $a \in \mathcal{A}(M)$ is a Hamilton vector field.

Let $\Lambda \subset M$ be an *isotropic submanifold*, $\omega|_\Lambda = 0$, $\dim \Lambda < \frac{1}{2}\dim M$.

By $T_\Lambda M$ we denote the symplectic vector bundle over Λ with fibers $T_\alpha M$, $\alpha \in \Lambda$. Consider the *isotropic subbundle* $T\Lambda \subset T_\Lambda M$ and the *coisotropic subbundle* $T\Lambda^\perp \subset T_\Lambda M$ whose fibers are the skew-orthogonal complements to the tangent spaces $T_\alpha \Lambda$,

$$T_\alpha \Lambda^\perp = \bigl\{u \in T_\alpha M : \Omega(u, v) = 0 \ \forall v \in T_\alpha \Lambda\bigr\}.$$

So, we have

$$T\Lambda \subset T\Lambda^\perp, \qquad \dim T\Lambda^\perp = \dim M - \dim \Lambda.$$

Consider two spaces $\mathcal{A}_\Lambda = \mathcal{A}_\Lambda(M)$ and $\mathcal{A}_\Lambda^\perp = \mathcal{A}_\Lambda^\perp(M)$ that consist of all infinitesimal symplectic automorphisms on M *tangent* to $T\Lambda$ and $T\Lambda^\perp$, respectively. Obviously, $\mathcal{A}_\Lambda \subset \mathcal{A}_\Lambda^\perp$ and $\mathcal{A}_\Lambda$ is a *Lie algebra*. The space of all Hamilton vector fields $\mathrm{ad}(H)$ on M that are *tangent* to Λ at each point form the Lie subalgebra $\mathrm{Ham}_\Lambda(M)$ of $\mathcal{A}_\Lambda(M)$. Furthermore, $\mathrm{Ham}_\Lambda(M)$ is generated by a subalgebra $\mathcal{F}_\Lambda(M)$ of the Poisson algebra $C^\infty(M)$,

$$\mathrm{Ham}_\Lambda(M) = \bigl\{\,\mathrm{ad}(H) \mid H \text{ runs over } \mathcal{F}_\Lambda(M)\bigr\}$$

In particular, any function $H \in \mathcal{F}_\Lambda(M)$ is constant along Λ.

Note that for any $a', a'' \in \mathcal{A}_\Lambda$ and $b \in \mathcal{A}_\Lambda^\perp$ we have

$$0 = (L_{a'}\Omega)(b, a'') = L_{a'}\bigl(\Omega(b, a'') - \Omega([a', b], a'') - \Omega(b, [a', a''])\bigr) = -\Omega([a', b], a'').$$

This means that

(1.1.11) $$[\mathcal{A}_\Lambda, \mathcal{A}_\Lambda^\perp] \subset \mathcal{A}^\perp.$$

Moreover,

(1.1.12) $$\text{if } b \in \mathcal{A}^\perp \text{ and } b\bigr|_\Lambda = 0, \text{ then } [a, b]\bigr|_\Lambda = 0 \text{ for any } a \in \mathcal{A}_\Lambda.$$

Now, following [**We**$_1$], let us consider the *symplectic normal bundle* E over an isotropic submanifold $\Lambda \subset M$ whose fibers are the quotient spaces

$$E_\alpha \stackrel{\text{def}}{=} T_\alpha \Lambda^\perp / T_\alpha \Lambda, \qquad \alpha \in \Lambda, \quad \dim E_\alpha = \dim M - 2 \dim \Lambda.$$

The symplectic structure along the fibers of E is defined by the vector bundle morphism $\mathcal{J} \colon E \to E^*$,

$$\langle \mathcal{J}(X), Y \rangle = \Omega(\widetilde{X}, \widetilde{Y}), \qquad X, Y \in E_\alpha,$$

where $\widetilde{X}, \widetilde{Y} \in T_\alpha \Lambda^\perp$ ($\alpha \in \Lambda$) are any two vectors such that $\nu(\widetilde{X}) = X$, $\nu(\widetilde{Y}) = Y$ under the natural projection

$$\nu \colon T\Lambda^\perp \to E.$$

We have the following exact sequence of $C^\infty(\Lambda)$-modules

$$0 \to \operatorname{Vect}(\Lambda) \to \Gamma(T\Lambda^\perp) \xrightarrow{\nu} \Gamma(E) \to 0.$$

For each vector field $a \in \mathcal{A}_\Lambda$, let us define an $\mathbb{R}$-linear operator $D_a \colon \Gamma(E) \to \Gamma(E)$ by the formula

(1.1.13) $$D_a(s) \stackrel{\text{def}}{=} \nu([a, b]\bigr|_\Lambda), \qquad s \in \Gamma(E),$$

where ν is the projection from (1.1.20) and $b \in \mathcal{A}_\Lambda^\perp$ is any vector field such that

(1.1.14) $$\nu(b\bigr|_\Lambda) = s.$$

It follows from (1.1.11) and (1.1.12) that the right-hand side of (1.1.13) is well defined, that is, independent of the choice of a vector field b satisfying (1.1.14) for a given section $s \in \Gamma(E)$.

THEOREM 1.1.1. *For each vector field $a \in \mathcal{A}_\Lambda$, formula (1.1.22) defines a covariant differential operator $D_a \in \mathfrak{D}(E)$ in the symplectic normal bundle E. The vector field $v \in \operatorname{Vect}(\Lambda)$ associated with the operator D_a (in condition (1.1.1)) is the restriction of a to Λ, $v = a\bigr|_\Lambda$. Moreover, the mapping*

$$a \mapsto D_a, \qquad a \in \mathcal{A}_\Lambda$$

gives a homomorphism of the Lie algebra $\mathcal{A}_\Lambda$ to $\mathfrak{D}(E)$,

(1.1.15) $$[D_{a'}, D_{a''}] = D_{[a', a'']} \quad \text{for all} \quad a', a'' \in \mathcal{A}_\Lambda.$$

Now we can introduce infinitesimal characteristics of (locally Hamilton) vector fields $a \in \mathcal{A}_\Lambda$ whose flows leave an isotropic submanifold Λ invariant.

DEFINITION 1.1.3. For each vector field $a \in \mathcal{A}_\Lambda$, the operator $D_a \in \mathfrak{D}(E)$ from (1.1.13) is called a *first variation operator* of a over an isotropic submanifold Λ. In particular, the first variation operator of a Hamilton vector field $\mathrm{ad}(H) \in \mathcal{A}_\Lambda$ ($H \in \mathcal{F}_\Lambda(M)$) is denoted by $D_H \stackrel{\mathrm{def}}{=} D_{\mathrm{ad}(H)}$.

Property (1.1.15) implies that the mapping $H \mapsto D_H$, $H \in \mathcal{F}_\Lambda(M)$, gives a homomorphism of the Lie algebra $\mathcal{F}_\Lambda$ to $\mathfrak{D}(E)$,

$$[D_{H'}, D_{H''}] = D_{\{H', H''\}}$$

for all $H', H'' \in \mathcal{F}_\Lambda(M)$.

If Λ is a compact isotropic submanifold, then the flow $\mathbb{S}_a^t \colon E \to E$ generated by $D_a \in \mathfrak{D}(E)$ is said to be the *linearized Hamilton flow* of $a \in \mathcal{A}$ (see Proposition 1.1.1). For a Hamilton vector field $\mathrm{ad}(H) \in \mathcal{A}_\Lambda$, this flow is denoted by $\mathbb{S}_H^t = \mathbb{S}_{\mathrm{ad}(H)}^t$.

Locally, the operator D_a is defined by the variation matrix $V_a(\alpha)$ ($\alpha \in \Lambda$) from (1.1.5), and the corresponding flow $\mathbb{S}_a^t$ is given by solutions of the *first variation equation* (1.1.10) over Λ.

A Hamilton system (M, Ω, H) is called *linear stable* over an invariant compact isotropic submanifold $\Lambda \subset M$ if the first variation operator D_H in the symplectic normal bundle E over Λ is metrically stable in the sense of Definition 1.1.2.

§1.2. Adapted symplectic connections and reducibility

First of all, we recall some facts from the theory of linear connections (for details, see [**GHV**]).

Let $(\pi \colon E \to B, \mathcal{J} \colon E \to E^*)$ be a real symplectic vector bundle. A family of $\mathbb{R}$-linear operators $\nabla_u \colon \Gamma(E) \to \Gamma(E)$ indexed by vector fields $u \in \mathrm{Vect}(B)$ and satisfying the following conditions:

(i) $u \to \nabla_u$ is a $C^\infty(B)$-linear mapping;

(ii) $\nabla_u(fs) = u(f)s + f\nabla_u s$ for all $u \in \mathrm{Vect}(B)$, $f \in C^\infty(B)$, $s \in \Gamma(E)$,

is called a *linear connection* in E. It is well known that in a given vector bundle E there always exists a linear connection.

By $A^q(B, E)$ we denote the space of q-forms on B with values in the vector bundle E. This space coincides with the space of sections $\Gamma(\bigwedge^q T^*B \otimes E)$, where $\bigwedge^q$ is the qth exterior product. In particular, $A^q(B)$ denotes the space of q-forms on B.

Then a linear connection in E can be viewed as a $\mathbb{R}$-linear mapping

$$\nabla \colon A^0(B, E) = \Gamma(E) \to \Gamma(T^*(B) \otimes E) = A^1(B, E)$$

such that $\nabla(fs) = f\nabla(s) + df \otimes s$ for all $s \in \Gamma(E)$, $f \in C^\infty(B)$. The *curvature 2-form* of ∇, $C \in A^2(B, \mathrm{End}(E))$, is defined as

$$(1.2.1) \qquad C(u', u'') = \nabla_{u'} \circ \nabla_{u''} - \nabla_{u''} \circ \nabla_{u'} - \nabla_{[u', u'']},$$

where $u', u'' \in \mathrm{Vect}(B)$. A linear connection ∇ is called *flat* if its curvature form is equal to zero, $C \equiv 0$.

A *dual connection* $\nabla^* \colon \Gamma(E^*) \to \Gamma(T^*B \otimes E^*)$ is defined by the relation

$$u(\langle \eta, s \rangle) = \langle \nabla_u^* \eta, s \rangle + \langle \eta, \nabla_u s \rangle$$

for all $u \in \mathrm{Vect}(B)$, $\eta \in \Gamma(E^*)$, and $s \in \Gamma(E)$.

A linear connection ∇ in a symplectic vector bundle $(E, \mathcal{J})$ is called *symplectic* if

(iii) the symplectic structure on E is *covariantly constant* or, equivalently, if

(1.2.2) $$\mathcal{J} \circ \nabla_u = \nabla_u^* \circ \mathcal{J} \quad \text{for any} \quad u \in \text{Vect}(B).$$

Condition (1.2.2) can be rewritten by the following way:

$$u(\langle \mathcal{J}s', s'' \rangle) = \langle \mathcal{J}\nabla_u s', s'' \rangle + \langle \mathcal{J}s', \nabla_u s'' \rangle,$$

for all $u \in \text{Vect}(B)$ and $s', s'' \in \Gamma(E)$.

Recall that two symplectic connections ∇ and $\widetilde{\nabla}$ in symplectic vector bundles $(E, \mathcal{J})$ and $(\widetilde{E}, \widetilde{\mathcal{J}})$ over Λ are called *gauge equivalent* if there exists a vector bundle symplectomorphism $L: E \to \widetilde{E}$, $L^* \circ \widetilde{\mathcal{J}} \circ L = \mathcal{J}$, such that $L^{-1} \circ \widetilde{\nabla} \circ L = \nabla$.

As is well known, each affine connection on a symplectic manifold induces a linear symplectic connection (see [**He, MRR**]). The same fact is valid for symplectic vector bundles.

PROPOSITION 1.2.1. *In a symplectic vector bundle $(E, \mathcal{J})$, each linear connection $\widetilde{\nabla}$ induces the symplectic connection ∇ by the formula*

$$\nabla_u = \widetilde{\nabla} u + \frac{1}{2}(\mathcal{J}^{-1} \circ \widetilde{\nabla}_u^* \circ \mathcal{J} - \widetilde{\nabla}_u), \qquad u \in \text{Vect}(B).$$

Thus, any symplectic vector bundle admits a symplectic connection.

PROOF. Let us look for a desired connection ∇ in the form

$$\nabla_u = \widetilde{\nabla}_u + \mathfrak{Q}_u, \qquad u \in \text{Vect}(B),$$

where $\mathfrak{Q}_u \in A^1(B, \text{End}(E))$ is an operator-valued 1-form. Condition (1.2.2) yields the following equation for $\mathfrak{Q}_u$:

(1.2.3) $$\mathcal{J}^{-1} \circ \mathfrak{Q}_u^* + \mathfrak{Q}_u \circ \mathcal{J}^{-1} = \mathfrak{P},$$

where $\mathfrak{P} = \mathcal{J}^{-1} \circ \widetilde{\nabla}_u^* - \widetilde{\nabla}_u \circ \mathcal{J}^{-1}$. Since $\mathcal{J}$ is antisymmetric, $\mathfrak{P}$ is also antisymmetric, $\langle \eta, \mathfrak{P}\eta' \rangle = -\langle \eta', \mathfrak{P}\eta \rangle$ for all $\eta', \eta \in \Gamma(E^*)$. This implies that the solution of (1.2.3) is given by the formula $\mathfrak{Q}_u = \frac{1}{2}\mathfrak{P} \circ \mathcal{J}$. □

Note that if ∇ is a symplectic connection in E, then for each vector field $v \in \text{Vect}(B)$, the operator ∇_v is a covariant differential operator in E, $\nabla_v \in \mathfrak{D}(E)$.

COROLLARY 1.2.1. *The Lie algebroid* CDO(E) *is transitive (see* §1.1*).*

Now we formulate the reducibility criteria for covariant differential operators via adapted connections.

Let ∇ be a symplectic connection on E. Then for each $D \in \mathfrak{D}(E)$, we have the decomposition

(1.2.4) $$D = \nabla_v + \Xi, \qquad \Xi \in \mathfrak{D}_0(E),$$

where the vector field v is defined by condition (1.1.1). It is clear that Ξ can be considered as the field of linear operators $\Xi = \{\Xi(\alpha): E_\alpha \to E_\alpha, \alpha \in B\}$ such that

(1.2.5) $$\Xi(\alpha) \in \text{sp}(E_\alpha; \mathbb{R}) \iff \mathcal{J}_\alpha \circ \Xi(\alpha) + \Xi^*(\alpha) \circ \mathcal{J}_\alpha = 0.$$

Define the pull-back $\nabla_v^\#$ of $v \in \mathrm{Vect}(B)$ by the formula like (1.1.9):

$$\nabla_v^\# = \sum_i v^i \frac{\partial}{\partial \alpha^i} - \sum_{\sigma,\sigma'} (v \lrcorner \theta_\sigma^{\sigma'}) x^\sigma \frac{\partial}{\partial x^{\sigma'}}. \tag{1.2.6}$$

Here $\theta = ((\theta_\sigma^{\sigma'}))$ is the matrix-valued connection 1-form on B,

$$\nabla X_\sigma = \sum_{\sigma'=1}^{2r} \theta_\sigma^{\sigma'} \otimes X_{\sigma'}, \qquad \sigma = 1, \ldots, 2r,$$

where $X_1(\alpha), \ldots, X_{2r}(\alpha)$ is a basis of local sections of E.

The connection ∇ induces the splitting of the tangent bundle TE,

$$TE = \mathcal{H} \oplus \mathcal{V}, \tag{1.2.7}$$

into the direct sum of *horizontal* $\mathcal{H}$ and *vertical* $\mathcal{V}$ subbundles. The fibers of $\mathcal{H}$ and $\mathcal{V}$ are defined by

$$\mathcal{H}_\xi = \mathrm{span}\{\nabla_v^\#(\xi), v \text{ runs over } \mathrm{Vect}(B)\},$$
$$\mathcal{V}_\xi = \ker(T_\xi \pi) \subset T_\xi E, \qquad \xi \in E.$$

Here $T\pi$ is the tangent mapping of the projection $\pi\colon E \to B$.

So, decompositions (1.2.4) and (1.2.7) imply

$$D^\# = \nabla_v^\# + \Xi^\#, \tag{1.2.8}$$

where $\nabla_v^\#$ and $\Xi^\#$ are the horizontal and vertical components, respectively.

Now we assume that the base B of the symplectic vector bundle E is *compact* and *connected*, and a symplectic connection ∇ in E is given. The parallel *transports* of ∇ along the trajectories of $v \in \mathrm{Vect}(B)$ are determined by the flow $\exp(t\nabla_v^\#)\colon E \to E$ of the vector field $\nabla_v^\#$ (see Proposition 1.1.1). Using this flow, we introduce a time-dependent field of operators $\Xi^t = \{\Xi^t(\alpha) \in \mathrm{sp}(E_\alpha;\mathbb{R}), \alpha \in B\}$ by the formula

$$\Xi^t = \exp(-t\nabla_v^\#) \circ \Xi \circ \exp(t\nabla_v^\#). \tag{1.2.9}$$

By $\mathfrak{D}_0(E)$ we denote the subset of the Lie algebra $\mathfrak{D}(E)$ consisting of all operators $D \in \mathfrak{D}(E)$ such that the vector field v in condition (1.1.1) corresponding to D vanishes, $v \equiv 0$. Thus, $\mathfrak{D}_0(E)$ is the kernel of the anchor mapping (1.1.3). Obviously, $\mathfrak{D}_0(E)$ is an ideal in $\mathfrak{D}(E)$,

$$[\mathfrak{D}(E), \mathfrak{D}_0(E)] \subset \mathfrak{D}_0(E).$$

Note that for each operator $D \in \mathfrak{D}_0(E)$, its flow $\mathbb{S}^t$ (defined in Proposition 1.1.1) is a family of fiberwise linear operators identical on the base B. Thus, $\mathbb{S}^t\big|_{E_\alpha} \in \mathrm{Sp}(E_\alpha, \mathbb{R})$ is a symplectic linear operator in the fiber E_α for all $\alpha \in B$, $t \in \mathbb{R}$.

Choose $\alpha^0 \in B$ and consider the flow $\mathbb{S}^t\colon E \to E$ generated by an operator $D \in \mathfrak{D}(E)$. It follows from the decomposition (1.2.8) that the orbit $\mathbb{S}^t\big|_{E_{\alpha^0}}$ has the representation

$$\mathbb{S}^t\big|_{E_{\alpha^0}} = \exp(t\nabla_v^\#)\big|_{E_{\alpha^0}} \circ \mathbb{T}(t). \tag{1.2.10}$$

Here $\mathbb{T}(t) \colon \mathbb{R} \to \mathrm{Sp}(E_{\alpha^0}; \mathbb{R})$ is the solution of the linear Hamilton system in the space E_{α^0},

(1.2.11) $$\frac{d\mathbb{T}}{dt} = \Xi^t(\alpha^0) \circ \mathbb{T}, \qquad \mathbb{T}(0) = I.$$

So, the symplectic connection ∇ allows us to reduce the dynamics (the flow) generated by $D \in \mathfrak{D}(E)$ to a time-dependent linear Hamilton system in the fiber E_{α^0} (see also [**BuNa, MRR**]).

We say that the vertical component of field Ξ in (1.2.4) is *covariantly constant* along the connection ∇ if

(1.2.12) $$\nabla_u \circ \Xi = \Xi \circ \nabla_u \qquad \forall u \in \mathrm{Vect}(B).$$

Locally, in terms of the connection 1-form $\theta = ((\theta_\sigma^{\sigma'}))$ and the matrix-valued function $\Xi = ((\Xi_\sigma^{\sigma'}(\alpha))$, condition (1.2.12) is equivalent to the equation on B

$$d\Xi + [\theta, \Xi] = 0.$$

On the other hand, (1.2.12) means that Ξ^t in (1.2.9) satisfies the condition

$$\frac{d\Xi^t}{dt} = 0,$$

or equivalently, $\Xi^t\big|_{E_\alpha}$ does not depend on $t \in \mathbb{R}$ for each $\alpha \in B$.

So, if Ξ is ∇-covariantly constant, then $\mathbb{T}(t)$ in (1.2.10) is the solution of the linear Hamilton system with constant coefficients, $\mathbb{T}(t) = \exp(t\Xi^0)$, $\Xi^0 = \Xi^t\big|_{E_{\alpha^0}}$.

DEFINITION 1.2.1. $D \in \mathfrak{D}(E)$ is ∇-*reducible in* E if there exists a flat symplectic linear connection in Ξ such that the vertical part E of D from the decomposition (1.2.4) is a ∇-covariantly constant. Then we say that the flow $\mathbb{S}^t$ generated by D is also ∇-reducible.

By $^\mathbb{C}E = E \otimes \mathbb{C}$ we denote the *complexification* of the real symplectic vector bundle E. Extending the vector bundle isomorphism $\mathcal{J} \colon E \to E^*$ to $^\mathbb{C}E$ by linearity, we obtain a complex symplectic vector bundle $(^\mathbb{C}E, \mathcal{J})$. All operators D, Ξ, and ∇ can also be extended to $\Gamma(^\mathbb{C}E)$.

We say that a covariant differential operator $D \in \mathfrak{D}(E)$ (or its flow $\mathbb{S}^t$) is ∇-reducible in $^\mathbb{C}E$ if the conditions in Definition 1.2.1 hold for a complex symplectic connection ∇ in $(^\mathbb{C}E, \mathcal{J})$ (that is, ∇ is a $\mathbb{C}$-linear mapping).

Note that a complex symplectic connection ∇ in $^\mathbb{C}E$ induces a real connection in E if $\overline{\nabla s} = \nabla \overline{s}$ for all $s \in \Gamma(^\mathbb{C}E)$ (here the bar denotes the *complex conjugation* in $^\mathbb{C}E$).

A symplectic vector bundle $(^\mathbb{C}E, \mathcal{J})$ is called (*symplectically*) *trivial* if there exists a set of global sections $Y_1(\alpha), \ldots, Y_{2r}(\alpha) \in \Gamma(^\mathbb{C}E)$ satisfying the conditions

$$\mathrm{span}\{Y_1(\alpha), \ldots, Y_{2r}(\alpha)\} = {}^\mathbb{C}E_\alpha \qquad \forall \alpha \in B,$$

$$J = ((\langle \mathcal{J}Y_{\sigma'}, Y_\sigma \rangle)) = \begin{bmatrix} 0 & -I \\ I & 0 \end{bmatrix}.$$

In this case, the set of sections $Y_1(\alpha), \ldots, Y_{2r}(\alpha)$ is said to be a *symplectic basis* (or a *symplectic trivialization*) of $^\mathbb{C}E$. In particular, the real symplectic vector bundle E is trivial if there is a real symplectic basis of E.

If $^{\mathbb{C}}E$ is trivial, then the flow $\mathbb{S}^t$ generated by $D \in \mathfrak{D}(E)$ on $^{\mathbb{C}}E$ is defined by solutions of the dynamical system on an open domain of $B \times \mathbb{C}^{2r}$:

$$\frac{d\alpha}{dt} = v(\alpha), \qquad \alpha = (\alpha^1, \ldots, \alpha^k) \in B.$$

$$\frac{dy}{dt} = V(\alpha)y, \qquad y = (y^1, \ldots, y^{2r}) \in \mathbb{C}^{2r}.$$

Here $V(\alpha) = ((V_\sigma^{\sigma'}(\alpha))) \in \mathrm{sp}(r;\mathbb{C})$ is the matrix-valued function on B represented with respect to the symplectic basis $Y_1(\alpha), \ldots, Y_{2r}(\alpha) \in \Gamma(^{\mathbb{C}}E)$,

(1.2.13) $$DY_\sigma(\alpha) = -\sum_{\sigma'=1}^{2r} V_\sigma^{\sigma'}(\alpha) Y_{\sigma'}(\alpha).$$

DEFINITION 1.2.2. A covariant differential operator $D \in \mathfrak{D}(E)$ is *reducible* in $^{\mathbb{C}}E$ (or in E) if the symplectic vector bundle $^{\mathbb{C}}E$ (or E) is trivial and there exists a symplectic basis $Y_1(\alpha), \ldots, Y_{2r}(\alpha)$ of $^{\mathbb{C}}E$ (or of E) such that the matrix-valued function $V(\alpha) = ((V_\sigma^{\sigma'}(\alpha)))$ from (1.2.13) is constant,

$$V_\sigma^{\sigma'}(\alpha) = \mathrm{const} \iff \langle \mathcal{J}DY_\sigma(\alpha), Y_{\sigma'}(\alpha) \rangle = \mathrm{const}$$

for all $\sigma', \sigma = 1, \ldots, 2r$, $\alpha \in B$. In this case, if B is compact, then the flow $\mathbb{S}^t$ is also called *reducible* in $^{\mathbb{C}}E$ (or in E).

Assume that B is compact. Then the homology group $H_1(B)$ is finitely generated and $H_1(B) = \widetilde{H}_1(B) \oplus \mathrm{Tor}\, H_1(B)$, where $\widetilde{H}_1(B)$ is the first *Betti group* (a free finitely generated abelian group) and $\mathrm{Tor}\, H_1(B)$ is the *torsion group* of B (see, for example, [**GHV**, **DNF**]).

THEOREM 1.2.1. *Let $D \in \mathfrak{D}(E)$ be a covariant differential operator in the real symplectic vector bundle E over B. If*
  (i) *the fundamental group $\pi_1(B)$ is commutative and the torsion group $\mathrm{Tor}\, H_1(B)$ is trivial, and*
  (ii) *the flow $\mathbb{S}^t$ generated by $D \in \mathfrak{D}(E)$ is ∇-reducible in E,*
then the flow $\mathbb{S}^t$ is reducible in $^{\mathbb{C}}E$.

PROOF. The proof is the same as in the classical Floquet theory [**YStz**], but it is given in terms of linear connections.

Let ∇ be a flat symplectic connection in E. Choose a point $\alpha^0 \in B$ and consider a smooth path $\alpha(t) : [0,1] \to B$ at α^0, $\alpha(0) = \alpha^0$. The linear symplectic isomorphism

$$g_{\alpha^0 \to \alpha(t)} \stackrel{\mathrm{def}}{=} \exp(t\nabla_{\dot\alpha(t)}^{\#})\big|_{E_{\alpha^0}} : E_{\alpha^0} \to E_{\alpha(t)}$$

is called the *parallel transport* along the path $\alpha(t)$ from α^0 to $\alpha(t)$. Locally, the parallel transport of a flat connection is given by the matrix-valued function $g = ((g_\sigma^{\sigma'}(\alpha)))$ on B satisfying the equation

(1.2.14) $$dg + \theta g = 0, \qquad g\big|_{\alpha^0} = I,$$

where θ is a connection 1-form. If $\Gamma = \{\alpha = \alpha(t), \alpha(0) = \alpha(1) = \alpha^0\}$ is a closed path in B, then the parallel transport $g_\Gamma = g(\alpha(t))\big|_{t=1}$ along Γ gives the mapping

(1.2.15) $$\Gamma \mapsto g_\Gamma \in \mathrm{Sp}(E_{\alpha^0}; \mathbb{R}).$$

Since the connection ∇ is flat, the parallel transport g_Γ is independent of a choice of Γ from a given homotopy class. In the general case, the correspondence (1.2.15) defines the homomorphism

$$\psi \colon \pi_1(B, \alpha^0) \mapsto \mathrm{Sp}(E_{\alpha^0}; \mathbb{R}).$$

The image under ψ is called a *holonomy group* of ∇ at $\alpha^0 \in B$,

$$\mathrm{Hol}^\nabla_{\alpha^0} = \psi\big(\pi_1(B, \alpha^0)\big) \subset \mathrm{Sp}(E_{\alpha^0}; \mathbb{R}).$$

So, the holonomy group $\mathrm{Hol}^\nabla_{\alpha^0}$ is a subgroup of the linear symplectic group $\mathrm{Sp}(r, \mathbb{R})$. If the holonomy group is Abelian (for instance, $\pi_1(B)$ is commutative), then the correspondence (1.2.15) defines the homomorphism $H_1(B) \to \mathrm{Hol}^\nabla_{\alpha^0}$.

If B is connected, then the holonomy groups related to distinct points of B are isomorphic.

In our case, by assumption (i) of Theorem 1.2.2, we have $H_1(B) \approx \pi_1(B)$, and the torsion of the group of $H_1(B)$ is trivial. This implies that [**GHV**]:

(a) there is a basis of closed 1-forms $\eta^1, \ldots, \eta^m$ on B dual to a basis of 1-cycles $\Gamma_1, \ldots, \Gamma_m$, $[\Gamma_j] \in H_1(B)$,

$$\oint_{\Gamma_j} \eta^i = \delta^i_j, \qquad i, j = 1, \ldots, m = \dim H_1(B);$$

(b) the elements $G_j = g_{\Gamma_j}$ of the holonomy group $\mathrm{Hol}^\nabla_{\alpha^0}$ are commuting symplectic operators,

$$G_j \in \mathrm{Sp}(E_{\alpha^0}; \mathbb{R}), \qquad [G_i, G_j] = 0, \quad i, j = 1, \ldots, m.$$

We introduce the fields of operators on B:

(1.2.16) $\qquad F_j(\alpha) = g(\alpha) \circ \ln(G_j) \circ g^{-1}(\alpha) \in \mathrm{Sp}(^{\mathbb{C}}E_\alpha), \quad \alpha \in B,$

where $j = 1, \ldots, m$; $g(\alpha)$ is defined by (1.2.14), and $\ln(G_j)$ is the principal branch of the logarithm.

Using (1.2.16), we define the complex connection on $^{\mathbb{C}}E$

$$\widetilde{\nabla} = \nabla + \sum_{j=1}^{m} F_j \otimes \eta^j.$$

The properties of ∇ and F_j imply that $\widetilde{\nabla}$ is a flat symplectic connection in $^{\mathbb{C}}E$ and its holonomy group is trivial, $\mathrm{Hol}^{\widetilde{\nabla}}_{\alpha^0} \approx \{I\}$.

Choose a symplectic basis $Y^0_1, \ldots, Y^0_{2r}$ in the fiber E_{α^0}. Then sections

(1.2.17) $\qquad Y_\sigma(\alpha) = g(\alpha) \circ \exp\left\{ -\int_{\alpha^0}^{\alpha} \sum_{j=1}^{m} \ln(G_j) \otimes \eta^j \right\} Y^0_\sigma,$

where $\sigma = 1, \ldots, 2r$, give the symplectic basis of $^{\mathbb{C}}E$ (the symplectic trivialization of $^{\mathbb{C}}E$). Moreover, $Y_\sigma(\alpha)$ are $\widetilde{\nabla}$-parallel,

(1.2.18) $\qquad \widetilde{\nabla} Y_\sigma = 0, \qquad \sigma = 1, \ldots, 2r.$

Finally, if the flow $\mathbb{S}^t$ generated by $D \in \mathfrak{D}(E)$ is ∇-reducible, then it is $\widetilde{\nabla}$-reducible. From decomposition (1.2.4) and condition (1.2.12) relative to $\widetilde{\nabla}$, we have
$$V = -v \,\lrcorner\, \widetilde{\theta} + \Xi, \qquad d\Xi + [\widetilde{\theta}, \Xi] = 0.$$
Here $V(\alpha)$ is a matrix-valued function of the operator D, and $\widetilde{\theta}$ is the connection 1-form of $\widetilde{\nabla}$ with respect to the basis (1.2.17). From (1.2.18), we have $\widetilde{\theta} \equiv 0$, and hence, $V = \mathrm{const}$ along B. This completes the proof of Theorem 1.2.2. $\square$

Note that if a flow $\mathbb{S}^t$ is reducible in $^\mathbb{C}E$ with respect to a symplectic basis $Y_1(\alpha), \ldots, Y_{2r}(\alpha) \in \Gamma(^\mathbb{C}E)$, then $\mathbb{S}^t$ is automatically ∇-reducible in $^\mathbb{C}E$ relative to the complex connection ∇ defined by
$$\nabla Y_\sigma = 0, \qquad \sigma = 1, \ldots, 2r.$$
So, we have the following statement.

COROLLARY 1.2.2. *Under assumption* (i) *of Theorem 1.2.2, the flow $\mathbb{S}^t$ generated by $D \in \mathfrak{D}(E)$ is reducible in $^\mathbb{C}E$ if and only if $\mathbb{S}^t$ is ∇-reducible in $^\mathbb{C}E$.*

Using well-known spectral properties of real linear symplectic operators (see, for example, [**AG, MeHl, YStz**]), we also deduce the next corollary.

COROLLARY 1.2.3. *If under assumptions* (i), (ii) *of Theorem 1.2.2, the spectra of the basic elements $G_1, \ldots, G_m$ of the holonomy group $\mathrm{Hol}_{\alpha^0}^\nabla$ do not contian -1, then the symplectic basis* (1.2.17) *is real, and thus, the flow $\mathbb{S}^t$ is reducible in E.*

Now let us recall the notion of *adapted symplectic connections*, which appeared in the context of the linearized Hamilton dynamics over isotropic submanifolds.[2]

DEFINITION 1.2.3. A symplectic connection ∇ in E (or in $^\mathbb{C}E$) is called *adapted* to a covariant differential operator $D \in \mathfrak{D}(E)$ if
$$\nabla_v = D \qquad (\text{or } \Xi \equiv 0), \tag{1.2.19}$$
where $v \in \mathrm{Vect}(B)$ is a vector field associated with the operator D (Ξ is determined by the decomposition (1.2.4)).

In terms of the flow $\mathbb{S}^t$ generated by D and the parallel transport of ∇, condition (1.2.19) means that for each α^0, the restriction $\mathbb{S}^t \big|_{E_{\alpha^0}} : E_{\alpha^0} \to E_{\alpha(t)}$ coincides with the parallel transport $\exp(t\nabla_{\dot{\alpha}(t)}^\#)$ along the trajectory $\alpha(t)$ of v issuing from α^0. In other words, the time-dependent field $\mathbb{T}^t$ in (1.2.9) is the identity operator $\mathbb{T}^t \equiv I$.

The following observation shows how the ∇-reducibility is related to the existence of flat adapted connections.

THEOREM 1.2.2. (a) *If there is a flat symplectic connection ∇ adapted to a covariant differential operator $D \in \mathfrak{D}(E)$, then the flow $\mathbb{S}^t$ generated by D is ∇-reducible.*

(b) *If a flow $\mathbb{S}^t$ is ∇-reducible and there exists a 1-form η on B such that*
$$v \,\lrcorner\, \eta = 1, \qquad d\eta = 0,$$

---

[2]Flat adapted connections over isotropic tori were exploited in the framework of the integrability problem and the construction of quasimodes via coherent states in [**Ka**₆]. Sufficient conditions for the flatness were found in [**Vo**₁], and the first systematic definitions and applications for general isotropic submanifolds appeared in [**KaVo**₁,₂].

where $v \in \mathrm{Vect}(B)$ is a vector field associated with the operator $D \in \mathfrak{D}(E)$, then the formula

$$\widetilde{\nabla} = \nabla + \Xi \otimes \eta,$$

where the field Ξ is defined in (1.2.4), gives a flat symplectic connection that is adapted to the operator D.

COROLLARY 1.2.4. *Let $B = \mathbb{T}^k \equiv \mathbb{R}^k/2\pi\mathbb{Z}^k$ be the k-torus. Let $D \in \mathfrak{D}(E)$, and let v be the image of D under the anchor mapping. Suppose that the flow of a vector field v on $\mathbb{T}^k$ is quasiperiodic, that is, v can be represented in the form*

$$v = \sum_{i=1}^{k} \omega_i \frac{\partial}{\partial \alpha_i}, \qquad \omega = (\omega_1, \ldots, \omega_k) \in \mathbb{R}^k,$$

with respect to some angle coordinates $\alpha_1, \ldots, \alpha_k$ on $\mathbb{T}^k$ and the frequency vector ω. Then the flow $\mathbb{S}^t$ generated by D is reducible in ${}^\mathbb{C}E$ if and only if there exists a flat symplectic connection ∇ in E which is adapted to the operator D.

In the periodic case $B = \mathbb{S}^1$ ($k = 1$), the flat adapted connection is uniquely defined as

$$\nabla_v = D, \qquad v \in \mathrm{Vect}(\mathbb{S}^1),$$

where v is a vector field on $\mathbb{S}^1$ without singular points. In this case, Corollary 1.2.4 is just the classical Floquet theorem.

REMARK 1.2.1. The above result gives a geometric reducibility criterion. Analytical reducibility criteria for 2-dimensional linear Hamilton systems with quasiperiodic coefficients were obtained in [**E, Jo, JoSe**]. These criteria lead to nonresonant conditions (of the Diophantine type) for frequencies of the quasiperiodic motion, which also arise in various applications of the KAM-theory [**AAv, AKN, Mel, Sev**].

§1.3. Flat symplectic connections in the normal bundle

Let (M, Ω) be a symplectic manifold, and let Λ be an isotropic submanifold in M. Our aim is to describe the conditions under which flat symplectic connections appear in the symplectic normal bundle $E = T\Lambda^\perp/T\Lambda$ over an isotropic submanifold Λ.

We say that a submanifold $N \subset M$ is *coisotropic on* an isotropic submanifold Λ [**KaVo**$_{1,2}$] if
 (a) $\Lambda \subset N$,
 (b) $\dim N = \mathrm{codim}\, \Lambda$,
 (c) $\ker \Omega|_{T_\Lambda N} = T\Lambda$ (or equivalently, $T_\Lambda N = T\Lambda^\perp$).

LEMMA 1.3.1. *Let a submanifold N be coisotropic on an isotropic submanifold Λ. In a neighborhood of each point of $\Lambda \in M$, there exists a set of C^∞-functions $f = (f_1, \ldots, f_k)$ and $\Phi = (\Phi_1, \ldots, \Phi_{2r})$ such that the submanifolds Λ and N are locally given as zero-level sets of these functions,*

(1.3.1) $$N = \{f_1 = 0, \ldots, f_k = 0\} \quad (k = \dim \Lambda),$$
(1.3.2) $$\Lambda = \{f_1 = 0, \ldots, f_k = 0\} \cap \{\Phi_1 = 0, \ldots, \Phi_{2r} = 0\} \quad (2r = \dim N - k).$$

Note that near Λ the pairwise Poisson brackets of functions f_i and Φ_σ in (1.3.1) and (1.3.2) satisfy the relations

$$\{f_i, \Phi_\sigma\}_{f=0} = O(|\Phi|), \qquad \{f_i, f_j\}\big|_{f=0} = O(|\Phi|^2) \qquad \text{as} \quad |\Phi| \to 0.$$

Here $|\Phi|^2 = \Phi_1^2 + \cdots + \Phi_{2r}^2$. The last property follows from the linearization of the Jacobi identity for the Poisson brackets of f_i, f_j, and Φ_σ near Λ.

Let us choose a set of functions f_i, Φ_σ from (1.3.1), (1.3.2) and introduce the following local objects associated with the pair (Λ, N).

It follows from conditions (a)–(c) and (1.3.1), (1.3.2) that the restrictions of the Hamilton vector fields $\operatorname{ad}(f_i)$ and $\operatorname{ad}(\Phi_\sigma)$ to Λ generate local trivializations (that is, local bases of sections) of the bundles $T\Lambda$ and E:

(1.3.3) $\qquad u_i = \operatorname{ad}(f_i)\big|_\Lambda \qquad (i = 1, \ldots, k),$

(1.3.4) $\qquad X_\sigma = \nu(\operatorname{ad}(\Phi_\sigma)\big|_\Lambda) \qquad (\sigma = 1, \ldots, 2r),$

where $\nu\colon T\Lambda^\perp \to E$ is the natural projection. The vector bundle morphism $\mathcal{J}\colon E \to E^*$, inducing the symplectic structure on E, is defined with respect to the local basis (1.3.4) by the matrix-valued function $J(\alpha) = ((J_{\sigma\sigma'}(\alpha)))$, $\alpha \in \Lambda$,

(1.3.5) $\qquad J_{\sigma\sigma'} = -J_{\sigma'\sigma} \stackrel{\text{def}}{=} \{\Phi_{\sigma'}, \Phi_\sigma\}\big|_\Lambda.$

It is clear that $\det J \neq 0$.

Let $\eta^1, \ldots, \eta^k$ be the basis of 1-forms dual to the local basis of vector fields $u_1, \ldots, u_k$ (1.3.9), $\eta^i(u_j) = \delta^i_j$.

We introduce a matrix-valued (local) 1-form on Λ:

(1.3.6) $\qquad \theta = \sum_{i=1}^k \theta_i \eta^i, \qquad \theta_i \stackrel{\text{def}}{=} J^{-1}\{\Phi \stackrel{\otimes}{,} \{\Phi, f_i\}\}\big|_\Lambda.$

Here we use the tensor notation $(\Phi \otimes \Psi)_{\gamma\sigma} = \Phi_\gamma \Psi_\sigma$. We also introduce a matrix-valued (local) 2-form on Λ:

(1.3.7) $\quad C = \frac{1}{2} \sum_{i,j=1}^k C_{ij} \eta^i \wedge \eta^j, \quad C_{ij} = -\sum_{s=1}^k \lambda_{ij}^s \theta_s + J^{-1}\{\Phi \stackrel{\otimes}{,} \{\Phi, \{f_i, f_j\}\}\}\big|_\Lambda,$

and the "structure" functions λ_{ij}^s on Λ are defined by the relations

(1.3.8) $\qquad [u_i, u_j] = \sum_{s=1}^k \lambda_{ij}^s u_s, \qquad \lambda_{ij}^s = \operatorname{ad}(\{f_i, f_j\}\big|_\Lambda) \lrcorner \eta^s.$

Let $C_N^\infty(M)$ denote the space of smooth functions on M that are constant along N. Since N is tangent to $T\Lambda^\perp$, the Hamilton vector field $\operatorname{ad}(H)$ of each $H \in C_N^\infty(M)$ is tangent to Λ, that is, $\operatorname{ad}(H) \in \mathcal{A}_\Lambda$. So, for each function $H \in C_N^\infty(M)$, we can consider the first variation operator $D_H\colon \Gamma(E) \to \Gamma(E)$ of the Hamilton vector field of $\operatorname{ad}(H)$ over Λ.

THEOREM 1.3.1 [**KaVo**$_{1,2}$, **Vo**$_1$]. *Let the submanifold N be coisotropic on Λ. Then there is a unique symplectic linear connection ∇ in the symplectic normal bundle E over Λ that satisfies the condition: for any $H \in C^\infty_N(M)$, the connection ∇ is adapted to the first variation operator D_H. Locally, the corresponding connection 1-form θ and the curvature 2-form C (with respect to the trivializations (1.3.3), (1.3.4)) are defined by formulas (1.3.6) and (1.3.7). The curvature of ∇ is equal to zero if and only if the Poisson bracket of any two functions $H', H'' \in C^\infty_N(M)$ has a third-order zero on Λ,*

$$(1.3.9) \qquad \{H', H''\}\big|_N = O(d_\Lambda^3) \quad \text{as } d_\Lambda \to 0.$$

Here d_Λ denotes the distance from a point on N to Λ. In general, if the curvature of ∇ is nonzero, the right-hand side of (1.3.9) is of order $O(d_\Lambda^2)$.

PROOF. The pairwise Poisson brackets of the functions $f_1, \ldots, f_k; \Phi_1, \ldots, \Phi_{2r}$ in (1.3.1), (1.3.2) satisfy the relations

$$(1.3.10) \qquad \{f_i, \Phi_\gamma\}\big|_{f=0} = \sum_{\sigma=1}^{2r} \theta^\sigma_{i\gamma} \Phi_\sigma + O(|\Phi|^2),$$

$$(1.3.11) \qquad \{f_i, f_j\}\big|_{f=0} = \frac{1}{2} \sum_{\gamma,\sigma=1}^{2r} \mathcal{R}^{\gamma\sigma}_{ij} \Phi_\gamma \Phi_\sigma + O(|\Phi|^3),$$

where θ^σ_{ij}, $\mathcal{R}^{\gamma\sigma}_{ij}$ are C^∞-functions in a neighborhood of a point in Λ. Applying the decompositions (1.3.12) and (1.3.13) to the Jacobi identities

$$(1.3.12) \qquad \underset{i,\gamma,\sigma}{\mathfrak{S}} \{f_i, \{\Phi_\gamma, \Phi_\sigma\}\} = 0, \qquad \underset{i,j,\sigma}{\mathfrak{S}} \{\{f_i, f_j\}, \Phi_\sigma\} = 0,$$

(where $\underset{i,\gamma,\sigma}{\mathfrak{S}}$ denotes the cyclic sum), we obtain the following equations for the matrix-valued functions $\theta_i = ((\theta^\sigma_{ij}))$, $\mathcal{R}_{ij} = ((\mathcal{R}^{\sigma\gamma}_{ij}))$:

$$(1.3.13) \qquad \theta_i^T J + J\theta_i - u_i(J) = 0,$$

$$(1.3.14) \qquad u_i(\theta_j) - u_j(\theta_i) - \sum_{s=1}^k \lambda^s_{ij} \theta_s + [\theta_i, \theta_j] = -\mathcal{R}_{ij} J.$$

Here we use the vector fields $u_1, \ldots, u_k$ and the matrix-valued function J defined by (1.3.3) and (1.3.5).

Comparing (1.3.14) and (1.3.7), we obtain $C_{ij} = -\mathcal{R}_{ij} J$.

So, formulas (1.3.13) and (1.3.14) can be rewritten in terms of the 1-form θ (1.3.6) and the 2-form C (1.3.7) as follows:

$$(1.3.15) \qquad d\theta + \theta \wedge \theta = C,$$
$$(1.3.16) \qquad dJ = \theta^T J + J\theta.$$

Equations (1.3.15), (1.3.16) show that locally θ defines the symplectic linear connection in E, and this connection is flat if and only if $\mathcal{R}_{ij} \equiv 0$.

Now let us consider two sets $\Phi = (\Phi_1, \ldots, \Phi_{2r})$ and $\widetilde{\Phi} = (\widetilde{\Phi}_1, \ldots, \widetilde{\Phi}_{2r})$ of C^∞-functions satisfying (1.3.1), (1.3.2) in a neighborhood of a point of Λ in M. It is clear that we have

$$\widetilde{\Phi}_\sigma = \sum_{\sigma'}^{2r} T_\sigma^{\sigma'} \Phi_{\sigma'} + O(|f|) + O(|\Phi|^2) \tag{1.3.17}$$

near Λ. Here $T = ((T_\sigma^{\sigma'}))$ is a transition matrix-valued function on Λ corresponding to two local bases $\{\widetilde{X}_\sigma = \nu(\mathrm{ad}(\widetilde{\Phi}_\sigma)|_\Lambda)\}$ and $\{X_\sigma\}$ from (1.3.4).

Relations (1.3.11), (1.3.14), and (1.3.17) give the following transformation rule between the forms $\widetilde{\theta}$, $\widetilde{C}$, and θ, C associated with the sets $\widetilde{\Phi}$ and Φ: $\widetilde{\theta} = dTT^{-1} + T\theta T^{-1}$, $\widetilde{C} = TCT^{-1}$. This shows that formula (1.3.6) defines a global connection in E.

Finally, for each function $H \in C_N^\infty(M)$ constant along N, we have

$$[\mathrm{ad}(H), \mathrm{ad}(\Phi_\sigma)]\Big|_\Lambda = \sum_{\gamma=1}^{2r} (v_H \lrcorner \theta_\sigma^\gamma) \mathrm{ad}(\Phi_\gamma)\Big|_\Lambda + (\text{a vector field on } \Lambda),$$

where $v_H = \mathrm{ad}(H)|_\Lambda$. This implies that the connection θ is adapted to the first variation operator D_H of $\mathrm{ad}(H)$. The variation matrix V_H of the Hamilton vector field $\mathrm{ad}(H)$ with respect to the basis (1.3.4) is given by

$$V_H = -J^{-1}\{\Phi\overset{\otimes}{,}\{\Phi,H\}\}\Big|_\Lambda. \tag{1.3.18}$$ □

Combining Theorem 1.3.1 and Theorem 1.2.1, we obtain the following statement.

THEOREM 1.3.2 (Reducibility criterion). *Let (M, Ω, H) be a Hamilton system, let $\Lambda \subset M$ be a compact connected isotropic submanifold, and let Λ be $\mathrm{ad}(H)$-invariant. Assume that*

(i) there exists a submanifold $N \subset M$ coisotropic on Λ and such that

$$H\big|_N = \mathrm{const} + O(d_\Lambda^3) \qquad (\text{as} \quad d_\Lambda \to 0). \tag{1.3.19}$$

Then the symplectic connection ∇ from (1.3.6), associated with N, is adapted to the first variation operator D_H of the Hamilton vector field $\mathrm{ad}(H)$. Let, in addition,

(ii) N satisfy the zero-curvature condition (1.3.9),

(iii) the torsion group $\mathrm{Tor}\, H_1(\Lambda)$ be trivial and the fundamental group $\pi_1(\Lambda)$ be commutative.

Then the linearized Hamilton flow $\mathbb{S}_H^t$ of $\mathrm{ad}(H)$ over Λ is reducible in $^\mathbb{C}E$.

Now let us state a ∇-reducibility criterion for linearized Hamilton flows over Λ in terms of the Poisson geometry.

Let a submanifold N be coisotropic on Λ. Suppose that the connection ∇ (1.3.6) generated by N is flat (that is, condition (1.3.9) holds). Then in a neighborhood of each point of $\Lambda \subset M$, there exists a set of C^∞-functions $\Phi = (\Phi_1, \ldots, \Phi_{2r})$ ($2r = \dim N - \dim \Lambda$) from (1.3.2) such that

$$[\mathrm{ad}(F), \mathrm{ad}(\Phi_\sigma)]\big|_\Lambda = 0 \qquad \forall F \in C_N^\infty(M), \quad \sigma = 1, \ldots, 2r, \tag{1.3.20}$$

$$J = ((\{\Phi_{\sigma'}, \Phi_\sigma\}|_\Lambda)) = \begin{bmatrix} 0 & -I \\ I & 0 \end{bmatrix}. \tag{1.3.21}$$

In this case, the Hamilton vector fields $\mathrm{ad}(\Phi_\sigma)$ give ∇-parallel sections defined by formula (1.3.4).

The set $\Phi = (\Phi_1, \ldots, \Phi_{2r})$ from (1.3.2), satisfying (1.3.20) and (1.3.21), is called a *parallel symplectic coordinates* over Λ associated with N.

THEOREM 1.3.3 (∇-Reducibility criterion). *Let a submanifold $N \subset M$ be coisotropic on an isotropic submanifold Λ. Suppose that the connection ∇ (1.3.6) associated with N is flat. If a function $H \in C^\infty(M)$ satisfies the following conditions*:

(a) $\{F, H\}\big|_\Lambda = 0$ *for any* $F \in C_N^\infty(M)$;

(b) *for arbitrary parallel symplectic coordinates* $\Phi = (\Phi_1, \ldots, \Phi_{2r})$ *over Λ associated with N,*

$$(1.3.22) \qquad \{\Phi_\sigma, \{\Phi_{\sigma'}, H\}\}\big|_\Lambda = \mathrm{const}, \qquad \sigma, \sigma' = 1, \ldots, 2r;$$

then Λ is $\mathrm{ad}(H)$-invariant and the first variation operator D_H of $\mathrm{ad}(H)$ over Λ is ∇-reducible. If the right hand side of (1.3.22) is equal to zero, then the connection ∇ (1.3.6) is adapted to D_H.

REMARK 1.3.1. Condition (1.3.22) does not depend on the choice of parallel symplectic coordinates Φ associated with a given submanifold N.

So, if Λ is compact and connected, then under the conditions (a) and (b) from Theorem 1.3.3, the flat connection ∇ associated with a submanifold N (that is, coisotropic on Λ) allows us to reduce the linearized Hamilton flow $\mathbb{S}_H^t$ to the following one (see §1.1):

$$\exp(t\Xi^0), \qquad \Xi^0 \stackrel{\mathrm{def}}{=} -J\{\Phi^\otimes\{\Phi, H\}\}\big|_{\alpha^0} \in \mathrm{Sp}(r, \mathbb{R}),$$

where $\alpha^0 \in \Lambda$ is a chosen point and $\Phi = (\Phi_1, \ldots, \Phi_{2r})$ are parallel symplectic coordinates over Λ associated with N.

Recall that a submanifold $N \subset M$ is called *coisotropic* if $TN^\perp \subset TN$. An equivalent condition is the following one: the space of functions $C_N^\infty(M)$ that are zero on N is the Lie algebra of the Poisson algebra $C^\infty(M)$,

$$f'\big|_N = 0, \quad f''\big|_N = 0 \quad \Rightarrow \quad \{f', f''\}\big|_N = 0.$$

A coisotropic submanifold N is called a *coisotropic extension* [**KaVo$_{1-3}$**] of an isotropic submanifold Λ if $N \supset \Lambda$ and $\mathrm{codim}\, N = \dim \Lambda$. In particular, each coisotropic extension of Λ is a coisotropic submanifold of M.

By Theorem 1.3.1, we obtain the following statement.

PROPOSITION 1.3.1. *The linear connection ∇ (1.3.6) generated by a coisotropic extension N of Λ is a flat symplectic connection in E.*

REMARK 1.3.2. The restriction of the symplectic form Ω to a given coisotropic extension N of Λ is a closed 2-form of constant rank. Thus, $\ker \Omega\big|_{TN}$ is an integrable distribution that defines the *characteristic foliation* $\mathcal{F}$ of N. Moreover, Λ is the leaf of $\mathcal{F}$ and E is the normal bundle of $\mathcal{F}$ over Λ. In terms of the foliation $\mathcal{F}$, the connection (1.3.6) can be characterized as follows:

$$\nabla_u s = \nu\big([a,b]\big|_\Lambda\big), \qquad u \in \mathrm{Vect}(\Lambda), \quad s \in \Gamma(E),$$

for all vector fields $a, b \in \mathrm{Vect}(N)$ such that a is tangent to $\mathcal{F}$, $a\big|_\Lambda = u$, and $\nu\big(b\big|_\Lambda\big) = s$ under the natural projection $\nu \colon T_\Lambda N \to E$. This shows that the connection (1.3.6) associated with a coisotropic extension N of Λ is a *Bott-type connection* [**Bo**$_{1,2}$]. For applications of Bott's connections to various situations we refer to [**Dz**$_{1-3}$, **L**$_1$, **Tab**, **We**$_1$, **Wo**].

Flat connections of type (1.3.6) play the crucial role in our further considerations. In particular, they allow us to introduce dynamical invariants associated with coisotropic extensions of isotropic submanifolds in the following way.

Let N be a coisotropic extension of a connected isotropic submanifold Λ. Let ∇ be a flat symplectic connection (1.3.6) generated by N. Choose a point $\alpha^0 \in \Lambda$. Recall that the holonomy group $\text{Hol}^{\nabla}_{\alpha^0}$ of the flat connection (1.3.6) at α^0 is the image under the homomorphism $\psi \colon \pi_1(\Lambda, \alpha^0) \mapsto \text{Sp}(E_{\alpha^0})$ (see §1.2). For each closed path $\Gamma \subset \Lambda$ (passing through α^0), we consider an element $G_{\Gamma,\alpha^0} \in \text{Hol}^{\nabla}_{\alpha^0}$ corresponding to the parallel transport of ∇ around Γ. Then the spectrum of G_{Γ,α^0} (with its multiplicity) is independent of a choice of Γ from a given homotopy class $[\Gamma] \in \pi_1(\Lambda)$ and of a point α^0,

$$\operatorname{spec} G_{\Gamma,\alpha^0} = \{\lambda^1(\Gamma), \ldots, \lambda^{2r}(\Gamma)\}, \quad \lambda^\sigma(\Gamma) \in \mathbb{C}, \quad (\operatorname{codim} N = \tfrac{1}{2}\dim M - r),$$
$$[\Gamma] = [\Gamma'] \in \pi_1(\Lambda) \implies \operatorname{spec} G_{\Gamma,\alpha^0} = \operatorname{spec} G_{\Gamma',\alpha^0}.$$

DEFINITION 1.3.1. Let N be a coisotropic extension of an isotropic submanifold Λ. The eigenvalues $\lambda^1(\Gamma), \ldots, \lambda^{2r}(\Gamma)$ of the symplectic operator $G_{\Gamma,\alpha^0} \in \text{Hol}^{\nabla}_{\alpha^0}$ are called *characteristic multipliers* of the characteristic distribution of the coisotropic extension $N \supset \Lambda$ at a closed path $\Gamma \subset \Lambda$.

For brevity, we omit the words "characteristic distribution" in Definition 1.3.1 and speak of the characteristic multipliers of a coisotropic extension $N \supset \Lambda$ at Γ.

In the case $\Lambda \approx \mathbb{S}^1$, $\operatorname{codim} N = 1$, Definition 1.3.1 coincides with the standard notion of the characteristic multipliers of a Hamilton vector field at a periodic orbit Λ lying in a regular energy surface N (see [**AbMar**]).

If the holonomy group $\text{Hol}^{\nabla}_{\alpha^0}$ of the flat connection ∇ (1.3.6) is commutative (for instance, $\pi_1(\Lambda)$ is commutative), then the characteristic multipliers $\lambda^\sigma(\Gamma)$ of a coisotropic extension N depend only on the *homology class* $[\Gamma] \in H_1(\Lambda)$ of a closed path $\Gamma \subset \Lambda$.

EXAMPLE 1.3.1 (*Partially integrable systems*). Let $\mathcal{Y}$ be a $2k$-dimensional symplectic submanifold of a symplectic manifold (M, Ω), that is, $\omega = \Omega\big|_{\mathcal{Y}}$ is nondegenerate. Then the restriction of the tangent bundle to $\mathcal{Y}$ splits into the skew-orthogonal direct sum of symplectic subbundles

(1.3.23) $$T_{\mathcal{Y}} M = T\mathcal{Y} \oplus T\mathcal{Y}^\perp.$$

Assume that there are C^∞-functions $f_1 \ldots, f_k$ on M, satisfying the following two conditions:

(i) $f_1 \ldots, f_k$ are functionally independent in a neighborhood of $\mathcal{Y}$ in M;
(ii) the pairwise Poisson brackets of f_j vanish along $\mathcal{Y}$,

(1.3.24) $$\{f_i, f_j\}\big|_{\mathcal{Y}} = 0 \quad (i, j = 1, \ldots, k).$$

Then $\mathcal{Y}$ is foliated by k-dimensional *Lagrangian* submanifolds:

(1.3.25) $$\Lambda_a = \{f_1 = a_1, \ldots, f_k = a_k\} \cap \mathcal{Y},$$

where $a = (a_1, \ldots, a_k) \in \mathbb{R}^k$. On the other hand, Λ_a considered as a submanifold of M is *isotropic*. The symplectic normal bundle over Λ_a is identified with $T_{\Lambda_a}\mathcal{Y}^\perp$, $E_a \approx T_{\Lambda_a}\mathcal{Y}^\perp$.

Moreover, the level set $N_a = \{f_1 = a_1, \ldots, f_k = a_k\} \supset \Lambda_a$ is a submanifold of M that is coisotropic on Λ_a. Hence, by Theorem 1.3.1, N_a induces the symplectic linear connection ∇^a (1.3.6) in E_a. The connection ∇^a is flat for each a if and only if the pairwise Poisson brackets of f_j satisfy the following condition near $\mathcal{Y}$:

  (iii) in a neighborhood of $\mathcal{Y}$ in M,

$$(1.3.26) \qquad \{f_i, f_j\} = O(d_\mathcal{Y}^3) \qquad (d_\mathcal{Y} \to 0),$$

where $d_\mathcal{Y}$ denotes the distance from a point to $\mathcal{Y}$.

We say that a Hamilton system (M, Ω, H) is *partially integrable* over a symplectic submanifold $\mathcal{Y} \subset M$ if there are functions $f_1, \ldots, f_k \in C^\infty(M)$ (infinitesimal integrals of motion) satisfying the above conditions (i)–(iii), and if the following condition holds as $d_\mathcal{Y} \to 0$:

  (iv) $\{H, f_j\} = O(d_\mathcal{Y}^3)$, $j = 1, \ldots, k$.

Assume also that

  (v) for a certain $a^0 \in \mathbb{R}^k$, the submanifold Λ_{a^0} is *compact* and *connected*.

Then under conditions (i)–(v), from the Liouville–Arnold theorem [**A**$_3$] (see also [**Jos**]) and Theorem 1.3.2, we obtain the following three facts.

  (a) There is a neighborhood U of a^0 in $\mathbb{R}^k$ such that for each $a \in U$, the submanifold Λ_a is diffeomorphic to the torus $\mathbb{T}^k$ and $\mathrm{ad}(H)$-invariant. The flow of $\mathrm{ad}(H)$ on Λ is quasiperiodic.

  (b) For each $a \in U$ the linearized Hamilton flow $\mathbb{S}_H^t$ of $\mathrm{ad}(H)$ over Λ_a is *reducible* in $^\mathbb{C}E_a$.

  (c) A certain neighborhood B of Λ_{a^0} in $\mathcal{Y}$ is diffeomorphic to the direct product $B \approx \mathbb{T}^k \times U$, where U satisfies properties (a), (b). Moreover, in B there are action-angle coordinates $s = (s^1, \ldots, s^k)$, $\alpha = (\alpha_1, \ldots, \alpha_k)$ ($\alpha_j \pmod{2\pi}$),

$$\omega = \sum_{i=1}^k ds^i \wedge d\alpha_i, \qquad \Lambda_a = \{s^1 = a^1, \ldots, s^k = a^k\}, \quad a \in U.$$

EXAMPLE 1.3.2 (*Systems with Poisson symmetry*). Let (M, Ω) be a symplectic manifold. Assume that we have two Poisson mappings

$$M \xrightarrow{A} P, \quad A = (A^1, \ldots, A^p); \qquad \{A^i, A^j\} = \Psi^{ij}(A);$$
$$M \xrightarrow{B} Q, \quad B = (B^1, \ldots, B^q); \qquad \{B^i, B^j\} = \Phi^{ij}(A),$$

where $\Psi = \frac{1}{2}\Psi^{ij} \frac{\partial}{\partial \xi^i} \wedge \frac{\partial}{\partial \xi^j}$ and $\Phi = \frac{1}{2}\Phi^{ij} \frac{\partial}{\partial \zeta^i} \wedge \frac{\partial}{\partial \zeta^j}$ are Poisson tensors on manifolds P ($\dim P = p$) and Q ($\dim Q = q$).

Suppose also that the following conditions hold:

  (i) $p + q = \dim M$;
  (ii) $\{A^i, B^j\} = 0$ ($i = 1, \ldots, p; j = 1, \ldots, q$);
  (iii) A and B are *surjective submersions*.

Conditions (i), (ii) mean that the tangent bundles over the fibers of B are *skew-orthogonal complements* to the tangent bundles over the fibers of A in TM, that is, B is *polar* to A [**KaMas**].

Choose a point $m^0 \in M$ and consider the subset

$$\Lambda_{m^0} \stackrel{\text{def}}{=} \{m \in M \colon A(m) = A(m^0), B(m) = B(m^0)\}.$$

It follows from (i)–(iii) that Λ_{m^0} is an *isotropic* submanifold of M and

$$\dim \Lambda_{m^0} = p - \operatorname{rank} \Psi(\xi^0) \equiv q - \operatorname{rank} \Phi(\zeta^0).$$

Here $\xi^0 = A(m^0)$ and $\zeta^0 = B(m^0)$. Let $\mathcal{O}_{\xi^0}$ and $\widetilde{\mathcal{O}}_{\zeta^0}$ denote the symplectic leaves in P and Q passing through the points ξ^0 and ζ^0, respectively. Assume that $\mathcal{O}_{\xi^0}$ and $\widetilde{\mathcal{O}}_{\zeta^0}$ are submanifolds of P and Q. Then it follows from conditions (i)–(iii) that $\operatorname{codim} \mathcal{O}_{\xi^0} = \operatorname{codim} \widetilde{\mathcal{O}}_{\zeta^0}$ and

$$N_{m^0} \stackrel{\text{def}}{=} A^{-1}(\mathcal{O}_{\xi^0}) \equiv B^{-1}(\widetilde{\mathcal{O}}_{\zeta^0})$$

is a *coisotropic* submanifold in M containing Λ_{m^0}.

So, by Proposition 1.3.1, we have that for each $m^0 \in M$ the pair (Λ_{m^0}, N_{m^0}) generates a *flat symplectic* connection ∇^{m^0} (1.3.6) in the symplectic normal bundle over Λ_{m^0}. Note that there is a global symplectic basis of ∇_{m^0}-parallel sections of $E_{\Lambda_{m^0}}$. Now let us discuss the structure of Λ_{m^0} in more detail. Consider the subspace

$$\mathfrak{g}_{\xi^0} \stackrel{\text{def}}{=} \ker \Psi(\xi^0) \subset T^*_{\xi^0} P.$$

As is well known (see, for example, [**KaMas**]), the Poisson tensor Ψ on P induces a *Lie algebra structure* on $\mathfrak{g}_{\xi^0}$, which can be described as follows. In a neighborhood W of $\xi^0 \in P$, there exists a set of C^∞-functions $k^1, \ldots, k^r$ such that, locally, $\mathcal{O}_{\xi^0}$ is the zero-level set,

$$\mathcal{O}_{\xi^0} = \{k^1 = 0, \ldots, k^r = 0\}, \qquad r = \operatorname{codim} \mathcal{O}_{\xi^0}.$$

This implies

(1.3.27) $$\{k^i, k^j\}\big|_{\mathcal{O}_{\xi^0}} = 0.$$

The property (1.3.27) and the Jacobi identity for Poisson brackets imply

$$[\operatorname{ad}(k^i), \operatorname{ad}(k^j)]\big|_{\xi^0} = \sum_{s=1}^r \lambda^{ij}_s \operatorname{ad}(k^s)\big|_{\xi^0},$$

where λ^{ij} are the structure constants of the Lie algebra $\mathfrak{g}$,

$$\underset{(i,j,l)}{\mathfrak{S}} \lambda^{ij}_s \lambda^{sl}_m = 0.$$

In terms of the Poisson bracket on P, the constants λ^{ij}_s can be written as

$$\lambda^{ij}_s = v_s(\{k^i, k^j\})(\xi^0),$$

where $v_1, \ldots, v_r$ are the vector fields on W dual to $dk^1, \ldots, dk^r$: $v_i(k^j) = \delta^j_i$. So, the Lie algebra structure on $\mathfrak{g}_{\xi^0}$ with respect to the basis $dk^1(\xi^0), \ldots, dk^r(\xi^0)$ is determined by the structure constants λ^{ij}_s.

By using the pull-back of the functions $k^1, \ldots, k^r$ to M via the submersion A, we can represent the coisotropic submanifold N_{m^0} as the zero-level set,

$$N_{m^0} = \{F^1 = 0, \ldots, F^r = 0\}, \qquad F^i \stackrel{\text{def}}{=} A^*(k^i).$$

Thus, the vector fields $u^i \stackrel{\text{def}}{=} \text{ad}\,(A^*(k^i))\big|_{\Lambda_{m^0}}$ span the tangent space $T_m \Lambda_{m^0}$ at each point m of Λ_{m^0}, and

(1.3.28) $$[u^i, u^j] = \sum_{s=1}^{r} \lambda_s^{ij} u^s.$$

Assume that Λ_{m^0} is *compact* and *connected*. Then it follows from (1.3.28) that $\Lambda_{m^0} \approx G/G_0$, where G is the connected Lie group corresponding to the Lie algebra $\mathfrak{g}$ (1.3.35) and G_0 is a discrete subgroup of G (see [**GHV**]). If $\lambda_s^{ij} = 0$, that is, $\mathfrak{g}$ is commutative (in particular, this is true if the symplectic leaf $\mathcal{O}_{\xi^0}$ is *regular*), then Λ_{m^0} is diffeomorphic to the torus $\mathbb{T}^r$.

Suppose that a Hamilton function $H \in C^\infty(M)$ satisfies the condition

(iv) $\quad \{H, A^i\}(m^0) = 0 \quad (i = 1, \ldots, p), \qquad \{H, B^j\}(m^0) = 0 \quad (i = 1, \ldots, q).$

Then the isotropic submanifold $\Lambda_{m^0} \subset M$ is $\text{ad}(H)$-invariant. Moreover, if

(v) $$H = (A \times B)^* f, \qquad f \in C^\infty(P \times Q),$$

then the first variation operator of $\text{ad}(H)$ over Λ_{m^0} is ∇_{m^0}-reducible in $E_{\Lambda_{m^0}}$.

§1.4. Invariant symplectic normal subbundles

Let (M, Ω) be a symplectic manifold. Assume that we have an $\text{ad}(H)$-invariant isotropic submanifold $\Lambda \subset M$, that is, $\text{ad}(H) \subset \mathcal{A}_\Lambda(M)$, $H \in C^\infty(M)$.

Recall that the isotropic embedding $\Lambda \subset M$ induces the following objects:
- the symplectic bundle $T_\Lambda M$ over Λ;
- the isotropic tangent bundle $T\Lambda$;
- the coisotropic bundle $T\Lambda^\perp$ (the skew-orthogonal complement to $T\Lambda$);

$$T\Lambda \subset T\Lambda^\perp \subset T_\Lambda M, \qquad \dim T\Lambda + \dim T\Lambda^\perp = \dim M,$$

- the symplectic normal bundle $E = T\Lambda^\perp / T\Lambda$ over Λ.

Let $W \subset T_\Lambda M$ be a subbundle. By $\text{Vect}_W(M)$ we denote the space of all vector fields on M that are tangent to W at each point of Λ.

We say that a subbundle $W \subset T_\Lambda M$ is $\text{ad}(H)$-*invariant* if

$$w \in \text{Vect}_W(M) \Rightarrow [\text{ad}(H), w] \in \text{Vect}_W(M).$$

In other words, the $\text{ad}(H)$-invariance of W means that the subbundle W is invariant with respect to the *first variation operator* of $\text{ad}(H)$ in $T_\Lambda M$, which is viewed as a covariant differential operator acting in $\Gamma(T_\Lambda M)$.

By assumption, the Hamilton vector field $\text{ad}(H)$ is tangent to $T\Lambda$, and hence, to $T\Lambda^\perp$. So, the subbundles $T\Lambda$ and $T\Lambda^\perp$ are $\text{ad}(H)$-invariant. We are interested in the existence of an $\text{ad}(H)$-*invariant splitting* of $T\Lambda^\perp$,

(1.4.1) $\qquad T\Lambda^\perp = T\Lambda \oplus W,$
(1.4.2) $\qquad W \quad \text{is ad}(H)\text{-invariant}.$

It follows from (1.4.1) that the restriction of the symplectic form Ω to W is nondegenerate, that is, W is a symplectic subbundle of $T_\Lambda M$.

The subbundle W satisfying (1.4.1), (1.4.2) will be called an *invariant symplectic normal subbundle* over Λ. Note that it follows from (1.4.1), (1.4.2) that there is an $\text{ad}(H)$-invariant splitting of the original symplectic bundle:

$$(1.4.3) \qquad T_\Lambda M = W^\perp \oplus W,$$

where $W^\perp$ is the skew-orthogonal complement to W that is also symplectic.

REMARK 1.4.1. Let $(M, \Omega, H; \Lambda \subset \mathcal{Y})$ be a *partially integrable* Hamilton system near an invariant isotropic submanifold Λ (see Example 1.3.1). Here $\mathcal{Y}$ is an $\text{ad}(H)$-invariant symplectic submanifold. If we set $W = (T_\Lambda, \mathcal{Y})^\perp$, then conditions (1.4.2), (1.4.3) hold automatically. So, possible obstructions to the existence of an invariant symplectic normal subbundle over Λ are infinitesimal obstructions to the partial integrability of a given Hamilton system.

We describe obstructions to the existence of an invariant symplectic normal subbundle over Λ in the following situation.

By the $\text{ad}(H)$-invariance of Λ, we have $H\big|_\Lambda = \text{const}$. Assume that $H\big|_\Lambda = 0$ and in a neighborhood of Λ in M, there is a set of smooth functions $f_1 = H, \ldots, f_k$ including the Hamilton function H and satisfying the following conditions:

(i) the functions $f_1, \ldots, f_k$ are functionally independent in a neighborhood of the zero-level set

$$N = \{f_1 = 0, \ldots, f_k = 0\},$$

that is, N is a submanifold of M;

(ii) $\Lambda \subset N$ and $\dim \Lambda = k$;

(iii) the pairwise Poisson brackets of f_j vanish along Λ,

$$\{f_i, f_j\}\big|_\Lambda = 0 \qquad (i, j = 1, \ldots, k).$$

This implies that the submanifold N is coisotropic on Λ, and Λ is parallelizable,

$$T_\alpha \Lambda = \text{span}\{u_1(\alpha), \ldots, u_k(\alpha)\} \qquad \forall \alpha \in \Lambda,$$

where

$$(1.4.4) \qquad u_j \stackrel{\text{def}}{=} \text{ad}(f_j)\big|_\Lambda.$$

Moreover, since $\text{ad}(f_i) \in \mathcal{A}_\Lambda(M)$, we can define the *first variation operator* $\widetilde{D}_{f_i}$: $\Gamma(T\Lambda^\perp) \to \Gamma(T\Lambda^\perp)$ of the Hamilton vector field $\text{ad}(f_i)$ in $T\Lambda^\perp$ for all $i = 1, \ldots, k$ (see §1.1).

A realization of such situation is given by the following proposition.

PROPOSITION 1.4.1. *If an isotropic submanifold $\Lambda \subset M$ is parallelizable, then in a neighborhood of Λ in M, there exists a set of functions $f_1, \ldots, f_k$ satisfying conditions* (i)–(iii). *If, in addition, Λ is $\text{ad}(H)$-invariant ($H\big|_\Lambda = 0$), then the Hamilton vector field $\text{ad}(H)$ has no singular points on Λ and the characteristic line bundle $\mathcal{L}_H$ of $\text{ad}(H)\big|_\Lambda$ satisfies the following property: if the quotient bundle $T\Lambda/\mathcal{L}_H$ is trivial, then the set $f_1, \ldots, f_k$ can be chosen so that $f_1 = H$.*

Let us choose a set of functions $f_1, \ldots, f_k$ satisfying conditions (i)–(iii) and a splitting (1.4.1), where $W \subset T\Lambda^\perp$ is a certain symplectic normal subbundle over Λ. In a coordinate neighborhood $\mathcal{U}$ of each point of Λ, let us choose a set of functions $\Phi_1, \ldots, \Phi_{2r}$ such that

$$(1.4.5) \qquad \Lambda = N \cap \{\Phi_1 = 0, \ldots, \Phi_{2r} = 0\}.$$

Thus, the sections

(1.4.6) $$u_1, \ldots, u_k, \qquad X = \mathrm{ad}(\Phi_1)\big|_\Lambda, \ldots, X_{2r} = \mathrm{ad}(\Phi_{2r})\big|_\Lambda$$

give a local trivialization of $T\Lambda^\perp$. Near Λ, the pairwise Poisson brackets of f_i and Φ_σ have the form (see §1.3)

$$\{f_i, \Phi_\sigma\} = \theta_{i\sigma}^{\sigma'} \Phi_{\sigma'} + \mathcal{T}_{i\sigma}^j f_j + O(|f|^2 + |\Phi|^2),$$
$$\{f_i, f_j\} = \lambda_{ij}^s f_s + O(|f|^2 + |\Phi|^2),$$

where $\theta_{i\sigma}^{\sigma'}$, $\mathcal{T}_{i\sigma}^j$, and λ_{ij}^s are (local) functions on Λ. Here and in the sequel, the sum is taken over repeated indices. Introduce the set of matrix-valued 1-forms on Λ:

(1.4.7) $$\lambda \circ \eta = ((\lambda_{sj}^i \eta^s)),$$

(1.4.8) $$\theta = ((\theta_{i\sigma}^{\sigma'} \eta^i)),$$

(1.4.9) $$\mathcal{T} \circ \eta = ((\mathcal{T}_{i\sigma}^j \eta^i)).$$

Here $\eta^1, \ldots, \eta^k$ are 1-forms on Λ dual to $u_1, \ldots, u_k$.

THEOREM 1.4.1. *The set of functions $f_1, \ldots, f_k$ satisfying the above conditions* (i)–(iii) *generates a linear connection $\widetilde{\nabla}$ in $T\Lambda^\perp$ defined by the relations*

(1.4.10) $$\widetilde{\nabla}_{u_i} = \widetilde{D}_{f_i} \qquad (i = 1, \ldots, k),$$

where $\widetilde{D}_{f_i}$ is the first variation operator of $\mathrm{ad}(f_i)$ in $T\Lambda^\perp$. The connection 1-form $\widetilde{\theta}$ of $\widetilde{\nabla}$ with respect to the trivialization (1.4.6) *can be written as*

(1.4.11) $$\widetilde{\theta} = \begin{bmatrix} \lambda \circ \eta & \mathcal{T} \circ \eta \\ 0 & \theta \end{bmatrix}.$$

The connection $\widetilde{\nabla}$ is flat if and only if the following conditions hold:

(1.4.12) $$\lambda_{ij}^s = \mathrm{const},$$

(1.4.13) $$\{f_i, f_j\} = \lambda_{ij}^s f_s + O(|f|^2 + |f||\Phi|^2 + |\Phi|^3).$$

Then the matrix-valued 1-forms from (1.4.7)–(1.4.9) *satisfy the zero-curvature equations*

$$d(\lambda \circ \eta) + \lambda \circ \eta \wedge \lambda \circ \eta = 0,$$
$$d(\mathcal{T} \circ \eta) + \lambda \circ \eta \wedge \mathcal{T} \circ \eta + \mathcal{T} \circ \eta \wedge \theta = 0,$$
$$d\theta + \theta \wedge \theta = 0.$$

REMARK 1.4.2. The subbundle $T\Lambda \subset T\Lambda^\perp$ is $\widetilde{\nabla}$-invariant. The restriction $\widetilde{\nabla}\big|_{T\Lambda}$ is flat if and only if conditions (1.4.12) are satisfied. Moreover, (1.4.12) implies that the vector fields $u_1, \ldots, u_k$ (1.4.4) form the Lie algebra $\mathfrak{g}$: $[u_i, u_j] = \sum_s \lambda_{ij}^s u_s$. Here λ_{ij}^s are the structure constants of $\mathfrak{g}$. If Λ is compact, then $\Lambda \approx G/G_0$, where G is the connected Lie group corresponding to $\mathfrak{g}$ and G_0 is a discrete subgroup of G.

REMARK 1.4.3. The subbundle W from (1.4.1) is isomorphic to the symplectic normal bundle E over Λ. Under this isomorphism, the restriction $\widetilde{\nabla}|_W$ gives the symplectic connection ∇ (1.3.6) in E associated with the submanifold $N = \{f_1 = 0, \ldots, f_k = 0\}$ that is coisotropic on Λ (see Theorem 1.3.1). So, condition (1.4.13) means that ∇ is flat.

Below we assume that conditions (1.4.12) and (1.4.13) hold. So, we have a flat linear connection $\widetilde{\nabla}$ (1.4.10) in $T\Lambda^\perp$. By $A^{q,p} = A^q(\Lambda, \wedge^p T\Lambda^\perp)$ we denote the space of q-forms on Λ with values in sections of the pth exterior power of the coisotropic subbundle $T\Lambda^\perp$. In particular, $A^{q,0} = A^q(\Lambda)$ is the space of q-forms on Λ.

Then the flat connection $\widetilde{\nabla}$ generates the *coboundary operator* $\partial_{\widetilde{\nabla}} \colon A^{q,p} \to A^{q+1,p}$, $\partial_{\widetilde{\nabla}}^2 = 0$, (locally) defined by the formula

$$(1.4.14) \qquad (\partial_{\widetilde{\nabla}} T)^{I_1,\ldots,I_p} = dT^{I_1,\ldots,I_p} + \sum_{(I_1,\ldots,I_p)} \widetilde{\theta}^{I_1}_J \wedge T^{JI_2,\ldots,I_p},$$

where $T = ((T^{I_1,\ldots,I_p})) \in A^{q,p}$ is a tensor field whose components $T^{I_1,\ldots,I_p}$ with respect to the trivialization (1.4.6) are q-forms on Λ, and $\widetilde{\theta} = ((\widetilde{\theta}^I_J))$ is the connection 1-form (1.4.11).

Using the matrix-valued 1-form (1.4.9), we define

$$(1.4.15) \qquad \boldsymbol{\tau} = ((\tau^{\sigma j})) \stackrel{\text{def}}{=} J\mathcal{T} \circ \eta = ((J^{\sigma\sigma'} T^j_{i\sigma'} \eta^i)),$$

where $J = ((J^{\sigma\gamma})) = ((\{\Phi^\gamma, \Phi^\sigma\}|_\Lambda))$.

LEMMA 1.4.1. *Formula* (1.4.15) *defines an element* $\boldsymbol{\tau} \in A^{1,1}$; *that is*, $\boldsymbol{\tau}$ *is a 1-form on* Λ *with values in* $\Gamma(T\Lambda^\perp)$. *Conditions* (1.4.12) *and* (1.4.13) *imply that* $\boldsymbol{\tau}$ *is a cocycle*, $\partial_{\widetilde{\nabla}} \boldsymbol{\tau} = 0$, *or equivalently, in a local notation,*

$$d\tau^{\sigma i} + \theta^\sigma_{\sigma'} \wedge \tau^{\sigma' i} + (\lambda \circ \eta)^i_j \wedge \tau^{\sigma j} = 0.$$

Here $\lambda \circ \eta$ *and* $\theta = ((\theta^\sigma_{\sigma'}))$ *are matrix-valued 1-forms from* (1.4.7) *and* (1.4.8). *The cohomology class* $[\boldsymbol{\tau}]$ *does not depend on the choice of a symplectic normal subbundle* W *from* (1.4.1).

The element $\boldsymbol{\tau}$ will be called the *torsion* of the pair $(f = (f_1, \ldots, f_k), W)$. If the torsion is equal to zero, then W is $\mathrm{ad}(f_i)$-invariant for all $i = 1, \ldots, k$.

Summarizing the above considerations, we obtain the following statement.

THEOREM 1.4.2. *Let* $(f = (f_1, \ldots, f_k), W)$ *be a set of functions satisfying conditions* (1.4.12), (1.4.13), *and let* W *be a symplectic normal subbundle. Then there exists an* $\mathrm{ad}(f_i)$*-invariant symplectic normal subbundle* $\widetilde{W}$ $(i = 1, \ldots, k)$ *if and only if the* $\partial_{\widetilde{\nabla}}$*-cohomology class of the torsion of the pair* (f, W) *vanishes,*

$$(1.4.16) \qquad [\boldsymbol{\tau}] = 0.$$

In this case the cocycle $\boldsymbol{\tau}$ *is a coboundary,* $\boldsymbol{\tau} = \partial_{\widetilde{\nabla}} b$, *or locally,*

$$(1.4.17) \qquad \tau^{\sigma i} = db^{\sigma i} + \theta^\sigma_{\sigma'} b^{\sigma' i} + (\lambda \circ \eta)^i_j b^{\sigma j},$$

where $b = ((b^{\sigma i})) \in A^{0,1}$ *is a primitive. The Hamilton vector fields of the functions* $\widetilde{\Phi}_\sigma = \Phi_\sigma - J_{\sigma\sigma'} b^{\sigma' i} f_i$ *(here* Φ_σ *satisfy* (1.4.5)*) give a local trivialization of* $\widetilde{W}$.

COROLLARY 1.4.1. *If Λ is a compact connected submanifold and*

(1.4.18) $$\lambda_{ij}^s = 0,$$

then $\Lambda \approx \mathbb{T}^k$ is a k-torus. In this case, condition (1.4.16) can be rewritten as

(1.4.19) $$[\tau^{(1)}] = \cdots = [\tau^{(k)}] = 0.$$

Here $\tau^{(i)} = ((\tau^{\sigma i})) \in A^{1,1}(\Lambda, E)$ is the set of cocycles of the coboundary operator ∂_∇ associated with the flat symplectic connection ∇ (1.3.6) in the normal symplectic bundle E over Λ.

If (1.4.18) holds, conditions (1.4.19) can be formulated more explicitly.

The case $\Lambda \approx \mathbb{T}^k$. Let $\Lambda \subset M$ be an isotropic k-torus,

$$\Lambda \approx \mathbb{T}^k = \{0 \leq \alpha^1 \leq 2\pi\} \times \cdots \times \{0 \leq \alpha^k \leq 2\pi\},$$

where $\alpha = (\alpha^1, \ldots, \alpha^k)$ are the angle coordinates. Let $N \supset \Lambda$ be a coisotropic extension of Λ in M. Then there exist functions $f_1, \ldots, f_k$ that are constant along N and satisfy conditions (i)–(iii). In the symplectic normal bundle E over Λ, we consider the flat symplectic connection ∇ (1.3.6) corresponding to N.

Let $\alpha^0 = (0, \ldots, 0) \in \Lambda$. As we know, the holonomy group $\mathrm{Hol}^\nabla_{\alpha^0} \subset \mathrm{Sp}(E_{\alpha^0})$ of ∇ at α^0 is commutative. Moreover, the spectra of elements $G \in \mathrm{Hol}^\nabla_{\alpha^0}$ do not depend on the choice of α^0.

We say that a coisotropic extension N of Λ in M is *nondegenerate* if there exists an element $G \in \mathrm{Hol}^\nabla_{\alpha^0}$ whose spectrum does not contain 1. In other words, there is a 1-cycle $\Gamma \subset \Lambda$ such that none of the characteristic multipliers $\lambda^1(\Gamma), \ldots, \lambda^{2r}(\Gamma)$ of N at Γ is equal to 1.

In particular, if $\Lambda \approx \mathbb{S}^1$ is a periodic orbit of a Hamilton vector field $\mathrm{ad}(H)$ and $N = \{H = \mathrm{const}\}$ is the energy surface, then Λ is said to be a *nondegenerate periodic orbit*.

THEOREM 1.4.3 (Infinitesimal version of the Poincaré theorem). *Let $N = \{f_1 = 0, \ldots, f_k = 0\}$ be a nondegenerate coisotropic extension of an isotropic k-torus Λ. Then there exists a unique symplectic normal subbundle $W \subset T\Lambda^\perp$ such that*

$$W \text{ is } \mathrm{ad}(f_i)\text{-invariant for all } i = 1, \ldots, k.$$

Thus, condition (1.4.19) is satisfied.

To prove this statement, we need the following algebraic fact. On a $(k+m)$-vector space $\mathcal{V}$, let us consider a set $\mathbf{K} = \mathbf{K}(k; GL(m))$ of linear nonsingular operators such that the restriction of each of them to a given k-vector subspace $\mathcal{L} \subset \mathcal{V}$ is the identity operator. It is easy to see that $\mathbf{K}$ is a matrix Lie group consisting of the matrices

(1.4.20) $$\mathcal{M} = \begin{pmatrix} I & B \\ 0 & G \end{pmatrix},$$

where I is the $k \times k$ identity matrix, $G \in GL(m)$, and B is a $k \times m$ matrix. By $\mathbf{K}_0$ we denote the subgroup of $\mathbf{K}$ generated by the elements (1.4.20) with $B = 0$. Note that for a given $\mathcal{M} \in \mathbf{K}$ satisfying the (nondegeneracy) condition

(1.4.21) $$(G - I) \text{ is invertible,}$$

the set of matrices $U \in \mathbf{K}$ such that

$$U \mathcal{M} U^{-1} \in \mathbf{K}_0$$

is described as follows:

$$U = \begin{pmatrix} I & B(I-G)^{-1} \\ 0 & S \end{pmatrix},$$

where S runs over $GL(m)$. This implies the following fact: if a linear operator $\mathcal{M} \in \mathbf{K}$ satisfies (1.4.21), then there exists a unique $\mathcal{M}$-invariant complement to $\mathcal{L}$ in $\mathcal{V}$. Moreover, for any $\mathcal{M}' \in \mathbf{K}$ commuting with $\mathcal{M}$, the $\mathcal{M}$-invariant complement to $\mathcal{L}$ is also $\mathcal{M}'$-invariant. This property follows from the formula

$$[\mathcal{M}, \mathcal{M}'] = \begin{pmatrix} I & B(G'-I) - B'(G-I) \\ 0 & [G, G'] \end{pmatrix}.$$

PROOF OF THEOREM 1.4.3. In $T\Lambda^\perp$, let us consider the flat connection $\widetilde{\nabla}$ (1.4.10) associated with the coisotropic extension $N = \{f_1 = 0, \ldots, f_k = 0\}$ of Λ. Since the vector fields $\operatorname{ad}(f_i)\big|_\Lambda$ ($i = 1, \ldots, k$) are pairwise commuting, we can choose the angle coordinates $\alpha = (\alpha^1, \ldots, \alpha^k)$ on Λ so that $\partial/\partial \alpha^i$ are $\widetilde{\nabla}$-parallel for all $i = 1, \ldots, k$.

Let us choose the basic 1-cycles $\Gamma_1, \ldots, \Gamma_k \subset \Lambda$ as follows:

$$\Gamma_j = \{\alpha^1 = 0\} \times \cdots \times \{0 \leq \alpha^j \leq 2\pi\} \times \cdots \times \{\alpha^k = 0\}.$$

Let $\operatorname{Hol}_{\alpha^0}^{\widetilde{\nabla}}$ be the holonomy group of $\widetilde{\nabla}$ at α^0. Consider the generators $\mathcal{M}_1, \ldots, \mathcal{M}_k$ of $\operatorname{Hol}_{\alpha^0}^{\widetilde{\nabla}}$ that correspond to the basis of 1-cycles $\Gamma_1, \ldots, \Gamma_k$. Then we have the following properties:

(a) $[\mathcal{M}_i, \mathcal{M}_j] = 0$ for all $i, j = 1, \ldots, k$.
(b) $\mathcal{M}_i$ is the identity operator on $T_{\alpha^0}\Lambda$ for all $i = 1, \ldots, k$.
(c) With respect to a certain basis of $T_{\alpha^0}\Lambda^\perp$, each $\mathcal{M}_i$ has the form (1.4.32), where $G_1, \ldots, G_k$ are just commuting generators of the holonomy group $\operatorname{Hol}_{\alpha^0}^{\nabla}$ at α^0; here ∇ is the flat connection (1.3.6) generated by N.
(d) At least one of the generators $G_1, \ldots, G_k$ satisfies the nondegeneracy condition (1.4.21).

This and the cited algebraic fact imply that there exists a unique complement W_0 to $T_{\alpha^0}\Lambda$ in $T_{\alpha^0}\Lambda^\perp$ such that W_0 is $\mathcal{M}_i$-invariant for all $i = 1, \ldots, k$.

Taking into account that the connection $\widetilde{\nabla}$ is compatible with the restriction ω of the symplectic form Ω to $T\Lambda^\perp$, we define the desired $\widetilde{\nabla}$-invariant symplectic normal subbundle $W \subset T\Lambda^\perp$ as $W_\alpha = \widetilde{g}_{\alpha^0 \to \alpha}(W_0)$. Here $\widetilde{g}_{\alpha^0 \to \alpha}$ is the $\widetilde{\nabla}$-parallel transport along a path from α^0 to $\alpha \in \Lambda$. $\square$

REMARK 1.4.4. In addition to the assumptions of Theorem 1.4.3, assume that the symplectic normal bundle E over Λ is trivial, $E \approx \mathbb{T}^k \times \mathbb{R}^{2r}$. Then the solvability of (1.4.17) is equivalent to the solvability of the following equation for a global smooth vector-function $b(\alpha) = (b^1(\alpha), \ldots, b^{2r}(\alpha))$ on Λ:

(1.4.22) $$db + \theta b = \tau.$$

Here the vector-valued 1-form $\tau = (\tau^1, \ldots, \tau^{2r})$ on Λ and the matrix-valued connection 1-form $\theta = ((\theta^\sigma_{\sigma'}))$ satisfy the equations

(1.4.23) $\quad\quad\quad d\tau + \theta \wedge \tau = 0 \quad$ (cocycle condition),

(1.4.24) $\quad\quad\quad d\theta + \theta \wedge \theta = 0 \quad$ (zero-curvature equation).

Note that the local solvability of (1.4.22) follows from (1.4.23) and (1.4.24). By (1.4.24), on the universal covering $\mathbb{R}^k \to \Lambda \approx \mathbb{R}^k/2\pi\mathbb{Z}^k$, there exists a unique matrix-valued solution $g = ((g^\sigma_{\sigma'}(\alpha)))$ $(\alpha = (\alpha^1, \ldots, \alpha^k) \in \mathbb{R}^k)$ of the Cauchy problem

$$dg + \theta g = 0, \quad\quad g\big|_{\alpha^0} = I.$$

Then the holonomy group $\operatorname{Hol}^\nabla_{\alpha^0}$ of the flat connection ∇ is generated by commuting (*monodromy*) matrices $G_1, \ldots, G_k$:

$$g(\alpha^1, \ldots, \alpha^j + 2\pi, \ldots, \alpha^k) = g(\alpha^1, \ldots, \alpha^k) G_j.$$

So, Theorem 1.4.3 implies that if one of the matrices $G_1, \ldots, G_k$, say G_1, satisfies the nondegeneracy condition: $\{1\} \cap \operatorname{spec} G_1 = \varnothing$, then equation (1.4.22) has a unique global solution $b(\alpha)$ that is defined by the formula

$$b(\alpha) = g(\alpha)\left(\int_{\alpha^0}^{\alpha} g^{-1}\tau + b_0\right), \quad\quad b_0 = (G_1^{-1} - I)^{-1} \int_{\alpha^0}^{\alpha^0 + 2\pi l^1} g^{-1}\tau.$$

Here $l^1 = (1, 0, \ldots, 0)$.

§1.5. Symplectic vector bundles as model phase spaces

Here we consider the following problem: in which case can a first variation operator given over an isotropic submanifold be considered as a Hamilton vector field in the symplectic normal bundle? Our approach to this problem is based on ideas of Sternberg's minimal coupling procedure [**St**] (see also [**GuSt**$_2$, **We**$_{3,4}$]) and on the geometric theory of linearized Hamilton systems along a given trajectory [**MRR**].

Let $(\pi\colon \mathcal{X} \to B, \mathcal{J}; \nabla)$ be a real symplectic vector bundle $\mathcal{X}$ over B equipped with a symplectic linear connection ∇. We assume that the base B is a symplectic manifold equipped with a nondegenerate closed 2-form ω.

Let X be a trivialization of $\mathcal{X}_U$ over an open subset $U \subseteq B$. Let us introduce the following notation and objects.

(a) Choose local coordinates $\xi = (\xi^1, \ldots, \xi^{2k})$ on $U \subseteq B$. The symplectic form ω on B has the local representation

$$\omega = \sum_{i<j} \omega_{ij}(\xi)\, d\xi^i \wedge d\xi^j \quad\quad (\omega_{ij} = -\omega_{ji}).$$

(b) Choose local coordinates $x = (x^1, \ldots, x^{2r})$ along the fibers of $\mathcal{X}_U$ associated with a basis of local sections $X_1(\xi), \ldots, X_{2r}(\xi)$. The vector bundle isomorphism $\mathcal{J}\colon \mathcal{X} \to \mathcal{X}^*$ inducing the symplectic structure along the fibers of $\mathcal{X}$ is given by the matrix-valued function $J = ((J_{\sigma\sigma'}(\xi)))$ on U: $J_{\sigma\sigma'}(\xi) = -J_{\sigma'\sigma}(\xi) = \langle \mathcal{J} X_{\sigma'}(\xi), X_\sigma(\xi)\rangle$.

(c) The symplectic connection ∇ in $\mathcal{X}$ is locally represented by the connection 1-form $\theta = ((\theta_\sigma^{\sigma'}))$ on U,

$$\nabla X_\sigma = \sum_{\sigma'=1}^{2r} \theta_\sigma^{\sigma'} \otimes X_{\sigma'}.$$

The curvature 2-form C of ∇ on $\mathcal{X}_U$ can be written as

$$C = \sum_{i<j} \mathcal{R}_{ij}\, d\xi^i \wedge d\xi^j, \qquad \mathcal{R}_{ij} = -\mathcal{R}_{ji} = ((\mathcal{R}_{ij,\sigma}^{\sigma'}(\xi))) \in \mathrm{sp}(r,\mathbb{R}) \otimes C^\infty(U).$$

(d) The connection ∇ splits the tangent bundle $T\mathcal{X}$ into the direct sum of the horizontal $\mathcal{H}$ and vertical $\mathcal{V}$ subbundles (see §1.2),

(1.5.1) $$T\mathcal{X} = \mathcal{H} \oplus \mathcal{V}.$$

Locally, at each point $p = (\xi, x)$, $\xi \in U$, $x \in \mathcal{X}_\xi$, we have

$$\mathcal{V}_p = \mathrm{span}\left\{\frac{\partial}{\partial x^\sigma}, \sigma = 1, \ldots, 2r\right\}, \qquad \mathcal{H}_p = \mathrm{span}\{\nabla_i^\#, i = 1, \ldots, 2k\},$$

where

(1.5.2) $$\nabla_i^\# = \frac{\partial}{\partial \xi^i} - \sum_{\sigma,\sigma'=1}^{2r} \theta_{\sigma,i}^{\sigma'}(\xi) x^\sigma \frac{\partial}{\partial x^{\sigma'}}, \qquad \theta_{i\sigma}^{\sigma'}(\xi) = \frac{\partial}{\partial \xi^i} \,\lrcorner\, \theta_\sigma^{\sigma'}.$$

Now let us introduce the 1-form on $\mathcal{X}_U$ by the formula

(1.5.3)
$$\mu^\nabla = \frac{1}{2}\langle Jx, dx\rangle + \frac{1}{2}\langle Jx, \theta x\rangle$$
$$= \frac{1}{2} \sum_{\alpha,\beta,\gamma=1}^{2r} J_{\alpha\gamma}(\xi)\left(x^\gamma\, dx^\alpha + x^\gamma x^\beta \sum_{\alpha,\beta,\gamma=1}^{2r} \theta_{i\beta}^\alpha(\xi)\, d\xi^i\right).$$

Using the transformation rules for J and θ, it is easy to see that formula (1.5.3) defines a global 1-form μ^∇ on $\mathcal{X}$. This 1-form has the following properties with respect to the horizontal $\mathcal{H}$ and vertical $\mathcal{V}$ subbundles:

(1.5.4) $$\mu^\nabla\Big|_{\mathcal{H}_p} = 0, \qquad \mu^\nabla\Big|_{\mathcal{V}_p} = \frac{1}{2}\langle Jx, dx\rangle,$$

(1.5.5) $$d\mu^\nabla(\nabla_i^\#, \nabla_j^\#) = \frac{1}{2}\langle Jx, \mathcal{R}_{ij} x\rangle.$$

We introduce the 2-form $\Omega^{\nabla,\omega}$ on $\mathcal{X}$,

(1.5.6) $$\Omega^{\nabla,\omega} = \pi^*\omega + d\mu^\nabla,$$

or, locally,

$$\Omega^{\nabla,\omega} = \frac{1}{2}\omega(\xi)\, d\xi \wedge d\xi + \frac{1}{2}\langle J\, dx \wedge dx\rangle + \langle J\, dx \wedge \theta x\rangle + \frac{1}{2}\langle Jx, Cx\rangle + \frac{1}{2}\langle J\theta x \wedge \theta x\rangle.$$

Here $C = d\theta + \theta \wedge \theta$ is a curvature 2-form. Obviously, $\Omega^{\nabla,\omega}$ is closed.

PROPOSITION 1.5.1. *The 2-form $\Omega^{\nabla,\omega}$ (1.5.6) is nondegenerate and closed in a neighborhood $\widetilde{\mathcal{X}}$ of B (as the zero-section) in $\mathcal{X}$, so that $(\widetilde{\mathcal{X}}, \Omega^{\nabla,\omega})$ is a symplectic manifold. This symplectic structure is compatible with the symplectic form ω on the base B and the symplectic structure along the fibers of $\mathcal{X}$:*

(a) The restriction of $\Omega^{\nabla,\omega}$ to B is equal to ω, and thus, B is a symplectic submanifold lying in $\mathcal{X}$.

(b) The restriction of $\Omega^{\nabla,\omega}$ to the fiber $\mathcal{X}_\xi$ is equal to $\frac{1}{2}\langle J dx \wedge dx\rangle$.

If the connection ∇ is flat ($C = 0$), then $\Omega^{\nabla,\omega}$ is nondegenerate everywhere on $\mathcal{X}$, and $(\mathcal{X}, \Omega^{\nabla,\omega})$ is a symplectic manifold.

This proof follows from the relations

(1.5.7)
$$\Omega^{\nabla,\omega}(\nabla_i^\#, \nabla_j^\#) = \omega_{ij} + \frac{1}{2}\langle Jx, \mathcal{R}_{ij}x\rangle,$$
$$\Omega^{\nabla,\omega}\left(\nabla_i^\#, \frac{\partial}{\partial x^\sigma}\right) = 0, \qquad \Omega^{\nabla,\omega}\left(\frac{\partial}{\partial x^\sigma}, \frac{\partial}{\partial x^{\sigma'}}\right) = -J_{\sigma\sigma'}. \qquad \square$$

So, in the general case, $\Omega^{\nabla,\omega}$ defines a symplectic structure on the open subset $\widetilde{\mathcal{X}} \subseteq \mathcal{X}$, where

(1.5.8) the matrix $\left(\!\left(\omega_{ij} + \frac{1}{2}\langle Jx, \mathcal{R}_{ij}x\rangle\right)\!\right)$ is nondegenerate,

that is, where the restriction of $\Omega^{\nabla,\omega}$ to the horizontal subbundle $\mathcal{H}$ is nondegenerate. The Poisson brackets on $(\widetilde{\mathcal{X}}, \Omega^{\nabla,\omega})$ with respect to some local coordinates are given by the relations

(1.5.9)
$$\{\xi^j, \xi^i\}_{\widetilde{\mathcal{X}}} = \omega^{ij} - \omega^{is}G_{ss'}\omega^{s'j},$$
$$\{\xi^j, x^\sigma\}_{\widetilde{\mathcal{X}}} = (\omega^{is}G_{ss'}\omega^{s'j} - \omega^{ij})\theta^\sigma_{i\sigma'}x^{\sigma'},$$
$$\{x^\sigma, x^\gamma\}_{\widetilde{\mathcal{X}}} = J^{\sigma\gamma} - \theta^\sigma_{j\sigma'}x^{\sigma'}(\omega^{ij} - \omega^{is}G_{ss'}\omega^{s'j})\theta^\gamma_{i\sigma''}x^{\sigma''},$$

where $\omega_{is}\omega^{sj} = \delta_i^j$, $J^{\sigma\gamma}J_{\gamma\sigma'} = \delta^\sigma_{\sigma'}$ and the sum is taken over repeating indices. The matrix $G = ((G_{ss'}))$ is determined by the formula

(1.5.10) $G = \omega - \omega \cdot \left(\omega + \frac{1}{2}\langle Jx, \mathcal{R}x\rangle\right)^{-1} \cdot \omega = \frac{1}{2}\langle Jx, \mathcal{R}x\rangle + O(|x|^3)$ as $|x| \to 0$.

Here $\omega = ((\omega_{ij}(\xi)))$, $\det \omega \neq 0$, $\mathcal{R} = ((\mathcal{R}_{ij}))$. If the connection ∇ is flat ($\mathcal{R} \equiv 0$), then $C \equiv 0$ and the Poisson bracket (1.5.9) is well defined on the entire manifold $\mathcal{X}$.

Thus, in general, the pair $(\mathcal{X}, \Omega^{\nabla,\omega})$ is a *presymplectic manifold*, that is, a manifold with a closed 2-form. We call $(\mathcal{X}, \Omega^{\nabla,\omega})$ the *model phase space* associated with $(\mathcal{X}, \mathcal{J}, \nabla, \omega)$. We say that a vector field X_F on $\mathcal{X}$ is a Hamilton vector field corresponding to a function $F \in C^\infty(\mathcal{X})$ if

(1.5.11) $$X_F \lrcorner \Omega^{\nabla,\omega} = -dF.$$

The *characteristic distribution* of $\Omega^{\nabla,\omega}$ is

(1.5.12) $\ker_{(\xi,x)} \Omega^{\nabla,\omega} = \left\{\nabla_u^\# \in \mathcal{H}_{(\xi,x)} \colon \nabla_u^\# \lrcorner \left(\pi^*\omega + \frac{1}{2}\mathcal{J}\langle x, \pi^*Cx\rangle\right)\right\} = 0.$

Thus, X_F is determined only up to vector fields tangent to the characteristic distribution of $\Omega^{\nabla,\omega}$, which are called *gauge vector fields*.

The Poisson algebra $C^\infty(\mathcal{X}, \Omega^{\nabla,\omega})$ associated with $\Omega^{\nabla,\omega}$ is given by the set of functions $F \in C^\infty(\mathcal{X})$ for which equation (1.5.11) has a solution X_F. The Poisson bracket on $C^\infty(\mathcal{X}, \Omega^{\nabla,\omega})$ is defined as

$$\{F', F''\}_{\mathcal{X}} = \Omega^{\nabla,\omega}(X_{F'}, X_{F''}). \tag{1.5.13}$$

The Jacobi and the Leibniz identities for this bracket follow from the fact that $\Omega^{\nabla,\omega}$ is closed. For the relationship between presymplectic manifolds and Poisson manifolds, see also [**Cou**$_{1,2}$, **DGMS**, **KaMas**, **L**$_2$].

Let $D \in \mathfrak{D}(\mathcal{X})$ be a covariant differential operator on $\mathcal{X}$ such that a vector field v on B associated with D from (1.1.1) is a Hamilton vector field,

$$v = \mathrm{ad}(f) \equiv -\sum_{i,j=1}^{2k} \omega^{ij}(\xi) \frac{\partial f}{\partial \xi^j} \frac{\partial}{\partial \xi^i}, \qquad f \in C^\infty(B). \tag{1.5.14}$$

Consider a vector field $D^\#$ on $\mathcal{X}$ induced by D and its decomposition into the horizontal and vertical parts (see (1.2.8)),

$$D^\# = \nabla^\#_{\mathrm{ad}(f)} + \Xi^\#. \tag{1.5.15}$$

Here $\Xi\colon \mathcal{X} \to \mathcal{X}$ is a vector bundle morphism with the identity base mapping (see (1.2.4), (1.2.5)).

THEOREM 1.5.1. *Let $D \in \mathfrak{D}(\mathcal{X})$ be a covariant differential operator whose image v under the anchor mapping satisfies (1.5.14).*

Then $D^\#$ is a Hamilton vector field on the model phase space $(\mathcal{X}, \Omega^{\nabla,\omega})$ if and only if the vertical part Ξ of D satisfies the following equation on B:

$$d\Xi + [\theta, \Xi] = \mathrm{ad}(f) \lrcorner\, C. \tag{1.5.16}$$

where θ is a connection 1-form and C is a curvature 2-form on B. In this case, the corresponding Hamilton function F is determined as follows:

$$D^\# = X_F, \qquad F(\xi, x) \stackrel{\mathrm{def}}{=} \pi^*(f(\xi)) + \frac{1}{2}\langle \mathcal{J} \circ \Xi x, x \rangle. \tag{1.5.17}$$

The subspace of operators $D \in \mathfrak{D}(\mathcal{X})$ that satisfy (1.5.14), (1.5.16) forms a Lie subalgebra. In particular, if the connection ∇ is flat, then condition (1.5.16) implies that Ξ is ∇-covariantly constant.

The proof readily follows from (1.5.7), (1.5.9), and (1.5.10).

COROLLARY 1.5.1. *Suppose that the base B is compact and the connection ∇ is flat. If under the assumption (1.5.14), the flow $\mathbb{S}^t$ of $D^\#$ is ∇-reducible (see Definition 1.2.1), then $\mathbb{S}^t$ is the flow of the Hamilton vector field X_F on $(\mathcal{X}, \Omega^{\nabla,\omega})$ defined by (1.5.17). Moreover, $B \subset \mathcal{X}$ (as the zero-section) is an X_F-invariant symplectic submanifold of $\mathcal{X}$. If $\Lambda \subset B$ is an $\mathrm{ad}(f)$-invariant Lagrangian submanifold of (B, ω), then $\Lambda \subset (\mathcal{X}, \Omega^{\nabla,\omega})$ is X_F-invariant and isotropic.*

Now let us consider the special case that appears in the context of symplectic geometry near isotropic submanifolds (for details, see [**GuSt**$_2$, **We**$_{3,4}$]).

We start with a symplectic vector-bundle $(E \to \Lambda, \mathcal{J})$ over a k-dimensional manifold Λ equipped with a symplectic linear connection ∇. Let us consider the Whitney sum of the cotangent bundle $T^*\Lambda$ and E over the base Λ:

$$\mathcal{X} = T^*\Lambda \oplus E, \qquad \pi_0\colon \mathcal{X} \to \Lambda. \tag{1.5.18}$$

Then the Whitney sum $\mathcal{X}$ can be viewed as the total space of the pull-back of the symplectic vector bundle $E \to \Lambda$ along the projection $j \colon T^*\Lambda \to \Lambda$,

(1.5.19) $$\pi \colon \mathcal{X} \to T^*\Lambda.$$

On the other hand, $\mathcal{X}$ is the total space of the pull-back of $T^*\Lambda$ to Λ,

(1.5.20) $$\pi_1 \colon \mathcal{X} \to E.$$

So, we have the commutative diagram

$$\begin{array}{ccc} \mathcal{X} & \xrightarrow{\pi_1} & E \\ \pi \downarrow & & \downarrow \\ T^*\Lambda & \xrightarrow{j} & \Lambda \end{array}$$

Consider also the pull-back $\widetilde{\mathcal{J}} \colon \mathcal{X} \to \mathcal{X}^*$ of the vector bundle morphism $\mathcal{J} \colon E \to E^*$ and the pull-back $\widetilde{\nabla}$ of the symplectic connection ∇ in E. Thus, the original triple $(E \to \Lambda, \mathcal{J}, \nabla)$ induces the symplectic vector bundle $(\pi \colon \mathcal{X} \to T^*\Lambda, \widetilde{\mathcal{J}}, \widetilde{\nabla})$ with symplectic connection $\widetilde{\nabla}$ over the base $B = T^*\Lambda$ equipped with the standard symplectic structure

(1.5.21) $$\omega = d\varrho, \qquad \varrho = \sum_{i=1}^{k} \eta_i \, d\alpha^i,$$

where $\alpha = (\alpha^1, \ldots, \alpha^k) \in \Lambda$, $\eta = (\eta_1, \ldots, \eta_k) \in T^*_\alpha \Lambda$ are local coordinates and ϱ is a canonical Liouville 1-form on $T^*\Lambda$. So, in this case, formula (1.5.6) defines a closed 2-form that depends only on the symplectic linear connection ∇ in E. In more detail, we have the following statement.

PROPOSITION 1.5.2. *Formula* (1.5.6) *defines the closed 2-form* $\Omega^\nabla \stackrel{\text{def}}{=} \Omega^{\widetilde{\nabla}, d\varrho}$ *that is nondegenerate everywhere on* $\mathcal{X} = T^*\Lambda \oplus E$. *Thus, the original symplectic vector bundle E over Λ with symplectic connection ∇ generates a symplectic manifold* $(\mathcal{X} = T^*\Lambda \oplus E, \Omega^\nabla)$ *with the properties:*
- *the embedding $\Lambda \hookrightarrow \mathcal{X}$ (as the zero-section of (1.5.18)) is an isotropic submanifold of $(\mathcal{X}, \Omega^\nabla)$;*
- *the embedding $T^*\Lambda \hookrightarrow \mathcal{X}$ (as the zero-section of (1.5.19)) is a symplectic submanifold of $(\mathcal{X}, \Omega^\nabla)$;*
- *the vertical subbundle $\ker d\pi_1 \subset T\mathcal{X}$ of the vector bundle (1.5.20) is isotropic.*

If ∇ is flat, then E is the coisotropic extension of Λ in $(\mathcal{X}, \Omega^\nabla)$.

PROOF. Let $\theta = \sum_i \theta_i(\alpha) \, d\alpha^i$, and let $C = \sum_{i<j} \mathcal{R}_{ij}$ be the connection 1-form and the curvature 2-form on Λ of the symplectic connection ∇ in E. Then the connection 1-form $\widetilde{\theta}$ and the curvature 2-form $\widetilde{C}$ on $T^*\Lambda$ of the induced connection $\widetilde{\nabla}$ in $\mathcal{X}$ are defined as the pull-backs under the projection $j \colon T^*\Lambda \to \Lambda$, $\widetilde{\theta} = j^*C$, $\widetilde{C} = j^*C$. Substituting these relations into formula (1.5.10) for the matrix G, we see that, with respect to the local coordinates $(\eta_1, \ldots, \eta_k, \alpha^1, \ldots, \alpha^k) \in T^*\Lambda$,

$(x^1, \ldots, x^{2r}) \in E_\alpha$ on $(\mathcal{X}, \Omega^\nabla)$, the Poisson brackets (1.5.9) take the form

(1.5.22)
$$\{\eta_j, \eta_i\}_\mathcal{X} = -\frac{1}{2}\langle J(\alpha)x, \mathcal{R}_{ji}(\alpha)x\rangle, \qquad \{\eta_j, \alpha^i\}_\mathcal{X} = \delta_j^i,$$
$$\{\alpha^i, \alpha^j\}_\mathcal{X} = 0, \qquad \{\alpha^i, x^\sigma\}_\mathcal{X} = 0,$$
$$\{\eta_i, x^\sigma\}_\mathcal{X} = -\sum_{\sigma'} \theta_{i\sigma'}^\sigma(\alpha) x^{\sigma'}, \qquad \{x^\sigma, x^{\sigma'}\}_\mathcal{X} = J^{\sigma\sigma'}(\alpha),$$

where $J^{\sigma\sigma'} J_{\sigma'\gamma} = \delta_\gamma^\sigma$, $J_{\sigma\sigma'} = \langle \mathcal{J}\frac{\partial}{\partial x^\sigma}, \frac{\partial}{\partial x^{\sigma'}}\rangle$. So, this implies that (1.5.22) defines the non-degenerate Poisson bracket on $\mathcal{X}$ corresponding to the symplectic structure

(1.5.23)
$$\Omega^\nabla = d(\pi^*\varrho + \pi_1^*\mu^\nabla).$$

Here μ^∇ is the 1-form on E defined by formula (1.5.3) via the symplectic connection ∇ on E. $\square$

Now we want to discuss the properties of the Poisson bracket (1.5.22) in the case where the symplectic connection ∇ is flat (that is, $C \equiv 0$). Let us consider the space $\mathrm{Lin}(E)$ of fiberwise linear functions on E. Then there is a natural isomorphism $l: \Gamma(E^*) \to \mathrm{Lin}(E)$.

Taking (1.5.22) into account, we deduce the following statement.

PROPOSITION 1.5.3. *On $\mathcal{X} = T^*\Lambda \oplus E$, each flat symplectic connection ∇ in $(E, \mathcal{J})$ generates a unique Poisson bracket $\{\,,\,\}_\mathcal{X}$ such that*
  (i) $\{\pi^*f, \pi^*g\}_\mathcal{X} = \pi^*(\{f, g\}_{T^*\Lambda})$,
  (ii) $\{\pi^*\varphi, \pi^*\psi\}_\mathcal{X} = \pi_0^*(\langle l^{-1}(\varphi), \mathcal{J}^{-1} \circ l^{-1}(\psi)\rangle)$,
  (iii) $\{\pi^*f, \pi^*\varphi\}_\mathcal{X} = \pi_1^*(l \circ \nabla_{u_f}^* \circ l^{-1}(\varphi))$,
for all $f, g \in C^\infty(T^*\Lambda)$, $\varphi, \psi \in \mathrm{Lin}(E)$. Here π and π_1 are the projections from (1.5.19), (1.5.20); $\pi_0: \mathcal{X} \to \Lambda$ is the Whitney sum (1.5.18); $u_f = \mathrm{ad}(f)\big|_\Lambda$ is the restriction to Λ of the Hamilton vector field $\mathrm{ad}(f)$ of $f \in C^\infty(T\Lambda^*)$. Conversely, each Poisson bracket on $\mathcal{X}$ satisfying conditions (i), (ii), *and*
  (iii) $\{\pi^*f, \pi_1^*\varphi\}_\mathcal{X} \subset \pi_1^*(\mathrm{Lin}(E))$
generates a unique flat symplectic connection ∇ in E with property (iii).

Let $(E, \mathcal{J}, \nabla)$ and $(E', \mathcal{J}', \nabla')$ be two symplectic vector bundles over Λ equipped with flat symplectic connections ∇ and ∇'. If ∇ and ∇' are gauge equivalent, then the corresponding Poisson structures $\{\,,\,\}_\mathcal{X}$ and $\{\,,\,\}_{\mathcal{X}'}$ are isomorphic.

Now let Λ be an isotropic submanifold of a symplectic manifold (M, Ω), and let E be the symplectic normal bundle over Λ equipped with a symplectic connection ∇. Then we have the model phase space $\mathcal{X} = T^*\Lambda \oplus E$, where the symplectic structure Ω^∇ is determined by (1.5.23). In this case the Hamiltonization problem for first variation operators over Λ can be solved in the following way.

Combining the results of §1.2 and of Theorem 1.5.1, we obtain the following statement.

THEOREM 1.5.2. *Let Λ be an isotropic submanifold of (M, Ω) and let E be the symplectic normal bundle over Λ endowed with a symplectic connection ∇. Let (M, Ω, H) be a Hamilton system whose flow leaves the isotropic submanifold Λ invariant, $H \in \mathcal{F}_\Lambda(M)$. Suppose that the vertical part Ξ of the first variation operator D_H of $\mathrm{ad}(H)$ over Λ is compatible with the curvature C of ∇ by the* *condition*

(1.5.24) $$d\Xi + [\theta, \Xi] = (\mathrm{ad}(H)|_\Lambda) \lrcorner\, C,$$

where θ is the connection 1-form corresponding to ∇. Then, on the model space $(\mathcal{X} = T^*\Lambda \oplus E, \Omega^\nabla)$, the operator D_H generates the Hamilton vector field X_F with the Hamilton function

(1.5.25) $$F = \sum_i \eta_i u^i + \frac{1}{2}\langle J\Xi x, x\rangle,$$

where $\mathrm{ad}(H)|_\Lambda = \sum_i u^i \partial/\partial \alpha^i$ and $\alpha = (\alpha^1, \ldots, \alpha^k) \in \Lambda$, $\eta = (\eta_1, \ldots, \eta_k) \in T^*_\alpha \Lambda$, and $x = (x^1, \ldots, x^{2r}) \in E_\alpha$ are local coordinates. The vector field X_F possesses the following properties:

(i) $\Lambda \hookrightarrow \mathcal{X}$ (as the zero-section of (1.5.18)) is an X_F-invariant isotropic submanifold.

(ii) $E \hookrightarrow \mathcal{X}$ (as the zero-section of (1.5.20)) is an X_F-invariant submanifold coisotropic on Λ, $E \supset \Lambda$.

(iii) The restriction of X_F to E coincides with the pull-back $D_H^\#$ (1.1.9) of D_H,

$$X_F|_E = D_H^\#.$$

In particular, if Λ is compact, then X_F is a complete vector field on $\mathcal{X}$ and the linearized Hamilton flow $\mathbb{S}_H^t$ of $\mathrm{ad}(H)$ over Λ coincides with a flow of the Hamilton vector field X_F along E. If ∇ is flat, then E is the X_F-invariant coisotropic extension of $\Lambda \subset \mathcal{X}$.

PROOF. To apply Theorem 1.5.1, let us use the following observation. Consider a natural isomorphism of the space of all fiberwise linear functions on $T^*\Lambda$ to the space of vector fields on Λ:

$$\lambda \colon \Gamma(T\Lambda) \to \mathrm{Lin}(T^*\Lambda),$$

$$\lambda(u) = \sum_i \eta_i u^i, \qquad u = \sum_i u^i \frac{\partial}{\partial \alpha^i}.$$

Then for each vector field u on Λ, we can define the pull-back to $T^*\Lambda$ as the Hamilton vector field of $\lambda(u)$ with respect to the standard symplectic structure on $T^*\Lambda$,

$$u \mapsto \mathrm{ad}\big(\lambda(u)\big) = \sum_i \left(\frac{\partial \lambda(u)}{\partial \eta_i} \frac{\partial}{\partial \alpha^i} - \frac{\partial \lambda(u)}{\partial \alpha^i} \frac{\partial}{\partial \eta_i}\right).$$

It is easy to see that $X_F = \widetilde{D}^\#$, where $\widetilde{D} \in \mathfrak{D}(\mathcal{X})$ is a covariant differential operator on the symplectic vector bundle $\pi \colon \mathcal{X} \to T^*\Lambda$ with the associated vector field $v = \mathrm{ad}\big(\lambda(u)\big)$ (see Definition 1.1.1). Moreover, from (1.5.24) and the property

$$\mathrm{ad}\big(\lambda(u)\big)(j^*f) = j^*\big(u(f)\big) \qquad \forall f \in C^\infty(\Lambda),$$

we have $\widetilde{D}|_E = D_H$. Now, applying Theorem 1.5.1 to $\widetilde{D}$, we obtain the desired result. $\square$

We say that the first variation operator D_H of $\mathrm{ad}(H)$ at Λ *admits a Hamiltonization* on the model phase space $(\mathcal{X}, \Omega^\nabla)$ if there exists a Hamilton vector field on $(\mathcal{X}, \Omega^\nabla)$ whose restriction to $E \subset \mathcal{X}$ coincides with $D^\#$. Thus, condition (1.5.24) implies that D_H admits a Hamiltonization on $\mathcal{X}$.

Geometrically, (1.5.24) means that the flow of $D_H^\#$ on E preserves the horizontal subbundle of TE associated with the connection ∇. There are nontrivial obstructions to the Hamiltonization of D_H on $\mathcal{X}$ with respect to symplectic structures of type (1.5.6). For instance, if $\Lambda \approx \mathbb{T}^2$ and E is a 2-dimensional trivial symplectic vector bundle over $\mathbb{T}^2$, then D_H admits a Hamiltonization on $(\mathcal{X}, \Omega^\nabla)$ relative to a certain connection ∇ if and only if D_H is *reducible* in $^\mathbb{C}E$.

Note that if a connection ∇ in E is flat, then the Hamiltonization condition (1.5.24) is equivalent to the ∇-reducibility of the first variation operator D_H in E.

COROLLARY 1.5.2. *Let a submanifold $N \subset M$ be coisotropic on Λ. Let $(\mathcal{X} = T^*\Lambda \oplus E, \Omega^\nabla)$ be the model phase space, where ∇ is the symplectic connection (1.3.6) associated with N. If the zero curvature condition (1.3.9) holds and a function $H \in C^\infty(M)$ satisfies assumptions (a), (b) of Theorem 1.3.3, then the pull-back $D_H^\#$ of the first variation operator D_H coincides with the Hamilton vector field X_F on E.*

Suppose that a flat symplectic connection ∇ in E is given. By $\mathcal{F}^\nabla_\Lambda(M)$ (see also Definition 1.1.2) we denote the set of all functions $H \in C^\infty(M)$ such that $\mathrm{ad}(H) \in \mathcal{A}_\Lambda(M)$ and D_H is ∇-reducible.

COROLLARY 1.5.3. *The set $\mathcal{F}^\nabla_\Lambda(M)$ is a Lie algebra with respect to the Poisson structure on M. The mapping*

$$H \mapsto F, \qquad H \in \mathcal{F}^\nabla_\Lambda(M),$$

given by (1.5.25) is a homomorphism of $\mathcal{F}^\nabla_\Lambda(M)$ to the Poisson algebra $C^\infty(\mathcal{X}, \Omega^\nabla)$,

$$\{H', H''\}_M \mapsto \{F', F''\}_\mathcal{X} \qquad \forall\, H', H'' \in \mathcal{F}^\nabla_\Lambda(M).$$

§1.6. Poisson brackets on vector bundles via Lie algebroids

Let $\pi \colon \mathcal{X} \to B$ be a vector bundle over B. Assume that the dual vector bundle $\pi^* \colon \mathcal{X}^* \to B$ is a *Lie algebroid*, that is, there is a bracket $[\,,\,] \colon \Gamma(\mathcal{X}^*) \times \Gamma(\mathcal{X}^*) \to \Gamma(\mathcal{X}^*)$ on the module of global sections of $\mathcal{X}^*$ and a vector bundle morphism $\mathbf{q} \colon \mathcal{X}^* \to TB$ such that:

(i) $[\,,\,]$ is a Lie bracket;
(ii) $[\eta', f\eta''] = f[\eta', \eta''] + \mathbf{q}(\eta')(f)\eta''$ for all $\eta', \eta'' \in \Gamma(\mathcal{X}^*)$, $f \in C^\infty(B)$;
(iii) $\mathbf{q}([\eta', \eta'']) = [\mathbf{q}(\eta'), \mathbf{q}(\eta'')]$ for all $\eta', \eta'' \in \Gamma(\mathcal{X}^*)$.

Following [**Mck**], we call $\mathbf{q}$ the *anchor mapping* of a Lie algebroid.

Let us consider the subspace $\mathrm{Lin}(\mathcal{X}) \subset C^\infty(\mathcal{X})$ of fiberwise linear functions on $\mathcal{X}$ and the natural isomorphism $l \colon \Gamma(\mathcal{X}^*) \to \mathrm{Lin}(\mathcal{X})$. The following result holds.

THEOREM 1.6.1 [**Cou**$_1$]. *There is a unique Poisson structure $\{\,,\,\}_\mathcal{X}$ on $\mathcal{X}$ satisfying the conditions*

(1.6.1) $$\begin{aligned}\{\varphi_1, \varphi_2\}_\mathcal{X} &= l\big([l^{-1}(\varphi_1), l^{-1}(\varphi_2)]\big), \\ \{\varphi, \pi^* f\}_\mathcal{X} &= \pi^*\big(\mathbf{q}(l^{-1}(\varphi))(f)\big), \qquad \{\pi^* f, \pi^* g\}_\mathcal{X} = 0,\end{aligned}$$

for all $\varphi_1, \varphi_2, \varphi \in \mathrm{Lin}(\mathcal{X})$ and $f, g \in C^\infty(B)$. Furthermore, any fiberwise linear Poisson bracket on $\mathcal{X}$ (that is, $\varphi_1, \varphi_2 \in \mathrm{Lin}(\mathcal{X}) \Rightarrow \{\varphi_1, \varphi_2\} \in \mathrm{Lin}(\mathcal{X})$) generates a Lie algebroid structure on $\mathcal{X}^*$.

Our goal is to investigate the question about the Hamilton structure of vector fields $D^{\#}$ ($D \in \mathfrak{D}(\mathcal{X})$) with respect to the Poisson bracket (1.6.1) under a given Lie algebra structure on the dual $\mathcal{X}^*$.

Let us choose local coordinates $\alpha = (\alpha^1, \ldots, \alpha^k)$ on B and local coordinates $x = (x^1, \ldots, x^{2r})$ along the fibers of $\mathcal{X}$ corresponding to the basis of local sections $X_1(\alpha), \ldots, X_{2r}(\alpha)$. By $\Gamma^1, \ldots, \Gamma^{2r}$ we denote the basis of local sections of $\mathcal{X}$ dual to $X_1, \ldots, X_{2r}$: $\langle \Gamma^\sigma, X_{\sigma'} \rangle = \delta^\sigma_{\sigma'}$. Then we can define the structure constants and the components of the anchor mapping $\mathbf{q}$ with respect to this basis:

$$[\Gamma^\sigma, \Gamma^{\sigma'}] = C^{\sigma\sigma'}_\gamma \Gamma^\gamma, \qquad \mathbf{q}(\Gamma^\sigma) = \mathbf{q}^{i\sigma}(\alpha) \frac{\partial}{\partial \alpha^i}.$$

We omit the sum over repeating indices.

The Poisson brackets (1.6.1) of the coordinate functions α^i and x^σ take the form

(1.6.2) $\qquad \{x^\sigma, x^{\sigma'}\}_{\mathcal{X}} = C^{\sigma\sigma'}_\gamma x^\gamma, \qquad \{x^\sigma, \alpha^i\}_{\mathcal{X}} = \mathbf{q}^{i\sigma}(\alpha), \qquad \{\alpha^i, \alpha^j\}_{\mathcal{X}} = 0.$

The Hamilton vector field X_F of a function $F \in C^\infty(\mathcal{X})$ corresponding to the bracket (1.6.1) can be written in the local coordinates as follows:

(1.6.3) $\qquad X_F = \sum_{\sigma,i} \left(\mathbf{q}^{i\sigma} \frac{\partial F}{\partial x^\sigma} \frac{\partial}{\partial \alpha^i} - \mathbf{q}^{i\sigma} \frac{\partial F}{\partial \alpha^i} \frac{\partial}{\partial x^\sigma} \right) + \sum_{\sigma,\sigma',\gamma} C^{\sigma\sigma'}_\gamma x^\gamma \frac{\partial F}{\partial x^\sigma} \frac{\partial}{\partial x^{\sigma'}}.$

It is clear that the Poisson bracket (1.6.1) is *degenerate* if and only if the kernel of the anchor mapping $\mathbf{q}$ is nontrivial. Consider the Lie algebra $\mathfrak{D}(\mathcal{X}^*)$ of covariant differential operators in the dual $\mathcal{X}^*$ of $\mathcal{X}$. The property (ii) says that the bracket $[,]$ acts on sections of $\mathcal{X}^*$ as covariant differential operators: for each $\eta \in \Gamma(\mathcal{X}^*)$, the $\mathbb{R}$-linear operator

$$\operatorname{ad}_\eta \stackrel{\text{def}}{=} [\eta, \cdot] \colon \Gamma(\mathcal{X}^*) \to \Gamma(\mathcal{X}^*)$$

belongs to $\mathfrak{D}(\mathcal{X}^*)$, where the vector field v (the anchor) on B associated to ad_η is defined by $v = \mathbf{q}(\eta)$. Property (iii) implies that the Lie algebroid $(\mathcal{X}^*, [,], \mathbf{q})$ induces the Lie subalgebra of $\mathfrak{D}(\mathcal{X}^*)$:

$$\widetilde{\mathfrak{D}}(\mathcal{X}^*) \stackrel{\text{def}}{=} \{\operatorname{ad}_\eta \mid \eta \text{ runs over } \Gamma(\mathcal{X}^*)\}.$$

THEOREM 1.6.2. *Let $\mathcal{X}^*$ be a Lie algebroid. Let $D \in \mathfrak{D}(\mathcal{X})$ be a covariant differential operator in $\mathcal{X}$. The pull-back $D^{\#}$ (1.1.9) of D to $\mathcal{X}$ is a Hamilton vector field with respect to the Poisson bracket (1.6.1) if and only if*

(1.6.4) $\qquad\qquad\qquad D^* \in \widetilde{\mathfrak{D}}(\mathcal{X}^*),$

that is, there exists a section $\eta \in \Gamma(\mathcal{X}^*)$ such that

(1.6.5) $\qquad\qquad\qquad D^* = \operatorname{ad}_\eta.$

In particular, the vector field v on B associated with D^* is the image of η under the anchor mapping,

(1.6.6) $\qquad\qquad\qquad v = \mathbf{q}(\eta).$

Thus, under the condition (1.6.4), $D^{\#}$ is a Hamilton vector field on $\mathcal{X}$, $D^{\#} = X_F$, where the Hamilton function F is a smooth fiberwise linear function on $\mathcal{X}$:

(1.6.7) $\qquad\qquad F = l(\eta) \equiv \langle \eta, x \rangle, \qquad x \in \mathcal{X}_\alpha, \quad \eta \in \Gamma(\mathcal{X}^*).$

PROOF. By using the local representation

$$D^{\#} = v_i(\alpha)\frac{\partial}{\partial \alpha^i} + V^{\sigma'}_\sigma(\alpha)x^\sigma \frac{\partial}{\partial x^{\sigma'}}$$

and (1.6.3), we show that $D^{\#}$ is Hamilton vector field if the following equations can be solved with respect to a function $F \in C^\infty(\mathcal{X})$:

(1.6.8) $\qquad v^i(\alpha) = \mathbf{q}^{i\sigma}(\alpha)\frac{\partial F}{\partial x^\sigma}, \qquad V^\sigma_{\sigma'}(\alpha)x^{\sigma'} = C^{\gamma\sigma}_{\sigma'}x^{\sigma'}\frac{\partial F}{\partial z^\gamma} - \mathbf{q}^{i\sigma}(\alpha)\frac{\partial F}{\partial \alpha^i}.$

On the other hand, conditions (1.6.5) and (1.6.6) may be written with respect to the components of the section $\eta = \eta_\sigma \Gamma^\sigma$ as

(1.6.9) $\qquad v^i(\alpha) = \mathbf{q}^{i\sigma}(\alpha)\eta_\sigma, \qquad V^\sigma_{\sigma'} = C^{\gamma\sigma}_{\sigma'}\eta_\gamma - \mathbf{q}^{i\sigma}\frac{\partial \eta_{\sigma'}}{\partial \alpha^i}.$

Comparing (1.6.8) and (1.6.9), we obtain the desired statement. $\square$

So, a Lie algebroid structure on the dual vector bundle $\mathcal{X}^*$ induces a Poisson manifold $(\mathcal{X}, \{\,,\,\}_{\mathcal{X}})$. Covariant differential operators in $\mathcal{X}$ whose dual operators form the Lie algebra $\widetilde{\mathfrak{D}}(\mathcal{X}^*)$, can be written as Hamilton vector fields on $\mathcal{X}$.

Let $\widetilde{\mathcal{X}}$ be the second vector bundle over the same base B, and let $\widetilde{\mathcal{X}}^*$ be a Lie algebroid with an anchor mapping $\widetilde{\mathbf{q}}$. Assume that a vector bundle morphism $\phi\colon \widetilde{\mathcal{X}} \to \mathcal{X}$ induces a *morphism of Lie algebroids* $\phi^*\colon \mathcal{X}^* \to \widetilde{\mathcal{X}}^*$, that is, $\widetilde{\mathbf{q}} \circ \phi^* = \mathbf{q}$ and

$$\phi^*[\eta', \eta''] = [\phi^*\eta', \phi^*\eta''] \quad \text{for all } \eta', \eta'' \in \Gamma(\mathcal{X}^*).$$

PROPOSITION 1.6.1. *The morphism* $\phi\colon \widetilde{\mathcal{X}} \to \mathcal{X}$ *is a Poisson mapping.*

This observation allows us to consider the reduction of the fiberwise linear Hamilton dynamics on $(\mathcal{X}, \{\,,\,\}_{\mathcal{X}})$ via morphisms of Lie algebroids.

If $D \in \mathfrak{D}(\mathcal{X})$ and $D^* \in \widetilde{\mathfrak{D}}(\mathcal{X})$, then it follows from Theorem 1.6.2 that $D^{\#} = X_F$ is a Hamilton vector field on $\mathcal{X}$. Using the morphism ϕ, we define the pull-back $\widetilde{D} \in \mathfrak{D}(\widetilde{\mathcal{X}})$ by the formula $\widetilde{D}^{\#} = \widetilde{X}_{F\circ\phi}$, so that the diagram

$$\begin{array}{ccc} \Gamma(\widetilde{\mathcal{X}}) & \xrightarrow{\phi} & \Gamma(\mathcal{X}) \\ \widetilde{D}\downarrow & & \downarrow D \\ \Gamma(\widetilde{\mathcal{X}}) & \xrightarrow{\phi} & \Gamma(\mathcal{X}) \end{array}$$

commutes. By Proposition 1.6.1, it follows that the projections under ϕ of the trajectories of the Hamilton vector field $\widetilde{D}^{\#}$ on $\widetilde{\mathcal{X}}$ are the trajectories of $D^{\#}$ on $\mathcal{X}$.

So, in the context of the Hamilton dynamics on $\mathcal{X}$, to study the Lie subalgebra $\widetilde{\mathfrak{D}}(\mathcal{X}) \subset \mathfrak{D}(\mathcal{X})$ is a crucial problem. Let us formulate this problem in more detail using the notion of *connections in Lie algebroids* [**Mck**].

Let $(\mathcal{X}^*, [\,,\,], \mathbf{q})$ be a *transitive* Lie algebroid (that is, let the anchor mapping $\mathbf{q}$ be a fiberwise submersion). The kernel $\mathfrak{g} = \ker \mathbf{g}$ of the anchor mapping is a subbundle of $\mathcal{X}^*$ and there is an exact sequence of vector bundles $\mathfrak{g} \longrightarrow \mathcal{X}^* \xrightarrow{\mathbf{q}} TB$. Furthermore, the kernel $\mathfrak{g}$ of $\mathbf{q}$ inherits the bracket structure of $\mathcal{X}^*$ such that $\mathfrak{g}$ is a *Lie algebra bundle*.

A vector bundle morphism $\gamma\colon TB \to \mathcal{X}^*$ is a *connection in the Lie algebroid* $\mathcal{X}^*$ if γ is anchor preserving (see [**Mck**]), $\mathbf{q} \circ \gamma = \mathrm{id}_{TB}$.

If, in addition, $(\mathcal{X}, \mathcal{J})$ is a symplectic vector bundle over B, then there is an induced symplectic bundle $(\mathcal{X}^*, \mathcal{J}^{-1})$. We say that a connection γ in $(\mathcal{X}^*, \mathcal{J}^{-1})$ is *symplectic* if $\mathcal{J}^{-1}$ is covariantly constant along $\mathrm{ad}_{\gamma(u)}$ for any $u \in \Gamma(TB)$,

$$\langle \mathrm{ad}_{\gamma(u)}(\eta'), \mathcal{J}^{-1}\eta'' \rangle + \langle \eta', \mathrm{ad}_{\gamma(u)}(\mathcal{J}^{-1}\eta'') \rangle = u(\langle \eta', \eta'' \rangle).$$

Let $D \in \mathfrak{D}(\mathcal{X})$ be a covariant differential operator in $\mathcal{X}$. We say that a (symplectic) connection γ in the Lie algebroid is *adapted* to D if the dual D^* can be represented as $D^* = \mathrm{ad}_{\gamma(v)}$, where v is a vector field on B associated with D.

As it follows from Theorem 1.6.2, the existence of a connection γ in $\mathcal{X}^*$ adapted to D implies that $D^\# = X_F$ is a Hamilton vector field with $F = \langle \gamma(u), x \rangle$.

Note that if the kernel of the anchor mapping $\mathbf{q}$ is trivial, then there is a unique connection γ in $\mathcal{X}^*$: $\gamma = \mathbf{q}^{-1}$.

As an application of the above results, let us consider Poisson structures on tangent bundles over symplectic manifolds [**A-S**, **MRR**] (see also [**Kle**, **Vais**]).

EXAMPLE 1.6.1. Let (M, Ω) be a symplectic manifold. Consider the antisymmetric nondegenerate 2-tensor field $\Psi \colon T^*M \to TM$ which gives the Poisson bracket on M,

$$\{f, g\} = \Omega\big(\mathrm{ad}(f), \mathrm{ad}(g)\big) = \langle df, \Psi dg \rangle = \sum_{i,j} \Psi^{ij} \frac{\partial f}{\partial \xi^i} \frac{\partial g}{\partial \xi^j},$$

where $\mathrm{ad}(f) = -\Psi df = ((-\Psi^{ij}\frac{\partial f}{\partial \xi^j}))$ is the Hamilton vector field of $f \in C^\infty(M)$ and ξ are local coordinates on M. As is well known, there is a Lie algebroid

(1.6.10) $$\big(\mathcal{X} = T^*M, \, [\,,\,] = \{\,,\,\}_\Psi, \, \mathbf{q} = -\Psi \big),$$

where $\{\,,\,\}_\Psi$ is the bracket of 1-forms on M [**AbMa**, **KaMas**]:

$$\{\beta', \beta''\}_\Psi = d(\langle \beta', \Psi\beta'' \rangle) + \Psi\beta'' \lrcorner d\beta' - \Psi'_\beta \lrcorner d\beta''.$$

By $\xi = (\xi^1, \ldots, \xi^{2n}) \in M$ and $y = (y^1, \ldots, y^{2n}) \in T_\xi M$, we denote local coordinates on TM. The structure constants C_s^{ij} and the components $\mathbf{q}^{ij}$ of the anchor mapping corresponding to the Lie algebroid (1.6.10) are defined by the components of the Poisson tensor Ψ,

$$\mathbf{q}^{ij} = -\Psi^{ij}, \qquad C_s^{ij} = \frac{\partial \Psi^{ij}}{\partial \xi^s}.$$

The Poisson brackets (1.6.2) generated by the Lie algebroid (1.6.10) on TM take the form

(1.6.11) $$\{y^i, y^j\}_{TM} = \frac{\partial \Psi^{ij}}{\partial \xi^s}(\xi)y^s, \qquad \{y^i, \xi^j\}_{TM} = \Psi^{ij}(\xi), \qquad \{\xi^i, \xi^j\}_{TM} = 0.$$

So, in this case, the anchor mapping $\mathbf{q}$ is an *isomorphism* and the Poisson bracket (1.6.11) on TM is *nondegenerate*. Moreover, the Lie algebroid (1.6.10) is transitive.

Let (M, Ω, H) be a Hamilton system. We consider the first variation equation of the Hamilton vector field $\mathrm{ad}(H)$ on M:

(1.6.12) $$\dot{\xi}^i = -\Psi^{is}(\xi)\frac{\partial H}{\partial \xi^s}, \qquad \dot{y}^j = V_s^j(\xi) y^s,$$

where $V_H = ((V_s^j(\xi)))$ is the variation matrix of $\mathrm{ad}(H)$ defined by the formula

$$V_i^j = -\frac{\partial \Psi^{js}}{\partial \xi^i}\frac{\partial H}{\partial \xi^s} - \Psi^{js}\frac{\partial^2 H}{\partial \xi^s \partial \xi^i}.$$

The first variation operator of $\mathrm{ad}(H)$ in TM is

$$D_H^{TM} \stackrel{\mathrm{def}}{=} [\mathrm{ad}(H), \cdot] \equiv -\Psi \circ \{dH, \cdot\}_\Psi.$$

So, condition (1.6.4) holds for any $H \in C^\infty(M)$.

Theorem 1.6.2 says that system (1.6.12) is the dynamical system of the vector field on TM

$$(D_H^{TM})^\# = -\Psi^{ij}\frac{\partial H}{\partial \xi^j}\frac{\partial}{\partial \xi^i} + \langle V_H(\xi)y, \frac{\partial}{\partial y}\rangle,$$

which is a Hamilton vector field with respect to the Poisson bracket (1.6.11),

(1.6.13) $\qquad (D_H^{TM})^\# = X_F, \qquad F(\xi, y) = \langle dH, y \rangle = \sum_{i=1}^{2n}\frac{\partial H}{\partial \xi^i}(\xi)y^i.$

Let us choose local Darboux coordinates $\xi = (p^1, \ldots, p^n, q_1, \ldots, q_n)$ on M, $\Omega = dp^i \wedge dq_i$. Then we have induced coordinates $y = (\delta p^1, \ldots, \delta p^n, \delta q_1, \ldots, \delta q_n)$ along the fibers of TM. In terms of these coordinates, the symplectic structure on TM corresponding to the Poisson bracket (1.6.11) can be written as

(1.6.14) $\qquad \Omega^\# = \sum_{i=1}^n dp^i \wedge d(\delta q^i) + d(\delta p^i) \wedge dq^i.$

REMARK 1.6.1. If, instead of a symplectic manifold (M, Ω), we consider a Poisson manifold (M, Ψ) (where a Poisson tensor Ψ may be degenerate), then the brackets of 1-forms on M associated with Ψ also define a Lie algebroid. This Lie algebroid also generates the Poisson brackets (1.6.11) in TM, and the first variation operator of any Hamilton vector field $\mathrm{ad}(H)$ on (M, Ψ) induces the Hamilton vector field (1.6.13) on TM. In the Poisson case, the Poisson structure (1.6.11) was introduced by Alvarez-Sanchez in [**A-S**] for applications to control theory. In the symplectic case, the symplectic structure (1.6.14) was used by Marsden, Ratiu, and Raugel [**MRR**] in the context of linearized Hamilton dynamics.

Following [**MRR**], let us describe the properties of the symplectic structure (1.6.14) and the vector field (1.6.13). Let us consider the *canonical involution* on M, $\sigma \colon TTM \to TTM$, locally given by $\sigma(\xi, y, \dot{\xi}, \dot{y}) \stackrel{\mathrm{def}}{=} (\xi, \dot{\xi}, y, \dot{y})$. Here $\dot{\xi}^i$ and $\dot{y}^i$ are the induced coordinates along the fibers of $T(TM)$. Thus, $\sigma \circ \sigma = $ identity and the diagram

$$\begin{array}{ccc} T(TM) & \xrightleftharpoons{\sigma} & T(TM) \\ {\scriptstyle \tau_{TM}}\downarrow & & \downarrow{\scriptstyle T\tau_M} \\ TM & \xrightarrow{\mathrm{id}} & TM \end{array}$$

commutes. Here τ is the natural projection, T is the tangent mapping. The canonical involution σ allows us to define a lift of vector fields on M to TM: if $u \colon M \to TM$ is a vector field on M, then $\sigma \circ Tu \colon TM \to TTM$ is a vector field on TM [**AbMa**]. In particular, the lift of a Hamilton vector field $\mathrm{ad}(H)$ on M is just the vector field $(D_H^{TM})^\#$ on TM,

$$(D_H^{TM})^\# = \sigma \circ T\bigl(\mathrm{ad}(H)\bigr).$$

Let $\varrho = \eta_i d\xi^i$ be the canonical Liouville 1-form on T^*M. Consider the pull-back of ϱ to TM via the vector bundle morphism $\Psi^{-1}\colon TM \to T^*M$: $\varrho_T = (\Psi^{-1})^*\varrho$. Then it is easy to see that the symplectic structure (1.6.14) is the pull-back of the standard symplectic structure $d\varrho$ on T^*M,

$$\Omega^\# = d\varrho_T \equiv (\Psi^{-1})^* d\varrho.$$

Note that $\Omega^\#$ has the following properties:

(i) $\Omega^\#$ is *homogeneous* in the sense that $L_Y \Omega^\# = -\Omega^\#$, where $Y = \sum_i y^i \frac{\partial}{\partial y^i}$ is the Liouville vector field on TM.

(ii) The vertical subbundle $VT(TM)$ is *Lagrangian*.

Using the above constructions, we can draw the following final conclusion about the linearized Hamilton dynamics. Let (M, Ω, H) be the original Hamilton system and D_H^{TM} the first variation operator of $\mathrm{ad}(H)$ in the tangent bundle TM. Then the pull-back $(D_H^{TM})^\#$ of D_H^{TM} to TM is the Hamilton vector field X_F (1.6.4) on the model phase space $(TM, \Omega^\#)$. If the original Hamilton system admits an $\mathrm{ad}(H)$-invariant isotropic submanifold $\Lambda \subset M$, then the model Hamilton system $(TM, \Omega^\#, F)$ also has the invariant isotropic submanifolds $T\Lambda \subset TM$. So, we obtain a different way (compared with the approach developed in §1.5) for the Hamilton description of the first variation equations over an invariant isotropic submanifold.

§1.7. Geometric stability and integrability

Let $(\pi\colon \mathcal{X} \to B, \mathcal{J}\colon \mathcal{X} \to \mathcal{X}^*)$ be a real symplectic vector bundle over a manifold B. The complexification $^{\mathbb{C}}\mathcal{X} = \mathcal{X} \otimes \mathbb{C}$ is viewed as a complex symplectic vector bundle over B whose symplectic structure is defined by the extension of the original structure with respect to linearity. In $^{\mathbb{C}}\mathcal{X}$, let us consider the form

(1.7.1) $$(X, Y)_+ = \frac{1}{2i}\langle \mathcal{J}X, \overline{Y}\rangle \qquad (X, Y \in {}^{\mathbb{C}}\mathcal{X}_\xi, \xi \in B),$$

where $\overline{Y}$ is the complex conjugate of Y.

Recall that a subbundle $\rho \subset {}^{\mathbb{C}}\mathcal{X}$ is said to be (see [**GL, YStz**])
- *positive* (or *negative*) if the form (1.7.1) is positive (or negative) definite in the fibers of ρ;
- *Kählerian* (or *anti-Kählerian*) if ρ is positive (or negative) and a Lagrangian subbundle of $^{\mathbb{C}}\mathcal{X}$. In this case, $^{\mathbb{C}}\dim \rho = \frac{1}{2} \dim \mathcal{X}$.

Note that the restriction of the form (1.7.1) to a given Kählerian subbundle defines a Hermitian structure.

Let $D \in \mathfrak{D}(\mathcal{X})$ be a covariant differential operator. Recall that a subbundle $\rho \subset {}^{\mathbb{C}}\mathcal{X}$ is called D-*invariant* (or, equivalently, invariant with respect to the flow $\mathbb{S}^t$ of D) if $s \in \Gamma(\rho) \Longrightarrow Ds \in \Gamma(\rho)$, or equivalently, $\mathbb{S}^t(\rho) = \rho$ for all $t \in \mathbb{R}$.

DEFINITION 1.7.1. A covariant differential operator $D \in \mathfrak{D}(\mathcal{X})$ is called *geometrically stable* if there exists a splitting of $^{\mathbb{C}}\mathcal{X}$ into the direct sum

(1.7.2) $$^{\mathbb{C}}\mathcal{X} = \rho \oplus \overline{\rho},$$

where ρ is a Kählerian D-invariant subbundle.

PROPOSITION 1.7.1. *If B is compact and $D \in \mathfrak{D}(\mathcal{X})$ is geometrically stable, then D is metrically stable (in the sense of Definition 1.1.2).*

PROOF. Let $\rho \subset {}^{\mathbb{C}}\mathcal{X}$ be a Kählerian D-invariant subbundle. Then the flow $\mathbb{S}^t$ of D preserves the Hermitian structure of ρ that is the restriction of the form (1.7.1) to ρ. This implies that $\mathbb{S}^t$ is uniformly bounded on ρ with respect to the above Hermitian structure. Using the invariance of ρ and the fact that any two Hermitian structures on ρ generate equivalent norms, we prove that D is metrically stable. □

REMARK 1.7.1. In the case $B \approx \mathbb{S}^1$, the geometric stability follows from metrical stability [**YStz**]. In general, these notions of stability are not equivalent. For the case $\dim \mathcal{X} = 2$, $B \approx \mathbb{T}^k$ ($k > 1$), the corresponding examples can be deduced from the Cameron–Johnson stability criterion [**Jo**].

DEFINITION 1.7.2. A symplectic linear connection ∇ in $\mathcal{X}$ is called *geometrically* stable if there is a Kählerian splitting (1.7.2) such that ρ is ∇_v-invariant for each vector field $v \in \mathrm{Vect}(B)$.

From Proposition 1.7.1, we see that the geometric stability of a symplectic connection in $\mathcal{X}$ implies the metric stability if B is compact.

Let us formulate criteria of geometric stability of a symplectic connection ∇ in $\mathcal{X}$ under the following assumptions:

(i) the base B is a compact connected manifold and the fundamental group $\pi_1(B)$ is commutative;

(ii) the symplectic connection ∇ is flat.

Choose a point $\xi^0 \in B$ and consider the holonomy group $\mathrm{Hol}_{\xi^0}^{\nabla}$ of ∇ at ξ^0. Recall that $\mathrm{Hol}_{\xi^0}^{\nabla}$ is independent of the choice of ξ^0 up to an isomorphism. Denote by $\widetilde{\mathrm{Hol}}_{\xi^0}^{\nabla} = \psi(\widetilde{H}_1(B))$ the image of the first Betti group $\widetilde{H}_1(B)$ under the homomorphism $\psi \colon H_1(B) \to \mathrm{Hol}_{\xi^0}^{\nabla}$. Then $\widetilde{\mathrm{Hol}}_{\xi^0}^{\nabla}$ is a finitely generated Abelian group which will be called the *reduced holonomy group* of ∇. Choosing a basis of 1-cycles $\Gamma_1, \ldots, \Gamma_m$ on B ($m = \dim \widetilde{H}_1(B)$), we obtain the pairwise commuting symplectic operators

$$(1.7.3) \qquad \begin{aligned} G_i &= \psi(\Gamma_i) \in \mathrm{Sp}(\mathcal{X}_{\xi^0}), & i &= 1, \ldots, m, \\ [G_i, G_j] &= 0, & i, j &= 1, \ldots, m, \end{aligned}$$

generating the reduced holonomy group $\widetilde{\mathrm{Hol}}_{\xi^0}^{\nabla}$.

PROPOSITION 1.7.2. *Under assumptions* (i) *and* (ii), *the following conditions are equivalent*:

(a) ∇ *is geometrically stable.*

(b) *There exists a splitting*

$$(1.7.4) \qquad {}^{\mathbb{C}}\mathcal{X} = \rho^1 \oplus \cdots \oplus \rho^r \oplus \overline{\rho}^1 \oplus \cdots \oplus \overline{\rho}^r, \qquad {}^{\mathbb{C}}\dim \rho^\nu = 1,$$

where the ρ^σ are ∇-invariant positive line subbundles of ${}^{\mathbb{C}}\mathcal{X}$ such that any two ρ^σ, $\rho^{\sigma'}$ are skew-orthogonal.

(c) *The generators $G_1, \ldots, G_m$* (1.7.3) *of the reduced holonomy group $\widetilde{\mathrm{Hol}}_{\xi^0}^{\nabla}$ satisfy the following conditions: each operator G_j is diagonalizable in ${}^{\mathbb{C}}\mathcal{X}_{\xi^0}$ and its* spectrum lies on the unit circle, $\operatorname{spec} G_j \subset \mathbb{S}^1 \subset \mathbb{C}$, that is, $G_1, \ldots, G_m$ are stable linear operators.

The proof is based on the Lyapunov–Poincaré criterion of stability [**GL, YStz**] and on the following facts.

(1) A symplectic operator $G \in \operatorname{Sp}(\mathcal{X}_{\xi^0})$ is stable if and only if there exists a G-invariant Kählerian subspace of $^{\mathbb{C}}\mathcal{X}_{\xi^0}$. Two commuting stable symplectic operators possess a common invariant Kählerian subspace.

(2) If there is a Kählerian subspace $\rho_{\xi^0} \in {}^{\mathbb{C}}\mathcal{X}_{\xi^0}$ invariant with respect to the holonomy group $\operatorname{Hol}_{\xi^0}^\nabla$, then there is a ∇-invariant Kählerian subbundle of $^{\mathbb{C}}\mathcal{X}$ whose fibers are defined as

$$(1.7.5) \qquad \rho_\xi = g_{\xi^0 \to \xi(t)}(\rho_{\xi^0}), \qquad \xi \in B,$$

where $g_{\xi^0 \to \xi(t)}$ is the parallel transport of ∇ along a path $\xi(t)$ such that $\xi(0) = \xi^0$, $\xi(1) = \xi$ (for more details, see [**KaVo$_2$, Vo$_1$**]).

(3) The elements of $\operatorname{Hol}_{\xi^0}^\nabla$ corresponding to the torsion elements of $H_1(B)$ are stable linear operators.

Now assume that we have a decomposition (1.7.4). It generates the projections

$$(1.7.6) \qquad \operatorname{pr}^\sigma \colon {}^{\mathbb{C}}\mathcal{X} \to \rho^\sigma \qquad (\sigma = 1, \ldots, r = \tfrac{1}{2}\dim \mathcal{X}),$$

which define a set of fiberwise quadratic functions $h^\sigma \colon \mathcal{X} \to \mathbb{R}$ by the formula

$$(1.7.7) \qquad h^\sigma(z) = \frac{1}{2i}\langle \mathcal{J} \circ \operatorname{pr}^\sigma(z), \overline{\operatorname{pr}^\sigma(z)} \rangle \qquad (\sigma = 1, \ldots, r)$$

for each $z \in \mathcal{X}$. Obviously, the functions h^σ are nonnegative. In more detail, these functions possess the following properties.

Let $\mathbb{R}^r_+$ be the positive octant,

$$\mathbb{R}^r_+ = \{c = (c^1, \ldots, c^r) \in \mathbb{R}^r : c^\sigma \geq 0, \sigma = 1, \ldots, r\}.$$

For each integer s, $0 \leq s \leq r$, we introduce an open domain

$$(1.7.8) \qquad \mathcal{C}^s = \bigcup_{0 \leq \sigma_1 < \cdots < \sigma_s \leq r} \{c \in \mathbb{R}^r_+ : c^{\sigma_1} > 0, \ldots, c^{\sigma_s} > 0; c^\sigma = 0 \,\forall\, \sigma \in \overline{\{\sigma_1, \ldots, \sigma_s\}}\}.$$

The set of functions (1.7.7) gives a *surjective* mapping

$$(1.7.9) \qquad \mathbf{h} = (h^1, \ldots, h^r) \colon \mathcal{X} \to \mathbb{R}^r_+.$$

Moreover, for each $c \in \mathcal{C}^s$, $0 \leq s \leq r$, the level set $\mathbf{h}^{-1}(c)$ is an $(s + \dim B)$-dimensional submanifold of $\mathcal{X}$, since the restriction of $\mathbf{h}$ to

$$\bigcup_{c \in \mathcal{C}^s} \mathbf{h}^{-1}(c)$$

is a surjective submersion.

Note that the splitting (1.7.4) induces the *symplectic splitting*

$$(1.7.10) \qquad \mathcal{X} = \mathcal{X}^1 \oplus \cdots \oplus \mathcal{X}^r, \qquad \dim \mathcal{X}^\sigma = 2,$$

where

$$(1.7.11) \qquad \mathcal{X}^\sigma = \mathcal{X} \cap (\rho^\sigma \oplus \overline{\rho}^\sigma)$$

is a 2-dimensional symplectic subbundle of $\mathcal{X}$ and any two subbundles $\mathcal{X}^\sigma$, $\mathcal{X}^{\sigma'}$ ($\sigma \neq \sigma'$) are skew-orthogonal. Then for each $\sigma = 1, \ldots, r$, we have

(1.7.12) $$h^\sigma(z) > 0 \quad \forall z \in \mathcal{X}^\sigma_\xi \setminus \{0\}, \quad \xi \in B.$$

This implies that the level set $\mathbf{h}^{-1}(c)$ is fibered by s-tori, that is, there is a fiber bundle over B

(1.7.13) $$\mathbb{T}^s \to \mathbf{h}^{-1}(c) \to B \quad (c \in \mathcal{C}^s)$$

whose fibers are defined as

(1.7.14) $$\mathbf{h}^{-1}_\xi(c) = \{z \in \mathcal{X}^{\sigma_1}_\xi : h^{\sigma_1}(z) = c^{\sigma_1}\} \times \cdots \times \{z \in \mathcal{X}^{\sigma_s}_\xi : h^{\sigma_s}(z) = c^{\sigma_s}\}.$$

So, we have the partition of $\mathcal{X}$ into submanifolds

(1.7.15) $$\mathcal{X} = \bigcup_{c \in \mathbb{R}^r_+} \mathbf{h}^{-1}(c) \quad \text{(disjoint union)},$$

which gives the *stratification* of $\mathcal{X}$ associated with a given splitting (1.7.4).

Now assume that under the conditions of Proposition 1.7.2 we have a symplectic connection ∇ in $\mathcal{X}$ that is geometrically stable. Choose a ∇-invariant splitting (1.7.4) and the corresponding set of functions $h^1, \ldots, h^r$ (1.7.7). In addition, assume that the base B of $\mathcal{X}$ is a symplectic manifold equipped with a nondegenerate closed 2-form ω. By Proposition 1.5.1, we can view $\mathcal{X}$ as the model phase space $(\mathcal{X}, \Omega^{\nabla,\omega})$, where the symplectic structure $\Omega^{\nabla,\omega}$ is defined by (1.5.6).

THEOREM 1.7.1. *If the base B of $\mathcal{X}$ is a compact connected symplectic manifold, $\pi_1(B)$ is commutative, and a flat symplectic connection ∇ in $\mathcal{X}$ is geometrically stable, then the following conditions* (a)–(c) *are satisfied:*
(a) *The functions $h^1, \ldots, h^r$ (1.7.7) are in involution on $\mathcal{X}$:*

(1.7.16) $$\{h^\sigma, h^{\sigma'}\}_\mathcal{X} = 0 \quad (\sigma, \sigma' = 1, \ldots, r).$$

For each $\sigma = 1, \ldots, r$, the open domain

$$\{z \in \mathcal{X} : h^\sigma(z) > 0\}$$

is fibered by 2π-periodic trajectories of the Hamilton vector field X_{h^σ} of h^σ.
(b) *Let $f^1, \ldots, f^k$ ($k = \frac{1}{2} \dim B$) be a set of functions independent on an open dense domain $U \subseteq B$ and pairwise Poisson commuting on (B, ω),*

$$\{f^i, f^j\}_B = 0 \quad (i, j = 1, \ldots, k).$$

Then the functions

(1.7.17) $$F^1 = \pi^* f^1, \ldots, F^k = \pi^* f^k; F^{k+1} = h^1, \ldots, F^{k+r} = h^r$$

(here $\pi \colon \mathcal{X} \to B$ is the projection) form a complete set of functions in involution on $(\mathcal{X}, \Omega^{\nabla,\omega})$ that are functionally independent on the dense domain

(1.7.18) $$\bigcup_{\xi \in U} \{z \in \mathcal{X}_\xi : h^\sigma(z) > 0, \sigma = 1, \ldots, r\}.$$

(c) If $\Lambda \subset B$ is a Lagrangian submanifold of (B,ω), then the zero section $\Lambda \subset \mathcal{X}$ and the restriction $\mathcal{X}_\Lambda \subset \mathcal{X}$ are isotropic and coisotropic submanifolds of $(\mathcal{X}, \Omega^{\nabla,\omega})$, respectively. For each $c \in \mathcal{C}^s$, the level set

$$\mathcal{L}(c) = \mathbf{h}^{-1}(c) \cap \mathcal{X}_\Lambda \tag{1.7.19}$$

is an $(s + \dim \Lambda)$-dimensional isotropic submanifold of $(\mathcal{X}, \Omega^{\nabla,\omega})$. The $\mathbb{T}^s$-action of 2π-periodic flows of the Hamilton vector fields $X_{h^{\sigma_1}}, \ldots, X_{h^{\sigma_s}}$ ($c^{\sigma_1} > 0, \ldots, c^{\sigma_s} > 0$) on $\mathcal{L}(c)$ induces the fiber bundle over Λ (see (1.7.13))

$$\mathbb{T}^s \to \mathcal{L}(c) \to \Lambda \qquad (c \in \mathcal{C}^s). \tag{1.7.20}$$

In particular, if $c \in \mathcal{C}^r$ or $c = 0$, then $\mathcal{L}(c)$ is Lagrangian or $\mathcal{L}(0) = \Lambda$. Thus, there is a stratification of the coisotropic submanifold $\mathcal{X}_\Lambda$ by isotropic submanifolds

$$\mathcal{X}_\Lambda = \bigcup_{\substack{c \in \mathcal{C}^s \\ 0 \leq s \leq r}} \mathcal{L}(c). \tag{1.7.21}$$

PROOF. Since the symplectic connection ∇ is flat and geometrically stable, we can choose a basis of local sections $X_1^+(\xi), \ldots, X_r^+(\xi), X_1^-(\xi), \ldots, X_r^-(\xi)$ of $\mathcal{X}$ over a coordinate chart of B such that

$$\begin{aligned} \langle \mathcal{J}X_\sigma^+, X_{\sigma'}^- \rangle &= \delta_{\sigma\sigma'}, & \langle \mathcal{J}X_\sigma^+, X_{\sigma'}^+ \rangle &= \langle \mathcal{J}X_\sigma^-, X_{\sigma'}^- \rangle = 0, \\ X_\sigma^+(\xi) - iX_\sigma^-(\xi) &\in \rho_\xi^c, & \nabla(X_\sigma^+ - iX_\sigma^-) &= 0, \end{aligned} \tag{1.7.22}$$

for all $\sigma, \sigma' = 1, \ldots, r$. Then we choose local coordinates $(\xi^i, x_+^\sigma, x_-^{\sigma'})$ of $\mathcal{X}$, where $d\xi^i$ vanish along the vertical subbundle $\mathcal{V}$ from (1.5.1) and $(x_+^\sigma, x_-^{\sigma'})$ are the coordinates along the fibers of $\mathcal{X}$ corresponding to the basis $(X_\sigma^+, X_{\sigma'}^-)$. The functions h^σ (1.7.7) and the Poisson brackets (1.5.22) on $\mathcal{X}$ take the form

$$h^\sigma = \frac{1}{2}[(x_+^\sigma)^2 + (x_-^\sigma)^2], \tag{1.7.23}$$

$$\begin{aligned} \{x_\pm^\sigma, \xi^i\}_\mathcal{X} &= 0, & \{x_+^\sigma, x_-^{\sigma'}\}_\mathcal{X} &= \delta^{\sigma\sigma'}, \\ \{x_+^\sigma, x_+^{\sigma'}\}_\mathcal{X} &= \{x_-^\sigma, x_-^{\sigma'}\}_\mathcal{X} = 0. \end{aligned} \tag{1.7.24}$$

This proves the theorem. $\square$

Note that functions (1.7.7) give a straightforward generalization of Lewis's invariant [**Lew**].

We assume that a symplectic connection ∇ in $\mathcal{X}$ is given and the assumptions of Theorem 1.7.1 are satisfied.

Let $D \in \mathfrak{D}(\mathcal{X})$ be a covariant differential operator such that the vector field v on B associated with D (see §1.1) is a Hamilton vector field on (B,ω),

$$v = \mathrm{ad}(f) \equiv -\sum_{i,j=1}^k \omega^{ij}(\xi) \frac{\partial f}{\partial \xi^i} \frac{\partial}{\partial \xi^j}. \tag{1.7.25}$$

Here $f \in C^\infty(B)$, $\omega = \omega_{ij} d\xi^i \wedge d\xi^j$, and $\omega^{is}\omega_{sj} = \delta_j^i$.

Let us consider the vector field $D^\#$ on $\mathcal{X}$ induced by D and the decomposition of $D^\#$ into the horizontal and vertical components (see (1.2.8)):

$$D^\# = \nabla_{\mathrm{ad}(f)}^\# + \Xi^\#, \tag{1.7.26}$$

where $\Xi = \{\Xi(\xi) \in \mathrm{sp}(\mathcal{X}_\xi), \xi \in B\}$ is the field of linear Hamilton operators in fibers of $\mathcal{X}$,
$$\mathcal{J}(\xi) \circ \Xi(\xi) + \Xi^*(\xi) \circ \mathcal{J}(\xi) = 0 \qquad \forall\, \xi \in B.$$
Assume that D is ∇-reducible,

(1.7.27) $\qquad\qquad\qquad D$ is ∇-covariantly constant.

Then by Theorem 1.5.1, $D^\#$ is a Hamilton vector field on $(\mathcal{X}, \Omega^{\nabla,\omega})$,

(1.7.28) $\qquad D^\# = X_F, \qquad F = \pi^*(f) + \dfrac{1}{2}\langle \mathcal{J} \circ \Xi z, z\rangle, \quad z \in \mathcal{X}.$

Let us choose a point $\xi^0 \in B$ and generators $G_1, \ldots, G_m$ (1.7.3) of the reduced holonomy group $\widetilde{\mathrm{Hol}}^\nabla_{\xi^0}$.

THEOREM 1.7.2. *Suppose that the linear Hamilton operator $\Xi(\xi^0) \in \mathrm{sp}(\mathcal{X}_{\xi^0})$ from (1.7.26) satisfies the following three conditions:*
  (i) *$\Xi(\xi^0)$ is diagonalizable and its spectrum is pure imaginary.*
  (ii) *$\Xi(\xi^0)$ commutes with the generators $G_1, \ldots, G_m \in \widetilde{\mathrm{Hol}}^\nabla_{\xi^0}$ of the reduced holonomy group of ∇ at ξ^0:*
$$[\Xi(\xi^0), G_i] = 0, \qquad i = 1, \ldots, m.$$
  (iii) *$\Xi(\xi^0)$ commutes with elements $\psi(\gamma)$, $\gamma \in \mathrm{Tor}\, H_1(B)$.*

  Then the following assertions (a) and (b) are true:
  (a) *There exists a ∇-invariant splitting (1.7.4) that is also D-invariant (in particular, D is geometrically stable). The functions $h^1, \ldots, h^r$ (1.7.7) generated by the D-invariant splitting (1.7.4) are the integrals of motion in involution of the Hamilton system $(\mathcal{X}, \Omega^{\nabla,\omega}, F)$ (1.7.28),*

(1.7.29) $\qquad\qquad \{F, h^\sigma\}_{\mathcal{X}} = 0, \qquad \{h^\sigma, h^{\sigma'}\} = 0, \qquad \sigma, \sigma' = 1, \ldots, r.$

If $\Lambda \subset B$ is an $\mathrm{ad}(f)$-invariant Lagrangian submanifold, the functions $h^1, \ldots, h^r$ generate the stratification (1.7.21) (of the coisotropic submanifold $\mathcal{X}_\Lambda$) by X_F-invariant isotropic submanifolds $\mathcal{L}(c)$ (1.7.19). Moreover, if Λ is a quasiperiodic torus of the system (B, ω, f), then for each $c \in \mathbb{R}^r$, the submanifold $\mathcal{L}(c)$ is the quasiperiodic torus of the Hamilton system $(\mathcal{X}, \Omega^{\nabla,\omega}, F)$.
  (b) *If the Hamilton system (B, ω, f) is completely integrable, then $(\mathcal{X}, \Omega^{\nabla,\omega}, F)$ is also completely integrable.*

REMARK 1.7.2. Assumptions (i), (ii) of Theorem 1.7.2 are independent of the choice of a point ξ^0 and of generators $G_1, \ldots, G_m \in \widetilde{\mathrm{Hol}}^\nabla_{\xi^0}$.

REMARK 1.7.3. If ∇ is adapted to D, then conditions (i)–(iii) are automatically satisfied.

To prove this theorem we use the following proposition.

PROPOSITION 1.7.3. *The functions $h^1, \ldots, h^r$ (1.7.7) that are generated by a ∇-invariant splitting (1.7.4) Poisson commute with the Hamilton function F (1.7.17) on $(\mathcal{X}, \Omega^{\nabla,\omega})$ if and only if the splitting (1.7.4) is D-invariant.*

PROOF. Using the local coordinates $\xi = (\xi^i)$, $x = (x_+^\sigma, x_-^{\sigma'})$ of $\mathcal{X}$ from (1.7.24) and condition (1.7.27), we obtain the following local representation of the Hamilton function (1.7.17):

$$(1.7.30) \qquad F(\xi, x) = f(\xi) + \frac{1}{2}\langle J\Xi^0 x, x\rangle,$$

where $\Xi^0 \in \text{sp}(r)$, $J = \begin{bmatrix} 0 & -I \\ I & 0 \end{bmatrix}$. If the Poisson brackets of $h^1, \ldots, h^r$ and F are equal to zero, then it follows from (1.7.23) and (1.7.24) that the second term in (1.7.30) has the form

$$(1.7.31) \qquad \frac{1}{2}\sum_{\sigma=1}^{r} \gamma_\sigma[(x_+^\sigma)^2 + (x_-^\sigma)^2], \qquad \gamma_\sigma = \text{const}.$$

Combining (1.7.22) and (1.7.31), we see that (1.7.4) is D-invariant. By the same arguments, we prove the inverse statement. □

PROOF OF THEOREM 1.7.2. Assumptions (i), (ii), (iii) imply that $\exp\big(\Xi(\xi^0)\big)$, $G_1, \ldots, G_m$, $\psi(\gamma)$, $\gamma \in \text{Tor}\big(H_1(B)\big)$, are stable commuting symplectic operators. The above arguments show that there exists a common invariant splitting (1.7.4) at a given point ξ^0. Using the parallel transport of ∇, by (1.7.5), we obtain the D-invariant splitting (1.7.4) over all of B. Applying Proposition 1.7.3, we complete the proof. □

Now let us formulate the above results for the situation when we start with an isotropic submanifold of a "nonlinear" phase space.

Let (M, Ω, Λ) be a symplectic manifold with isotropic submanifold $\Lambda \subset M$. By Theorem 1.3.1, a coisotropic extension N of Λ induces a flat symplectic connection ∇ (1.3.6) in the symplectic normal bundle E over Λ.

DEFINITION 1.7.3. A coisotropic extension N of an isotropic submanifold $\Lambda \subset M$ is called *stable* if the flat symplectic connection ∇ (1.3.6) in E is geometrically stable.

If the connection ∇ (1.3.6) in E is geometrically stable and Λ satisfies the assumptions of Proposition 1.7.2, then there is a ∇-invariant splitting (1.7.4) of E which generates the functions $\tilde{h}^1, \ldots, \tilde{h}^r$ (1.7.7). Moreover, the pull-back $\tilde{\nabla}$ of ∇ to the Whitney sum $\mathcal{X} = T^*\Lambda \oplus E$ under the projection $j: T^*\Lambda \to \Lambda$ will also be a flat geometrically stable connection in $\mathcal{X}$. The $\tilde{\nabla}$-invariant splitting (1.7.4) of $^{\mathbb{C}}\mathcal{X}$ is induced by the pull-back of the ∇-invariant splitting of $^{\mathbb{C}}E$. If a function H is constant along the coisotropic extension N of Λ, which corresponds to ∇, then the first variation operator D_H automatically satisfies assumptions (i), (ii), (iii) of Theorem 1.7.2. So, we obtain the following statement.

THEOREM 1.7.3. *Let $\Lambda \subset M$ be an isotropic submanifold satisfying the conditions*:
  (i) *Λ is compact and connected, and $\pi_1(\Lambda)$ is commutative*;
  (ii) *there exists a stable coisotropic extension N of Λ.*
Then, for each function $H \in C^\infty(M)$ constant along N, $H\big|_N = \text{const}$, the first variation operator $D_H \in \mathfrak{D}(E)$ of the Hamilton vector field $\text{ad}(H)$ possesses the following two properties.

(a) The flow of $D_H^{\#}$ on E coincides with the flow of the Hamilton system $(\mathcal{X} = T^*\Lambda \oplus E, \Omega^\nabla, F)$, where ∇ is the symplectic connection in E associated with the stable coisotropic extension N of Λ, Ω^∇ is the symplectic structure from (1.5.23), and

(1.7.32) $$F = \pi^*\bigl(\langle \varrho, \mathrm{ad}(H)\rangle\big|_\Lambda\bigr)$$

(here $\pi\colon \mathcal{X} \to T^*\Lambda$ is the projection and ϱ is the Liouville 1-form on $T^*\Lambda$). There is a stratification of E (as a coisotropic submanifold of $\mathcal{X}$) into $D_H^{\#}$-invariant isotropic submanifolds:

(1.7.33) $$E = \bigcup_{c \in \mathbb{R}_+^r} \widetilde{\mathbf{h}}^{-1}(c),$$

where the surjective mapping $\widetilde{\mathbf{h}}\colon E \to \mathbb{R}_+^r$ is defined by (1.7.7) and by the ∇-invariant splitting (1.7.4) of $^{\mathbb{C}}E$.

(b) If Λ is an $\mathrm{ad}(H)$-invariant quasiperiodic torus, then the Hamilton system $(\mathcal{X} = T^*\Lambda \oplus E, \Omega^\nabla, F)$ is completely integrable.

§1.8. Variations of holonomy and characteristic multipliers

Let $(\pi\colon \mathcal{X} \to B, \mathcal{J})$ be a symplectic vector bundle over a connected base B. Assume that there is a fiber bundle

(1.8.1) $$\Lambda \to B \xrightarrow{p} \mathfrak{P} \qquad (\Lambda \text{ is a typical fiber})$$

whose fibers $\Lambda_a = p^{-1}(a) \approx \Lambda$ are compact connected manifolds. Denote by $\mathcal{X}_{\Lambda_a}$ the restriction $\mathcal{X}\big|_{\Lambda_a}$ of $\mathcal{X}$ to the fiber Λ_a, that is, the pull-back of $\mathcal{X}$ to Λ_a via the inclusion $\Lambda_a \subset B$.

Assume also that for each $a \in \mathfrak{P}$, we have a symplectic connection ∇^a in $\mathcal{X}_{\Lambda_a}$ such that the collection $\{\nabla^a, a \in \mathfrak{P}\}$ forms a *smooth* family, that is, ∇^a varies smoothly with respect to $a \in \mathfrak{P}$. The question is: how does the holonomy group of ∇^a vary if one varies a?

We answer this question under the following assumptions:

(i) ∇^a is flat for each $a \in \mathfrak{P}$.

(ii) There is a global section of the fiber bundle $p\colon B \to \mathfrak{P}$, $e\colon \mathfrak{P} \to B$, $p \circ e = \mathrm{id}$.

(iii) The fiber bundle $p\colon B \to \mathfrak{P}$ admits an atlas of charts such that the corresponding transition functions belong to the group $\mathrm{Diff}_0(\Lambda)$ consisting of all diffeomorphisms homotopic to the identity mapping of the typical fiber Λ.

Note that condition (iii) holds if $B \approx \Lambda \times \mathfrak{P}$ is a trivial fiber bundle. Also note that conditions similar to (ii) and (iii) appear in the construction of *global action-angle variables* [**Ne**$_1$, **DzDl**, **Du**]).

Assume that the fundamental group $\pi_1(\Lambda)$ is finitely generated, that is, there are 1-cycles $\gamma_1, \ldots, \gamma_m$ of Λ that generate $\pi_1(\Lambda)$. For each $a \in \mathfrak{P}$, consider the holonomy group $\mathrm{Hol}_{e(a)}^{\nabla^a} \subset \mathrm{Sp}(\mathcal{X}_{e(a)})$ of the flat connection ∇^a at $e(a) \in \Lambda_a$. This holonomy group is also finitely generated. Let

$$\mathrm{Hol}_{\mathfrak{P}} = \bigcup_{a \in \mathfrak{P}} \mathrm{Hol}_{e(a)}^{\nabla^a}.$$

PROPOSITION 1.8.1. (a) *Under assumptions* (i)–(iii), *we have the following mappings* (*sections*):

(1.8.2) $\quad G_j \colon \operatorname{Hol}_{\mathfrak{P}} \to \mathfrak{P}, \qquad \mathfrak{P} \ni a \mapsto G_j(a) \in \operatorname{Hol}^{\nabla^a}_{e(a)}, \qquad j = 1, \ldots, m,$

such that for each $a \in \mathfrak{P}$, *the elements* $G_1(a), \ldots, G_m(a)$ *generate the holonomy group* $\operatorname{Hol}^{\nabla^a}_{e(a)}$ *and smoothly depend on* a.

(b) *If there exists a flat connection* ∇ *in* $\mathcal{X}$ *such that*

(1.8.3) $\qquad\qquad\qquad \nabla\big|_{\mathcal{X}_{\Lambda_a}} = \nabla^a \qquad \forall\, a \in \mathfrak{P},$

then the spectra of $G_1(a), \ldots, G_m(a)$ *do not depend on* $a \in \mathfrak{P}$,

(1.8.4) $\qquad \operatorname{spec} G_j(a) = \{\lambda_j^1, \ldots, \lambda_j^{2r}\} \qquad \forall\, j = 1, \ldots, m, \quad a \in \mathfrak{P},$

where λ_j^σ *are constants.*

This proposition allows us to obtain the following statements. Assume that, in addition to conditions (i)–(iii), there is a line subbundle $\rho \subset \mathcal{X}$ possessing the property

(1.8.5) $\qquad\qquad\qquad \rho_{\Lambda_a}$ is ∇^a-invariant for all $a \in \mathfrak{P}$.

Then it follows from Proposition 1.8.1 that there are *smooth functions* $\lambda_j^\rho = \lambda_j^\rho(a) \in C^\infty(\mathfrak{P}) \otimes \mathbb{C}$ $(j = 1, \ldots, m)$ that are eigenvalues of $G_1(a), \ldots, G_m(a)$,

(1.8.6) $\qquad\qquad \lambda_j^\rho(a) \in \operatorname{spec}(G_j(a)) \qquad (j = 1, \ldots, m) \quad \forall\, a \in \mathfrak{P}$

corresponding to the common eigenspace $\rho_{e(a)}$. Assume that there exists a connection ∇ in $\mathcal{X}$ (not necessarily flat) satisfying (1.8.3) and such that

(1.8.7) $\qquad\qquad\qquad\qquad \rho$ is ∇-invariant.

Denote by C^ρ the curvature 2-form of $\nabla\big|_\rho$ on B. By conditions (ii) and (iii), there is an isomorphism

(1.8.8) $\qquad\qquad\qquad \imath_a \colon \pi_1(\Lambda) \mapsto \pi_1(\Lambda_a), \qquad a \in \mathfrak{P}.$

Choose a basis of 1-cycles $\gamma_1, \ldots, \gamma_m \in \Lambda$. Then for each $a \in \mathfrak{P}$ we can define basic cycles $\Gamma_1(a), \ldots, \Gamma_m(a) \in \Lambda_a$ such that

(1.8.9) $\qquad\qquad [\Gamma_i(a)] = [\imath_a(\gamma_i)] \qquad \forall\, i = 1, \ldots, m, \quad a \in \mathfrak{P}.$

Choosing a point $a^0 \in \mathfrak{P}$ and using the curvature 2-form C^ρ, we obtain the "variation" formula

(1.8.10) $\qquad\qquad \lambda_j^\rho(a) = \lambda_j^\rho(a^0) \exp\left(\int_{\Sigma_j(a)} C^\rho \right).$

Here $\Sigma_j(a)$ is a membrane (surface) in B such that $\partial \Sigma_j(a) = \Gamma_j(a) \cup \Gamma_j(a^0)$.

If the connection ∇ is flat, then $C^\rho = 0$, and (1.8.10) implies

$$\lambda_j^\rho(a) = \lambda_j^\rho(a^0) \quad \forall\, a \in \mathfrak{P} \implies \lambda_j^\rho(a) = \operatorname{const}.$$

Moreover, if ρ is trivial, then $\lambda_j^\rho(a^0)$ can be written as

(1.8.11) $\qquad\qquad \lambda_j^\rho(a^0) = \exp\left(-\oint_{\Gamma_j(a^0)} \theta^\rho \right).$

Here θ^ρ is the closed connection 1-form of $\nabla^{a^0}\big|_\rho$ on Λ_{a^0} corresponding to a global section Z of ρ, $\nabla^{a^0} Z = \theta^\rho \otimes Z$.

EXAMPLE 1.8.1. Let us consider the same situation as in Example 1.3.1.

Let $(\mathcal{Y}, f)$ be a pair that consists of a symplectic submanifold $\mathcal{Y}$ of (M, Ω) and of a set of functions $f = (f_1, \ldots, f_k)$ on M satisfying conditions (i)–(iii) of Example 1.3.1. In this case, the level sets of $f_1, \ldots, f_k$ on $\mathcal{Y}$ define k-dimensional Lagrangian submanifolds Λ_a, $a \in U$. By assumption, $\Lambda_{a^0} \approx \mathbb{T}^k$ for a certain $a^0 \in U$. Choose a neighborhood $B \approx \mathbb{T}^k \times U$ of Λ_{a^0} in $\mathcal{Y}$, satisfying properties (a)–(c) from Example 1.3.1. In particular, B is equipped with action-angle coordinates $s = (s^1, \ldots, s^k)$, $\alpha = (\alpha^1, \ldots, \alpha^k)$. Then we have the following objects:
- a symplectic vector bundle over B: $\mathcal{X} = (T_B \mathcal{Y})^\perp$;
- a fiber bundle over an open domain $U \subset \mathbb{R}^k$: $\mathbb{T}^k \to B \xrightarrow{p} U$, associated with the action-angle coordinates (s, α);
- a flat symplectic connection ∇^a (1.3.6) in $\mathcal{X}_{\Lambda_a}$ for each $a \in U$.

In addition, assume that the symplectic submanifold $B \subset M$ is the zero-level set of some functions $\Phi = (\Phi_1, \ldots, \Phi_{2r})$, $B = \{\Phi_1 = 0, \ldots, \Phi_{2r} = 0\}$.

In particular, since

$$(1.8.12) \qquad X_1(s, \alpha) = \mathrm{ad}(\Phi_1)\big|_B, \ldots, X_{2r}(s, \alpha) = \mathrm{ad}\,\Phi_{2r}\big|_B$$

form the basis of global sections of $\mathcal{X}$, it follows that $\mathcal{X}$ is trivial. Moreover, we have

$$\det J(s, \alpha) \neq 0 \quad \forall (s, \alpha) \in B, \qquad J(s, \alpha) \stackrel{\text{def}}{=} ((\{\Phi_{\sigma'}, \Phi_\sigma\}\big|_B)).$$

Using the set of functions $f = (f_1, \ldots, f_k)$, $\Phi = (\Phi_1, \ldots, \Phi_{2r})$, we define $2r \times 2r$ matrix-valued *smooth* functions $\theta_1(s, \alpha), \ldots, \theta_k(s, \alpha) \in C^\infty(B) \otimes \mathrm{Sp}(r)$ via formula (1.3.6). Then for each $s = a \in U$, the matrix-valued functions $\theta_i(s, \alpha)$ give the connection 1-form of ∇^a. Hence, $\theta_i(s, \alpha)$ satisfy the following equations on $B \approx \mathbb{T}^k \times U$:

$$\frac{\partial \theta_j(s, \alpha)}{\partial \alpha^i} - \frac{\partial \theta_i(s, \alpha)}{\partial \alpha^j} + [\theta_i(s, \alpha), \theta_j(s, \alpha)] = 0,$$

$$\theta_i^*(s, \alpha) J(s, \alpha) + J(s, \alpha) \theta_i(s, \alpha) = \frac{\partial J(s, \alpha)}{\partial \alpha^i}$$

for all $i, j = 1, \ldots, k$.

Thus, the connection ∇ in $\mathcal{X}$, satisfying the compatibility conditions (1.8.3), can be represented by the following matrix-valued 1-form on B (with respect to the trivialization (1.8.12)):

$$(1.8.13) \qquad \theta = \sum_{i=1}^k \left(\theta_i(s, \alpha) \, d\alpha^i + \mu_i(s, \alpha) \, ds^i \right),$$

where $\mu_i(s, \alpha)$ are any smooth $2r \times 2r$ matrix-valued functions. The curvature 2-form C of (1.8.13) is given by the formula

$$C = \sum_{i,j=1}^k \left[\left(\frac{\partial \theta_i}{\partial s^j} - \frac{\partial \mu_j}{\partial \alpha^i} + [\mu_j, \theta_i] \right) ds^j \wedge d\alpha^i + \left(\frac{\partial \mu_i}{\partial s^j} + \mu_j \mu_i \right) ds^j \wedge ds^i \right].$$

This implies that the problem of the existence of a flat connection ∇ in $\mathcal{X}$ satisfying (1.8.3) is reduced to solving the following equations on B for $\mu_1(s,\alpha),\ldots,\mu_k(s,\alpha)$:

$$\frac{\partial \mu_j}{\partial \alpha^i} + [\theta_i, \mu_j] = \frac{\partial \theta_i}{\partial s^j}, \qquad \frac{\partial \mu_j}{\partial s^i} - \frac{\partial \mu_i}{\partial s^j} + [\mu_i, \mu_j] = 0$$

for all $i,j = 1,\ldots,k$.

§1.9. Action-angle variables and the quasiperiodic motion

Let $(\mathcal{X}, \Omega^{\nabla,\omega})$ be the model phase space generated by a flat symplectic connection ∇ in a symplectic vector bundle $\pi\colon \mathcal{X} \to B$ over a manifold B endowed with a symplectic form ω.

We make the following assumptions.

(i) On B there are *action-angle variables* $s = (s^1,\ldots,s^k)$, $\alpha = (\alpha_1,\ldots,\alpha_k)$ (mod 2π) such that $\omega = ds \wedge d\alpha$, and B is trivially fibered by the Lagrangian k-tori

$$\Lambda(a) = \{s^1 = a^1,\ldots, s^k = a^k\}, \qquad a = (a^1,\ldots,a^k) \in U,$$

over a connected open bounded domain $U \subset \mathbb{R}^k$. So, smoothly varying $a \in U$, we can choose the basis of 1-cycles on $\Lambda(a)$:

$$\Gamma_i(a) = \{\alpha_1 = 0,\ldots, 0 \leq \alpha_i \leq 2\pi,\ldots, \alpha_k = 0\} \qquad (i = 1,\ldots, k).$$

Then

$$s^i = \frac{1}{2\pi} \oint_{\Gamma_i(a)} \varrho, \qquad \omega = d\varrho.$$

(ii) The connection ∇ in $\mathcal{X}$ is geometrically stable.

Then, by Proposition 1.7.2, there is a ∇-invariant splitting (1.7.4) over B. Consider the surjective mapping $\mathbf{h} = (h^1,\ldots,h^r)$ defined by (1.7.7), and denote

$$\mathcal{X}^{\mathrm{reg}} = \{z \in \mathcal{X}\colon h^\sigma(z) > 0, \sigma = 1,\ldots,r\}.$$

It follows from Theorem 1.7.1 that for each $a \in U$, the restriction $\mathcal{X}^{\mathrm{reg}}_{\Lambda(a)}$ is a coisotropic submanifold of $(\mathcal{X}, \Omega^{\nabla,\omega})$.

PROPOSITION 1.9.1. *There is a trivial fiber bundle*

(1.9.1) $$\mathbb{T}^{k+r} \to \mathcal{X}^{\mathrm{reg}} \to U \times C^r$$

whose fibers are $(k+r)$-dimensional Lagrangian tori

(1.9.2) $$\mathcal{L}(a,c) = \mathcal{X}^{\mathrm{reg}}_{\Lambda(a)} \cap \mathbf{h}^{-1}(c), \qquad a \in U, \quad c \in C^r,$$

and 2π-periodic flows of the Hamilton vector fields $X_{s^1},\ldots,X_{s^k}, X_{h^1},\ldots,X_{h^r}$ generate the Hamilton action of $\mathbb{T}^{k+r}$ on $\mathcal{X}^{\mathrm{reg}}$ whose orbits coincide with $\mathcal{L}(a,c)$.

Now let us introduce action-angle variables on $\mathcal{X}^{\mathrm{reg}}$ that are associated with the action-angle variables (s,α) on B and with the Lagrangian fiber bundle (1.9.1).

Recall that the ∇-invariant splitting (1.7.4) induces the real symplectic splitting (1.7.10) by 2-dimensional ∇-invariant symplectic subbundles:

(1.9.3) $$\mathcal{X}^\sigma = \mathcal{X} \cap (\rho^\sigma \oplus \overline{\rho}^\sigma) \qquad (\sigma = 1,\ldots,r).$$

PROPOSITION 1.9.2. *Symplectic subbundles $\mathcal{X}^1, \ldots, \mathcal{X}^r$ from (1.9.3) are trivial; that is, there are global sections $Y_\sigma^+, Y_\sigma^- \in \Gamma(\mathcal{X}^\sigma)$ such that*

$$(1.9.4) \quad \langle JY_\sigma^+, Y_{\sigma'}^+ \rangle = \langle JY_\sigma^-, Y_{\sigma'}^- \rangle = 0, \quad \langle JY_\sigma^+, Y_{\sigma'}^- \rangle = \delta_{\sigma\sigma'}$$

for all $\sigma, \sigma' = 1, \ldots, r$.

Choose a symplectic trivialization (1.9.4) and introduce *Floquet 1-forms* of ∇ on B by the formula

$$\beta_\sigma = \langle JY_\sigma^+, \nabla Y_\sigma^+ \rangle \quad (\sigma = 1, \ldots, r).$$

Since ∇ is flat, the β_σ are *closed* on B and

$$\nabla(Y_\sigma^+ - iY_\sigma^-) = i\beta_\sigma \otimes (Y_\sigma^+ - iY_\sigma^-).$$

Then the *Floquet exponents* $\beta_\sigma^1, \ldots, \beta_\sigma^k$ related to the basic 1-cycles $\Gamma_1(a), \ldots, \Gamma_k(a)$ are defined as the periods of Floquet 1-forms:

$$\beta_\sigma^i = \frac{1}{2\pi} \oint_{\Gamma_i(a)} \beta_\sigma \quad (i = 1, \ldots, k).$$

It follows from Proposition 1.9.1 that the $\beta_{\sigma,i}^i$ do not depend on $a \in U$. If $G_1, \ldots, G_k$ are the generators of the holonomy group $\text{Hol}_{\xi^0}^\nabla$, related to 1-cycles $\Gamma_1(a), \ldots, \Gamma_k(a)$, then

$$\text{spec}\, G_j = \{\exp(\mp 2\pi i \beta_\sigma^j) \mid \sigma = 1, \ldots, r\}.$$

THEOREM 1.9.1. *Under assumptions (i), (ii), the action-angle variables (s, α) on B and the Lagrangian fiber bundle (1.9.1) induce the action-angle variables $I = (I^1, \ldots, I^k, I^{k+1}, \ldots, I^{k+r})$, $\varphi = (\varphi_1, \ldots, \varphi_k, \varphi_{k+1}, \ldots, \varphi_{k+r})$ (mod 2π) on $\mathcal{X}^{\text{reg}}$:*

$$\Omega^{\nabla,\omega} = dI \wedge d\varphi.$$

These coordinates are defined by the formulas

$$(1.9.5) \quad I^i = \pi^* s^i + \sum_{\sigma=1}^r \beta_\sigma^i h^\sigma \quad (i = 1, \ldots, k), \quad I^{k+\sigma} = h^\sigma \quad (\sigma = 1, \ldots, r),$$

and

$$\varphi_i = \pi^* \alpha_i, \quad \varphi_{k+\sigma} = t_\sigma - \sum_{i=1}^k \beta_\sigma^i \alpha_i,$$

where t_σ is "time" along the trajectories of the Hamilton vector field X_{h^σ}.

PROOF. By $y_+^1, \ldots, y_+^r, y_-^1, \ldots, y_-^r$ we denote the global coordinates on $\mathcal{X}$ that correspond to the symplectic trivialization (1.9.4). Then the symplectic 2-form $\Omega^{\nabla,\omega}$ takes the form

$$\Omega^{\nabla,\omega} = d(\pi^* \rho + \mu^\nabla),$$

where

$$(1.9.6) \quad \mu^\nabla = \langle y_+, dy_- \rangle - \frac{1}{2} d(\langle y_+, y_- \rangle) + \frac{1}{2} \sum_{\sigma=1}^r \beta_\sigma \big[(y_+^\sigma)^2 + (y_-^\sigma)^2\big].$$

Using the global sections $Y_1^+, \ldots, Y_r^+ \in \Gamma(\mathcal{X})$, we can choose the basis of $\mathcal{L}(a,c)$, consisting of 1-cycles $\widetilde{\Gamma}_1(a,c), \ldots, \widetilde{\Gamma}_k(a,c), \widetilde{\Gamma}_{k+1}(a,c), \ldots, \widetilde{\Gamma}_{k+r}(a,c)$ that smoothly depend on $(a,c) \in U \times C^r$ and satisfy the relations

$$\pi\big(\widetilde{\Gamma}_i(a,c)\big) = \Gamma_i(a) \quad (i = 1, \ldots, k), \qquad \pi\big(\widetilde{\Gamma}_{k+\sigma}(a,c)\big) = \xi^0 \quad (\sigma = 1, \ldots, r).$$

Here $\xi^0 \in B$ is a given point. Defining the action variables $I^1, \ldots, I^{k+r}$ on $\mathcal{X}^{\text{reg}}$ by the standard formula

$$I^j = \frac{1}{2\pi} \oint_{\widetilde{\Gamma}_j(a,c)} (\pi^*\varrho + \mu^\nabla)$$

and using (1.9.6), we obtain (1.9.5). □

PROPOSITION 1.9.3. *If a covariant differential operator $D \in \mathfrak{D}(\mathcal{X})$ is ∇-reducible, then the spectrum of the linear operator $\Xi(\xi) \in \text{sp}(\mathcal{X}_\xi)$ from (1.7.26) is independent of a point $\xi \in B$.*

DEFINITION 1.9.1. The eigenvalues of $\Xi(\xi)$ associated with a ∇-reducible covariant differential operator $D \in \mathfrak{D}(\mathcal{X})$ are called ∇-*relative Floquet exponents* of D.

Assume that a ∇-reducible operator $D \in \mathfrak{D}(\mathcal{X})$ satisfies (1.7.25) and assumptions (1), (2) of Theorem 1.7.2 at a point ξ^0. Then there exists a Kählerian subspace $\rho(\xi^0) \in {}^{\mathbb{C}}\mathcal{X}_{\xi^0}$ such that
- $\rho(\xi^0)$ is G-invariant $\forall G \in \text{Hol}_{\xi^0}^\nabla$;
- $\rho(\xi^0)$ is $\Xi(\xi^0)$-invariant.

So, the restriction $\Xi(\xi^0)\big|_{\rho(\xi^0)}$ is an Hermitian operator, and we can introduce "positive" ∇-relative Floquet exponents of D as follows:

$$\{i\gamma_1, \ldots, i\gamma_r\} = \text{spec}\, \Xi(\xi^0)\big|_{\rho(\xi^0)}, \qquad \gamma \in \mathbb{R}.$$

Under the cited assumptions on $D \in \mathfrak{D}(\mathcal{X})$, Theorem 1.7.2 and Theorem 1.9.1 imply the following statement.

THEOREM 1.9.2. *Let $(\mathcal{X}, \Omega^{\nabla, \omega}, F)$ be the Hamilton system (1.7.28) associated with $D \in \mathfrak{D}(\mathcal{X})$. Let the Hamilton function $f \in C^\infty(B)$ from (1.7.28) depend only on the action variables $s = (s^1, \ldots, s^k)$, $f = f(s)$, and let $\omega(a) = (\omega_1(a), \ldots, \omega_k(a))$, $\omega_i(a) = \partial f(a)/\partial s_i$, be the frequencies of the field $\text{ad}(f)$ on the quasiperiodic tori $\Lambda(a)$.*

Then $\mathcal{L}(a,c)$ (1.9.2) are quasiperiodic tori of the Hamilton system $(\mathcal{X}, \Omega^{\nabla,\omega}, F)$ with frequency vector $\widetilde{\omega} = (\widetilde{\omega}_1, \ldots, \widetilde{\omega}_k, \widetilde{\omega}_{k+1}, \ldots, \widetilde{\omega}_{k+r})$ defined by

(1.9.7)
$$\widetilde{\omega}_i = \omega_i(a) \qquad (i = 1, \ldots, k),$$
$$\widetilde{\omega}_{k+\sigma} = -\sum_{i=1}^k \omega_i(a)\beta_\sigma^i + \gamma_\sigma \qquad (\sigma = 1, \ldots, r).$$

Note that the first term in (1.9.7) is the so-called *rotation number* [**JM**] of the flow $\mathbb{S}^t$ of D on $\mathcal{X}^\sigma$, which measures the mean rotation of $\mathbb{S}^t$ on the $\{Y_\sigma^+, Y_\sigma^-\}$-plane. The rotation number can be expressed via *averaging* over the k-torus $\Lambda(a)$:

$$-\sum_{i=1}^k \omega_i(a)\beta_\sigma^i = -\int_{\Lambda(a)} (\text{ad}(f) \lrcorner \beta_\sigma) \frac{d\alpha_1 \wedge \cdots \wedge d\alpha_k}{(2\pi)^k}.$$

In the next section we discuss in more detail the properties of the rotation numbers in the context of linearized Hamilton dynamics over isotropic submanifolds.

§1.10. Invariants of stable symplectic connections. Generalized rotation numbers and Gelfand–Lidskii indices

In this section, we discuss some aspects of the classification theory of geometrically stable symplectic connections and introduce generalized versions of the dynamical invariants such as the *Johnson–Moser rotation number* [**JM**] and the *Gelfand–Lidskii index* [**GL**]. Conceptually, we shall follow the approach suggested in [**KaVo**$_{1-3}$].

Let $(E, \mathcal{J})$ be a real symplectic vector bundle over a manifold Λ. We assume that E admits a Lagrangian subbundle. In this case, E possesses the following properties:

- The first Chern class $c_1(E) \in H^2(\Lambda, \mathbb{R})$ is trivial (this shows that a symplectic vector bundle with nontrivial first Chern class does not admit Lagrangian subbundles [**We**$_2$]).
- The complexification $^{\mathbb{C}}E$ admits a Kählerian subbundle (this property is based on the fact [**We**$_2$, **LiMr**, **Wo**] that in a given symplectic vector bundle E there exists a complex structure compatible with the symplectic structure on E).
- There exists a geometrically stable symplectic connection in E.

(I) *Gauge forms and the Maslov class.* Fixing a Lagrangian subbundle $\mathcal{L} \subset E$, we can associate a gauge 1-form $\varkappa$ on the base Λ with a given geometrically stable symplectic connection ∇ in E in the following way. Let us choose a ∇-invariant Kählerian subbundle $\rho \subset {}^{\mathbb{C}}E$, and let us consider in ρ the induced Hermitian connection

$$(1.10.1) \qquad \nabla^+ = \nabla\Big|_\rho.$$

Recall that the Hermitian structure on ρ is determined by the form $(\cdot, \cdot)_+ = \frac{1}{2i}\langle \mathcal{J}\cdot, \overline{\cdot}\rangle$. Let $Z_1, \ldots, Z_r$ and $u_1, \ldots, u_r$ be two bases of local sections of ρ and $\mathcal{L}$, respectively (where $r = \dim \rho$). Define the function

$$(1.10.2) \qquad \mathbf{j} = \frac{\det\langle \mathcal{J}u \overset{\otimes}{,} Z\rangle}{\det\langle \mathcal{J}u \overset{\otimes}{,} \overline{Z}\rangle}, \qquad |\mathbf{j}| > 0.$$

Here we use the notation $((\langle \mathcal{J}u\overset{\otimes}{,}Z\rangle))_{\sigma\sigma'} = \det\langle \mathcal{J}u_\sigma, Z_{\sigma'}\rangle$. Note that $\mathbf{j}$ is the Jacobian of the projection of ρ to $\overline{\rho}$ along $^{\mathbb{C}}\mathcal{L}$. Now we define the gauge 1-form $\varkappa = \varkappa(L, \nabla)$ of a connection ∇ relative to a Lagrangian subbundle $\mathcal{L}$ as follows:

$$(1.10.3) \qquad \varkappa = \varkappa(L, \nabla) \overset{\mathrm{def}}{=} \mathrm{Im}\left(-\frac{1}{2}\mathbf{j}^{-1}d\mathbf{j} + \mathrm{tr}\,\theta^+\right),$$

where θ^+ is the matrix-valued connection form of ∇^+ taken with respect to the basis $\{Z_\sigma\}$ in (1.10.2). Obviously, under the change of the basis $\{Z_\sigma\}$, the first and the second terms in (1.10.3) have the same transition rules up to the sign. So, $\varkappa$ is a well-defined 1-form on Λ satisfying the relation

$$(1.10.4) \qquad i\,d\varkappa = \mathrm{tr}\,\mathcal{C}^+,$$

where $\mathcal{C}^+$ is the curvature 2-form of ∇^+. In particular, it follows from (1.10.4) that $c_1(E) = 0$.

PROPOSITION 1.10.1. (a) *For any geometrically stable symplectic connection ∇ in E formula* (1.10.3) *determines the global 1-form $\varkappa(L, \nabla)$ on Λ. If ∇ satisfies the weak flatness condition*

(1.10.5) $$\operatorname{tr} \mathcal{C}^+ = 0,$$

then $\varkappa(\mathcal{L}, \nabla)$ is a closed 1-form.

(b) *The functorial property holds. That is, if $f\colon \widetilde{\Lambda} \to \Lambda$ is a C^∞-map, then*

$$f^*\varkappa(\mathcal{L}, \nabla) = \varkappa(f^*\mathcal{L}, f^*\nabla),$$

where $f^*\mathcal{L}$ is the Lagrangian subbundle in the pull-back f^*E of E via f.

(c) *If $\mathcal{L}$ and $\widetilde{\mathcal{L}}$ are two Lagrangian subbundles of E, then*

(1.10.6) $$\varkappa(\mathcal{L}, \nabla) = \varkappa(\widetilde{\mathcal{L}}, \nabla) - \pi\mu(\mathcal{L}, \widetilde{\mathcal{L}}).$$

Here $\mu(\mathcal{L}, \widetilde{\mathcal{L}})$ is the Arnold 1-form on Λ determining the Maslov class for Lagrangian subbundles $\mathcal{L}$ and $\widetilde{\mathcal{L}}$.

Statements (a) and (b) readily follow from the definition of the gauge form.

To justify statement (c), we recall the definition of the Arnold form and properties of the Maslov class (see also [**A**$_1$, **KaVo**$_5$, **KaMas**]).

Let $\mathcal{L}$ and $\widetilde{\mathcal{L}}$ be two Lagrangian subbundles of E, and let ρ be a Kählerian subbundle of $^\mathbb{C}E$. Consider the decomposition $^\mathbb{C}E = {}^\mathbb{C}\widetilde{\mathcal{L}} \oplus \rho$ and the projection

(1.10.7) $$^\mathbb{C}\mathcal{L} \to {}^\mathbb{C}\widetilde{\mathcal{L}} \qquad \text{along} \quad \rho.$$

In local terms, the Jacobian $\mathbb{J}$ of projection (1.10.7) can be written as follows:

(1.10.8) $$\mathbb{J} = \frac{\det\langle \mathcal{J}u^{\otimes}_?, Z\rangle}{\det\langle \mathcal{J}\widetilde{u}^{\otimes}_?, Z\rangle}.$$

Here $\{u_\sigma\}$, $\{\widetilde{u}_\sigma\}$, and $\{Z_\sigma\}$ are bases of local sections of the subbundles $\mathcal{L}$, $\widetilde{\mathcal{L}}$, and ρ, respectively.

Then the Arnold 1-form on Λ associated with a pair $(\mathcal{L}, \widetilde{\mathcal{L}})$ is defined by

(1.10.9) $$\mu(\mathcal{L}, \widetilde{\mathcal{L}}) = \left(\frac{\mathbb{J}^2}{|\mathbb{J}|^2}\right)^* \phi,$$

where $\phi = -dz/2\pi i z$ is the fundamental 1-form on the unit circle $\mathbb{S}^1 = \{|z| = 1\}$.

Comparing (1.10.8) and (1.10.9) with (1.10.2) and (1.10.3), we deduce (1.10.6).

The Maslov class of a pair $(\mathcal{L}, \widetilde{\mathcal{L}})$ is determined by the cohomology class of $\mu(\mathcal{L}, \widetilde{\mathcal{L}})$, which is integral, $[\mu(\mathcal{L}, \widetilde{\mathcal{L}})] \in H^1(\Lambda; \mathbb{Z})$, and independent of the choice of a Kählerian subbundle ρ.

Now assume that we have a geometrically stable symplectic connection ∇ in E satisfying the weak flatness condition (1.10.5) relative to a ∇-invariant Kählerian subbundle ρ. Then the determinant bundle $\det(\rho)$ of ρ is flat, and there is a flat connection in $\det(\rho)$ induced by the Hermitian connection ∇^+ (1.10.1). For a closed path γ through a point $\alpha^0 \in \Lambda$, consider the parallel transport G^+_γ of ∇^+ around γ.

Then the map $\gamma \mapsto \det G^+_\gamma$ induces a homomorphism of the homology group $H_1(\Lambda)$ into the Abelian group $\mathbb{S}^1$. Using the gauge 1-form $\varkappa(\mathcal{L}, \nabla)$ (1.10.3), we obtain

$$\det G^+_\gamma = \exp\left(-i \oint_\gamma \varkappa(\mathcal{L}, \nabla)\right) \cdot \exp\left(-i\pi w_1^L(\gamma)\right),$$

where w_1^L is the first Stiefel–Whitney class of a Lagrangian subbundle L.

(II) *Rotation numbers.* Here we assume that the base Λ of a symplectic vector bundle E is a compact connected manifold.

Let $D \in \mathfrak{D}(E)$ be a covariant differential operator in E, and let $v \in \mathrm{Vect}(\Lambda)$ be the anchor of D (see §1.1). Assume that D is geometrically stable and choose a D-invariant Kählerian subbundle ρ of $^{\mathbb{C}}E$. Consider the restriction D^+ of D to ρ, $D^+ = D\big|_\rho$. Then D^+ is a covariant differential operator in ρ preserving the Hermitian structure on ρ. Introduce a global real C^∞-function $\mathbf{r}(\mathcal{L}, D)$ on Λ associated with D and a Lagrangian subbundle $\mathcal{L} \subset E$ by the formula

$$(1.10.10) \qquad \mathbf{r}(\mathcal{L}, D) \stackrel{\text{def}}{=} \mathrm{Im}\left(\frac{1}{2}\mathbf{j}^{-1}v(\mathbf{j}) + \mathrm{tr}\, W\right).$$

Here $\mathbf{j}$ is determined by (1.10.2) relative to a basis of local sections $\{Z_\sigma\}$ of ρ, and W is the matrix-valued function of D^+ taken with respect to the same basis:

$$D^+ Z_\sigma = -\sum_{\sigma'} W_\sigma^{\sigma'} Z_{\sigma'}.$$

Denote the "space" mean value of $\mathbf{r}(\mathcal{L}, D)$ relative to a smooth measure $d\sigma$ on Λ by

$$(1.10.11) \qquad \langle \mathbf{r}(\mathcal{L}, D)\rangle_\sigma \stackrel{\text{def}}{=} \int_\Lambda \mathbf{r}(\mathcal{L}, D)\, d\sigma.$$

Assume that the flow of the anchor v of D is *uniquely ergodic*, that is, it has precisely one normalized invariant (smooth) measure $d\sigma$ [**CFS**]. Then $\langle \mathbf{r}(\mathcal{L}, D)\rangle_\sigma$ is an *invariant of the dynamical system* $D^\#$ on E generated by a geometrically stable covariant differential operator D. Moreover, formula (1.10.11) gives a generalization of the Johnson–Moser rotation number that was introduced in [**JM**] for two-dimensional linear Hamilton systems with almost periodic coefficients. More precisely, in the quasiperiodic case, this observation may be formulated in the following way.

Let E be a trivial *two-dimensional* symplectic vector bundle over Λ, where $\Lambda \approx \mathbb{T}^k = \mathbb{R}^k / 2\pi\mathbb{Z}^k$ is a k-torus. Suppose that the anchor v of $D \in \mathfrak{D}(E)$ has a *quasiperiodic flow* on Λ,

$$v = \omega^1 \frac{\partial}{\partial \alpha^1} + \cdots + \omega^k \frac{\partial}{\partial \alpha^k},$$

where $\alpha = (\alpha^1, \ldots, \alpha^k)$ ($\alpha^j \bmod 2\pi$) are angular coordinates on $\mathbb{T}^k$ and $\omega = (\omega^1, \ldots, \omega^k)$ is the *frequency vector* of v. Assume that ω satisfies the nonresonant condition

$$(1.10.12) \qquad \omega \cdot n = 0, \quad n \in \mathbb{Z}^k \quad \Longrightarrow \quad n = 0.$$

Then the flow of v is uniquely ergodic and the normalized invariant measure $d\sigma$ on Λ is defined by the volume k-form on $\mathbb{T}^k$,

$$(1.10.13) \qquad d\sigma = \frac{1}{(2\pi)^k} d\alpha^1 \wedge \cdots \wedge d\alpha^k.$$

Let $X^+ = X^+(\alpha)$, $X^- = X^-(\alpha)$ be a symplectic trivialization of E, $E_\alpha = \mathrm{span}\{X^+(\alpha), X^-(\alpha)\}$ and $\langle \mathcal{J}X^+, X^-\rangle = 1$. Then the flow of the vector field $D^\#$

on E (generated by D) is described by the solutions of the two-dimensional linear Hamilton system

$$\dot{x} = V(\alpha(t))x, \qquad x = (x^+, x^-) \in \mathbb{R}^2. \tag{1.10.14}$$

Here $\alpha(t) = t\omega + \alpha^0$ is the trajectory of v starting at $\alpha^0 \in \mathbb{T}^k$, and $V(\alpha) \in \mathrm{sp}(1,\mathbb{R}) \otimes C^\infty(\mathbb{T}^k)$ is a matrix-valued function of D relative to the basis $\{X^+, X^-\}$.

Choose a Lagrangian subbundle $\mathcal{L}$ of E as follows:

$$\mathcal{L}(\alpha) = \mathrm{span}\{X^+(\alpha)\}. \tag{1.10.15}$$

PROPOSITION 1.10.2. *If the nonresonant condition* (1.10.12) *holds and system* (1.10.14) *is geometrically stable, then the Johnson–Moser rotation number of system* (1.10.14) *coincides with the "space" average* $\langle \mathbf{r}(\mathcal{L}, D) \rangle_\sigma$ (1.10.11), *where the measure* $d\sigma$ *and the Lagrangian subbundle* $\mathcal{L}$ *are taken from* (1.10.13) *and* (1.10.15).

The proof of this statement is based on an interpretation of the rotation number in terms of solutions of Riccati's equation on a torus [**KaVo**$_5$, **VoIt**].

REMARK 1.10.1. According to the original definition [**JM**], the rotation number for system (1.10.14) is given by the limit

$$\lim_{t\to\infty} \frac{1}{t} \arg\left(x^+(t) - ix^-(t)\right),$$

where $x(t) = \left(x^+(t), x^-(t)\right)$ is a real nonzero solution of (1.10.4). This limit exists and is independent of the choice of a solution $x(t)$ (this fact is based on ergodic theory arguments [**JM**]). Thus, the rotation number measures the average rotation of nonzero solutions of system (1.10.14). In the general case (in which an operator $D \in \mathfrak{D}(E)$ need not be geometrically stable and $\dim E \geq 2$), the rotation number of D may be also defined via "time" averaging of a function of type (1.10.11) along a trajectory of the anchor v of D. For interpretations and applications of rotation numbers, see also [**E, JoNer, CFS**].

In the following we consider only the stable case. For a given geometrically stable covariant differential operator $D \in \mathfrak{D}(E)$ we call the quantity $\langle \mathbf{r}(\mathcal{L}, D) \rangle_\sigma$ the *rotation number of D relative to a Lagrangian subbundle $\mathcal{L} \subset E$* (under a given v-invariant normalized measure $d\sigma$).

Comparing the rotation numbers of D taken relative to two Lagrangian subbundles $\mathcal{L}$ and $\widetilde{\mathcal{L}}$, we obtain

$$\langle \mathbf{r}(\mathcal{L}, D) \rangle_\sigma - \langle \mathbf{r}(\widetilde{\mathcal{L}}, D) \rangle_\sigma = -\pi \langle v \,\lrcorner\, \mu(\mathcal{L}, \widetilde{\mathcal{L}}) \rangle_\sigma. \tag{1.10.16}$$

Here the right-hand side of (1.10.16) includes the "space" average of the interior product of the anchor v and the Arnold 1-form for $\mathcal{L}$ and $\widetilde{\mathcal{L}}$. Using the v-invariance of the measure $d\sigma$, we deduce from (1.10.6) that if the Maslov class for Lagrangian subbundles $\mathcal{L}$ and $\widetilde{\mathcal{L}}$ is trivial, then the rotation numbers of D relative to $\mathcal{L}$ and $\widetilde{\mathcal{L}}$ coincide; in particular, this is true if $\mathcal{L}$ is transverse to $\widetilde{\mathcal{L}}$, $\mathcal{L} \cap \widetilde{\mathcal{L}} = \{0\}$. For instance, to obtain the Johnson–Moser rotation number for system (1.10.4), one can take the Lagrangian subbundle $\mathcal{L} = \mathrm{span}\{X^-(\alpha)\}$ instead of (1.10.15).

The next property describes the behavior of rotation numbers under the reduction of symplectic vector bundles. Assume that a geometrically stable operator

$D \in \mathfrak{D}(E)$ admits an invariant symplectic subbundle $E' \subset E$. Then the skew-orthogonal complement of E' in E,

$$E'' = (E')^\perp$$

is also a D-invariant symplectic subbundle of E. Moreover, the restrictions

$$D' = D\Big|_{E'} \quad \text{and} \quad D'' = D\Big|_{E''}$$

are geometrically stable covariant differential operators in E' and E'', respectively. From (1.10.16), the definition of the function (1.10.10), and the *additivity property* of the Arnold form [**KaVo**$_5$], we deduce a relation between the rotation numbers of D and D', D''.

PROPOSITION 1.10.3. *For any Lagrangian subbundles $\mathcal{L} \subset E$, $\mathcal{L}' \subset E'$, $\mathcal{L}'' \subset E''$, we have*

(1.10.17) $\quad \langle \mathbf{r}(\mathcal{L}, D) \rangle_\sigma = \langle \mathbf{r}(\mathcal{L}', D') \rangle_\sigma + \langle \mathbf{r}(\mathcal{L}'', D'') \rangle_\sigma - \pi \langle v \,\lrcorner\, \mu(\mathcal{L}, \mathcal{L}' \oplus \mathcal{L}'') \rangle_\sigma.$

Now we shall discuss some properties of rotation numbers in the context of adapted connections.

DEFINITION 1.10.1. A *symplectic connection* ∇ in E is said to be *weakly adapted* to a geometrically stable covariant differential operator $D \in \mathfrak{D}(E)$ if

(i) there exists a D-invariant Kählerian subbundle $\rho \subset {}^\mathbb{C}E$ that is also ∇-invariant,

$$\rho \text{ is } (D, \nabla)\text{-invariant};$$

(ii) $\operatorname{tr}(\nabla_v^+ - D^+) = 0$, where ∇^+ and D^+ are the restrictions of ∇ and D to ρ and v is the anchor of D.

Assume that a geometrically stable operator $D \in \mathfrak{D}(E)$ admits a weakly adapted symplectic connection ∇ satisfying condition (1.10.5). Then the gauge 1-form $\varkappa(\mathcal{L}, \nabla)$ associated with a (D, ∇)-invariant Kählerian subbundle ρ is closed and

(1.10.18) $\quad\quad\quad\quad\quad\quad v \,\lrcorner\, \varkappa(\mathcal{L}, \nabla) = -\mathbf{r}(\mathcal{L}, \nabla).$

Choose a basis of 1-cycles $\Gamma_1, \ldots \Gamma_m$ on Λ and consider the closed 1-forms $\eta^1, \ldots, \eta^m$ on Λ dual to these 1-cycles,

$$\frac{1}{2\pi} \oint_{\Gamma_i} \eta^j = \delta_i^j.$$

Assume that $\eta^1, \ldots, \eta^m$ can be chosen so that the anchor v of D satisfies the condition

(1.10.19) $\quad\quad\quad v \,\lrcorner\, \eta^i = \omega^i, \quad \omega^i = \operatorname{const} \quad (i = 1, \ldots, m).$

Then we claim that for a given v-invariant measure $d\sigma$ and a Lagrangian subbundle $\mathcal{L} \subset E$, the rotation number of D relative to $\mathcal{L}$ may be written in terms of "frequencies" $\omega_1, \ldots, \omega_m$ from (1.10.19) and the periods of the gauge form,

(1.10.20) $\quad\quad \langle \mathbf{r}(\mathcal{L}, D) \rangle_\sigma = -\sum_{i=1}^m \omega^i \varkappa_i, \quad \varkappa_i \stackrel{\text{def}}{=} \frac{1}{2\pi} \oint_{\Gamma_i} \varkappa(\mathcal{L}, \nabla).$

Indeed, using (1.10.18), the fact that

$$\varkappa(\mathcal{L}, \nabla) - \sum_{i=1}^{m} \varkappa_i \eta^i = \text{(exact 1-form)}$$

and the relation

$$\int_\Lambda v(f)\,d\sigma = -\int_\Lambda f\,{}^\sigma\!\operatorname{div}(v)\,d\sigma = 0, \qquad f \in C^\infty(\Lambda)$$

(where ${}^\sigma\!\operatorname{div}(v)$ is the divergence of v with respect to the measure $d\sigma$), we deduce (1.10.20).

(III) *Strongly stable connections and the relative Gelfand–Lidskii index.* Here we demonstrate elements of the classification theory for a special class of flat symplectic connections of stable type that is based on the Gelfand–Krein–Lidskii theory of parametric resonance for linear periodic Hamilton systems [**GL**] (see also [**YStz**]).

We shall assume that the base Λ of a symplectic vector bundle E is a compact connected manifold, the fundamental group $\pi_1(\Lambda)$ is commutative, and the torsion $\operatorname{Tor} H_1(\Lambda)$ is trivial.

Let $\{\Gamma_1, \ldots, \Gamma_m\}$ be a basis of 1-cycles on Λ. Denote by $\boldsymbol{\Gamma} = \{[\Gamma_1], \ldots, [\Gamma_m]\}$ the set of the corresponding homology classes in $H_1(\Lambda)$.

Using the notion of *strong stability* for linear symplectic operators due to Gelfand, Krein, and Lidskii, we can introduce the following class of symplectic connections.

DEFINITION 1.10.2. A *flat symplectic connection* ∇ in E is said to be $\boldsymbol{\Gamma}$-*strongly stable* if there exists a basis $\{\Gamma_1, \ldots, \Gamma_m\}$ of 1-cycles on Λ through a point $\alpha^0 \in \Lambda$ such that the parallel transports $G_{\Gamma_1}, \ldots, G_{\Gamma_m}$ of ∇ around $\Gamma_1, \ldots, \Gamma_m$ (generating the holonomy group $\operatorname{Hol}^\nabla_{\alpha^0}$) satisfy the following conditions: each $G_{\Gamma_j} \subset \operatorname{Sp}(E_{\alpha^0})$ is a strongly stable symplectic operator, that is, G_{Γ_j} is diagonalizable in ${}^\mathbb{C}E_{\alpha^0}$, its spectrum lies on the unit circle in $\mathbb{C}$, and each eigenspace of G_{Γ_j} is positive or negative (with respect to the form $(\cdot, \cdot)_+$).

For a given $\boldsymbol{\Gamma}$, we denote by $\mathcal{R}(E; \boldsymbol{\Gamma})$ the set of all $\boldsymbol{\Gamma}$-strongly stable symplectic connections in E. In general, $\mathcal{R}(E; \boldsymbol{\Gamma})$ depends on the choice of $\boldsymbol{\Gamma}$.

PROPOSITION 1.10.4. *If* $\nabla \in \mathcal{R}(E; \boldsymbol{\Gamma})$, *then* ∇ *is geometrically stable and there exists a unique ∇-invariant Kählerian subbundle* $\rho \subset {}^\mathbb{C}E$.

The proof of this statement follows from the same arguments as in §1.7, Krein's criterion of strong stability [**GL, YStz**] and its geometric interpretation: a linear symplectic operator $G \in \operatorname{Sp}(E_{\alpha^0})$ is strongly stable if and only if there exists a unique G-invariant Kählerian subspace of ${}^\mathbb{C}E_{\alpha^0}$.

So, it follows from Proposition 1.10.4 that under a given Lagrangian subbundle $\mathcal{L} \subset E$, to any $\nabla \in \mathcal{R}(E; \boldsymbol{\Gamma})$ we can associate the *unique gauge 1-form* $\varkappa(\mathcal{L}; \nabla)$ (1.10.3) which is taken with respect to the unique ∇-invariant Kählerian subbundle $\rho \subset {}^\mathbb{C}E$. In other words, there is a mapping

(1.10.21)
$$\begin{aligned}\mathcal{R}(E; \boldsymbol{\Gamma}) &\to \{\text{closed 1-forms on } \Lambda\}, \\ \nabla &\mapsto \varkappa(\mathcal{L}, \nabla) = (\text{gauge 1-form of } \nabla),\end{aligned}$$

which is of class C^∞ in the following sense. If $\{\nabla^t \in \mathcal{R}(E; \boldsymbol{\Gamma}),\ t \in (a, b)\}$ is a smooth t-parameter family, then $\{\varkappa^t = \varkappa(\mathcal{L}, \nabla^t)\}$ is a smooth family of closed 1-forms.

Let $\nabla \in \mathcal{R}(E;\mathbf{\Gamma})$ be a $\mathbf{\Gamma}$-strongly stable symplectic connection. Consider the flat Hermitian connection $\nabla^+ = \nabla|_\rho$, where ρ is the unique ∇-invariant Kählerian subbundle. Denote by $G^+_{\Gamma_1}, \ldots, G^+_{\Gamma_m} \in U(\rho_{\alpha^0})$ the parallel transports of ∇^t around the basic 1-cycles $\Gamma_1, \ldots, \Gamma_m$ on Λ generating $\mathbf{\Gamma}$. It follows from the Krein criterion of strong stability that the spectra

$$\operatorname{spec} G^+_{\Gamma_i} = \{\lambda^+_{1,i}, \ldots, \lambda^+_{r,i}\}$$

of unitary operators $\{G^+_{\Gamma_i}\}$ have the properties

(1.10.22) $\qquad |\lambda^+_{\sigma,i}| = 1, \qquad \lambda^+_{\sigma,i} \neq \pm 1;$

(1.10.23) $\qquad \lambda^+_{\sigma,i} \cdot \lambda^+_{\sigma',i} \neq 1$

for all $\sigma, \sigma' = 1, \ldots, r = \dim E/2$ and $i = 1, \ldots, m = \dim H_1(\Lambda)$.

Conditions (1.10.22), (1.10.23) imply the following representation for the eigenvalues $\lambda^+_{\sigma,i}$:

(1.10.24) $\qquad \lambda^+_{\sigma,i} = \exp(-2\pi i \beta^\#_{\sigma,i}), \qquad \beta^\#_{\sigma,i} \in \mathbb{R},$

where the real numbers $\beta^\#_{\sigma,i}$ (the *normalized Floquet exponents*) are uniquely determined by the conditions

(1.10.25) $\qquad 0 < |\beta^\#_{\sigma,j}| < \frac{1}{2} \qquad (\sigma = 1, \ldots, r;\ j = 1, \ldots, m).$

Now we define

(1.10.26) $\qquad \nu_i(\mathcal{L}, \nabla) \stackrel{\text{def}}{=} \frac{1}{2\pi} \oint_{\Gamma_i} \varkappa(\mathcal{L}, \nabla) - \sum_{\sigma=1}^r \beta^\#_{\sigma,i}.$

From (1.10.10) we deduce that for any Lagrangian subbundle $\mathcal{L} \subset E$ and $\mathbf{\Gamma}$-strongly stable symplectic connection ∇, formula (1.10.26) defines a set of integers,

$$\nu_i(\mathcal{L}, \nabla) \in \mathbb{Z}, \qquad i = 1, \ldots, m,$$

associated with basic 1-cycles $\Gamma_1, \ldots, \Gamma_m$ generating $\mathbf{\Gamma}$. Formula (1.10.26) represents a direct analog of the Gelfand–Lidskii index for strongly stable linear periodic Hamilton systems with a fixed period [**GL**].

Indeed, let $\Lambda \approx \mathbb{S}^1 = \mathbb{R}/2\pi\mathbb{Z}$ be a circle and E a trivial symplectic vector bundle over $\mathbb{S}^1$. Then $\pi_1(\mathbb{S}^1) \approx H_1(\mathbb{S}^1) = \mathbb{Z}$. Consider a closed path $\Gamma \colon [0, 2\pi] \to \mathbb{S}^1$ whose homology class $\mathbf{\Gamma} = [\Gamma]$ generates $H_1(\mathbb{S}^1)$. Choose a vector field $v = \partial/\partial\alpha$ on $\mathbb{S}^1$ such that Γ is a 2π-periodic orbit of v (where α is an angular coordinate on $\mathbb{S}^1$).

We choose a symplectic trivialization $\{X^+_1(\alpha), \ldots, X^+_r(\alpha); X^-_1(\alpha), \ldots, X^-_r(\alpha)\}$ on E such that $\langle \mathcal{J} X^+_\sigma, X^-_{\sigma'}\rangle = \delta_{\sigma\sigma'}$, $\langle \mathcal{J} X^+_\sigma, X^+_{\sigma'}\rangle = \langle \mathcal{J} X^-_\sigma, X^-_{\sigma'}\rangle = 0$.

Each symplectic connection ∇ in $E \to \mathbb{S}^1$ is flat and uniquely determined by the covariant derivative ∇_v along v. We can view ∇_v as a 2π-periodic linear Hamilton system

(1.10.27) $\qquad \dot{x} = V(t)x, \qquad x \in \mathbb{R}^{2r},$

where $V(t) = V(t + 2\pi)$ is a $\mathrm{sp}(r)$-valued function of ∇_v with respect to a given symplectic basis $\{X^+_\sigma(\alpha), X^-_{\sigma'}(\alpha)\}$. So, in this case the set $\mathcal{R}(E;\mathbf{\Gamma})$ coincides with the set of all strongly stable 2π-periodic linear Hamilton systems on $\mathbb{R}^{2r}$ (see [**GL**]).

Choosing a Lagrangian subbundle $\mathcal{L}$ as

$$\mathcal{L}(\alpha) = \operatorname{span}\{X_1^+(\alpha), \ldots, X_r^+(\alpha)\},$$

we arrive at the following fact.

PROPOSITION 1.10.5. *The Gelfand–Lidskii index of the strongly stable linear 2π-periodic Hamilton system (1.10.27) is given by (1.10.26).*

In the general case, the integer $\nu_i(\mathcal{L}, \nabla)$ (1.10.26) will be called the *relative Gelfand–Lidskii index* of a $\mathbf{\Gamma}$-strongly stable symplectic connection $\nabla \in \mathcal{R}(E; \mathbf{\Gamma})$ at the basic 1-cycle Γ_i.

Just as in the classical theory [**GL**], the relative Gelfand–Lidskii index is one obstruction to the homotopy of connections in the class $\mathcal{R}(E; \mathbf{\Gamma})$. We say that two $\mathbf{\Gamma}$-strongly stable connections ∇ and $\widetilde{\nabla}$ are *homotopic* if there is a smooth family of $\mathbf{\Gamma}$-strongly stable connections

$$\nabla^t \in \mathcal{R}(E; \mathbf{\Gamma}), \qquad 0 \le t \le 1,$$

such that $\nabla^0 = \nabla$, $\nabla^1 = \widetilde{\nabla}$.

PROPOSITION 1.10.6. (a) *Let us choose a Lagrangian subbundle $\mathcal{L}$ and a set of homology classes $\mathbf{\Gamma} = \{[\Gamma_1], \ldots, [\Gamma_m]\} \subset H_1(\Lambda)$. If two $\mathbf{\Gamma}$-strongly stable symplectic connections ∇ and $\widetilde{\nabla}$ are homotopic, then the relative Gelfand–Lidskii indices of ∇ and $\widetilde{\nabla}$ coincide at the basic 1-cycles $\Gamma_1, \ldots, \Gamma_m$ (generating $\mathbf{\Gamma}$), that is,*

$$\nu_i(\mathcal{L}, \nabla) = \nu_i(\mathcal{L}, \widetilde{\nabla}) \qquad (i = 1, \ldots, m).$$

(b) *If $\mathcal{L}$ and $\widetilde{\mathcal{L}}$ are two Lagrangian subbundles of E, then for each $\nabla \in \mathcal{R}(E; \mathbf{\Gamma})$ we have*

(1.10.28) $$\nu_i(\mathcal{L}, \nabla) - \nu_i(\widetilde{\mathcal{L}}, \nabla) = \frac{1}{2}\mu_i,$$

where $\mu_i = \oint_{\Gamma_i} \mu(\widetilde{\mathcal{L}}, \mathcal{L})$ is the Maslov index of the basic 1-cycle Γ_i.

The proof of this statement follows from the above properties of gauge forms and Arnold's form.

(IV) *Phase space invariants of the linearized Hamilton dynamics over isotropic submanifolds.* Now we would like to discuss the above constructions in the context of the Hamilton dynamics over isotropic submanifolds. Consider the standard phase space

$$\left(M = (\mathbb{R}_p^n)^* \times \mathbb{R}_q^n, \ \omega = dp \wedge dq\right).$$

Then there is a natural Lagrangian subbundle of TM:

$$\mathcal{P} \approx \mathbb{R}_p^n \qquad \text{(the vertical polarization)}.$$

Let Λ be an isotropic submanifold of M. Consider the skew-orthogonal complement $T\Lambda^\perp$ of $T\Lambda$ in $T_\Lambda M$ and the projection $\operatorname{pr}\colon T\Lambda^\perp \to E$, where $E = T\Lambda^\perp/T\Lambda$ is the symplectic bundle over Λ. Then we can define the *reduced Lagrangian subbundle* $\mathcal{P}^{\text{red}}$ of E associated with $\mathcal{P}$ as follows [**We$_2$**]:

(1.10.29) $$\mathcal{P}^{\text{red}} \stackrel{\text{def}}{=} \operatorname{pr}(\mathcal{P} \cap T\Lambda^\perp), \qquad \dim \mathcal{P}^{\text{red}} = \frac{1}{2}\dim E.$$

Let (M, H, Λ) be a Hamilton system with a compact invariant isotropic submanifold Λ. Let D_H be the first variant operator of the Hamilton vector field $\operatorname{ad}(H)$

over Λ. We say that Λ is an *invariant isotropic submanifold of stable type* if the first variation operator D_H is geometrically stable.

We assume that Λ is an isotropic submanifold of stable type. Then using the reduced Lagrangian subbundle $\mathcal{P}^{\mathrm{red}}$ (1.10.29) in the symplectic normal bundle E over Λ, we can introduce the function $\mathbf{r}(\mathcal{P}^{\mathrm{red}}, D_H)$ given by (1.10.10). Let $v_H = \mathrm{ad}(H)\big|_\Lambda$ be the restriction of the Hamilton field to Λ.

DEFINITION 1.10.3. If (M, H, Λ) is a Hamilton system with an invariant compact isotropic submanifold Λ of stable type and if the flow of v_H is uniquely ergodic, then the *rotation number of the first variation operator* D_H (or the linearized Hamilton flow of $\mathrm{ad}(H)$ over Λ) is the space average $\langle \mathbf{r}(\mathcal{P}^{\mathrm{red}}, D_H)\rangle_\sigma$ (1.10.11) (where $d\sigma$ is the v_H-invariant measure on Λ).

Now let N be a coisotropic extension of an isotropic submanifold Λ in M. Assume that Λ satisfies the conditions from item (III). Consider the flat symplectic connection ∇ (1.3.6) associated with N.

We say that a *coisotropic extension* N of Λ is $\boldsymbol{\Gamma}$-*strongly stable* if the connection ∇ (1.3.6) is $\boldsymbol{\Gamma}$-strongly stable.

DEFINITION 1.10.4. If N is a $\boldsymbol{\Gamma}$-strongly stable coisotropic extension of Λ in the phase space M, then the *Gelfand–Lidskii indices of* N at basic 1-cycles $\Gamma_1, \ldots, \Gamma_m$ (generating $\boldsymbol{\Gamma}$) are defined as the relative Gelfand–Lidskii indices $\nu_i(\mathcal{P}^{\mathrm{red}}, \nabla)$ (see (1.10.26)) taken with respect to the reduced Lagrangian subbundle (1.10.29) and Bott's connection (1.3.6) associated with N.

EXAMPLE 1.10.1 (Gelfand–Lidskii indices for periodic orbits of strongly stable type). Let $\Lambda = \Gamma = \{p = p(t), q = q(t)\}$ be a T-periodic orbit of a Hamilton vector field $\mathrm{ad}(H)$ on M. Then $v_H = \omega \partial/\partial t$ (where $\omega = 2\pi/T$ is the frequency), and the energy surface of H containing Γ defines the natural coisotropic extension of Γ,

$$(1.10.30) \qquad N = \{(p,q) \in M \mid H(p,q) = \mathrm{const}\} \supset \Gamma.$$

The symplectic connection (1.3.6) associated with the energy surface (1.10.30) is uniquely determined by the first variation operator D_H over Γ, $\nabla_{v_H} = D_H$. We observe that this connection is $\boldsymbol{\Gamma}$-strongly stable if and only if the monodromy operator of the first variation operator D_H is a strongly stable symplectic operator (see also [**GL**]).

In this case, we say that Γ is a *periodic orbit* of $\mathrm{ad}(H)$ of *strongly stable type*.

Thus we define the *Gelfand–Lidskii index* for a *periodic orbit* Γ of stable type as $\nu_\Gamma = \nu_\Gamma(\mathcal{P}^{\mathrm{red}}, \nabla)$, where ∇ is the connection (1.3.6) associated with the energy surface (1.10.30).

Note that for a given periodic orbit Γ of strongly stable type, the symplectic normal bundle E over Γ is trivial. One can show that there is a trivial Lagrangian subbundle $\mathcal{L}$ of E such that $\nu_\Gamma(\mathcal{L}, D_H) = 0$. Such Lagrangian subbundles are called *minimal*; we denote them by $\mathcal{L}^{\mathrm{min}}$. Using (1.10.28), we obtain

$$(1.10.31) \qquad \nu_\Gamma = \frac{1}{2}\mu_\Gamma,$$

where $\mu_\Gamma = \oint_\Gamma \mu(\mathcal{L}^{\mathrm{min}}, \mathcal{P}^{\mathrm{red}})$ is the Maslov index of Γ associated with the minimal Lagrangian subbundle $\mathcal{L}^{\mathrm{min}}$ and the reduced Lagrangian subbundle $\mathcal{P}^{\mathrm{red}}$. Although formula (1.10.31) relates the Gelfand–Lidskii index and the Maslov index of a periodic orbit Γ, we would like to stress that the nature of these invariants is different.

Part 2. Symplectic Geometry and Nonlinear Dynamics near Isotropic Submanifolds

§2.1. Symplectic and coisotropic extensions. General properties

In this section we formulate some general facts about symplectic geometry near isotropic submanifolds, which are based on the isotropic embedding theorem due to Weinstein [**We**$_{1,2}$], the coisotropic embedding theorem due to Gotay [**Go**], and the minimal coupling procedure due to Sternberg [**St**].

Let (M, Ω) and $(\widetilde{M}, \widetilde{\Omega})$ be two symplectic manifolds with isotropic embeddings of a manifold Λ,
$$\imath\colon \Lambda \to M, \qquad \widetilde{\imath}\colon \Lambda \to \widetilde{M}.$$
Suppose that there is a diffeomorphism $g\colon U \to \widetilde{U}$ from a neighborhood U of $\imath(\Lambda)$ in M onto a neighborhood $\widetilde{U}$ of $\widetilde{\imath}(\Lambda)$ in $\widetilde{M}$ that satisfies the conditions

(2.1.1) $$\widetilde{\imath} = g \circ \imath,$$

(2.1.2) $$g^*\widetilde{\Omega} = \Omega.$$

In this case, the isotropic embeddings $\imath$ and $\widetilde{\imath}$ of Λ are called *equivalent* (or *neighborhood equivalent*), see [**We**$_3$].

The symplectomorphism g from (2.1.1), (2.1.2) determines a symplectic vector bundle isomorphism

(2.1.3) $$L_g\colon E \to \widetilde{E}$$

from the symplectic normal bundle E over $\imath(\Lambda)$ onto the symplectic normal bundle $\widetilde{E}$ over $\widetilde{\imath}(\Lambda)$, which implies the following commutative diagram:

$$\begin{array}{ccc} T_{\imath(\Lambda)}U & \xrightarrow{dg} & T_{\widetilde{\imath}(\Lambda)}\widetilde{U} \\ \downarrow & & \downarrow \\ E & \xrightarrow{L_g} & \widetilde{E} \end{array}$$

Conversely, we have the following proposition.

PROPOSITION 2.1.1 (the isotropic embedding theorem, [**We**$_{1,2}$]). *For a given symplectic isomorphism $L\colon E \to \widetilde{E}$ of symplectic normal bundles over $\imath(\Lambda)$ and $\widetilde{\imath}(\Lambda)$, there exists a symplectomorphism g, satisfying (2.1.1), (2.1.2), such that $L = L_g$.*

Thus, there is a one-to-one correspondence between the neighborhood equivalence classes of isotropic embeddings and the isomorphism classes of symplectic vector bundles.

Note that the construction of a "canonical model" for an isotropic embedding with a given symplectic normal bundle was described in [**We**$_{3,4}$]. In the presence of a compact group of automorphisms, there is an equivalent version of the isotropic embedding theorem. For details, see [**GuSt**$_2$].

Let $\Lambda \subset M$ be an isotropic submanifold of a symplectic manifold (M, Ω). In the symplectic normal bundle E over Λ, let us choose a symplectic connection ∇ and consider the model phase space $(\mathcal{X} = T^*\Lambda \oplus E, \Omega^\nabla)$ determined by (1.5.23). Then the zero section $\Lambda \subset \mathcal{X}$ is an isotropic submanifold and its symplectic normal bundle is naturally isomorphic to E. Applying Proposition 2.1.1, we obtain the following statement.

PROPOSITION 2.1.2. *The isotropic embeddings $\Lambda \subset M$ and $\Lambda \subset \mathcal{X}$ are equivalent, that is, there exists a diffeomorphism $g\colon U \to \widetilde{U}$ of a neighborhood U of $\Lambda \subset M$ onto a neighborhood U of the zero section $\Lambda \subset \mathcal{X}$ such that*

(2.1.4) $$g\big|_{\Lambda} = \mathrm{id}_{\Lambda},$$
(2.1.5) $$g^{*}\Omega^{\nabla} = \Omega.$$

Note that the symplectic form Ω^{∇} (1.5.23) on the Whitney sum $\mathcal{X} = T^{*}\Lambda \oplus E$ can be obtained using the Sternberg minimal coupling procedure (see [**GuSt**$_2$, **St**, **We**$_4$]). The symplectic structure (1.5.23) depends on the choice of a symplectic connection ∇ in E. Any two such symplectic structures are equivalent in a neighborhood of the zero section of $\mathcal{X}$. So, in this sense, we have no symplectic geometry near isotropic submanifolds if there is no additional geometric structure. For example, note that there is an interesting problem of deformation of isotropic submanifolds in Kähler manifolds [**ChMo**].

In our case, such an additional geometric structure is a coisotropic extension of Λ.

Let us consider an isotropic submanifold Λ of (M, Ω). Recall that
- a symplectic submanifold $\mathcal{Y} \subset M$ is a *symplectic extension* of Λ in M if
    (i) $\Lambda \subset M$,
    (ii) $\dim \mathcal{Y} = 2 \dim \Lambda$;
- a coisotropic submanifold $N \subset M$ is a *coisotropic extension* of Λ in M if
    (i) $\Lambda \subset N$,
    (ii) $\mathrm{codim}\, N = \dim \Lambda$.

THEOREM 2.1.1. *There always exists a symplectic extension $\mathcal{Y}$ of a given isotropic submanifold Λ in M. Any two symplectic extensions $\mathcal{Y}$ and $\widetilde{\mathcal{Y}}$ of Λ in M are equivalent, that is, there are neighborhoods U and $\widetilde{U}$ of Λ in M and a symplectomorphism $g\colon U \to \widetilde{U}$, satisfying (2.1.1), such that $g(\mathcal{Y} \cap U) = \widetilde{\mathcal{Y}} \cap \widetilde{U}$.*

PROOF. By Theorem 1.2.1, there always exists a symplectic connection ∇ in the symplectic normal bundle E over Λ. Let us choose ∇ and consider the isotropic embedding $\Lambda \subset \mathcal{X}$ into the model phase space $(\mathcal{X} = T^{*}\Lambda \oplus E, \Omega^{\nabla})$. Then $T^{*}\Lambda$ (as the zero section of the vector bundle $\mathcal{X} \to T^{*}\Lambda$) is a symplectic extension of Λ in $\mathcal{X}$. Using a symplectomorphism $g\colon U \to \widetilde{U}$ from (2.1.4), (2.1.5), we can set $\mathcal{Y} = g^{-1}(T^{*}\Lambda \cap \widetilde{U})$.

The equivalence of two symplectic extensions of a given Λ follows from the Darboux–Weinstein theorem [**GuSt**$_2$, **M**, **We**$_2$] (see also [**SLe**]). □

THEOREM 2.1.2. *There exists a coisotropic extension N of a given isotropic submanifold Λ in M if and only if the symplectic normal bundle E over Λ admits a flat symplectic linear connection ∇.*

PROOF. If N is a coisotropic extension of Λ, then there is a flat symplectic connection ∇ in E given by (1.3.6). Conversely, if ∇ is a flat symplectic connection in E, then $E \subset \mathcal{X}$ (as the zero section of the vector bundle $\mathcal{X} \to E$) is the coisotropic extension of Λ in the model phase space $(\mathcal{X} = T^{*}\Lambda \oplus E, \Omega^{\nabla})$ (see Theorem 1.5.2).

Thus, using a diffeomorphism g from (2.1.4), (2.1.5), we define the coisotropic extension N of Λ in M by the formula $N = g^{-1}(E \cap \widetilde{U})$. □

COROLLARY 2.1.1. *There is no coisotropic extension of Λ in M if the first Chern class $c_1(E) \in H_1(\Lambda, \mathbb{R})$ of E is nontrivial.*

EXAMPLE 2.1.1. Let us consider the Whitney sum $\mathcal{X} = T^*\mathbb{S}^2 \oplus T\mathbb{S}^2$ as the symplectic manifold equipped with symplectic structure Ω^∇ of type (1.5.23), where ∇ is the Riemann connection in $E = T\mathbb{S}^2$. Then the isotropic submanifold $\Lambda = \mathbb{S}^2 \subset \mathcal{X}$ admits no coisotropic extension, since $c_1(T\mathbb{S}^2) \neq 0$.

COROLLARY 2.1.2. *If the symplectic normal bundle E over Λ is trivial, then there exists a coisotropic extension of Λ in M. In particular, this is always true if $\dim \Lambda = 1$ (that is, $\Lambda = \mathbb{R}$ or $\Lambda = \mathbb{S}^1$).*

The next statement illustrates, from the dynamical viewpoint, the role played by coisotropic extensions.

PROPOSITION 2.1.3. *Let $\Lambda \subset M$ be a closed isotropic orbit of a Lie group action in M, that is,*

$$T_\alpha \Lambda = \mathrm{span}\{v_1(\alpha), \ldots, v_k(\alpha)\}, \qquad \alpha \in \Lambda,$$

$$[v_i, v_j] = \sum_{s=1}^k \lambda_{ij}^s v_s, \qquad \lambda_{ij}^s = \mathrm{const},$$

where $v_1, \ldots, v_k$ are vector fields on Λ generating a Lie algebra, and $k = \dim \Lambda$. Then the existence of smooth functions $f_1, \ldots, f_k \in C^\infty(U)$ (in a neighborhood of $\Lambda \subset M$) with the properties

$$\{f_i, f_j\} = \sum_{s=1}^k \lambda_{ij}^s f_s, \qquad \mathrm{ad}(f_j)\big|_\Lambda = v_j, \qquad f_j\big|_\Lambda = 0,$$

is equivalent to the existence of a coisotropic extension of Λ in M.

COROLLARY 2.1.3. *If $\Lambda \approx \mathbb{T}^k$ and E is trivial, then in a neighborhood of Λ in M, there are Darboux coordinates $p = (p^1, \ldots, p^k)$, $\alpha = (\alpha^1, \ldots, \alpha^k)$ ($\alpha^j \mod (2\pi)$), $x = (x^1, \ldots, x^{2r})$,*

$$\Omega = dp \wedge d\alpha + \frac{1}{2}\langle J dx \wedge dx \rangle$$

such that

$$\Lambda = \{p^1 = \cdots = p^k = 0\} \cap \{x^1 = \cdots = x^{2r} = 0\}.$$

In particular, this is true if Λ admits a flat stable symplectic connection in E.

The proof follows from Theorem 1.5.2 and Proposition 2.1.2.

To classify coisotropic extensions of a given isotropic submanifold, we need the following proposition.

PROPOSITION 2.1.4 (the coisotropic embedding theorem, [**Go**]). *Let $k: N \to M$ and $\widetilde{k}: N \to \widetilde{M}$ be two coisotropic embeddings of a manifold N into the symplectic manifolds (M, Ω) and $(\widetilde{M}, \widetilde{\Omega})$. Assume that $k^*\Omega = \widetilde{k}^*\widetilde{\Omega}$. Then there exist* neighborhoods U of $\boldsymbol{k}(N)$ in M and $\widetilde{U}$ of $\widetilde{k}(N)$ in $\widetilde{M}$ and a symplectomorphism $g\colon U \to \widetilde{U}$ such that $g \circ k = \widetilde{k}$.

Let $\imath\colon \Lambda \to M$ and $\widetilde{\imath}\colon \Lambda \to M$ be two isotropic embeddings of a manifold Λ into the symplectic manifolds (M, Ω) and $(\widetilde{M}, \widetilde{\Omega})$.

DEFINITION 2.1.1. A coisotropic extension N of $\imath(\Lambda)$ in M is *equivalent* to a coisotropic extension $\widetilde{N}$ of $\widetilde{\imath}(\Lambda)$ in $\widetilde{M}$ if there exists a symplectomorphism $g\colon U \to \widetilde{U}$ of a neighborhood U of $\imath(\Lambda)$ in M onto a neighborhood $\widetilde{U}$ of $\widetilde{\imath}(\Lambda)$ in $\widetilde{M}$ such that

(2.1.6) $$g \circ \imath = \widetilde{\imath}, \qquad g(N \cap U) = \widetilde{N} \cap \widetilde{U}.$$

In this case, we see that the restriction $g_0\colon N \cap U \to \widetilde{N} \cap \widetilde{U}$ of g identifies the presymplectic manifolds $(N, \Omega_N = \Omega|_N)$ and $(\widetilde{N}, \widetilde{\Omega}_{\widetilde{N}} = \widetilde{\Omega}|_{\widetilde{N}})$ in neighborhoods of $\imath(\Lambda)$ and $\widetilde{\imath}(\Lambda)$.

Conversely, Proposition 2.1.4 implies the following proposition.

PROPOSITION 2.1.5. *If there exists a diffeomorphism* $g_0\colon W \to \widetilde{W}$ *of a neighborhood* W *of* $\imath(\Lambda)$ *in* N *onto a neighborhood* $\widetilde{W}$ *of* $\widetilde{\imath}(\Lambda)$ *in* $\widetilde{N}$ *such that*

(2.1.7) $$g_0 \circ \imath = \widetilde{\imath}, \qquad g_0^*(\widetilde{\Omega}_{\widetilde{N}}) = \Omega_N,$$

then the coisotropic extensions N *and* $\widetilde{N}$ *are equivalent.*

The equivalence problem for a special class of coisotropic extensions will be considered in §2.3.

§2.2. Poincaré extension theorem for tori

Assume that a Hamilton system (M, Ω, H) has a periodic orbit γ lying on the energy surface $N_{E_0} = \{H = E_0\}$. If none of the characteristic multipliers of the Hamilton vector field $\mathrm{ad}(H)$ at γ is equal to 1, then by the Poincaré theorem, there exists a family of periodic orbits γ_E of $\mathrm{ad}(H)$ lying on the energy surfaces N_E for $|E - E_0|$ sufficiently small (see [**AbMa**]). Geometrically, this means that γ_{E_0} admits a 2-dimensional $\mathrm{ad}(H)$-invariant symplectic extension in M (a cylinder of closed orbits).

Our goal is to discuss a quasiperiodic analog of the Poincaré theorem. The basic point is to formulate this theorem in terms of characteristic multipliers of a coisotropic extension (Definition 1.3.1). We shall use standard notions of the theory of foliations [**Tam**] and the qualitative theory of dynamical systems [**AbMa**].

Let (M, Ω) be a $2n$-dimensional symplectic manifold. Assume that

(i) there is a surjective submersion $\boldsymbol{f}\colon M \to \mathcal{B}$, where $\mathcal{B} \subset \mathbb{R}^k$ is an open neighborhood of the origin in $\mathbb{R}^k$, $k < n$; and

(ii) the components of the map $\boldsymbol{f} = (f_1, \ldots, f_k)$ are Poisson commutative with each other on M,
$$\{f_i, f_j\} = 0 \qquad (i, j = 1, \ldots, k).$$

It follows from (i), (ii) that the level sets

(2.2.1) $$N_b = \boldsymbol{f}^{-1}(b), \qquad b \in \mathcal{B},$$

are $(2n - k)$-dimensional coisotropic submanifolds of M. We suppose that the N_b are connected.

We also assume that

(iii) there is a k-dimensional isotropic submanifold $\Lambda \subset M$ lying on the zero-level set $N_0 = \mathbf{f}^{-1}(0)$, $\Lambda \subset N_0$, that is diffeomorphic to the k-torus, $\Lambda \approx \mathbb{T}^k$.

In this situation, we have the following objects.

- The isotropic plane distribution $\mathcal{H} \subset TM$ with fibers

$$\mathcal{H}_m = \ker_m \Omega\big|_{T_m N_b}, \qquad m \in N_b, \quad b \in \mathcal{B}.$$

Obviously, $\mathcal{H}$ is integrable. By $\mathcal{F}$ we denote the corresponding isotropic foliation of M, $\mathcal{H} = T(M, \mathcal{F})$. The space of smooth functions on M that are constant along the fibers of $\mathcal{F}$ is denoted by $C_\mathcal{F}^\infty(M)$.

- Since the fundamental group $\pi_1(\Lambda)$ is commutative, the holonomy group of the Bott-type flat connection ∇ (1.3.6) generated by the coisotropic extension N_0 is also commutative. For each 1-cycle Γ of Λ, $[\Gamma] \in H_1(\Lambda)$, we define the characteristic multipliers $\{\lambda^1(\Gamma), \ldots, \lambda^{2(n-k)}(\Gamma)\}$ of the coisotropic extension N_0 of Λ at Γ.

THEOREM 2.2.1. *Under the above conditions* (i)–(iii), *suppose that there is a 1-cycle Γ of Λ such that the characteristic multipliers $\{\lambda^1(\Gamma), \ldots, \lambda^{2(n-k)}(\Gamma)\}$ of the coisotropic extension N_0 of Λ at Γ satisfy the condition*

$$(2.2.2) \qquad \lambda^\sigma(\Gamma) \neq 1 \qquad \forall \sigma = 1, \ldots, 2(n-k).$$

Then there exists a unique symplectic extension $\mathcal{Y}$ of Λ in M that is invariant under the flows of the Hamilton vector fields $\mathrm{ad}(f_1), \ldots, \mathrm{ad}(f_k)$. Moreover, there is a trivial fibration $\mathcal{Y} \to \mathcal{B}_\varepsilon$ over an open ball $\mathcal{B}_\varepsilon \subset \mathcal{B}$ (of radius $\varepsilon > 0$ centered at $\{0\}$ in $\mathbb{R}^k$) whose fibers

$$(2.2.3) \qquad \Lambda_b = \mathcal{Y} \cap N_b, \qquad b \in \mathcal{B}_\varepsilon,$$

are $\mathrm{ad}(f_i)$-invariant isotropic k-tori with quasiperiodic motion (for all $i = 1, \ldots, k$).

COROLLARY 2.2.1. *If integrals of motion $f_1, \ldots, f_k$ that are in involution and satisfy the conditions* (i)–(iii) *are admitted by a Hamilton system (M, Ω, H), then this system has a cylinder filled by quasiperiodic k-tori*

$$\mathcal{Y} = \bigcup_{b \in \mathcal{B}_\varepsilon} \Lambda_b \approx \mathbb{T}^k \times \mathcal{B}_\varepsilon.$$

Here the uniqueness means that if we have two $\mathrm{ad}(f_i)$-invariant symplectic extensions $\mathcal{Y}$ and $\mathcal{Y}'$ of Λ, then $\mathcal{Y} \cap \mathcal{U} = \mathcal{Y}' \cap \mathcal{U}$ in some neighborhood $\mathcal{U}$ of Λ in M.

REMARK 2.2.1. A theorem similar to Theorem 2.2.1 was first announced by Nekhoroshev in [**Ne**$_2$], where an analog of condition (2.2.2) was formulated in a different manner (via the "Poincaré mappings" associated with the isotropic foliation $\mathcal{F}$). Here we present the complete proof of Theorem 2.2.1.

First, we need some auxiliary facts.

We say that a $(2n-k)$-dimensional submanifold Σ of M is a local *cross-section* of the isotropic foliation $\mathcal{F}$ at a point $\alpha_0 \in \Lambda$ if

$$\Sigma \cap \Lambda = \{\alpha_0\}, \qquad T_m \Sigma \oplus \mathcal{H}_m = T_m M \quad \forall m \in \Sigma.$$

Choose a point $\alpha_0 \in \Lambda$ and a local cross-section Σ of $\mathcal{F}$ at α_0. Then there exists an $\varepsilon > 0$ such that for each $b \in \mathcal{B}_\varepsilon$, the intersection $S_b = \Sigma \cap N_b$ is a $2(n-k)$-dimensional symplectic submanifold of M and $\alpha_0 \in S_0$. We define a "corrected" local cross-section of $\mathcal{F}$ at α_0 as

$$\Sigma_\varepsilon = \bigcup_{b \in \mathcal{B}_\varepsilon} S_b.$$

DEFINITION 2.2.1. An open neighborhood $\mathcal{U}$ of α_0 in M is said to be *admissible* for the pair $(\mathcal{F}, \Sigma_\varepsilon)$ if
- $\mathcal{U}$ is a (*distinguished*) *foliated chart* of $\mathcal{F}$, and
- for any $b \in \mathcal{B}_\varepsilon$, $m \in S_b$, and for the *connected* component $\mathcal{L}_m^{(i)}$ in $\mathcal{U}$ of a leaf $\mathcal{L}_m$ containing m, the intersection $\mathcal{L}_m^{(i)} \cap S_b$ consists exactly one point.

We choose angle coordinates $\alpha = (\alpha^1, \ldots, \alpha^k)$ ($\alpha^i \mod 2\pi$) on the torus Λ and the corresponding basis of 1-cycles $\Gamma_1, \ldots, \Gamma_k \subset \Lambda$:

$$\Gamma_i = \{\alpha^1 = 0, \ldots, 0 \leq \alpha^i \leq 2\pi, \ldots, \alpha^k = 0\}.$$

Choose $\alpha_0 = (0, \ldots, 0)$. Assumptions (i)–(iii) imply that the Hamilton vector fields $\mathrm{ad}(f_1), \ldots, \mathrm{ad}(f_k)$ generate an isotropic subbundle $\mathcal{H} \subset TM$ and the commuting flows $g_{f_i}^t$ of $\mathrm{ad}(f_i)$ are quasiperiodic on Λ. Thus, we can choose angle coordinates $\alpha = (\alpha^1, \ldots, \alpha^k)$ on Λ and components $f_1, \ldots, f_k$ of the map $\boldsymbol{f} : M \to \mathcal{B}$ such that

$$\mathrm{ad}(f_i)\big|_\Lambda = \partial/\partial\alpha^i \qquad (i = 1, \ldots, k),$$

that is, the restrictions of the flows $g_{f_i}^t$ to Λ are 2π-periodic, and

$$g_{f_i}^{2\pi}(\alpha_0) = \alpha_0 \qquad (i = 1, \ldots, k).$$

It follows from the flow box theorem [**AbMa**] that there is a neighborhood $\mathcal{U}$ of α_0 in M admissible for $(\mathcal{F}, \Sigma_\varepsilon)$ that possesses the following property: for each $i = 1, \ldots, k$, the image under the flow $g_{f_i}^{2\pi}$

(2.2.4) $\qquad \widetilde{\mathcal{U}}_i = g_{f_i}^{2\pi}(\mathcal{U}) \qquad$ is also an admissible neighborhood of α_0 in M.

So, choosing an admissible neighborhood $\mathcal{U}$ of α_0 in M with property (2.2.4), we see that the map $\boldsymbol{f}$ induces the following fibrations over $\mathcal{B}_\varepsilon$:

$$e : \Sigma_\varepsilon \cap \mathcal{U} \to \mathcal{B}_\varepsilon,$$
$$\widetilde{e}_i : \Sigma_\varepsilon \cap \widetilde{\mathcal{U}}_i \to \mathcal{B}_\varepsilon \qquad (i = 1, \ldots, k),$$

whose fibers $S_b \cap \mathcal{U}$, $S_b \cap \widetilde{\mathcal{U}}_i$ ($b \in \mathcal{B}_\varepsilon$) are $2(n-k)$-dimensional symplectic submanifolds of M.

Now we are ready to define *Poincaré mappings*

$$P^i : \Sigma_\varepsilon \cap \mathcal{U} \to \Sigma_\varepsilon \cap \widetilde{\mathcal{U}}_i \qquad (i = 1, \ldots, k)$$

of the isotropic foliation $\mathcal{F}$ at the basic 1-cycles $\Gamma_1, \ldots, \Gamma_k \subset \Lambda$ in the following way.

For a given $m \in \Sigma_\varepsilon \cap \mathcal{U}$, consider the leaf $\mathcal{L}_m$ of the foliation $\mathcal{F}$ that contains m. Since the flows $g_{f_i}^t$ preserve the foliation $\mathcal{F}$, all points $g_{f_1}^{2\pi}(m), \ldots, g_{f_k}^{2\pi}(m)$ also belong to $\mathcal{L}_m$. On the other hand, by our assumptions, we have

$$g_{f_i}^{2\pi}(m) \in \widetilde{\mathcal{U}}_i \qquad (i = 1, \ldots, k),$$

where $\widetilde{\mathcal{U}}_i$ are admissible neighborhoods of α_0 in M. For each $1 \leq i \leq k$, in $\widetilde{\mathcal{U}}_i$, consider the connected component $\mathcal{L}_m^{(j)}$ of the leaf $\mathcal{L}_m$ that contains $g_{f_i}^{2\pi}(m)$,

$$g_{f_i}^{2\pi}(m) \in \mathcal{L}_m^{(j)} \equiv \mathcal{L}_{g_{f_i}^{2\pi}(m)}^{(j)}.$$

Then the intersection $\mathcal{L}_m^{(j)} \cap S_{e(m)}$ consists of exactly one point, and we define

(2.2.5) $$P^i(m) \stackrel{\text{def}}{=} \mathcal{L}_{g_{f_i}^{2\pi}(m)}^{(j)} \cap S_{e(m)} \qquad (m \in \Sigma_\varepsilon \cap \mathcal{U}).$$

Now let us formulate the properties of the *Poincaré mappings* (2.2.5).

In the symplectic normal bundle E over Λ we consider the flat symplectic connection ∇ (1.3.6) associated with the coisotropic extension N_0 of Λ in M. Consider the generators $G_1, \ldots, G_k \in \mathrm{Sp}(E_{\alpha_0})$ of the holonomy group $\mathrm{Hol}_{\alpha_0}^\nabla$ of ∇ at $\alpha_0 \in \Lambda$ that corresponds to the basis of 1-cycles $\Gamma_1, \ldots, \Gamma_k \subset \Lambda$.

Note that for the symplectic cross-section $S_0 = N_0 \cap \Sigma$ at $\alpha_0 \in \Lambda$, we have

$$T_{\alpha_0} S_0 \subset (T_{\alpha_0} \Lambda)^\perp,$$

and hence there is the natural symplectomorphism $\widehat{\tau} \colon E_{\alpha_0} \to T_{\alpha_0} S_0$.

LEMMA 2.2.1. *There exist an $\varepsilon > 0$ and a neighborhood $\mathcal{U}$ of α_0 in M admissible for the pair $(\mathcal{F}, \Sigma_\varepsilon)$ such that the Poincaré mappings P^i (2.2.5) are smooth mappings possessing the following properties:*

(a) $P^i(\alpha_0) = \alpha_0$.

(b) *The diagram*

$$\begin{array}{ccc} \Sigma_\varepsilon & \xrightarrow{P^i} & \Sigma_\varepsilon \cap \widetilde{\mathcal{U}}_i \\ e \downarrow & & \downarrow \widetilde{e}_i \\ \mathcal{B}_\varepsilon & \xrightarrow{\text{id}} & \mathcal{B}_\varepsilon \end{array}$$

commutes; moreover, for each $b \in \mathcal{B}_\varepsilon$ the restriction maps

$$P_b^i \equiv P^i\big|_{S_b \cap \mathcal{U}} \colon S_b \cap \mathcal{U} \to S_b \cap \widetilde{\mathcal{U}}_i$$

are symplectomorphisms.

(c) *The tangent maps $T_{\alpha_0} P_b^i \colon T_{\alpha_0} S_0 \to T_{\alpha_0} S_0$ of P_b^i at $m = \alpha_0$ and $b = 0$ are conjugate to the generators $G_1, \ldots, G_k$ of the holonomy group $\mathrm{Hol}_{\alpha_0}^\nabla$,*

$$(T_{\alpha_0} P_b^i)\big|_{b=0} = \widehat{\tau}^{-1} \circ G_i \circ \widehat{\tau} \qquad (i = 1, \ldots, k).$$

(d) *The Poincaré mappings commute pairwise,*

$$P^i \circ P^j(m) = P^j \circ P^i(m) \qquad (i, j = 1, \ldots, k)$$

for any $m \in (P^i)^{-1}(\Sigma_\varepsilon \cap \widetilde{\mathcal{U}}_i \cap \widetilde{\mathcal{U}}_j) \cap (P^j)^{-1}(\Sigma_\varepsilon \cap \widetilde{\mathcal{U}}_i \cap \widetilde{\mathcal{U}}_j)$.

PROOF. By assumption, the commuting Hamilton vector fields $\mathrm{ad}(f_1), \ldots,$ $\mathrm{ad}(f_k)$ form the integrable plane distribution $\mathcal{H}$. This implies that we can choose an $\varepsilon > 0$ and a neighborhood $\mathcal{U}$ of α_0 in M admissible for $(\mathcal{F}, \Sigma_\varepsilon)$ such that each connected component in $\widetilde{\mathcal{U}}_i = g_{f_i}^{2\pi}(\mathcal{U})$ of a leaf of $\mathcal{F}$ is an orbit of the k-flow

$$g^{\mathbf{t}} = g_{f_1}^{t_1} \circ \cdots \circ g_{f_k}^{t_k} \qquad (\mathbf{t} = (t_1, \ldots, t_k) \in \mathbb{R}^k).$$

Taking such a neighborhood $\mathcal{U}$, we see that there are smooth mappings

(2.2.6) $$\mathbf{t}^{(i)}: \mathcal{U} \cap \Sigma_\varepsilon \to \mathbb{R}^k \qquad (i = 1, \ldots, k)$$

defined by the relations

(2.2.7) $$g^{\mathbf{t}^{(i)}(m)}(g_{f_i}^{2\pi}(m)) = P^i(m) \qquad (m \in \Sigma_\varepsilon \cap \mathcal{U})$$

and satisfying the property

$$\mathbf{t}^{(i)}(m) = 0 \iff m = \alpha_0.$$

These arguments and Theorem 1.3.1 imply properties (a)–(c). Part (d) follows from the commutativity of the flows $g_{f_i}^t$ and from (2.2.7). □

LEMMA 2.2.2. *Assume that there is a 1-cycle* $\Gamma \subset \Lambda$ *whose multipliers* $\{\lambda^\sigma(\Gamma)\}$ *satisfy condition* (2.2.2), *say,* $\Gamma = \Gamma_1$. *Then there are* $\varepsilon > 0$ *and an admissible neighborhood* $\mathcal{U}$ *of* α_0 *in* M *such that the Poincaré mappings* P^i *with properties* (a)–(d) *from Lemma* 2.2.1 *are well defined, and there is a smooth map*

$$\widehat{\delta}: \mathcal{B}_\varepsilon \to \widetilde{\mathcal{U}}_1 \cap \mathcal{U} \cap \Sigma_\varepsilon, \qquad e \circ \widehat{\delta} = \mathrm{id}_{\mathcal{B}_\varepsilon}, \qquad \widehat{\delta}(0) = \alpha_0 \in \Lambda,$$

which gives the fixed points of the map P^1,

(2.2.8) $$P_b^1(s) = s \iff s = \widehat{\delta}(b),$$

where $s \in S_b \cap \widetilde{\mathcal{U}}_1 \cap \mathcal{U}$, $b \in \mathcal{B}_\varepsilon$.

PROOF. By (2.2.2) and property (c) of Lemma 2.2.1, the matrix $TP^1\big|_{(0,\alpha_0)} - I$ has no zero eigenvalues. Applying the implicit mapping theorem [**AbMa**] to P^1, we prove the existence of the map $\widehat{\delta}$ with desired properties. □

COROLLARY 2.2.1. *The map*

(2.2.9) $$\widehat{\delta}: \mathcal{B}_\varepsilon \mapsto \widetilde{\mathcal{U}}_1 \cap \cdots \cap \widetilde{\mathcal{U}}_k \cap \mathcal{U} \cap \Sigma_\varepsilon$$

can be chosen in such a way that

(2.2.10) $$P^i\big(b, \widehat{\delta}(b)\big) = \widehat{\delta}(b) \qquad \forall i = 1, \ldots, k, \quad b \in \mathcal{B}_\varepsilon.$$

PROOF. Since the Poincaré mappings P^i commute pairwise, we have

$$P_b^i\big(P_b^1(\widehat{\delta}(b))\big) = P_b^1\big(P_b^i(\widehat{\delta}(b))\big), \qquad b \in \mathcal{B}_\varepsilon.$$

Taking into account (2.2.8) and denoting $s = P_b^i(\widehat{\delta}(b))$, we obtain $P_b^1(s) = s$. Using again (2.2.8), we deduce that

$$s = \widehat{\delta}(b) \implies P_b^i\big(\widehat{\delta}(b)\big) = \widehat{\delta}(b). \quad \square$$

PROOF OF THEOREM 2.2.1. Using the map $\widehat{\delta}$ (2.2.9) and (2.2.10), let us show that the leaf
$$\Lambda_b = \mathcal{L}_{\widehat{\delta}(b)}, \qquad b \in \mathcal{B}_\varepsilon,$$
of the foliation $\mathcal{F}$ containing $\widehat{\delta}(b) \in \widetilde{\mathcal{U}}_1 \cap \cdots \cap \widetilde{\mathcal{U}}_k \cap S_b \subset N_b$ is an embedded isotropic submanifold of M that is diffeomorphic to the k-torus.

We define the smooth maps
$$c^{(i)} \stackrel{\text{def}}{=} \mathbf{t}^{(i)} \circ \widehat{\delta} \colon \mathcal{B}_\varepsilon \to \mathbb{R}^k \qquad (i = 1, \ldots, k)$$
$$c^{(i)}(b) = \bigl(c_1^{(i)}(b), \ldots, c_k^{(i)}(b)\bigr), \qquad b \in \mathcal{B}_\varepsilon,$$
where the $\mathbf{t}^{(i)}$ are defined by (2.2.6) and (2.2.7). In a neighborhood of Λ in M, we introduce the set of vector fields

$$(2.2.11) \quad u^i(m) = \sum_{j=1}^k \frac{1}{2\pi} c_j^{(i)}\bigl(\boldsymbol{f}(m)\bigr) \operatorname{ad}(f_j)(m) + \operatorname{ad}(f_i)(m) \qquad (i = 1, \ldots, k).$$

Note that $c_j^i(0) = 0$, and hence $u^i\big|_\Lambda = \operatorname{ad}(f_i)\big|_\Lambda \equiv \partial/\partial \alpha^i$. Moreover, we have:
- $[u^i, u^j] = 0$ for all $i, j = 1, \ldots, k$;
- $\operatorname{span}\{u^1(m), \ldots, u^k(m)\} = T_m \Lambda_b$ for all $m \in \Lambda_b$, $b \in \mathcal{B}_\varepsilon$;
- the trajectories $\gamma^1_{\widehat{\delta}(b)}, \ldots, \gamma^k_{\widehat{\delta}(b)}$ of the vector fields $u^1, \ldots, u^k$ passing through $\widehat{\delta}(b) \in \Lambda_b$ are 2π-periodic. This follows from the identity

$$g_{u^i}^{2\pi}\bigl(\widehat{\delta}(b)\bigr) = g_{f_i}^{2\pi} \circ g^{\mathbf{t}^{(i)} \widehat{\delta}(b)}\bigl(\widehat{\delta}(b)\bigr) = P_b^i\bigl(\widehat{\delta}(b)\bigr) = \widehat{\delta}(b).$$

As is easy to see, all trajectories of the vector fields $u^1, \ldots, u^k$ (2.2.11) on Λ_b are 2π-periodic. Hence, similarly to the Liouville–Arnold theorem [**A**$_3$, **AKN**], by standard arguments we prove that the leaf $\Lambda_b = \mathcal{L}_{\widehat{\delta}(b)}$ is a k-torus.

The uniqueness follows from Theorem 1.4.3. $\square$

§2.3. Flat coisotropic extensions

Here we describe a class of flat coisotropic extensions of a given isotropic submanifold that come from coisotropic extensions of an isotropic submanifold in a model phase space. As we shall see in §2.5, such coisotropic extensions generate a class of Hamilton systems that are partially integrable near an isotropic submanifold. First, we define flat coisotropic extensions in terms of symplectic geometry over an isotropic submanifold (via flat Darboux atlases) and give its interpretation by means of Bott-type connections. Finally, we shall see that the space of equivalence classes of flat coisotropic extensions is isomorphic to the moduli space of flat symplectic connections in the symplectic normal bundle over an isotropic submanifold.

Let $\Lambda \subset M$ be a connected isotropic submanifold of a symplectic manifold (M, Ω), $\dim M = 2n$, $\dim \Lambda = k < n$.

DEFINITION 2.3.1. A Darboux chart $\bigl(\mathcal{U}, (p, \tau; z)\bigr)$ over Λ is an open connected domain $\mathcal{U} \subset M$ equipped with a set of smooth coordinate functions $(p, \tau; z)$ on $\mathcal{U}$:
$$\tau = (\tau^1, \ldots, \tau^k), \qquad p = (p^1, \ldots, p^k),$$
$$z = (z^1, \ldots, z^{2r}), \qquad r = n - k,$$

satisfying the following conditions:
 (i) $\mathcal{U} \cap \Lambda = \{p^1 = 0, \ldots, p^k = 0\} \cap \{z^1 = 0, \ldots, z^{2r} = 0\} \neq \varnothing$;
 (ii) $\Omega|_{\mathcal{U}} = \frac{1}{2}\langle J dz \wedge dz\rangle$, $J = \begin{pmatrix} 0 & -I \\ I & 0 \end{pmatrix}$.

Note that by Proposition 2.1.2, there always exists a Darboux chart containing a given point of Λ (see also [**KaVo**$_3$]).

DEFINITION 2.3.2. A *flat Darboux atlas* $\mathbb{A} = \mathbb{A}_\Lambda$ over Λ is a family of Darboux charts $\{\mathcal{U}_I; (p_I, \tau_I; z_I)\}$ such that

(2.3.1) $\qquad U_\mathbb{A} = \bigcup_I \mathcal{U}_I \quad$ is a neighborhood of Λ in M,

and on the intersection of two charts $\mathcal{U}_I$ and $\mathcal{U}_J$, the corresponding coordinate functions $(p_I, \tau_I; z_I)$ and $(p_J, \tau_J; z_J)$ satisfy the relations

(2.3.2) $\qquad p_J = C^{JI} p_I, \qquad C^{JI} \in \mathrm{GL}(k) \otimes C^\infty(\mathcal{U}_I \cap \mathcal{U}_J),$

(2.3.3) $\qquad z_J = T^{JI} z_I, \qquad T^{JI} \in \mathrm{Sp}(2r),$

where the transition matrix T_{JI} is constant,

(2.3.4) $\qquad T^{JI} = \mathrm{const} \quad \text{on } \mathcal{U}_I \cap \mathcal{U}_J.$

Proposition 2.1.2 implies the following statement.

PROPOSITION 2.3.1. *There exists a flat Darboux atlas over a given isotropic submanifold $\Lambda \subset M$ if and only if the symplectic normal bundle E of Λ is flat, that is, E admits a flat symplectic connection.*

Note that each Darboux atlas $\mathbb{A}$ over Λ generates the coisotropic extension $N_\mathbb{A} \subset U_\mathbb{A}$ of Λ, which is defined in a Darboux chart $(\mathcal{U}_I, (p_I, \tau_I; z_I))$ by the equations

(2.3.5) $\qquad N_\mathbb{A} = \{m \in \mathcal{U}_I \mid p_I^1(m) = 0, \ldots, p_I^k(m) = 0\}.$

Two flat Darboux atlases $\mathbb{A}$ and $\mathbb{A}'$ over Λ are said to be *equivalent* if the union $\mathbb{A} \cup \mathbb{A}'$ is also a flat Darboux atlas over Λ. Obviously, coisotropic extensions generated by equivalent flat Darboux atlases coincide in a neighborhood of Λ in M.

Using the Darboux atlas $\mathbb{A}$ over Λ, we can compute the holonomy group of the flat connection ∇ (1.3.6) associated with the coisotropic extension $N_\mathbb{A}$ (2.3.5) in the following way. Let us choose a point $\alpha_0 \in \Lambda$. For a given closed path $\Gamma \subset \Lambda$ containing α_0, we choose a distinguished chain

$$\mathcal{U}_0, \mathcal{U}_1, \ldots, \mathcal{U}_{\mathcal{N}-1}, \mathcal{U}_\mathcal{N} = \mathcal{U}_0 \ni \alpha_0$$

of Darboux charts $(\mathcal{U}_i, (p_i, \tau_i; z_i))$ that cover Γ,

$$\Gamma \cap \mathcal{U}_j \cap \mathcal{U}_{j+1} \neq \varnothing, \qquad \Gamma \subset \bigcup_{j=0}^{\mathcal{N}} \mathcal{U}_j.$$

By

(2.3.6) $\qquad G(\Gamma) = T_{\mathcal{U}_0 \cap \mathcal{U}_{\mathcal{N}-1}} \cdot \ldots \cdot T_{\mathcal{U}_2 \cap \mathcal{U}_1} \cdot T_{\mathcal{U}_1 \cap \mathcal{U}_0} \in \mathrm{Sp}(2r),$

we denote the product of transition matrices from (2.3.3), (2.3.4). Since the restrictions $Z_1 = \partial/\partial z^1|_\Lambda, \ldots, Z_{2r} = \partial/\partial z^{2r}|_\Lambda$ define the local parallel sections of the flat connection ∇ (1.3.6) associated with $N_{\mathbb{A}}$, we see that the matrices $G(\Gamma)$ (2.3.6) generate the holonomy group $\operatorname{Hol}^{\nabla}_{\alpha_0}$.

DEFINITION 2.3.3. A coisotropic extension N of Λ in M is said to be *flat* if there exists a Darboux atlas $\mathbb{A} = \{\mathcal{U}_I, (p_I, \tau_I; z_I)\}$ adapted to N in the sense that $N_{\mathbb{A}} \subseteq N$, that is, N is locally defined in the neighborhood $U_{\mathbb{A}}$ of Λ as follows:

$$N = \{p_I^1 = 0, \ldots, p_I^k = 0\}.$$

For a given coisotropic extension N of Λ, by $\mathcal{H} \subset TN$ we denote the *characteristic* subbundle of the symplectic structure Ω with fibers

(2.3.7) $$\mathcal{H}_m = \ker{}_m \Omega\big|_N, \qquad m \in N.$$

Since the restriction of Ω to N is a closed 2-form of constant rank, the characteristic subbundle $\mathcal{H}$ is *integrable* and

(2.3.8) $$\mathcal{H}\big|_\Lambda = T\Lambda, \qquad \dim \mathcal{H} = \dim \Lambda.$$

We can single out the following properties of flat coisotropic extensions in terms of characteristic subbundles.

PROPOSITION 2.3.2. *Let $N \subset M$ be a flat coisotropic extension of Λ with adapted flat Darboux atlas $\mathbb{A} = \{(\mathcal{U}_I; p_I, \tau_I; z_I)\}$. Then in the neighborhood $U = U_{\mathbb{A}}$ of Λ in M, there is a 1-form $\gamma = \gamma_{\mathbb{A}}$ satisfying the properties*

(2.3.9) $$\Omega = d\gamma \qquad \text{on } U;$$
(2.3.10) $$u \,\lrcorner\, \gamma = 0 \qquad \text{for each } u \in \mathcal{H}_m, \, m \in U \cap N.$$

In each Darboux chart $\mathcal{U}_I$, the 1-form γ takes the form

(2.3.11) $$\gamma = \frac{1}{2}\langle Jz_I, dz_I\rangle.$$

REMARK 2.3.1. Conditions (2.3.9), (2.3.10) mean that the restriction $\overset{\circ}{\gamma}$ of the 1-form γ to $N \cap U$ is an *absolute integral invariant* of the characteristic subbundle $\mathcal{H}$ (see [**LiMr**]), namely, for each closed curve $\sigma \subset U \cap N$ and each one-parameter smooth deformation $\{\sigma_t, t \in [0,1]\}$ of σ on U along the leaves of $\mathcal{H}$, we have

$$\oint_{\sigma_t} \overset{\circ}{\gamma} = \oint_\sigma \overset{\circ}{\gamma} \qquad \text{for all } t \in [0,1].$$

REMARK 2.3.2. Conditions (2.3.9), (2.3.10) imply that there is a vector field Z on U satisfying the relations

(2.3.12) $$L_Z \Omega = \Omega \qquad \text{on } U,$$
(2.3.13) $$Z(m) \in T_m N \qquad \forall m \in N \cap U.$$

In a Darboux chart $(\mathcal{U}_I, (p, \tau, z))$ of $\mathbb{A}$, the vector field Z has the local representation

(2.3.14) $$Z = \langle z_I, \partial/\partial z_I\rangle.$$

On the contrary, from the existence of a vector field Z with properties (2.3.12), (2.3.13), it follows that the 1-form

(2.3.15) $$\gamma = Z \,\lrcorner\, \Omega$$

satisfies (2.3.9), (2.3.10). A vector field Z satisfying (2.3.12) is said to be the *Liouville vector field* (see [**LiMr**]).

DEFINITION 2.3.3. Let N be a coisotropic extension of an isotropic submanifold Λ in (M, Ω). If in a neighborhood of Λ in N there is a *Liouville vector field* $\overset{\circ}{Z}$,

$$(2.3.16) \qquad L_{\overset{\circ}{Z}}\omega = \omega, \qquad \omega = \Omega\big|_N,$$

then the triple $(N, \omega, \overset{\circ}{Z})$ is called an *exact coisotropic extension* of Λ.

So, Proposition 2.3.2 says that any flat coisotropic extension is exact. Also note that equivalent flat Darboux atlases $\mathbb{A}$ and $\widetilde{\mathbb{A}}$ over Λ generate the same Liouville vector fields,

$$\mathbb{A} \sim \widetilde{\mathbb{A}} \implies \overset{\circ}{Z}_\mathbb{A} = \overset{\circ}{Z}_{\widetilde{\mathbb{A}}} \quad \text{and} \quad \overset{\circ}{\gamma}_\mathbb{A} = \overset{\circ}{\gamma}_{\widetilde{\mathbb{A}}}.$$

Now we discuss the properties of flat coisotropic extensions in terms of Bott-type connections.

Let $N \subset M$ be a coisotropic extension of an isotropic submanifold Λ. Consider the characteristic subbundle $\mathcal{H}$ of Ω on N and its normal bundle

$$(2.3.17) \qquad \nu\mathcal{H} \overset{\text{def}}{=} TN/\mathcal{H} \equiv TN/TN^\perp.$$

The symplectic structure Ω induces a nondegenerate skew-symplectic bilinear form in $\nu\mathcal{H}$ via the natural projection

$$(2.3.18) \qquad \nu\colon TN \to \nu\mathcal{H}.$$

Thus, $\nu\mathcal{H}$ is the symplectic bundle over N whose restriction to Λ is the symplectic normal bundle over Λ, $\nu\mathcal{H}\big|_\Lambda = E$. In the normal bundle $\nu\mathcal{H}$, there is a *partial connection* (an $\mathcal{H}$-connection) $\nabla^h \colon \Gamma(\mathcal{H}) \oplus \Gamma(\nu\mathcal{H}) \to \Gamma(\nu\mathcal{H})$, which is defined as follows:

$$(2.3.19) \qquad \nabla^h_u s = \nu([u, X]), \qquad s \in \Gamma(\nu\mathcal{H}),$$

where $u \in \Gamma(\mathcal{H})$ and $X \in \text{Vect}(N)$ is such that $\nu(X) = s$.

We choose the decomposition

$$(2.3.20) \qquad TN = \mathcal{H} \oplus \mathcal{V},$$

where the subbundle $\mathcal{V}$ is the complement to $\mathcal{H}$. We call $\mathcal{H}$ a *horizontal* subbundle and $\mathcal{V}$ a *vertical* subbundle. Since the restriction of Ω to $\mathcal{V}$ is nondegenerate, we have the identification $\nu\mathcal{H} \approx \mathcal{V}$. Then the partial connection ∇^h in $\mathcal{V}$ can be determined by the relation (see also Remark 1.3.2)

$$(2.3.21) \qquad (\nabla^h_u X) \lrcorner \omega = u \lrcorner (L_X \omega)$$

for $u \in \Gamma(\mathcal{H})$, $X \in \Gamma(\mathcal{V})$. Indeed, it is easy to see that the kernel of the 1-form on the right-hand side of (2.3.21) belongs to $\mathcal{H}$ for each $u \in \Gamma(\mathcal{H})$. Since Ω is nondegenerate along the fibers of $\mathcal{V}$, we can deduce that there is a unique section $\nabla^h_u X$ that resolves (2.3.21). So, the partial connection ∇^h is a characteristic of the coisotropic extension N that possesses the following properties:

- ∇^h is flat;
- ∇^h is symplectic;

- the restriction of ∇^h to the symplectic normal bundle E over Λ coincides with the connection ∇ (1.3.6),

$$\nabla^h\big|_\Lambda = \nabla. \qquad (2.3.22)$$

Now we can formulate the flatness criterion of a coisotropic extension N in terms of the extension of the partial connection ∇^h.

THEOREM 2.3.1. *A coisotropic extension N of a connected isotropic submanifold $\Lambda \subset M$ is flat if and only if there exists a linear connection*

$$\nabla^{\mathrm{ex}}\colon \mathrm{Vect}(N') \oplus \Gamma(\nu\mathcal{H}) \to \Gamma(\nu\mathcal{H})$$

in the normal bundle $\nu\mathcal{H}$ over a neighborhood N' of Λ in N, and this connection has the following properties:
  (a) ∇^{ex} *is flat;*
  (b) ∇^{ex} *is symplectic;*
  (c) *for any $u, v \in \mathrm{Vect}(N')$*

$$\nabla^{\mathrm{ex}}_u \nu(v) - \nabla^{\mathrm{ex}}_v \nu(u) = \nu\big([u,v]\big), \qquad (2.3.23)$$

where the linear map $\nu\colon \mathrm{Vect}(N') \to \Gamma(\nu\mathcal{H})$ is induced by the natural projection (2.3.18).

REMARK 2.3.3. Condition (2.3.23) implies that the covariant derivative of ∇^{ex} along the horizontal vector fields coincides with the partial connection ∇^h,

$$\nabla^{\mathrm{ex}}_u = \nabla^h_u, \qquad u \in \Gamma(\mathcal{H}),$$

that is, the connection ∇^{ex} is canonical in the sense of Bott with respect to the foliation $\mathcal{H}$ (see [**Bo$_2$**, **L$_1$**]).

PROOF. *Necessity.* Suppose that we have a flat coisotropic extension N with an adapted flat Darboux atlas $\mathbb{A} = \{(\mathcal{U}_I; p_I, \tau_I; z_I)\}$. Then the desired connection ∇^{ex} in $\nu\mathcal{H}$ over $N' = N \cap U_\mathbb{A}$ is uniquely defined by the relations

$$\nabla^{\mathrm{ex}}\nu\big(\partial/\partial z_I^\sigma\big|_{N'}\big) = 0 \qquad \text{on } \mathcal{U}_I \cap N \qquad (2.3.24)$$

for $\sigma = 1, \ldots, 2r$.

Before we prove the sufficiency, we need some auxiliary lemmas.

LEMMA 2.3.1. *Let N be a coisotropic extension of Λ. Then in a neighborhood N' of Λ in N, there is a fiber bundle over Λ*

$$\mathbf{n}\colon N' \to \Lambda \qquad (2.3.25)$$

that induces the decomposition

$$TN = \mathcal{H} \oplus \mathcal{V} \qquad \text{on } N', \qquad (2.3.26)$$

where $\mathcal{H}$ is the characteristic subbundle of Ω on N' and $\mathcal{V}$ is the characteristic subbundle of $\mathbf{n}$, $\mathcal{V}_m = \ker_m \mathbf{n}$, $m \in N'$.

This lemma follows from (2.3.8) and the tubular neighborhood theorem (see, for example, [**LiMr**]).

Let us choose a fiber bundle (2.3.25) over Λ and consider the cotangent bundle $j\colon T^*\Lambda \to \Lambda$. Then we can define the *fibered product* of N' and $T^*\Lambda$ (see [**LiMr**]):

$$\widetilde{M} = N' \times_\Lambda T^*\Lambda \qquad (2.3.27)$$

It can be viewed as the pullback of (2.3.25) via j, $\mathbf{n}_1 \colon \widetilde{M} \to T^*\Lambda$, or as the pullback of j via (2.3.25),

(2.3.28) $$\mathbf{n}_2 \colon \widetilde{M} \to N'.$$

We introduce the following 2-form on $\widetilde{M}$:

(2.3.29) $$\widetilde{\Omega} = \mathbf{n}_1^*(d\rho) + \mathbf{n}_2^*(\Omega|_{N'}).$$

It is clear that $(\widetilde{M}, \widetilde{\Omega})$ is a symplectic manifold. Moreover, N' (as the zero section of (2.3.28)) is the coisotropic extension of Λ (as the zero section of $\mathbf{n}_1 \circ j$) in $\widetilde{M}$.

We obtain the following lemma by applying Proposition 2.1.4 to the coisotropic embeddings (M, Ω, N') and $(\widetilde{M}, \widetilde{\Omega}, N')$.

LEMMA 2.3.2. *A neighborhood N' of Λ in M from Lemma 2.3.1 can be chosen so that there is a symplectomorphism g of a neighborhood U of N' in M onto a neighborhood $\widetilde{U}$ of N' in $\widetilde{M}$ that satisfies the condition $g|_{N'} = \mathrm{id}$.*

Sufficiency. We choose a neighborhood N' of Λ in M that satisfies the assumptions of Lemma 2.3.2. Then we have the decomposition (2.3.26) of TN' into the Whitney sum of two integrable subbundles: horizontal $\mathcal{H}$ and vertical $\mathcal{V}$. In this case, the spaces of smooth sections $\Gamma(\mathcal{H})$ and $\Gamma(\mathcal{V})$ are the Lie algebras of vector fields on N'. Assume that we have a linear connection ∇^{ex} in $\mathcal{V}$ over N' that satisfies the above assumptions (a)–(c). It follows from these assumptions that

(α) if $u \in \Gamma(\mathcal{H})$, $X \in \Gamma(\mathcal{V})$, then

(2.3.30) $$\nabla_u^{\mathrm{ex}} X = \nu([u, X]),$$

where $\nu \colon \mathrm{Vect}(N') \to \Gamma(\mathcal{V})$ is the linear map induced by the decomposition (2.3.26);

(β) if $X', X'' \in \Gamma(\mathcal{V})$, then

(2.3.31) $$\nabla_{X'}^{\mathrm{ex}} X'' - \nabla_{X''}^{\mathrm{ex}} X' = [X', X''].$$

In a simply connected neighborhood W of a given point of Λ in N', we choose a basis of local vector fields $u_1, \ldots, u_k \in \Gamma(\mathcal{H})$, $X_1, \ldots, X_{2r} \in \Gamma(\mathcal{V})$ with the properties:

(i) $u_i = \partial/\partial \alpha_i$, where $\alpha = (\alpha^1, \ldots, \alpha^k)$ are local coordinates along the fibers of $\mathcal{H}$ from a Frobenius chart of the foliation of $\mathcal{H}$; and

(ii) X_σ are ∇^{ex}-parallel sections,

$$\nabla^{\mathrm{ex}} X_\sigma = 0 \quad (\sigma = 1, \ldots, 2r) \quad \text{on } W;$$

$$\Omega(X_\sigma, X_{\sigma'}) = J_{\sigma\sigma'}, \quad J = ((J_{\sigma\sigma'})) = \begin{pmatrix} 0 & -I \\ I & 0 \end{pmatrix}, \quad \text{on } W.$$

Properties (i), (ii) and conditions (2.3.30), (2.3.31) imply that u_i, X_σ pairwise commute.

Define 1-forms $\gamma^1, \ldots, \gamma^{2r}$ on W by the relations

$$X_\sigma \lrcorner \gamma^{\sigma'} = \delta_\sigma^{\sigma'}, \qquad u_i \lrcorner \gamma^\sigma = 0.$$

Then it is clear that γ^σ are closed 1-forms, $d\gamma^\sigma = 0$ ($\sigma = 1, \ldots, 2r$). Since W is simply connected, there are coordinate functions $x^1, \ldots, x^{2r} \in C^\infty(W)$ such that

$$\gamma^\sigma = dx^\sigma, \qquad x^\sigma|_{\Lambda \cap W} = 0.$$

By $\widetilde{U}$ we denote the fibered product of the fiber bundle (2.3.25) and $T^*\Lambda$ over $W \cap \Lambda$; by $\eta = (\eta^1, \ldots, \eta^k)$ we denote the local coordinates in $T^*\Lambda$ ($\alpha \in W \cap \Lambda$) dual to $\partial/\partial \alpha^1, \ldots, \partial/\partial \alpha^k$. Then it is easy to see that the coordinate chart $(\widetilde{\mathcal{U}}(\eta, \alpha, x))$ is the Darboux chart over Λ in the model phase space $(\widetilde{M}, \widetilde{\Omega})$ defined by (2.3.27), (2.3.29). Moreover, it follows from the above constructions that the transformation rules (2.3.2)–(2.3.4) hold on the intersections of two such Darboux charts. So, there exists a flat Darboux atlas over Λ in $(\widetilde{M}, \widetilde{\Omega})$ adapted to the coisotropic extension N'. The pullback of this atlas via the symplectomorphism g from Lemma 2.3.2 gives the desired flat Darboux atlas adapted to the original coisotropic extension N of Λ in (M, Ω). $\square$

EXAMPLE 2.3.1. Starting from an isotropic submanifold Λ of a symplectic manifold (M, Ω), we can consider the model phase space $(\mathcal{X} = T^*\Lambda \oplus E, \Omega^\nabla)$ from (1.5.23) generated by the flat symplectic connection ∇ (1.3.6) in the symplectic normal bundle E over Λ. We claim that the coisotropic extension E of Λ in $\mathcal{X}$ is flat. Indeed, the pullback $\nabla^\#$ of the connection ∇ via $j : T^*\Lambda \to \Lambda$ satisfies the assumptions (a)–(c) of Theorem 2.3.1; that is, we can choose $\nabla^{\text{ex}} = \nabla^\#$.

On the other hand, the flat Darboux atlas over Λ adapted to the coisotropic extension E in $\mathcal{X}$ consists of Darboux charts $(\mathcal{U}, (\eta, \alpha; x))$ such that E is trivial over $\mathcal{U} \cap \Lambda$, $x = (x^1, \ldots, x^{2r})$ are local coordinates along the fibers of E corresponding to the basis of ∇-parallel sections, and (η, α) are local canonical coordinates on $T^*_{\mathcal{U} \cap \Lambda}\Lambda$.

So, this example shows that if the symplectic normal E over Λ admits a flat symplectic connection, then the set of flat coisotropic extensions of Λ in M is not empty.

The next natural problem is to classify flat coisotropic extensions of a given isotropic submanifold with accuracy up to the equivalence relation (in the sense of Definition 2.1.1).

THEOREM 2.3.2. *Two coisotropic extensions N and $\widetilde{N}$ of a given isotropic submanifold Λ in a symplectic manifold (M, Ω) are equivalent if and only if the corresponding flat connections ∇ and $\widetilde{\nabla}$ defined by (1.3.6) are gauge equivalent in the symplectic normal bundle E over Λ.*

PROOF. The necessity is obvious.

Sufficiency. By Proposition 2.1.5 and the tubular neighborhood theorem, it suffices to consider the following situation.

Let Λ be a connected submanifold of a manifold N. Let ω and $\widetilde{\omega}$ be closed 2-forms on N of constant rank with characteristic subbundles

$$\mathcal{H} = \ker \omega, \qquad \widetilde{\mathcal{H}} = \ker \widetilde{\omega}$$

such that $\mathcal{H}|_\Lambda = \widetilde{\mathcal{H}}|_\Lambda = T\Lambda$.

By Lemma 2.3.1 in a neighborhood of Λ in N, there exists an integrable subbundle $\mathcal{V} \subset TN$ that is simultaneously the complement to $\mathcal{H}$ and to $\widetilde{\mathcal{H}}$,

(2.3.32) $$TN = \mathcal{H} \oplus \mathcal{V} = \widetilde{\mathcal{H}} \oplus \mathcal{V}.$$

Then the 2-forms ω and $\widetilde{\omega}$ are nondegenerate along $\mathcal{V}$. Assume that

$$\omega\big|_{\mathcal{V}_\Lambda} = \widetilde{\omega}\big|_{\mathcal{V}_\Lambda}.$$

Let $(N, \omega, \nabla^{\text{ex}})$ and $(N, \widetilde{\omega}, \widetilde{\nabla}^{\text{ex}})$ be two *flat* presymplectic manifolds equipped with linear connections ∇^{ex} and $\widetilde{\nabla}^{\text{ex}}$ in $\mathcal{V}$ that satisfy assumptions (a)–(c) of Theorem 2.3.1. If the corresponding flat connections (1.3.6) are gauge equivalent under a symplectomorphism $L\colon E \to E$, then by the tubular neighborhood theorem, there exists a diffeomorphism that is identical on Λ and covers L. So, we may assume that

(2.3.33) $$\nabla^{\text{ex}}\big|_{\mathcal{V}_\Lambda} = \widetilde{\nabla}^{\text{ex}}\big|_{\mathcal{V}_\Lambda},$$

that is, under the restriction to Λ, the partial connections associated with $\mathcal{H}$ and $\widetilde{\mathcal{H}}$ coincide. Let us prove that these presymplectic submanifolds are isomorphic in neighborhoods of Λ in N.

By the same arguments as in the proof of Theorem 2.3.1, we obtain that there exists an open covering of a neighborhood U of Λ in N,

$$U = \bigcup_I \mathcal{U}_I, \qquad \Lambda \cap \mathcal{U}_I \neq \varnothing,$$

by coordinate connected charts $(\mathcal{U}_I, \alpha_I, x_I)$ that are *adapted* to the triple $(N, \omega, \nabla^{\text{ex}})$ in the following sense:

(i) $\partial/\partial \alpha_I^i \in \Gamma(\mathcal{H})$ for all $i = 1, \ldots, \varkappa = \dim \Lambda$;

(ii) $\partial/\partial x_I^\sigma \in \Gamma(\mathcal{V})$ are ∇^{ex}-parallel sections for all $\sigma = 1, \ldots, 2r = \text{codim}\, \Lambda$;

(iii) $\omega\big|_U = d\gamma$, where γ is a 1-form on U that has the following local representation in each chart $(\mathcal{U}_I, \alpha_I, x_I)$:

(2.3.34) $$\gamma\big|_{\mathcal{U}_I} = \langle J x_I, dx_I \rangle, \qquad x_I\big|_{\mathcal{U}_I \cap \Lambda} = 0.$$

It follows from properties (i)–(iii) that the change of coordinates on the intersection of two charts $\mathcal{U}_I \cap \mathcal{U}_J \neq \varnothing$ is given by the relations

(2.3.35) $$\alpha_J = \alpha_J(\alpha_I), \qquad x_J = T^{JI} x_I, \qquad T^{JI} \in \text{Sp}(2r),$$

where T^{JI} is a constant matrix. Note that the compatibility conditions (2.3.27), (2.3.33) allow us to choose a neighborhood $\widetilde{U}$ of Λ and a covering $\{\widetilde{\mathcal{U}}_I\}$ such that

$$\widetilde{\mathcal{U}}_I \subset \mathcal{U}_I \quad \text{and} \quad \Lambda \cap \mathcal{U}_I = \Lambda \cap \widetilde{\mathcal{U}}_I \neq \varnothing \qquad \forall I.$$

Moreover, each chart $\widetilde{\mathcal{U}}_I$ is equipped with coordinates $(\widetilde{\alpha}_I, \widetilde{x}_I)$ adapted to the second triple $(N, \widetilde{\omega}, \widetilde{\nabla}^{\text{ex}})$ with the following transformation rule on the intersection $\mathcal{U}_I \cap \mathcal{U}_J$:

(2.3.36) $$\widetilde{\alpha}_J = \widetilde{\alpha}_J(\widetilde{\alpha}_I), \qquad \widetilde{x}_J = \widetilde{T}^{JI} \widetilde{x}_I,$$

where the transition matrix $\widetilde{T}^{JI}$ is the same as in (2.3.35),

(2.3.37) $$\widetilde{T}^{JI} = T^{JI} \qquad \forall J, I.$$

Moreover, we may set

(2.3.38) $$\widetilde{\alpha}_I = \alpha_I\big|_{\widetilde{\mathcal{U}}_I} \qquad \forall I.$$

So, under this choice of coverings $\{\mathcal{U}_I\}$ and $\{\widetilde{\mathcal{U}}_I\}$, the corresponding coordinate functions (α_I, x_I) and $(\widetilde{\alpha}_I, \widetilde{x}_I)$ define the diffeomorphisms

(2.3.39) $$\begin{aligned}\varphi_I &\colon \mathcal{U}_I \to V_I \times W_I \subset \mathbb{R}^k \times \mathbb{R}^{2r}, \\ \widetilde{\varphi}_I &\colon \widetilde{\mathcal{U}}_I \to V_I \times \widetilde{W}_I \subset \mathbb{R}^k \times \mathbb{R}^{2r},\end{aligned}$$

where $V_I \subset \mathbb{R}^k$ are open subsets and W_I, $\widetilde{W}_I$ are open neighborhoods of $\{0\}$ in $\mathbb{R}^{2r}$ such that

(2.3.40) $$\widetilde{W}_I \subset W_I \quad \forall I.$$

Then the primitives γ_I and $\widetilde{\gamma}_I$ (2.3.34) of the presymplectic structures ω and $\widetilde{\omega}$ are pull-backs of the Liouville 1-form $\eta = \langle Jy, dy \rangle$ ($y \in \mathbb{R}^{2r}$) that define the standard presymplectic structure $d\eta$ on $\mathbb{R}^k \times \mathbb{R}^{2r}$, $\gamma_I = \varphi_I^*(\eta)$, $\widetilde{\gamma}_I = \widetilde{\varphi}_I^*(\eta)$. Since conditions (2.3.33) and (2.3.40) are compatible, the mappings (2.3.39) define the diffeomorphisms

(2.3.41) $$f_I = \varphi_I^{-1} \circ \widetilde{\varphi}_I : \widetilde{\mathcal{U}}_I \to \mathcal{U}'_I, \qquad \mathcal{U}'_I = \varphi_I^{-1}(V_I \times \widetilde{W}_I) \subset \mathcal{U}_I.$$

The properties (2.3.37) and (2.3.38) imply the following transformation rules for the mappings (2.3.39):

$$\varphi_J = (\widehat{\alpha}_{JI} \times \widehat{T}^{JI}) \circ \varphi_I \quad \text{on} \quad \mathcal{U}_I \cap \mathcal{U}_J \neq \varnothing,$$
$$\widetilde{\varphi}_J = (\widehat{\alpha}_{JI} \times \widehat{T}^{JI}) \circ \widetilde{\varphi}_I \quad \text{on} \quad \widetilde{\mathcal{U}}_I \cap \widetilde{\mathcal{U}}_J \neq \varnothing,$$

where $\widehat{\alpha}_{JI}$ are local diffeomorphisms of $\mathbb{R}^k$ and $\widehat{T}^{JI} : \mathbb{R}^{2r} \to \mathbb{R}^{2r}$ are linear symplectic mappings of $(\mathbb{R}^{2r}, d\eta)$. This implies

$$f_I = f_J \quad \text{on} \quad \widetilde{\mathcal{U}}_I \cap \widetilde{\mathcal{U}}_J \neq \varnothing,$$

and hence, the family $\{f_I\}$ defines a smooth mapping

(2.3.42) $$f : \widetilde{U} \to U', \qquad U' = \bigcup_I \mathcal{U}'_I \subset U.$$

Obviously, f is a local diffeomorphism such that $f|_\Lambda = \mathrm{id}_\Lambda$ and $f^*\gamma = \widetilde{\gamma}$. Now one can easily see that under a contraction of the neighborhood $\widetilde{U}$, the mapping f becomes a diffeomorphism. $\square$

COROLLARY 2.3.2. *Let $\mathbb{A}$ and $\widetilde{\mathbb{A}}$ be two flat Darboux atlases over a connected isotropic submanifold Λ, and let $N_\mathbb{A}$ and $N_{\widetilde{\mathbb{A}}}$ be the corresponding flat coisotropic extensions of Λ. Let ∇ and $\widetilde{\nabla}$ be flat connections (1.3.6) generated by $N_\mathbb{A}$ and $N_{\widetilde{\mathbb{A}}}$. If the holonomy groups $\mathrm{Hol}^\nabla_{\alpha_0}$ and $\mathrm{Hol}^{\widetilde{\nabla}}_{\alpha_0}$ of ∇ and $\widetilde{\nabla}$ are conjugate at a given point $\alpha_0 \in \Lambda$,*

(2.3.43) $$\mathrm{Hol}^\nabla_{\alpha_0} = \{T^{-1} \circ G \circ T, G \text{ runs over } \mathrm{Hol}^{\widetilde{\nabla}}_{\alpha_0}\},$$

then the pairs $(N_\mathbb{A}, \overset{\circ}{Z}_\mathbb{A})$ and $(N_{\widetilde{\mathbb{A}}}, \overset{\circ}{Z}_{\widetilde{\mathbb{A}}})$ (where $\overset{\circ}{Z}_\mathbb{A}$ and $\overset{\circ}{Z}_{\widetilde{\mathbb{A}}}$ are the Liouville vector fields (2.3.14)) are isomorphic, that is, there exists a symplectomorphism g from (2.1.7) such that

(2.3.44) $$g_* \overset{\circ}{Z}_\mathbb{A} = \overset{\circ}{Z}_{\widetilde{\mathbb{A}}}.$$

As an important consequence of Theorem 2.3.2 and Proposition 2.1.2, we also obtain the following result.

THEOREM 2.3.3. *A flat coisotropic extension N of a connected isotropic submanifold Λ in a symplectic manifold (M, Ω) is equivalent to the coisotropic extension E of Λ in the model phase space $(\mathcal{X} = T^*\Lambda \oplus E, \Omega^\nabla)$ generated by the flat connection ∇ (1.3.6) associated with N in the symplectic normal bundle E over $\Lambda \subset M$.*

We conclude this section with a discussion of how the above results are related to the classification of coisotropic extensions.

Let $\Lambda \subset N$ be an isotropic submanifold. The equivalence class $[N]$ of a coisotropic extension N of Λ will be called a *coisotropic germ* of Λ in M. A coisotropic germ $[N]$ is flat if N is a flat coisotropic extension. By $\mathbb{G}(\Lambda, M)$ we denote the set of all coisotropic germs of Λ in M and by $\mathbb{G}^{\text{flat}}(\Lambda, M)$ the subset of flat coisotropic germs. Let E be the symplectic normal bundle over Λ. Consider the *moduli space* [**AtBo**]

$$\mathcal{M}(E) = \{\text{flat symplectic connections in } E\}/\sim,$$

where $\sim$ denotes the gauge equivalence relation. Then we have the *surjective* map

$$\lambda \colon \mathbb{G}(\Lambda, M) \to \mathcal{M}(E), \qquad \lambda([N]) = [\nabla],$$

where ∇ is the flat connection (1.3.6) in E associated with a coisotropic extension N of Λ. Theorem 2.3.2 says that the restriction of λ to the set of flat coisotropic germs

$$\lambda_0 \colon \mathbb{G}^{\text{flat}}(\Lambda, M) \to \mathcal{M}(E)$$

is a *bijective* map.

§2.4. Inflations and torus actions

As is well known, the convexity theorem ([**At**], [**GuTs**$_1$]) describes properties of Hamilton torus actions. In the case of a compact symplectic manifold, the theory of normal forms for effective torus actions is given by results of [**Dl**] (see also [**Gu**$_1$]).

Let $\Lambda \subset M$ be an isotropic submanifold, and let $\mathfrak{a}$ be a Hamilton torus action on M such that Λ is the set of fixed points of $\mathfrak{a}$. By the equivariant version of the Darboux–Weinstein theorem [**G, GuSt**$_2$], in a neighborhood of Λ, an action $\mathfrak{a}$ is uniquely defined by its linearized action on the symplectic normal bundle over Λ. We are interested in the situation where $\mathfrak{a}$ possesses the following additional property: $\mathfrak{a}$ preserves a given coisotropic extension N of Λ. Our goal is to investigate such torus actions for stable flat coisotropic extensions.

THEOREM 2.4.1. *Suppose that a Hamilton system (M, Ω, H) has an invariant isotropic submanifold $\Lambda \subset M$ with the following three properties:*

(i) Λ is a connected compact manifold with commutative fundamental group $\pi_1(\Lambda)$;

(ii) there exists a flat coisotropic extension N of Λ in M such that $H\big|_N = \text{const}$;

(iii) N is stable (that is, the flat connection ∇ (1.3.6) associated with N is geometrically stable).

Then there is a surjective map

(2.4.1) $\qquad \mathbb{F} \colon U \to S, \qquad \mathbb{F} = (F^1, \ldots, F^r) \qquad (r = (\dim N - \dim \Lambda)/2)$

from a neighborhood U of Λ in M onto a semineighborhood S of $\{0\}$ in $\mathbb{R}_+^r$ whose components $F^1, \ldots, F^r \in C^\infty(U)$ are smooth nonnegative functions satisfying the following conditions (a)–(d):

(a) $F^1, \ldots, F^r$ *Poisson commute,*

$$\{F^\sigma, F^{\sigma'}\} = 0 \quad on\ U \qquad (\sigma, \sigma' = 1, \ldots, r),$$

and the functions F^σ and its differentials dF^σ vanish only on Λ,

$$F^\sigma|_\Lambda = 0, \qquad dF^\sigma|_{T_\Lambda M} = 0 \qquad (\sigma = 1, \ldots, r).$$

(b) *The Poisson bracket of the Hamilton function H and each F^σ vanish on N,*

$$\{H, F^\sigma\} = 0 \qquad on\ N \cap U \qquad (\sigma = 1, \ldots, r).$$

(c) *For each $s = 0, 1, \ldots, r$ and $c = (c^1, \ldots, c^r) \in C^s \cap S$ (where the open domains C^s are defined by* (1.7.8)*), the level set*

$$(2.4.2) \qquad \Lambda(c) = \mathbb{F}^{-1}(c) \cap N$$

is an $(s + \dim \Lambda)$-dimensional isotropic submanifold of M that is invariant under the flow of the Hamilton vector field $\mathrm{ad}(H)$. *In particular, $\Lambda(0) = \Lambda$, and $\Lambda(c)$ is Lagrangian for $c \in C^r \cap S$.*

(d) *For each $c = (c^1, \ldots, c^r) \in C^s \cap S$ (where $c^{\sigma_1} > 0, \ldots, c^{\sigma_s} > 0$), the flows of the Hamilton vector fields* $\mathrm{ad}(F^{\sigma_1}), \ldots, \mathrm{ad}(F^{\sigma_s})$ *are 2π-periodic on $\Lambda(c)$ and induce a $\mathbb{T}^s$-principle bundle over Λ*

$$(2.4.3) \qquad \mathbb{T}^s \to \Lambda(c) \to \Lambda.$$

PROOF. Consider the model phase space $(\mathcal{X} = T^*\Lambda \oplus E, \Omega^\nabla)$ generated by the flat connection ∇ (1.3.6) associated with N. Since ∇ is geometrically stable, we can define nonnegative functions $h^1, \ldots, h^r \in C^\infty(\mathcal{X})$ by (1.7.7). By Theorem 2.3.3, there exists a symplectomorphism g of a neighborhood U of Λ in M onto a neighborhood of Λ in $\mathcal{X}$ such that $g|_\Lambda = \mathrm{id}$, and that sends N to E.

Then we define the functions (2.4.1) as follows: $F^\sigma = h^\sigma \circ g$ ($\sigma = 1, \ldots, r$). □

DEFINITION 2.4.1. *The submanifolds $\Lambda(c)$ from Theorem 2.4.1 are called inflations of the isotropic submanifold Λ.*

For this terminology, see also [**KaVo**₃].

COROLLARY 2.4.1. *There exists a flat Darboux atlas over Λ such that at each Darboux chart $(\mathcal{U}, (p, \tau; z = (z^1, \ldots, z^{2r})))$, the Hamilton function H and "partial" integrals of motion F^σ* (2.4.2) *take the form*

$$H\big|_\mathcal{U} = \sum_{i=1}^k a_i p^i + \mathrm{const}, \qquad a_i = a_i(p, \tau, z) \in C^\infty(\mathcal{U}),$$

$$F^\sigma\big|_\mathcal{U} = \frac{1}{2}\big[(z^\sigma)^2 + (z^{\sigma+r})^2\big],$$

where $k = \dim \Lambda$ and the a_i are smooth functions on $\mathcal{U}$.

COROLLARY 2.4.2. *Suppose that an isotropic submanifold Λ of a symplectic manifold (M, Ω) and a coisotropic extension N of Λ satisfy the assumptions of Theorem 2.4.1. Then in a neighborhood W of Λ in N, there exists an action of the r-torus $\mathbb{T}^r = \mathbb{S}^1 \times \cdots \times \mathbb{S}^1$,*

$$\phi \colon \mathbb{T}^r \times W \to W, \qquad r = \frac{1}{2}(\dim N - \dim \Lambda),$$

with the following properties:
 (I) ϕ *is effective on* W;
 (II) Λ *is the fixed point set of* ϕ;
 (III) *the $\mathbb{T}^r$-action ϕ is transverse to the characteristic subbundle $\mathcal{H} = \ker \omega$ of the presymplectic 2-form $\omega = \Omega\big|_N$ on W,*

$$T_m \mathcal{O} \cap \mathcal{H}_m = \{0\} \qquad \forall m \in W,$$

where $\mathcal{O}$ is the orbit of ϕ that passes through $m \in W$;
 (IV) *the presymplectic 2-form ω is invariant under the $\mathbb{T}^r$-action ϕ,*

$$\phi_a^* \omega = \omega \qquad \forall a \in \mathbb{T}^r.$$

Moreover, if $\mathbb{F} \colon U \to S$ is a map satisfying assumptions (a)–(d) of Theorem 2.4.1, then the restrictions of the Hamilton vector fields $\mathrm{ad}(F^1), \ldots, \mathrm{ad}(F^r)$ to $W = N \cap U$ define infinitesimal generators of a $\mathbb{T}^r$-action with properties (I)–(IV).

Below we assume that a given isotropic submanifold Λ satisfies the conditions of Theorem 2.4.1.

If N is a coisotropic extension of Λ in (M, Ω), then we denote by (N, ω) the corresponding presymplectic manifold with a closed 2-form $\omega = \Omega\big|_N$ of constant rank.

DEFINITION 2.4.2. A $\mathbb{T}^r$-action ϕ on a coisotropic extension (N, ω) of Λ is called a *transverse presymplectic $\mathbb{T}^r$-action* if ϕ satisfies conditions (I)–(IV) of Corollary 2.4.2. If, in addition, infinitesimal generators of ϕ are Hamilton vector fields on N (as on a presymplectic manifold), then ϕ is called a *transverse Hamilton $\mathbb{T}^r$-action* on (N, ω).

In the Hamilton case, ϕ is described by a *moment map* [**AbMa**].

Let $(N, \omega, \overset{\circ}{Z})$ be an exact coisotropic extension of Λ with Liouville vector field $\overset{\circ}{Z}$ ($L_{\overset{\circ}{Z}} \omega = \omega$). We say that $(N, \omega, \overset{\circ}{Z})$ is *invariant* under a $\mathbb{T}^r$-action ϕ on N, if ϕ preserves $\overset{\circ}{Z}$, $(\phi_a)_* \overset{\circ}{Z} = \overset{\circ}{Z}$ for any $a \in \mathbb{T}^r$.

PROPOSITION 2.4.1. *Let $(N, \omega, \overset{\circ}{Z})$ be an exact coisotropic extension of Λ that is invariant under a transverse presymplectic $\mathbb{T}^r$-action ϕ on N. Then ϕ is the transverse Hamilton $\mathbb{T}^r$-action on (N, ω) with the moment map $\mathbb{F}_0 \colon N \to \mathbb{R}^r$ whose components are defined as follows:*

(2.4.4) $$F_0^\sigma = \omega(\overset{\circ}{Z}, v^\sigma) \qquad (\sigma = 1, \ldots, r),$$

where $v^1, \ldots, v^r$ are the infinitesimal generators of ϕ corresponding to the group product $\mathbb{T}^r = \mathbb{S}^1 \times \cdots \times \mathbb{S}^1$. The moment map $\mathbb{F}_0$ possesses the following three properties:

(a) $\mathbb{F}_0(m) = 0 \Longrightarrow m \in \Lambda$.

(b) $F_0^1, \ldots, F_0^r$ are constant along the leaves of the characteristic distribution $\mathcal{H} = \ker \omega$.

(c) There is a $\mathbb{T}^r$-invariant open dense subset N^{reg} in N such that for each $c \in \mathbb{F}_0(N^{\mathrm{reg}}) \subset \mathbb{R}^r$, the level set $\Lambda(c) = \mathbb{F}_0^{-1}(c)$ is a Lagrangian submanifold of M on which ϕ acts freely. So, $\Lambda(c)$ is a $\mathbb{T}^r$-principle bundle.

Moreover, the $\mathbb{T}^r$-action ϕ on N can be uniquely (up to equivalence) extended to a Hamilton $\mathbb{T}^r$-action on a neighborhood U of N in M so that the corresponding moment map $\mathbb{F}_0 \colon U \to \mathbb{R}^r$ coincides with $\mathbb{F}_0$ on N, $\mathbb{F}_0 = \mathbb{F}|_N$.

The proof of the existence of a $\mathbb{T}^r$-action ϕ follows from the equivariant version of the coisotropic embedding theorem [**GuSt**$_2$].

Note that if there is a smooth *retraction* of N onto Λ, then the moment map $\mathbb{F}_0$ (2.4.4) does not depend on the choice of the Liouville vector field $\overset{\circ}{Z}$ on (N, ω) (for example, this is true if N is a tubular neighborhood of Λ in M).

Two transverse Hamilton $\mathbb{T}^r$-actions ϕ and $\widetilde{\phi}$ on coisotropic extensions (N, ω) and $(\widetilde{N}, \widetilde{\omega})$ of Λ are said to be equivalent if there are $\mathbb{T}^r$-invariant neighborhoods W and $\widetilde{W}$ of Λ in N and $\widetilde{N}$ and a diffeomorphism $g \colon W \to \widetilde{W}$ such that

$$(2.4.5) \qquad g\big|_\Lambda = \mathrm{id}_\Lambda, \qquad g^*\widetilde{\omega} = \omega,$$

$$(2.4.6) \qquad g \circ \phi_a = \widetilde{\phi}_a \circ g \qquad \forall a \in \mathbb{T}^r.$$

In terms of the moment maps $\mathbb{F}_0$ and $\widetilde{\mathbb{F}}_0$ (corresponding to ϕ and $\widetilde{\phi}$), the equivariance condition (2.4.6) can be written as $\mathbb{F}_0 = \widetilde{\mathbb{F}}_0 \circ g$.

Now we will describe torus actions on stable flat coisotropic extensions in terms of flat Darboux atlases. Recall that a given isotropic submanifold $\Lambda \subset M$ is connected and compact, and $\pi_1(\Lambda)$ is commutative. Assume that
- Λ admits a flat Darboux atlas $\mathbb{A}$, and
- the coisotropic extension $N_\mathbb{A}$ of Λ generated by $\mathbb{A}$ is stable.

Let $(E, \mathcal{J})$ be the symplectic normal bundle over Λ. By assumption, the flat connection ∇ (1.3.6) in E associated with $N_\mathbb{A}$ is geometrically stable. Choose a point $\alpha_0 \in \Lambda$ and consider the holonomy group $\mathrm{Hol}^\nabla_{\alpha_0} \subset \mathrm{Sp}(E_{\alpha_0})$ of ∇ at α_0. By Proposition 1.7.2, there is a decomposition

$$(2.4.7) \qquad {}^\mathbb{C}E_{\alpha_0} = \rho^1_{\alpha_0} \oplus \cdots \oplus \rho^r_{\alpha_0} \oplus \overline{\rho}^1_{\alpha_0} \oplus \cdots \oplus \overline{\rho}^r_{\alpha_0}$$

where
- the $\rho^\sigma_{\alpha_0}$ are positive 1-dimensional subspaces;
- any two subspaces $\rho^\sigma_{\alpha_0}$ and $\rho^{\sigma'}_{\alpha_0}$ are skew-orthogonal; and
- $\rho^\sigma_{\alpha_0}$ is G-invariant for all $G \in \mathrm{Hol}^\nabla_{\alpha_0}$ and for all $\sigma = 1, \ldots, r$.

The decomposition (2.4.7) induces the linear Hamilton $\mathbb{T}^r$-action on the symplectic vector space E_{α_0},

$$(2.4.8) \qquad \ell \colon \mathbb{T}^r \times E_{\alpha_0} \to E_{\alpha_0},$$

whose moment map $\mathbf{h}_0 \colon E_{\alpha_0} \to \mathbb{R}^r$ is defined by the quadratic forms

$$h^\sigma_0(x) = \frac{1}{i}\langle \mathcal{J}(\alpha_0)\,\mathrm{pr}^\sigma_{\alpha_0}(x), \overline{\mathrm{pr}^\sigma_{\alpha_0}(x)}\rangle, \qquad x \in E_{\alpha_0},$$

where $\sigma = 1, \ldots, r$ and $\text{pr}^\sigma_{\alpha_0} : {}^{\mathbb{C}}E_{\alpha_0} \to \rho^\sigma_{\alpha_0}$ is the projection defined by the decomposition (2.4.7).

Note that linear actions of type (2.4.8) generated by different decompositions (2.4.7) of ${}^{\mathbb{C}}E_{\alpha_0}$ are equivalent (see, for instance, [**GuSt₂**]).

We claim that a pair $(\mathbb{A}, \oplus)$ consisting of a stable flat Darboux atlas $\mathbb{A}$ and a decomposition (2.4.7) induces a transverse Hamilton $\mathbb{T}^r$-action on $N_{\mathbb{A}}$,

$$\phi_{\mathbb{A},\oplus} : \mathbb{T}^r \times N_{\mathbb{A}} \to N_{\mathbb{A}}, \tag{2.4.9}$$

which is defined in the following way. By Proposition 1.7.2, there is a *unique* ∇-invariant splitting

$$^{\mathbb{C}}E = \rho^1 \oplus \cdots \oplus \rho^r \oplus \overline{\rho}^1 \oplus \cdots \oplus \overline{\rho}^r \tag{2.4.10}$$

(where ρ^σ are ∇-invariant positive line bundles) that coincides with the decomposition (2.4.13) at $\alpha = \alpha_0$. Consider the corresponding projections

$$\text{pr}^\sigma : {}^{\mathbb{C}}E \to \rho^\sigma \qquad (\sigma = 1, \ldots, r).$$

At each Darboux chart $(\mathcal{U}, p, \tau, z = (z^1, \ldots, z^{2r}))$ of $\mathbb{A}$, we define smooth functions $F^1, \ldots, F^r$:

$$F^\sigma = F^\sigma_{\mathbb{A},\oplus} = \frac{1}{4} \langle z, (JK^\sigma + K^\sigma J)z \rangle, \tag{2.4.11}$$

where $J = \begin{pmatrix} 0 & -I \\ I & 0 \end{pmatrix}$ and $K^\sigma = ((K^\sigma_{\sigma'\sigma''}))$ are antisymmetric $2r \times 2r$ matrices of the form

$$K^\sigma = ((\Omega(\text{pr}^\sigma(Z_{\sigma'}), \text{pr}^\sigma(Z_{\sigma''})))), \qquad Z_{\sigma'} =: \text{ad}(z^{\sigma'})\Big|_{\Lambda \cap \mathcal{U}}.$$

From the properties of flat Darboux atlases and the splitting (2.4.10), we deduce that (2.4.11) gives smooth functions $F^\sigma_{\mathbb{A},\oplus}$ that are well defined in a neighborhood $U_{\mathbb{A}}$ of Λ in M. So, to each pair $(\mathbb{A}, \oplus)$ we can assign the smooth map

$$\mathbb{F}_{\mathbb{A},\oplus} = (F^1_{\mathbb{A},\oplus}, \ldots, F^r_{\mathbb{A},\oplus}) : U_{\mathbb{A}} \to \mathbb{R}^r. \tag{2.4.12}$$

Note that equivalent flat Darboux atlases over Λ define the same mapping (2.4.12).

THEOREM 2.4.2. *Let $\Lambda \subset M$ be an isotropic submanifold satisfying the assumptions of Theorem 2.4.1. Let $(\mathbb{A}, \oplus)$ be a pair consisting of a flat Darboux atlas $\mathbb{A}$ over Λ and a decomposition (2.4.7) at a point $\alpha_0 \in \Lambda$. Suppose that the coisotropic extension $N_{\mathbb{A}}$ of Λ (associated with $\mathbb{A}$) is stable. Then:*

(a) *The pair $(\mathbb{A}, \oplus)$ induces the Hamilton $\mathbb{T}^r$-action on a neighborhood $U_{\mathbb{A}}$ of Λ in M:*

$$\mathbb{T}^r \times U_{\mathbb{A}} \to U_{\mathbb{A}} \tag{2.4.13}$$

whose moment map is just $\mathbb{F}_{\mathbb{A},\oplus}$ (2.4.12). The moment map $\mathbb{F}_{\mathbb{A},\oplus}$ possesses the properties (a)–(d) of Theorem 2.4.1 and is independent of the choice of data $(\mathbb{A}, \oplus, \alpha_0)$ up to equivalence. The linearized action of (2.4.13) on E_{α_0} coincides with ℓ (2.4.8).

(b) *The coisotropic extension $N_{\mathbb{A}}$ is invariant under the action (2.4.13) and the restriction of this action to $N_{\mathbb{A}}$ defines the transverse Hamilton $\mathbb{T}^r$-action*

$$\phi_{\mathbb{A},\oplus} : \mathbb{T}^r \times N_{\mathbb{A}} \to N_{\mathbb{A}}.$$

Moreover, if $(\widetilde{\mathbb{A}}, \oplus)$ is another pair consisting of a flat Darboux atlas $\widetilde{\mathbb{A}}$ (with stable coisotropic extension $N_{\widetilde{\mathbb{A}}}$) and a decomposition of type (2.4.7) at $\alpha_0 \in \Lambda$, then the induced $\mathbb{T}^r$-actions $\phi_{\mathbb{A},\oplus}$ and $\phi_{\widetilde{\mathbb{A}},\oplus}$ are equivalent if and only if the conjugation condition (2.3.43) is satisfied.

PROOF. The proof of the first part of Theorem 2.4.2 follows from our constructions. To show an equivalence of the $\mathbb{T}^r$-actions $\phi_{\mathbb{A},\oplus}$ and $\phi_{\widetilde{\mathbb{A}},\oplus}$ under conditions (2.3.49), let us examine the proof of Theorem 2.3.2 in the case of a compact group action. Let G be a compact Lie group acting on N with ω and $\widetilde{\omega}$ both invariant under this action. First, let us choose a G-invariant Riemann metric on N (see [**GuSt**$_2$]). Using a G-invariant exponential map, we can arrange a vertical subbundle $\mathcal{V}$ in (2.3.37) to be also G-invariant. Assume that the flat connections ∇^{ex} and $\widetilde{\nabla}^{\text{ex}}$ in $\mathcal{V}$ are such that the G-action preserves ∇^{ex} and $\widetilde{\nabla}^{\text{ex}}$. This implies that local diffeomorphisms (2.3.41) and, hence, (2.3.42) are G-equivariant. In our case, $G = \mathbb{T}^r$, and the $\mathbb{T}^r$-actions $\phi_{\mathbb{A},\oplus}$ and $\phi_{\widetilde{\mathbb{A}},\oplus}$ preserve the flat connections ∇^{ex} and $\widetilde{\nabla}^{\text{ex}}$ from (2.3.24). This completes the proof. □

COROLLARY 2.4.3. *Let $(\mathbb{A}, \oplus)$ be a pair satisfying the assumptions of Theorem 2.4.2. Consider the flat connection ∇ (1.3.6) in the symplectic normal bundle E over Λ, where ∇ is determined by the coisotropic extension $N_\mathbb{A}$ of Λ. Let $(\mathcal{X} = T^*\Lambda \oplus E, \Omega^\nabla)$ be the corresponding model phase space, where E is viewed as the coisotropic extension of the zero section Λ. Then by Theorem 1.7.1, the decomposition (2.4.10) induces the mapping $\mathbf{h}\colon \mathcal{X} \to \mathbb{R}^r_+$ (1.7.9), which is the moment map of the transverse Hamilton $\mathbb{T}^r$-action $\ell^\#$ on E. Moreover, the $\mathbb{T}^r$-action $\phi_{\mathbb{A},\oplus}$ on $N_\mathbb{A}$ with moment map $\mathbb{F}_{\mathbb{A},\oplus}$ (2.4.12) is isomorphic to the $\mathbb{T}^r$-action $\ell^\#$ on $E \subset \mathcal{X}$, that is, there is a diffeomorphism $g\colon W \to \widetilde{W}$ of a neighborhood W of Λ in $N_\mathbb{A}$ onto a neighborhood $\widetilde{W}$ of Λ in E satisfying (2.4.5) and (2.4.6),*

(2.4.14) $$\mathbb{F}_{\mathbb{A},\oplus} = \mathbf{h} \circ g.$$

As an application of Theorem 2.4.2, let us derive the following useful fact. Assume that we have a flat Darboux atlas $\mathbb{A}$ over Λ satisfying the conditions of Theorem 2.4.2. Then the $\mathbb{T}^r$-action $\phi_{\mathbb{A},\oplus}$ determines the stratification of N by isotropic submanifolds $\Lambda(c)$ (2.4.2). Consider the regular set of the momentum map $\mathbb{F}_{\mathbb{A},\oplus}$:

(2.4.15) $$N^{\text{reg}} := F_{\mathbb{A},\oplus}^{-1}(\mathcal{C}^r \cap \mathcal{S}) \cap N_\mathbb{A}.$$

The $\mathbb{T}^r$-action ϕ is free on N^{reg}, and N^{reg} is fibered by Lagrangian submanifolds $\Lambda(c)$ ($c \in \mathcal{C}^r \cap \mathcal{S}$),

(2.4.16) $$\Lambda(c) \to N^{\text{reg}} \xrightarrow{\mathbb{F}} \mathcal{C}^r \cap \mathcal{S}.$$

By Corollary 2.4.3, an open domain $\mathcal{S} \subset \mathbb{R}^r$ can be chosen so that there is an isomorphism g from (2.4.14). In this case, $\mathcal{S}$ will be called *admissible*. So, the $\mathbb{T}^r$-action ϕ and a diffeomorphism g from (2.4.24) induce a $\mathbb{T}^r$-principle bundle over Λ:

(2.4.17) $$\mathbb{T}^r \to \Lambda(c) \to \Lambda$$

for each $c \in \mathcal{C}^r \cap \mathcal{S}$. Thus (2.4.16). (2.2.17) define the $\mathbb{T}^r$-principle bundle

(2.4.18) $$\mathbb{T}^r \to N^{\text{reg}} \to \Lambda.$$

Multiplying (2.4.16) by (2.4.18), we obtain the fibration

$$(2.4.19) \qquad N^{\text{reg}} \to (\mathcal{C}^r \cap \mathcal{S}) \times \Lambda.$$

Assume that

$$(2.4.20) \qquad \text{Tor } H_1(\Lambda) = 0.$$

Then, taking into account (2.4.20) and the geometric stability of ∇, we see that the line subbundles $\rho^1, \ldots, \rho^r$ from the ∇-invariant splitting (2.4.10) are trivial. This implies that the fibration (2.4.19) is trivial, and hence there is a global section

$$(2.4.21) \qquad e \colon (\mathcal{C}^r \cap \mathcal{S}) \times \Lambda \to N^{\text{reg}}.$$

For any $c \in \mathcal{C}^r \cap \mathcal{S}$, by

$$(2.4.22) \qquad e_c \colon \Lambda \to N^{\text{reg}}$$

we denote the restriction of e to the slice $\{c\} \times \Lambda$. Thus, for a given closed path $\Gamma \subset \Lambda$, we can consider the lift of $\Gamma \colon e_c(\Gamma) \subset \Lambda(c)$. Note also that the ∇-invariant splitting (2.4.10) allows us to consider the $\mathbb{T}^r$-principle bundle (2.4.17) as the fibered product of $\mathbb{S}^1$-principle bundles over Λ,

$$(2.4.23) \qquad \Lambda(c) = \mathcal{O}(c^1) \times \cdots \times \mathcal{O}(c^r).$$

PROPOSITION 2.4.2. *For a given closed path $\Gamma \subset \Lambda$, consider a 2-surface $\Sigma(c)$ of $N_\mathbb{A}$ such that $\partial \Sigma(c) = e_c(\Gamma)$. Then the formula*

$$(2.4.24) \qquad \beta_\sigma(\Gamma) = \frac{1}{2\pi} \frac{\partial}{\partial c^\sigma} \left(\oint_{e_c(\Gamma)} \gamma_\mathbb{A} \right) \equiv \frac{1}{2\pi} \frac{\partial}{\partial c^\sigma} \left(\int_{\Sigma(c)} \Omega \right)$$

determines the Floquet exponents of $N_\mathbb{A}$ at Γ, that is,

$$\lambda_\sigma(\Gamma) = \exp\left(\mp 2\pi i \beta_\sigma(\Gamma) \right) \qquad (\sigma = 1, \ldots, r)$$

are the characteristic multipliers of $N_\mathbb{A}$ at $\Gamma \subset \Lambda$. Here the primitive 1-form $\gamma_\mathbb{A}$ is defined by a given flat Darboux atlas $\mathbb{A}$ over Λ from (2.3.11). If $\tilde{e}$ is another section of the fibration (2.4.11), then $\beta_\sigma(\Gamma) - \tilde{\beta}_\sigma(\Gamma) = \mu_\Gamma \in \mathbb{Z}$, where μ_Γ is the Maslov index of Γ, that is, the Γ-period of the Arnold form for Lagrangian subbundles of E generated by the global sections e and $\tilde{e}$. Furthermore, the periods of $\gamma_\mathbb{A}$ on the fibers of $\mathbb{S}^1$-principle bundles from (2.4.23) are equal to

$$\frac{1}{2\pi} \oint_{\mathcal{O}_\alpha(c^\sigma)} \gamma_\mathbb{A} = c^\sigma \qquad (\sigma = 1, \ldots, r).$$

The proof follows from Corollary 2.4.3 and Theorem 1.9.1. For the properties of the Arnold form determining the Maslov class, see §1.10.

§2.5. Liouville tori

Let Λ be an isotropic submanifold of a symplectic manifold (M, Ω). Assume that Λ is diffeomorphic to the standard k-torus,

(2.5.1) $$\Lambda \approx \mathbb{T}^k.$$

Let us also assume that
(i) there is a flat Darboux atlas $\mathbb{A}$ over Λ, and
(ii) the flat coisotropic extension $N_\mathbb{A}$ associated with $\mathbb{A}$ is stable.

Let us choose a decomposition (2.4.7). Then by Theorem 2.4.2, we have a Hamilton $\mathbb{T}^r$-action $\mathbb{T}^r \times U_\mathbb{A} \to U_\mathbb{A}$ on a neighborhood $U_\mathbb{A} \supset N_\mathbb{A}$ of Λ in M with moment map $\mathbb{F}_{\mathbb{A},\oplus}$ (2.4.12). Here $k+r = \frac{1}{2}\dim M$. Thus, we have the stratification of $N_\mathbb{A}$ by isotropic submanifolds:

(2.5.2) $$\Lambda(c) = \mathbb{F}_{\mathbb{A},\oplus}^{-1}(c) \cap N_\mathbb{A}, \qquad c \in \mathcal{C}^r \cap \mathcal{S} \qquad (0 \le s \le r).$$

In what follows, we assume that an open domain $\mathcal{S}$ is admissible. As we know from Theorem 2.4.1, each $\Lambda(c)$ is a $\mathbb{T}^r$-principle bundle over Λ. It follows from (2.5.1) and the stability assumption that the principal bundle $\Lambda(c)$ from (2.4.17) is trivial, and hence $\Lambda(c) \approx \mathbb{T}^s \times \mathbb{T}^k$. In particular, if $c \in \mathcal{C}^r \cap \mathcal{S}$, then $\Lambda(c)$ is a Lagrangian torus ($\dim \Lambda(c) = k+r$). Let us consider the regular set N^{reg} (2.4.15) of the moment map $\mathbb{F}_{\mathbb{A},\oplus}$.

THEOREM 2.5.1. *Under the listed assumptions, fibration (2.4.19) by Lagrangian tori $\Lambda(c)$ is trivial,*

$$N^{\mathrm{reg}} \approx \mathbb{T}^{r+k} \times (\mathcal{C}^r \cap \mathcal{S}).$$

In a neighborhood of N^{reg} in M, each global section e (2.4.21) determines the action-angle variables

(2.5.3) $$\begin{aligned} I &= (I^1, \ldots, I^k, I^{k+1}, \ldots, I^{k+r}), \\ \varphi &= (\varphi^1, \ldots, \varphi^k, \varphi^{k+1}, \ldots, \varphi^{k+r}) \quad (\varphi^i \bmod 2\pi) \\ \Omega &= \sum_{i=1}^{k+r} dI^i \wedge d\varphi_i \end{aligned}$$

such that for each $c \in \mathcal{C}^r \cap \mathcal{S}$, the Lagrangian torus (2.5.2) is the level set of the action variables,

(2.5.4) $$\Lambda(c) = \left\{ I^1 = \sum_{\sigma=1}^r c^\sigma \beta_\sigma^1, \ldots, I^k = \sum_{\sigma=1}^r c^\sigma \beta_\sigma^k, \ I^{k+1} = c^1, \ldots, I^{k+r} = c^r \right\}.$$

Here β_σ^i are the Floquet exponents (2.4.24) of $N_\mathbb{A}$ at the 1-cycles $\Gamma^1, \ldots, \Gamma^k \subset \Lambda$:

(2.5.5) $$e_c(\Gamma^i) = \{\varphi^1 = 0, \ldots, 0 \le \varphi^i \le 2\pi, \ldots, \varphi^k = 0; \varphi^{k+1} = 0, \ldots, \varphi^{k+r} = 0\}$$

for $i = 1, \ldots, k$.

The proof follows from Theorem 2.4.2, Corollary 2.4.3, and Theorem 1.9.1.

REMARK 2.5.1. If instead of the regular set N^{reg}, we consider

$$N^s = \mathbb{F}_{\mathbb{A},\oplus}^{-1}(\mathcal{C}^s \cap \mathcal{S}) \cap N_{\mathbb{A}}, \qquad (0 \leq s < r),$$

then we obtain the same fact: the fibration of N^s by $(k+s)$-dimensional isotropic tori $\Lambda(c)$ is trivial. In a neighborhood of N^s in M, there are generalized action-angle variables [**Ne$_1$**].

Under the same assumptions as in Theorem 2.5.1, we obtain the following theorem.

THEOREM 2.5.2. *Let $H \in C^\infty(M)$ be a Hamilton function with the properties:*
(i) $H\big|_{N_{\mathbb{A}}} = \text{const}$;
(ii) *there are "partial" integrals of motion $H^1, \ldots, H^k \in C^\infty(M)$,*

$$H^i\big|_{N_{\mathbb{A}}} = \text{const}, \qquad d(\{H^i, H^j\})\big|_{TN_{\mathbb{A}}} = d(\{H^i, H\})\big|_{TN_{\mathbb{A}}} = 0,$$

and the differentials $dH^1, \ldots, dH^k$ are linearly independent on $N_{\mathbb{A}}$;
(iii) *the $\mathbb{T}^r$-action with moment map $\mathbb{F}_{\mathbb{A},\oplus}$ (2.4.12) preserves the Hamilton vector field $\mathrm{ad}(H)$ on $N_{\mathbb{A}}$,*

$$d(\{F^\sigma_{\mathbb{A},\oplus}, H\})\big|_{TN_{\mathbb{A}}} = 0 \qquad (\sigma = 1, \ldots, r).$$

Then for each $c \in \mathcal{C}^r \cap \mathcal{S}$, the Lagrangian torus $\Lambda(c)$ is invariant under the flow of the Hamilton vector field $\mathrm{ad}(H)$. Each $\Lambda(c)$ carries a quasiperiodic motion along the trajectories of $\mathrm{ad}(H)$ with the frequency vector

$$\widetilde{\omega}(c) = \big(\widetilde{\omega}_1(c), \ldots, \widetilde{\omega}_k(c); \widetilde{\omega}_{k+1}(c), \ldots, \widetilde{\omega}_{k+r}(c)\big)$$

whose components are defined as follows:

(2.5.6) $$\widetilde{\omega}_i(c) = \frac{\partial H}{\partial I^i}\bigg|_{\Lambda(c)} \qquad (i = 1, \ldots, k),$$

(2.5.7) $$\widetilde{\omega}_{k+\sigma}(c) = -\sum_{i=1}^k \beta^i_\sigma \widetilde{\omega}_i(c) \qquad (\sigma = 1, \ldots, r).$$

The same fact is true for low-dimensional isotropic tori $\Lambda(c)$, $c \in \mathcal{C}^s \cap \mathcal{S}$, $0 \leq s \leq r$.

COROLLARY 2.5.1. *Suppose that, in addition to the assumptions of Theorem 2.5.1, the characteristic multipliers of $N_{\mathbb{A}}$ at Λ satisfy the resonance conditions*

(2.5.8) $$\lambda_1(\Gamma) = \cdots = \lambda_{2r}(\Gamma) = 1$$

for each closed path $\Gamma \subset \Lambda$. Let $\mathcal{F}$ be the foliation on $N_{\mathbb{A}}$ associated with the characteristic subbundle $\mathcal{H} = \ker \Omega\big|_{N_{\mathbb{A}}}$. Then Theorem 2.5.2 says that the leaves of $\mathcal{F}$ are diffeomorphic to a k-torus. Moreover, in a neighborhood of Λ in M, there are functionally independent functions $H^1, \ldots, H^k$, $H^i\big|_{N_{\mathbb{A}}} = \text{const}$, such that all trajectories of $\mathrm{ad}(H^i)\big|_{N_{\mathbb{A}}}$ are 2π-periodic for each $i = 1, \ldots, k$.

Condition (2.5.8) means that the linear holonomy group of the (compact) leaf Λ of the foliation $\mathcal{F}$ is trivial. So, in the resonant case, we obtain the "coisotropic" version of the Reeb stability theorem [**GuSt$_2$, Tam**].

EXAMPLE 2.5.1. Let $f^1, \ldots, f^k, \Phi^1, \ldots, \Phi^{2r}$ be a set of smooth functions on a symplectic manifold (M, Ω) ($\dim M = 2(k+r)$) satisfying the following conditions.

(i) The zero-level set
$$\Lambda = \{f^1 = 0, \ldots, f^k = 0\} \cap \{\Phi^1 = 0, \ldots, \Phi^{2r} = 0\}$$
is compact and connected. On $T_\Lambda M$, the differentials $df^1, \ldots, df^k, d\Phi^1, \ldots, d\Phi^{2r}$ are linearly independent.

(ii) The pairwise Poisson brackets of f^i and Φ^i satisfy the following relations as $|f| = [(f^1)^2 + \cdots + (f^k)^2]^{1/2} \to 0$:
$$\{f^i, f^j\} = O(|f|^2),$$
$$\{f^i, \Phi^\sigma\} = -\sum_{\sigma'=1}^{r} \overset{\circ}{V}^{i\sigma}_{\sigma'} \Phi^{\sigma'} + O(|f|),$$
for all $i = 1, \ldots, k$ and $\sigma = 1, \ldots, 2r$. Here $\overset{\circ}{V}^i = ((\overset{\circ}{V}^{i\sigma}_{\sigma'})) = \text{const}$ are the same $2r \times 2r$ matrices.

(iii) $\det((\{\Phi^\sigma, \Phi^{\sigma'}\}|_\Lambda)) \neq 0$.

Then we claim that
- Λ is a k-dimensional isotropic torus of M;
- there is a neighborhood U of Λ in M such that the zero-level set
$$N = \{f^1 = 0, \ldots, f^k = 0\} \cap U$$
is a coisotropic extension of Λ;
- N is flat;
- if, in addition, $\overset{\circ}{V}^i$ is diagonalizable for all $i = 1, \ldots, k$ and its eigenvalues are pure imaginary, then N is stable.

So, we can apply Theorem 2.5.1 and Theorem 2.5.2 to this situation.

In particular, if $\overset{\circ}{V}^i \equiv 0$ ($i = 1, \ldots, k$), then the resonance conditions (2.5.8) hold. It follows from Corollary 2.5.1 that the trajectories of Hamilton vector fields $\text{ad}(f^1), \ldots, \text{ad}(f^k)$ on N are quasiperiodic and belong to isotropic k-tori. This is just the "coisotropic" version of the Nekhoroshev theorem [**Ne**[1]].

PART 3. GEOMETRIC QUANTIZATION OVER ISOTROPIC SUBMANIFOLDS

§3.1. Hermitian operators satisfying Dirac axioms

The goal of this section is to give an isotropic (non-Lagrangian) version of the quantization formula proposed in [**Ka**[2-4]] for the case of Lagrangian submanifolds. As we shall see later, this quantization formula naturally correlates with the global formula for quasimodes within the framework of the approach developed in [**Ka**[1-5], **KaKz**[1,2], **KaVo**[1,6,8]].

Let (M, Ω) be a symplectic manifold and P a *polarization* on M, that is, an integrable plane distribution of Lagrangian subspaces $P_m \subset {}^{\mathbb{C}}T_m M$, $m \in M$. In the same way as in the Kostant–Souriau theory of geometric quantization (see, for instance, [**Ki**]), we can define a Lie algebra (with respect to the Poisson bracket on M) of classical *observables*:

$$C_P^\infty(M) = \{H \in C^\infty(M)\colon \text{the flow of the Hamilton vector field } \operatorname{ad}(H) \text{ preserves } P\}.$$

Quantum observables should represent this algebra. We construct them as first-order differential operators acting on functions on the normal bundle over the isotropic submanifolds. But first we recall the quantization formula in the Lagrangian case.

Let $(\Lambda, d\sigma)$ be a Lagrangian submanifold of M,

$$\Omega\big|_\Lambda = 0, \qquad \dim \Lambda = \frac{1}{2} \dim M,$$

equipped with a smooth measure $d\sigma$. Assume that a given polarization P is *transversal* to Λ at each point,

(3.1.1) $$P_m \cap {}^\mathbb{C}T_m\Lambda = \{0\} \qquad \text{for any} \quad m \in \Lambda.$$

For example, the transversality condition (3.1.1) holds automatically if P is a *Kählerian* or *anti-Kählerian polarization*. Following [**Ka**$_7$], let us define the first-order differential operator on Λ by the formula

(3.1.2) $$\check{H} = H\big|_\Lambda - i\hbar\left(v_H + \frac{1}{2}{}^\sigma\!\operatorname{div}(v_H)\right),$$

where $H \in C_P^\infty(M)$, the (complex) vector field v_H on Λ is defined as the projection of the Hamilton vector field $\operatorname{ad}(H)$ to ${}^\mathbb{C}T\Lambda$ along the polarization P, and ${}^\sigma\!\operatorname{div}$ denotes the divergence with respect to the measure $d\sigma$.

Let $\mathcal{F} \subseteq C_P^\infty(M)$ be a Lie subalgebra. Then (see [**Ka**$_7$]) formula (3.12) defines the linear correspondence

$$H \to \check{H}, \qquad H \in \mathcal{F},$$

satisfying the Dirac axioms

(3.1.3) $$1 \to \mathbb{I} \qquad \text{(identity operator)},$$

(3.1.4) $$\{H_1, H_2\}^{\check{}} = \frac{i}{\hbar}[\check{H}_1, \check{H}_2].$$

Now let Λ be an isotropic submanifold of (M, Ω),

$$\dim \Lambda = k < \frac{1}{2}\dim M.$$

Let $(E, \mathcal{J})$ be the symplectic normal bundle over Λ. Assume that E is endowed with a symplectic connection ∇ and there is a decomposition

(3.1.5) $${}^\mathbb{C}E = \tau \oplus \rho,$$

where τ and ρ are Lagrangian subbundles, ${}^\mathbb{C}\!\dim \tau = {}^\mathbb{C}\!\dim \rho = r = (1/2)\dim E$.

We make the following assumptions:
(i) The Lagrangian subbundle τ is ∇-invariant.
(ii) The curvature 2-form $C \in A^2(\Lambda, \operatorname{End}(E))$ of the connections ∇ is compatible with the second Lagrangian subbundle ρ,

(3.1.6) $$C(u_1, u_2)(\rho) \subset \rho,$$

for any vector fields $u_1, u_2 \in \text{Vect}(\Lambda)$.

Conditions (i), (ii) can be locally expressed as follows. Choose local coordinates $\alpha = (\alpha^1, \ldots, \alpha^k)$, $y = (y^1, \ldots, y^k)$, and $z = (z^1, \ldots, z^k)$ on the complexification $^{\mathbb{C}}E$ associated with the decomposition (3.15): $\{y^\sigma\}$ are local coordinates along the fibers of τ, and $\{z^\sigma\}$ are local coordinates along the fibers of ρ. Let $\theta = ((\theta_\sigma^{\sigma'}))$ be the $(2r \times 2r)$-matrix-valued connection 1-form of ∇ corresponding to the local coordinates $\{y^\sigma, z^{\sigma'}\}$. Then condition (i) means that the connection 1-form θ has the form

$$(3.1.7) \qquad \theta = \begin{pmatrix} ^\tau\theta & m \\ 0 & ^\rho\theta \end{pmatrix}.$$

Here $(r \times r)$-matrix-valued 1-forms $^\tau\theta$ and $^\rho\theta$ represent linear connections $^\tau\nabla$ and $^\rho\nabla$ in the subbundles τ and ρ that are induced by the original connection ∇. Condition (ii) implies that the $(r \times r)$-matrix-valued 1-form m from (3.1.7) satisfies the following equation on the manifold Λ:

$$(3.1.8) \qquad dm + {}^\tau\theta \wedge m + m \wedge {}^\rho\theta = 0.$$

Note that if the Lagrangian subbundle τ is positive or negative and

$$(3.1.9) \qquad \rho = \overline{\tau},$$

then $\rho \cap \tau = \{0\}$, and from the ∇-invariance of τ it follows that ρ is also ∇-invariant, that is, $m = 0$.

Let us consider the subspace $\mathcal{F}_\Lambda(M)$ of smooth functions H on M such that a given isotropic submanifold Λ is invariant under the flow of the Hamilton vector field $\text{ad}(H)$. From §§1.1 and 1.2 we know that for each $H \in \mathcal{F}_\Lambda(M)$, the first variation operator D_H of the Hamilton vector field $\text{ad}(H)$ over Λ can be considered as a covariant differential operator on E (hence on $^{\mathbb{C}}E$). The given symplectic connection ∇ in E allows us to split D_H into two components,

$$(3.1.10) \qquad D_H = \nabla_{v_H} + \Xi, \qquad v_H = \text{ad}(H)\Big|_\Lambda,$$

where $\Xi = \{\Xi(\alpha) \in \text{sp}(E_\alpha; \mathbb{R}) \mid \alpha \in \Lambda\}$ is a field of linear Hamilton operators on E (see (1.2.4), (1.2.5)).

Let us introduce the subspace $\mathcal{F}_\Lambda^{\nabla,\tau}(M) \subset \mathcal{F}_\Lambda(M)$ consisting of smooth functions $H \in \mathcal{F}_\Lambda(M)$ satisfying the following conditions:

(a) The first variation operator D_H is compatible with the given symplectic connection ∇ in E in the following sense:

$$(3.1.11) \qquad [\nabla_u, D_H] - \nabla_{[u, v_H]} = 0$$

for any vector field $u \in \text{Vect}(\Lambda)$.

(b) The Lagrangian subbundle τ from (1.3.5) is D_H-invariant,

$$(3.1.12) \qquad s \in \Gamma(\tau) \Longrightarrow D_H(s) \in \Gamma(\tau).$$

In terms of the field Ξ in (3.1.10), the connection 1-form θ, and the curvature 2-form C, condition (3.1.11) can be written as follows (see (1.5.16)):

$$(3.1.13) \qquad d\Xi + [\theta, \Xi] = v_H \lrcorner C.$$

Since τ is ∇-invariant, condition (3.1.12) is equivalent to the condition that τ is Ξ-invariant. This implies that locally we have

$$\Xi = \begin{pmatrix} {}^\tau\Xi & S \\ 0 & {}^\rho\Xi \end{pmatrix}. \tag{3.1.14}$$

Moreover, the first variation operator D_H induces covariant differential operators ${}^\tau D_H$ and ${}^\rho D_H$ acting on sections of the Lagrangian subbundles τ and ρ, respectively.

From Theorem 1.5.1, we deduce the following statement.

PROPOSITION 3.1.1. *The subspace $\mathcal{F}_\Lambda^{\nabla,\tau}(M)$ is a Lie subalgebra of the Poisson algebra $C^\infty(M)$.*

We call $\mathcal{F}_\Lambda^{\nabla,\tau}(M)$ *an algebra of observables* associated with a triple (Λ, ∇, τ).

If a connection ∇ is flat, then the observables H are described by the following conditions:
- an isotropic submanifold Λ is invariant under the flow of the Hamilton vector field $\mathrm{ad}(H)$;
- the first variation operator D_H of $\mathrm{ad}(H)$ is ∇-reducible in E (in particular, this is true if the symplectic connection ∇ is adapted to D_H);
- D_H preserves the Lagrangian subbundle τ.

We would like to extend the quantization formula (3.1.2) to observables $H \in \mathcal{F}_\Lambda^{\nabla,\tau}(M)$.

So, we introduce the following objects associated with the Lagrangian subbundle ρ from (3.1.5):
- a smooth measure $d\sigma$ on Λ;
- a $\mathbb{C}$-bilinear nondegenerate form $\mathbf{b}$ on ρ that can be considered as an isomorphism of vector bundles $\mathbf{b} \colon \rho \to \rho^*$, $\det \mathbf{b} \neq 0$.

For each observable $H \in \mathcal{F}_\Lambda^{\nabla,\tau}(M)$ and a triple $(\rho, d\sigma, \mathbf{b})$, let us define a first-order differential operator $\check{H}$ on ρ by the formula

$$\check{H} = \left[H\Big|_\Lambda + \frac{1}{2}\langle Sz, z\rangle\right] - i\hbar\left[v_H - \left\langle (v_H \lrcorner {}^\rho\theta)z, \frac{\partial}{\partial z}\right\rangle - \left\langle {}^\rho\Xi z, \frac{\partial}{\partial z}\right\rangle\right] \\ - \frac{i\hbar}{2}\left[\sigma \mathrm{div}\, v_H + \left(\frac{1}{2}\mathrm{tr}\,(b^{-1}v_H(b)) - \mathrm{tr}(v_H \lrcorner {}^\rho\theta)\right) - \mathrm{tr}\,\Xi\right], \tag{3.1.15}$$

where v_H is the restriction of the Hamilton vector field $\mathrm{ad}(H)$ to Λ, $b = ((b_{\sigma\sigma'}))$ is the symmetric $(r \times r)$-matrix of the form $\mathbf{b}$ in local coordinates $z = (z^1, \ldots, z^r)$ on ρ, the matrix-valued 1-form ${}^\rho\theta$, and the functions ${}^\rho\Xi$, S are defined by (3.1.7) and (3.1.14).

THEOREM 3.1.1. *Formula (3.1.15) determines the linear map*

$$H \mapsto \check{H}, \qquad H \in \mathcal{F}_\Lambda^{\nabla,\tau}(M), \tag{3.1.16}$$

satisfying the Dirac axioms (3.1.3), (3.1.4).

The proof of Theorem 3.1.1 is based on the following interpretation of the prequantization formula (3.1.15).

Consider the model phase space $(\mathcal{X} = T^*\Lambda \oplus E, \Omega^\nabla)$ associated with a given symplectic connection ∇ in E. Here the symplectic 2-form Ω^∇ on $\mathcal{X}$ is defined by (1.5.23). From the results of §1.5 we deduce the following facts.

(i) Condition (3.1.11) implies that for each function $H \in \mathcal{F}_\Lambda^{\nabla,\tau}(M)$, the pullback $D_H^\#$ (1.1.9) of the first variation operator D_H coincides on E with the Hamilton vector field X_F (see Theorem 1.5.1); i.e.,

$$(3.1.17) \qquad D_H^\# = X_F \qquad \text{on} \quad E.$$

Here the Hamilton function F is given by

$$(3.1.18) \qquad F = \sum_{i=1}^k v_H^i(\alpha)\eta_i + \frac{1}{2}\langle \mathcal{J}\Xi x, x\rangle,$$

where $\eta = (\eta_1, \ldots, \eta_k)$, $\alpha = (\alpha^1, \ldots, \alpha^k)$ are canonical coordinates on $T^*\Lambda$, $v_H = \sum_i v_H^i(\alpha)\partial/\partial\alpha_i$, and $x = (x^1, \ldots, x^{2r})$ are local coordinates along the fibers of E. Denote by $Q_\Lambda^\nabla(\mathcal{X})$ the space of functions F on $\mathcal{X}$ of the form (3.1.18), where $v = v_H$ is a vector field on Λ and Ξ is a tensor field of type (1.1) on E satisfying condition (3.1.13). Then $Q_\Lambda^\nabla(\mathcal{X})$ is the Lie algebra with respect to the Poisson bracket on $\mathcal{X}$ (see Corollary 1.5.2). And the linear map

$$(3.1.19) \qquad \mathcal{F}_\Lambda^{\nabla,\tau}(M) \to Q_\Lambda^\nabla(\mathcal{X}), \qquad H \mapsto F,$$

actually is a *homomorphism* of Lie algebras.

(ii) Consider the plane distribution on $\mathcal{X}$

$$P \subset {}^\mathbb{C}T\mathcal{X}, \qquad {}^\mathbb{C}\dim P = \frac{1}{2}\dim \mathcal{X},$$

induced by the Lagrangian subbundle τ from (3.1.5). The fibers of P are defined as follows:

$$(3.1.20) \qquad P_\xi = T^*_{\pi_0(\xi)}\Lambda \oplus \tau_{\pi_0(\xi)}, \qquad \xi \in \mathcal{X},$$

where $\pi_0 : \mathcal{X} \to \Lambda$ is the natural projection. Then P is a *polarization* on $\mathcal{X}$. Consider the space $Q_\Lambda^{\nabla,\tau}(\mathcal{X})$ of functions $F \in Q_\Lambda^\nabla(\mathcal{X})$ such that the flow of the Hamilton vector field X_F preserves the polarization P on $\mathcal{X}$. Then it is easy to see that $Q_\Lambda^{\nabla,\tau}(\mathcal{X})$ is a Lie subalgebra of $Q_\Lambda^\nabla(\mathcal{X})$, and the image under the homomorphism (3.1.19) is just $Q_\Lambda^{\nabla,\tau}(\mathcal{X})$. The space $Q_\Lambda^{\nabla,\tau}(\mathcal{X})$ will be called an *algebra of observables* over the model phase space $(\mathcal{X}, \Omega^\nabla)$ associated with a triple (Λ, ∇, τ).

(iii) Let ρ be the second Lagrangian subbundle from the decomposition (3.1.5). Consider the complexification ${}^\mathbb{C}\mathcal{X}$ of the vector bundle $\pi : \mathcal{X} \to T^*\Lambda$ (see §1.5). Then ${}^\mathbb{C}\mathcal{X}$ inherits the complex-valued symplectic structure $\Omega_\mathbb{C}^\nabla$. The pair $({}^\mathbb{C}\mathcal{X}, \Omega_\mathbb{C}^\nabla)$ will be called the *complexified model space*. The inclusion map $\rho \hookrightarrow {}^\mathbb{C}E \hookrightarrow {}^\mathbb{C}\mathcal{X}$ allows us to consider the subbundle ρ as a smooth submanifold of ${}^\mathbb{C}\mathcal{X}$. Condition (3.1.6) implies that ρ is a *Lagrangian submanifold* of $({}^\mathbb{C}\mathcal{X}, \Omega_\mathbb{C}^\nabla)$ (see also (1.5.5)). Moreover, the polarization P (3.1.20) is transversal to ρ. Thus, for each $F \in Q_\Lambda^{\nabla,\tau}(\mathcal{X})$, we can define the complex-valued vector field ${}^\rho X_F$ on ρ as the projection of the Hamilton vector field X_F to ρ along P,

$${}^\rho X_F = \mathrm{pr}(X_F) \equiv \mathrm{pr}(D_H^\#),$$

where $\mathrm{pr}: {}^\mathbb{C}T_\rho\mathcal{X} \to T\rho$ is the projection. It is clear that ${}^\rho X_F = {}^\rho D_H^\#$, ${}^\rho D_H \in \mathfrak{D}(\rho)$.

(iv) Let $(d\sigma, \mathbf{b})$ be a pair consisting of a smooth measure on Λ and a bilinear nondegenerate form on ρ. Consider the Lie algebra of smooth vector fields $D^\#$ on ρ, where D runs over the space of covariant differential operators $\mathfrak{D}(\rho)$. For each $D^\#$, let us define the divergence with respect to the pair $(d\sigma, \mathbf{b})$,

$$D^\# \mapsto \mathrm{div}(D^\#) \in C^\infty(\Lambda) \otimes \mathbb{C},$$

by the formula

$$\mathrm{div}(D^\#) = {}^\sigma\mathrm{div}(v) + \left(\frac{1}{2} \mathrm{tr}\left(b^{-1} \mathbf{L}_v(b) \right) + \mathrm{tr}\, V \right),$$

where the vector field v on Λ and the matrix-valued function V are defined by (1.1.9). For any $D_1, D_2 \in \mathfrak{D}(\rho)$, we have the standard property

$$D_1^\#\left(\mathrm{div}(D_2^\#) \right) - D_2^\#\left(\mathrm{div}(D_2^\#) \right) = \mathrm{div}\left([D_1^\#, D_2^\#] \right).$$

Here $D^\#(\cdot) = L_{D^\#}(\cdot)$ denotes the Lie derivative along the vector field $D^\#$.

So, summarizing these facts, we see that the right-hand side of the quantization formula (3.1.15) consist of the following parts:

$$[\text{first term}] = F\big|_\rho \equiv (\text{restriction of the Hamilton function } F \quad (3.1.18)$$
$$\text{to the Lagrangian submanifold } \rho \subset {}^\mathbb{C}\mathcal{X});$$
$$[\text{second term}] = {}^\rho X_F \equiv (\text{projection of the Hamilton vector field } X_F$$
$$\text{to } \rho \text{ along the polarization } \widetilde{P});$$
$$[\text{third term}] = \mathrm{div}({}^\rho X_F) \equiv (\text{divergence of } {}^\rho X_F \text{ with respect to } (d\sigma, \mathbf{b})).$$

Thus, we have a complete analogy with the "Lagrangian" quantization formula (3.1.2). So, the proof of Theorem 3.1.1 follows the same arguments as in the Lagrangian case [**Ka**[7], **KaKz**[1,2]].

Note that a version of the quantization formula (3.1.15) over an isotropic submanifold for the case of Kählerian polarization P was proposed in [**KaVo**[7]].

Now we show that operators of type (3.1.15) can be viewed as self-adjoint operators in a certain Hilbert space. Then, for these operators, we can consider the spectral problem in this Hilbert space.

Our starting point is the following. Suppose that Λ be an isotropic submanifold of (M, Ω) and the complexification ${}^\mathbb{C}E$ of the symplectic normal bundle E over Λ splits into a direct sum,

(3.1.21) $$\qquad\qquad {}^\mathbb{C}E = \bar\rho \oplus \rho,$$

where ρ is a *positive* Lagrangian subbundle. Thus, the form

(3.1.22) $$\qquad (X, Y)_+ = \frac{1}{2i} \langle \mathcal{J}X, \overline{Y} \rangle, \qquad X, Y \in {}^\mathbb{C}E_\alpha,$$

is positive definite on ρ, and hence the restriction of $(\,,\,)_+$ to ρ defines the *Hermitian structure*.

Let us define by $\mathcal{F}_\Lambda^{\bar\rho \oplus \rho}(M)$ the space of functions $H \in \mathcal{F}_\Lambda(M)$ such that the first variation operator D_H of the Hamilton vector field $\mathrm{ad}(H)$ preserves the invariant subbundle ρ. Since D_H is a real operator, the complex conjugate subbundle $\bar\rho$ is also D_H-invariant.

It is clear that the space $\mathcal{F}_\Lambda^{\bar{\rho}\oplus\rho}(M)$ is a Lie subalgebra of the Poisson algebra $C^\infty(M)$.

Let us choose an Hermitian basis of local sections $Z_1, \ldots, Z_r$ of ρ,

$$(3.1.23) \qquad (Z_\sigma, Z_{\sigma'})_+ = \delta_{\sigma\sigma'},$$

and denote by $z = (z^1, \ldots, z^r)$ the local coordinates along the fibers of ρ corresponding to the basis (3.1.23).

Then for each $H \in \mathcal{F}_\Lambda^{\bar{\rho}\oplus\rho}(M)$, the first variation matrix V_H of D_H (defined by (1.1.5)) with respect to local coordinates $z^1, \ldots, z^r, \bar{z}^1, \ldots, \bar{z}^r$ on $^\mathbb{C}E$ takes the form

$$(3.1.24) \qquad V_H = \begin{pmatrix} iW_H & 0 \\ 0 & -iW_H \end{pmatrix},$$

where W_H is a real $r \times r$ symmetric matrix $W_H^T = W_H$.

We assume that we have a pair $(d\sigma, \mathbf{b})$ consisting of a smooth measure $d\sigma$ on Λ and a $\mathbb{C}$-bilinear nondegenerate form $\mathbf{b}$ on ρ. For each $H \in \mathcal{F}_\Lambda^{\bar{\rho}\oplus\rho}(M)$, we define the first-order differential operator $\check{H}$ by the formula

$$(3.1.25) \qquad \begin{aligned} \check{H} = H\big|_\Lambda - i\hbar\left[v_H + i\left\langle W_H z, \frac{\partial}{\partial z}\right\rangle\right] - \left[\frac{i\hbar}{2}\sigma \operatorname{div} v_H\right] \\ - \frac{i\hbar}{2}\left[\frac{1}{2}\operatorname{tr}\left(b^{-1} v_H(b)\right) + i \operatorname{tr} W_H\right]. \end{aligned}$$

PROPOSITION 3.1.2. *Formula (3.1.25) determines a correspondence $H \to \check{H}$ for which the Dirac axioms (3.1.3), (3.1.4) hold.*

Note that if we have a symplectic connection ∇ in E leaving the subbundle ρ invariant, then we can choose $\tau = \bar{\rho}$ in (3.1.5), and conditions (i), (ii) from §3.1 hold automatically. The algebra of observables $\mathcal{F}_\Lambda^{\nabla,\bar{\rho}}(M)$ associated with the triple $(\Lambda, \nabla, \bar{\rho})$ belongs to the space $\mathcal{F}_\Lambda^{\bar{\rho}\oplus\rho}(M)$,

$$(3.1.26) \qquad \mathcal{F}_\Lambda^{\nabla,\bar{\rho}}(M) \subset \mathcal{F}_\Lambda^{\bar{\rho}\oplus\rho}(M).$$

Moreover, for each $H \in \mathcal{F}_\Lambda^{\nabla,\bar{\rho}}(M)$, the quantization formula (3.1.15) coincides with (3.1.25).

Now we would like to define an inner product on the space of fiberwise holomorphic functions on ρ, which is compatible with the prequantization formula (3.1.25). Let us introduce the following objects.

(1) By using the form (3.1.22), we define the function on $^\mathbb{C}E$:

$$(3.1.27) \qquad K(\alpha, z, \bar{z}) = \sum_{\sigma,\sigma'=1}^r \left(Z_\sigma(\alpha), Z_{\sigma'}(\alpha)\right)_+ z^\sigma \bar{z}^{\sigma'},$$

where $\{Z_\sigma(\alpha)\}$ is a basis of local sections of ρ, $\alpha \in \Lambda$, and $z = \{z^\sigma\}$ are the corresponding coordinates. This function possesses the following two properties:

- $K(\alpha, z, \bar{z})$ is positive definite on ρ,

$$K(\alpha, z, \bar{z}) > 0, \qquad z \in \rho_\alpha, \quad \alpha \in \Lambda, \quad z \neq 0.$$

- For each $H \in \mathcal{F}_\Lambda^{\bar{\rho}\oplus\rho}(M)$, the function K is the integral of motion of the vector field $D_H^\#$, $L_{D_H^\#}(K) = 0$. The function K is an analog of the so-called "Kähler scalar" for Kähler manifolds (see, for example, [**Che**]).

(2) Let $\mathbf{g}\colon \rho \to \overline{\rho}^*$ be a vector bundle morphism which defines the Hermitian structure (3.1.22) on ρ. Assume that we have a $\mathbb{C}$-bilinear nondegenerate form on ρ, $\mathbf{b}\colon \rho \to \rho^*$. Then it is possible to introduce a smooth real function on Λ:

$$\Delta_{\mathbf{b}} = \frac{\det g}{|\det b|}, \tag{3.1.28}$$

where $g = ((g_{\sigma\sigma'}))$ and $b = ((b_{\sigma\sigma'}))$ are matrices of Hermitian structure (3.1.22) and of the bilinear form $\mathbf{b}$ in a certain basis of local sections of ρ. It is clear that $\Delta > 0$ uniformly on Λ.

Now we assume that Λ is compact and a smooth measure $d\sigma$ on Λ satisfies the normalization condition

$$\int_\Lambda d\sigma = 1.$$

Let us define the Hilbert space $\mathcal{H}(\rho)$ of fiberwise holomorphic functions $\varphi(\alpha, z)$ on ρ ($\alpha \in \Lambda$, $z = (z^1, \ldots, z^r) \in \rho_\alpha$) with the inner product

$$(\varphi_1, \varphi_2)_\mathcal{H} = \left(\frac{1}{4\pi}\right)^r \int_\rho \varphi_1(\alpha, z)\overline{\varphi_2(\alpha, z)} e^{-K(\alpha, z, \bar{z})} \frac{d\sigma}{\sqrt{\Delta_\mathbf{b}}} \sqrt{\det g}\, dz\, d\bar{z}. \tag{3.1.29}$$

With respect to the inner product (3.1.29), the square-integrable functions on ρ form a Hilbert space of the Fock–Bargmann–Segal type. Note that formula (3.1.29) in our case contains the integration along the isotropic submanifold Λ with respect to the measure $d\sigma/\sqrt{\Delta_\mathbf{b}}$.

So, to each triple $(d\sigma, \mathbf{b}, \mathbf{g})$, we have assigned the Hilbert space $\mathcal{H}(\rho)$ and the Lie algebra of observables $\mathcal{F}_\Lambda^{\overline{\rho}\oplus\rho}(M)$.

THEOREM 3.1.2. *To each real function $H \in \mathcal{F}_\Lambda^{\overline{\rho}\oplus\rho}(M)$, the quantization formula (3.2.25) assigns the essentially self-adjoint operator $\check{H}$ acting in the Hilbert space $\mathcal{H}(\rho)$.*

This statement is proved by straightforward computations in suitable local coordinates on ρ.

§3.2. Model spectral problem

Now a natural further step is to investigate the spectral problem for the operator $\check{H}$ (3.1.25), where $H \in \mathcal{F}_\Lambda^{\overline{\rho}\oplus\rho}(M)$:

$$\check{H}\varphi = \lambda\varphi, \qquad \varphi \in \mathcal{H}(\rho). \tag{3.2.1}$$

First, let us take some preparatory steps.

We assume that in the symplectic normal bundle E of Λ, there is a symplectic connection ∇ such that

$$\rho \text{ is } \nabla\text{-invariant}. \tag{3.2.2}$$

Then the restriction

$$\nabla^+ = \nabla\Big|_\rho \tag{3.2.3}$$

of ∇ to the Hermitian bundle $(\rho,(\,,\,)_+)$ is an *Hermitian connection*. For the given pair $(\mathbf{b}, \nabla^+)$ consisting of a $\mathbb{C}$-bilinear nondegenerate form $\mathbf{b}$ on ρ and the Hermitian connection ∇^+, we define the 1-*form* on Λ by the formula

$$(3.2.4) \qquad \theta_{\mathbf{b}}^+ = -\frac{1}{2}\operatorname{tr}(b^{-1}\,db) + \operatorname{tr}\theta^+.$$

Here $b = ((b_{\sigma\sigma''}))$ and $\theta^+ = (((\theta^+)_{\sigma'}^{\sigma}))$ are matrices of the bilinear form $\mathbf{b}$ and the Hermitian connection ∇^+ relative to a basis of local sections of ρ. We also define the 1-form $\varkappa = \varkappa_{\mathbf{b}}$ on Λ as the imaginary part of the form $\theta_{\mathbf{b}}^+$,

$$(3.2.5) \qquad \varkappa = \varkappa_{\mathbf{b}} = \operatorname{Im}\theta_{\mathbf{b}}^+.$$

It is easy to see that the 1-form $\varkappa$ and the trace of the curvature 2-form C^+ of the connection ∇^+ are related by the formula $i\,d\varkappa = \operatorname{tr} C^+$. So, if the 2-form $\operatorname{tr} C^+$ vanishes,

$$(3.2.6) \qquad \operatorname{tr} C^+ = 0,$$

then $\varkappa$ is a closed real 1-form on Λ.

Below we suppose that (3.2.6) holds.

Let us consider the splitting of the homology group of Λ,

$$H_1(\Lambda) = \operatorname{Tor} H_1(\Lambda) \oplus \widetilde{H}_1(\Lambda).$$

Here $\operatorname{Tor} H_1(\Lambda)$ is the torsion group of $H_1(\Lambda)$, and $\widetilde{H}_1(\Lambda)$ is the group without torsion dual to the cohomology group $H^1(\Lambda, \mathbb{R})$. Then there is a basis of closed 1-forms $\eta^1, \ldots, \eta^{k'}$ on Λ dual to a basis of 1-cycles $\Gamma_1, \ldots, \Gamma_{k'}$, $[\Gamma_i] \subset \widetilde{H}_1(\Lambda)$,

$$(3.2.7) \qquad \frac{1}{2\pi}\oint_{\Gamma_i} \eta^j = \delta_i^j, \qquad i,j = 1,\ldots,k' = \dim \widetilde{H}_1(\Lambda).$$

THEOREM 3.2.1 ("vacuum states"). *Let $(\Lambda, d\sigma, \rho, \mathbf{b}, \nabla)$ be initial data consisting of a compact connected isotropic submanifold Λ with smooth measure $d\sigma$, a positive Lagrangian subbundle $\rho \subset {}^{\mathbb{C}}E$ equipped with a $\mathbb{C}$-bilinear nondegenerate form $\mathbf{b}$, and a symplectic connection ∇ in E. Let ρ and ∇ satisfy conditions (3.2.2) and (3.2.6). Suppose that a function $H \in C^\infty(M)$ is compatible with these data in the following way:*

(a) *$H \in \mathcal{F}_\Lambda(M)$, and the measure $d\sigma$ is invariant under the flow of the restriction v_H of the Hamilton vector field $\operatorname{ad}(H)$ to Λ;*

(b) *the given symplectic connection ∇ is adapted to the first variation operator D_H of $\operatorname{ad}(H)$ at Λ;*

(c) *there is a basis of closed 1-forms $\eta^1, \ldots, \eta^{k'}$ on Λ satisfying (3.2.7) and such that*

$$(3.2.8) \qquad v_H \lrcorner \eta^i = \omega^i = \mathrm{const} \qquad (i = 1, \ldots, k').$$

Then $H \in \mathcal{F}_\Lambda^{\nabla,\rho}(M)$, and the self-adjoint operator $\check{H}$ (3.1.25) has the following eigenfunctions $\varphi_m \in \mathcal{H}(\rho)$ and corresponding eigenvalues $\lambda_m \in \mathbb{R}$:

$$(3.2.9) \quad \varphi_m(\alpha) = (\Delta_{\mathbf{b}})^{1/4} \exp\left[\frac{i}{2}\int_{\alpha^0}^{\alpha} \varkappa - i\sum_{i=1}^{k'}\int_{\alpha^0}^{\alpha}(\varkappa_i/2 + m_i)\eta^i\right], \quad \alpha \in \Lambda,$$

$$(3.2.10) \qquad \lambda_m = H\Big|_\Lambda + \hbar\lambda_m^{(1)}, \qquad \lambda_m^{(1)} = -\sum_{i=1}^{k'}(\varkappa_i/2 + m_i)\omega^i.$$

Here the function $\Delta_{\mathbf{b}}$ is defined by (3.1.28), $m = (m_1, \ldots, m_{k'}) \in \mathbb{Z}^{k'}$ is any integer vector, $\alpha^0 \in \Lambda$ is an arbitrary fixed point, $\varkappa_1, \ldots, \varkappa_{k'}$ are the periods of the 1-form $\varkappa$ (3.2.5) along the basic 1-cycles $\Gamma_1, \ldots, \Gamma_{k'}$ from (3.2.7), and

$$(3.2.11) \qquad \varkappa_i = \frac{1}{2\pi}\oint_{\Gamma_i}\varkappa.$$

Moreover, if the nonresonance condition holds,

$$(3.2.12) \qquad \langle \omega, s\rangle \neq 0 \quad \forall s \in \mathbb{Z}^{k'}, \quad s \neq 0,$$

then all eigenvalues λ_m are simple (of multiplicity one) and

$$(\varphi_m, \varphi_{m'})_\mathcal{H} = \delta_{mm'}.$$

Using the definitions of $\varkappa_{\mathbf{b}}$ and $\Delta_{\mathbf{b}}$, we verify formulas (3.2.9) and (3.2.10) by straightforward computations.

REMARK 3.2.1. Instead of condition (b), it suffices to assume that the Hermitian connection ∇^+ (3.2.3) in ρ is *weakly adapted* to the first variation operator D_H of $\mathrm{ad}(H)$ at Λ in the sense that ρ is D_H-invariant and

$$(3.2.13) \qquad \mathrm{tr}\, W_H = -v_H \,\lrcorner\, \mathrm{tr}\,\theta^+,$$

where W_H is the first variation matrix of the covariant differential operator ${}^\rho D_H$ locally defined by (3.1.24), and the traces in (3.2.13) are taken with respect to the same basis of local sections of ρ. See also Definition 1.10.1.

Now we describe excitations of the "vacuum states" (3.2.9), (3.2.10) by using fiberwise polynomial eigenfunctions of the spectral problem (3.2.11).

Let us make the following additional assumptions. Suppose that there are line subbundles $\rho^1, \ldots, \rho^{r'}$ of the Hermitian bundle ρ ($r' \leq r = {}^\mathbb{C}\dim \rho$) that are invariant with respect to the given connection ∇^+,

$$(3.2.14) \qquad {}^\mathbb{C}\dim \rho^\sigma = 1, \qquad \rho^\sigma \text{ is } \nabla^+\text{-invariant} \quad (\sigma = 1, \ldots, r').$$

Then the restriction of ∇^+ to each line bundle ρ^σ is also a Hermitian connection. We assume that

$$(3.2.15) \qquad \nabla^+\Big|_{\rho^\sigma} \text{ is flat} \quad (\sigma = 1, \ldots, r')$$

and

$$(3.2.16) \qquad \mathrm{Tor}\, H_1(\Lambda) = 0.$$

By conditions (3.2.15) and (3.2.16), it follows that each line subbundle ρ^σ is *trivial*. So, we can choose global sections $s_1, \ldots, s_{r'}$,

$$s_\sigma \in \Gamma(\rho^\sigma), \qquad (s_\sigma, s_\sigma)_+ = 1,$$

and introduce the global Floquet 1-forms $\beta^1, \ldots, \beta^{r'}$ on Λ associated with these trivializations by the formula

$$(3.2.17) \qquad \beta^\sigma = \frac{1}{i}(\nabla^+ s_\sigma, s_\sigma)_+ \qquad (\sigma = 1, \ldots, r').$$

It is clear that β^σ are closed forms. We define the Floquet exponents along the basic 1-cycles $\Gamma_1, \ldots, \Gamma_{k'}$ from (3.2.7) as the periods of the 1-forms $\beta^1, \ldots, \beta^{r'}$:

$$\beta_i^\sigma = \frac{1}{2\pi} \oint_{\Gamma_i} \beta^\sigma \qquad (\sigma = 1, \ldots, r'; i = 1, \ldots, k').$$

For each integer vector $n = (n^1, \ldots, n^{r'}) \in \mathbb{Z}^{r'}$, let us define the fiberwise polynomial functions $\mathbb{P}_{n^1}(\alpha, z), \ldots, \mathbb{P}_{n^{r'}}(\alpha, z) \in \mathcal{H}(\rho)$ by the formula

$$(3.2.18) \qquad \mathbb{P}_{n^\sigma}(\alpha, z) = \left[\sum_{\sigma'=1}^{r'} z^{\sigma'}(Z_{\sigma'}, s_\sigma)_+ \right]^{n^\sigma}.$$

Here $Z_1, \ldots, Z_r$ is a basis of local sections of ρ, and $z = (z^1, \ldots, z^r)$ are local coordinates along the fibers of ρ corresponding to this basis.

THEOREM 3.2.2. *Under the assumptions of Theorem* 3.2.1 *and conditions* (3.2.14)–(3.2.15), *the operator $\check{H}$ (3.1.25) has the eigenfunctions*

$$(3.2.19) \qquad \varphi_{m,n}(\alpha, z) = \frac{2^{(n^1 + \cdots + n^{r'})/2}}{\sqrt{n^1! \ldots n^{r'}!}} \varphi_m(\alpha) \mathbb{P}_{n^1}(\alpha, z) \cdot \ldots \cdot \mathbb{P}_{n^{r'}}(\alpha, z)$$

corresponding to the eigenvalues

$$(3.2.20) \qquad \lambda_{m,n} = \lambda_m - \hbar \sum_{\sigma=1}^{r'} n^\sigma \left(\sum_{i=1}^{k'} \beta_i^\sigma \omega^i \right).$$

Moreover,

$$(\varphi_{m,n}, \varphi_{m',n'})_\mathcal{H} = \delta_{mm'} \delta_{nn'}.$$

The proof of this statement follows the same argument as in [**KaVo$_1$**].

REMARK 3.2.2. For $r' = r$, the system $\{\varphi_{m,n}\}$ (3.2.19) is a total orthogonal basis of the Hilbert space $\mathcal{H}(\rho)$. In particular, this is true if a symplectic connection ∇ in E (adapted to D_H) is geometrically stable and the fundamental group $\pi_1(\Lambda)$ is commutative.

If condition (3.2.16) does not hold, then the line vector bundles ρ^σ from (3.2.14), (3.2.15) may be nontrivial. In this case, a natural problem is to find Floquet type solutions of the spectral problem (3.2.1).

Let $\mathfrak{f}\colon \widetilde{\Lambda} \to \Lambda$ be the *universal covering* of the manifold Λ. Denote by $\widetilde{\rho}$ the pullback of the original Hermitian vector bundle ρ to the universal covering manifold $\widetilde{\Lambda}$. We would like to investigate the Bloch spectral problem (3.2.1) in the space $\text{Pol}(\widetilde{\rho})$ of smooth fiberwise polynomial functions $\varphi(\widetilde{\alpha}, z)$ ($\widetilde{\alpha} \in \widetilde{\Lambda}$, $z \in \widetilde{\rho}_{\widetilde{\alpha}}$) on the pull-back bundle $\widetilde{\rho}$. By $\mathfrak{M}(\pi_1(\Lambda, \alpha))$ ($\alpha \in \Lambda$) we denote the *monodromy group* of the universal covering of Λ.

Instead of condition (3.2.16), we assume that the Hermitian line subbundles $\rho^1, \ldots, \rho^{r'}$ from (3.2.14), (3.2.15) satisfy the following condition: there are Euclidean real line subbundles $\varepsilon^1, \ldots, \varepsilon^{r'}$ of ρ such that

$$(3.2.21) \qquad \rho^\sigma = {}^{\mathbb{C}}\varepsilon^\sigma \qquad (\sigma = 1, \ldots, r').$$

Note that each ε^σ is defined by a *cocycle of the Hermitian bundle ρ^σ with values in the orthogonal group $O(r)$*.

Condition (3.2.21) allows us to define the global Floquet 1-forms $\beta^1, \ldots, \beta^{r'}$ on Λ by formula (3.2.17), where $s^1, \ldots, s^{r'}$ are any unit (local) sections of the Euclidean line subbundles $\varepsilon^1, \ldots, \varepsilon^{r'}$,

$$(3.2.22) \qquad s^\sigma \in \Gamma(\varepsilon^\sigma), \qquad (s^\sigma, s^\sigma)_+ = 1 \qquad (\sigma = 1, \ldots, r').$$

Let us choose a basis of unit vectors $\overset{\circ}{s}{}^1 \in \varepsilon^1_{\alpha^0}, \ldots, \overset{\circ}{s}{}^{r'} \in \varepsilon^{r'}_{\alpha^0}$ and define the global sections $\widetilde{s}^1(\widetilde{\alpha}), \ldots, \widetilde{s}^{r'}(\widetilde{\alpha})$ of the pull-back line vector bundles $\widetilde{\rho}^1, \ldots, \widetilde{\rho}^{r'}$ by the formula

$$(3.2.23) \qquad \widetilde{s}_\sigma(\widetilde{\alpha}) = \exp\left(-i \int_{\alpha^0}^{\mathfrak{f}(\widetilde{\alpha})} \beta^\sigma\right) g_{\alpha^0 \to \mathfrak{f}(\widetilde{\alpha})}(\overset{\circ}{s}{}^\sigma), \qquad \widetilde{\alpha} \in \widetilde{\Lambda},$$

for a fixed point $\alpha^0 \in \Lambda$. Here $g_{\alpha^0 \to \mathfrak{f}(\widetilde{\alpha})}$ denotes the parallel transport of the connection ∇^+ along a path on Λ connecting α^0 and $\mathfrak{f}(\widetilde{\alpha}) \in \Lambda$. We introduce the set of fiberwise polynomial functions on $\widetilde{\rho}$ as in (3.2.18),

$$(3.2.24) \qquad \mathbb{P}_{n^\sigma}(\widetilde{\alpha}, z) = \left[\sum_{\sigma'=1}^{r'} z^{\sigma'} \left(Z_{\sigma'}(\mathfrak{f}(\widetilde{\alpha})), \widetilde{s}_\sigma(\widetilde{\alpha})\right)_+\right]^{n^\sigma}.$$

Substituting the functions (3.2.14) into formula (3.2.19), we obtain Bloch eigenfunctions $\varphi_{m,n}(\widetilde{\alpha}, z)$ of the operator (3.1.25) corresponding to the eigenvalues (3.2.20) and such that, for each point $\alpha \in \Lambda$ and any element $\Gamma \subset \pi_1(\Lambda, \alpha)$, the function $\varphi_{m,n}(\widetilde{\alpha}, z)$ is changed under the action of the monodromy group $\mathfrak{M}(\pi_1(\Lambda, \alpha))$ according to the rule

$$(3.2.25) \qquad \varphi_{m,n}\bigl(\mathbf{m}_\Gamma(\widetilde{\alpha}), z\bigr) = \exp\left[i\pi \sum_{\sigma=1}^{r'} n^\sigma w_1^{\varepsilon^\sigma}(\Gamma')\right] \varphi_{m,n}(\widetilde{\alpha}, z) \qquad (\widetilde{\alpha} \in \widetilde{\Lambda}).$$

Here $\Gamma' \in H_1(\Lambda)$ is the image of Γ under the homomorphism $\pi_1(\Lambda, \alpha) \to H_1(\Lambda)$, $w_1^{\varepsilon^\sigma} \in H^1(\Lambda, \mathbb{Z}_2)$ is the Stiefel–Whitney class of the real vector bundle ε^σ, and $\mathbf{m}_\Gamma \in \mathfrak{M}(\pi_1(\Lambda, \alpha))$ is the element of the monodromy group corresponding to Γ.

Note that there is another interesting situation in which we can construct "unusual" fiberwise polynomial solutions of the spectral problem (3.2.5). Namely, an original symplectic connection ∇ adapted to D_H admits no invariant line subbundles, and the holonomy group of the connection ∇ is noncommutative. Such a case was analyzed in [**KaVo**$_1$], for instance, for the Klein bottle.

§3.3. Geometric phases and integer cohomology classes

Let us again recall what our basic geometric objects are:
• a compact connected isotropic submanifold Λ of a symplectic manifold (M,Ω); Λ is equipped with a smooth measure $d\sigma$;
• a *Kählerian prepolarization* over Λ; that is, a Lagrangian subbundle Π of the complexification ${}^{\mathbb{C}}T_\Lambda M$,

$$\Pi \subset {}^{\mathbb{C}}T_\Lambda M, \qquad {}^{\mathbb{C}}\dim \Pi = \frac{1}{2}\dim M,$$

which is positive definite with respect to the form

$$(3.3.1) \qquad (X,Y)_+ = \frac{1}{2i}\Omega(X,\overline{Y}) \qquad (X,Y \in {}^{\mathbb{C}}T_m M);$$

• a decomposition of the real symplectic vector bundle $T_\Lambda M$ into the Whitney sum

$$(3.3.2) \qquad T_\Lambda M = \mathcal{P} \oplus \mathcal{Q}$$

of two real Lagrangian subbundles $\mathcal{P}$ and $\mathcal{Q}$, where $\mathcal{Q}$ is equipped with a *Riemannian metric* $\langle\,,\,\rangle_\mathcal{Q}$.

Such subbundles $\mathcal{P}$ and $\mathcal{Q}$ will be called *"vertical"* and *"horizontal" prepolarizations* over Λ, respectively.

It follows from these assumptions that we have the decomposition

$$(3.3.3) \qquad {}^{\mathbb{C}}T_\Lambda M = \Pi \oplus \overline{\Pi},$$

where Π and $\overline{\Pi}$ are skew-orthogonal. Moreover, the complexifications ${}^{\mathbb{C}}\mathcal{P}$ and ${}^{\mathbb{C}}\mathcal{Q}$ are *transversal* to Π (and hence to $\overline{\Pi}$) at each point of Λ,

$$(3.3.4) \qquad \Pi \cap {}^{\mathbb{C}}\mathcal{P} = \Pi \cap {}^{\mathbb{C}}\mathcal{Q} = \{0\}.$$

Let $(E, \mathcal{J}\colon E \to E^*)$ be the symplectic normal bundle over the given isotropic submanifold Λ. Consider the skew-orthogonal complement $T\Lambda^\perp$ to $T\Lambda$ in $T_\Lambda M$ and the natural projection

$$(3.3.5) \qquad \mathrm{pr}\colon T\Lambda^\perp \to E.$$

Then we have the following short exact sequence of complex vector bundles:

$$(3.3.6) \qquad 0 \to {}^{\mathbb{C}}T\Lambda \to {}^{\mathbb{C}}(T\Lambda^\perp) \xrightarrow{\mathrm{pr}} {}^{\mathbb{C}}E \to 0.$$

The given Kählerian prepolarization Π induces the *positive isotropic* subbundle

$$(3.3.7) \qquad \Pi'' = \Pi \cap {}^{\mathbb{C}}(T\Lambda^\perp),$$

where ${}^{\mathbb{C}}\dim \Pi'' = (1/2)(\dim E)$.

Recall that the form $(\,,\,)_+$ (3.3.1) determines the Hermitian structure on Π. Consider the $(\,,\,)_+$-orthogonal complement to Π'' in Π,

$$(3.3.8) \qquad \Pi' = \mathrm{orth}_{(\,,\,)_+} \Pi''.$$

Then Π' is the positive isotropic subbundle such that

$$(3.3.9) \qquad {}^{\mathbb{C}}\dim \Pi' = \dim \Lambda, \qquad {}^{\mathbb{C}}T\Lambda \subset \Pi' \oplus \overline{\Pi}'.$$

Furthermore, under the projection (3.3.5), the subbundle Π'' induces the *positive Lagrangian subbundle* ρ of the complexification ${}^{\mathbb{C}}E$,

$$(3.3.10) \qquad \rho = \mathrm{pr}\left(\Pi \cap {}^{\mathbb{C}}(T\Lambda^{\perp})\right) \subset {}^{\mathbb{C}}E,$$

which will be called a *reduced* almost Kählerian polarization over Λ.

Now we shall introduce several objects associated with the symbolic data:

$$(3.3.11) \qquad \{(\Lambda, d\sigma), \Pi, \mathcal{P}, (Q, \langle\,,\,\rangle), (\rho, \mathbf{b})\},$$

where $\mathbf{b}$ is a $\mathbb{C}$-bilinear nondegenerate form on ρ.

(I) *Induced symmetric bilinear forms.* Consider the $\mathbb{C}$-linear map

$$(3.3.13) \qquad \mathfrak{G}\colon \overline{\Pi} \to \Pi^{*}$$

that determines the Hermitian structure (3.3.1) on the Kählerian prepolarization Π,

$$\langle \mathfrak{G}(v), w \rangle = (w, \overline{v})_{+}, \qquad v \in \overline{\Pi}_{\alpha}, \quad w \in \Pi_{\alpha}, \quad \alpha \in \Lambda.$$

The decomposition

$$(3.3.14) \qquad T_{\Lambda}M = \overline{\Pi} \oplus {}^{\mathbb{C}}\mathcal{P}$$

allows us to define the isomorphism $\mathfrak{J}\colon \Pi \to \overline{\Pi}$ as the projection of Π on $\overline{\Pi}$ along ${}^{\mathbb{C}}\mathcal{P}$. Introduce the $\mathbb{C}$-linear map $\mathfrak{B}\colon \Pi \to \Pi^{*}$ as follows:

$$(3.3.15) \qquad \mathfrak{B} = \mathfrak{G} \circ \mathfrak{J}.$$

It is easy to see that (3.3.15) defines a $\mathbb{C}$-linear nondegenerate symmetric form on the bundle Π.

By the above constructions, we have the $\mathfrak{G}$-orthogonal splitting

$$(3.3.16) \qquad \Pi = \Pi' \oplus \Pi''.$$

This implies that the restrictions of the $\mathbb{C}$-bilinear form (3.3.15) to the subbundles Π' and Π'' also define the $\mathbb{C}$-bilinear symmetric nondegenerate forms $\mathfrak{B}'_{\mathcal{P}}$ and $\mathfrak{B}''_{\mathcal{P}}$:

$$(3.3.17) \qquad \mathfrak{B}'_{\mathcal{P}}\colon \Pi' \to (\Pi')^{*}, \qquad \mathfrak{B}'_{\mathcal{P}} = \mathfrak{B}\big|_{\Pi'},$$

$$(3.3.18) \qquad \mathfrak{B}''_{\mathcal{P}}\colon \Pi'' \to (\Pi'')^{*}, \qquad \mathfrak{B}''_{\mathcal{P}} = \mathfrak{B}\big|_{\Pi''},$$

associated with the given vertical prepolarization $\mathcal{P}$. Moreover, we have the $\mathbb{C}$-bilinear symmetric nondegenerate form on the subbundle ρ (3.3.10):

$$(3.3.19) \qquad \mathfrak{b} = \mathfrak{b}_{\mathcal{P}} = \imath^{*} \circ \mathfrak{B}''_{\mathcal{P}} \circ \imath.$$

Here the vector bundle isomorphism $\imath\colon \rho \to \Pi''$ is defined by the projection (3.3.5), $\mathrm{pr} \circ \imath = \mathrm{id}_{\rho}$. In local terms, the matrix of the form $\mathfrak{b}$ (3.3.19) can be written as

$$\mathfrak{b} = \left(\imath(Z) \overset{\otimes}{,} X\right)_{+} \cdot \left(X \overset{\otimes}{,} X\right)_{+} \cdot \left(\imath(Z) \overset{\otimes}{,} X\right)_{+}^{t}.$$

Here $Z = (Z_1, \ldots, Z_r)$ is a basis of local sections of ρ, and $X = (X_1, \ldots, X_r)$ is the basis of local sections of Π generated by a basis of local sections $e = (e_1, \ldots, e_n)$ of $\mathcal{P}$ in the sense that

$$(3.3.21) \qquad X_i + \overline{X}_i = e_i.$$

On the other hand, let us define the second $\mathbb{C}$-bilinear symmetric nondegenerate form $\mathfrak{B}'_{T\Lambda}$ on Π' associated with the real isotropic subbundle $^\mathbb{C}T\Lambda \subset \Pi' \oplus \overline{\Pi}'$ by the formula

$$(3.3.22) \qquad \mathfrak{B}'_{T\Lambda} = \mathfrak{G}\Big|_{\overline{\Pi}'} \circ \mathfrak{l},$$

where

$$(3.3.23) \qquad \mathfrak{l}: \Pi' \to \overline{\Pi}'$$

is the projection of Π' onto $\overline{\Pi}'$ along $^\mathbb{C}T\Lambda$.

(II) *Topological phases and integral classes over isotropic submanifolds.* Using bilinear forms (3.3.17), (3.3.22), (3.3.19), and the given $\mathbb{C}$-bilinear nondegenerate form $\mathbf{b}: \rho \to \rho^*$, let us define the following smooth functions:

$$(3.3.24) \qquad \mathbf{j}' = \frac{\det \mathfrak{B}'_{T\Lambda}}{\det \mathfrak{B}'_{\mathcal{P}}}: \Lambda \to \mathbb{C},$$

$$(3.3.25) \qquad \mathbf{j}''_\mathbf{b} = \frac{\det \mathbf{b}}{\det \mathfrak{b}_\mathcal{P}}: \Lambda \to \mathbb{C}.$$

Obviously, $|\mathbf{j}'| > 0$ and $|\mathbf{j}''_\mathbf{b}| > 0$ uniformly on Λ. Thus, these mappings define the Arnold type 1-forms on Λ

$$(3.3.26) \qquad \mu' = -\left(\frac{\mathbf{j}'}{|\mathbf{j}'|}\right)^* \phi,$$

$$(3.3.27) \qquad \mu''_\mathbf{b} = -\left(\frac{\mathbf{j}''_\mathbf{b}}{|\mathbf{j}''_\mathbf{b}|}\right)^* \phi,$$

where $\phi = \frac{dz}{2\pi i z}$ is the fundamental 1-form on the unit circle $\mathbb{S}^1 = \{|z| = 1\} \subset \mathbb{C}$, $\oint_{\mathbb{S}^1} \phi = 1$.

PROPOSITION 3.3.1. *Formulas* (3.3.26), (3.3.27) *determine the closed 1-forms μ' and $\mu''_\mathbf{b}$ on Λ with integer cohomology classes,*

$$(3.3.28) \qquad [\mu'] \subset H^1(\Lambda, \mathbb{Z}), \qquad [\mu''_\mathbf{b}] \subset H^1(\Lambda, \mathbb{Z}).$$

These forms possess the following properties. The cohomology class of μ' is independent of the choice of the Kählerian prepolarization Π. If the vertical prepolarization $\mathcal{P}$ and the tangent bundle $T\Lambda$ of the isotropic submanifold Λ satisfy

$$\mathcal{P}_\alpha \cap T_\alpha \Lambda = \{0\}, \qquad \alpha \in \Lambda,$$

then the cohomology class $[\mu']$ vanishes, $[\mu'] = 0$. If the $\mathbb{C}$-bilinear form $\mathbf{b}$ coincides with $\mathfrak{b}_\mathcal{P}$ (3.3.19), $\mathbf{b} = \mathfrak{b}_\mathcal{P}$, then $[\mu''_\mathbf{b}] = 0$.

The proof follows from standard arguments [$\mathbf{A}_1$, **KaMas**, **KaVo**$_5$].

If Λ is a Lagrangian submanifold, $\dim \Lambda = \dim M/2$ and $\Pi = \Pi'$, $\rho = \{0\}$, then the cohomology class of the form μ' is just the Maslov class of the Lagrangian submanifold Λ. To prove this fact, we need the following relations between the *Jacobian* $\mathbb{J}$ of the projection (see [**KaVo**$_{1,5,8}$])

$$(3.3.29) \qquad {}^\mathbb{C}T\Lambda \oplus \bar{\rho} \to {}^\mathbb{C}\mathcal{P}$$

along Π and the function (3.3.24):

$$\mathbf{j}' = \frac{\det \mathfrak{b}}{|\det \mathfrak{b}|} \left(\frac{\mathbb{J}}{|\mathbb{J}|}\right)^2. \tag{3.3.30}$$

In the Lagrangian case, the function $\mathbf{j}'$ is equal to $(\mathbb{J}/|\mathbb{J}|)^2$, and hence, in this case, the form $[\mu']$ coincides with the standard form representing the Maslov class.

In general, the cohomology class of the form μ' (3.3.26) can be viewed as an *analog of the Maslov class for the isotropic submanifold* Λ.

The integrals

$$-\frac{1}{2}\int_{\alpha^0}^{\alpha}\mu', \qquad -\frac{1}{2}\int_{\alpha^0}^{\alpha}\mu''_{\mathfrak{b}}, \qquad (\alpha^0, \alpha \in \Lambda)$$

will be called *"tangential"* and *"normal" geometric phases over* an isotropic submanifold Λ.

Let us formulate some properties of the cohomology class $[\mu']$. Let $\mathcal{Y}$ be a *symplectic extension* of an isotropic submanifold Λ in M. Introduce the *tangential vertical prepolarization* $\mathcal{P}^{\text{tan}}$ as follows:

$$\mathcal{P}^{\text{tan}} = \mathcal{P} \cap T_\Lambda \mathcal{Y}.$$

This is a Lagrangian subbundle of the symplectic bundle $T_\Lambda \mathcal{Y}$; its dimension is equal to $(2 \dim \Lambda)$.

PROPOSITION 3.3.2. *For any symplectic extension $\mathcal{Y}$ of Λ in M, the cohomology class of the Arnold form $\mu(T\Lambda, \mathcal{P}^{\text{tan}})$ associated with Lagrangian subbundles $T\Lambda$ and $\mathcal{P}^{\text{tan}}$ of $T_\Lambda\mathcal{Y}$ coincides with the cohomology class $[\mu']$ of the 1-form (3.3.26):*

$$[\mu(T\Lambda, \mathcal{P}^{\text{tan}})] = [\mu'].$$

The proof of this statement follows from the definition of μ', the properties of the Arnold form [**KaVo**[5]] and the following fact: for any two symplectic extensions $\mathcal{Y}$ and $\widetilde{\mathcal{Y}}$ of Λ in M the subbundles $T_\Lambda \mathcal{Y}$ and $T_\Lambda \widetilde{\mathcal{Y}}$ are homotopic in the class of symplectic subbundles of $T_\Lambda M$ containing $T\Lambda$ as a Lagrangian subbundle.

Thus, we conclude that the value of the cohomology class $[\mu']$ at an element $[\gamma] \in H_1(\Lambda)$ measures the twisting of the intersection of the vertical prepolarization $\mathcal{P}$ and the isotropic subbundle $T\Lambda$ around a closed path γ.

Let Λ be an isotropic submanifold of the standard phase space $M = (\mathbb{R}_p^n)^* \times \mathbb{R}_q^n$ equipped with the natural polarization $\mathcal{P} \approx \mathbb{R}_p^n$. In this case, one can study the invariance of the cohomology class $[\mu']$ with respect to the near identity unitary transformations $g \in \text{Sp}(n) \cap O(2n)$ of the phase space $\mathbb{R}^{2n}$.

It is known that in the Lagrangian case, the Maslov class is invariant with respect to such transformations; this leads to the computation of the Maslov indices in terms of the *cycle of singularities* of the Lagrangian submanifold [**A**[1]]. In the isotropic case, the cohomology class $[\mu']$ is invariant only with respect to small *"tangential"* (relative to a given submanifold Λ) *unitary deformations* of Λ in $\mathbb{R}^{2n}$. This property agrees with the following fact [**DKrV, NVo**]: the codimension of the cycle of singularities of an isotropic submanifold $\Lambda \subset \mathbb{R}^{2n}$ in general position is larger than 1.

(III) *Second-order differential operators.* Note that the $\mathbb{C}$-bilinear symmetric form $\mathfrak{b}$ (3.3.19) induces the second-order differential operator $\check{\mathfrak{b}} = \mathcal{H}(\rho) \to \mathcal{H}(\rho)$

acting on the Hilbert space of smooth fiberwise holomorphic functions on ρ,

$$\text{(3.3.31)} \qquad \check{\mathfrak{b}} = \frac{1}{4} \sum_{\sigma\sigma'=1}^{r} \mathfrak{b}^{\sigma\sigma'}(\alpha) \frac{\partial^2}{\partial z^\sigma \partial z^{\sigma'}}.$$

Here $\mathfrak{b}^{-1} = ((\mathfrak{b}^{\sigma\sigma'}(\alpha)))$ is the matrix of the inverse map $\mathfrak{b}^{-1}\colon \rho^* \to \rho$ with respect to some local coordinates $\{z^\sigma\}$ along the fibers of ρ.

(IV) *Volume elements and modular function.* It follows from property (3.3.9) that there is an isomorphism

$$\text{(3.3.32)} \qquad \mathbf{f}\colon {}^{\mathbb{C}}T\Lambda \to \Pi',$$

which is defined as the projection of ${}^{\mathbb{C}}T\Lambda$ to Π' along $\overline{\Pi}'$.

PROPOSITION 3.3.3. *There is an Euclidean structure on $T\Lambda$ (or a Riemannian metric on Λ) defined by the formula*

$$\text{(3.3.33)} \qquad \langle u, v \rangle_{T\Lambda} = \bigl(\mathbf{f}(u), \mathbf{f}(v)\bigr)_+ \qquad (u, v \in T_\alpha \Lambda).$$

Here $(\,,\,)_+$ *is the form* (3.3.1).

By $d\sigma'$ we denote the measure on Λ generated by the Euclidean structure (3.3.33). Note that the density of $d\sigma'$ is equal to $|\det \mathfrak{B}'_{T\Lambda}|^{1/2}$, where the $\mathbb{C}$-bilinear form $\mathfrak{B}'_{T\Lambda}$ is defined by (3.3.17). Consider the ratio of volume elements corresponding to a given measure $d\sigma$ on Λ and the induced measure $d\sigma'$,

$$\text{(3.3.34)} \qquad \Delta_\Lambda = d\sigma/d\sigma'.$$

This is a smooth global function on Λ.

Now let us consider the triple $(\Pi, Q, \langle\,,\,\rangle_Q)$ consisting of an almost Kählerian polarization Π over Λ and a horizontal almost polarization Q over Λ with a Riemannian metric.

Using the decomposition ${}^{\mathbb{C}}T_\Lambda M = \Pi \oplus {}^{\mathbb{C}}\mathcal{P}$, we consider the projection

$$\text{(3.3.35)} \qquad Q \to \Pi \quad \text{along} \quad {}^{\mathbb{C}}\mathcal{P}.$$

Then the image $\mathcal{E}$ under the projection (3.3.35) is a Euclidean subbundle with respect to the form $(\,,\,)_+$ (3.3.1). On the other hand, under the projection (3.3.35), a given Riemannian metric $\langle\,,\,\rangle_Q$ on Q induces the Riemannian metric on $\mathcal{E}$. The ratio of volume elements on $\mathcal{E}$ induced by a given Riemannian metric $\langle\,,\,\rangle_Q$ and by the form $(\,,\,)_+$ is a smooth function on Λ. Let us denote this ratio by Δ_Q. In local terms, Δ_Q can be written as follows:

$$\text{(3.3.36)} \qquad \Delta_Q(\alpha) = \left[\frac{\det G(\alpha)}{\det \operatorname{Im} A(\alpha)}\right]^{1/2}, \qquad \alpha \in \Lambda.$$

Here $G(\alpha) = ((G_{ij}(\alpha)))$ is the Riemann tensor of $\langle\,,\,\rangle_Q$ with respect to a basis of local sections $\widetilde{e}_1, \ldots, \widetilde{e}_n$ of Q, the symmetric matrix $A(\alpha)$ is defined by the formula

$$\text{(3.3.37)} \qquad A = \bigl[\Omega(e\overset{\otimes}{,}X)\bigr]^{-1} \cdot \Omega(X\overset{\otimes}{,}\widetilde{e}),$$

where $X = (X_\sigma)$ is a basis of local sections of Π, and $e = (e^1, \ldots, e^n)$ is the basis of $\mathcal{P}$ dual to $\widetilde{e} = (\widetilde{e}^1, \ldots, \widetilde{e}^n)$,

$$\Omega(e^i, \widetilde{e}_j) = \delta^i_j.$$

It is clear that $\operatorname{Im} A(\alpha) > 0$. Finally, we define the positive *modular* function on Λ as follows:

(3.3.38) $$\Delta = \Delta_\Lambda \cdot \Delta_Q > 0 \quad \text{on} \quad \Lambda.$$

§3.4. Deformed coherent states

Following [**KaVo**$_{1,4,8}$], we describe deformed coherent states and an integral operator over an isotropic submanifold, which can be exploited either as an intertwining homomorphism in the representation theory, or as a simple global ansatz in the theory of semiclassical approximation.

For simplicity we assume that the original symplectic manifold (M, Ω) is the Euclidean space, $M = (\mathbb{R}^n_p)^* \times \mathbb{R}^n_q$ with coordinates $p = (p_1, \ldots, p_n)$, $q = (q^1, \ldots, q^n)$ and with standard symplectic structure $\Omega = dp \wedge dq$.

We choose the vertical and horizontal polarizations on M in the natural way,

$$\mathcal{P} \approx \mathbb{R}^n_p, \quad \mathcal{Q} \approx \mathbb{R}^n_q.$$

The Riemannian metric $\langle \, , \, \rangle_Q$ on Q is induced by the Euclidean structure on $\mathbb{R}^n_q$, so that the Riemann tensor G with respect to the basis

$$\widetilde{e}_1 = \partial/\partial q^1, \ldots, \widetilde{e}_n = \partial/\partial q^n$$

takes the form $G_{ij} = \delta_{ij}$. A given almost Kählerian polarization Π over Λ can be defined as

(3.4.1) $$\Pi = \operatorname{span}\{X_1(\alpha), \ldots, X_n(\alpha)\}, \quad \alpha \in \Lambda,$$

where $\{X_i(\alpha)\}$ is the global basis of Π:

(3.4.2) $$X_i(\alpha) = \sum_{j=1}^n A_{ij}(\alpha) e^j + \widetilde{e}_i.$$

Here $e^1 = \partial/\partial p_1, \ldots, e^n = \partial/\partial p_n$, and the global matrix-valued function $A(\alpha) = ((A_{ij}(\alpha)))$ has the properties

$$A^t(\alpha) = A(\alpha), \quad \operatorname{Im} A(\alpha) > 0 \quad \text{on} \quad \Lambda.$$

The matrix $\mathfrak{b} = ((\mathfrak{b}_{\sigma\sigma'}))$ of the bilinear form $\mathfrak{b}$ (3.3.19) with respect to a basis $Z = (Z_1, \ldots, Z_r)$ of ρ is given by the formula

(3.4.3) $$\mathfrak{b}^{-1} = \left[(X \overset{\otimes}{,} Z)^t_+ \cdot (\operatorname{Im} A)^{-1} \cdot (X \overset{\otimes}{,} Z)_+ \right].$$

Now we introduce coherent states over Λ.

Suppose that the embedding of the isotropic submanifolds Λ into $M = \mathbb{R}^{2n}$ is described by the equations

$$\Lambda = \{p = p(\alpha), q = q(\alpha)\}.$$

For any $\alpha, \alpha^0 \in \Lambda$, we define *deformed coherent states* over Λ by the formula

(3.4.4) $$\mathbb{I}(\alpha, \alpha^0 \mid q) = \frac{1}{\sqrt{\Delta(\alpha)}} \exp\left\{ \frac{i}{\hbar} \Phi(\alpha, \alpha^0 \mid q) \right\} \cdot \exp\left\{ -\frac{i n}{2} \int_{\alpha^0}^\alpha (\mu' + \mu''_\mathfrak{b}) \right\},$$

where the phase $\Phi(\alpha, \alpha^0 \mid q)$ has the form

(3.4.5) $$\Phi(\alpha, \alpha^0 \mid q) = \int_{\alpha^0}^\alpha \langle p(\alpha), dq(\alpha) \rangle + \langle p(\alpha), x(\alpha, q) \rangle + \frac{1}{2} \langle A(\alpha) x(\alpha, q), x(\alpha, q) \rangle.$$

The closed 1-forms μ' and $\mu''_{\mathfrak{b}}$ and the smooth function Δ are defined by (3.3.26), (3.3.27), and (3.3.38), respectively. In (3.4.5) we set

$$x(\alpha, q) = q - Q(\alpha), \qquad \alpha \in \Lambda, \quad q \in \mathbb{R}^n_q.$$

Note that
$$\operatorname{Im} \Phi \geq 0 \text{ and } \operatorname{Im} \Phi = 0 \text{ if and only if } q = q(\alpha).$$

Now, using the projection (3.3.25) and the projection

$$\Pi \to \Pi'' \text{ along } \Pi',$$

we can define the sequence of vector bundle morphisms

(3.4.6) $$\mathbb{R}^n_q \to \Pi \to \Pi'' \to \rho.$$

From (3.4.6) we obtain the morphisms

$$\mathfrak{k} \colon \mathbb{R}^n_q \to \rho, \qquad \mathfrak{k}^* \colon \operatorname{Pol}(\rho) \to C^\infty(\mathbb{R}^n_q),$$

where $\operatorname{Pol}(\rho)$ is the subspace of smooth fiberwise polynomial functions on ρ. By $\operatorname{Pol}_0(\rho)$ we also denote the subspace of $\operatorname{Pol}(\rho)$ that consists of functions whose supports are contained in a certain simply connected domain of Λ.

Now we can defined the following integral operator $\mathbb{I}_\Lambda \colon \operatorname{Pol}_0(\rho) \to C^\infty(\mathbb{R}^n_q)$ associated with symbolic data (3.3.11):

(3.4.7) $$\mathbb{I}_\Lambda(\varphi)(q) = \frac{1}{c} \int_\Lambda \mathbb{I}(\alpha, \alpha^0 \mid q) \cdot \widetilde{\varphi}\left(\alpha; \frac{x(\alpha, q)}{\sqrt{\hbar}}\right) d\sigma,$$

where
$$\widetilde{\varphi} = \mathfrak{k}^*(\exp(-\check{\mathfrak{b}})\varphi), \qquad \varphi(\alpha, z) \in \operatorname{Pol}_0(\rho).$$

Here the second order differential operator $\check{\mathfrak{b}}$ is given by (3.3.31), and the morphism $\mathfrak{k}^* \colon \operatorname{Pol}(\rho) \to C^\infty(\mathbb{R}^n_q)$ is defined by (3.4.6).

If the normalizing constant c from (3.4.7) and the smooth measure $d\sigma$ are chosen so that

$$c = 4^{k/4} \cdot (\pi\hbar)^{(k+n)/4} \qquad (k = \dim \Lambda, \quad n = \frac{1}{2} \dim M), \qquad \int_\Lambda d\sigma = 1,$$

then for each $\varphi(\alpha) \in C^\infty(\Lambda) \subset \operatorname{Pol}_0(\rho)$ we obtain the following asymptotic formula as $\hbar \to 0$:

$$\bigl(\mathbb{I}_\Lambda(\varphi), \mathbb{I}_\Lambda(\varphi)\bigr)_{L^2(\mathbb{R}^n_q)} = \int_\Lambda |\varphi(\alpha)|^2 d\sigma + O(\hbar).$$

Note that if the quantization rule

(3.4.8) $$\left[\frac{1}{2\pi\hbar} p\, dq - \frac{1}{4}\mu' - \frac{1}{4}\mu''_{\mathfrak{b}}\right] \in H^1(\Lambda, \mathbb{Z})$$

is satisfied, then the integral operator (3.4.7) is well defined on the entire space $\operatorname{Pol}(\rho)$. Obviously, $\operatorname{Pol}_0(\rho) \subset \operatorname{Pol}(\rho) \subset \mathcal{H}(\rho)$. Let us consider the Lie algebra $\mathcal{F}^{\bar\rho \oplus \rho}_\Lambda(\mathbb{R}^{2n})$ consisting of smooth functions $H = H(p, q) \in C^\infty(\mathbb{R}^{2n})$ such that
- Λ is $\operatorname{ad}(H)$-invariant,
- ρ is D_H-invariant, where D_H is the first variation operator of the Hamilton vector field $\operatorname{ad}(H)$.

Let $H = H(p,q) \in \mathcal{F}_\Lambda^{\bar{\rho} \oplus \rho}(\mathbb{R}^{2n})$ be a real smooth function whose derivatives increase at infinity not faster than a polynomial. Then the *Weyl operator*

$$\widehat{H} = H(-i\hbar\partial/\partial q, q)$$

is well defined [**Shub**].

THEOREM 3.4.1. *The following intertwining property holds on* Pol_0 *as* $\hbar \to 0$:

(3.4.9) $$\widehat{H} \circ \mathbb{I}_\Lambda = \mathbb{I}_\Lambda \circ \check{H} \qquad (\mathrm{mod} O(\hbar^{3/2})),$$

where $\check{H}$ *is the self-adjoint operator (in the Hilbert space* $\mathcal{H}_0(\rho)$*) defined by* (3.1.25). *If the quantization condition* (3.4.8) *is satisfied, then we have the commutation formula* (3.4.9) *on the entire space* $\mathrm{Pol}(\rho)$.

The complete proof of Theorem 3.4.1 can be found in [**KaVo$_1$**]. For the construction of the integral operator $\mathbb{I}_\Lambda$ in the case $M = T^*\mathcal{M}$ (where $\mathcal{M}$ is a Riemannian manifold), see also [**KaVo$_1$**].

REMARK 3.4.1. If the quantization rule (3.4.8) is satisfied, the integral operator (3.4.7) is also well defined on the entire Hilbert space $\mathcal{H}(p)$ of fiberwise holomorphic functions

$$\mathbb{I}_\Lambda : \mathcal{H}(\rho) \to L^2(\mathbb{R}_q^n).$$

This fact can be deduced from the relation between the "Kähler scalar" $K(\alpha, z, \bar{z})$ (3.1.27) and the complex phase function $\Phi(\alpha, \alpha^0 \mid q)$ (3.4.5):

$$K(\alpha, \mathfrak{k}(q), \bar{\mathfrak{k}}(q)) \leq \mathrm{Im}\, \Phi(\alpha, \alpha^0 \mid q) \qquad \forall \alpha \in \Lambda, \; q \in \mathbb{R}^n,$$

where the map $\mathfrak{k}$ is defined by (3.4.6).

REMARK 3.4.2. Let $\widetilde{\rho}$ be the pull-back of the vector bundle ρ to the universal covering manifold $\widetilde{\Lambda}$ of Λ. The integral operator $\mathbb{I}_\Lambda$ can also be defined on functions on $\widetilde{\rho}$ of the form

$$\exp\left(-\frac{i}{2} \int_{\alpha^0}^\alpha \varkappa \right) \varphi(\widetilde{\alpha}, z) \qquad (\widetilde{\alpha} \in \widetilde{\rho},\; z \in \widetilde{\rho}_{\widetilde{\alpha}}),$$

where $\varkappa$ is a real closed 1-form on Λ and $\varphi(\widetilde{\alpha}, z)$ is a fiberwise polynomial function on $\widetilde{\rho}$ of type (3.1.25). Then the operator $\mathbb{I}_\Lambda$ is well defined on such functions if the corrected quantization rule holds. Here corrections to (3.4.8) are given by the cohomology class of the form $\varkappa$ and by the Stiefel–Whitney classes.

§3.5. Quantization conditions

Recall that a pair $(\lambda = \lambda(\hbar), \psi = \psi(q;\hbar))$ is called a *quasimode* for the pseudodifferential Weyl operator

(3.5.1) $$\widehat{H} = H(-i\hbar\partial/\partial q, q)$$

if the following estimates hold as $\hbar \to 0$:

$$\|(\widehat{H} - \lambda)\psi\|_{L^2(\mathbb{R}^n)} = O(\hbar^N), \qquad \|\psi\|_{L^2(\mathbb{R}^n)} = 1 + O(\hbar)$$

for a certain $N > 1$. It is well known [**MasF**] that if the operator $\widehat{H}$ is essentially self-adjoint in $L^2(\mathbb{R}^n)$, then the distance from the point $\lambda(\hbar) \in \mathbb{R}$ to the spectrum of $\widehat{H}$ is of order $O(\hbar^N)$,

$$\operatorname{spec} \widehat{H} \cap (\lambda - C\hbar^N, \lambda + C\hbar^N) \neq \varnothing,$$

where C is a constant independent of $\hbar$.

The main idea of the semiclassical method for constructing quasimodes for the operator $\widehat{H}$ is to use invariant Lagrangian or isotropic submanifolds $\Lambda \subset \mathbb{R}^{2n}$ that lie on the energy surface of the symbol H, $H\big|_\Lambda = E$.

A quasimode (λ, ψ) associated with a classical object Λ must satisfy the following *a priori* hypothesis: λ is close to the classical energy level E, and the *frequency set* (or the *oscillation front*) [**GuSt$_3$, KaMas**] of the function ψ lies in Λ,

$$\lambda = E + O(\hbar), \qquad \operatorname{osc} \psi \subseteq \Lambda.$$

The usual well-known and well-developed approaches to construction of quasimodes are based on matching local WKB–type asymptotic expressions. The construction, due to Voros [**Vor**] (in the case $\dim \Lambda = 1$, $\dim M = 2$), was based on the Bargmann transform (integration over the complex plane). The closest to what follows below is Heller's idea of Gaussian packet integration over a trajectory [**Hel**]. We consider a scheme for constructing quasimodes using the integral operator (3.4.7), which appeared in [**Ka$_{1,2,5,6}$**] for the case of Lagrangian submanifolds and in [**KaVo$_{1,4,6,8}$**] for the isotropic case (and then was studied in a number of other works; see [**GrPa, PU**] and the references therein).

To formulate the main results, let us take some preparatory steps.

Assume that the symbol $H = H(p,q)$ of a given Weyl operator (3.5.1) is a smooth real function on the phase space $(M = \mathbb{R}^{2n}, \Omega = dp \wedge dq)$.

We say that a triple $(\Lambda, d\sigma, \rho)$ consisting of a compact connected submanifold $\Lambda \subset M$ with smooth measure $d\sigma$ and a positive Lagrangian subbundle $\rho \subset {}^\mathbb{C}E$ (E is the symplectic normal bundle of Λ) is *compatible* with the symbol H if

- Λ and $d\sigma$ are invariant under the flow of the Hamilton vector field $\operatorname{ad}(H)$, and
- the subbundle ρ is invariant with respect to the first variation operator D_H of $\operatorname{ad}(H)$ of Λ.

These conditions imply that $H \in \mathcal{F}_\Lambda^{\bar\rho \oplus \rho}(M)$, the function H is constant along Λ,

(3.5.2) $$H\big|_\Lambda = E = \mathrm{const},$$

and the first variation operator D_H is geometrically stable.

PROPOSITION 3.5.1. *For a given positive Lagrangian subbundle $\rho \subset {}^\mathbb{C}E$, there exists a Kählerian prepolarization $\Pi \subset {}^\mathbb{C}T_\Lambda M$ such that ρ is related to Π by condition* (3.3.10).

The proof of Proposition 3.5.1 is based on the following well-known fact: in a symplectic vector bundle, there always exists an almost complex structure (see, for instance, [**LiMr, We$_2$, Wo**]).

PROOF. Let ρ be a positive Lagrangian subbundle of $^{\mathbb{C}}E$ and let $\langle\,,\,\rangle$ be a metric on $T_\Lambda M$. Then we have the decomposition of $T_\Lambda M$ into the direct sum of two skew-orthogonal symplectic subbundles E' and E'', $T_\Lambda M = E' \oplus E''$, which are defined in the following unique way. The subbundle E'' is defined as the orthogonal complement of $T\Lambda$ in $T\Lambda^\perp$ (relative to a given metric $\langle\,,\,\rangle$) and $E' = (E'')^\perp$ is the skew-orthogonal complement of E' in $T_\Lambda M$. Thus we obtain $\dim E' = 2\dim \Lambda$ and $T\Lambda$ is a Lagrangian subbundle of E'. Taking the almost complex structure $g\colon E' \to E'$, $g^2 = -\operatorname{id}$, which is uniquely defined by the symplectic structure and the metric on E' (see [**We**$_2$]), we introduce the positive Lagrangian subbundle $\Pi' = (I - ig)(^{\mathbb{C}}T\Lambda) \subset {}^{\mathbb{C}}E'$. Finally, we choose $\Pi = \Pi' \oplus \operatorname{pr}_0^{-1}(\rho)$, where pr_0 is the restriction of the projection (3.3.5) to E''. $\square$

So, Proposition 3.5.1 says that the given initial triple $(\Lambda, d\sigma, \rho)$ can be included into the symbolic data (3.3.11). In what follows, we suppose that such symbolic data are fixed.

DEFINITION 3.5.1. A real closed 1-form $\varkappa^{\text{gauge}}$ on Λ (depending on the parameter $\hbar \in (0,1]$) is called a *gauge form* of the symbol H if for each $\hbar \in (0,1]$,

$$\varkappa_H^{\text{gauge}} =: \operatorname{ad}(H) \lrcorner \varkappa^{\text{gauge}} = \operatorname{const} \text{ on } \Lambda,$$
$$\varkappa_H^{\text{gauge}} = O(1) \qquad \text{as } \hbar \to 0.$$

A gauge form $\varkappa^{\text{gauge}}$ is called *admissible* if the following corrected quantization rule holds on Λ:

$$\left[\frac{1}{2\pi\hbar} p\,dq - \frac{1}{4}\mu' - \frac{1}{4}\mu_b''\right] - \left[\frac{\varkappa^{\text{gauge}}}{4\pi}\right] \in H^1(\Lambda, \mathbb{Z}).$$

THEOREM 3.5.1. *Let $(\Lambda, d\sigma, \rho)$ be a triple compatible with the symbol H of the Weyl operator (3.5.1). Assume that there is an admissible gauge form $\varkappa^{\text{gauge}}$ of H. If φ is an eigenfunction of the operator $\check{H}$ (3.1.25) corresponding to an eigenvalue $\widetilde{\lambda}$,*

$$\check{H}\varphi = \widetilde{\lambda}\varphi, \qquad \|\varphi\|_{\mathcal{H}(\rho)} = 1,$$

then the pair (λ, ψ),

(3.5.3) $$\lambda = \widetilde{\lambda} - \frac{\hbar}{2}\varkappa_H^{\text{gauge}},$$

(3.4.4) $$\psi(q, \hbar) = \mathbb{I}_\Lambda\left(\exp\left(-\frac{i}{2}\int_{\alpha^0}^\alpha \varkappa^{\text{gauge}}\right) \cdot \varphi\right)$$

is the quasimode for the operator $\widehat{H}$ (3.5.1),

$$\|(H - \lambda)\psi\|_{L^2} = O(\hbar^{3/2}), \qquad \|\psi\|_{L^2} = 1 + O(\hbar).$$

The proof of this statement readily follows from Theorem 3.4.1.

So, Theorem 3.5.1 shows that the problem of constructing the quasimodes associated with a given triple $(\Lambda, d\sigma, \rho)$ can be reduced to the following problems:
- to justify the existence of an admissible gauge form (or the resolution of the quantization conditions of Bohr–Sommerfeld type); and
- to solve the model spectral problem (3.2.1).

EXAMPLE 3.5.1. Let us assume that the restriction v_H of the Hamilton vector field ad(H) to Λ satisfies condition (3.2.8). Then the admissible gauge form of H can be chosen as follows:

$$\varkappa^{\text{gauge}} = \sum_{i=1}^{k'} \delta_i \eta^i, \tag{3.5.5}$$

$$\boldsymbol{\delta}(\hbar) = \big(\delta_1(\hbar), \ldots, \delta(\hbar)\big) =: \left(\frac{1}{\pi\hbar} \oint p\,dq - 2m(\hbar)\right) - \frac{1}{2}\oint(\mu' + \mu''_{\mathbf{b}}). \tag{3.5.6}$$

Here we use the vector notation

$$\oint p\,dq = \left(\oint_{\Gamma_1} p\,dq, \ldots, \oint_{\Gamma_{k'}} p\,dq\right),$$

where $\Gamma_1, \ldots, \Gamma_{k'}$ are the basic 1-cycles on Λ dual to the closed 1-forms $\eta^1, \ldots, \eta^{k'}$ from (3.2.7). An integer vector $m = m(\hbar) \in \mathbb{Z}^{k'}$ in (3.5.6) must satisfy the estimate

$$\left\|\frac{1}{2\pi\hbar}\oint p\,dq - m\right\| = O(1), \quad \text{as} \quad \hbar \to 0. \tag{3.5.7}$$

(I) *Vacuum quasimodes assigned to an individual isotropic submanifold*. Combining Theorem 3.2.1 and Theorem 3.5.1 with Example 3.5.1, we obtain the following statement.

PROPOSITION 3.5.2 ("vacuum quasimodes"). *Let $(\Lambda, d\sigma, \rho)$ be a triple compatible with the symbol H of the Weyl operator (3.5.1). Assume that the following conditions* (a) *and* (b) *hold*:
(a) *There are basic closed 1-forms $\eta^1, \ldots, \eta^{k'}$ from (3.2.7) such that*

$$v_H \,\lrcorner\, \eta^i = \omega^i = \text{const} \qquad \left(v_H = \text{ad}(H)\big|_\Lambda\right) \tag{3.5.8}$$

for all $i = 1, \ldots, k'$.
(b) *There is a symplectic connection ∇ in the symplectic normal bundle E of Λ such that the subbundle ρ is ∇-invariant,*

$$\text{tr}\left(\text{curvature of } \nabla\big|_\rho\right) = 0, \tag{3.5.9}$$

and the connection ∇ is weakly adapted to the first variation operator D_H of the Hamilton vector field ad(H) (that is, condition 3.2.13 holds).

Then to the Weyl operator $\widehat{H}$ there corresponds a set of quasimodes $\big(\lambda_m = \lambda_m(\hbar), \psi_m = \psi_m(q,\hbar)\big)$, where the functions ψ_m are determined by (3.5.4) (the gauge form $\varkappa^{\text{gauge}}$ is taken from (3.5.5)), and

$$\lambda_m = E + \hbar\big[\lambda^{(1)}_{\text{dyn}} + \lambda^{(1)}_{\text{geom}} + \lambda^{(1)}_{\text{sym}} + \omega \cdot m\big], \tag{3.5.10}$$

$$\lambda^{(1)}_{\text{dyn}} = -\frac{1}{4\pi}\omega \cdot \left(\oint \varkappa_{\mathbf{b}} - \pi \oint \mu''_{\mathbf{b}}\right), \tag{3.5.11}$$

$$\lambda^{(1)}_{\text{geom}} = \frac{1}{4}\omega \cdot \oint \mu', \tag{3.5.12}$$

$$\lambda^{(1)}_{\text{sym}} = -\frac{1}{2\pi\hbar}\omega \cdot \oint p\,dq. \tag{3.5.13}$$

Here $m = (m_1, \ldots, m_{k'}) \in \mathbb{Z}^{k'}$ is an integer vector satisfying condition (3.5.7), $\omega = (\omega^1, \ldots, \omega^{k'})$ is the "frequency" vector with components defined by (3.5.8), and the 1-forms $\varkappa_{\mathfrak{b}}$, μ', and $\mu''_{\mathfrak{b}}$ are given in (3.2.5), (3.3.26), and (3.3.27).

Note that the components (3.5.11)–(3.5.13) are independent of the choice of basic 1-forms $\eta^1, \ldots, \eta^{k'}$ and the corresponding "frequency" vector ω from condition (3.5.8), and of a $\mathbb{C}$-bilinear form $\mathfrak{b}$ entering the symbolic data (3.3.11). To verify the last statement, let us investigate the nature of each summand in the formula for approximate eigenvalues (3.5.10).

Note that we have the identity

$$(3.5.14) \qquad \varkappa_{\mathfrak{b}} - \pi\mu''_{\mathfrak{b}} = \varkappa_{\mathfrak{b}},$$

where $\mathfrak{b}$ is a $\mathbb{C}$-bilinear symmetric nondegenerate form on ρ defined by (3.3.19). In terms of the 1-form θ^+ corresponding to the Hermitian connection $\nabla^+ = \nabla\big|_\rho$, the form $\varkappa_{\mathfrak{b}}$ can be written as follows:

$$(3.5.15) \qquad \varkappa_{\mathfrak{b}} = \left[-\frac{1}{2}\operatorname{Im}\left(\operatorname{tr}(\mathfrak{b}^{-1}\,d\mathfrak{b})\right) + \operatorname{Im}(\operatorname{tr}\theta^+) \right].$$

Using condition (3.5.8) and the invariance of the measure $d\sigma$, we obtain the identity

$$(3.5.16) \qquad \frac{1}{2\pi}\omega \cdot \oint \varkappa_{\mathfrak{b}} = \int_\Lambda v_H \,\lrcorner\, \varkappa_{\mathfrak{b}}\, d\sigma.$$

The right-hand side of (3.5.16) determines a dynamical characteristic of the flow of the vector field $D_H^\#$ on E, which is the *rotation number* of the first variation operator D_H (see §1.10).

So, we obtain

$$(3.5.17) \qquad \lambda_{\mathrm{dyn}}^{(1)} = -\frac{1}{2}\left(\int_\Lambda v_H \,\lrcorner\, \varkappa_{\mathfrak{b}}\, d\sigma\right) \equiv \frac{1}{2}(\text{rotation number of } D_H).$$

As we saw in §3.3, the cohomology of the closed 1-form μ' (3.3.26) gives an analog of the Maslov class for the pair $(T\Lambda, \mathcal{P})$ consisting of the isotropic subbundle $T\Lambda$ and the vertical real Lagrangian subbundle $\mathcal{P}$. The second term (3.5.12) can be also written as the average over Λ,

$$(3.5.18) \qquad \lambda_{\mathrm{geom}}^{(1)} = \frac{\pi}{2}\int_\Lambda (v_H \,\lrcorner\, \mu')\, d\sigma.$$

Since the restriction of the symplectic potential $p\,dq$ (the Liouville form) to Λ is a closed 1-form, the last term (3.5.13) is

$$(3.5.19) \qquad \lambda_{\mathrm{sym}}^{(1)} = -\frac{1}{\hbar}\int_\Lambda \left(v_H \,\lrcorner\, (p\,dq)\right) d\sigma.$$

Resuming our discussion, we conclude that formula (3.5.10) for approximate eigenvalues includes the following phase space invariants of a Hamilton system H with an invariant isotropic submanifold Λ: the average of the Liouville form, the average of Arnold's form along Λ, and the rotation number of the first variation operator of the Hamilton vector field $\operatorname{ad}(H)$ over Λ.

There is an interesting special case in which the "normal" Maslov class can be calculated via the Gelfand-Lidskii indices of stable cycles.

Assume that the symplectic normal bundle E of Λ admits a Lagrangian subbundle $\mathcal{L}$. Then for each positive Lagrangian subbundle $\rho \in {}^{\mathbb{C}}E$, the given $\mathcal{L}$ induces a $\mathbb{C}$-bilinear form $\mathbf{b}$ on ρ (defined similarly to (3.3.22)), and hence, there is a naturally determined closed 1-form $\mu''_{\mathbf{b}}$ (3.3.27). In this case, it follows from the results of §1.10 that the form $\mu''_{\mathbf{b}}$ coincides with the Arnold 1-form associated with the pair of Lagrangian subbundles $\mathcal{L}$ and $\mathcal{P}^{\text{red}}$ (1.10.29). So, the cohomology class of $\mu''_{\mathbf{b}}$ represents the "normal" Maslov class of Λ for the pair $(\mathcal{L}, \mathcal{P}^{\text{red}})$.

To specify the choice of $\mathcal{L}$, we consider the following situation. Assume that the symplectic connection ∇ from condition (a) of Proposition 3.5.2 and a Lagrangian subbundle $\mathcal{L}$ satisfy the following two conditions:

(i) There are basic cycles $\Gamma_1, \ldots, \Gamma_m$ of Λ such that the connection ∇ is $\boldsymbol{\Gamma}$-strongly stable (Definition 1.10.2), where $\boldsymbol{\Gamma}$ is determined by 1-cycles $\Gamma_1, \ldots, \Gamma_m$.

(ii) The Lagrangian subbundle $\mathcal{L}$ is "minimal", that is, the Gelfand–Lidskii index of ∇ relative to $\mathcal{L}$ (see §1.10) vanishes at all basic 1-cycles $\Gamma_1, \ldots, \Gamma_m$.

COROLLARY 3.3.1. *Assume that assumptions* (i), (ii), *as well as condition* (3.5.8), *hold relative to some basic 1-cycles* $\Gamma_1, \ldots, \Gamma_m$ *on* Λ. *Then the "dynamical" term* (3.5.11) *consists of two parts: the normalized Floquet exponents* (*see* §1.10) *and the double Gelfand–Lidskii indices of* ∇ *relative to the reduced polarization* $\mathcal{P}^{\text{red}}$ (1.10.29) (*see Proposition* 1.10.5).

Moreover the quantization rule (3.4.8) *taken with respect to the above basic cycles* $\Gamma_1, \ldots, \Gamma_m$ *includes the following "index" part:*

$$(\text{tangential Maslov index of } \Lambda \text{ at } \Gamma_i/4)$$
$$+ \frac{1}{2} \left(\text{Gelfand–Lidskii index of } \nabla \text{ relative to } \mathcal{P}^{\text{red}} \text{ at } \Gamma_i \right).$$

Some observations concerning the role of phase space invariants of Hamilton systems in semiclassical asymptotic expansions can be also found in [**Au, CRLj, Vo₂, Vor, Mas, BeDo₁, DMasN**].

(II) *Excited quasimodes.* Now under the assumptions of §3.2, we want to describe "excitations" of the quasimodes (λ_m, ψ_m) from Proposition 3.5.2.

As in Example 3.5.1, let us define the "excited" gauge 1-form $\varkappa^{\text{gauge}}$ by formula (3.5.5), where the components of the vector $\delta(\hbar)$ are given by the right-hand side of (3.5.6) plus the correction

$$\sum_{\sigma=1}^{r'} n^\sigma w_1^{\varepsilon^\sigma}(\Gamma_i) \qquad (i = 1, \ldots, k').$$

Here $w_1^{\varepsilon^\sigma}(\Gamma_i)$ is the value of the Stiefel–Whitney class of the Euclidean subbundle ε^σ from condition (3.2.21) at the basic 1-cycle $\Gamma_i \in \widetilde{H}_1(\Lambda, \mathbb{R})$ from (3.2.7), and the components of an integer vector $n = (n^1, \ldots, n^{r'}) \in \mathbb{Z}^{r'}$ satisfy the condition

(3.5.20) $\qquad n^\sigma = n^\sigma(\hbar) = O(1) \qquad \text{as } \hbar \to 0.$

Using this "excited" gauge 1-form and the set of fiberwise polynomial functions $\varphi_{m,n}$ defined by (3.2.19) and (3.3.34), we introduce the functions

(3.5.21) $\qquad \psi_{m,n} = \mathbb{I}_\Lambda \left(\exp\left(-\frac{i}{2} \int_{\alpha^0}^{\alpha} \varkappa^{\text{gauge}} \right) \varphi_{m,n} \right).$

Let us denote

$$(3.5.22) \qquad \lambda_{m,n} = \lambda_m - \hbar\left[\sum_{\sigma=1}^{r'} n^\sigma \left(\int_\Lambda v_H \lrcorner \beta^\sigma \, d\sigma + \sum_{i=1}^{k'} \frac{\omega^i}{2} w_1^{\varepsilon^\sigma}(\Gamma_i)\right)\right].$$

Here λ_m is defined by (3.5.10), β^σ are the Floquet 1-forms (3.2.17) associated with the Euclidean subbundles ε^σ from (3.2.21), and ω^i are the components of the "frequency" vector from condition (3.2.17).

PROPOSITION 3.5.3 ("excited quasimodes"). *Let $(\Lambda, d\sigma, \rho)$ be a triple compatible with the symbol H of the Weyl operator (3.5.1). Assume that conditions (a) and (b) of Proposition 3.5.2 and the following additional assumptions hold:*

(c) *There exist line subbundles $\rho^1, \ldots, \rho^{r'}$ ($r' \leq r$) of ρ satisfying conditions (3.2.14), (3.2.15), (3.2.21).*

(d) *The Euclidean line subbundles $\varepsilon^1, \ldots, \varepsilon^{r'}$ from (3.2.21) satisfy the conditions*

$$(3.5.23) \qquad w_1^{\varepsilon^\sigma}(\Gamma) = 0 \qquad (\sigma = 1, \ldots, r')$$

for each element $\Gamma \in \operatorname{Tor} H_1(\Lambda)$.

Then formulas (3.5.21) and (3.5.22) define the quasimodes $(\lambda_{m,n}, \psi_{m,n})$ for the operator (3.5.1) modulo $O(\hbar^{3/2})$.

Note that if the symbol H of the Weyl operator (3.5.1) is constant along a *stable coisotropic extension* N of an isotropic submanifold Λ, then it follows from the results of §1.3 that there is the flat symplectic connection ∇ (1.3.6) of Bott type adapted to the first variation operator D_H. If the fundamental group $\pi_1(\Lambda)$ is commutative, then there is a ∇-invariant positive Lagrangian subbundle $\rho \subset {}^\mathbb{C}E$. So, we conclude that if $\Lambda \approx \mathbb{T}^k$ is an isotropic torus and the flow of the Hamilton vector field $\operatorname{ad}(H)$ on Λ is quasiperiodic, then all assumptions of Proposition 3.5.3 are satisfied, and formulas (3.5.21) and (3.5.22) give the quasimodes associated with the pair $(\Lambda \approx \mathbb{T}^k, N)$.

EXAMPLE 3.5.2 *Quasimodes associated with a flat coisotropic extension.* We suppose that the assumptions of Theorem 2.4.1 are satisfied for an isotropic k-dimensional submanifold Λ of the phase space $(M = \mathbb{R}^{2n}, \Omega = dp \wedge dq, \mathcal{P})$ with vertical polarization $\mathcal{P} \approx \mathbb{R}_p^n$. In particular, Λ admits a flat stable coisotropic extension $N \subset \mathbb{R}^{2n}$. We also assume that conditions (3.2.16), (3.5.8) hold and the symbol H of the operator (3.4.1) is constant along the coisotropic extension N, $H|_N = E$. Moreover, let there be an $\operatorname{ad}(H)$-invariant smooth measure $d\sigma$ on Λ.

Then by Theorem 2.4.1, an open domain N^{reg} (near Λ) is fibered by the Lagrangian submanifolds $\Lambda(c)$ (2.4.2), $c = (c^1, \ldots, c^r) \in \mathcal{C}^r \cap S$, $r = n - k$. For each $c \in \mathcal{C}^r \cap \rho$, the submanifolds $\Lambda(c)$ are invariant with respect to the flow of the Hamilton vector field $\operatorname{ad}(H)$, and there exists the $\mathbb{T}^r$-principle bundle (2.4.17) over Λ.

Let us choose basic 1-cycles $\Gamma_1, \ldots, \Gamma_k \in H_1(\Lambda)$ and a global section e (see (2.4.22)), $e_c \colon \Lambda \to \Lambda(c)$. Then for each c, we can define induced 1-cycles

$$\widetilde{\Gamma}_1(c), \ldots, \widetilde{\Gamma}_k(c); \widetilde{\Gamma}_{k+1}(c), \ldots, \widetilde{\Gamma}_{k+r}(c) \in H_1(\widetilde{\Lambda}(c)).$$

Here $\widetilde{\Gamma}_i(c) = e_c(\Gamma_i)$, $i = 1, \ldots, k$, each $\widetilde{\Gamma}_{k+\sigma}(c)$, $\sigma = 1, \ldots, r$, lies in a fiber of the $\mathbb{T}^r$-principle bundle (2.4.17), and $\widetilde{\Gamma}_{k+\sigma}(c)$ is uniquely defined by a global section e and the fibered product of $\mathbb{S}^1$-principle bundles over Λ (2.4.23).

For any $m \in \mathbb{Z}^k$, $n \in \mathbb{Z}^r$, let us define

$$(3.5.24) \qquad \lambda_{m,n} = E - \hbar \sum_{\sigma=1}^{r} (n^\sigma + \tfrac{1}{4}\mu^\sigma) \cdot \sum_{i=1}^{k} \omega^i \beta_i^\sigma + \hbar \sum_{i=1}^{k} (m^i + \tfrac{1}{4}\mu^i)\omega^i,$$

where $\omega^1, \ldots, \omega^k$ are given by (3.4.8), $\beta_i^\sigma = \beta^\sigma(\Gamma_i)$ are the Floquet exponents of the coisotropic extension N at the basic 1-cycles Γ_i defined by (2.4.24), and $\mu^i = [\mu]\big(\widetilde{\Gamma}_i(c)\big)$, $i = 1, \ldots, k$, $\mu^\sigma = [\mu]\big(\widetilde{\Gamma}_\sigma(c)\big)$, $\sigma = 1, \ldots, r$, are values of the Maslov class. Here $\mu = \mu(T\Lambda(c), \mathcal{P})$ is the Arnold 1-form on the Lagrangian submanifold $\Lambda(c)$ that defines the Maslov class.

We claim that if integer vectors m and n satisfy conditions (3.5.7) and (3.5.20), then formula (3.5.24) gives the semiclassical eigenvalues $\lambda_{m,n}$ of the operator (3.5.1) modulo $O(\hbar^2)$. This fact follows from the Lagrangian version [**Ka**$_1$] of Proposition 3.5.3.

On the other hand, under the above assumptions we can apply Proposition 3.5.3 directly to the triple $(\Lambda, d\sigma, \rho)$, where the positive Lagrangian subbundle $\rho \subset {}^{\mathbb{C}}E$ from (2.4.10) and a flat Darboux atlas $\mathbb{A}$ over Λ define the $\mathbb{T}^r$-principle bundle (2.4.18).

Now we can prove that formula (3.5.22) gives the same asymptotic behavior of eigenvalues $\lambda_{m,n}$ as formula (3.5.24). Let us choose a $\mathbb{C}$-bilinear form **b** on ρ such that for a given global section e the 1-forms μ', $\mu''_{\mathbf{b}}$ from (3.3.26), (3.3.27) and the Arnold 1-form $\mu = \mu(T\Lambda(c), \mathcal{P})$ are related by the condition

$$[\mu''_{\mathbf{b}}] = [e_c^*(\mu) - \mu'] \in H^1(\Lambda, \mathbb{Z}).$$

Then the 1-form $\varkappa_{\mathbf{b}}$ (3.2.14) has the representation

$$\varkappa_{\mathbf{b}} = \frac{1}{2} \sum_{\sigma=1}^{r} \mu^\sigma \frac{\partial}{\partial c^\sigma}\big(e_c^*(\gamma_{\mathbb{A}})\big) + 4\pi\nu,$$

where μ^σ are the Maslov indices of 1-cycles $\widetilde{\Gamma}_{k+1}(c), \ldots, \widetilde{\Gamma}_{k+r}(c)$ on $\Lambda(c)$, $\gamma_{\mathbb{A}}$ is the Liouville 1-form on N from (2.4.24) associated with a flat Darboux atlas $\mathbb{A}$ $(d\gamma_{\mathbb{A}} = (dp \wedge dq)\big|_N)$, and ν is a closed 1-form on Λ with integer cohomology class, $[\nu] \in H^1(\Lambda, \mathbb{Z})$. Then one can easily see that formulas (3.5.22) and (3.5.24) coincide under an appropriate choice of integer vectors m.

References

[AbMa]   R. Abraham and J. E. Marsden, *Foundations of Mechanics*, 2nd Ed., Addison Wesley, Redwood City, 1985.

[A-S]   G. Alvarez-Sanchez, *Geometric Methods of Classical Mechanics Applied to Control Theory*, PhD Thesis, Univ. of California, Berkeley, 1986.

[A$_1$]   V. I. Arnold, *On a characteristic class entering into conditions of quantization*, Funktsional. Anal. i Prilozhen. **1** (1967), no. 1, 1–13; English transl. in Functional Anal. Appl. **1** (1967).

[A$_2$]   _____, *Modes and quasimodes*, Funktsional. Anal. i Prilozhen. **6** (1972), no. 2, 12–20; English transl. in Functional Anal. Appl. **6** (1972).

[A$_3$]   _____, *On a theorem of Liouville concerning integrable problems of dynamics*, Trans. Amer. Math. Soc. **61** (1967), 292–296.

[A$_4$]   _____, *Sturm theorems and symplecitci geometry*, Funktsional. Anal. i Prilozhen. **19** (1985), no. 4, 1–11; English transl. in Functional Anal. Appl. **19** (1985).

[AAv]   V. I. Arnold and A. Avez, *Ergodic Problems of Classical Mechanics*, Benjamin, New York, 1968.

[AG]   V. I. Arnold and A. B. Givental', *Symplectic geometry*, Itogi Nauki i Tekhniki: Sovremennye Problemy Mat.: Fundamental'nye Napravleniya, vol. 4, VINITI, Moscow, 1985, pp. 7–139; English transl. in Encyclopedia of Math. Sci., vol. 4 (Dynamical Systems, IV), Springer-Verlag, Berlin and New York, 1990.

[AKN]   V. I. Arnold, V. V. Kozlov, and A. I. Neistadt, *Mathematical aspects of classical and celestial mechanics*, Itogi Nauki i Tekhniki: Sovremennye Problemy Mat.: Fundamental'nye Napravleniya, vol. 3, VINITI, Moscow, 1985, pp. 5–302; English transl. in Encyclopedia of Math. Sci., vol. 3 (Dynamical Systems, III), Springer-Verlag, Berlin and New York, 1988.

[At]   M. Atiyah, *Convexity and commuting Hamiltonians*, Bull. London Math. Soc. **14** (1982), 1–15.

[AtBo] M. Atiyah and R. Bott, *The Yang–Mills equation over Riemann surfaces*, Philos. Trans. Roy. Soc. London Ser. A **308** (1982), 523–615.

[Ba]   V. M. Babich, *Eigenfunctions concentrated near closed geodesics*, Zap. Nauchn. Sem. LOMI **9** (1968); English transl., Sem. Math. V. A. Steklov Math. Inst. Leningrad, vol. 9, Plenum Press, New York, 1968.

[BaLa] V. M. Babich and V. F. Lazutkin, *On eigenfunctions concentrated near closed geodesics*, Problemy Mat. Fiz., vyp. 2, Izdat. Leningrad. Univ., Leningrad, 1967, pp. 15–25; English transl., Problems of Math. Phys., no. 2, Plenum Press, New York, 1968, pp. 9–18.

[BeDo$_1$] V. V. Belov and S. Yu. Dobrokhotov, *Canonical Maslov operator on isotropic manifolds with a complex germ and its applications to spectral problems*, Dokl. Akad. Nauk SSSR **298** (1988), 1037–1042; English transl. in Soviet Math. Dokl. **37** (1988).

[BeDo$_2$] _____, *Maslov's semiclassical asymptotics with complex phases. I: General approach*, Teoret. Mat. Fiz. **92** (1992), 215–254; English transl. in Theoret. and Math. Phys. **92** (1992).

[Bo$_1$]   R. Bott, *On a topological obstruction to integrability*, Global Analysis, Proc. Sympos. Pure Math., vol. 16, Amer. Math. Soc., Providence, RI, 1970, pp. 127-131.

[Bo$_2$]   R. Bott, *Lectures on Characteristic Classes and Foliations*, Lectures on Algebraic and Differential Topology (Second Latin Amer. School in Math., Mexico City, 1971), Lecture Notes in Math., vol. 279, Springer-Verlag, Berlin and New York, 1972, pp. 1–94.

[Bu]   V. S. Buslaev, *Quantization and WKB-method*, Trudy Mat. Inst. Steklov **110** (1970), 5–28; English transl. in Proc. Steklov Inst. Math. **110** (1972).

[BuNa] V. S. Buslaev and E. A. Nalimova, *Trace formula in general Hamiltonian mechanics*, Teoret. Mat. Fiz. **60** (1984), 344–355; English transl. in Theoret. and Math. Phys. **60** (1984).

[ChMo] B. Chern and J.M. Morvan, *Deformations of isotropic submanifolds in Kähler manifolds*, J. Geom. Phys. **13** (1994), 79–104.

[Che]   S. S. Chern, *Complex Manifolds without Potential Theory*, Van Nostrand Math. Studies, vol. 15, Van Nostrand, Princeton, NJ, 1967.

[CdV]   Y. Colin de Verdière, *Quasi-modes sur les variétés Riemanniennes*, Invent. Math. **43** (1977), 15–42.

[CFS]   I. P. Cornfeld, S. V. Fomin, and Ya. G. Sinai, *Ergodic Theory*, Springer-Verlag, Berlin and New York, 1982.

[Cou$_1$] T. J. Courant, *Dirac manifolds*, Trans. Amer. Math. Soc. **319** (1990), 631–661.

[Cou$_2$] _____, *Tangent Dirac structures*, J. Phys. A **23** (1990), 5153–5168.

[CRLj] S. C. Creagh, J. M. Robbins, and R. G. Littlejohn, *Geometrical properties of Maslov indices in the semiclassical trace formula for the density of states*, Phys. Rev. A **42** (1990), 1907–1922.

[Dz$_1$]   P. Dazord, *Sur la géométrie des sous-fibrés et des feuilletages lagrangiens*, Ann. Sci. École Norm. Sup. (4) **13** (1981), 465–480.

[Dz$_2$]   _____, *Feuilletages en géométrie symplectique*, C. R. Acad. Sci. Paris Sér. I Math. **294** (1982), 489–491.

[Dz$_3$]   _____, *Stabilité et linéarisation dans les variétés de Poisson*, Séminaire Sud-rhodanien de Géométrie, Rencontre de Balarué. I, Travaux en Cours, Hermann, Paris, 1985, pp. 59–75.

[DzDl] P. Dazord and T. Delzant, *Le problème général des variables actions-angles*, J. Differential Geom. **26** (1987), 223–251.

[Dl] T. Delzant, *Hamiltoniens périodiques et images convexes de l'application moment*, Bull. Soc. Math. France **116** (1988), 315–339.

[DKrV] S. Yu. Dobrokhotov, A. I. Krakhnov, and Yu. M. Vorobjev, *On topological quantization conditions for isotropic submanifolds of noncomplete dimension*, Internat. Topological Conf. Baku, Abstracts of Reports (1987), 75.

[DMasN] V. L. Dubnov, V. P. Maslov, and V. E. Nazaikinskii, *The complex Lagrangioan germ and the canonical operator*, Russian J. Math. Phys. **3** (1995), 141–190.

[DGMS] B. A. Dubrovin, M. Giordano, G. Marmo, and A. Simoni, *Poisson brackets on presymplectic manifolds*, Internat. J. Modern Phys. A **8** (1993), 3747–3771.

[DNF] B. A. Dubrovin, S. P. Novikov, and A. T. Fomenko, *Modern Geometry*, 2nd ed., Nauka, Moscow, 1986; English transl. of 1st ed., Parts I, II, Springer-Verlag, Berlin and New York, 1984, 1985.

[Du] J. J. Duistermaat, *On global action angle coordinates*, Comm. Pure Appl. Math. **33** (1980), 687–706.

[DuGu] J. J. Duistermaat and V. Guillemin, *The spectrum of positive elliptic operators and periodic bicharacteristics*, Invent. Math. **29** (1975), 39–79.

[E] L. H. Eliasson, *Floquet solutions for the 1-dimensional quasi-periodic Schrödinger equation*, Comm. Math. Phys. **146** (1992), 447–482.

[GL] I. M. Gelfand and V. B. Lidskii, *On the structure of stability domains of linear canonical systems of differential equations with periodic coefficients*, Uspekhi Mat. Nauk **10** (1955), no. 1, 3–40; English transl., Amer. Math. Soc. Transl. (2) **8** (1958), 143–181.

[G] A. B. Givental, *Periodic mappings in symplectic topology*, Funktsional Anal. i Prilozhen. **23** (1989), no. 4, 37–52; English transl. in Functional Anal. Appl. **23** (1989).

[Go] M. J. Gotay, *On coisotropic embeddings of presymplectic manifolds*, Proc. Amer. Math. Soc. **84** (1982), 111–114.

[GrPa] S. Graffi and A. Parmeggiani, *Quantum evolution and classical flow in complex phase space*, Comm. Math. Phys. **128** (1990), 393–409.

[GHV] W. Greub, S. Halperin, and R. Vanstone, *Connections, Curvature, and Cohomology*. Vol. II, Academic Press, New York–London, 1973.

[Gu$_1$] V. Guillemin, *Moment Maps and Combinatorial Invariants of Hamiltonian T^n-Spaces*, Progress in Mathematics, vol. 122, Birkhäuser, Boston, 1994.

[Gu$_2$] _____, *Symplectic spinors and partial differential equations*, Géométrie Symplectique et Physique Mathématique, Colloques Internat. du C.N.R.S., vol. 237, Éditions CNRS, Paris, 1975, pp. 217–252.

[GuSt$_1$] V. Guillemin and S. Sternberg, *Convexity properties of the moment mapping*, Invent. Math. **67** (1982), 491–513.

[GuSt$_2$] _____, *Symplectic Technique in Physics*, Cambridge Univ. Press, Cambridge, 1984.

[GuSt$_3$] _____, *Geometric Asymptotics*, Math. Surveys and Monographs, vol. 14, Amer. Math. Soc., Providence, RI, 1977.

[GuWe] V. Guillemin and A. Weinstein, *Eigenvalues associated with closed geodesics*, Bull. Amer. Math. Soc. **82** (1976), 92–94.

[Gut] M. Gutzwiller, *Periodic orbits and classical quantization conditions*, J. Math. Phys. **12** (1971), 343–358.

[Hel] E. J. Heller, *Time dependent approach to semiclassical dynamics*, J. Chem. Phys. **62** (1975), 1544–1555.

[He] H. Hess, *Connections on symplectic manifolds and geometric quantization*, Differential-Geometric Methods in Math. Phys. (Proc. Conf., Aix-en-Provence/Salamanca, 1979), Lecture Notes in Math., vol. 836, Springer-Verlag, Berlin and New York, 1980, pp. 153–166.

[Jo] R. A. Johnson, *On a Floquet theory for almost periodic two-dimensional linear systems*, J. Differential Equations **37** (1980), 184–205.

[JM] R. A. Johnson and J. Moser, *The rotation number for almost periodic potentials*, Comm. Math. Phys. **84** (1982), 403–438.

[JoNer] R. A. Johnson and M. Nerurkar, *Exponential dichotomy and rotation number for linear Hamiltonian systems*, J. Differential Equations **108** (1994), 201–216.

[JoSe] R. A. Johnson and G. R. Sell, *Smoothness of spectral subbundles and reducibility of quasiperiodic linear differential systems*, J. Differential Equations **41** (1981), 262–288.

[Jos] R. Jost, *Winkel- und Wirkungsvariable für allgemeine mechanische Systeme*, Helv. Phys. Acta **41** (1968), 965–968.

[Ka₁] M. V. Karasev, *Connections on Lagrangian submanifolds and certain problems of the semiclassical approximation theory*, Zap. Nauchn. Sem. LOMI **172** (1989), 41–54; English transl., J. Soviet Math. **10** (1992), no. 5, 1053–1062.

[Ka₂] _____, *Simple quantization formula*, Symplectic Geometry and Mathematical Physics, Actes du colloque en l'honneur de J.-M. Souriau (P. Donato and others, eds.), Birkhäuser, Basel–Boston, 1991, pp. 234–243.

[Ka₃] _____, *Quantization and coherent states over Lagrangian submanifolds*, Russian J. Math. Phys. **3** (1995), 393–400.

[Ka₄] _____, *Quantization by means of two-dimensional surfaces (membranes). Geometrical formulas for wave-functions*, Contemp. Math. **179** (1994), 83–113.

[Ka₅] _____, *New global asymptotics and anomalies in the problem of quantization of the adiabatic invariant*, Funktsional. Anal. i Prilozhen. **24** (1990), no. 2, 24–36; English transl., Functional Anal. Appl. **24** (1990), 104–114.

[Ka₆] _____, *To the Maslov theory of quasiclassical asymptotics. Examples of new global quantization formula applications*, Preprint No. ITP-89-78E, Inst. Theoret. Phys., Kiev, 1989.

[Ka₇] _____, *Integrals over membranes, transition amplitudes and quantization*, Russian J. Math. Phys. **1** (1993), 523–526.

[KaKz₁] M. V. Karasev and M. V. Kozlov, *Exact and semiclassical representation over Lagrangian submanifolds in $su(2)^*$, $so(4)^*$, and $su(1,1)^*$*, J. Math. Phys. **34** (1993), 4986–5006.

[KaKz₂] _____, *Representations of compact semisimple Lie algebras over Lagrangian submanifolds*, Funktsional. Anal. i Prilozhen. **28** (1994), no. 4, 16–27; English transl., Functional Anal. Appl. **28** (1994), 238–246.

[KaMas] M. V. Karasev and V. P. Maslov, *Nonlinear Poisson Brackets. Geometry and Quantization*, Nauka, Moscow, 1991; English transl., Transl. Math. Monographs, vol. 119, Amer. Math. Soc., Providence, RI, 1993.

[KaVo₁] M. V. Karasev and Yu. M. Vorobjev, *Integral representations over isotropic submanifolds and equations of zero curvature invariant*, preprint MIEM, A Math-QDS-92-01, Moscow (1992); Adv. Math. (to appear).

[KaVo₂] _____, *Linear connections for Hamiltonian dynamics over isotropic submanifolds*, Progress in Nonlinear Differential Equations and their Applications, vol. 12, Birkhäuser, 1994, pp. 235–252.

[KaVo₃] _____, *Symplectic curvature and Arnold form over isotropic submanifolds*, J. Math. Sci. **82** (1996), 3789–3799.

[KaVo₄] _____, *Quasimodes generated by characters of a dynamical group and deformation of the Kirillov form*, Funktsional Anal. i Prilozhen. **26** (1992), no. 1, 71–73; English transl., Functional Anal. Appl. **26** (1992), 57–59.

[KaVo₅] _____, *On an analog of Maslov's class in the non-Lagrangian case*, Problems of Geometry, Topology, and Mathematical Physics, Novoe v Global. Anal., vol. 12, Voronezh Gos. Univ., Voronezh, 1992, pp. 37–48. (Russian)

[KaVo₆] _____, *Connections and excited wavepackets over invariant isotropic tori*, Quantization and Coherent States Methods (A. T. Ali, I. Mladenov, and A. Odzijewicz, eds.), World Scientific, Singapore, 1993, pp. 179–189.

[KaVo₇] _____, *Symplectic connections, quantization over isotropic submanifolds, and Dirac axioms*, Quantization, Coherent States, and Poisson Structures (A. Strasburger et al., eds.), PWN, Warsaw, to appear.

[KaVo₈] _____, *Hermitian bundles over isotropic submanifolds and correction to Kostant–Souriau quantization rule*, Preprint No. ITP-90-85E, Inst. Theoret. Phys., Kiev, 1991.

[Kel] J. B. Keller, *Corrected Bohr–Sommerfeld quantum conditions for nonseparable systems*, Ann. of Physics **4** (1958), 180–188.

[Ki] A. A. Kirillov, *Geometric quantization*, Itogi Nauki i Tekhniki: Sovremennye Problemy Mat.: Fundamental′nye Napravleniya, vol. 4, VINITI, Moscow, 1985, pp. 141–178; English transl. in Encyclopedia of Math. Sci., vol. 4 (Dynamical Systems, IV), Springer-Verlag, Berlin and New York, 1990.

[Kle]   J. Klein, *Structures symplectiques ou J-symplectiques homogènes sur l'espace tangent à une variété*, Sympos. Math (INDAM), vol. XIV, Academic Press, London, 1974, pp. 181–192.

[Kra$_1$] A. D. Krakhnov, *Eigenfunctions concentrated near a conditionally periodic geodesic*, Methods of the Qualitative Theory of Differential Equations, vol. 1, Izd. Gor'kov. Univ., Gor'ki, 1975, pp. 75–87. (Russian)

[Kra$_2$] _____, *Asymptotic eigenvalues of pseudodifferential operators and invariant tori*, Uspekhi Mat. Nauk **31** (1976), no. 3, 217–218. (Russian)

[Lew]   H. R. Lewis, *Class of exact invariants for classical and quantum time-dependent harmonic oscillators*, J. Math. Phys. **9** (1968), 1976–1986.

[Li]   P. Libermann, *Sur le problème d'équivalence de certaines structures infinitésimales*, Ann. Mat. Pure Appl. (4) **36** (1954), 27–120.

[LiMr]   P. Libermann and C.-M. Marle, *Symplectic Geometry and Analytical Mechanics*, Reidel, Dordrecht, 1987.

[L$_1$]   A. Lichnerowicz, *Feuilletages, géométrie riemannienne et géométrie symplectique*, Riv. Mat. Univ. Parma (4) **10** (1984), 81–90.

[L$_2$]   _____, *Les variétés de Poisson et leurs algèbres de Lie associées*, J. Differential Geom. **12** (1977), 253–300.

[Lj]   R. G. Littlejohn, *Symplectically invariant WKB wave functions*, Phys. Rev. Lett. **54** (1985), 1742–1745.

[Mck]   K. C. H. Mackenzie, *Lie Groupoids and Lie Algebroids in Differential Geometry*, London Math. Soc. Lecture Note Ser., vol. 124, Cambridge Univ. Press, Cambridge, 1987.

[MRR]   J. Marsden, T. Ratiu, and G. Raugel, *Symplectic connections and the linearization of Hamiltonian systems*, Proc. Roy. Soc. Edinburgh Sect. A **117** (1991), 329–380.

[Mas]   V. P. Maslov, *The Complex WKB Method for Nonlinear Equations*, Nauka, Moscow, 1977; English transl., Birkhäuser, Basel–Boston, 1994.

[MasF]   V. P. Maslov and M. V. Fedoryuk, *Semiclassical Approximation in Quantum Mechanics*, Nauka, Moscow, 1976; English transl., Reidel, Dordrecht, 1981.

[Mel]   V. K. Melnikov, *On some cases of conservation of quasi-periodic motion under a small change of the Hamiltonian function*, Soviet Math. Dokl. **6** (1965), 1592–1596.

[MeHl]   K. R. Meyer and G. R. Hall, *Introduction to Hamiltonian Dynamical Systems and the N-Body Problem*, Springer-Verlag, Berlin and New York, 1992.

[MiFm]   A. S. Mishchenko and A. T. Fomenko, *Generalized Liouville integration method for Hamiltonian systems*, Funktsional. Anal. i Prilozhen. **12** (1978), no. 2, 46–56; English transl. in Functional Anal. Appl. **12** (1978).

[MiSSh]   A. S. Mishchenko, B. Yu. Sternin, and V. E. Shatalov, *The geometry of complex phase space and Maslov's canonical operator*, Itogi Nauki i Tekhniki: Sovremennye Problemy Mat., vol. 8, VINITI, Moscow, 1977, pp. 1–40; English transl., J. Soviet Math. **13** (1980), 1–23.

[M]   J. Moser, *On the volume element on a manifold*, Trans. Amer. Math. Soc. **120** (1965), 286–294.

[NVo]   V. E. Nazaikinskii and Yu. M. Vorobjev, *On the cycle of singularities of isotropic submanifolds*, Russian J. Math. Phys. **5** (1997), 131–135.

[Ne$_1$]   N. N. Nekhoroshev, *Action-angle variables and their generalizations*, Trudy Moskov. Mat. Ob. **26** (1972), 180–198; English transl. in Trans. Moscow Math. Soc. **26** (1972).

[Ne$_2$]   _____, *The Poincaré–Lyapunov–Liouville–Arnol'd theorem*, Funktsional. Anal. i Prilozhen. **28** (1994), no. 2, 128–129; English transl. in Functional Anal. Appl. **28** (1994).

[Nov]   S. P. Novikov, *Hamiltonian formalism and many-valued analog of the Morse theory*, Uspekhi Mat. Nauk **37** (1982), no. 5, 3–49; English transl. in Russian Math. Surveys **37** (1982).

[PU]   T. Paul and A. Uribe, *The semi-classical trace formula and propagation of wave packets*, J. Funct. Anal. **132** (1995), 192–242.

[Pr]   J. Pradines, *Théorie de Lie pour les groupoïdes des différentiables*, C. R. Acad. Sci. Paris Sér. A **264** (1967), 245–248.

[Ral]   J. V. Ralston, *On the construction of quasimodes associated with stable periodic orbits*, Comm. Math. Phys. **51** (1976), 219–242.

[Sev]   M. V. Sevryuk, *Lower-dimensional tori in reversible systems*, Chaos **1** (1991), 160–167.

[Shub] M. A. Shubin, *Pseudodifferential Operators and Spectral Theory*, Springer-Verlag, Berlin and New York, 1987.
[SLe] R. Sjamaar and E. Lerman, *Stratified symplectic spaces and reduction*, Ann. of Math. **134** (1991), 375–422.
[St] S. Sternberg, *Minimal coupling and the symplectic mechanics of a classical particle in the presence of a Yang–Mills field*, Proc. Nat. Acad. Sci. U.S.A. **74** (1977), 5253–5254.
[Tab] S. Tabachnikov, *Geometry of Lagrangian and Legendrian 2-web*, Diff. Geom. Appl. **3** (1993), 265–284.
[Tam] I. Tamura, *Topology of Foliations: An Introduction*, Amer. Math. Soc., Providence, RI, 1992.
[Vais] I. Vaisman, *Second-order Hamiltonian vector fields on tangent bundles*, Diff. Geom. Appl. **5** (1995), 153–170.
[Vo$_1$] Yu. M. Vorobjev, *The Maslov complex germ generated by linear connections*, Mat. Zametki **48** (1990), no. 6, 29–37; English transl., Math. Notes **48** (1990), 1191–1197.
[Vo$_2$] _____, *The Riccati equation over a torus and semiclassical quantization of multiperiodic motion*, Quantization and Infinite-Dimensional Systems (J.-P. Antoine et al., eds.), Plenum, New York, 1994, pp. 205–212.
[VoIt] Yu. M. Vorobjev and V. M. Itskov, *On quasimodes associated with quasiperiodic motion of stable type*, Mat. Zametki **55** (1994), no. 5, 36–42; English transl., Math. Notes **55** (1994), 461–468.
[Vor] A. Voros, *Wentzel–Kramers–Brillouin method in the Bargmann representation*, Phys. Rev. A **40** (1989), 6814–6825.
[We$_1$] A. Weinstein, *Symplectic manifolds and their Lagrangian submanifolds*, Adv. in Math. **6** (1971), 329–346.
[We$_2$] _____, *Lectures on Symplectic Manifolds*, Amer. Math. Soc., Providence, RI, 1977.
[We$_3$] _____, *Neighborhood classification of isotropic embeddings*, J. Differential Geom. **16** (1981), 125–128.
[We$_4$] _____, *A universal phase space for particles in Yang–Mills fields*, Lett. Math. Phys. **2** (1978), 117–120.
[Wo] N. M. J. Woodhouse, *Geometric Quantization*, Clarendon Press, Oxford, 1992.
[YStz] V. A. Yakubovich and V. M. Starzhiuskii, *Linear Differential Equations with Periodic Coefficients*. Vols. I, II, Wiley, New York, 1975.

DEPARTMENT OF APPLIED MATHEMATICS, MOSCOW STATE INSTITUTE OF ELECTRONICS AND MATHEMATICS, B.TREKHSVYATITEL'SKII, 3/12, MOSCOW 109028, RUSSIA
E-mail address: vorob@amath.msk.ru

Translated by the authors

Infinitesimal Poisson Cohomology

Vladimir Itskov, Mikhail Karasev, and Yurii Vorobjev

ABSTRACT. The cohomology spaces arising in the infinitesimal Poisson geometry are considered in the Lie algebroids framework. It is shown that Lie algebroid structure can be reduced to the vector bundle over an integral leaf of the anchor image. In particular, the Schouten–Lichnerowicz complex can be reduced to a complex associated with the normal vector bundle over each symplectic leaf. The reduced Lie algebroid is transitive and its cohomology is finite-dimensional.

The calculation of cohomology with trivial coefficients for transitive Lie algebroids is carried out following the general approach due to Mackenzie. Simple formulas are given in some particular cases. In the case of Abelian algebroids, a classification theorem and formulas for some cohomology spaces are given.

A formal equivalence problem is considered for two Poisson brackets having the same symplectic leaf. Sufficient conditions for the formal equivalence are given in terms of certain cohomology spaces of a transitive Lie algebroid.

Contents

Introduction
§1. Coboundary Lichnerowicz operator over a symplectic leaf
§2. Reduction of Lie algebroids
§3. Connections, curvature, and characteristic classes in transitive Lie algebroids
§4. Calculation of cohomology of a transitive Lie algebroid
§5. Abelian transitive algebroids
§6. Cohomology of homogeneous algebroids, examples
§7. Formal equivalence
Appendix: De Rham cohomology of flat connections
References

1991 *Mathematics Subject Classification*. Primary 58F05, 58H15; Secondary 53C05.

This research was partially supported by the Russian Foundation for Basic Research under grant No. 96-01-01426, and by the Ministry of General and Professional Education of the Russian Federation under grant No. 95-0-1.7-91

©1998 American Mathematical Society

Introduction

The Poisson cohomology introduced by A. Lichnerowicz [**Li**$_2$] occurs in many problems of Hamiltonian mechanics and quantization [**BFFLS, Hu, KM, Li**$_2$, **Va, VK**$_{1,2}$]. For regular Poisson manifolds, the method for computation of this cohomology was first developed in [**VK**$_{1,2}$]; see also [**KM, Va, Xu**]. In the case of an action of a compact Poisson group, the Poisson cohomology was computed in [**Gi**]. The problem of calculating the Poisson cohomology for general (irregular) Poisson manifolds seems to be very attractive but rather difficult [**VK**$_3$].

It turns out that the Lichnerowicz differential can be restricted to the complex associated with the normal vector bundle over each symplectic leaf of the Poisson manifold. The relationship between reduced cohomology and the Poisson cohomology of the whole manifold is the first interesting problem in this framework (it is not studied in the present paper). We are interested in another problem: how to calculate the reduced infinitesimal cohomology itself. Note that the infinitesimal Poisson cohomology appears in investigation of linearized Hamiltonian dynamics, deformation of Poisson bracket, and normal forms over a given symplectic leaf.

As shown in [**BV, Hu**], the Poisson calculus fits in the more general context of Lie algebroids. The reduction process and the calculation of the restricted cohomology have no specific features of the Poisson setting. The restricted Poisson calculus falls in the general scheme of transitive Lie algebroids, and the restricted Poisson cohomology is a cohomology with trivial coefficients of a certain transitive Lie algebroid.

The general procedure for calculating the cohomology of transitive Lie algebroids was developed by Mackenzie [**Mz**]. In the present paper we use a slightly modified approach and develop it for the case of trivial coefficients. For a special class of *quasiparallelizable* Lie algebroids we obtain simple formulas that are similar to formulas for the spectral sequence of a fiber bundle. We also pay special attention to Abelian algebroids because they correspond to the nondegenerate symplectic leaves and play a significant role in calculations related to non-Abelian algebroids. Since in some applications it is also significant to know the space of cocycles, we describe this space (of various dimensions) in the case of transitive Lie algebroids.

In the present paper we also treat the problem of equivalence of Poisson brackets having the same symplectic leaf. The first step is the classification up to formal equivalence, i.e., equivalence of Poisson tensors considered as formal power series in coordinates along the directions normal to the symplectic leaf. This problem has been investigated in detail only in the case when the symplectic leaf is a single point (see [**Co**$_{1,2}$, **Du, Ly**]). We consider a formal equivalence problem for brackets having the same transitive Lie algebroid structure over the symplectic leaf. We give a sufficient condition for equivalence of such Poisson brackets in terms of cohomology of that transitive Lie algebroid with coefficients in a certain vector bundle.

The paper is organized as follows.

In §1 we give the necessary facts from Poisson calculus.

In §2 we give definitions for Lie algebroids and prove the restriction theorem.

In §3 we discuss some facts from the theory of connections in transitive Lie algebroids and introduce characteristic classes.

In §4 we calculate the cohomology of non-Abelian transitive Lie algebroids. For the quasiparallelizable Lie algebroids we present a simple formula for the second term of the spectral sequence in terms of the cohomology of a finite-dimensional normal algebra and the cohomology of a certain Abelian algebroid or the de Rham cohomology of the base. We show some applications of this formula and give sufficient conditions for the quasiparallelizability of a transitive Lie algebroid. We also describe the one- and two-dimensional spaces of cocycles.

In §5 we study the structure and the cohomology of Abelian algebroids. We present a classification theorem and calculate the space of cocycles and the cohomology space.

In §6 we discuss some properties of homogeneous Lie algebroids. By using the methods developed in §§ 4 and 5, we also calculate the cohomology for homogeneous algebroids over two different types of orbits of the coadjoint action of $E(3)$.

In §7 we consider a formal equivalence problem for Poisson brackets inducing the same transitive Lie algebroid structure over a symplectic leaf. We give a sufficient condition for formal equivalence in terms of cohomology classes of the transitive Lie algebroid with coefficients in a symmetric power of the normal algebra bundle. We also discuss calculation of these cohomology spaces.

§1. Coboundary Lichnerowicz operator over a symplectic leaf

All geometric objects considered in the present paper are infinitely differentiable, and we also assume that all linear spaces are over real numbers (although all the results remain valid over any field of zero characteristic).

Let $\mathcal{N}$ be a manifold equipped with the *Poisson bracket*

$$\{f, g\} = \Psi(df, dg), \qquad f, g, \in C^\infty(\mathcal{N}),$$

where Ψ is the corresponding antisymmetric tensor field, called sometimes a Poisson tensor [**MR**]. The tensor Ψ defines a morphism of vector bundles $q\colon T^*\mathcal{N} \to T\mathcal{N}$,

(1.1) $$\langle \beta_2, q\beta_1 \rangle \stackrel{\text{def}}{=} \Psi(\beta_1, \beta_2), \qquad \beta_1, \beta_2 \in \Gamma(T^*\mathcal{N}).$$

(Here $\langle \cdot, \cdot \rangle$ is the pairing of 1-forms and vector fields.) It is well known [**Do, Ka, Ko**] that the Poisson bracket on functions can be extended to a bracket on 1-forms $\{,\}\colon \Gamma(T^*\mathcal{N}) \times \Gamma(T^*\mathcal{N}) \to \Gamma(T^*\mathcal{N})$,

(1.2) $$\{\beta_1, \beta_2\} \stackrel{\text{def}}{=} \mathcal{L}_{q\beta_1} \beta_2 - \mathcal{L}_{q\beta_2} \beta_1 - d\Psi(\beta_1, \beta_2),$$

where $\mathcal{L}_{q\beta_1}\beta_2$ denotes the Lie derivative of the differential form β_2 along the vector field $q\beta_1$. The bracket (1.2) satisfies the Jacobi identity and endows the space of 1-forms with a Lie algebra structure; the mapping q is a homomorphism of this algebra into the Lie algebra of vector fields:

(1.3) $$q\{\beta_1, \beta_2\} = [q\beta_1, q\beta_2].$$

Moreover, we have the identities

(1.4) $$d\{f, g\} = \{df, dg\},$$
(1.5) $$\{\beta_1, f\beta_2\} = f\{\beta_1, \beta_2\} + (\mathcal{L}_{q\beta_1}(f))\beta_2, \qquad f, g, \in C^\infty(\mathcal{N}).$$

By $\mathcal{V}^k(\mathcal{N})$ we denote the space of contravariant antisymmetric tensor fields of the type $(k, 0)$ on the manifold $\mathcal{N}$. The *Schouten bracket* [**Sn**]

$$[\![\,,\,]\!]\colon \mathcal{V}^k(\mathcal{N}) \times \mathcal{V}^l(\mathcal{N}) \to \mathcal{V}^{k+l-1}(\mathcal{N})$$

defines the structure of a Lie superalgebra on $\mathcal{V}^{\cdot}(\mathcal{N}) \stackrel{\text{def}}{=} \bigoplus_{l=0}^{\dim \mathcal{N}} \mathcal{V}^k(\mathcal{N})$. It is well known [Li$_2$] that the Poisson tensor satisfies the condition $[\![\Psi, \Psi]\!] = 0$ and defines a *coboundary operator* $D: \mathcal{V}^k(\mathcal{N}) \to \mathcal{V}^{k+1}(\mathcal{N})$,

(1.6) $$DQ \stackrel{\text{def}}{=} -[\![\Psi, Q]\!], \qquad Q \in \mathcal{V}^k(\mathcal{N}), \qquad D^2 = 0.$$

Note (see, for example, [KSM]) that the operator (1.6) is a standard coboundary operator [F] associated with the representation of the Lie algebra of differential forms in the ring $C^\infty(\mathcal{N})$:

$$DQ(\beta_0, \beta_1, \ldots, \beta_k) = \sum_{j=0}^{k} (-1)^j \mathcal{L}_{q\beta_j}\big(Q(\beta_0, \beta_1, \ldots, \widehat{\beta}_j, \ldots, \beta_k)\big)$$
$$+ \sum_{0 \le i < j \le k} (-1)^{i+j} Q(\{\beta_i, \beta_j\}, \beta_0, \beta_1, \ldots, \widehat{\beta}_i, \ldots, \widehat{\beta}_j, \ldots, \beta_k).$$

(Here $Q \in \mathcal{V}^k(\mathcal{N})$, $\beta_j \in \Gamma(T^*\mathcal{N})$, and the "hat" stands over omitted terms.)

Let $\mathcal{O}$ be a *symplectic leaf* of the Poisson bracket (1.0), i.e., an integral leaf of the characteristic distribution $\operatorname{Ran}(q)$ [We]. The embedding $\mathcal{O} \hookrightarrow \mathcal{N}$ defines a vector bundle $T^*_{\mathcal{O}}\mathcal{N}$ over a symplectic leaf $\mathcal{O}$ with fiber $T^*_x\mathcal{N}$ ($x \in \mathcal{O}$). A restriction of differential forms to the leaf $\mathcal{O}$ is denoted by $r_{\mathcal{O}}: \Gamma(T^*\mathcal{N}) \to \Gamma(T^*_{\mathcal{O}}\mathcal{N})$. Similarly, we define the vector bundles $\bigwedge^k T_{\mathcal{O}}\mathcal{N}$ and the restriction operation $r^*_{\mathcal{O}}: \mathcal{V}^k(\mathcal{N}) \to \bigwedge^k T_{\mathcal{O}}\mathcal{N}$.

PROPOSITION 1.1. *The operation* $\{,\}_{\mathcal{O}}$ *is defined on sections of the bundle* $T^*_{\mathcal{O}}\mathcal{N}$, *and this operation gives a natural restriction of the bracket* (1.2),

$$r_{\mathcal{O}}\{\beta_1, \beta_2\} = \{r_{\mathcal{O}}\beta_1, r_{\mathcal{O}}\beta_2\}_{\mathcal{O}}.$$

COROLLARY 1.2. *The coboundary operator* $D_{\mathcal{O}}: \Gamma(\bigwedge^k T_{\mathcal{O}}\mathcal{N}) \to \Gamma(\bigwedge^{k+1} T_{\mathcal{O}}\mathcal{N})$ *that is a restriction of the Lichnerowicz operator* (1.6),

$$r^*_{\mathcal{O}} D = D_{\mathcal{O}} r^*_{\mathcal{O}}, \qquad D^2_{\mathcal{O}} = 0,$$

is defined on the sections of the bundle $\bigwedge^k T_{\mathcal{O}}\mathcal{N}$.

REMARK 1.3. This restriction can be performed over any Poisson submanifold.

The proof of Proposition 1.1 follows from the more general Theorem 2.1 in the next section.

§2. Reduction of Lie algebroids

A *Lie algebroid* [Mz, Pr] is a triple $(A \to B, q, \{,\})$, where $A \to B$ is a vector bundle, $q: A \to TB$ is a morphism of vector bundles, and $\{,\}: \Gamma(A) \times \Gamma(A) \to \Gamma(A)$ is a bilinear operation on sections that satisfies the following properties:

• the space of sections of the bundle A forms a Lie algebra under the operation $\{,\}$;

• the mapping $q: \Gamma(A) \to \mathcal{V}^1(B)$ is a homomorphism of Lie algebras:

(2.1) $$q\{\alpha_1, \alpha_2\} = [q\alpha_1, q\alpha_2], \qquad \alpha_1, \alpha_2 \in \Gamma(A);$$

• any two sections $\alpha_1, \alpha_2 \in \Gamma(A)$ and each function $\varphi \in C^\infty(B)$ satisfy the relation

(2.2) $$\{\alpha_1, \varphi\alpha_2\} = \varphi\{\alpha_1, \alpha_2\} + \bigl(\mathcal{L}_{q\alpha_1}(\varphi)\bigr)\alpha_2.$$

A Lie algebroid is called *transitive* if the morphism q is a surjection.

EXAMPLE 1. Obviously, the triple $(T^*\mathcal{N} \to \mathcal{N}, q, \{\,,\,\})$, where $\mathcal{N}$ is a Poisson manifold, q is defined by (1.1), and $\{\,,\,\}$ is defined by (1.2), is a Lie algebroid.

EXAMPLE 2 (Transformation algebroid [**Mz**]). Suppose that an action of a finite dimensional Lie algebra $(G, [\,]_G)$ is given on the manifold B, that is, a homomorphism $q\colon G \to \mathcal{V}^1(B)$ is defined. The triplet $(G \times B \to B, q, \{\,,\,\})$ is a Lie algebroid with commutator on sections defined as

(2.3) $$\{\alpha_1, \alpha_2\} = [\alpha_1, \alpha_2]_G + \nabla^0_{q\alpha_1}\alpha_2 - \nabla^0_{q\alpha_2}\alpha_1.$$

(Here ∇^0 is a flat connection in the vector bundle $G \times B \to B$.)

An equivalent of the Poisson cohomology for a Lie algebroid is the cohomology of an infinite-dimensional Lie algebra $\Gamma(A)$ with coefficients in the $\Gamma(A)$-module $C^\infty(B)$ (i.e., the cohomology that corresponds to the representation $\alpha \mapsto \mathcal{L}_{q\alpha}$ in a one-dimensional trivial bundle).

Let us consider a standard complex [**F, Mz**] $C^k(A) \stackrel{\text{def}}{=} \Gamma(\wedge^k A^*)$, $k \geq 0$, and the differential operator $D\colon C^k(A) \to C^{k+1}(A)$,

(2.4) $$Df(\alpha_0, \alpha_1, \ldots, \alpha_k) \stackrel{\text{def}}{=} \sum_{j=0}^k (-1)^j \mathcal{L}_{q\alpha_j}\bigl(f(\alpha_0, \alpha_1, \ldots, \widehat{\alpha}_j, \ldots, \alpha_k)\bigr)$$
$$+ \sum_{0 \leq i < j \leq k} (-1)^{i+j} f\bigl(\{\alpha_i, \alpha_j\}, \alpha_0, \alpha_1, \ldots, \widehat{\alpha}_i, \ldots, \widehat{a}_j, \ldots, \alpha_k\bigr).$$

(Here $f \in C^k(A)$, $\alpha_j \in \Gamma(A)$, and the hat stands over omitted terms.) We denote the cocycles and the coboundaries of the operator (2.4) by $Z^k(A)$ and $B^k(A)$, respectively. The cohomology of the Lie algebroid [**Mz**] is defined as

$$\mathcal{H}^k(A) = Z^k(A)/B^k(A).$$

Note that on the cochains $C^k(A)$ the standard Grassmann multiplication is defined, with respect to which the operator (2.4) is antiderivation and the cohomology $\mathcal{H}^\cdot(A) \stackrel{\text{def}}{=} \bigoplus_k \mathcal{H}^k(A)$ inherits the structure of the Grassmann algebra [**F**]. The mapping $q^*\colon \Gamma(\wedge^k T^*B) \to C^k(A)$ is a homomorphism of Grassmann differential algebras (i.e., $q^*(\omega_1 \wedge \omega_2) = (q^*\omega_1) \wedge (q^*\omega_2)$ and $Dq^* = q^*d$), and thus, defines a homomorphism of Grassmann algebras $q^\dagger\colon H^\cdot(B) \to \mathcal{H}^\cdot(A)$.

For each submanifold $M \hookrightarrow B$, by $A_M \to M$ we denote the vector bundle obtained by the restriction of the bundle $A \to B$ to the base M; by $r_M\colon \Gamma(A) \to \Gamma(A_M)$ and $r_M^*\colon C^k(A) \to C^k(A_M)$ we denote the restrictions of sections of the corresponding bundles.

Let us consider a distribution $\mathcal{D}_x \stackrel{\text{def}}{=} (\operatorname{Ran} q_{|x}) \subset T_xB$. An *integral leaf* is a submanifold $\mathcal{O} \hookrightarrow B$ such that $T_y\mathcal{O} = \mathcal{D}_y$ for any point $y \in \mathcal{O}$. We shall say that an integral leaf $\mathcal{O}$ is *tame* if each section $\alpha \in \Gamma(A_\mathcal{O})$ of the restricted vector bundle has an extension $\widetilde{\alpha} \in \Gamma(A)$, $\alpha = r_\mathcal{O}\widetilde{\alpha}$.

Obviously, each compact leaf is tame. The orbit of a coadjoint action in a "wild" Lie algebra [**Ki**$_1$] is an example of an integral leaf of the distribution $\operatorname{Ran} q$ that is not tame.

THEOREM 2.1. *Let the triple $(A \to B, q, \{\, ,\, \})$ be a Lie algebroid and let $\mathcal{O}$ be a tame integral leaf of the distribution $\operatorname{Ran} q$. Then there is a natural structure of a transitive Lie algebroid $(A_\mathcal{O} \to \mathcal{O}, q, \{\, ,\, \}_\mathcal{O})$ such that*

$$(2.5) \qquad r_\mathcal{O}\{\alpha_1, \alpha_2\} = \{r_\mathcal{O}\alpha_1, r_\mathcal{O}\alpha_2\}_\mathcal{O}.$$

COROLLARY 2.2. *Let $D_\mathcal{O}$ be the coboundary operator (2.4) in the Lie algebroid $A_\mathcal{O}$. Then $r_\mathcal{O}^* D = D_\mathcal{O} r_\mathcal{O}^*$, and the homomorphism of Grassmann algebras $r_\mathcal{O}^\dagger : \mathcal{H}^\cdot(A) \to \mathcal{H}^\cdot(A_\mathcal{O})$ is well defined.*

REMARK 2.3. *The theorem remains true for each tame submanifold $M \hookrightarrow B$ such that $\mathcal{D}_x \subseteq T_x M$ for each $x \in M$.*

To prove Theorem 2.1 we need the following observation.

LEMMA 2.4. *Suppose that a submanifold $M \hookrightarrow B$ satisfies the condition $\mathcal{D}_x \subseteq T_x M$ for each point $x \in M$. Then the linear subspace $J_M \stackrel{\text{def}}{=} \operatorname{Ker} r_M$ is an ideal in the Lie algebra $\Gamma(A)$. Moreover, the linear subspace $C_0^\cdot(A, M) \stackrel{\text{def}}{=} \operatorname{Ker} r_M^*$ is invariant under the action of the differential operator (2.4).*

PROOF. It suffices to prove this statement for a small open domain $U \subset B$ such that $U \cap M \neq \varnothing$. In the domain U there is a basis of sections $X_1, \ldots, X_n \in \Gamma(A_U)$; the section $\alpha_0 = \sum_{i=1}^n a_i(x) X_i$ belongs to the space J_M if and only if $a_i(x)\big|_{x \in M} = 0$. For any section $\alpha \in \Gamma(A_U)$

$$r_M\{\alpha, \alpha_0\} = \sum_{i=1}^n \left(r_M a_i(x)\{\alpha, X_i\} + r_M \mathcal{L}_{q\alpha}(a_i) X_i \right) = 0$$

(since the vector field $q\alpha$ is tangent to the manifold M). Hence, $\{\alpha, \alpha_0\} \in J_M$. The second part of this lemma can be proved similarly. $\square$

PROOF OF THEOREM 2.1. Since $\mathcal{O}$ is tame, for each section $\beta \in \Gamma(A_\mathcal{O})$ there exists its extension $\widetilde{\beta} \in \Gamma(A)$, $\beta = r_\mathcal{O} \widetilde{\beta}$. We define the operation $\{\alpha, \beta\}_\mathcal{O} \stackrel{\text{def}}{=} r_\mathcal{O}\{\widetilde{\alpha}, \widetilde{\beta}\}$, where $\widetilde{\alpha}$ and $\widetilde{\beta}$ are the extensions of the sections $\alpha, \beta \in \Gamma(A_\mathcal{O})$.

Now we show that this operation is well defined. Let $\widetilde{\beta}_1, \widetilde{\beta}_2$ and $\widetilde{\alpha}_1, \widetilde{\alpha}_2$ be extensions of the sections β and α, respectively,

$$r_\mathcal{O}\{\widetilde{\alpha}_1, \widetilde{\beta}_1\} - r_\mathcal{O}\{\widetilde{\alpha}_2, \widetilde{\beta}_2\} = \frac{1}{2} r_\mathcal{O}\left(\{\widetilde{\alpha}_1 - \widetilde{\alpha}_2, \widetilde{\beta}_1 + \widetilde{\beta}_2\} + \{\widetilde{\alpha}_1 + \widetilde{\alpha}_2, \widetilde{\beta}_1 - \widetilde{\beta}_2\}\right) = 0$$

(since the sections $(\widetilde{\alpha}_1 - \widetilde{\alpha}_2)$ and $(\widetilde{\beta}_1 - \widetilde{\beta}_2)$ belong to the ideal $J_\mathcal{O} = \operatorname{Ker}(r_\mathcal{O})$). Obviously, the triple $(A_\mathcal{O} \to \mathcal{O}, q, \{\, ,\, \}_\mathcal{O})$ is a transitive Lie algebroid. $\square$

§3. Connections, curvature, and characteristic classes in transitive Lie algebroids

Let $(A \to B, q, \{\, ,\, \})$ be a transitive Lie algebroid. We consider a vector bundle $\mathfrak{g} \to B$, whose fiber is the kernel of the morphism q:

$$(3.0) \qquad \mathfrak{g}_x \stackrel{\text{def}}{=} \operatorname{Ker} q_{|x}$$

(since the morphism q is surjective, $\mathfrak{g} = \bigcup_x \mathfrak{g}_x$ is a subbundle of the bundle $A \to B$). We have the following statement [**Mz**].

PROPOSITION 3.1. (a) *The space of sections* $\Gamma(\mathfrak{g})$ *is an ideal in the Lie algebra* $\Gamma(A)$.

(b) *The restriction of the bracket* $\{\, ,\, \}$ *to the sections of the subbundle* $\mathfrak{g}$

$$(3.1) \qquad [\theta_1, \theta_2] \stackrel{\text{def}}{=} \{\theta_1, \theta_2\}, \qquad \theta_1, \theta_2 \in \Gamma(\mathfrak{g}),$$

is linear over the ring of functions

$$(3.2) \qquad [\theta_1, \varphi \theta_2] = \varphi [\theta_1, \theta_2], \qquad \varphi \in C^\infty(B),$$

and generates a Lie algebra structure in each fiber $\mathfrak{g}_x$; *the structures corresponding to different points of the base are isomorphic.*

PROOF. Statement (a) follows from formula (2.1). Formula (3.2) follows from (2.2). The algebras $\mathfrak{g}_x$ are isomorphic because there exists a Lie connection in the Lie algebroid (see below). □

COROLLARY 3.2. *The adjoint action of the Lie algebroid on the vector bundle* $\mathfrak{g}$

$$(3.3) \qquad \operatorname{ad}_\beta \theta \stackrel{\text{def}}{=} \{\beta, \theta\}, \qquad \beta \in \Gamma(A), \quad \theta \in \Gamma(\mathfrak{g}),$$

has the following properties:

$$(3.4) \qquad \operatorname{ad}_{\varphi \beta} \theta = \varphi \operatorname{ad}_\beta \theta, \qquad \varphi \in C^\infty(B),$$

$$(3.5) \qquad \operatorname{ad}_\beta(\varphi \theta) = \varphi \operatorname{ad}_\beta \theta + \mathcal{L}_{q\beta}(\varphi)\theta,$$

$$(3.6) \qquad \operatorname{ad}_\beta[\theta_1, \theta_2] = [\operatorname{ad}_\beta \theta_1, \theta_2] + [\theta_1, \operatorname{ad}_\beta \theta_2].$$

Moreover, let us note that the subbundle $z\mathfrak{g}$ ($z\mathfrak{g}_x$ is the center of the Lie algebra $\mathfrak{g}_x$) and $[\mathfrak{g}, \mathfrak{g}]$ are invariant under the adjoint action (3.3).

A transitive Lie algebroid is called *Abelian*, if the Lie algebra $\mathfrak{g}_x$ is commutative. The algebroid $(T^*_\mathcal{O} \mathcal{N} \to \mathcal{O}, q, \{\, ,\, \}_\mathcal{O})$ over a *nondegenerate* symplectic leaf $\mathcal{O}$ of the Poisson manifold $\mathcal{N}$ is an example of an Abelian algebroid (see [**We**]).

A linear connection ∇ in the vector bundle $\mathfrak{g} \to B$, satisfying the condition

$$(3.7) \qquad \nabla_v [\theta_1, \theta_2] = [\nabla_v \theta_1, \theta_2] + [\theta_1, \nabla_v \theta_2], \qquad v \in \mathcal{V}^1(B).$$

is called a *Lie connection* [**Mz**] in a transitive Lie algebroid.

Clearly, this condition is equivalent to the fact that the parallel transport of the connection ∇, $T^\nabla_\gamma \colon \mathfrak{g}_{\gamma(0)} \to \mathfrak{g}_{\gamma(1)}$, along each path $\gamma \colon [0,1] \to B$ is an isomorphism of Lie algebras.

The *adjoint connection* [**Mz**] is the major example of the Lie connection. Let $\mathcal{P} \subset A$ be some subbundle transversal to the subbundle $\mathfrak{g}$:

$$(3.8) \qquad A_x = \mathcal{P}_x \oplus \mathfrak{g}_x.$$

Since $\operatorname{Ker} q_{|x} = \mathfrak{g}_x$ and $\operatorname{Ran} q_{|x} = T_x B$, we have the isomorphism of vector bundles $p\colon TB \to \mathcal{P}$ that is inverse to the morphism q:

(3.9) $$q \circ p = id_{\mathcal{V}^1(B)}.$$

The adjoint connection is defined via the adjoint action (3.3):

(3.10) $$\nabla_v^{\mathcal{P}} \theta \stackrel{\text{def}}{=} \operatorname{ad}_{pv} \theta, \quad v \in \mathcal{V}^1(B), \quad \theta \in \Gamma(\mathfrak{g}).$$

(Formulas (3.4), (3.5), and (3.9) imply that (3.10) actually defines a linear connection in the vector bundle $\mathfrak{g}$; that is, it satisfies (A.1) and (A.2).)

Now let us calculate the curvature (A.3) of the adjoint connection (3.10).

PROPOSITION 3.3. (a). *The curvature of the adjoint connection has the form*

(3.11) $$K(v_1, v_2)\theta = [R^{\mathcal{P}}(v_1, v_2), \theta],$$

where the antisymmetric morphism of the vector bundles $R^{\mathcal{P}}\colon TB \otimes TB \to \mathfrak{g}$ *is defined as*

(3.11a) $$R^{\mathcal{P}}(v_1, v_2) \stackrel{\text{def}}{=} \{pv_1, pv_2\} - p[v_1, v_2], \quad v_1, v_2 \in \mathcal{V}^1(B).$$

(b) *The adjoint connection has zero curvature if and only if*

(3.12) $$\{\Gamma(\mathcal{P}), \Gamma(\mathcal{P})\} \subseteq \Gamma(\mathcal{P} \oplus z\mathfrak{g}).$$

PROOF. We have
$$K(v_1, v_2)\theta = \operatorname{ad}_{pv_1}(\operatorname{ad}_{pv_2} \theta) - \operatorname{ad}_{pv_2}(\operatorname{ad}_{pv_1} \theta) - \operatorname{ad}_{p[v_1,v_2]} \theta$$
$$= \operatorname{ad}_{(\{pv_1, pv_2\} - p[v_1,v_2])} \theta = \operatorname{ad}_{R^{\mathcal{P}}(v_1,v_2)} \theta.$$

By (2.1) and (3.9), $qR^{\mathcal{P}}(v_1, v_2) = 0$. Thus, statement (a) is proved.

To prove statement (b), we note that curvature (3.11) is identically equal to zero if and only if

(3.13) $$R^{\mathcal{P}}(v_1, v_2) \in \Gamma(z\mathfrak{g}) \quad \forall v_1, v_2 \in \mathcal{V}^1(B).$$

If condition (3.12) holds, then $\{pv_1, pv_2\} \in \Gamma(\mathcal{P} \oplus z\mathfrak{g})$ and (3.13) is satisfied because $\mathcal{P} \cap z\mathfrak{g} = \{0\}$. Conversely, the sections $\beta_1, \beta_2 \in \Gamma(\mathcal{P})$ can be represented as $\beta_j = pq\beta_j$; therefore,

$$\{\beta_1, \beta_2\} = \{pq\beta_1, pq\beta_2\} = \left(R^{\mathcal{P}}(q\beta_1, q\beta_2) + p[q\beta_1, q\beta_2]\right) \in \Gamma(\mathcal{P} \oplus z\mathfrak{g}). \quad \square$$

COROLLARY 3.4. *The restriction of the adjoint connection* $\nabla^{\mathcal{P}}$ *to the invariant subbundle* $z\mathfrak{g}$ *has zero curvature and is independent of the choice of a transversal subbundle* $\mathcal{P}$. *Thus it defines a universal flat connection in* $z\mathfrak{g}$. *In particular, in an Abelian algebroid the bundle* $\mathfrak{g} \to B$ *is flat.*

PROOF. By (3.11), the curvature is equal to zero. Now let us consider two subbundles $\mathcal{P}$ and $\widetilde{\mathcal{P}}$ that are transversal to $\mathfrak{g}$, and the corresponding morphisms p and $\widetilde{p}$ (3.9). For each vector field $v \in \mathcal{V}^1(B)$ we have $(pv - \widetilde{p}v) \in \Gamma(\mathfrak{g})$; therefore, $\nabla_v^{\mathcal{P}} \theta - \nabla_v^{\widetilde{\mathcal{P}}} \theta = [pv - \widetilde{p}v, \theta] = 0$ for each section $\theta \in \Gamma(z\mathfrak{g})$. Hence, the adjoint connection in the subbundle $z\mathfrak{g}$ is independent of the choice of $\mathcal{P}$. $\quad \square$

REMARK 3.5. In the book by Mackenzie [**Mz**] the morphism $p\colon TB \to \mathcal{P}$ from (3.9) is called a connection in the Lie algebroid. A curvature of this connection is $R^{\mathcal{P}}$, and this connection is called *flat* if $R^{\mathcal{P}}(v_1, v_2) \equiv 0$ (i.e., $\{\Gamma(\mathcal{P}), \Gamma(\mathcal{P})\} \subseteq \Gamma(\mathcal{P})$).

REMARK 3.6. In the context of Poisson geometry the adjoint connection appeared in [**VK**$_3$].

Characteristic classes. To complete this section we would like to show that each transitive Lie algebroid structure over a base B gives a subring of $H^*(B)$ which is a generalization of characteristic classes for vector bundles.

Let $P^k(\mathfrak{g}_x) \stackrel{\text{def}}{=} S^k \mathfrak{g}_x^*$ denote the space of polynomials of degree k on the Lie algebra (3.0) (here S^k denotes the kth symmetric power). Consider a vector bundle $\text{inv}^k(\mathfrak{g}) = \bigcup_{x \in B} \text{inv}^k(\mathfrak{g}_x)$ whose fiber
(3.14)
$$\text{inv}^k(\mathfrak{g}_x) \stackrel{\text{def}}{=} \Big\{ f \in P^k(\mathfrak{g}_x), (\ell_{\theta_0} f)(\theta_1, \theta_2, \ldots, \theta_k)$$
$$= \sum_{j=1}^k f(\theta_1, \ldots, \theta_{j-1}, [\theta_0, \theta_j], \theta_{j+1}, \ldots, \theta_k) = 0, \ \theta_0, \theta_1, \ldots, \theta_k \in \mathfrak{g}_x \Big\}$$

is a space of ad-invariant polynomials on $\mathfrak{g}_x$.

PROPOSITION 3.7. *Every adjoint connection* (3.10) *induces a flat connection in* $\text{inv}^k(\mathfrak{g})$; *this connection does not depend on the choice of a transversal subbundle* $\mathcal{P}$.

PROOF. For each adjoint connection (3.10), a connection in a vector bundle $P^k(\mathfrak{g}) = \bigcup_{x \in B} P^k(\mathfrak{g}_x)$ is defined by the formula

(3.15) $\quad (\nabla_u^{\mathcal{P}} s)(\theta_1, \ldots, \theta_k) = \mathcal{L}_u \big(s(\theta_1, \ldots, \theta_k) \big) - \sum_{j=1}^k s(\theta_1, \ldots, \nabla_u^{\mathcal{P}} \theta_j, \ldots \theta_k)$

(here $s \in \Gamma(P^k(\mathfrak{g}))$ and $\theta_i \in \Gamma(\mathfrak{g})$). Now observe that for each $\theta \in \Gamma(\mathfrak{g})$ and $s \in \Gamma(P^k(\mathfrak{g}))$,
$$\ell_\theta \nabla_u^{\mathcal{P}} s = \nabla_u^{\mathcal{P}} \ell_\theta s - \ell_{\nabla_u \theta} s$$
(here ℓ_θ is defined as in (3.14); $u \in \mathcal{V}^1(B)$). Therefore, $\text{inv}^k(\mathfrak{g})$ is an invariant subbundle in $P^k(\mathfrak{g})$ with respect to any adjoint connection (3.15), and thus we may regard the connection (3.15) as a connection on $\text{inv}^k(\mathfrak{g})$. To see that this connection on $\text{inv}^k(\mathfrak{g})$ has zero curvature, notice that
$$\big(K^{\nabla^{\mathcal{P}}}(u_1, u_2) \big) s = \ell_{R^{\mathcal{P}}(u_1, u_2)} s = 0 \quad \text{for each } s \in \Gamma\big(\text{inv}^k(\mathfrak{g}) \big).$$

Now let $\mathcal{P}$ and $\mathcal{P}'$ be two different transversal subbundles (3.8), and let p and p' be the corresponding morphisms (3.9). Then
$$\nabla_u^{\mathcal{P}} s - \nabla_u^{\mathcal{P}'} s = \ell_{p'(u)-p(u)} s = 0 \quad \forall s \in \Gamma(\text{inv}^k(\mathfrak{g})), \quad u \in \mathcal{V}^1(B).$$

Therefore, the connection in $\text{inv}^k(\mathfrak{g})$ does not depend on the choice of a transversal subbundle $\mathcal{P}$. □

Now consider a space $\text{Inv}^k(\mathfrak{g}_x)$ of $\text{Aut}(\mathfrak{g}_x)$-invariant polynomials on $\mathfrak{g}_x$:

(3.16) $\quad \text{Inv}^k(\mathfrak{g}_x) \stackrel{\text{def}}{=} \big\{ f \in P^k(\mathfrak{g}_x), \ f(g \cdot \theta_1, g \cdot \theta_2, \ldots, g \cdot \theta_k) = f(\theta_1, \theta_2 \ldots, \theta_k),$
$$\theta_1, \theta_2, \ldots, \theta_k \in \mathfrak{g}_x, \forall g \in \text{Aut}(\mathfrak{g}_x) \big\}$$

(here $\mathrm{Aut}(\mathfrak{g}_x)$ is the group of all automorphisms of the Lie algebra $\mathfrak{g}_x$, and $g \cdot \theta$ denotes the action of $g \in \mathrm{Aut}(\mathfrak{g}_x)$ on $\theta \in \mathfrak{g}_x$).

It is easy to see that $\mathrm{Inv}^k(\mathfrak{g}_x)$ is naturally imbedded into the space of parallel sections of the vector bundle $\mathrm{inv}^k(\mathfrak{g})$; therefore, for each $f \in \mathrm{Inv}^k(\mathfrak{g}_x)$ we can define a differential form $c_f \in \Gamma(\Lambda^{2k} T^* B)$,
(3.17)
$$c_f(u_1, u_2, \ldots, u_{2k}) = f(R^{\mathcal{P}}, \ldots, R^{\mathcal{P}})$$
$$\stackrel{\mathrm{def}}{=} \sum_{\sigma \in S_{2k}} (-1)^\sigma \frac{1}{(2k)!} f\big(R^{\mathcal{P}}(u_{\sigma(1)}, u_{\sigma(2)}), \ldots, R^{\mathcal{P}}(u_{\sigma(2k-1)}, u_{\sigma(2k)})\big),$$

(here the sum is taken over all permutations of $\{1, \ldots, 2k\}$, $u_1, \ldots, u_{2k} \in \mathcal{V}^1(B)$, and $R^{\mathcal{P}}$ is a "curvature" (3.11a) of some adjoint connection).

THEOREM 3.8. *For each $f \in \mathrm{Inv}^k(\mathfrak{g}_x)$, a differetial form c_f (3.17) is closed and its cohomology class in $H^{2k}(B)$ does not depend on the choice of a transversal subbundle $\mathcal{P}$.*

PROOF. Since each $f \in \mathrm{Inv}^k(\mathfrak{g}_x)$ is a parallel section of $\mathrm{inv}^k(\mathfrak{g})$ and $\nabla^{\mathcal{P}} R^{\mathcal{P}} = 0$ (see [**Mz**]), we have
$$dc_f = df(R^{\mathcal{P}}, R^{\mathcal{P}}, \ldots, R^{\mathcal{P}})$$
$$= \sum_{j=1}^k f(R^{\mathcal{P}}, \ldots, \nabla R^{\mathcal{P}}, \ldots, R^{\mathcal{P}}) + (\nabla^{\mathcal{P}} f)(R^{\mathcal{P}}, \ldots, R^{\mathcal{P}}) = 0.$$

Now we show that the cohomology class of c_f does not depend on the choice of a transversal subbundle $\mathcal{P}$. Let $\mathcal{P}$ and $\mathcal{P}'$ be distinct subbundles in A transversal to $\mathfrak{g}$ (3.8), and let p and p' be the corresponding morphisms (3.9). Define
$$p^t(u) \stackrel{\mathrm{def}}{=} (1-t)p(u) + tp'(u).$$

It is easy to see that for each $t \in \mathbb{R}$, $p^t(u)$ satisfies (3.9) and thus defines a transversal subbundle $\mathcal{P}^t$ for each $t \in \mathbb{R}$.

Let R^t denote the curvature (3.11a) of the transversal subbundle $\mathcal{P}^t$, and for each $f \in \mathrm{Inv}^k(\mathfrak{g}_x)$ define $c_f^t = f(R^{\mathcal{P}^t}, \ldots, R^{\mathcal{P}^t})$, where the right-hand side is defined as the right-hand side of (3.17) with $R^{\mathcal{P}}$ replaced by $R^{\mathcal{P}^t}$. This differential form is closed for each $t \in \mathbb{R}$.

In order to show that the cohomology class of c_f^t does not depend on t, consider another transitive algebroid $(A_1 \to B \times \mathbb{R}, q_1, \{\}_1)$, where $A_1 = \pi_1^* A \oplus \pi_2^* T\mathbb{R}$ (here π_1 and π_2 are natural projections of $B \times \mathbb{R}$ onto B and $\mathbb{R}$ respectively),
$$q_1(\alpha \oplus v) = q(\alpha) + v, \qquad \alpha \in \Gamma(A), \quad v \in \mathcal{V}^1(\mathbb{R}),$$
$$\{\alpha_1 \oplus v_1, \alpha_2 \oplus v_2,\}_1 = \{\alpha_1, \alpha_2\} \oplus [v_1, v_2], \quad \forall \alpha_i \in \Gamma(A), \quad v_i \in \mathcal{V}^1(\mathbb{R}), \quad i = 1, 2.$$

Introduce a vector bundle morphism $p_1 \colon T(B \times \mathbb{R}) \to A_1$,
$$p_1(u + v) = \big((1-t)p(u) + tp'(u)\big) \oplus v,$$

where t is the coordinate on $\mathbb{R}$, $u \in \mathcal{V}^1(B)$, and $v \in \mathcal{V}^1(\mathbb{R})$. It is easy to see that $q_1 \circ p_1 = \mathrm{id}_{T(B \times \mathbb{R})}$; thus p_1 defines a subbundle in A_1 transversal to $\mathfrak{g}_1 = \ker q_1 =$

$\pi_1^* \mathfrak{g}$. Direct calculation shows that the curvature $R^{\mathcal{P}_1}$ (3.11a) of this transversal subbundle satisfies

$$\text{(3.18)} \quad R^{\mathcal{P}_1}(u_1 + v_1, u_2 + v_2) = R^{\mathcal{P}^t}(u_1, u_2) + \mathcal{L}_{v_1}(t)(p'(u_2) - p(u_2)) \\ - \mathcal{L}_{v_2}(t)(p'(u_1) - p(u_1))$$

(here $u_i \in \mathcal{V}^1(B)$ and $v_i \in \mathcal{V}^1(\mathbb{R})$, $i = 1, 2$).

Since $\mathfrak{g}_1 = \ker q_1 = \pi_1^* \mathfrak{g}$, it follows that $\text{Inv}^k(\mathfrak{g}_{1x}) = \text{Inv}^k(\mathfrak{g}_x)$; therefore for each $f \in \text{Inv}^k(\mathfrak{g}_x)$ we can define $\omega_f = f(R_1, \ldots, R_1)$ using formula (3.17). This is a closed differential form on $B \times \mathbb{R}$, and using (3.18) it is easy to see that $c_f^t = i_t^* \omega_f$, where $i_t \colon B \to B \times \mathbb{R}$ is defined by the formula

$$i_t(x) = (x, t), \qquad x \in B, \quad t \in \mathbb{R}.$$

Since all the i_t are homotopic, the cohomology classes of $i_t^* \omega_f$ coincide for all $t \in \mathbb{R}$; in particular, the cohomology classes of $c_f^0 = f(R^{\mathcal{P}}, \ldots, R^{\mathcal{P}})$ and $c_f^1 = f(R^{\mathcal{P}'}, \ldots, R^{\mathcal{P}'})$ coincide. This proves that the cohomology class of (3.17) does not depend on the choice of a transversal subbundle. □

EXAMPLE (Chern characteristic classes of a vector bundle). Let $\mathcal{E} \to B$ be a vector bundle. Consider the transitive algebroid $A = \text{CDO}(\mathcal{E})$ of covariant differential operators [**Mz**]. Every section α of A acts on $\Gamma(\mathcal{E})$ as a first-order differential operator $\alpha(fs) = \mathcal{L}_{q(\alpha)}(f)s + f\alpha(s)$. It is easy to see that each morphism $p \colon TB \to \text{CDO}(\mathcal{E})$ satisfying (3.9) defines a linear connection ∇^p on $\mathcal{E}$ ($\nabla_u^p s = p(u)(s)$) and the curvature (3.11a) corresponding to p is exactly the curvature (A.3) of the connection ∇^p. This proves that the cohomology classes c_f are exactly characteristic classes of the vector bundle $\mathcal{E}$ (see, for example, [**NS**], Appendix C).

§4. Calculation of cohomology of a transitive Lie algebroid

Let $(A \to B, q, \{\,,\,\})$ be a transitive Lie algebroid. We will always assume that the bundle $\mathfrak{g}$ (3.0) can be decomposed into the direct sum of subbundles $\mathfrak{g} = \mathfrak{g}_0 \oplus \mathfrak{h}$, which satisfies two conditions:
- $\mathfrak{g}_0$ and $\mathfrak{h}$ are invariant under the adjoint action (3.3);
- $\mathfrak{g}_0$ is a subbundle of the center $\mathfrak{g}_0 \subset z\mathfrak{g}$.

In what follows, we are interested in two extreme cases: (1) $\mathfrak{h} = \mathfrak{g}$ and (2) $\mathfrak{g}_0 = z\mathfrak{g}$ (though, generally speaking, not every transitive Lie algebroid has an ad-invariant subbundle $\mathfrak{h}$ that is transversal to $z\mathfrak{g}$).

Let us choose some transversal subbundle $\mathcal{P}$ (3.8). We consider a vector bundle

$$\text{(4.0)} \quad A_0 \stackrel{\text{def}}{=} \mathcal{P} \oplus \mathfrak{g}_0 \simeq TB \oplus \mathfrak{g}_0,$$

on which we define the structure of an Abelian algebroid $(A_0 \to B, q, \{\,,\,\}_0)$. A bracket on sections of the subbundle $A_0 \subset A$ is defined by the formula

$$\text{(4.1)} \quad \{\alpha_1, \alpha_2\}_0 \stackrel{\text{def}}{=} \{\alpha_1, \alpha_2\} - \pi\{\alpha_1, \alpha_2\},$$

where $\pi \colon A \to \mathfrak{h}$ is the projection on the subbundle $\mathfrak{h}$ along the subbundle A_0.

By $\mathcal{K}^{s,t}$ we denote the space of s-cochains of the Lie algebroid A_0 with values in the space of sections of the t-th external power of the subbundle dual to $\mathfrak{h}$:

$$\text{(4.2)} \quad \mathcal{K}^{s,t} \stackrel{\text{def}}{=} \Gamma(\bigwedge\nolimits^t \mathfrak{h}^* \otimes \bigwedge\nolimits^s A_0^*).$$

The space $\mathcal{K}^\cdot = \bigoplus_k \mathcal{K}^k$, where $\mathcal{K}^k = \bigoplus_{s+t=k} \mathcal{K}^{s,t}$ is a Grassmann algebra with respect to the multiplication

(4.3) $$(\eta_1 \otimes \omega_1) \square (\eta_2 \otimes \omega_2) = (-1)^{s_1+t_2}(\eta_1 \wedge \eta_2) \otimes (\omega_1 \wedge \omega_2),$$

where $\eta_i \in \Gamma(\wedge^{t_i}\mathfrak{h}^*)$, $\omega_i \in C^{s_i}(A_0)$, $i = 1, 2$.

We define the mappings $p^{s,t} \colon C^{t+s}(A) \to \mathcal{K}^{s,t}$ and $e^{s,t} \colon \mathcal{K}^{s,t} \to C^{t+s}(A)$ as follows:

(4.4) $$p^{s,t}f(\theta_1,\ldots,\theta_t)(\beta_1,\ldots,\beta_s) \stackrel{\text{def}}{=} f(\theta_1,\ldots,\theta_t,\beta_1,\ldots,\beta_s)$$

(here $f \in C^{t+s}(A)$, $\theta_j \in \Gamma(\mathfrak{h})$, and $\beta_l \in \Gamma(A_0)$) and

$$e^{s,t}\eta(\alpha_1,\ldots,\alpha_{t+s}) \stackrel{\text{def}}{=} \frac{1}{t!s!} \sum_{\sigma \in S_{t+s}} (-1)^\sigma \eta\big(\pi\alpha_{\sigma(1)},\ldots,\pi\alpha_{\sigma(t)}\big)\big(\alpha_{\sigma(t+1)}$$
$$- \pi\alpha_{\sigma(t+1)},\ldots,\alpha_{\sigma(t+s)} - \pi\alpha_{\sigma(t+s)}\big)$$

(here $\eta \in \mathcal{K}^{s,t}$ and $\alpha_l \in \Gamma(A)$, $l = 1,\ldots,t+s$, and the summation is taken over the set of permutations).

We write $p^k \stackrel{\text{def}}{=} \sum_{t+s=k} p^{s,t}$ and $e^k \stackrel{\text{def}}{=} \sum_{t+s=k} e^{s,t}$.

PROPOSITION 4.1. *The mapping* $p^k \colon C^k(A) \to \mathcal{K}^k$ *is an isomorphism of Grassmann algebras*

(4.5) $$(p^k)^{-1} = e^k, \quad p^{k_1+k_2}(f_1 \wedge f_2) = p^{k_1}f_1 \square p^{k_2}f_2 \quad (f_i \in C^{k_i}(A),\ i=1,2),$$

satisfying the commutation relation

(4.6) $$p^{k+1}D = (d_0 + d_1 + \delta)p^k,$$

where the operators $d_0 \colon \mathcal{K}^{s,t} \to \mathcal{K}^{s,t+1}$, $d_1 \colon \mathcal{K}^{s,t} \to \mathcal{K}^{s+1,t}$, *and* $\delta \colon \mathcal{K}^{s,t} \to \mathcal{K}^{s+2,t-1}$ *are defined as follows:*

(4.7)
$$d_0\eta(\theta_0,\theta_1,\ldots,\theta_t)(\alpha_1,\ldots,\alpha_s)$$
$$\stackrel{\text{def}}{=} \sum_{0\leq i<j\leq t} (-1)^{i+j}\eta\big([\theta_i,\theta_j],\theta_0,\theta_1,\ldots,\widehat{\theta_i},\ldots,\widehat{\theta_j},\ldots,\theta_t\big)(\alpha_1,\ldots,\alpha_s),$$

(4.8) $$d_1 \stackrel{\text{def}}{=} \sum_{t,s}(-1)^t \overline{\nabla}^{s,t},$$

(4.9)
$$\overline{\nabla}^{s,t}\eta(\theta_1,\ldots,\theta_t)(\alpha_0,\alpha_1,\ldots,\alpha_s)$$
$$\stackrel{\text{def}}{=} \sum_{j=0}^{s}(-1)^j\bigg(\mathcal{L}_{q\alpha_j}\big(\eta(\theta_1,\ldots,\theta_t)(\alpha_0,\ldots,\widehat{\alpha}_j,\ldots,\alpha_s)\big)$$
$$- \sum_{\tau=1}^{t}\eta(\theta_1,\ldots,\theta_{\tau-1},\nabla^{\mathcal{P}}_{q\alpha_j}\theta_\tau,\theta_{\tau+1}\ldots,\theta_t)(\alpha_0,\ldots,\widehat{\alpha}_j,\ldots,\alpha_s)\bigg)$$
$$+ \sum_{0\leq l<m\leq s}(-1)^{l+m}\eta(\theta_1,\ldots,\theta_t)(\{\alpha_l,\alpha_m\}_0,\alpha_0,\alpha_1,\ldots,\widehat{\alpha}_l,\ldots,\widehat{\alpha}_m,\ldots,\alpha_s),$$

(4.10)
$$\delta\eta(\theta_1,\ldots,\theta_{t-1})(\alpha_0,\alpha_1,\ldots,\alpha_s,\alpha_{s+1})$$
$$\stackrel{\text{def}}{=} \sum_{0\leq l<m\leq s+1} (-1)^{l+m}$$
$$\times \eta\big(\pi R^{\mathcal{P}}(q\alpha_l,q\alpha_m),\theta_1,\ldots,\theta_{t-1}\big)(\alpha_0,\alpha_1,\ldots,\widehat{\alpha}_l,\ldots,\widehat{\alpha}_m,\ldots,\alpha_s,\alpha_{s+1}),$$

(here π is the projection on the subbundle $\mathfrak{h}$ along $\mathfrak{g}_0$) and satisfy the following commutation relations:

(4.11) $$d_0^2 = 0,$$
(4.12) $$d_0 d_1 + d_1 d_0 = 0,$$
(4.13) $$d_1^2 + d_0 \delta + \delta d_0 = 0,$$
(4.14) $$d_1 \delta + \delta d_1 = 0,$$
(4.15) $$\delta^2 = 0.$$

PROOF. Formulas (4.5)–(4.10) can be verified by direct calculations. Each formula (4.1i), $i = 1, 2, \ldots, 5$, follows from the identity $p^{s+i-1,t+3-i} D^2 e^{s,t} = 0$. □

REMARK 4.2. The operator d_0 is a standard coboundary operator for the zero representation of the Lie algebra $\Gamma(\mathfrak{h})$ in the space $C^{\cdot}(A_0) = \bigoplus C^s(A_0)$. The operator d_1 coincides up to the sign with the differential operator (A.4) (see Appendix) that corresponds to the connection in the bundle $\bigoplus_t \bigwedge^t \mathfrak{h}^*$ induced by the adjoint connection $\nabla^{\mathcal{P}}$. In particular, for any section $\eta \in \mathcal{K}^{s,0}$ we have $d_1\eta = D_0\eta$, where D_0 is the coboundary operator (2.4) of the Abelian algebroid $(A_0 \to B, q, \{\,,\,\}_0)$.

REMARK 4.3. Each of the operators d_0, d_1, and δ is antidifferential with respect to Grassmann multiplication (4.3):

$$(d_0 + d_1 + \delta)(\eta_1 \,\square\, \eta_2) = \big((d_0 + d_1 + \delta)\eta_1\big) \,\square\, \eta_2 + (-1)^{s_1+t_1} \eta_1 \,\square\, (d_0 + d_1 + \delta)\eta_2,$$

where $\eta_i \in \mathcal{K}^{s_i,t_i}$, $i = 1, 2$.

Let us consider the Hochschild–Serre spectral sequence $(E_r^{s,t}, d_r^{s,t})$ [**HS**], related to the ideal $\Gamma(\mathfrak{h})$ of the Lie algebra $\Gamma(A)$. We have the following result (see [**Mz**]).

PROPOSITION 4.4. *The spectral sequence $(E_r^{s,t}, d_r^{s,t})$ converges to the cohomology space $\mathcal{H}^{\cdot}(A)$:*

(4.16) $$\mathcal{H}^k(A) \simeq \bigoplus_{t+s=k} E_r^{s,t} \quad \text{for} \quad r \geq \max(t+1, s) + 1,$$

and

(4.17) $$\begin{aligned}(E_0^{s,t}, d_0^{s,t}) &\simeq (\mathcal{K}^{s,t}, d_0), \\ (E_1^{s,t}, d_1^{s,t}) &\simeq (H_{d_0}\mathcal{K}^{s,t}, d_1),\end{aligned}$$
(4.18) $$(E_2^{s,t}, d_2^{s,t}) \simeq (H_{d_1}H_{d_0}\mathcal{K}^{s,t}, \delta - d_1 d_0^{-1} d_1),$$

where the operators d_0, d_1, and δ are defined by (4.7)–(4.10).

The proof is standard [**HS**].

Let us consider a vector bundle $H^t\mathfrak{h} \to B$, whose fiber is the cohomololgy space $H^t(\mathfrak{h}_x)$ of the Lie algebra $\mathfrak{h}_x$. It is easy to see that the space of sections of this bundle is isomorphic to the space $H_{d_0}\mathcal{K}^{0,t}$ (the isomorphism can be established by some scalar product in the bundle $\bigwedge^t \mathfrak{h}^*$; in this case the fiber $H^t\mathfrak{h}_{|x}$ is realized as the orthogonal complement to the subbundle of coboundaries in the subbundle of cocycles). Similarly, one can obtain the isomorphism

$$(4.19) \qquad E_1^{s,t} \simeq H_{d_0}\mathcal{K}^{s,t} \simeq \Gamma(H^t\mathfrak{h} \otimes \bigwedge^s A_0^*).$$

By ∇^t we denote the connection in the vector bundle $\bigwedge^t \mathfrak{h}^*$ induced by the adjoint connection. If we restrict formula (4.12) to the space $\mathcal{K}^{0,t}$, then we see that the connection ∇^t satisfies the formula

$$\nabla_v^{t+1} d_0 \eta = d_0 \nabla_v^t \eta \qquad \forall \eta \in \mathcal{K}^{0,t}, \quad v \in \mathcal{V}^1(B).$$

Hence, the connection ∇^t is invariant over the subbundle of cocycles and the subbundle of coboundaries of the operator d_0. Thus this connection induces a connection in the vector bundle $H^t\mathfrak{h}$, which we denote by the same symbol ∇^t.

The following statement is a reformulation of the result obtained in [**Mz**].

PROPOSITION 4.5. (a) *In the bundle $H^t\mathfrak{h}$ the connection ∇^t, which is induced by the adjoint connection $\nabla^\mathcal{P}$, is flat and independent of the choice of the transversal subbundle $\mathcal{P}$.*

(b) *The second term of the spectral sequence coincides with the cohomology space of the operator $\overline{\nabla}^t$ defined as in (A.6) associated with the flat connection ∇^t:*

$$(4.20) \qquad E_2^{s,t} \simeq H^s_{\overline{\nabla}^t}(A_0, H^t\mathfrak{h}).$$

PROOF. We define a morphism of vector bundles $\ell \colon \mathfrak{h} \otimes \bigwedge^t \mathfrak{h}^* \to \bigwedge^t \mathfrak{h}^*$ by the formula

$$\ell_\theta \eta(\theta_1, \ldots, \theta_t) \stackrel{\text{def}}{=} \sum_{\tau=1}^t \eta(\theta_1, \ldots, \theta_{\tau-1}, [\theta, \theta_\tau], \theta_{\tau+1}, \ldots, \theta_t).$$

For each cohomology class $[\eta]_{d_0} \in H_{d_0}\mathcal{K}^{0,t}$ we have $[\ell_\theta \eta]_{d_0} = 0$ (see, for example, [**CE**]). One can easily verify that

$$\left(\nabla_{v_1}^t \nabla_{v_2}^t - \nabla_{v_2}^t \nabla_{v_1}^t - \nabla_{[v_1,v_2]}^t\right)\eta = -\ell_{\pi R^\mathcal{P}(v_1,v_2)}(\eta) \qquad \forall \eta \in \mathcal{K}^{0,t}.$$

Therefore, in the bundle $H^t\mathfrak{h}$ the curvature of the connection ∇^t is equal to zero.

Let $\mathcal{P}$ and $\mathcal{P}'$ be two subbundles (3.8) transversal to $\mathfrak{g}$, let p and p' be morphisms (3.9), and let ∇^t and $\nabla^{t'}$ be the corresponding connections in the bundle $\bigwedge^t \mathfrak{h}^*$. Obviously, $(pv - p'v) \in \Gamma(\mathfrak{g})$ for any vector field $v \in \mathcal{V}^1(B)$. It is easy to see that

$$(\nabla_v^t - \nabla_v^{t'})\eta = \ell_{\pi(pv-p'v)}\eta \qquad \forall \eta \in \Gamma(\bigwedge^t \mathfrak{h}^*).$$

Therefore, the adjoint connections $\nabla^\mathcal{P}$ and $\nabla^{\mathcal{P}'}$ induce the same connection in the bundle $H^t\mathfrak{h}$. Statement (a) is thereby proved.

Statement (b) follows from (4.17), (4.19), (4.9), and statement (a). □

A simple consequence of formula (4.20) is the following generalization of the theorem stating that the de Rham cohomology is finite-dimensional.

PROPOSITION 4.6. *Suppose the base B of a transitive Lie algebroid is a finite-type manifold. Then the cohomology space $\mathcal{H}^{\cdot}(A)$ is finite-dimensional.*

PROOF. Let us consider the spectral sequence corresponding to the invariant subbundle $\mathfrak{h} = \mathfrak{g}$. Then we have $E_2^{s,t} \simeq H_{\nabla^t}^s(H^t\mathfrak{g})$ (see the definition for the cohomology of the flat connection in the Appendix). By applying Lemma A.3, we prove that the second term of this spectral sequence is finite-dimensional. Hence, the space $\mathcal{H}^k(A) \simeq \bigoplus_{s+t=k} E_\infty^{s,t}$ is also finite-dimensional. $\square$

A transitive Lie algebroid is called *quasiparallelizable* if there exists a decomposition $\mathfrak{g} = \mathfrak{g}_0 \oplus \mathfrak{h}$ into the direct sum of ad-invariant subbundles such that $\mathfrak{g}_0 \subset z\mathfrak{g}$, and for any $t \geq 1$ the canonical connection ∇^t has a trivial holonomy group in the bundle $H^t\mathfrak{h}$.

Obviously, each transitive Lie algebroid over a simply connected base B is quasiparallelizable ($\mathfrak{h} = \mathfrak{g}$). It also should be mentioned that we can ensure the quasiparallelizability of transitive Lie algebroids by imposing some conditions on the Lie algebra $\mathfrak{g}_x$.

PROPOSITION 4.7. *For a Lie algebroid to be quasiparallelizable, it suffices that the following two conditions hold simultaneously.*

(i) *The Lie algebra $\mathfrak{g}_x$ can be decomposed into the direct sum of its center and its commutant, $\mathfrak{g}_x = z\mathfrak{g}_x \oplus [\mathfrak{g}_x, \mathfrak{g}_x]$.*

(ii) *The Lie algebra $\mathfrak{h}_x = [\mathfrak{g}_x, \mathfrak{g}_x]$ is complete (that is, each derivation of the Lie algebra $\mathfrak{h}_x$ is interior) and the group $\mathrm{Aut}(\mathfrak{h}_x)$ of automorphisms of the Lie algebra $\mathfrak{h}_x$ is connected.*

PROOF. For each path $\gamma \colon [0,1] \to B$, $\gamma(0) = \gamma(1) = x_0$, the parallel transport $T_\gamma \colon H^t(\mathfrak{h}_{x_0}) \to H^t(\mathfrak{h}_{x_0})$ has the form $T_\gamma[C]_{d_0} = [T_\gamma^t C]_{d_0}$, where the parallel transport $T_\gamma^t \colon C^t(\mathfrak{h}_{x_0}) \to C^t(\mathfrak{h}_{x_0})$ satisfies the formula

$$(T_\gamma^t C)(\theta_1, \ldots, \theta_t) = C(T_\gamma^{\mathcal{P}} \theta_1, \ldots, T_\gamma^{\mathcal{P}} \theta_t), \qquad \theta_1, \ldots, \theta_t \in \mathfrak{h}_{x_0},$$

and $T_\gamma^{\mathcal{P}} \colon \mathfrak{h}_{x_0} \to \mathfrak{h}_{x_0}$ is the parallel transport of the adjoint connection $\nabla^{\mathcal{P}}$ in the invariant subbundle $\mathfrak{h} = [\mathfrak{g}, \mathfrak{g}]$.

Obviously, $T_\gamma^{\mathcal{P}}$ is an automorphism of the Lie algebra $\mathfrak{h}_x$. It follows from condition (ii) that each automorphism $a \in \mathrm{Aut}(\mathfrak{h}_x)$ can be represented in the form of a finite product of automorphisms of the form $a = \exp(\mathrm{ad}\,\theta)$, where $\theta \in \mathfrak{h}_x$ (see, e.g., Proposition 4.6.1 in [**GG**]). Hence, to complete the proof, it remains to show that the group of interior automorphisms acts *trivially* on the space $H^t(\mathfrak{h}_{x_0})$.

We consider the one-parametric group of linear isomorphisms $a_\tau \colon H^t(\mathfrak{h}_{x_0}) \to H^t(\mathfrak{h}_{x_0})$, $a_\tau[C]_{d_0} \stackrel{\mathrm{def}}{=} [C \cdot \mathrm{Ad}(\exp(\tau\theta))]_{d_0}$, where $\theta \in \mathfrak{h}_{x_0}$ and

$$\bigl(C \cdot \mathrm{Ad}(\exp(\tau\theta))\bigr)(\theta_1, \ldots, \theta_t) \stackrel{\mathrm{def}}{=} C(e^{\mathrm{ad}\,\tau\theta}\theta_1, \ldots, e^{\mathrm{ad}\,\tau\theta}\theta_t).$$

As is easy to see,

$$\frac{d}{d\tau}\bigl(C \cdot \mathrm{Ad}(\exp(\tau\theta))\bigr)(\theta_1, \ldots, \theta_t)$$
$$= \sum_{l=1}^{t} C\bigl(e^{\mathrm{ad}\,\tau\theta}\theta_1, \ldots, e^{\mathrm{ad}\,\tau\theta}\theta_{l-1}, [\theta, e^{\mathrm{ad}\,\tau\theta}\theta_l], e^{\mathrm{ad}\,\tau\theta}\theta_{l+1}, \ldots, e^{\mathrm{ad}\,\tau\theta}\theta_t\bigr)$$
$$= \ell_\theta\bigl(C \cdot \mathrm{Ad}(\exp(\tau\theta))\bigr)(\theta_1, \ldots, \theta_t).$$

Since $[\ell_\theta C]_{d_0} = 0$ for each cocycle $C \in Z_{d_0}(\mathfrak{h}_x)$ (see [**CE**]), we have $\frac{d}{d\tau}a_\tau = 0$; hence, $a_\tau \equiv a_0 = \mathrm{id}$. The proposition is proved. □

THEOREM 4.8. *Let $(A \to B, q, \{\,,\,\})$ be a quasiparallelizable Lie algebroid. Then the second term of the spectral sequence, corresponding to the subbundle $\mathfrak{h}$, has the form*

(4.21) $$E_2^{s,t} \simeq \mathbf{h}^t \otimes \mathcal{H}^s(A_0),$$

where $\mathcal{H}^s(A_0)$ is the cohomology space of the Abelian algebroid $(\mathcal{P} \oplus \mathfrak{g}_0 \to B, q, \{\,,\,\}_0)$ and $\mathbf{h}^t \cong H^t \mathfrak{h}_{|x}$ is the cohomology space of a finite-dimensional Lie algebra $\mathfrak{h}_x$. In this case for any $\zeta \in \mathbf{h}^t$, $\varkappa \in \mathcal{H}^s(A_0)$, we have

(4.22) $$d_2^{s,t}(\zeta \otimes \varkappa) = d_2^{0,t}\zeta \boxtimes \varkappa,$$

where the operation of Grassmann multiplication $\boxtimes$ in the space $E_2^{s,t}$ is induced by the Grassmann multiplication (4.3) in the space $E_0^{s,t}$.

PROOF. By applying Lemma A.4. (see the Appendix) to (4.20), we obtain (4.21). Formula (4.22) follows from (4.18) and Remark 4.3. □

COROLLARY 4.9. *Let $d_k^{s,t}[\zeta \otimes \varkappa]_k = 0$ for any $k = 2, 3, \ldots, r - 1$. Then*

$$d_r^{s,t}[\zeta \otimes \varkappa]_r = (d_r^{0,t}[\zeta]_r) \boxtimes [\varkappa]_r,$$

where $[\cdot]_r$ is the cohomology class in the space $E_r^{s,t} = \mathrm{Ker}\, d_{r-1}^{s,t} / \mathrm{Ran}\, d_{r-1}^{s-r,t+r-1}$.

REMARK 4.10. *If $\mathfrak{h} = \mathfrak{g}$, i.e., $A_0 = \mathcal{P} \simeq TB$, then the mapping $q^\dagger : \mathcal{H}^k(A_0) \to H^k(B)$ is an isomorphism of Grassmann algebras.*

COROLLARY 4.11. *Let the base B of a transitive Lie algebroid be a simply connected finite-type manifold. Then the Euler characteristic*

$$\chi(A) \stackrel{\mathrm{def}}{=} \sum_{k=0}^{\infty} (-1)^k \dim \mathcal{H}^k(A)$$

has the form

$$\chi(A) = \chi_B \left(\sum_{t=0}^{\infty} (-1)^t \dim H^t(\mathfrak{h}_x) \right),$$

where χ_B is the Euler characteristic of the manifold B.

Let us consider some simple examples.

EXAMPLE 4.1. Suppose that the Lie algebra $\mathfrak{g}_x$ is *reductive*, that is, $\mathfrak{g}_x = z\mathfrak{g}_x \oplus [\mathfrak{g}_x, \mathfrak{g}_x]$, where $\mathfrak{h}_x = [\mathfrak{g}_x, \mathfrak{g}_x]$ is a semisimple Lie algebra. By (3.6), each summand in the direct sum of vector bundles $\mathfrak{g} = z\mathfrak{g} \oplus [\mathfrak{g}, \mathfrak{g}]$ is ad-invariant. Therefore, an Abelian algebroid $(A_0 \to B, q, \{\,,\,\}_0)$ with $A_0 = \mathcal{P} \oplus z\mathfrak{g} \simeq TB \oplus z\mathfrak{g}$ is defined. Since the Lie algebra $\mathfrak{h}_x$ is semisimple, by the Whitehead lemma [**CE**] we have $H^1(\mathfrak{h}_x) = H^2(\mathfrak{h}_x) = 0$. Hence, if we consider the spectral sequence related to the subbundle $\mathfrak{h} = [\mathfrak{g}, \mathfrak{g}]$, we obtain $E_2^{s,1} = E_2^{s,2} = 0$. Thus, we have proved (see formula (4.16)) that the first cohomology spaces have the form

$$\mathcal{H}^1(A) \simeq \mathcal{H}^1(A_0),$$
$$\mathcal{H}^2(A) \simeq \mathcal{H}^2(A_0),$$
$$\mathcal{H}^3(A) \simeq \mathcal{H}^3(A_0) \oplus \mathrm{Ker}\, d_4^{0,3},$$

where $\mathcal{H}^k(A_0)$ is the cohomology space of an Abelian algebroid. We calculate the cohomology of an Abelian algebroid in the next section.

EXAMPLE 4.2. Let the base B of a quasiparallelizable Lie algebroid have the de Rham cohomology of spherical type, i.e.,

$$\dim H^s(B) = \begin{cases} 1, & s = 0, n, \\ 0, & s \neq 0, n, \end{cases} \quad n \geq 2.$$

Consider the spectral sequence related to the subbundle $\mathfrak{h} = \mathfrak{g}$. Formula (4.21) has the form

$$E_2^{s,t} = \begin{cases} H^t(\mathfrak{g}_x), & s = 0, n, \\ 0, & s \neq 0, n. \end{cases}$$

Hence $E_n^{s,t} = E_2^{s,t}$, and the operator $d_n^{0,k} \colon E_n^{0,k} \to E_n^{k,k-n+1}$ defines a linear mapping

(4.23) $$\varpi_k \colon \mathbf{h}^k \to \mathbf{h}^{k-n+1},$$

where for $k < 0$ the cohomology spaces $\mathbf{h}^k \stackrel{\text{def}}{=} H^k(\mathfrak{g}_x)$ of the Lie algebra $\mathfrak{g}_x$ are supplemented by zeros. Obviously, the spectral sequence degenerates in the $(n+1)$th term $E_{n+1}^{s,t} = \operatorname{Ker} d_n^{s,t} / \operatorname{Ran} d_n^{s+n,t-n+1}$. Therefore, we have the following formulas for the cohomology of the considered Lie algebroid:

(4.24) $$\mathcal{H}^k(A) \simeq \mathbf{h}^k \quad \forall k = 0, 1, \ldots, n-2,$$

(4.25) $$\mathcal{H}^{n-1}(A) \simeq \operatorname{Ker} \varpi_{n-1},$$

(4.26) $$\mathcal{H}^k(A) \simeq (\operatorname{Ker} \varpi_k) \oplus (\mathbf{h}^{k-n}/\operatorname{Ran} \varpi_{k-1}), \quad k \geq n.$$

EXAMPLE 4.3. Suppose that the base of the considered transitive Lie algebroid is simply connected ($\pi_1(B) = 0$). In this case A is quasiparallelizable and $E_2^{1,t} = 0$. Therefore, we have a linear mapping

$$\mathcal{E}_k = d_2^{0,k} \colon \mathbf{h}^k \to \mathbf{h}^{k-1} \otimes H^2(B),$$

and $\mathcal{H}^1(A) \simeq \operatorname{Ker} \mathcal{E}_1$. If in addition $H^3(B) = 0$, then

$$\mathcal{H}^2(A) \simeq \operatorname{Ker} \mathcal{E}_2 \oplus \left(H^2(B)/\operatorname{Ran} \mathcal{E}_1 \right).$$

EXAMPLE 4.4. Suppose that the Lie algebroid A admits a flat adjoint connection $\nabla^{\mathcal{P}}$. Also suppose that there exists an ad-invariant subbundle $\mathfrak{h}$, transversal to $z\mathfrak{g}$ ($\mathfrak{g} = z\mathfrak{g} \oplus \mathfrak{h}$). Consider operators (4.7)–(4.10) in $K^{s,t}$ (4.2), corresponding to the decomposition $A = \mathcal{P} \oplus z\mathfrak{g} \oplus \mathfrak{h}$. Then the following equalities hold:

$$\delta = 0, \quad d_1^2 = 0.$$

In other words, the coboundary operator (4.6) is a sum of two coboundary operators d_0 and d_1; the calculation of cohomology for such an operator falls in the well-known procedure for the spectral sequence of a double complex (see, e.g., [**BT**]).

If in addition we assume that the holonomy group of this flat connection is trivial (for example, B is simply connected), then it is easy to see that the spectral sequence degenerates at $E_2^{s,t}$ (i.e. $d_r^{s,t} = 0 \ \forall r \geq 2$), and

$$\mathcal{H}^k(A) \simeq \bigoplus_{t+s=k} \mathbf{h}^t \otimes \mathcal{H}^s(A_0).$$

In some applications (see, for example, [**KM**]) it is important to describe the cocycles $Z^k(A)$ of the operator D. We give some formulas for $Z^1(A)$ and $Z^2(A)$.

By $Z^{s,t}_{d_0}$ ($B^{s,t}_{d_0}$) and $Z^{s,t}_{d_1}$ ($B^{s,t}_{d_1}$) we denote the kernels (images) of the operators d_0 and d_1 respectively. Let us fix a linear mapping $d_1^{-1} \colon B^{s+1,t}_{d_1} \to \mathcal{K}^{s,t}$ such that $d_1 d_1^{-1} = id_{B^{s+1,t}_{d_1}}$. Let us consider the following linear spaces:

(4.27) $$L_0^{s,t} \stackrel{\text{def}}{=} \{\eta \in Z^{s,t}_{d_0} \mid d_1\eta \in B^{s,t}_{d_0}\},$$

(4.28) $$L_1^{s,t} \stackrel{\text{def}}{=} \{\eta \in L_0^{s,t} \mid \delta\eta \in B^{s+2,t-1}_{d_1} \text{ and } d_0 d_1^{-1}\delta\eta = d_1\eta\},$$

(4.29) $$L_2^{0,2} \stackrel{\text{def}}{=} \{\eta \in L_1^{0,2} \mid \delta d_1^{-1}\delta\eta \in B^{3,0}_{d_1}\}.$$

PROPOSITION 4.12. *There exist isomorphisms*
$$i_1 \colon L_1^{0,1} \oplus L_0^{1,0} \xrightarrow{\simeq} Z^1(A),$$
$$i_2 \colon L_2^{0,2} \oplus L_1^{1,1} \oplus L_0^{2,0} \xrightarrow{\simeq} Z^2(A)$$

that are given by the formulas

(4.29) $$i_1(\eta^{0,1} + \eta^{1,0}) = e^{0,1}\eta^{0,1} + e^{1,0}(\eta^{1,0} - d_1^{-1}\delta\eta^{0,1}),$$

(4.30) $$\begin{aligned} i_2(\eta^{0,2} + \eta^{1,1} + \eta^{2,0}) &= e^{0,2}\eta^{0,2} + e^{1,1}(\eta^{1,1} - d_1^{-1}\delta\eta^{0,2}) \\ &\quad + e^{2,0}(\eta^{2,0} - d_1^{-1}\delta\eta^{1,1} + d_1^{-1}\delta d_1^{-1}\delta\eta^{0,2}) \end{aligned}$$

(here $\eta^{s,t} \in L_t^{s,t}$).

The proof is a straightforward verification.

§5. Abelian transitive algebroids

We shall say that two Lie algebroids are isomorphic: $(A_1 \to B, q_1, \{\,,\,\}_1) \stackrel{f}{\simeq} (A_2 \to B, q_2, \{\,,\,\}_2)$ if there exists an isomorphism of vector bundles $f \colon A_1 \to A_2$ such that $q_2 \circ f = q_1$ and

(5.0) $$f\{\alpha, \beta\}_1 = \{f\alpha, f\beta\}_2 \qquad \forall \alpha, \beta \in \Gamma(A_1).$$

Let $(A \to B, q, \{\,,\,\})$ be an Abelian transitive algebroid. As pointed out in Corollary 3.4, the adjoint connection in the Abelian algebroid is flat and independent of the choice of the transversal subbundle $\mathcal{P}$; we denote the adjoint connection by ∇. Note that if we obtain an Abelian transitive algebroid by restriction of the algebroid for a regular Poisson manifold, then the adjoint connection is a Bott-type connection [**Bo**$_1$, **Li**$_3$] associated to the symplectic foliation.

By choosing a transversal subbundle $\mathcal{P}$ (3.8), we obtain an isomorphism of Lie algebroids $(A \to B, q, \{\,,\,\}) \stackrel{f_\mathcal{P}}{\simeq} (TB \oplus \mathfrak{g} \to B, q_1, \{\,,\,\}_{R^\mathcal{P}})$, where q_1 is the projection of $TB \oplus \mathfrak{g}$ on the first summand, $f_\mathcal{P}\alpha = q\alpha \oplus (\alpha - pq\alpha)$, and the bracket $\{\,,\,\}_{R^\mathcal{P}}$ is defined as follows:

(5.1) $$\{u_1 \oplus \theta_1, u_2 \oplus \theta_2\}_{R^\mathcal{P}} = [u_1, u_2] \oplus \bigl(R^\mathcal{P}(u_1, u_2) + \nabla_{u_1}\theta_2 - \nabla_{u_2}\theta_1\bigr).$$

By $Z^s_\nabla(\mathfrak{g})$ and $H^s_\nabla(\mathfrak{g})$ we denote the cocycles and cohomology of the differential (A.4) corresponding to the adjoint connection in the vector bundle $\mathfrak{g}$ (see §7).

LEMMA 5.1. (a) $R^{\mathcal{P}} \in Z^2_{\nabla}(\mathfrak{g})$.

(b) *The cohomology class $[R^{\mathcal{P}}] \in H^2_{\nabla}(\mathfrak{g})$ is independent of the choice of a transversal subbundle $\mathcal{P}$.*

PROOF. (a). The equality $\nabla R^{\mathcal{P}} = 0$ follows from the Jacoby identity for the bracket (5.1) and coincides with the Bianchi identity for the connection in the Lie algebroid in the sense of Mackenzie.

(b). Suppose that $\mathcal{P}$ and $\mathcal{P}'$ are two transversal subbundles (3.8), and p and p' are the corresponding morphisms (3.9). We define $T \in \Gamma(T^*B \otimes \mathfrak{g})$ as $T(u) \stackrel{\text{def}}{=} pu - p'u$ ($u \in \mathcal{V}^1(B)$). Then

$$\begin{aligned} R^{\mathcal{P}}(u_1, u_2) - R^{\mathcal{P}'}(u_1, u_2) &= \{pu_1, pu_2\} - p[u_1, u_2] - \{p'u_1, p'u_2\} + p'[u_1, u_2] \\ &= \{pu_1, pu_2 - p'u_2\} + \{pu_1 - p'u_1, p'u_2\} - T([u_1, u_2]) \\ &= \nabla_{u_1} T(u_2) - \nabla_{u_2} T(u_1) - T([u_1, u_2]) = \nabla T(u_1, u_2). \end{aligned}$$

Hence, we have proved that $R^{\mathcal{P}} = R^{\mathcal{P}'} + \nabla T$. The lemma is proved. □

The above lemma states that the structure of an Abelian algebroid uniquely determines a certain cohomology class in $H^2_{\nabla}(\mathfrak{g})$. The converse statement also holds. Suppose that the vector bundle $\mathcal{E} \to B$ has a flat connection ∇. By $\text{Iso}_\nabla(\mathcal{E})$ we denote the group of isomorphisms of the vector bundle $\mathcal{E}$, preserving the connection ∇:

$$\text{Iso}_\nabla(\mathcal{E}) = \{F \in \Gamma(GL(\mathcal{E})) \mid \nabla_u F\theta = F\nabla_u \theta \ \forall u \in \mathcal{V}^1(B), \theta \in \Gamma(\mathcal{E})\}.$$

The group $\text{Iso}_\nabla(\mathcal{E})$ acts on the space $H^s_\nabla(\mathcal{E})$ in the natural way: $F[R] = [FR]$.

THEOREM 5.2. (a) *Each section $R \in Z^2_\nabla(\mathcal{E})$ determines the structure of an Abelian algebroid $(TB \oplus \mathcal{E} \to B, q, \{\,,\,\}_R)$, where q is the projection on the first summand and the bracket $\{\,,\,\}_R$ is defined by (5.1).*

(b) *The space of nonisomorphic structures of Abelian algebroids on $A = TB \oplus \mathcal{E}$ with adjoint connection ∇ coincides with the space of orbits for the action of the group $\text{Iso}_\nabla(\mathcal{E})$ in the space $H^2_\nabla(\mathcal{E})$.*

PROOF. Statement (a) is verified by direct calculations; it is an Abelian version of a more general statement that can be found in [**Mz**].

(b) Let us consider two Abelian algebroids $(TB \oplus \mathcal{E} \to B, q, \{\,,\,\}_1)$ and $(TB \oplus \mathcal{E} \to B, q, \{\,,\,\}_2)$, where q is the projection on the first summand and the brackets $\{\,,\,\}_1, \{\,,\,\}_2$ are defined by the formula

(5.1i) $$\{u_1 \oplus \theta_1, u_2 \oplus \theta_2\}_i = [u_1, u_2] \oplus (R_i(u_1, u_2) + \nabla_{u_1} \theta_2 - \nabla_{u_2} \theta_1),$$

where $R_i \in Z^2_\nabla(\mathcal{E})$, $i = 1, 2$. Let f be an isomorphism of the vector bundle $TB \oplus \mathcal{E}$ such that $qf = q$. We write

(5.2) $$f(u \oplus \theta) = u \oplus (T(u) + F\theta), \qquad T \in C^1(\mathcal{E}), \quad F \in \Gamma(GL(\mathcal{E})).$$

It is easy to see that, by (5.0), f in (5.2) is an isomorphism of the algebroids (5.1i) if and only if the following two conditions are satisfied:

(5.3) $$F\nabla_u \theta = \nabla_u F\theta \qquad \forall u \in \mathcal{V}^1(B), \quad \theta \in \Gamma(\mathcal{E});$$
(5.4) $$R_2(u_2, u_1) = FR_1(u_2, u_1) + \nabla T(u_1, u_2).$$

Formula (5.3) means that $F \in \mathrm{Iso}_\nabla(\mathcal{E})$. Formula (5.4) means that the cohomology classes $[R_1], [R_2] \in H^2_\nabla(\mathcal{E})$ lie on the same orbit of the action of the group $\mathrm{Iso}_\nabla(\mathcal{E})$. The theorem is proved. $\square$

COROLLARY 5.3. *Let the base B be simply connected. Then the space of non-isomorphic structures of an Abelian algebroid on the bundle $TB \oplus \mathbb{R}^n$ is isomorphic to the space*

$$(5.5) \quad (H^2(B) \otimes \mathbb{R}^n)/GL(n) \simeq G_0(H^2(B)) \sqcup G_1(H^2(B)) \sqcup \cdots \sqcup G_n(H^2(B)),$$

where $G_k(V)$ is the space of k-dimensional planes in the linear space V and $\sqcup$ is the disjoint union.

REMARK 5.4. In fact, Corollary 5.3 says that up to an isomorphism the structure of an Abelian algebroid that has a trivial holonomy group is uniquely determined by a linear subspace in $H^2(B)$.

Now we calculate the cohomology of the Abelian algebroid. Consider the spectral sequence $(E_r^{s,t}, d_r^{s,t})$ corresponding to the subbundle $\mathfrak{g}$ (see Section 4). Then the algebroid $(A_0 \to B, q, \{,\}_0)$ is isomorphic to the trivial algebroid $(TB \to B, \mathrm{id}_{TB}, [\,])$, and $K^{s,t} = \Gamma(\Lambda^t \mathfrak{g}^* \otimes \Lambda^s T^* B)$.

By $Z_\nabla^{s,t}$, $B_\nabla^{s,t}$, and $H_\nabla^{s,t}$, we denote the cocycles, coboundaries, and cohomology of the operator $\nabla^{s,t}$ (A.4), (4.9), corresponding to the flat connection in the bundle $\Lambda^t \mathfrak{g}^*$, induced by the coadjoint connection. By Proposition 4.4, we have

$$(E_2^{s,t}, d_2^{s,t}) \simeq (H_\nabla^{s,t}, \widetilde{\delta}),$$

where the operator $\widetilde{\delta}\colon H_\nabla^{s,t} \to H_\nabla^{s+2,t-1}$ generated by the operator δ in (4.10) is the pairing of the cohomology class from $H_\nabla^{s,t} = H_\nabla^s(\Lambda^t \mathfrak{g}^*)$ with the cohomology class $[R^\mathcal{P}] \in H_\nabla^2(\mathfrak{g})$ (cf. Theorem 8 in [**HS**]). As a corollary, we obtain the following proposition.

PROPOSITION 5.5. *If $[R^\mathcal{P}] = 0$, then $\mathcal{H}^k(A) = \bigoplus_{s+t=k} H_\nabla^{s,t}$.*

For each linear subspace $M \subset Z_\nabla^{s,t}$ we define

$$(\nabla^{-1}\delta)M \stackrel{\mathrm{def}}{=} \{\zeta \in \mathcal{K}^{s+1,t-1} \mid \nabla^{s+1,t-1}\zeta \in \delta M\}.$$

Let us define the sequence of embedded linear spaces

$$Z_\nabla^{s,t} = L_0^{s,t} \supset L_1^{s,t} \supset \cdots \supset L_t^{s,t} = L_\infty^{s,t},$$

$$L_r^{s,t} \stackrel{\mathrm{def}}{=} \{\eta \in L_{r-1}^{s,t} \mid \big(\delta(\nabla^{-1}\delta)^{r-1}\{\eta\}\big) \cap B_\nabla^{s+r+1,t-r} \neq \varnothing\}, \qquad r \geq 1.$$

Since $\delta \nabla^{s,t} = \nabla^{s+2,t-1}\delta$ (see (4.14)), we obviously have $B_\nabla^{s,t} \subset L_\infty^{s,t}$ and $\delta Z_\nabla^{s-2,t+1} \subset L_\infty^{s,t}$.

THEOREM 5.6. *There exists an isomorphism of the following spaces:*

$$(5.6) \qquad Z^k(A) \simeq \bigoplus_{s=0}^{k} L_{k-s}^{s,k-s},$$

$$(5.7)\ \mathcal{H}^k(A) \simeq L_k^{0,k} \oplus \big(L_{k-1}^{1,k-1}/B_\nabla^{1,k-1}\big) \oplus \bigoplus_{s=2}^{k} L_{k-s}^{s,k-s}/(B_\nabla^{s,k-s} + \delta Z_\nabla^{s-2,k-s+1}).$$

The proof of this theorem follows from the standard consideration (see, e.g., [**BT**]) of the spectral sequence for the double complex $(K^{s,t}, d = (-1)^t \nabla^{s,t} + \delta)$; in particular, $d_r^{s,t} = \delta(\nabla^{-1}\delta)^{r-1}$ and the isomorphism $i_k \colon \bigoplus_{s=0}^{k} L_{k-s}^{s,k-s} \xrightarrow{\simeq} Z^k(A)$ is given by the formula

$$i_k\left(\sum_{s=0}^{k} \eta^{s,k-s}\right) = \sum_{s=0}^{k} e^{s,k-s} \sum_{l=0}^{s} (-d_1^{-1}\delta)^{s-l} \eta^{l,k-l},$$

where $\eta^{s,k-s} \in L_{k-s}^{s,k-s}$ and to each $\eta^{s,k-s}$ the operator $(-d_1^{-1}\delta)^r$ assigns the last element from the chain of equalities:

$$\delta \eta^{s,k-s} = -d_1\bigl((-d_1^{-1}\delta)\eta^{s,k-s}\bigr),$$
$$\delta\bigl((-d_1^{-1}\delta)\eta^{s,k-s}\bigr) = -d_1\bigl((-d_1^{-1}\delta)^2 \eta^{s,k-s}\bigr),$$
$$\dots\dots\dots\dots\dots\dots\dots\dots\dots\dots\dots\dots\dots\dots\dots$$
$$\delta\bigl((-d_1^{-1}\delta)^{r-1}\eta^{s,k-s}\bigr) = -d_1\bigl((-d_1^{-1}\delta)^r \eta^{s,k-s}\bigr)$$

REMARK. Note that as a corollary of Theorem 5.2 we obtain that the cohomology space of an Abelian transitive algebroid can be calculated using only the cohomology class $[R^{\mathcal{P}}] \in H^2_\nabla(\mathcal{E})$. In the case of a trivial holonomy group this means that the cohomology of an algebroid is calculated in terms of multiplicative properties of some subspace in $H^2(B)$. This approach has been used in a number of papers. Formulas similar to (5.6) and (5.7) for the dimensions $k = 1, 2, 3$ were obtained in [**VK**$_{1,2}$] for some kinds of regular Poisson manifolds (see the algebroid version of these formulas in Theorem 5.7). Formulas similar to (5.7) were obtained in [**Xu**] for any k. For the spectral sequence of regular Poisson manifold see also [**Va**].

The case of a trivial holonomy group. Let us consider the case in which the holonomy group of the adjoint connection is trivial. This happens if the base of an Abelian algebroid is simply connected. In the case of an Abelian algebroid over a nondegenerate symplectic leaf $\mathcal{O}$ of a Poisson manifold, for a holonomy group to be trivial it also suffices that there exists a set of Casimir functions $k_1, k_2, \ldots, k_n$ ($n = \mathrm{codim}\,\mathcal{O} = \dim \mathfrak{g}_x$) in a tubular neighborhood of this leaf such that their differentials are linearly independent at each point of this leaf.

Let us choose a basis of parallel sections $X_1, X_2, \ldots, X_n \in Z^{0,1}_\nabla$, $n = \dim \mathfrak{g}_x$, such that the closed differential forms

(5.8) $$\omega_j \stackrel{\text{def}}{=} \delta X_j, \qquad j = 1, 2, \ldots, n,$$

satisfy the conditions

(5.9) $\quad [\omega_1], \ldots, [\omega_q]$ are linearly independent in $H^2(B)$,
$\quad \omega_{q+1} = d\beta_{q+1}, \omega_{q+2} = d\beta_{q+2}, \ldots, \omega_n = d\beta_n, \qquad \beta_j \in \Gamma(T^*B)$.

Obviously, $\mathrm{Span}\{[\omega_j]\} \subset H^2(B)$ is a subspace that determines the structure of the considered Abelian algebroid (see Remark 5.4). In particular, if all the forms ω_j are exact ($q = 0$), i.e., if $[R^{\mathcal{P}}] = 0$, then we have $d_2^{s,t} = \delta = 0$ and $\mathcal{H}^k(A) \simeq \bigoplus_{s+t=k} (\bigwedge^t \mathfrak{g}_z) \otimes H^s(B)$.

Condition (5.9) means that $L_1^{0,1} = \mathrm{Span}\{X_j\}_{j=q+1,\ldots,n}$.

THEOREM 5.7. *Let the holonomy group of the adjoint connection in an Abelian transitive algebroid be trivial. Then*

$$\mathcal{H}^1(A) \simeq L_1^{0,1} \oplus H^1(B),$$
$$\mathcal{H}^2(A) \simeq \bigwedge^2 L_1^{0,1} \oplus W^1 \oplus (L_1^{0,1} \otimes H^1(B)) \oplus \widetilde{H}^2(B),$$

and

$$H^1(B) = 0 \implies \mathcal{H}^3(A) \simeq \bigwedge^3 L_1^{0,1} \oplus (\widetilde{H}^2 \otimes L_1^{0,1}) \oplus \widetilde{W}^2 \oplus H^3(B),$$

where

$$\widetilde{H}^s(B) \stackrel{\text{def}}{=} H^s(B) / \big(\operatorname{Span}\{[\omega_j]\}_{j=1,\ldots,q} \wedge H^{s-2}(B) \big),$$
$$W^s \subset (Z_\nabla^{0,1} / L_1^{0,1}) \otimes H^s(B),$$
$$W^s \stackrel{\text{def}}{=} \Big\{ \sum_{j=1}^q X_j \otimes [\varkappa_j] \mid [\varkappa_j] \in H^s(B), \sum_{j=1}^q \omega_j \wedge \varkappa_j = d\beta, \ \beta \in \Gamma(\wedge^{s+1} T^* B) \Big\},$$
$$\widetilde{W}^2 \stackrel{\text{def}}{=} W^2 / \operatorname{Span} \{ (X_j \otimes [\omega_i] - X_i \otimes [\omega_j]) \mid i,j \leq q \}.$$

The proof of this theorem follows from (5.7) and, in fact, is presented in [**VK**$_1$] for the case of regular Poisson manifolds.

§6. Cohomology of homogeneous algebroids, examples

Let $\mathbb{G}$ a Lie group, $\mathbb{H}$ a closed connected subgroup, and $B \stackrel{\text{def}}{=} \mathbb{H} \setminus \mathbb{G}$ a *right homogeneous manifold*.

Let us consider a mapping $\tau^* \colon C^\infty(B) \to C^\infty(\mathbb{G})$, $\tau^* \varphi \stackrel{\text{def}}{=} \varphi \circ \tau$, where $\tau \colon \mathbb{G} \to B$ is the natural projection. Obviously, the image of τ^* is the space of functions invariant under the left action $L_h g = hg$ of $\mathbb{H}$ on the Lie group $\mathbb{G}$.

By G we denote a Lie algebra of left-invariant vector fields on the Lie group $\mathbb{G}$. We consider a homomorphism $q' \colon G \to \mathcal{V}^1(B)$,

(6.0) $$\mathcal{L}_{q'u} \varphi = (\tau^*)^{-1} \mathcal{L}_u \tau^* \varphi, \quad u \in G, \quad \varphi \in C^\infty(B).$$

This homomorphism can be extended in a natural way to the morphism of vector bundles $q \colon B \times G \to TB$, where $q\imath u = q'u$, and $\imath \colon G \to \Gamma(B \times G)$ is an isomorphism between the space of constant sections of a trivial vector bundle $A \stackrel{\text{def}}{=} B \times G \to B$ and G.

A transitive Lie algebroid $(A \to B, q, \{\,,\,\})$, where $A = B \times G$, q is defined as above, and $\{\,,\,\}$ is defined by (2.3) (see Example 2 in section 2), will be called a *homogeneous algebroid*.

In fact, the cohomology of a homogeneous algebroid is exactly the cohomology of the Lie algebra G with coefficients in the G-module $C^\infty(\mathbb{H} \setminus \mathbb{G})$. Nevertheless, we prefer to use the algebroid approach.

The following theorem gives a "lower bound" for the cohomology of a homogeneous algebroid in the case when $\mathbb{H} \setminus \mathbb{G}$ is compact.

PROPOSITION 6.1. *Suppose that $B = \mathbb{H} \setminus \mathbb{G}$ is compact and possesses a $\mathbb{G}$-invariant measure μ. Then the homomorphism $\imath \colon G \to \Gamma(A)$ induces the inclusion of the cohomology of a Lie algebra into the cohomology of a homogeneous algebroid*

(6.1) $$H^{\cdot}(G) \hookrightarrow \mathcal{H}^{\cdot}(A).$$

To prove this proposition we need the following simple algebraic lemma.

LEMMA 6.2. *Let $\mathcal{K}$ be a linear space endowed with the differential d, $d^2 = 0$, and with a chain projection $I\colon \mathcal{K} \to \mathcal{K}$,*

(6.2) $$I^2 = I, \qquad dI - Id = 0.$$

Then there is an exact sequence

$$0 \to H_d \operatorname{Ran} I \to H_d \mathcal{K} \to H_d \operatorname{Ker} I \to 0,$$

where H_d denotes the cohomology of d.

PROOF OF PROPOSITION 6.1. Let us consider an operator $\imath_*\colon C^k(G) \to C^k(A)$ such that $(\imath_* c)(\imath u_1, \ldots, \imath u_k) = c(u_1, u_2, \ldots, u_k)$. Obviously, $\imath_*$ is a chain mapping: $D\imath_* = \imath_* d_G$, where d_G is the standard differential in $C^{\cdot}(G) \oplus_k \bigwedge^k G^*$.

Let us also introduce an averaging operator $I_\mu \colon C^k(A) \to C^k(A)$ by the formula

$$(I_\mu f)(\imath u_1, \ldots, \imath u_k) = \int_B f(\imath u_1, \ldots, \imath u_k)\mu, \qquad f \in C^k(A), \quad u_j \in G.$$

The image of I_μ is exactly the image of the inclusion $\imath_*$, so that the operator $\imath_*$ induces an isomorphism of $H^k(G)$ and the cohomology of $\operatorname{Ran} I_\mu$ with the differential D (2.4).

Since B is compact, we may suppose that $\int_B \mu = 1$, and therefore $I_\mu^2 = I_\mu$. Since μ is invariant under the action of $\mathbb{G}$, we have

$$\int_B \mathcal{L}_{q'u}(\varphi)\mu = 0 \qquad \forall u \in G, \quad \forall \varphi \in C^\infty(B),$$

and therefore

$$(DI_\mu f - I_\mu D f)(\imath u_1, \ldots, \imath u_k)$$
$$= -\sum_{j=0}^{k}(-1)^j \int_B \mathcal{L}_{q'u_j}\bigl(f(\imath u_o, \imath u_1, \ldots, \widehat{\imath u_j}, \ldots \imath u_k)\bigr)\mu = 0.$$

By applying Lemma 6.2 to the chain projection I_μ, we complete the proof. $\square$

The following proposition is somewhat dual to Proposition 6.2.

PROPOSITION 6.3. *Suppose that $\mathbb{H}$ is compact. Then there is an isomorphism of Grassmann algebras*

(6.3) $$\mathcal{H}^{\cdot}(A) \simeq H^{\cdot}(\mathbb{G}),$$

where $H^{\cdot}(\mathbb{G})$ is the de Rham cohomology of $\mathbb{G}$.

Although the proof of this proposition can be derived from [**Mz**] or, probably, from another source unknown to the authors, we give a brief sketch of the proof.

PROOF. Let us define the inclusion mapping $\tau_k^*\colon C^k(A) \to \Omega^k(\mathbb{G})$ of $C^k(A)$ into the space of differential forms on the group $\mathbb{G}$ by the formula

$$(\tau_k^* f)(u_1, \ldots, u_k) = \tau^*\bigl(f(\imath u_1, \ldots, \imath u_k)\bigr), \qquad u_j \in G, \quad f \in C^k(A).$$

The image of τ_k^* is the space of differential forms $\Omega_L^k(\mathbb{G}, \mathbb{H})$ that are invariant under the left action of the subgroup $\mathbb{H}$

$$(6.4) \qquad \operatorname{Ran} \tau_k^* = \Omega_L^k(\mathbb{G}, \mathbb{H}) \stackrel{\text{def}}{=} \{\omega \in \Omega^k(\mathbb{G}) \mid L_h^*\omega = \omega \ \forall h \in \mathbb{H}\}.$$

The inclusion mapping τ_k^* satisfies the conditions

$$\tau_{k+1}^* D = d\tau_k^*,$$
$$\tau_{k_1+k_2}^* f_1 \wedge f_2 = \tau_{k_1}^* f_1 \wedge \tau_{k_2}^* f_2, \qquad f_i \in C^{k_i}(A), \quad i=1,2,$$

and thus, defines an isomorphism between Grassmann algebras $\mathcal{H}^k(A)$ and the de Rham cohomology $H_L^k(\mathbb{G}, \mathbb{H})$ of $\mathbb{H}$-left-invariant differential forms.

By μ_h we denote the Haar measure on a compact group $\mathbb{H}$ such that $\int \mu_h = 1$. Introduce an operator $I_\mathbb{H} \colon \Omega^k(\mathbb{G}) \to \Omega^k(\mathbb{G})$:

$$I_\mathbb{H}\omega \stackrel{\text{def}}{=} \int L_h^*\omega \mu_h.$$

The image of $I_\mathbb{H}$ is exactly the space of $\mathbb{H}$-left-invariant differential forms (6.4). It can be easily verified that $I_\mathbb{H}$ satisfies conditions (6.2). Applying Lemma 6.2, it remains to prove that each closed differential form that lies in $\operatorname{Ker} I_\mathbb{H}$ is exact.

Let $[\omega] \in H^k(\mathbb{G})$, $I_\mathbb{H}\omega = 0$. Since the subgroup $\mathbb{H}$ is connected, the diffeomorphism $L_h \colon \mathbb{G} \to \mathbb{G}$ is homotopic to the identity, and thus, generates an identical mapping in $H^k(\mathbb{G})$:

$$L_h^*\omega = \omega + d\widetilde{\omega}_h.$$

Since $\omega = \operatorname{Ker} I_\mathbb{H}$, we have

$$0 = \int L_h^*\omega \mu_h = \int (\omega + d\widetilde{\omega}_h)\mu_h = \omega + d\int \widetilde{\omega}_h \mu_h;$$

hence, $\omega = -d\int \widetilde{\omega}_h \mu_h$. This completes the proof. $\square$

REMARK 6.4. Suppose that the group $\mathbb{G}$ is compact. If we consider the subgroup $\mathbb{H} = \mathbb{G}$ (i.e., the case in which B is a point), we obtain the classical result $H^{\cdot}(\mathbb{G}) = H^{\cdot}(G)$. For another closed subgroup $\mathbb{H} \subset \mathbb{G}$ we obtain the transitive case of the result proved in [**GW**]: $\mathcal{H}^{\cdot}(A) = H^{\cdot}(G)$.

Let us consider the coadjoint action of a Lie group $\mathbb{G}$ on its coalgebra G^*. On the Euclidean space G^* a *linear Poisson bracket* is defined (see, e.g., [**Ki**$_1$]) and the Lie algebroid $(G \times G^* \to G^*, q, \{\,,\,\})$, which can be constructed by this Poisson bracket, falls into both examples of Section 2. For a tame orbit $\mathcal{O}$ let us consider a reduced homogeneous algebroid $(G \times \mathcal{O} \to \mathcal{O}, q, \{\,,\,\})$. Each orbit $\mathcal{O}$ of a coadjoint action possess the invariant measure $\mu = \underbrace{\omega_\mathcal{O} \wedge \omega_\mathcal{O} \wedge \cdots \wedge \omega_\mathcal{O}}_{\frac{1}{2}(\dim \mathcal{O}) \text{ times}}$ which is the symplectic volume corresponding to the Kirillov symplectic form $\omega_\mathcal{O}$ (see [**Ki**$_1$]). Therefore, for each homogeneous algebroid over the compact orbit of coadjoint action the assumptions of Proposition 6.1 are satisfied.

Poisson cohomology of transitive Lie algebroids over $e(3)^*$. Below we calculate the cohomology for homogeneous algebroids over the orbits of the coadjoint action of the group $E(3)$. As will be shown, without the assumption that the group $\mathbb{G}$ is compact, homogeneous algebroids over orbits of different types may have different cohomology. This example also shows that without any additional assumptions the inclusion (6.1) need not be an isomorphism.

Let us consider a Lie group $\mathbb{G} = E(3)$ of motions in the Euclidean space $\mathbb{R}^3$. In $\mathcal{N} = e(3)^*$ we introduce coordinates $(x_1, x_2, x_3, y_1, y_2, y_3)$ so that they satisfy the following commutation relations under the linear Poisson bracket:

$$\{y_i, y_j\} = 0, \qquad i,j = 1,2,3;$$
$$\{x_1, x_2\} = x_3, \qquad \{x_2, x_3\} = x_1, \qquad \{x_3, x_1\} = x_2,$$
$$\{x_1, y_2\} = y_3, \qquad \{x_2, y_3\} = y_1, \qquad \{x_3, y_1\} = y_2,$$

The linear Poisson bracket in $\mathcal{N} = e(3)^*$ has two Casimir functions:

$$k_1 = y_1^2 + y_2^2 + y_3^2, \qquad k_2 = x_1 y_1 + x_2 y_2 + x_3 y_3.$$

There are two kinds of orbits of coadjoint action in $e(3)^*$ (see, e.g., [**MR**]):
  (i) *nondegenerate orbits* $\mathcal{O}_{k_1,k_2} = \{k_1 = \text{const} \neq 0, k_2 = \text{const}\} \simeq TS^2$;
  (ii) *degenerate orbits* $\mathcal{O}_\rho = \{x_1^2 + x_2^2 + x_3^2 = \rho^2 \neq 0, y_1 = y_2 = y_3 = 0\} \simeq S^2$.

Nondegenerate case. Let us calculate the cohomology of the homogeneous algebroid over a nondegenerate orbit $\mathcal{O}_{k_1,k_2}$. Following [**VK**$_2$] we introduce vector fields $u_1, u_2 \in \mathcal{V}^1(e(3)^*)$ such that

(6.5) $$\mathcal{L}_{u_j}(k_i) = \delta_{ji}$$

(here δ_{ij} is the Kronecker symbol). By (6.5) we have $X_j \stackrel{\text{def}}{=} r^*_{\mathcal{O}_{k_1,k_2}} u_j \in \Gamma(\mathfrak{g}^*)$ of vector fields u_j to the base $\mathcal{O}_{k_1,k_2}$ gives the basis of parallel sections in the bundle $\mathfrak{g}^*$. The differential forms ω_j (5.8) can be determined from the relation $Du_j = q^*\widetilde{\omega}_j$, where D is the Lichnerowicz differential (1.6) on the Poisson manifold $\mathcal{N} = e(3)^*$, $q^* : \bigwedge^2 T^*\mathcal{N} \to \bigwedge^2 T\mathcal{N}$ is defined by the morphism q (1.1), and the natural restriction $r : \Gamma(\bigwedge^k T^*\mathcal{N}) \to \Gamma(\bigwedge^k T\mathcal{O}_{k_1,k_2})$ gives the closed differential forms $\omega_j = r\widetilde{\omega}_j$.

Since the homogeneous algebroid over a nondegenerate orbit is Abelian, to calculate its cohomology we only need to determine the subspace $\mathrm{Span}\{[\omega_j]\}$ in the one-dimensional space $H^2(\mathcal{O}_{k_1,k_2})$. The vector fields u_1 and u_2 (6.5) are given by the formulas

$$u_1 = \sum_{j=1}^{3} \left(\frac{1}{2k_1} x_j - \frac{k_2}{(k_1)^2} y_j \right) \frac{\partial}{\partial x_j} + \frac{1}{2k_1} \sum_{i=1}^{3} y_i \frac{\partial}{\partial y_i},$$

$$u_2 = \frac{1}{k_1} \sum_{j=1}^{3} y_j \frac{\partial}{\partial x_j}.$$

It is easy to calculate that

$$Du_2 = \frac{2}{k_1}\left(y_1 \frac{\partial}{\partial x_2} \wedge \frac{\partial}{\partial x_3} + y_2 \frac{\partial}{\partial x_3} \wedge \frac{\partial}{\partial x_1} + y_3 \frac{\partial}{\partial x_1} \wedge \frac{\partial}{\partial x_2} \right) = q^*\widetilde{\omega}_2,$$

where $\widetilde{\omega}_2 = \frac{2}{k_1^2}(y_1 \, dy_2 \wedge dy_3 + y_2 \, dy_3 \wedge dy_1 + y_3 \, dy_1 \wedge dy_2)$ and the restriction $\omega_2 = r\widetilde{\omega}_2 \in \Gamma(\wedge^2 \mathcal{O}_{k_1,k_2})$ gives the basic form in $H^2(\mathcal{O}_{k_1,k_2})$. Therefore, the considered Lie algebroid falls in the nontrivial type of the only two possible types of Abelian algebroids over TS^2. The cohomology of this Lie algebroid is easily calculated by using either Theorems 5.7 and 5.6 or Example 4.2. The *Betti numbers*

$$b_k \stackrel{\text{def}}{=} \dim \mathcal{H}^k(A)$$

of this Lie algebroid are

(6.6) $\qquad b_1 = b_3 = b_4 = 1, \qquad b_2 = b_5 = b_6 = 0.$

Degenerate case. Let us consider a homogeneous algebroid over the degenerate orbit $\mathcal{O}_\rho = \{x_1^2 + x_2^2 + x_3^2 = \rho^2 \neq 0, \ y_1 = y_2 = y_3 = 0\}$. It should be noted that despite the fact that the vector bundle $\mathfrak{g}$ is trivial ($\mathfrak{g}$ has a basis of parallel sections $\{dy_j, \ j=1,2,3, \ \sum_{j=1}^3 x_j \, dx_j\}$), $\mathfrak{g}$ does not admit a flat adjoint connection. This follows from a fact that an invariant subbundle $\mathfrak{g}^1 = [\mathfrak{g}, \mathfrak{g}]$ is isomorphic (as a vector bundle) to $TS^2 \to S^2$, and thus, cannot have a flat linear connection.

We introduce a transversal subbundle $\mathcal{P}$ (3.8),

(6.7) $\qquad \mathcal{P}_x \stackrel{\text{def}}{=} \left\{ \beta_x \in T_x^* \mathcal{N} \ \Big| \ \left\langle \beta_x, \frac{\partial}{\partial y_j} \right\rangle = 0, \ j=1,2,3; \quad \left\langle \beta_x, \sum_{j=1}^3 x_j \frac{\partial}{\partial x_j} \right\rangle = 0 \right\}$

for $x \in \mathcal{O}_\rho$. The curvature of this transversal subbundle has the form

$$R^{\mathcal{P}}(u_1, u_2) = -\frac{1}{\rho^2} \omega_\rho(u_1, u_2) \, r_{\mathcal{O}_\rho}^* \left(\sum_{j=1}^3 x_j \, dx_j \right),$$

where $\omega_\rho = \frac{1}{\rho^2}(x_1 \, dx_2 \wedge dx_3 + x_2 \, dx_3 \wedge dx_1 + x_3 \, dx_1 \wedge dx_2)$ is the Kirillov symplectic form of the orbit $\mathcal{O}_\rho$; it has a nonzero cohomology class in $H^2(\mathcal{O}_\rho)$. We also introduce sections $Y_0, Y_1 \in \Gamma(\mathfrak{g}^*)$, $\eta \in \Gamma(\wedge^2 \mathfrak{g}^*)$,

(6.8) $\qquad Y_0 \stackrel{\text{def}}{=} r_{\mathcal{O}_\rho}^* \sum_{j=1}^3 x_j \frac{\partial}{\partial x_j}, \qquad Y_1 \stackrel{\text{def}}{=} r_{\mathcal{O}_\rho}^* \sum_{j=1}^3 x_j \frac{\partial}{\partial y_j},$

(6.9) $\qquad \eta \stackrel{\text{def}}{=} r_{\mathcal{O}_\rho}^* \left(x_1 \frac{\partial}{\partial y_2} \wedge \frac{\partial}{\partial y_3} + x_2 \frac{\partial}{\partial y_3} \wedge \frac{\partial}{\partial y_1} + x_3 \frac{\partial}{\partial y_1} \wedge \frac{\partial}{\partial y_2} \right).$

It is easy to calculate that these sections are parallel with respect to the connection induced by the adjoint connection $\nabla^{\mathcal{P}}$ (6.7).

We consider the spectral sequence that corresponds to the subbundle $\mathfrak{g}$ ($K^{s,t} = \Gamma(\wedge^t \mathfrak{g}^* \otimes \wedge^s T^* \mathcal{O}_\rho)$). The Lie algebra $\mathfrak{g}_x$ is isomorphic to $e(2) \oplus \mathbb{R}^1$ [**MR**], and it turns out that the cohomology space $E_1^{0,t} = H_{d_0} \Gamma(\wedge^t \mathfrak{g})$ can be written via parallel sections (6.8), (6.9):

$$E_1^{0,1} = \text{Span}(Y_0, Y_1),$$
$$E_1^{0,2} = \text{Span}([\eta]_{d_0}, [Y_0 \wedge Y_1]_{d_0}),$$
$$E_1^{0,3} = \text{Span}([\eta \wedge Y_0]_{d_0}, [\eta \wedge Y_1]_{d_0}),$$
$$E_1^{0,4} = \text{Span}([\eta \wedge Y_1 \wedge Y_0]_{d_0}).$$

This Lie algebroid falls in the case of Example 4.2. Therefore, $E_2^{s,t} = E_1^{s,t}$, and to calculate the cohomology space $\mathcal{H}^k(A)$ it remains to obtain the linear mapping ϖ (4.23) that is defined by $d_2^{0,t}$. Since all cohomology classes of $E_2^{s,t}$ can be represented via parallel sections, the mapping $d_2^{s,t}$ (4.18) coincides with δ (4.10) on such sections. It is easy to calculate that the following equalities hold:

$$d_2^{0,1}[Y_0] = [\omega_\rho], \qquad d_2^{0,1}[Y_1] = 0,$$
$$d_2^{0,2}[\eta] = 0, \qquad d_2^{0,2}[Y_0 \wedge Y_1] = -Y_1 \boxtimes [\omega_\rho],$$
$$d_2^{0,3}[Y_0 \wedge \eta] = [\eta] \boxtimes [\omega_\rho], \qquad d_2^{0,3}[Y_1 \wedge \eta] = 0,$$
$$d_2^{0,4}[\eta \wedge Y_1 \wedge Y_0] = -[\eta \wedge Y_1] \boxtimes [\omega_\rho].$$

These equalities give us the mapping (4.23), and thus, the cohomology spaces (4.24)–(4.26). We have the following Betti numbers of the Lie algebroid over $\mathcal{O}_\rho$:

$$b_1 = b_2 = b_4 = b_5 = b_6 = 1, \qquad b_3 = 2.$$

§7. Formal equivalence

Let $\mathcal{O}$ be a symplectic manifold, $\mathcal{E} \to \mathcal{O}$ a vector bundle, $p\colon \mathcal{E}^* \to \mathcal{O}$ its dual, and $o\colon \mathcal{O} \to \mathcal{E}^*$ the zero section.

We say that a Poisson bracket on the total manifold $\mathcal{E}^*$ is *proper* if the zero section $o\colon \mathcal{O} \to \mathcal{E}^*$ is a Poisson mapping. In other words, a Poisson bracket on $\mathcal{E}^*$ is proper if the image of the zero section $\mathcal{O} \hookrightarrow \mathcal{E}^*$ is a symplectic leaf, and the induced Poisson bracket on $\mathcal{O}$ coincides with the Poisson bracket of the symplectic structure on $\mathcal{O}$.

Proper Poisson brackets arise when we consider the geometry in the vicinity of a (possibly degenerate) symplectic leaf $\mathcal{O}$ of a Poisson manifold $\mathcal{N}$. We can consider a diffeomorphism between a tubular neighborhood of $\mathcal{O}$ in $\mathcal{N}$ and the normal vector bundle $\mathcal{E}^* = T_\mathcal{O}\mathcal{N}/T\mathcal{O}$ (note that this diffeomorphism is not unique). The pullback of the initial Poisson bracket to $\mathcal{E}^*$ is a proper Poisson bracket.

By μ we denote an ideal of the commutative ring $C^\infty(\mathcal{E}^*)$ of functions that vanish on the submanifold $\mathcal{O} \hookrightarrow \mathcal{E}^*$. Consider a filtration of ideals

$$(7.1) \qquad C^\infty(\mathcal{E}^*) = \mu^0 \supset \mu^1 \supset \mu^2 \supset \mu^3 \supset \cdots \supset \mu^\infty = \bigcap_{k=0}^\infty \mu^k,$$

where $\mu^k = \mu\mu^{k-1}$ is an ideal of functions that have k zero terms in the Taylor expansion along the fiber coordinates on $\mathcal{E}^*$. Considering $\mathcal{V}(\mathcal{E}^*) = \bigoplus_k \mathcal{V}^k(\mathcal{E}^*)$ as a module over the ring $C^\infty(\mathcal{E}^*)$, we also can establish a filtration

$$\mathcal{V}^k(\mathcal{E}^*) \supset \mu^1 \mathcal{V}^k(\mathcal{E}^*) \supset \mu^2 \mathcal{V}^k(\mathcal{E}^*) \supset \cdots \supset \mu^\infty \mathcal{V}^k(\mathcal{E}^*).$$

In what follows we are interested only in the behavior of all considered objects in an open small neighborhood of the symplectic leaf $\mathcal{O}$. We shall call $g\colon U \to g(U)$ a *local diffeomorphism* if U is an open neighborhood in $\mathcal{E}^*$ that contains $\mathcal{O}$ and the diffeomorphism g satisfies the condition $g(\mathcal{O}) = \mathcal{O}$. Local diffeomorphisms form a group.

We shall say that two proper Poisson brackets given by Poisson tensors $\Psi, \Psi_1 \in \mathcal{V}^2(\mathcal{E}^*)$ are *formally equivalent* if for any $p \geq 2$, there exists a local diffeomorphism

g such that $g|_{\mathcal{O}} = \mathrm{id}_{\mathcal{O}}$ and

(7.2) $$g_*\Psi - \Psi_1 \in \mu^p \mathcal{V}^2(\mathcal{E}^*).$$

Let $\Psi \in \mathcal{V}^2(\mathcal{E}^*)$ be a Poisson tensor of a proper Poisson bracket on $\mathcal{E}^*$. Let $(T^*_{\mathcal{O}}\mathcal{E}^* \to \mathcal{O}, \{\,,\,\}, q)$ be the corresponding transitive Lie algebroid. Note that the zero section of $\mathcal{E}^*$ induces a natural isomorphism $T_{\mathcal{O}}\mathcal{E}^* = T\mathcal{O} \oplus \mathcal{E}^*$, where $\mathcal{E}^*$ is identified with its vertical subbundle in $T\mathcal{E}^*$ restricted on $\mathcal{O}$. Therefore, we also have a natural isomorphism $T^*_{\mathcal{O}}\mathcal{E}^* = T^*\mathcal{O} \oplus \mathcal{E}$. It is easy to see that the properness of the considered Poisson bracket implies that $\mathfrak{g}_x = \mathrm{Ker}\, q|_x = \mathcal{E}_x$ ($x \in \mathcal{O}$), and in what follows we shall always identity $\mathcal{E}$ and $\mathfrak{g}$.

Let $\rho^p_\alpha \colon \Gamma(S^p\mathcal{E}) \to \Gamma(S^p\mathcal{E})$ be a representation of the algebroid $(T^*_{\mathcal{O}}\mathcal{E}^*, \{\,,\,\}, q)$ in the symmetric product of the vector bundle $\mathcal{E}$. This representation can be given by the following formulas (here we consider sections of $\mathcal{E}$ as sections of $\mathfrak{g} \subset T^*_{\mathcal{O}}\mathcal{E}^*$):

$$\rho^0_\alpha f = \mathcal{L}_{q\alpha} f, \qquad \alpha \in \Gamma(T^*_{\mathcal{O}}\mathcal{E}^*), \quad f \in C^\infty(\mathcal{O}) = \Gamma(S^0\mathcal{E});$$
$$\rho^1_\alpha \theta = \{\alpha, \theta\}, \qquad \theta \in \Gamma(\mathcal{E});$$
$$\rho^{p_1+p_2}_\alpha(s_1 \cdot s_2) = (\rho^{p_1}_\alpha s_1) \cdot s_2 + s_1 \cdot \rho^{p_2}_\alpha s_2, \qquad s_i \in \Gamma(S^{p_i}\mathcal{E})$$

(here the dot stands for symmetric multiplication in $S^p\mathcal{E}$). Define the standard coboundary operator $D_p \colon C^k_p \to C^{k+1}_p$ in $C^k_p = \Gamma(\bigwedge^k T_{\mathcal{O}}\mathcal{E}^* \otimes S^p\mathcal{E})$ associated to the representation ρ^p_α:

$$D_p\eta(\alpha_0, \ldots, \alpha_k) = \sum_{j=0}^k (-1)^j \rho^p_{\alpha_j} \eta(\alpha_0, \ldots, \widehat{\alpha_j}, \ldots, \alpha_k)$$
$$+ \sum_{i<j} (-1)^{i+j} \eta(\{\alpha_i, \alpha_j\}, \alpha_0, \ldots, \widehat{\alpha_i}, \ldots, \widehat{\alpha_j}, \ldots, \alpha_k), \quad \eta \in C^k_p.$$

Denote by $\mathcal{H}^k_p$ the cohomology space of the operator D_p.

THEOREM 7.1. *Let the symplectic manifold $\mathcal{O}$ be compact. If two proper Poisson brackets on $\mathcal{E}^*$ induce the same transitive Lie algebroid's structure on $T^*_{\mathcal{O}}\mathcal{E}^*$ such that*

(7.3) $$\mathcal{H}^2_p = 0 \quad \text{for any} \quad p \geq 2,$$

then these Poisson brackets are formally equivalent.

To give a sketch of the proof, we need the following lemmas.

LEMMA 7.1. *Two proper Poisson brackets with Poisson tensors $\Psi, \Psi_1 \in \mathcal{V}^2(\mathcal{E}^*)$ define the same Lie algebroid structure in $T^*_{\mathcal{O}}\mathcal{E}^*$ if and only if $\Psi - \Psi_1 \in \mu^2 \mathcal{V}^2(\mathcal{E}^*)$.*

Note that $\mu^p \mathcal{V}^k(\mathcal{E}^*)/\mu^{p+1}\mathcal{V}^k(\mathcal{E}^*) \simeq C^k_p$ and denote the natural projection by $\pi_p \colon \mu^p(\mathcal{E}^*) \to C^k_p$. Let $\Psi \in \mathcal{V}^2(\mathcal{E}^*)$ be the Poisson tensor of a proper Poisson bracket, and let $D_\Psi \colon \mathcal{V}^k(\mathcal{E}^*) \to \mathcal{V}^{k+1}(\mathcal{E}^*)$ be the corresponding Lichnerowicz operator (1.6).

LEMMA 7.2. *The following relations hold:*

$$[\mu^{p_1}\mathcal{V}^{k_1}(\mathcal{E}^*), \mu^{p_2}\mathcal{V}^{k_2}(\mathcal{E}^*)] \subseteq \mu^{p_1+p_2-1}\mathcal{V}^{k_1+k_2-1}(\mathcal{E}^*);$$

(7.4) $$D_\Psi \mu^p \mathcal{V}^k(\mathcal{E}^*) \subseteq \mu^p \mathcal{V}^{k+1}(\mathcal{E}^*);$$

(7.5) $$\pi_p D_\Psi Q = D_p \pi_p Q \quad \text{for all} \quad Q \in \mu^p \mathcal{V}^k(\mathcal{E}^*).$$

LEMMA 7.3. *Let $\mathcal{O}$ be compact and $X \in \mu^2 \mathcal{V}^1(\mathcal{E}^*)$. Then there is a one-parameter family of local diffeomorphisms $\exp(tX)$ such that*

$$\exp(tX)\big|_{\mathcal{O}} = \mathrm{id}_{\mathcal{O}} \quad \text{and} \quad \frac{d}{dt}\exp(tX)_*\Phi = \mathcal{L}_X \exp(tX)_*\Phi$$

for any tensor field Φ.

(The last lemma follows from more general facts in the theory of ordinary differential equations (see, for example, [**NS**]).)

The decomposition of $T^*_{\mathcal{O}}\mathcal{E}^*$ induced by the zero section gives a transversal subbundle (3.8), hence naturally defines an adjoint connection ∇ (3.10). Recall that a linear connection in $\mathcal{E} \to \mathcal{O}$ defines a linear connection in $\mathcal{E}^* \to \mathcal{O}$, hence a decomposition of $T\mathcal{E}^*$ into horizontal and vertical subbundles. This decomposition defines a mapping $\chi^\nabla_{p,k}: C^k_p \to \mu^p \mathcal{V}^k(\mathcal{E}^*)$ such that $\chi^\nabla_{0,1}: \Gamma(T_{\mathcal{O}}\mathcal{E}^*) \to \mu^1(\mathcal{E}^*)$ maps two summands of the decomposition $T_{\mathcal{O}}\mathcal{E}^* = T\mathcal{O} \oplus \mathcal{E}^*$ into the horizontal and vertical subbundles, respectively, and $\chi^\nabla_{p,k}$ satisfies the following properties:

$$\pi_p \circ \chi^\nabla_{p,k} = \mathrm{id}_{C^k_p},$$
$$\chi^\nabla_{(p_1+p_2),(k_1+k_2)}(\eta_1 \overline{\wedge} \eta_2) = \chi^\nabla_{p_1,k_1}\eta_1 \wedge \chi^\nabla_{p_2,k_2}\eta_2$$

(here $\eta_i \in C^{k_i}_{p_i}$, $i = 1, 2$, and $\overline{\wedge}: C^{k_1}_{p_1} \times C^{k_2}_{p_2} \to C^{k_1+k_2}_{p_1+p_2}$ is the exterior multiplication),

$$\chi^\nabla_{0,0} = p^* \quad (\text{here} \quad p: \mathcal{E}^* \to \mathcal{O}),$$

and $\chi^\nabla_{1,0}$ provides the isomorphism between the sections of $\mathcal{E}$ and fiberwise linear functions of $\mathcal{E}^*$.

LEMMA 7.4. *Let $\Psi \in \mathcal{V}^2(\mathcal{E}^*)$ be a Poisson tensor of a proper Poisson bracket, and let $\Psi_1, \Psi_2 \in \mathcal{V}^2(\mathcal{E}^*)$ satisfy the conditions*

(7.6) $$\sigma \overset{\text{def}}{=} \Psi_2 - \Psi_1 \in \mu^{p-1}\mathcal{V}^2(\mathcal{E}^*), \quad p \geq 3,$$
(7.7) $$\Psi_2 - \Psi \in \mu^2 \mathcal{V}^2(\mathcal{E}^*).$$

Assume that a section $\eta \in C^1_{p-1}$ is a solution of the homological equation

(7.8) $$D_{p-1}\eta = \pi_{p-1}\sigma.$$

Then the vector field

(7.9) $$X \overset{\text{def}}{=} \chi^\nabla_{p-1,1}\eta$$

satisfies the conditions

(7.10) $$\exp(tX)_*\Psi_2 - \Psi \in \mu^2 \mathcal{V}^2(\mathcal{E}^*),$$
(7.11) $$\frac{d}{dt}\big(\exp(tX)_*\Psi_2\big) + \sigma \in \mu^p \mathcal{V}^2(\mathcal{E}^*).$$

(*Here $\exp(tX)$ is a one-parameter group of local diffeomorphisms for the vector field X.*)

Using the above lemmas, we can give the sketch of the proof of Theorem 7.1.

PROOF. Let Ψ be a Poisson tensor of a proper Poisson bracket on $\mathcal{E}^*$ and let $\Psi_1 \in \mathcal{V}^2(\mathcal{E}^*)$ be such that $\Psi - \Psi_1 \in \mu^2 \mathcal{V}^2(\mathcal{E}^*)$. To prove that for any $p \geq 2$ there exists a local diffeomorphism g such that its restriction to $\mathcal{O}$ is the identity and (7.2) holds, we shall use induction on p. If $p = 2$, then $g = \mathrm{id}$. Assume that there exists a local diffeomorphism g_1 such that $g_1\big|_{\mathcal{O}} = \mathrm{id}_{\mathcal{O}}$ and $g_{1*}\Psi - \Psi_1 \in \mu^{p-1}\mathcal{V}^2(\mathcal{E}^*)$.

Let us denote $\Psi_2 \stackrel{\mathrm{def}}{=} g_{1*}\Psi$ and σ as in (7.6). Applying π_{p-1} to the equality $[\![\Psi_2, \Psi_2]\!] = 0$ and using formulas (7.6) and (7.7), we get $D_{p-1}\pi_{p-1}\sigma = 0$. Since $\mathcal{H}^2_{p-1} = 0$, the homological equation (7.8) has a solution; therefore, we can define a one-parameter group of local diffeomorphisms $\exp(tX)$ of the vector field (7.9).

Define a local diffeomorphism $g \stackrel{\mathrm{def}}{=} \exp(tX)\big|_{t=1} \circ g_1$. Since $X \in \mu^{p-1}\mathcal{V}^1(\mathcal{E}^*) \subset \mu^2\mathcal{V}^1(\mathcal{E}^*)$, the restriction of g on $\mathcal{O}$ is identity. To prove that $g_*\Psi - \Psi_1 \in \mu^p\mathcal{V}^2(\mathcal{E}^*)$, we show that

$$(7.12) \qquad \Phi_t = \exp(tX)_*\Psi_2 - (\Psi_2 - t\sigma) \in \mu^p\mathcal{V}^2(\mathcal{E}^*), \qquad \forall t \in \mathbb{R}.$$

Note that $\Phi_0 = 0$, and $\frac{d}{dt}\Phi_t = \frac{d}{dt}\exp(tX)_*\Psi_2 + \sigma \in \mu^p\mathcal{V}^2(\mathcal{E}^*)$ (by Lemma 7.3). Therefore, (7.2) holds. $\square$

REMARK. The proof of Theorem 7.1 is an improved version of that for the case in which $\mathcal{O}$ is a point. A theorem similar to Theorem 7.1 for the case of a single point was proved by O. Lychagina [**Ly**].

Now we would like to discuss briefly some facts about the cohomology space $\mathcal{H}^k_p$ and to give a sufficient condition for vanishing of $\mathcal{H}^2_p$. All facts given in §4 can be easily generalized to the case of the cochain complex C^k_p and the operator D_p.

Consider the representation ρ^p of the Lie algebra $\Gamma(T^*_{\mathcal{O}}\mathcal{E}^*)$ in $\Gamma(S^p\mathcal{E})$. Let $(E^{s,t}_{(p)r}, d^{s,t}_{(p)r})$ denote the Hochshild–Serre spectral sequence [**HS**] related to the ideal $\Gamma(\mathcal{E})$ in the Lie algebra $\Gamma(T^*_{\mathcal{O}}\mathcal{E}^*)$.

THEOREM 7.2. *The spectral sequence $(E^{s,t}_{(p)r}, d^{s,t}_{(p)r})$ converges to the cohomology space $\mathcal{H}^k_p$, i.e.,*

$$\mathcal{H}^k_p \simeq \bigoplus_{s+t=k} E^{s,t}_{(p)r}, \qquad \text{for} \quad r \geq \max(t+1, s) + 1.$$

Moreover, if the base $\mathcal{O}$ is simply connected, then

$$E^{s,t}_{(p)2} \simeq H^t(\mathfrak{g}_x, S^p\mathfrak{g}_x) \otimes H^s(\mathcal{O}),$$

where $H^s(\mathcal{O})$ is the de Rham cohomology of the symplectic leaf $\mathcal{O}$ and $H^t(\mathfrak{g}_x, S^p\mathfrak{g}_x)$ is the cohomology of the finite-dimensional Lie algebra $\mathfrak{g}_x$ with coefficients in its pth symmetric product.

REMARK 7.3. Note that if the "curvature" (3.11a) of the adjoint connection in $\mathcal{E}$ induced by the decomposition of $T_{\mathcal{O}}\mathcal{E}^*$ vanishes ($R \equiv 0$), then the spectral sequence $(E^{s,t}_{(p)r}, d^{s,t}_{(p)r})$ degenerates in the second term, and given that $\mathcal{O}$ is simply connected, we have $\mathcal{H}^k_p = \bigoplus_{s+t=k} H^t(\mathfrak{g}_x, S^p\mathfrak{g}_x) \otimes H^s(\mathcal{O})$.

COROLLARY 7.4. *Let the symplectic leaf $\mathcal{O}$ be simply connected and*

(7.15) $$H^2(\mathcal{O}) = 0;$$

also assume that the normal algebra generated in each fiber $\mathcal{E}_x = \mathfrak{g}_x$ by a proper Poisson bracket satisfies the condition

(7.16) $$H^2(\mathfrak{g}_x, S^p \mathfrak{g}_x) = 0.$$

Then $\mathcal{H}_p^2 = 0$.

Finally, observe that condition (7.16) is satisfied for each $p \geq 0$ if the Lie algebra $\mathfrak{g}_x$ is isomorphic to the direct sum of its one-dimensional center and a semisimple Lie algebra. Together with (7.15) and the assumption that $\mathcal{O}$ is simply connected, this implies (7.3); however, in this case, the symplectic leaf cannot be compact. So, to ensure a formal equivalence of two proper Poisson brackets, the vector fields (7.9) must be complete.

Appendix. The de Rham cohomology of flat connections

In this section we collect well-known facts [**BT, GHV, Li**$_1$] about flat connections and cohomology of locally constant vector bundles, which we use in the present paper.

Let $\mathcal{E} \to B$ be a vector bundle. The operation ∇ that to each vector field $v \in \mathcal{V}^1(B)$ assigns a mapping of sections $\nabla_v \colon \Gamma(\mathcal{E}) \to \Gamma(\mathcal{E})$ and satisfies the axioms

(A.1) $\quad \nabla_{\varphi v_1 + v_2} \eta = \varphi \nabla_{v_1} \eta + \nabla_{v_2} \eta, \qquad \eta \in \Gamma(\mathcal{E}), \quad \varphi \in C^\infty(B),$

(A.2) $\quad \nabla_v(\varphi \eta_1 + \eta_2) = \varphi \nabla_v \eta_1 + \nabla_v \eta_2 + \mathcal{L}_v(\varphi) \eta_1, \qquad \eta_1, \eta_2 \in \Gamma(\mathcal{E}).$

is called a *connection in the vector bundle*.

The connection ∇ in the vector bundle $\mathcal{E}$ naturally generates connections in the bundles $\mathcal{E}^*$, $\bigwedge^t \mathcal{E}$, and $\mathcal{E}/\mathcal{E}_1$, where $\mathcal{E}_1 \subset \mathcal{E}$ is a ∇-invariant subbundle.

A section $K \in \Gamma(\mathrm{Hom}(\mathcal{E}) \otimes \bigwedge^2 T^*B)$, defined by the formula

(A.3) $\quad K(v_1, v_2)\eta \stackrel{\mathrm{def}}{=} \nabla_{v_1} \nabla_{v_2} \eta - \nabla_{v_2} \nabla_{v_1} \eta - \nabla_{[v_1, v_2]} \eta, \quad v_1, v_2 \in \mathcal{V}^1(B), \quad \eta \in \Gamma(\mathcal{E}),$

is called the *curvature* of the connection ∇.

The connection ∇ is called *flat* if it has zero curvature. The parallel transport of a flat connection depends only on the homotopy class of the path [**Li**$_1$]; therefore, in any simply connected domain U there is a basis of parallel sections, obtained by the parallel transport of the basis in some fiber $\mathcal{E}_{x_0}$ ($x_0 \in U$), and the holonomy group of the connection ∇ is a homomorphism of the fundamental group $\pi_1(B)$ into the group $GL(\mathcal{E}_{x_0})$. In particular, if the base is simply connected, then the holonomy group of a flat connection is trivial and the bundle $\mathcal{E}$ is trivial ($\mathcal{E} = \mathcal{E}_{x_0} \times B$).

Let us define a space of cochains $C^k(\mathcal{E}) \stackrel{\mathrm{def}}{=} \Gamma(\mathcal{E} \otimes \bigwedge^k T^*B)$ and an operator $\nabla \colon C^k(\mathcal{E}) \to C^{k+1}(\mathcal{E})$, which we denote by the same symbol as the connection,

(A.4) $$\nabla \eta(v_0, v_1, \ldots, v_k) \stackrel{\mathrm{def}}{=} \sum_{j=0}^{k} (-1)^j \nabla_{v_j} \eta(v_0, v_1, \ldots, \widehat{v}_j, \ldots, v_k)$$
$$+ \sum_{0 \leq i < j \leq k} (-1)^{i+j} \eta([v_i, v_j], v_0, v_1, \ldots, \widehat{v}_i, \ldots, \widehat{v}_j, \ldots, v_k).$$

It is easy to see that the operator ∇^2 has the form

$$\nabla^2 \eta(v_0, v_1, \ldots, v_k, v_{k+1})$$
$$= -\sum_{0 \leq i < j \leq k+1} (-1)^{i+j} K(v_i, v_j) \eta(v_0, \ldots, \widehat{v}_i, \ldots, \widehat{v}_j, \ldots, v_{k+1}).$$

Therefore, we have the following statement.

LEMMA A.1. *A connection is flat if and only if $\nabla^2 = 0$.*

For each flat connection ∇, by $Z^k_\nabla(\mathcal{E})$, $B^k_\nabla(\mathcal{E})$, and $H^k_\nabla(\mathcal{E}) = Z^k_\nabla/B^k_\nabla(\mathcal{E})$ we denote cocycles, coboundaries, and the cohomology of the operator (A.4). The cohomology space $H_\nabla(\mathcal{E}) = \bigoplus_k H^k_\nabla(\mathcal{E})$ depends on the choice of the flat connection ∇^k. We have the following statement [**BT**, Prop. 7.4].

LEMMA A.2. *The connection ∇ has a trivial holonomy group if and only if*

(A.5) $$H^k_\nabla(\mathcal{E}) \simeq H^k(B) \otimes \mathcal{E}_{x_0} \qquad \forall k \geq 0$$

(here $H^k(B)$ is the de Rham cohomology of the base).

LEMMA A.3. *Suppose that the base B is a finite-type manifold. Then for any flat connection ∇ the space $H_\nabla(\mathcal{E})$ is finite-dimensional.*

To prove Lemma A.3 we apply the Mayer–Vietoris sequence and repeat the proof of [**BT**, Prop. 5.3.1] for the fact that the de Rham cohomology is finite-dimensional.

In Section 4 we have used a slightly generalized version of Lemma A.2. Suppose that $(A \to B, q, \{,\})$ is a transitive Lie algebroid, and $\mathcal{E} \to B$ is a vector bundle. We define the space of cochains $C^k(A, \mathcal{E}) = \Gamma(\mathcal{E} \otimes \bigwedge^k A^*)$. For each connection ∇ in the bundle $\mathcal{E}$ we define the operator $\overline{\nabla} \colon C^k(A, \mathcal{E}) \to C^{k+1}(A, \mathcal{E})$ as follows:

(A.6)
$$\overline{\nabla}\eta(\alpha_0, \ldots, \alpha_k) \stackrel{\text{def}}{=} \sum_{j=0}^{k} (-1)^j \nabla_{q\alpha_j} \eta(\alpha_0, \ldots, \widehat{\alpha}_j, \ldots, \alpha_k)$$
$$+ \sum_{0 \leq i < j \leq k} (-1)^{i+j} \eta(\{\alpha_i, \alpha_j\}, \alpha_0, \ldots, \widehat{\alpha}_i, \ldots, \widehat{\alpha}_j, \ldots, \alpha_k)$$

(here $\eta \in C^k(A, \mathcal{E})$, $\alpha_j \in \Gamma(A)$). Obviously, the connection ∇ is flat if and only if $\overline{\nabla}^2 = 0$. By $H^k_\nabla(A, \mathcal{E})$ we denote the cohomology of the operator (A.6).

LEMMA A.4. *The holonomy group of the flat connection ∇ is trivial if and only if*

(A.7) $$H^k_\nabla(A, \mathcal{E}) \simeq \mathcal{H}^k(A) \otimes \mathcal{E}_{x_0} \qquad \forall k \geq 0,$$

where $\mathcal{H}^k(A)$ is the cohomology of a transitive Lie algebroid A and $\mathcal{E}_{x_0}$ is a fiber of the bundle $\mathcal{E}$.

The proof of this lemma is similar to that of Lemma A.2.

References

[AM] D. V. Alekseevsky and P. W. Michor, *Differential geometry of g-manifolds*, Diff. Geom. and Appl. **5** (1995), 371–403.

[BFFLS] F. Bayen, M. Flato, C. Fronsdal, A. Lichnerowicz, and D. Sternheimer, *Deformation theory and quantization*, Ann. of Physics **111** (1978), 51–161.

[BV] K. H. Bhaskara and K. Viswanath, *Calculus on Poisson manifolds*, Bull. London Math. Soc. **20** (1988), 62–72.

[Bo$_1$] R. Bott, *On a topological obstruction to integrability*, Proc. Sympos. Pure Math., vol. 16, Amer. Math. Soc., Providence, RI, 1970, pp. 127–131.

[BT] R. Bott and L. W. Tu, *Differential Forms in Algebraic Topology*, Springer-Verlag, Berlin and New York, 1982.

[Br] J.-L. Brylinski, *A differential complex for Poisson manifolds*, J. Differential Geom. **28** (1988), 93–114.

[CE] C. Chevalley and S. Elenberg, *Cohomology theory of Lie groups and Lie algebras*, Trans. Amer. Math. Soc. **63** (1948), 85–124.

[Co$_1$] J. F. Conn, *Normal forms for analytic Poisson structures*, Ann. of Math. **119** (1984), 577–601.

[Co$_2$] _____, *Normal forms for smooth Poisson structures*, Ann. of Math. **121** (1985), 565–593.

[Cu] T. J. Courant, *Dirac manifolds*, Trans. Amer. Math. Soc. **319** (1990), 631–661.

[Do] I. Ya. Dorfman, *Deformations of Hamiltonian structures and integrable systems*, Nonlinear and Turbulent Processes in Physics, Vol. 3 (Kiev, 1983), Harwood Academic Publ., Chur, 1984, pp. 1313–1318.

[Du] J. P. Dufour, *Linéarisation de certaines structures de Poisson*, J. Differential Geom. **32** (1990), 415–428.

[F] D. B. Fuks, *Cohomology of Infinite-Dimensional Lie Algebras*, Consultants Bureau, New York, 1986.

[Gi] V. L. Ginzburg, *Equivariant Poisson cohomology and a spectral sequence associated with a moment map*, Preprint dg–ga/9611002v2.

[GW] V. L. Ginzburg and A. Weinstein, *Lie–Poisson structures on some Poisson–Lie groups*, J. Amer. Math. Soc. **5** (1992), 445–453.

[GG] M. Goto and F. D. Grosshans, *Semisimple Lie Algebras*, Lecture Notes in Pure and Appl. Math., vol. 38, Marcel Dekker, New York, 1978.

[GHV] W. Greub, S. Halperin, and R. Vanstone, *Connections, Curvature, and Cohomology*. Vol. 2, Academic Press, New York, 1973.

[Hu] J. Huebschmann, *Poisson cohomology and quantization*, J. Reine Angew. Math. **408** (1990), 57–113.

[HS] G. P. Hochschild and J. P. Serre, *Cohomology of Lie algebras*, Ann. of Math. (2) **57** (1953), 591–603.

[Ka] M. V. Karasev, *Analogues of objects of the Lie group theory for nonlinear Poisson brackets*, Izv. Akad. Nauk SSSR Ser. Mat. **50** (1986), 508–538; English transl., Math. USSR-Izv. **28** (1987), 497–527.

[KaMas] M. V. Karasev and V. P. Maslov, *Nonlinear Poisson Brackets. Geometry and Quantization*, Nauka, Moscow, 1991; English transl., Transl. Math. Monographs, vol. 119, Amer. Math. Soc., Providence, RI, 1993.

[Ki$_1$] A. A. Kirillov, *Elements of the Theory of Representations*, Nauka, Moscow, 1972; English transl., Springer-Verlag, Berlin and New York, 1976.

[Ki$_2$] _____, *Local Lie algebras*, Uspekhi Mat. Nauk **31** (1976), no. 4, 57–76; English transl. in Russian Math. Surveys **31** (1976).

[Ko] J.-L. Koszul, *Crochet de Schouten–Nijenhuis et cohomologie*, Elie Cartan et les mathématiques d'aujourd'hui, Astérisque, Numéro Hors Série, Soc. Math. France, Paris, 1985, pp. 257–271.

[KSM] Y. Kosmann–Schwarzbach and F. Magri, *Poisson–Nijenhuis structures*, Ann. Inst. H. Poincaré Phys. Théor. **53** (1990), 35–81.

[Li$_1$] A. Lichnerowicz, *Théorie globale des connexions et groups d'holonomie*, Edizioni Cremonese, Rome, 1955.

[Li$_2$] _____, *Les variétés de Poisson et leurs algèbres de Lie associées*, J. Differential Geom. **12** (1977), 253–300.

[Li₃] _____, *Feuilletages, géométrie riemannienne et géométrie symplectique*, Riv. Mat. Univ. Parma (4) **10** (1984), 81–90.

[Ly] O. V. Lychagina, *Normal forms of Poisson structures*, Mat. Zametki **61** (1997), 220–235; English transl. in Math. Notes **61** (1997).

[Mz] K. Mackenzie, *Lie Groupoids and Lie Algebroids in Differential Geometry*, London Math. Soc. Lecture Note Ser., vol. 124, Cambridge Univ. Press, Cambridge, 1987.

[MR] J. Marsden and T. Ratiu, *Introduction to Mechanics and Symmetry*. Vol. 1, Springer-Verlag, Berlin and New York, 1994.

[MS] J. W. Milnor and J. D. Stasheff, *Characteristic Classes*, Princeton Univ. Press, Princeton, New Jersey, 1974.

[NS] V. V. Nemytski and V. V. Stepanov, *Qualitative Theory of Differential Equations*, Dover, New York, 1989.

[Pr] J. Pradines, *Théorie de Lie pour les groupoïdes différentiables. Calcul différentiel dans la catégorie des groupoïdes infinitésimaux*, C. R. Acad. Sci. Paris Sér. A **264** (1967), 245–248.

[Sn] J. A. Schouten, *Über Differentialkomitanten zweier kontravarianter Grössen*, Nederl. Acad. Wetensch. Proc. Ser. A **43** (1940), 449–452.

[Va] I. Vaisman, *Lectures on the Geometry of Poisson Manifolds*, Birkhäuser, Basel and Boston, 1994.

[VK₁] Yu. M. Vorobjev and M. V. Karasev, *About Poisson manifolds and Schouten brackets*, Funktsional. Anal. i Prilozhen. **22** (1988), no. 1, 1–11; English transl. in Functional Anal. Appl. **22** (1988).

[VK₂] _____, *Corrections to classical dynamics and quantization conditions arising in the deformation of Poisson brackets*, Dokl. Akad. Nauk SSSR **297** (1987), 1294–1298; English transl. in Soviet Math. Dokl. **36** (1988).

[VK₃] _____, *Deformation and cohomology of Poisson brackets*, Topological and Geometrical Methods of Analysis (Novoe v Global. Anal., vyp. 9), Izdat. Voronezh. Gos. Univ., Voronozh, 1989, pp. 75–89; English transl., Global Analysis—Studies and Applications, Lecture Notes in Math., vol. 1453, Springer-Verlag, Berlin and New York, 1990, pp. 271–289.

[We] A. Weinstein, *The local structure of Poisson manifolds*, J. Differential Geom. **18** (1983), 523–557.

[Xu] P. Xu, *Poisson cohomology of regular Poisson manifolds*, Ann. Inst. Fourier (Grenoble) **42** (1992), 967–988.

V. I.: Department of Mathematics, University of Minnesota, 127 Vincent Hall, 206 Church St., Minneaplolis, MN 55455, USA
E-mail address: itskov@math.umn.edu

M. K and Yu. V.: Department of Applied Mathematics, Moscow State Institute of Electronics and Mathematics, B.Trekhsvyatitel'skii, 3/12, Moscow 109028, Russia
E-mail address: vorob@amath.msk.ru

Translated by the authors

Selected Titles in This Subseries

(*Continued from the front of this publication*)

10 **A. V. Babin and M. I. Vishik, Editors,** Properties of global attractors of partial differential equations, 1992

9 **A. M. Vershik, Editor,** Representation theory and dynamical systems, 1992

8 **E. B. Vinberg, Editor,** Lie groups, their discrete subgroups, and invariant theory, 1992

7 **M. Sh. Birman, Editor,** Estimates and asymptotics for discrete spectra of integral and differential equations, 1991

6 **A. T. Fomenko, Editor,** Topological classification of integrable systems, 1991

5 **R. A. Minlos, Editor,** Many-particle Hamiltonians: spectra and scattering, 1991

4 **A. A. Suslin, Editor,** Algebraic K-theory, 1991

3 **Ya. G. Sinaĭ, Editor,** Dynamical systems and statistical mechanics, 1991

2 **A. A. Kirillov, Editor,** Topics in representation theory, 1991

1 **V. I. Arnold, Editor,** Theory of singularities and its applications, 1990